Algebra
Second Edition

Michael Artin

Massachusetts Institute of Technology

Prentice Hall

Boston   Columbus   Indianapolis   New York   San Fransisco   Upper Saddle River
Amsterdam   Cape Town   Dubai   London   Madrid   Milan   Munich   Paris   Montréal   Toronto
Delhi   Mexico City   São Paulo   Sydney   Hong Kong   Seoul   Singapore   Taipei   Tokyo

Editor-in-Chief: *Deirdre Lynch*
Senior Acquisitions Editor: *William Hoffman*
Sponsoring Editor: *Caroline Celano*
Senior Managing Editor: *Karen Wernholm*
Production Project Manager: *Beth Houston*
Cover Designer: *Christina Gleason*
Executive Marketing Manager: *Jeff Weidenaar*
Marketing Assistant: *Kendra Bassi*
Manufacturing Manager: *Evelyn Beaton*
Production Coordination, Technical Illustrations, and Composition: *Laserwords*

Library of Congress Cataloging-in-Publication Data

Artin, Michael.
  Algebra : applications, and algorithms / by Michael Artin. -- 2nd ed.
    p. cm.
  ISBN-13: 978-0-13-241377-0
  ISBN-10: 0-13-241377-9
  1. Algebra. I. Title.
   QA154.2.A77 2011
   512.9--dc22

                                          2010017573

Many of the designations used by manufacturers and sellers to distinguish their products are claimed as trademarks. Where those designations appear in this book, and Pearson was aware of a trademark claim, the designations have been printed in initial caps or all caps.

Prentice Hall
is an imprint of

www.pearsonhighered.com

1 2 3 4 5 6 7 8 9 10 CW 14 13 12 11 10

ISBN-13: 978-0-13-241377-0

ISBN-10: 0-13-241377-9

To my wife Jean

Contents

Preface xi

1 Matrices
1.1 The Basic Operations 1
1.2 Row Reduction 10
1.3 The Matrix Transpose 17
1.4 Determinants 18
1.5 Permutations 24
1.6 Other Formulas for the Determinant 27
Exercises 31

2 Groups
2.1 Laws of Composition 37
2.2 Groups and Subgroups 40
2.3 Subgroups of the Additive Group of Integers 43
2.4 Cyclic Groups 46
2.5 Homomorphisms 47
2.6 Isomorphisms 51
2.7 Equivalence Relations and Partitions 52
2.8 Cosets 56
2.9 Modular Arithmetic 60
2.10 The Correspondence Theorem 61
2.11 Product Groups 64
2.12 Quotient Groups 66
Exercises 69

3 Vector Spaces
3.1 Subspaces of $\mathbb{R}^n$ 78
3.2 Fields 80
3.3 Vector Spaces 84
3.4 Bases and Dimension 86
3.5 Computing with Bases 91
3.6 Direct Sums 95
3.7 Infinite-Dimensional Spaces 96
Exercises 98

4 Linear Operators
4.1 The Dimension Formula 102
4.2 The Matrix of a Linear Transformation 104

4.3	Linear Operators	108
4.4	Eigenvectors	110
4.5	The Characteristic Polynomial	113
4.6	Triangular and Diagonal Forms	116
4.7	Jordan Form	120
	Exercises	125

5 Applications of Linear Operators

5.1	Orthogonal Matrices and Rotations	132
5.2	Using Continuity	138
5.3	Systems of Differential Equations	141
5.4	The Matrix Exponential	145
	Exercises	150

6 Symmetry

6.1	Symmetry of Plane Figures	154
6.2	Isometries	156
6.3	Isometries of the Plane	159
6.4	Finite Groups of Orthogonal Operators on the Plane	163
6.5	Discrete Groups of Isometries	167
6.6	Plane Crystallographic Groups	172
6.7	Abstract Symmetry: Group Operations	176
6.8	The Operation on Cosets	178
6.9	The Counting Formula	180
6.10	Operations on Subsets	181
6.11	Permutation Representations	181
6.12	Finite Subgroups of the Rotation Group	183
	Exercises	188

7 More Group Theory

7.1	Cayley's Theorem	195
7.2	The Class Equation	195
7.3	p-Groups	197
7.4	The Class Equation of the Icosahedral Group	198
7.5	Conjugation in the Symmetric Group	200
7.6	Normalizers	203
7.7	The Sylow Theorems	203
7.8	Groups of Order 12	208
7.9	The Free Group	210
7.10	Generators and Relations	212
7.11	The Todd-Coxeter Algorithm	216
	Exercises	221

8 Bilinear Forms

8.1	Bilinear Forms	229
8.2	Symmetric Forms	231

8.3   Hermitian Forms 232
8.4   Orthogonality 235
8.5   Euclidean Spaces and Hermitian Spaces 241
8.6   The Spectral Theorem 242
8.7   Conics and Quadrics 245
8.8   Skew-Symmetric Forms 249
8.9   Summary 252
       Exercises 254

9   Linear Groups
9.1   The Classical Groups 261
9.2   Interlude: Spheres 263
9.3   The Special Unitary Group SU_2 266
9.4   The Rotation Group SO_3 269
9.5   One-Parameter Groups 272
9.6   The Lie Algebra 275
9.7   Translation in a Group 277
9.8   Normal Subgroups of SL_2 280
       Exercises 283

10 Group Representations
10.1 Definitions 290
10.2 Irreducible Representations 294
10.3 Unitary Representations 296
10.4 Characters 298
10.5 One-Dimensional Characters 303
10.6 The Regular Representation 304
10.7 Schur's Lemma 307
10.8 Proof of the Orthogonality Relations 309
10.9 Representations of SU_2 311
       Exercises 314

11 Rings
11.1 Definition of a Ring 323
11.2 Polynomial Rings 325
11.3 Homomorphisms and Ideals 328
11.4 Quotient Rings 334
11.5 Adjoining Elements 338
11.6 Product Rings 341
11.7 Fractions 342
11.8 Maximal Ideals 344
11.9 Algebraic Geometry 347
       Exercises 354

12 Factoring

12.1	Factoring Integers	359
12.2	Unique Factorization Domains	360
12.3	Gauss's Lemma	367
12.4	Factoring Integer Polynomials	371
12.5	Gauss Primes	376
	Exercises	378

13 Quadratic Number Fields

13.1	Algebraic Integers	383
13.2	Factoring Algebraic Integers	385
13.3	Ideals in $\mathbb{Z}[\sqrt{-5}]$	387
13.4	Ideal Multiplication	389
13.5	Factoring Ideals	392
13.6	Prime Ideals and Prime Integers	394
13.7	Ideal Classes	396
13.8	Computing the Class Group	399
13.9	Real Quadratic Fields	402
13.10	About Lattices	405
	Exercises	408

14 Linear Algebra in a Ring

14.1	Modules	412
14.2	Free Modules	414
14.3	Identities	417
14.4	Diagonalizing Integer Matrices	418
14.5	Generators and Relations	423
14.6	Noetherian Rings	426
14.7	Structure of Abelian Groups	429
14.8	Application to Linear Operators	432
14.9	Polynomial Rings in Several Variables	436
	Exercises	437

15 Fields

15.1	Examples of Fields	442
15.2	Algebraic and Transcendental Elements	443
15.3	The Degree of a Field Extension	446
15.4	Finding the Irreducible Polynomial	449
15.5	Ruler and Compass Constructions	450
15.6	Adjoining Roots	456
15.7	Finite Fields	459
15.8	Primitive Elements	462
15.9	Function Fields	463
15.10	The Fundamental Theorem of Algebra	471
	Exercises	472

16 Galois Theory
16.1   Symmetric Functions 477
16.2   The Discriminant 481
16.3   Splitting Fields 483
16.4   Isomorphisms of Field Extensions 484
16.5   Fixed Fields 486
16.6   Galois Extensions 488
16.7   The Main Theorem 489
16.8   Cubic Equations 492
16.9   Quartic Equations 493
16.10 Roots of Unity 497
16.11 Kummer Extensions 500
16.12 Quintic Equations 502
       Exercises 505

APPENDIX

Background Material
A.1   About Proofs 513
A.2   The Integers 516
A.3   Zorn's Lemma 518
A.4   The Implicit Function Theorem 519
      Exercises 521

Bibliography   523

Notation   525

Index   529

Preface

Important though the general concepts and propositions may be with which the modern and industrious passion for axiomatizing and generalizing has presented us, in algebra perhaps more than anywhere else, nevertheless I am convinced that the special problems in all their complexity constitute the stock and core of mathematics, and that to master their difficulties requires on the whole the harder labor.

—Herman Weyl

This book began many years ago in the form of supplementary notes for my algebra classes. I wanted to discuss some concrete topics such as symmetry, linear groups, and quadratic number fields in more detail than the text provided, and to shift the emphasis in group theory from permutation groups to matrix groups. Lattices, another recurring theme, appeared spontaneously.

My hope was that the concrete material would interest the students and that it would make the abstractions more understandable – in short, that they could get farther by learning both at the same time. This worked pretty well. It took me quite a while to decide what to include, but I gradually handed out more notes and eventually began teaching from them without another text. Though this produced a book that is different from most others, the problems I encountered while fitting the parts together caused me many headaches. I can't recommend the method.

There is more emphasis on special topics here than in most algebra books. They tended to expand when the sections were rewritten, because I noticed over the years that, in contrast to abstract concepts, with concrete mathematics students often prefer more to less. As a result, the topics mentioned above have become major parts of the book.

In writing the book, I tried to follow these principles:

1. The basic examples should precede the abstract definitions.

2. Technical points should be presented only if they are used elsewhere in the book.

3. All topics should be important for the average mathematician.

Although these principles may sound like motherhood and the flag, I found it useful to have them stated explicitly. They are, of course, violated here and there.

The chapters are organized in the order in which I usually teach a course, with linear algebra, group theory, and geometry making up the first semester. Rings are first introduced in Chapter 11, though that chapter is logically independent of many earlier ones. I chose this arrangement to emphasize the connections of algebra with geometry at the start, and because, overall, the material in the first chapters is the most important for people in other fields. The first half of the book doesn't emphasize arithmetic, but this is made up for in the later chapters.

About This Second Edition

The text has been rewritten extensively, incorporating suggestions by many people as well as the experience of teaching from it for 20 years. I have distributed revised sections to my class all along, and for the past two years the preliminary versions have been used as texts. As a result, I've received many valuable suggestions from the students. The overall organization of the book remains unchanged, though I did split two chapters that seemed long.

There are a few new items. None are lengthy, and they are balanced by cuts made elsewhere. Some of the new items are an early presentation of Jordan form (Chapter 4), a short section on continuity arguments (Chapter 5), a proof that the alternating groups are simple (Chapter 7), short discussions of spheres (Chapter 9), product rings (Chapter 11), computer methods for factoring polynomials and Cauchy's Theorem bounding the roots of a polynomial (Chapter 12), and a proof of the Splitting Theorem based on symmetric functions (Chapter 16). I've also added a number of nice exercises. But the book is long enough, so I've tried to resist the temptation to add material.

NOTES FOR THE TEACHER

This book is designed to allow you to choose among the topics. Don't try to cover the book, but do include some of the interesting special topics such as symmetry of plane figures, the geometry of SU_2, or the arithmetic of imaginary quadratic number fields. If you don't want to discuss such things in your course, then this is not the book for you.

There are relatively few prerequisites. Students should be familiar with calculus, the basic properties of the complex numbers, and mathematical induction. An acquaintance with proofs is obviously useful. The concepts from topology that are used in Chapter 9, Linear Groups, should not be regarded as prerequisites.

I recommend that you pay attention to concrete examples, especially throughout the early chapters. This is very important for the students who come to the course without a clear idea of what constitutes a proof.

One could spend an entire semester on the first five chapters, but since the real fun starts with symmetry in Chapter 6, that would defeat the purpose of the book. Try to get to Chapter 6 as soon as possible, so that it can be done at a leisurely pace. In spite of its immediate appeal, symmetry isn't an easy topic. It is easy to be carried away and leave the students behind.

These days most of the students in my classes are familiar with matrix operations and modular arithmetic when they arrive. I've not been discussing the first chapter on matrices in class, though I do assign problems from that chapter. Here are some suggestions for Chapter 2, Groups.

1. Treat the abstract material with a light touch. You can have another go at it in Chapters 6 and 7.

2. For examples, concentrate on matrix groups. Examples from symmetry are best deferred to Chapter 6.

3. Don't spend much time on arithmetic; its natural place in this book is in Chapters 12 and 13.

4. De-emphasize the quotient group construction.

Quotient groups present a pedagogical problem. While their construction is conceptually difficult, the quotient is readily presented as the image of a homomorphism in most elementary examples, and then it does not require an abstract definition. Modular arithmetic is about the only convincing example for which this is not the case. And since the integers modulo n form a ring, modular arithmetic isn't the ideal motivating example for quotients of groups. The first serious use of quotient groups comes when generators and relations are discussed in Chapter 7. I deferred the treatment of quotients to that point in early drafts of the book, but, fearing the outrage of the algebra community, I eventually moved it to Chapter 2. If you don't plan to discuss generators and relations for groups in your course, then you can defer an in-depth treatment of quotients to Chapter 11, Rings, where they play a central role, and where modular arithmetic becomes a prime motivating example.

In Chapter 3, Vector Spaces, I've tried to set up the computations with bases in such a way that the students won't have trouble keeping the indices straight. Since the notation is used throughout the book, it may be advisable to adopt it.

The matrix exponential that is defined in Chapter 5 is used in the description of one-parameter groups in Chapter 10, so if you plan to include one-parameter groups, you will need to discuss the matrix exponential at some point. But you must resist the temptation to give differential equations their due. You will be forgiven because you are teaching algebra.

Except for its first two sections, Chapter 7, again on groups, contains optional material. A section on the Todd-Coxeter algorithm is included to justify the discussion of generators and relations, which is pretty useless without it. It is fun, too.

There is nothing unusual in Chapter 8, on bilinear forms. I haven't overcome the main pedagogical problem with this topic – that there are too many variations on the same theme, but have tried to keep the discussion short by concentrating on the real and complex cases.

In the chapter on linear groups, Chapter 9, plan to spend time on the geometry of SU_2. My students complained about that chapter every year until I expanded the section on SU_2, after which they began asking for supplementary reading, wanting to learn more. Many of our students aren't familiar with the concepts from topology when they take the course, but I've found that the problems caused by the students' lack of familiarity can be managed. Indeed, this is a good place for them to get an idea of a manifold.

I resisted including group representations, Chapter 10, for a number of years, on the grounds that it is too hard. But students often requested it, and I kept asking myself: If the chemists can teach it, why can't we? Eventually the internal logic of the book won out and group representations went in. As a dividend, hermitian forms got an application.

You may find the discussion of quadratic number fields in Chapter 13 too long for a general algebra course. With this possibility in mind, I've arranged the material so that the end of Section 13.4, on ideal factorization, is a natural stopping point.

It seemed to me that one should mention the most important examples of fields in a beginning algebra course, so I put a discussion of function fields into Chapter 15. There is always the question of whether or not Galois theory should be presented in an undergraduate course, but as a culmination of the discussion of symmetry, it belongs here.

Some of the harder exercises are marked with an asterisk.

Though I've taught algebra for years, various aspects of this book remain experimental, and I would be very grateful for critical comments and suggestions from the people who use it.

ACKNOWLEDGMENTS

Mainly, I want to thank the students who have been in my classes over the years for making them so exciting. Many of you will recognize your own contributions, and I hope that you will forgive me for not naming you individually.

Acknowledgments for the First Edition

Several people used my notes and made valuable suggestions – Jay Goldman, Steve Kleiman, Richard Schafer, and Joe Silverman among them. Harold Stark helped me with the number theory, and Gil Strang with the linear algebra. Also, the following people read the manuscript and commented on it: Ellen Kirkman, Al Levine, Barbara Peskin, and John Tate. I want to thank Barbara Peskin especially for reading the whole thing twice during the final year.

The figures which needed mathematical precision were made on the computer by George Fann and Bill Schelter. I could not have done them by myself. Many thanks also to Marge Zabierek, who retyped the manuscript annually for about eight years before it was put onto the computer where I could do the revisions myself, and to Mary Roybal for her careful and expert job of editing the manuscript.

I haven't consulted other books very much while writing this one, but the classics by Birkhoff and MacLane and by van der Waerden from which I learned the subject influenced me a great deal, as did Herstein's book, which I used as a text for many years. I also found some good exercises in the books by Noble and by Paley and Weichsel.

Acknowledgments for the Second Edition

Many people have commented on the first edition – a few are mentioned in the text. I'm afraid that I will have forgotten to mention most of you.

I want to thank these people especially: Annette A' Campo and Paolo Maroscia made careful translations of the first edition, and gave me many corrections. Nathaniel Kuhn and James Lepowsky made valuable suggestions. Annette and Nat finally got through my thick skull how one should prove the orthogonality relations.

I thank the people who reviewed the manuscript for their suggestions. They include Alberto Corso, Thomas C. Craven, Sergi Elizade, Luis Finotti, Peter A. Linnell, Brad Shelton, Hema Srinivasan, and Nik Weaver. Toward the end of the process, Roger Lipsett read and commented on the entire revised manuscript. Brett Coonley helped with the many technical problems that arose when the manuscript was put into TeX.

Many thanks, also, to Caroline Celano at Pearson for her careful and thorough editing of the manuscript and to Patty Donovan at Laserwords, who always responded graciously to my requests for yet another emendation, though her patience must have been tried at times.

And I talk to Gil Strang and Harold Stark often, about everything.

Finally, I want to thank the many MIT undergraduates who read and commented on the revised text and corrected errors. The readers include Nerses Aramyan, Reuben Aronson, Mark Chen, Jeremiah Edwards, Giuliano Giacaglia, Li-Mei Lim, Ana Malagon, Maria Monks, and Charmaine Sia. I came to rely heavily on them, especially on Nerses, Li-Mei, and Charmaine.

"One, two, three, five, four..."
"No Daddy, it's one, two, three, four, five."
"Well if I want to say one, two, three, five, four, why can't I?"
"That's not how it goes."

—Carolyn Artin

CHAPTER 1

Matrices

Erftlich wird alles dasjenige eine Größe genennt,
welches einer Vermehrung oder einer Verminderung fähig ist,
oder wozu sich noch etwas hinzusetzen oder davon wegnehmen läßt.

—Leonhard Euler[1]

Matrices play a central role in this book. They form an important part of the theory, and many concrete examples are based on them. Therefore it is essential to develop facility in matrix manipulation. Since matrices pervade mathematics, the techniques you will need are sure to be useful elsewhere.

1.1 THE BASIC OPERATIONS

Let m and n be positive integers. An $m \times n$ *matrix* is a collection of mn numbers arranged in a rectangular array

(1.1.1)
$$
\begin{array}{c}
n \text{ columns} \\
m \text{ rows} \begin{bmatrix} a_{11} & \cdots & a_{1n} \\ \vdots & & \vdots \\ a_{m1} & \cdots & a_{mn} \end{bmatrix}
\end{array}
$$

For example, $\begin{bmatrix} 2 & 1 & 0 \\ 1 & 3 & 5 \end{bmatrix}$ is a 2×3 matrix (two rows and three columns). We usually introduce a symbol such as A to denote a matrix.

The numbers in a matrix are the *matrix entries*. They may be denoted by a_{ij}, where i and j are indices (integers) with $1 \leq i \leq m$ and $1 \leq j \leq n$, the index i is the *row index*, and j is the *column index*. So a_{ij} is the entry that appears in the ith row and jth column of the matrix:

$$
i \begin{bmatrix} & & j & \\ & & \vdots & \\ \cdots & & a_{ij} & \cdots \\ & & \vdots & \end{bmatrix}
$$

---

[1] This is the opening sentence of Euler's book *Algebra*, which was published in St. Petersburg in 1770.

In the above example, $a_{11} = 2$, $a_{13} = 0$, and $a_{23} = 5$. We sometimes denote the matrix whose entries are a_{ij} by (a_{ij}).

An $n \times n$ matrix is called a *square* matrix. A 1×1 matrix $[a]$ contains a single number, and we do not distinguish such a matrix from its entry.

A $1 \times n$ matrix is an n-dimensional *row vector*. We drop the index i when $m = 1$ and write a row vector as

$$[a_1 \cdots a_n], \text{ or as } (a_1, \ldots, a_n).$$

Commas in such a row vector are optional. Similarly, an $m \times 1$ matrix is an

m-dimensional *column vector*:

$$\begin{bmatrix} b_1 \\ \vdots \\ b_m \end{bmatrix}.$$

In most of this book, we won't make a distinction between an n-dimensional column vector and the point of n-dimensional space with the same coordinates. In the few places where the distinction is useful, we will state this clearly.

Addition of matrices is defined in the same way as vector addition. Let $A = (a_{ij})$ and $B = (b_{ij})$ be two $m \times n$ matrices. Their sum $A + B$ is the $m \times n$ matrix $S = (s_{ij})$ defined by

$$s_{ij} = a_{ij} + b_{ij}.$$

Thus

$$\begin{bmatrix} 2 & 1 & 0 \\ 1 & 3 & 5 \end{bmatrix} + \begin{bmatrix} 1 & 0 & 3 \\ 4 & -3 & 1 \end{bmatrix} = \begin{bmatrix} 3 & 1 & 3 \\ 5 & 0 & 6 \end{bmatrix}.$$

Addition is defined only when the matrices to be added have the same shape – when they are $m \times n$ matrices with the same m and n.

Scalar multiplication of a matrix by a number is also defined as with vectors. The result of multiplying an $m \times n$ matrix A by a number c is another $m \times n$ matrix $B = (b_{ij})$, where $b_{ij} = ca_{ij}$ for all i, j. Thus

$$2 \begin{bmatrix} 2 & 1 & 0 \\ 1 & 3 & 5 \end{bmatrix} = \begin{bmatrix} 4 & 2 & 0 \\ 2 & 6 & 10 \end{bmatrix}.$$

Numbers will also be referred to as *scalars*. Let's assume for now that the scalars are real numbers. In later chapters other scalars will appear. Just keep in mind that, except for occasional reference to the geometry of real two- or three-dimensional space, everything in this chapter continues to hold when the scalars are complex numbers.

The complicated operation is *matrix multiplication*. The first case to learn is the product AB of a row vector A and a column vector B, which is defined when both are the same size, say m. If the entries of A and B are denoted by a_i and b_i, respectively, the product AB is the 1×1 matrix, or scalar,

(1.1.2)
$$a_1 b_1 + a_2 b_2 + \cdots + a_m b_m.$$

Thus

$$[1 \quad 3 \quad 5] \begin{bmatrix} 1 \\ -1 \\ 4 \end{bmatrix} = 1 - 3 + 20 = 18.$$

The usefulness of this definition becomes apparent when we regard A and B as vectors that represent indexed quantities. For example, consider a candy bar containing m ingredients. Let a_i denote the number of grams of $(ingredient)_i$ per bar, and let b_i denote the cost of $(ingredient)_i$ per gram. The matrix product AB computes the cost per bar:

$$(\text{grams/bar}) \cdot (\text{cost/gram}) = (\text{cost/bar}).$$

In general, the product of two matrices $A = (a_{ij})$ and $B = (b_{ij})$ is defined when the number of columns of A is equal to the number of rows of B. If A is an $\ell \times m$ matrix and B is an $m \times n$ matrix, then the product will be an $\ell \times n$ matrix. Symbolically,

$$(\ell \times m) \cdot (m \times n) = (\ell \times n).$$

The entries of the product matrix are computed by multiplying all rows of A by all columns of B, using the rule (1.1.2). If we denote the product matrix AB by $P = (p_{ij})$, then

(1.1.3)
$$p_{ij} = a_{i1} b_{1j} + a_{i2} b_{2j} + \cdots + a_{im} b_{mj}.$$

This is the product of the ith row of A and the jth column of B.

$$\begin{bmatrix} & & & \\ a_{i1} & \cdots & \cdots & a_{im} \\ & & & \end{bmatrix} \begin{bmatrix} b_{1j} \\ \cdot \\ \cdot \\ \cdot \\ b_{mj} \end{bmatrix} = \begin{bmatrix} & \cdot & \\ \cdot & p_{ij} & \cdot \\ & \cdot & \end{bmatrix}$$

For example,

(1.1.4)
$$\begin{bmatrix} 2 & 1 & 0 \\ 1 & 3 & 5 \end{bmatrix} \begin{bmatrix} 1 \\ -1 \\ 4 \end{bmatrix} = \begin{bmatrix} 1 \\ 18 \end{bmatrix}.$$

This definition of matrix multiplication has turned out to provide a very convenient computational tool. Going back to our candy bar example, suppose that there are ℓ candy bars. We may form the $\ell \times m$ matrix A whose ith row measures the ingredients of $(bar)_i$. If the cost is to be computed each year for n years, we may form the $m \times n$ matrix B whose jth column measures the cost of the ingredients in $(year)_j$. Again, the matrix product $AB = P$ computes cost per bar: $p_{ij} = \text{cost of } (bar)_i \text{ in } (year)_j$.

One reason for matrix notation is to provide a shorthand way of writing linear equations. The system of equations

$$
\begin{array}{rcl}
a_{11}x_1 + \cdots + a_{1n}x_n &=& b_1 \\
a_{21}x_1 + \cdots + a_{2n}x_n &=& b_2 \\
\vdots \qquad\qquad \vdots & & \vdots \\
a_{m1}x_1 + \cdots + a_{mn}x_n &=& b_m
\end{array}
$$

can be written in matrix notation as

(1.1.5) $$AX = B$$

where A denotes the matrix of coefficients, X and B are column vectors, and AX is the matrix product:

$$
\begin{bmatrix} \\ A \\ \end{bmatrix}
\begin{bmatrix} x_1 \\ \cdot \\ \cdot \\ \cdot \\ x_n \end{bmatrix}
=
\begin{bmatrix} b_1 \\ \cdot \\ \cdot \\ b_m \end{bmatrix}
$$

We may refer to an equation of this form simply as an "equation" or as a "system." The matrix equation

$$
\begin{bmatrix} 2 & 1 & 0 \\ 1 & 3 & 5 \end{bmatrix}
\begin{bmatrix} x_1 \\ x_2 \\ x_3 \end{bmatrix}
=
\begin{bmatrix} 1 \\ 18 \end{bmatrix}
$$

represents the following system of two equations in three unknowns:

$$
\begin{array}{rcl}
2x_1 + x_2 &=& 1 \\
x_1 + 3x_2 + 5x_3 &=& 18.
\end{array}
$$

Equation (1.1.4) exhibits one solution, $x_1 = 1$, $x_2 = -1$, $x_3 = 4$. There are others.

The sum (1.1.3) that defines the product matrix can also be written in summation or "sigma" notation as

(1.1.6) $$p_{ij} = \sum_{v=1}^{m} a_{iv}b_{vj} = \sum_{v} a_{iv}b_{vj}.$$

Each of these expressions for p_{ij} is a shorthand notation for the sum. The large sigma indicates that the terms with the indices $\nu = 1, \ldots, m$ are to be added up. The right-hand notation indicates that one should add the terms with all possible indices ν. It is assumed that the reader will understand that, if A is an $\ell \times m$ matrix and B is an $m \times n$ matrix, the indices should run from 1 to m. We've used the greek letter "nu," an uncommon symbol elsewhere, to distinguish the index of summation clearly.

Our two most important notations for handling sets of numbers are the summation notation, as used above, and matrix notation. The summation notation is the more versatile of the two, but because matrices are more compact, we use them whenever possible. One of our tasks in later chapters will be to translate complicated mathematical structures into matrix notation in order to be able to work with them conveniently.

Various *identities* are satisfied by the matrix operations. The *distributive laws*

(1.1.7) $$A(B + B') = AB + AB', \quad \text{and} \quad (A + A')B = AB + A'B$$

and the *associative law*

(1.1.8) $$(AB)C = A(BC)$$

are among them. These laws hold whenever the matrices involved have suitable sizes, so that the operations are defined. For the associative law, the sizes should be $A = \ell \times m$, $B = m \times n$, and $C = n \times p$, for some ℓ, m, n, p. Since the two products (1.1.8) are equal, parentheses are not necessary, and we will denote the triple product by ABC. It is an $\ell \times p$ matrix. For example, the two ways of computing the triple product

$$ABC = \begin{bmatrix} 1 \\ 2 \end{bmatrix} \begin{bmatrix} 1 & 0 & 1 \end{bmatrix} \begin{bmatrix} 2 & 0 \\ 1 & 1 \\ 0 & 1 \end{bmatrix}$$

are

$$(AB)C = \begin{bmatrix} 1 & 0 & 1 \\ 2 & 0 & 2 \end{bmatrix} \begin{bmatrix} 2 & 0 \\ 1 & 1 \\ 0 & 1 \end{bmatrix} = \begin{bmatrix} 2 & 1 \\ 4 & 2 \end{bmatrix} \quad \text{and} \quad A(BC) = \begin{bmatrix} 1 \\ 2 \end{bmatrix} \begin{bmatrix} 2 & 1 \end{bmatrix} = \begin{bmatrix} 2 & 1 \\ 4 & 2 \end{bmatrix}.$$

Scalar multiplication is compatible with matrix multiplication in the obvious sense:

(1.1.9) $$c(AB) = (cA)B = A(cB).$$

The proofs of these identities are straightforward and not very interesting.

However, the *commutative law* does not hold for matrix multiplication, that is,

(1.1.10) $$AB \neq BA, \quad \text{usually.}$$

Even when both matrices are square, the two products tend to be different. For instance,

$$\begin{bmatrix} 1 & 1 \\ 0 & 0 \end{bmatrix}\begin{bmatrix} 2 & 0 \\ 1 & 1 \end{bmatrix} = \begin{bmatrix} 3 & 1 \\ 0 & 0 \end{bmatrix}, \quad \text{while} \quad \begin{bmatrix} 2 & 0 \\ 1 & 1 \end{bmatrix}\begin{bmatrix} 1 & 1 \\ 0 & 0 \end{bmatrix} = \begin{bmatrix} 2 & 2 \\ 1 & 1 \end{bmatrix}.$$

If it happens that $AB = BA$, the two matrices are said to *commute*.

Since matrix multiplication isn't commutative, we must be careful when working with matrix equations. We can multiply both sides of an equation $B = C$ on the left by a matrix A, to conclude that $AB = AC$, provided that the products are defined. Similarly, if the products are defined, we can conclude that $BA = CA$. We cannot derive $AB = CA$ from $B = C$.

A matrix all of whose entries are 0 is called a *zero matrix*, and if there is no danger of confusion, it will be denoted simply by 0.

The entries a_{ii} of a matrix A are its *diagonal entries*. A matrix A is a *diagonal matrix* if its only nonzero entries are diagonal entries. (The word *nonzero* simply means "different from zero." It is ugly, but so convenient that we will use it frequently.)

The diagonal $n \times n$ matrix all of whose diagonal entries are equal to 1 is called the $n \times n$ *identity matrix*, and is denoted by I_n. It behaves like the number 1 in multiplication: If A is an $m \times n$ matrix, then

(1.1.11) $AI_n = A \quad \text{and} \quad I_m A = A.$

We usually omit the subscript and write I for I_n.

Here are some shorthand ways of depicting the identity matrix:

$$I = \begin{bmatrix} 1 & & 0 \\ & \ddots & \\ 0 & & 1 \end{bmatrix} = \begin{bmatrix} 1 & & \\ & \ddots & \\ & & 1 \end{bmatrix}.$$

We often indicate that a whole region in a matrix consists of zeros by leaving it blank or by putting in a single 0.

We use $*$ to indicate an arbitrary undetermined entry of a matrix. Thus

$$\begin{bmatrix} * & \cdots & * \\ & \ddots & \vdots \\ & & * \end{bmatrix}$$

may denote a square matrix A whose entries below the diagonal are 0, the other entries being undetermined. Such a matrix is called *upper triangular*. The matrices that appear in (1.1.14) below are upper triangular.

Let A be a (square) $n \times n$ matrix. If there is a matrix B such that

(1.1.12) $AB = I_n \quad \text{and} \quad BA = I_n,$

then B is called an *inverse* of A and is denoted by A^{-1}:

(1.1.13)
$$A^{-1}A = I = AA^{-1}.$$

A matrix A that has an inverse is called an *invertible* matrix.

For example, the matrix $A = \begin{bmatrix} 2 & 1 \\ 5 & 3 \end{bmatrix}$ is invertible. Its inverse is $A^{-1} = \begin{bmatrix} 3 & -1 \\ -5 & 2 \end{bmatrix}$, as can be seen by computing the products AA^{-1} and $A^{-1}A$. Two more examples:

(1.1.14)
$$\begin{bmatrix} 1 & \\ & 2 \end{bmatrix}^{-1} = \begin{bmatrix} 1 & \\ & \frac{1}{2} \end{bmatrix} \quad \text{and} \quad \begin{bmatrix} 1 & 1 \\ & 1 \end{bmatrix}^{-1} = \begin{bmatrix} 1 & -1 \\ & 1 \end{bmatrix}.$$

We will see later that a square matrix A is invertible if there is a matrix B such that either one of the two relations $AB = I_n$ or $BA = I_n$ holds, and that B is then the inverse (see (1.2.20)). But since multiplication of matrices isn't commutative, this fact is not obvious. On the other hand, an inverse is unique if it exists. The next lemma shows that there can be only one inverse of a matrix A:

Lemma 1.1.15 Let A be a square matrix that has a *right inverse*, a matrix R such that $AR = I$ and also a *left inverse*, a matrix L such that $LA = I$. Then $R = L$. So A is invertible and R is its inverse.

Proof. $R = IR = (LA)R = L(AR) = LI = L.$ $\qquad\qquad\qquad\qquad\qquad\qquad\qquad\qquad$ □

Proposition 1.1.16 Let A and B be invertible $n \times n$ matrices. The product AB and the inverse A^{-1} are invertible, $(AB)^{-1} = B^{-1}A^{-1}$ and $(A^{-1})^{-1} = A$. If $A_1, \ldots, A_m$ are invertible $n \times n$ matrices, the product $A_1 \cdots A_m$ is invertible, and its inverse is $A_m^{-1} \cdots A_1^{-1}$.

Proof. Assume that A and B are invertible. To show that the product $B^{-1}A^{-1} = Q$ is the inverse of $AB = P$, we simplify the products PQ and QP, obtaining I in both cases. The verification of the other assertions is similar. $\qquad\qquad\qquad\qquad\qquad\qquad$ □

The inverse of $\begin{bmatrix} 1 & \\ & 2 \end{bmatrix}\begin{bmatrix} 1 & 1 \\ & 1 \end{bmatrix} = \begin{bmatrix} 1 & 1 \\ & 2 \end{bmatrix}$ is $\begin{bmatrix} 1 & -1 \\ & 1 \end{bmatrix}\begin{bmatrix} 1 & \\ & \frac{1}{2} \end{bmatrix} = \begin{bmatrix} 1 & -\frac{1}{2} \\ & \frac{1}{2} \end{bmatrix}.$

• It is worthwhile to memorize the inverse of a 2×2 matrix:

(1.1.17)
$$\begin{bmatrix} a & b \\ c & d \end{bmatrix}^{-1} = \frac{1}{ad - bc}\begin{bmatrix} d & -b \\ -c & a \end{bmatrix}.$$

The denominator $ad - bc$ is the *determinant* of the matrix. If the determinant is zero, the matrix is not invertible. We discuss determinants in Section 1.4.

Though this isn't clear from the definition of matrix multiplication, we will see that most square matrices are invertible, though finding the inverse explicitly is not a simple problem when the matrix is large. The set of all invertible $n \times n$ matrices is called the n-dimensional *general linear group*. It will be one of our most important examples when we introduce the basic concept of a group in the next chapter.

For future reference, we note the following lemma:

Lemma 1.1.18 A square matrix that has either a row of zeros or a column of zeros is not invertible.

Proof. If a row of an $n \times n$ matrix A is zero and if B is any other $n \times n$ matrix, then the corresponding row of the product AB is zero too. So AB is not the identity. Therefore A has no right inverse. A similar argument shows that if a column of A is zero, then A has no left inverse. $\qquad\square$

Block Multiplication

Various tricks simplify matrix multiplication in favorable cases; block multiplication is one of them. Let M and M' be $m \times n$ and $n \times p$ matrices, and let r be an integer less than n. We may decompose the two matrices into blocks as follows:

$$M = [A|B] \quad \text{and} \quad M' = \left[\frac{A'}{B'}\right],$$

where A has r columns and A' has r rows. Then the matrix product can be computed as

(1.1.19) $$MM' = AA' + BB'.$$

Notice that this formula is the same as the rule for multiplying a row vector and a column vector.

We may also multiply matrices divided into four blocks. Suppose that we decompose an $m \times n$ matrix M and an $n \times p$ matrix M' into rectangular submatrices

$$M = \left[\begin{array}{c|c} A & B \\ \hline C & D \end{array}\right], \quad M' = \left[\begin{array}{c|c} A' & B' \\ \hline C' & D' \end{array}\right],$$

where the number of columns of A and C are equal to the number of rows of A' and B'. In this case the rule for block multiplication is the same as for multiplication of 2×2 matrices:

(1.1.20) $$\left[\begin{array}{c|c} A & B \\ \hline C & D \end{array}\right]\left[\begin{array}{c|c} A' & B' \\ \hline C' & D' \end{array}\right] = \left[\begin{array}{c|c} AA' + BC' & AB' + BD' \\ \hline CA' + DC' & CB' + DD' \end{array}\right]$$

These rules can be verified directly from the definition of matrix multiplication.

Please use block multiplication to verify the equation

$$\left[\begin{array}{cc|c} 1 & 0 & 5 \\ 0 & 1 & 3 \end{array}\right] \left[\begin{array}{cc|cc} 2 & 3 & 1 & 1 \\ 4 & 8 & 0 & 0 \\ \hline 1 & 0 & 1 & 0 \end{array}\right] = \left[\begin{array}{cc|cc} 7 & 3 & 6 & 1 \\ 7 & 8 & 3 & 0 \end{array}\right].$$

Besides facilitating computations, block multiplication is a useful tool for proving facts about matrices by induction.

Matrix Units

The matrix units are the simplest nonzero matrices. The $m \times n$ matrix unit e_{ij} has a 1 in the i, j position as its only nonzero entry:

(1.1.21)
$$e_{ij} = \begin{array}{c} \\ i \end{array} \overset{j}{\left[\begin{array}{ccc} & \vdots & \\ \cdots & 1 & \cdots \\ & \vdots & \end{array}\right]}$$

We usually denote matrices by uppercase (capital) letters, but the use of a lowercase letter for a matrix unit is traditional.

• The set of matrix units is called a *basis* for the space of all $m \times n$ matrices, because every $m \times n$ matrix $A = (a_{ij})$ is a *linear combination* of the matrices e_{ij}:

(1.1.22)
$$A = a_{11}e_{11} + a_{12}e_{12} + \cdots = \sum_{i,j} a_{ij} e_{ij}.$$

The indices i, j under the sigma mean that the sum is to be taken over all $i = 1, \ldots, m$ and all $j = 1, \ldots, n$. For instance,

$$\left[\begin{array}{cc} 3 & 2 \\ 1 & 4 \end{array}\right] = 3 \left[\begin{array}{cc} 1 & \\ & \end{array}\right] + 2 \left[\begin{array}{cc} & 1 \\ & \end{array}\right] + 1 \left[\begin{array}{cc} & \\ 1 & \end{array}\right] + 4 \left[\begin{array}{cc} & \\ & 1 \end{array}\right] = 3e_{11} + 2e_{12} + 1e_{21} + 4e_{22}.$$

The product of an $m \times n$ matrix unit e_{ij} and an $n \times p$ matrix unit $e_{j\ell}$ is given by the formulas

(1.1.23)
$$e_{ij} e_{j\ell} = e_{i\ell} \quad \text{and} \quad e_{ij} e_{k\ell} = 0 \text{ if } j \neq k$$

• The column vector e_i, which has a single nonzero entry 1 in the position i, is analogous to a matrix unit, and the set $\{e_1, \ldots, e_n\}$ of these vectors forms what is called the *standard basis* of the n-dimensional space $\mathbb{R}^n$ (see Chapter 3, (3.4.15)). If X is a column vector with entries $(x_1, \ldots, x_n)$, then

(1.1.24)
$$X = x_1 e_1 + \cdots + x_n e_n = \sum_i x_i e_i.$$

The formulas for multiplying matrix units and standard basis vectors are

(1.1.25) $$e_{ij}\,e_j = e_i, \quad \text{and} \quad e_{ij}\,e_k = 0 \ \text{ if } \ j \neq k.$$

1.2 ROW REDUCTION

Left multiplication by an $n \times n$ matrix A on $n \times p$ matrices, say

(1.2.1) $$AX = Y,$$

can be computed by operating on the rows of X. If we let X_i and Y_i denote the ith rows of X and Y, respectively, then in vector notation,

(1.2.2) $$Y_i = a_{i1}X_1 + \cdots + a_{in}X_n,$$

$$
A
\begin{bmatrix}
-X_1- \\
-X_2- \\
\vdots \\
-X_n-
\end{bmatrix}
=
\begin{bmatrix}
-Y_1- \\
-Y_2- \\
\vdots \\
-Y_n-
\end{bmatrix}.
$$

For instance, the bottom row of the product

$$
\begin{bmatrix} 0 & 1 \\ -2 & 3 \end{bmatrix}
\begin{bmatrix} 1 & 2 & 1 \\ 1 & 3 & 0 \end{bmatrix}
=
\begin{bmatrix} 1 & 3 & 0 \\ 1 & 5 & -2 \end{bmatrix}
$$

can be computed as $-2[1 \ \ 2 \ \ 1] + 3[1 \ \ 3 \ \ 0] = [1 \ \ 5 \ \ -2]$.

Left multiplication by an invertible matrix is called a *row operation*. We discuss these row operations next. Some square matrices called *elementary matrices* are used. There are three types of elementary 2×2 matrices:

(1.2.3) (i) $\begin{bmatrix} 1 & a \\ 0 & 1 \end{bmatrix}$ or $\begin{bmatrix} 1 & 0 \\ a & 1 \end{bmatrix}$, (ii) $\begin{bmatrix} 0 & 1 \\ 1 & 0 \end{bmatrix}$, (iii) $\begin{bmatrix} c & \\ & 1 \end{bmatrix}$ or $\begin{bmatrix} 1 & \\ & c \end{bmatrix}$,

where a can be any scalar and c can be any nonzero scalar.

There are also three types of elementary $n \times n$ matrices. They are obtained by splicing the elementary 2×2 matrices symmetrically into an identity matrix. They are shown below with a 5×5 matrix to save space, but the size is supposed to be arbitrary.

(1.2.4)

Type (i):

$$
i \begin{bmatrix} 1 & & & & \\ & 1 & a & & \\ & & 1 & & \\ & & & 1 & \\ & & & & 1 \end{bmatrix}
\quad \text{or} \quad
j \begin{bmatrix} 1 & & & & \\ & 1 & & & \\ & & 1 & 1 & \\ & a & & 1 & \\ & & & & 1 \end{bmatrix}
\quad (i \neq j).
$$

One nonzero off-diagonal entry is added to the identity matrix.

Type (ii):

$$
\begin{matrix} & i & j \end{matrix}
$$
$$
\begin{matrix} i \\ \\ j \end{matrix}
\begin{bmatrix} 1 & & & & \\ & 0 & & 1 & \\ & & 1 & & \\ & 1 & & 0 & \\ & & & & 1 \end{bmatrix} .
$$

The ith and jth diagonal entries of the identity matrix are replaced by zero, and 1's are added in the (i, j) and (j, i) positions.

Type (iii):

$$
i \begin{bmatrix} 1 & & & \\ & 1 & & \\ & & c & \\ & & & 1 \\ & & & & 1 \end{bmatrix}
\quad (c \neq 0).
$$

One diagonal entry of the identity matrix is replaced by a nonzero scalar c.

• The elementary matrices E operate on a matrix X this way: To get the matrix EX, you must:

(1.2.5)   Type(i): with a in the i, j position, "add $a \cdot$(*row j*) of X to (*row i*), "
       Type(ii): "interchange (*row i*) and (*row j*) of X,"
       Type(iii): "multiply (*row i*) of X by a nonzero scalar c."

These are the *elementary row operations*. Please verify the rules.

Lemma 1.2.6 Elementary matrices are invertible, and their inverses are also elementary matrices.

Proof. The inverse of an elementary matrix is the matrix corresponding to the inverse row operation: "subtract $a \cdot$(*row j*) from (row i)," "interchange (*row* i) and (*row* j)" again, or "multiply (*row i*) by c^{-1}." $\qquad\square$

We now perform elementary row operations (1.2.5) on a matrix M, with the aim of ending up with a simpler matrix:

$$M \xrightarrow{\text{sequence of operations}} \longrightarrow \cdots \longrightarrow M'.$$

Since each elementary operation is obtained by multiplying by an elementary matrix, we can express the result of a sequence of such operations as multiplication by a sequence $E_1, \ldots, E_k$ of elementary matrices:

$$(1.2.7) \qquad\qquad M' = E_k \cdots E_2 E_1 M.$$

This procedure to simplify a matrix is called *row reduction*.

As an example, we use elementary operations to simplify a matrix by clearing out as many entries as possible, working from the left.

$$(1.2.8) \qquad M = \begin{bmatrix} 1 & 1 & 2 & 1 & 5 \\ 1 & 1 & 2 & 6 & 10 \\ 1 & 2 & 5 & 2 & 7 \end{bmatrix} \to \to \begin{bmatrix} 1 & 1 & 2 & 1 & 5 \\ 0 & 0 & 0 & 5 & 5 \\ 0 & 1 & 3 & 1 & 2 \end{bmatrix} \to$$

$$\begin{bmatrix} 1 & 1 & 2 & 1 & 5 \\ 0 & 1 & 3 & 1 & 2 \\ 0 & 0 & 0 & 5 & 5 \end{bmatrix} \to \to \begin{bmatrix} 1 & 0 & -1 & 0 & 3 \\ 0 & 1 & 3 & 1 & 2 \\ 0 & 0 & 0 & 1 & 1 \end{bmatrix} \to \begin{bmatrix} 1 & 0 & -1 & 0 & 3 \\ 0 & 1 & 3 & 0 & 1 \\ 0 & 0 & 0 & 1 & 1 \end{bmatrix} = M'.$$

The matrix M' cannot be simplified further by row operations.

Here is the way that row reduction is used to solve systems of linear equations. Suppose we are given a system of m equations in n unknowns, say $AX = B$, where A is an $m \times n$ matrix, B is a given column vector, and X is an unknown column vector. To solve this system, we form the $m \times (n + 1)$ block matrix, sometimes called the *augmented matrix*

$$(1.2.9) \qquad M = [A|B] = \begin{bmatrix} a_{11} & \cdots & a_{1n} & b_1 \\ & \vdots & \vdots & \vdots \\ a_{m1} & \cdots & a_{mn} & b_n \end{bmatrix},$$

and we perform row operations to simplify M. Note that $EM = [EA|EB]$. Let

$$M' = [A'|B']$$

be the result of a sequence of row operations. The key observation is this:

Proposition 1.2.10 The systems $A'X = B'$ and $AX = B$ have the same solutions.

Proof. Since M' is obtained by a sequence of elementary row operations, there are elementary matrices $E_1, \ldots, E_k$ such that, with $P = E_k \cdots E_1$,

$$M' = E_k \cdots E_1 M = PM.$$

The matrix P is invertible, and $M' = [A'|B'] = [PA|PB]$. If X is a solution of the original equation $AX = B$, we multiply by P on the left: $PAX = PB$, which is to say, $A'X = B'$. So X also solves the new equation. Conversely, if $A'X = B'$, then $P^{-1}A'X = P^{-1}B'$, that is, $AX = B$.   □

For example, consider the system

$$(1.2.11) \qquad \begin{aligned} x_1 + x_2 + 2x_3 + x_4 &= 5 \\ x_1 + x_2 + 2x_3 + 6x_4 &= 10 \\ x_1 + 2x_2 + 5x_3 + 2x_4 &= 7. \end{aligned}$$

Its augmented matrix is the matrix whose row reduction is shown above. The system of equations is equivalent to the one defined by the end result M' of the reduction:

$$\begin{aligned} x_1 \phantom{{}+x_2} - x_3 \phantom{{}+3x_3} &= 3 \\ x_2 + 3x_3 &= 1 \\ x_4 &= 1. \end{aligned}$$

We can read off the solutions of this system easily: If we choose $x_3 = c$ arbitrarily, we can solve for x_1, x_2, and x_4. The general solution of (1.2.11) can be written in the form

$$x_3 = c, \quad x_1 = 3 + c, \quad x_2 = 1 - 3c, \quad x_4 = 1,$$

where c is arbitrary.

We now go back to row reduction of an arbitrary matrix. It is not hard to see that, by a sequence of row operations, any matrix M can be reduced to what is called a *row echelon matrix*. The end result of our reduction of (1.2.8) is an example. Here is the definition: A *row echelon matrix* is a matrix that has these properties:

(1.2.12)

(a) If (*row i*) of M is zero, then (*row j*) is zero for all $j > i$.

(b) If (*row i*) isn't zero, its first nonzero entry is 1. This entry is called a *pivot*.

(c) If (*row* $(i+1)$) isn't zero, the pivot in (*row* $(i+1)$) is to the right of the pivot in (*row i*).

(d) The entries above a pivot are zero. (The entries below a pivot are zero too, by **(c)**.)

The pivots in the matrix M' of (1.2.8) and in the examples below are shown in boldface.

To make a row reduction, find the first column that contains a nonzero entry, say m. (If there is none, then M is zero, and is itself a row echelon matrix.) Interchange rows using an elementary operation of Type (ii) to move m to the top row. Normalize m to 1 using an operation of Type (iii). This entry becomes a pivot. Clear out the entries below this pivot by a sequence of operations of Type (i). The resulting matrix will have the block form

$$\begin{bmatrix} 0\cdots 0 & 1 & * & \cdots & * \\ 0\cdots 0 & 0 & * & \cdots & * \\ \vdots & \vdots & \vdots & & \vdots \\ 0\cdots 0 & 0 & * & \cdots & * \end{bmatrix} , \text{ which we write as } \begin{bmatrix} & 1 & B_1 \\ \hline & & D_1 \end{bmatrix} = M_1.$$

We now perform row operations to simplify the smaller matrix D_1. Because the blocks to the left of D_1 are zero, these operations will have no effect on the rest of the matrix M_1. By induction on the number of rows, we may assume that D_1 can be reduced to a row echelon matrix, say to D_2, and M_1 is thereby reduced to the matrix

$$\begin{bmatrix} & 1 & B_1 \\ \hline & & D_2 \end{bmatrix} = M_2.$$

This matrix satisfies the first three requirements for a row echelon matrix. The entries in B_1 above the pivots of D_2 can be cleared out at this time, to finish the reduction to row echelon form. □

It can be shown that the row echelon matrix obtained from a matrix M by row reduction doesn't depend on the particular sequence of operations used in the reduction. Since this point will not be important for us, we omit the proof.

As we said before, row reduction is useful because one can solve a system of equations $A'X = B'$ easily when A' is in row echelon form. Another example: Suppose that

$$[A'|B'] = \begin{bmatrix} 1 & 6 & 0 & 1 & | & 1 \\ 0 & 0 & 1 & 2 & | & 3 \\ 0 & 0 & 0 & 0 & | & 1 \end{bmatrix}.$$

There is no solution to $A'X = B'$ because the third equation is $0 = 1$. On the other hand,

$$[A'|B'] = \begin{bmatrix} 1 & 6 & 0 & 1 & | & 1 \\ 0 & 0 & 1 & 2 & | & 3 \\ 0 & 0 & 0 & 0 & | & 0 \end{bmatrix}$$

has solutions. Choosing $x_2 = c$ and $x_4 = c'$ arbitrarily, we can solve the first equation for x_1 and the second for x_3. The general rule is this:

Proposition 1.2.13 Let $M' = [A'|B']$ be a block row echelon matrix, where B' is a column vector. The system of equations $A'X = B'$ has a solution if and only if there is no pivot in the last column B'. In that case, arbitrary values can be assigned to the unknown x_i, provided that (*column i*) does not contain a pivot. When these arbitrary values are assigned, the other unknowns are determined uniquely.   □

Every *homogeneous* linear equation $AX = 0$ has the *trivial* solution $X = 0$. But looking at the row echelon form again, we conclude that if there are more unknowns than equations then the homogeneous equation $AX = 0$ has a *nontrivial* solution.

Corollary 1.2.14 Every system $AX = 0$ of m homogeneous equations in n unknowns, with $m < n$, has a solution X in which some x_i is nonzero.

Proof. Row reduction of the block matrix $[A|0]$ yields a matrix $[A'|0]$ in which A' is in row echelon form. The equation $A'X = 0$ has the same solutions as $AX = 0$. The number, say r, of pivots of A' is at most equal to the number m of rows, so it is less than n. The proposition tells us that we may assign arbitrary values to $n - r$ variables x_i.   □

We now use row reduction to characterize invertible matrices.

Lemma 1.2.15 A square row echelon matrix M is either the identity matrix I, or else its bottom row is zero.

Proof. Say that M is an $n \times n$ row echelon matrix. Since there are n columns, there are at most n pivots, and if there are n of them, there has to be one in each column. In this case, $M = I$. If there are fewer than n pivots, then some row is zero, and the bottom row is zero too.   □

Theorem 1.2.16 Let A be a square matrix. The following conditions are equivalent:

(a) A can be reduced to the identity by a sequence of elementary row operations.

(b) A is a product of elementary matrices.

(c) A is invertible.

Proof. We prove the theorem by proving the implications **(a)** $\Rightarrow$ **(b)** $\Rightarrow$ **(c)** $\Rightarrow$ **(a)**. Suppose that A can be reduced to the identity by row operations, say $E_k \cdots E_1 A = I$. Multiplying both sides of this equation on the left by $E_1^{-1} \cdots E_k^{-1}$, we obtain $A = E_1^{-1} \cdots E_k^{-1}$. Since the inverse of an elementary matrix is elementary, **(b)** holds, and therefore **(a)** implies **(b)**. Because a product of invertible matrices is invertible, **(b)** implies **(c)**. Finally, we prove the implication **(c)** $\Rightarrow$ **(a)**. If A is invertible, so is the end result A' of its row reduction. Since an invertible matrix cannot have a row of zeros, Lemma 1.2.15 shows that A' is the identity.   □

Row reduction provides a method to compute the inverse of an invertible matrix A: We reduce A to the identity by row operations: $E_k \cdots E_1 A = I$ as above. Multiplying both sides of this equation on the right by A^{-1},

$$E_k \cdots E_1 I = E_k \cdots E_1 = A^{-1}.$$

Corollary 1.2.17 Let A be an invertible matrix. To compute its inverse, one may apply elementary row operations $E_1, \ldots, E_k$ to A, reducing it to the identity matrix. The same sequence of operations, when applied to the identity matrix I, yields A^{-1}.   □

Example 1.2.18 We invert the matrix $A = \begin{bmatrix} 1 & 5 \\ 2 & 6 \end{bmatrix}$. To do this, we form the 2×4 block matrix

$$[A|I] = \begin{bmatrix} 1 & 5 & | & 1 & 0 \\ 2 & 6 & | & 0 & 1 \end{bmatrix}.$$

We perform row operations to reduce A to the identity, carrying the right side along, and thereby end up with A^{-1} on the right.

$$[A|I] = \begin{bmatrix} 1 & 5 & | & 1 & 0 \\ 2 & 6 & | & 0 & 1 \end{bmatrix} \longrightarrow \begin{bmatrix} 1 & 5 & | & 1 & 0 \\ 0 & -4 & | & -2 & 1 \end{bmatrix} \longrightarrow$$

(1.2.19)
$$\begin{bmatrix} 1 & 5 & | & 1 & 0 \\ 0 & 1 & | & \frac{1}{2} & -\frac{1}{4} \end{bmatrix} \longrightarrow \begin{bmatrix} 1 & 0 & | & -\frac{3}{2} & \frac{5}{4} \\ 0 & 1 & | & \frac{1}{2} & -\frac{1}{4} \end{bmatrix} = [I|A^{-1}]. \qquad \square$$

Proposition 1.2.20 Let A be a square matrix that has either a left inverse or a right inverse, a matrix B such that either $BA = I$ or $AB = I$. Then A is invertible, and B is its inverse.

Proof. Suppose that $AB = I$. We perform row reduction on A. Say that $A' = PA$, where $P = E_k \cdots E_1$ is the product of the corresponding elementary matrices, and A' is a row echelon matrix. Then $A'B = PAB = P$. Because P is invertible, its bottom row isn't zero. Then the bottom row of A' can't be zero either. Therefore A' is the identity matrix (1.2.15), and so P is a left inverse of A. Then A has both a left inverse and a right inverse, so it is invertible and B is its inverse.

   If $BA = I$, we interchange the roles of A and B in the above reasoning. We find that B is invertible and that its inverse is A. Then A is invertible, and its inverse is B. $\qquad \square$

   We come now to the main theorem about square systems of linear equations:

Theorem 1.2.21 Square Systems. The following conditions on a square matrix A are equivalent:

(a) A is invertible.

(b) The system of equations $AX = B$ has a unique solution for every column vector B.

(c) The system of homogeneous equations $AX = 0$ has only the trivial solution $X = 0$.

Proof. Given the system $AX = B$, we reduce the augmented matrix $[A|B]$ to row echelon form $[A'|B']$. The system $A'X = B'$ has the same solutions. If A is invertible, then A' is the identity matrix, so the unique solution is $X = B'$. This shows that **(a)** $\Rightarrow$ **(b)**.

   If an $n \times n$ matrix A is not invertible, then A' has a row of zeros. One of the equations making up the system $A'X = 0$ is the trivial equation. So there are fewer than n pivots.

The homogeneous system $A'X = 0$ has a nontrivial solution (1.2.13), and so does $AX = 0$ (1.2.14). This shows that if **(a)** fails, then **(c)** also fails, hence that **(c)** $\Rightarrow$ **(a)**.

Finally, it is obvious that **(b)** $\Rightarrow$ **(c)**. $\qquad\qquad\qquad\qquad\qquad\qquad\qquad$ □

We want to take particular note of the implication **(c)** $\Rightarrow$ **(b)** of the theorem:

If the homogeneous equation $AX = 0$ has only the trivial solution,
then the general equation $AX = B$ has a unique solution for every column vector B.

This can be useful because the homogeneous system may be easier to handle than the general system.

Example 1.2.22 There exists a polynomial $p(t)$ of degree n that takes prescribed values, say $p(a_i) = b_i$, at $n + 1$ distinct points $t = a_0, \ldots, a_n$ on the real line.[2] To find this polynomial, one must solve a system of linear equations in the undetermined coefficients of $p(t)$. In order not to overload the notation, we'll do the case $n = 2$, so that

$$p(t) = x_0 + x_1 t + x_2 t^2.$$

Let a_0, a_1, a_2 and b_0, b_1, b_2 be given. The equations to be solved are obtained by substituting a_i for t. Moving the coefficients x_i to the right, they are

$$x_0 + a_i x_1 + a_i^2 x_2 = b_i$$

for $i = 0, 1, 2$. This is a system $AX = B$ of three linear equations in the three unknowns x_0, x_1, x_2, with

$$\begin{bmatrix} 1 & a_0 & a_0^2 \\ 1 & a_1 & a_1^2 \\ 1 & a_2 & a_2^2 \end{bmatrix}$$

The homogeneous equation, in which $B = 0$, asks for a polynomial with 3 roots a_0, a_1, a_2. A nonzero polynomial of degree 2 can have at most two roots, so the homogeneous equation has only the trivial solution. Therefore there is a unique solution for every set of prescribed values b_0, b_1, b_2.

By the way, there is a formula, the *Lagrange Interpolation Formula*, that exhibits the polynomial $p(t)$ explicitly. $\qquad\qquad\qquad\qquad\qquad\qquad\qquad\qquad\qquad$ □

1.3   THE MATRIX TRANSPOSE

In the discussion of the previous section, we chose to work with rows in order to apply the results to systems of linear equations. One may also perform column operations to simplify a matrix, and it is evident that similar results will be obtained.

---

[2]Elements of a set are said to be *distinct* if no two of them are equal.

Rows and columns are interchanged by the transpose operation on matrices. The *transpose* of an $m \times n$ matrix A is the $n \times m$ matrix A^t obtained by reflecting about the diagonal: $A^t = (b_{ij})$, where $b_{ij} = a_{ji}$. For instance,

$$\begin{bmatrix} 1 & 2 \\ 3 & 4 \end{bmatrix}^t = \begin{bmatrix} 1 & 3 \\ 2 & 4 \end{bmatrix} \quad \text{and} \quad \begin{bmatrix} 1 & 2 & 3 \end{bmatrix}^t = \begin{bmatrix} 1 \\ 2 \\ 3 \end{bmatrix}.$$

Here are the rules for computing with the transpose:

(1.3.1) $(AB)^t = B^t A^t, \quad (A+B)^t = A^t + B^t, \quad (cA)^t = cA^t, \quad (A^t)^t = A.$

Using the first of these formulas, we can deduce facts about *right multiplication* from the corresponding facts about left multiplication. The elementary matrices (1.2.4) act by right multiplication AE as the following *elementary column operations*

(1.3.2)   "with a in the i, j position, add $a \cdot (column\ i)$ to $(column\ j)$";

"interchange $(column\ i)$ and $(column\ j)$";

"multiply $(column\ i)$ by a nonzero scalar c."

Note that in the first of these operations, the indices i, j are the reverse of those in (1.2.5a).

1.4   DETERMINANTS

Every square matrix A has a number associated to it called its determinant, and denoted by $\det A$. We define the determinant and derive some of its properties here.

The determinant of a 1×1 matrix is equal to its single entry

(1.4.1) $$\det [a] = a,$$

and the determinant of a 2×2 matrix is given by the formula

(1.4.2) $$\det \begin{bmatrix} a & b \\ c & d \end{bmatrix} = ad - bc.$$

The determinant of a 2×2 matrix A has a geometric interpretation. Left multiplication by A maps the space $\mathbb{R}^2$ of real two-dimensional column vectors to itself, and the area of the parallelogram that forms the image of the unit square via this map is the absolute value of the determinant of A. The determinant is positive or negative, according to whether the orientation of the square is preserved or reversed by the operation. Moreover, $\det A = 0$ if and only if the parallelogram degenerates to a line segment or a point, which happens when the columns of the matrix are proportional.

A picture of this operation, in which the matrix is $\begin{bmatrix} 3 & 2 \\ 1 & 4 \end{bmatrix}$, is shown on the following page. The shaded region is the image of the unit square under the map. Its area is 10.

This geometric interpretation extends to higher dimensions. Left multiplication by a 3×3 real matrix A maps the space $\mathbb{R}^3$ of three-dimensional column vectors to itself, and the absolute value of its determinant is the volume of the image of the unit cube.

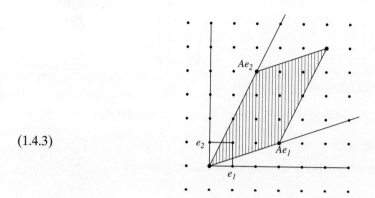

(1.4.3)

The set of all real $n \times n$ matrices forms a space of dimension n^2 that we denote by $\mathbb{R}^{n \times n}$. We regard the determinant of $n \times n$ matrices as a function from this space to the real numbers:

$$\det : \mathbb{R}^{n \times n} \to \mathbb{R}.$$

The determinant of an $n \times n$ matrix is a function of its n^2 entries. There is one such function for each positive integer n. Unfortunately, there are many formulas for these determinants, and all of them are complicated when n is large. Not only are the formulas complicated, but it may not be easy to show directly that two of them define the same function.

We use the following strategy: We choose one of the formulas, and take it as our definition of the determinant. In that way we are talking about a particular function. We show that our chosen function is the *only* one having certain special properties. Then, to show that another formula defines the same determinant function, one needs only to check those properties for the other function. This is often not too difficult.

We use a formula that computes the determinant of an $n \times n$ matrix in terms of certain $(n-1) \times (n-1)$ determinants by a process called *expansion by minors*. The determinants of submatrices of a matrix are called *minors*. Expansion by minors allows us to give a recursive definition of the determinant.

The word *recursive* means that the definition of the determinant for $n \times n$ matrices makes use of the determinant for $(n-1) \times (n-1)$ matrices. Since we have defined the determinant for 1×1 matrices, we will be able to use our recursive definition to compute 2×2 determinants, then knowing this, to compute 3×3 determinants, and so on.

Let A be an $n \times n$ matrix and let A_{ij} denote the $(n-1) \times (n-1)$ submatrix obtained by crossing out the ith row and the jth column of A:

(1.4.4)

$$i \;\begin{array}{c} j \\ \boxed{} \end{array}\; = A_{ij}.$$

For example, if

$$A = \begin{bmatrix} 1 & 0 & 3 \\ 2 & 1 & 2 \\ 0 & 5 & 1 \end{bmatrix}, \quad \text{then} \quad A_{21} = \begin{bmatrix} 0 & 3 \\ 5 & 1 \end{bmatrix}.$$

- Expansion by minors on the first column is the formula

(1.4.5) $\det A = a_{11} \det A_{11} - a_{21} \det A_{21} + a_{31} \det A_{31} - \cdots \pm a_{n1} \det A_{n1}.$

The signs alternate, beginning with $+$.

It is useful to write this expansion in summation notation:

(1.4.6) $$\det A = \sum_{\nu} \pm a_{\nu 1} \det A_{\nu 1}.$$

The alternating sign can be written as $(-1)^{\nu+1}$. It will appear again. We take this formula, together with (1.4.1), as a *recursive definition of the determinant*.

For 1×1 and 2×2 matrices, this formula agrees with (1.4.1) and (1.4.2). The determinant of the 3×3 matrix A shown above is

$$\det A = 1 \cdot \det \begin{bmatrix} 1 & 2 \\ 5 & 1 \end{bmatrix} - 2 \cdot \det \begin{bmatrix} 0 & 3 \\ 5 & 1 \end{bmatrix} + 0 \cdot \det \begin{bmatrix} 0 & 3 \\ 1 & 2 \end{bmatrix} = 1 \cdot (-9) - 2 \cdot (-15) = 21.$$

Expansions by minors on other columns and on rows, which we define in Section 1.6, are among the other formulas for the determinant.

It is important to know the many special properties satisfied by determinants. We present some of these properties here, deferring proofs to the end of the section. Because we want to apply the discussion to other formulas, the properties will be stated for an unspecified function δ.

Theorem 1.4.7 Uniqueness of the Determinant. There is a unique function δ on the space of $n \times n$ matrices with the properties below, namely the determinant (1.4.5).

(i) With I denoting the identity matrix, $\delta(I) = 1$.

(ii) δ is linear in the rows of the matrix A.

(iii) If two adjacent rows of a matrix A are equal, then $\delta(A) = 0$.

The statement that δ is linear in the rows of a matrix means this: Let A_i denote the ith row of a matrix A. Let A, B, D be three matrices, all of whose entries are equal, except for those in the rows indexed by k. Suppose furthermore that $D_k = cA_k + c'B_k$ for some scalars c and c'. Then $\delta(D) = c\,\delta(A) + c'\delta(B)$:

(1.4.8) $$\delta \begin{bmatrix} \vdots \\ cA_i + c'B_i \\ \vdots \end{bmatrix} = c\,\delta \begin{bmatrix} \vdots \\ \text{---}\,A_i\,\text{---} \\ \vdots \end{bmatrix} + c'\,\delta \begin{bmatrix} \vdots \\ \text{---}\,B_i\,\text{---} \\ \vdots \end{bmatrix}.$$

This allows us to operate on one row at a time, the other rows being left fixed. For example, since $[0 \quad 2 \quad 3] = 2[0 \quad 1 \quad 0] + 3[0 \quad 0 \quad 1]$,

$$\delta \begin{bmatrix} 1 & & \\ & 2 & 3 \\ & & 1 \end{bmatrix} = 2\delta \begin{bmatrix} 1 & & \\ & 1 & \\ & & 1 \end{bmatrix} + 3\delta \begin{bmatrix} 1 & & \\ & & 1 \\ & & 1 \end{bmatrix} = 2 \cdot 1 + 3 \cdot 0 = 2.$$

Perhaps the most important property of the determinant is its compatibility with matrix multiplication.

Theorem 1.4.9 Multiplicative Property of the Determinant. For any $n \times n$ matrices A and B, $\det(AB) = (\det A)(\det B)$.

The next theorem gives additional properties that are implied by those listed in (1.4.7).

Theorem 1.4.10 Let δ be a function on $n \times n$ matrices that has the properties (1.4.7)**(i,ii,iii)**. Then

(a) If A' is obtained from A by adding a multiple of (*row j*) of A to (*row i*) and $i \neq j$, then $\delta(A') = \delta(A)$.

(b) If A' is obtained by interchanging (*row i*) and (*row j*) of A and $i \neq j$, then $\delta(A') = -\delta(A)$.

(c) If A' is obtained from A by multiplying (*row i*) by a scalar c, then $\delta(A') = c\,\delta(A)$. If a row of a matrix A is equal to zero, then $\delta(A) = 0$.

(d) If (*row i*) of A is equal to a multiple of (*row j*) and $i \neq j$, then $\delta(A) = 0$.

We now proceed to prove the three theorems stated above, in reverse order. The fact that there are quite a few points to be examined makes the proofs lengthy. This can't be helped.

Proof of Theorem 1.4.10. The first assertion of **(c)** is a part of linearity in rows (1.4.7)**(ii)**. The second assertion of **(c)** follows, because a row that is zero can be multiplied by 0 without changing the matrix, and it multiplies $\delta(A)$ by 0.

Next, we verify properties **(a),(b),(d)** when i and j are adjacent indices, say $j = i+1$. To simplify our display, we represent the matrices schematically, denoting the rows in question by $R = (row\ i)$ and $S = (row\ j)$, and suppressing notation for the other rows. So $\begin{bmatrix} R \\ S \end{bmatrix}$ denotes our given matrix A. Then by linearity in the ith row,

(1.4.11) $$\delta \begin{bmatrix} R + cS \\ S \end{bmatrix} = \delta \begin{bmatrix} R \\ S \end{bmatrix} + c\delta \begin{bmatrix} S \\ S \end{bmatrix}.$$

The first term on the right side is $\delta(A)$, and the second is zero (1.4.7). This proves **(a)** for adjacent indices. To verify **(b)** for adjacent indices, we use **(a)** repeatedly. Denoting the rows by R and S as before:

(1.4.12)

$$\delta \begin{bmatrix} R \\ S \end{bmatrix} = \delta \begin{bmatrix} R - S \\ S \end{bmatrix} = \delta \begin{bmatrix} R - S \\ S + (R - S) \end{bmatrix} = \delta \begin{bmatrix} R - S \\ R \end{bmatrix} = \delta \begin{bmatrix} -S \\ R \end{bmatrix} = -\delta \begin{bmatrix} S \\ R \end{bmatrix}.$$

Finally, **(d)** for adjacent indices follows from **(c)** and (1.4.7)**(iii)**.

To complete the proof, we verify **(a)**,**(b)**,**(d)** for an arbitrary pair of distinct indices. Suppose that (*row i*) is a multiple of (*row j*). We switch adjacent rows a few times to obtain a matrix A' in which the two rows in question are adjacent. Then **(d)** for adjacent rows tells us that $\delta(A') = 0$, and **(b)** for adjacent rows tells us that $\delta(A') = \pm \delta(A)$. So $\delta(A) = 0$, and this proves **(d)**. At this point, the proofs of that we have given for **(a)** and **(b)** in the case of adjacent indices carry over to an arbitrary pair of indices.   □

The rules (1.4.10)**(a)**,**(b)**,**(c)** show how multiplication by an elementary matrix affects δ, and they lead to the next corollary.

Corollary 1.4.13 Let δ be a function on $n \times n$ matrices with the properties (1.4.7), and let E be an elementary matrix. For any matrix A, $\delta(EA) = \delta(E)\delta(A)$. Moreover,

  (i) If E is of the first kind (*add a multiple of one row to another*), then $\delta(E) = 1$.
  (ii) If E is of the second kind (*row interchange*), then $\delta(E) = -1$.
  (iii) If E is of the third kind (*multiply a row by c*), then $\delta(E) = c$.

Proof. The rules (1.4.10)**(a)**,**(b)**,**(c)** describe the effect of an elementary row operation on $\delta(A)$, so they tell us how to compute $\delta(EA)$ from $\delta(A)$. They tell us that $\delta(EA) = \epsilon \, \delta(A)$, where $\epsilon = 1, -1$, or c according to the type of elementary matrix. By setting $A = I$, we find that $\delta(E) = \delta(EI) = \epsilon\delta(I) = \epsilon$.   □

Proof of the multiplicative property, Theorem 1.4.9. We imagine the first step of a row reduction of A, say $EA = A'$. Suppose we have shown that $\delta(A'B) = \delta(A')\delta(B)$. We apply Corollary 1.4.13: $\delta(E)\delta(A) = \delta(A')$. Since $A'B = E(AB)$ the corollary also tells us that $\delta(A'B) = \delta(E)\delta(AB)$. Thus

$$\delta(E)\delta(AB) = \delta(A'B) = \delta(A')\delta(B) = \delta(E)\delta(A)\delta(B).$$

Canceling $\delta(E)$, we see that the multiplicative property is true for A and B as well. This being so, induction shows that it suffices to prove the multiplicative property after row-reducing A. So we may suppose that A is row reduced. Then A is either the identity, or else its bottom row is zero. The property is obvious when $A = I$. If the bottom row of A is zero, so is the bottom row of AB, and Theorem 1.4.10 shows that $\delta(A) = \delta(AB) = 0$. The property is true in this case as well.   □

Proof of uniqueness of the determinant, Theorem 1.4.7. There are two parts. To prove uniqueness, we perform row reduction on a matrix A, say $A' = E_k \cdots E_1 A$. Corollary 1.4.13 tells us how to compute $\delta(A)$ from $\delta(A')$. If A' is the identity, then $\delta(A') = 1$. Otherwise the bottom row of A' is zero, and in that case Theorem 1.4.10 shows that $\delta(A') = 0$. This determines $\delta(A)$ in both cases.

Note: It is a natural idea to try defining determinants using compatibility with multiplication and Corollary 1.4.13. Since we can write an invertible matrix as a product of elementary matrices, these properties determine the determinant of every invertible matrix. But there are many ways to write a given matrix as such a product. Without going through some steps as we have, it won't be clear that two such products will give the same answer. It isn't easy to make this idea work.

To complete the proof of Theorem 1.4.7, we must show that the determinant function (1.4.5) we have defined has the properties (1.4.7). This is done by induction on the size of the matrices. We note that the properties (1.4.7) are true when $n = 1$, in which case $\det [a] = a$. So we assume that they have been proved for determinants of $(n-1) \times (n-1)$ matrices. Then all of the properties (1.4.7), (1.4.10), (1.4.13), and (1.4.9) are true for $(n-1) \times (n-1)$ matrices. We proceed to verify (1.4.7) for the function $\delta = \det$ defined by (1.4.5), and for $n \times n$ matrices. For reference, they are:

(i) With I denoting the identity matrix, $\det (I) = 1$.

(ii) det is linear in the rows of the matrix A.

(iii) If two adjacent rows of a matrix A are equal, then $\det (A) = 0$.

(i) If $A = I_n$, then $a_{11} = 1$ and $a_{\nu 1} = 0$ when $\nu > 1$. The expansion (1.4.5) reduces to $\det (A) = 1 \det (A_{11})$. Moreover, $A_{11} = I_{n-1}$, so by induction, $\det (A_{11}) = 1$ and $\det (I_n) = 1$.

(ii) To prove linearity in the rows, we return to the notation introduced in (1.4.8). We show linearity of each of the terms in the expansion (1.4.5), i.e., that

(1.4.14) $$d_{\nu 1} \det (D_{\nu 1}) = c \, a_{\nu 1} \det (A_{\nu 1}) + c' \, b_{\nu 1} \det (B_{\nu 1})$$

for every index ν. Let k be as in (1.4.8).

Case 1: $\nu = k$. The row that we operate on has been deleted from the minors A_{k1}, B_{k1}, D_{k1} so they are equal, and the values of det on them are equal too. On the other hand, a_{k1}, b_{k1}, d_{k1} are the first entries of the rows A_k, B_k, D_k, respectively. So $d_{k1} = c \, a_{k1} + c' \, b_{k1}$, and (1.4.14) follows.

Case 2: $\nu \neq k$. If we let A'_k, B'_k, D'_k denote the vectors obtained from the rows A_k, B_k, D_k, respectively, by dropping the first entry, then A'_k is a row of the minor $A_{\nu 1}$, etc. Here $D'_k = c \, A'_k + c' \, B'_k$, and by induction on n, $\det (D'_{\nu 1}) = c \det (A'_{\nu 1}) + c' \det (B'_{\nu 1})$. On the other hand, since $\nu \neq k$, the coefficients $a_{\nu 1}, b_{\nu 1}, d_{\nu 1}$ are equal. So (1.4.14) is true in this case as well.

(iii) Suppose that rows k and $k + 1$ of a matrix A are equal. Unless $\nu = k$ or $k + 1$, the minor $A_{\nu 1}$ has two rows equal, and its determinant is zero by induction. Therefore, at most two terms in (1.4.5) are different from zero. On the other hand, deleting either of the equal rows gives us the same matrix. So $a_{k1} = a_{k+11}$ and $A_{k1} = A_{k+11}$. Then

$$\det (A) = \pm a_{k1} \det (A_{k1}) \mp a_{k+11} \det (A_{k+11}) = 0.$$

This completes the proof of Theorem 1.4.7. $\qquad\qquad\qquad\qquad\qquad\qquad\qquad\square$

Corollary 1.4.15

(a) A square matrix A is invertible if and only if its determinant is different from zero. If A is invertible, then $\det(A^{-1}) = (\det A)^{-1}$.

(b) The determinant of a matrix A is equal to the determinant of its transpose A^t.

(c) Properties (1.4.7) and (1.4.10) continue to hold if the word *row* is replaced by the word *column* throughout.

Proof. **(a)** If A is invertible, then it is a product of elementary matrices, say $A = E_1 \cdots E_r$ (1.2.16). Then $\det A = (\det E_1) \cdots (\det E_k)$. The determinants of elementary matrices are nonzero (1.4.13), so $\det A$ is nonzero too. If A is not invertible, there are elementary matrices $E_1, \ldots, E_r$ such that the bottom row of $A' = E_1 \cdots E_r A$ is zero (1.2.15). Then $\det A' = 0$, and $\det A = 0$ as well. If A is invertible, then $\det(A^{-1})\det A = \det(A^{-1}A) = \det I = 1$, therefore $\det(A^{-1}) = (\det A)^{-1}$.

(b) It is easy to check that $\det E = \det E^t$ if E is an elementary matrix. If A is invertible, we write $A = E_1 \cdots E_k$ as before. Then $A^t = E_k^t \cdots E_1^t$, and by the multiplicative property, $\det A = \det A^t$. If A is not invertible, neither is A^t. Then both $\det A$ and $\det A^t$ are zero.

(c) This follows from **(b)**.   □

1.5   PERMUTATIONS

A *permutation* of a set S is a bijective map p from a set S to itself:

(1.5.1) $$p : S \to S.$$

The table

(1.5.2)

i	1	2	3	4	5
p(i)	3	5	4	1	2

exhibits a permutation p of the set $\{1, 2, 3, 4, 5\}$ of five indices: $p(1) = 3$, etc. It is bijective because every index appears exactly once in the bottom row.

The set of all permutations of the indices $\{1, 2, \ldots, \mathbf{n}\}$ is called the *symmetric group*, and is denoted by S_n. It will be discussed in Chapter 2.

The benefit of this definition of a permutation is that it permits composition of permutations to be defined as composition of functions. If q is another permutation, then doing first p then q means composing the functions: $q \circ p$. The composition is called the *product permutation*, and will be denoted by qp.

Note: People sometimes like to think of a permutation of the indices $1, \ldots, \mathbf{n}$ as a list of the same indices in a different order, as in the bottom row of (1.5.2). This is not good for us. In mathematics one wants to keep track of what happens when one performs two or more permutations in succession. For instance, we may want to obtain a permutation by repeatedly switching pairs of indices. Then unless things are written carefully, keeping track of what has been done becomes a nightmare.   □

The tabular form shown above is cumbersome. It is more common to use *cycle notation*. To write a cycle notation for the permutation p shown above, we begin with an arbitrary index, say **3**, and follow it along: $p(3) = 4$, $p(4) = 1$, and $p(1) = 3$. The string of three indices forms a *cycle* for the permutation, which is denoted by

(1.5.3) $$(341).$$

This notation is interpreted as follows: the index **3** is sent to **4**, the index **4** is sent to **1**, and the parenthesis at the end indicates that the index **1** is sent back to **3** at the front by the permutation:

Because there are three indices, this is a 3-cycle.

Also, $p(2) = 5$ and $p(5) = 2$, so with the analogous notation, the two indices **2, 5** form a 2-cycle (**25**). 2-cycles are called *transpositions*.

The complete cycle notation for p is obtained by writing these cycles one after the other:

(1.5.4) $$p = (341)(25).$$

The permutation can be read off easily from this notation.

One slight complication is that the cycle notation isn't unique, for two reasons. First, we might have started with an index different from **3**. Thus

$$(341), \quad (134) \quad \text{and} \quad (413)$$

are notations for the same 3-cycle. Second, the order in which the cycles are written doesn't matter. Cycles made up of disjoint sets of indices can be written in any order. We might just as well write

$$p = (52)(134).$$

The indices (which are **1, 2, 3, 4, 5** here) may be grouped into cycles arbitrarily, and the result will be a cycle notation for some permutation. For example, $(34)(2)(15)$ represents the permutation that switches two pairs of indices, while fixing **2**. However, 1-cycles, the indices that are left fixed, are often omitted from the cycle notation. We might write this permutation as $(34)(15)$. The 4-cycle

(1.5.5) $$q = (1452)$$

is interpreted as meaning that the missing index **3** is left fixed. Then in a cycle notation for a permutation, every index appears at most once. (Of course this convention assumes that the set of indices is known.) The one exception to this rule is for the identity permutation. We'd rather not use the empty symbol to denote this permutation, so we denote it by 1.

To compute the product permutation qp, with p and q as above, we follow the indices through the two permutations, but we must remember that qp means $q \circ p$, "first do p, then q." So since p sends $3 \rightarrow 4$ and q sends $4 \rightarrow 5$, qp sends $3 \rightarrow 5$. Unfortunately, we read cycles from left to right, but we have to run through the permutations from right to left, in a zig-zag fashion. This takes some getting used to, but in the end it is not difficult. The result in our case is a 3-cycle:

$$\overset{\text{then this}\qquad\qquad\text{first do this}}{qp \;=\; [(1452)] \circ [(341)(25)] \;=\; (135),}$$

the missing indices **2** and **4** being left fixed. On the other hand,

$$pq = (234).$$

Composition of permutations is not a commutative operation.

There is a *permutation matrix P* associated to any permutation p. Left multiplication by this permutation matrix permutes the entries of a vector X using the permutation p.

For example, if there are three indices, the matrix P associated to the cyclic permutation $p = (123)$ and its operation on a column vector are as follows:

$$(1.5.6) \qquad\qquad PX = \begin{bmatrix} 0 & 0 & 1 \\ 1 & 0 & 0 \\ 0 & 1 & 0 \end{bmatrix} \begin{bmatrix} x_1 \\ x_2 \\ x_3 \end{bmatrix} = \begin{bmatrix} x_3 \\ x_1 \\ x_2 \end{bmatrix}.$$

Multiplication by P shifts the first entry of the vector X to the second position and so on.

It is essential to write the matrix of an arbitrary permutation down carefully, and to check that the matrix associated to a product pq of permutations is the product matrix PQ. The matrix associated to a transposition (25) is an elementary matrix of the second type, the one that interchanges the two corresponding rows. This is easy to see. But for a general permutation, determining the matrix can be confusing.

• To write a permutation matrix explicitly, it is best to use the $n \times n$ matrix units e_{ij}, the matrices with a single 1 in the i, j position that were defined before (1.1.21). The matrix associated to a permutation p of S_n is

$$(1.5.7) \qquad\qquad P = \sum_i e_{pi,i}.$$

(In order to make the subscript as compact as possible, we have written pi for $p(\mathbf{i})$.)

This matrix acts on the vector $X = \sum e_j x_j$ as follows:

$$(1.5.8) \qquad PX = \Big(\sum_i e_{pi,i}\Big)\Big(\sum_j e_j x_j\Big) = \sum_{i,j} e_{pi,i} e_j \, x_j = \sum_i e_{pi,i} e_i x_i = \sum_i e_{pi} x_i.$$

This computation is made using formula (1.1.25). The terms $e_{pi,i} e_j$ in the double sum are zero when $i \neq j$.

To express the right side of (1.5.8) as a column vector, we have to reindex so that the standard basis vectors on the right are in the correct order, $e_1, \ldots, e_n$ rather than in the permuted order $e_{p1}, \ldots, e_{pn}$. We set $pi = k$ and $i = p^{-1}k$. Then

(1.5.9)
$$\sum_i e_{pi} x_i = \sum_k e_k x_{p^{-1}k}.$$

This is a confusing point: Permuting the entries x_i of a vector by p permutes the indices by p^{-1}.

For example, the 3×3 matrix P of (1.5.6) is $e_{21} + e_{32} + e_{13}$, and

$$PX = (e_{21} + e_{32} + e_{13})(e_1 x_1 + e_2 x_2 + e_3 x_3) = e_1 x_3 + e_2 x_1 + e_3 x_2.$$

Proposition 1.5.10

(a) A permutation matrix P always has a single 1 in each row and in each column, the rest of its entries being 0. Conversely, any such matrix is a permutation matrix.

(b) The determinant of a permutation matrix is ± 1.

(c) Let p and q be two permutations, with associated permutation matrices P and Q. The matrix associated to the permutation pq is the product PQ.

Proof. We omit the verification of **(a)** and **(b)**. The computation below proves **(c)**:

$$PQ = \left(\sum_i e_{pi,i} \right)\left(\sum_j e_{qj,j} \right) = \sum_{i,j} e_{pi,i}\, e_{qj,j} = \sum_j e_{pqj,qj}\, e_{qj,j} = \sum_j e_{pqj,j}.$$

This computation is made using formula (1.1.23). The terms $e_{pi,i} e_{qj,j}$ in the double sum are zero unless $i = qj$. So PQ is the permutation matrix associated to the product permutation pq, as claimed. $\qquad\square$

• The determinant of the permutation matrix associated to a permutation p is called the *sign* of the permutation :

(1.5.11)
$$\operatorname{sign} p = \det P = \pm 1.$$

A permutation p is *even* if its sign is $+1$, and *odd* if its sign is -1. The permutation (123) has sign $+1$. It is even, while any transposition, such as (12), has sign -1 and is odd.

Every permutation can be written as a product of transpositions in many ways. If a permutation p is equal to the product $\tau_1 \cdots \tau_k$, where τ_i are transpositions, the number k will always be even if p is an even permutation and it will always be odd if p is an odd permutation.

This completes our discussion of permutations and permutation matrices. We will come back to them in Chapters 7 and 10.

1.6   OTHER FORMULAS FOR THE DETERMINANT

There are formulas analogous to our definition (1.4.5) of the determinant that use expansions by minors on other columns of a matrix, and also ones that use expansions on rows.

Again, the notation A_{ij} stands for the matrix obtained by deleting the ith row and the jth column of a matrix A.

Expansion by minors on the jth column:

$$\det A = (-1)^{1+j} a_{1j} \det A_{1j} + (-1)^{2+j} a_{2j} \det A_{2j} + \cdots + (-1)^{n+j} a_{nj} \det A_{nj},$$

or in summation notation,

(1.6.1) $$\det A = \sum_{\nu=1}^{n} (-1)^{\nu+j} a_{\nu j} \det A_{\nu j}.$$

Expansion by minors on the ith row:

$$\det A = (-1)^{i+1} a_{i1} \det A_{i1} + (-1)^{i+2} a_{i2} \det A_{i2} + \cdots + (-1)^{i+n} a_{in} \det A_{in},$$

(1.6.2) $$\det A = \sum_{\nu=1}^{n} (-1)^{i+\nu} a_{i\nu} \det A_{i\nu}.$$

For example, expansion on the second row gives

$$\det \begin{bmatrix} 1 & 1 & 2 \\ 0 & 2 & 1 \\ 1 & 0 & 2 \end{bmatrix} = -0 \det \begin{bmatrix} 1 & 2 \\ 0 & 2 \end{bmatrix} + 2 \det \begin{bmatrix} 1 & 2 \\ 1 & 2 \end{bmatrix} - 1 \det \begin{bmatrix} 1 & 1 \\ 1 & 0 \end{bmatrix} = 1.$$

To verify that these formulas yield the determinant, one can check the properties (1.4.7). The alternating signs that appear in the formulas can be read off of this figure:

(1.6.3) $$\begin{bmatrix} + & - & + & \cdots \\ - & + & - & \\ + & - & + & \\ \vdots & & & \ddots \end{bmatrix}$$

The notation $(-1)^{i+j}$ for the alternating sign may seem pedantic, and harder to remember than the figure. However, it is useful because it can be manipulated by the rules of algebra.

We describe one more expression for the determinant, the complete expansion. The complete expansion is obtained by using linearity to expand on all the rows, first on (*row 1*), then on (*row 2*), and so on. For a 2×2 matrix, this expansion is made as follows:

$$\det \begin{bmatrix} a & b \\ c & d \end{bmatrix} = a \det \begin{bmatrix} 1 & 0 \\ c & d \end{bmatrix} + b \det \begin{bmatrix} 0 & 1 \\ c & d \end{bmatrix}$$

$$= ac \det \begin{bmatrix} 1 & 0 \\ 1 & 0 \end{bmatrix} + ad \det \begin{bmatrix} 1 & 0 \\ 0 & 1 \end{bmatrix} + bc \det \begin{bmatrix} 0 & 1 \\ 1 & 0 \end{bmatrix} + bd \det \begin{bmatrix} 0 & 1 \\ 0 & 1 \end{bmatrix}.$$

The first and fourth terms in the final expansion are zero, and

$$\det \begin{bmatrix} a & b \\ c & d \end{bmatrix} = ad \det \begin{bmatrix} 1 & 0 \\ 0 & 1 \end{bmatrix} + bc \det \begin{bmatrix} 0 & 1 \\ 1 & 0 \end{bmatrix} = ad - bc.$$

Carrying this out for $n \times n$ matrices leads to the *complete expansion* of the determinant, the formula

(1.6.4)
$$\det A = \sum_{\text{perm}\, p} (\text{sign } p)\, a_{1,p1} \cdots a_{n,pn},$$

in which the sum is over all permutations of the n indices, and (sign p) is the sign of the permutation.

For a 2×2 matrix, the complete expansion gives us back Formula (1.4.2). For a 3×3 matrix, the complete expansion has six terms, because there are six permutations of three indices:

(1.6.5) $\det A =$

$$a_{11}a_{22}a_{33} + a_{12}a_{23}a_{31} + a_{13}a_{21}a_{32} - a_{11}a_{23}a_{32} - a_{12}a_{21}a_{33} - a_{13}a_{22}a_{31}.$$

As an aid for remembering this expansion, one can display the block matrix $[A|A]$:

(1.6.6)
$$\begin{bmatrix} a_{11} & a_{12} & a_{13} & a_{11} & a_{12} & a_{13} \\ a_{21} & a_{22} & a_{23} & a_{21} & a_{22} & a_{23} \\ a_{31} & a_{32} & a_{33} & a_{31} & a_{32} & a_{33} \end{bmatrix}$$

The three terms with positive signs are the products of the terms along the three diagonals that go downward from left to right, and the three terms with negative signs are the products of terms on the diagonals that go downward from right to left.

Warning: The analogous method will not work with 4×4 determinants.

The complete expansion is more of theoretical than of practical importance. Unless n is small or the matrix is very special, it has too many terms to be useful for computation. Its theoretical importance comes from the fact that determinants are exhibited as polynomials in the n^2 variable matrix entries a_{ij}, with coefficients ± 1. For example, if each matrix entry a_{ij} is a differentiable function of a variable t, then because sums and products of differentiable functions are differentiable, $\det A$ is also a differentiable function of t.

The Cofactor Matrix

The *cofactor matrix* of an $n \times n$ matrix A is the $n \times n$ matrix $\text{cof}(A)$ whose i, j entry is

(1.6.7)
$$\text{cof}(A)_{ij} = (-1)^{i+j} \det A_{ji},$$

where, as before, A_{ji} is the matrix obtained by crossing out the jth row and the ith column. So the cofactor matrix is the *transpose* of the matrix made up of the $(n-1) \times (n-1)$ minors of A, with signs as in (1.6.3). This matrix is used to provide a formula for the inverse matrix.

   If you need to compute a cofactor matrix, it is safest to make the computation in three steps: First compute the matrix whose i, j entry is the minor $\det A_{ij}$, then adjust signs, and finally transpose. Here is the computation for a particular 3×3 matrix:

(1.6.8)

$$A = \begin{bmatrix} 1 & 1 & 2 \\ 0 & 2 & 1 \\ 1 & 0 & 2 \end{bmatrix} : \begin{bmatrix} 4 & -1 & -2 \\ 2 & 0 & -1 \\ -3 & 1 & 2 \end{bmatrix}, \begin{bmatrix} 4 & 1 & -2 \\ -2 & 0 & 1 \\ -3 & -1 & 2 \end{bmatrix}, \begin{bmatrix} 4 & -2 & -3 \\ 1 & 0 & -1 \\ -2 & 1 & 2 \end{bmatrix} = \mathrm{cof}(A).$$

Theorem 1.6.9   Let A be an $n \times n$ matrix, let $C = \mathrm{cof}(A)$ be its cofactor matrix, and let $\alpha = \det A$. If $\alpha \neq 0$, then A is invertible, and $A^{-1} = \alpha^{-1} C$. In any case, $CA = AC = \alpha I$.

Here αI is the diagonal matrix with diagonal entries equal to α. For the inverse of a 2×2 matrix, the theorem gives us back Formula 1.1.17. The determinant of the 3×3 matrix A whose cofactor matrix is computed in (1.6.8) above happens to be 1, so for that matrix, $A^{-1} = \mathrm{cof}(A)$.

Proof of Theorem 1.6.9.   We show that the i, j entry of the product CA is equal to α if $i = j$ and is zero otherwise. Let A_i denote the ith column of A. Denoting the entries of C and A by c_{ij} and a_{ij}, the i, j entry of the product CA is

(1.6.10)
$$\sum_\nu c_{i\nu} a_{\nu j} = \sum_\nu (-1)^{\nu+i} \det A_{\nu i} a_{\nu j}.$$

When $i = j$, this is the formula (1.6.1) for the determinant by expansion by minors on column j. So the diagonal entries of CA are equal to α, as claimed.

   Suppose that $i \neq j$. We form a new matrix M in the following way: The entries of M are equal to the entries of A, except for those in column i. The ith column M_i of M is equal to the jth column A_j of A. Thus the ith and the jth columns of M are both equal to A_j, and $\det M = 0$.

   Let D be the cofactor matrix of M, with entries d_{ij}. The i, i entry of DM is

$$\sum_\nu d_{i\nu} m_{\nu i} = \sum_\nu (-1)^{\nu+i} \det M_{\nu i} m_{\nu i}.$$

This sum is equal to $\det M$, which is zero.

   On the other hand, since the ith column of M is crossed out when forming $M_{\nu i}$, that minor is equal to $A_{\nu i}$. And since the ith column of M is equal to the jth column of A, $m_{\nu i} = a_{\nu j}$. So the i, i entry of DM is also equal to

$$\sum_\nu (-1)^{\nu+i} \det A_{\nu i} a_{\nu j},$$

which is the i, j entry of CA that we want to determine. Therefore the i, j entry of CA is zero, and $CA = \alpha I$, as claimed. It follows that $A^{-1} = \alpha^{-1}\operatorname{cof}(A)$ if $\alpha \neq 0$. The computation of the product AC is done in a similar way, using expansion by minors on rows. $\qquad \square$

A general algebraical determinant in its developed form may be likened to a mixture of liquids seemingly homogeneous, but which, being of differing boiling points, admit of being separated by the process of fractional distillation.

—James Joseph Sylvester

EXERCISES

Section 1 The Basic Operations

1.1. What are the entries a_{21}, and a_{23} of the matrix $A = \begin{bmatrix} 1 & 2 & 5 \\ 2 & 7 & 8 \\ 0 & 9 & 4 \end{bmatrix}$?

1.2. Determine the products AB and BA for the following values of A and B:

$$A = \begin{bmatrix} 1 & 2 & 3 \\ 3 & 3 & 1 \end{bmatrix}, \quad B = \begin{bmatrix} -8 & -4 \\ 9 & 5 \\ -3 & -2 \end{bmatrix}; \quad A = \begin{bmatrix} 1 & 4 \\ 1 & 2 \end{bmatrix}, \quad B = \begin{bmatrix} 6 & -4 \\ 3 & 2 \end{bmatrix}.$$

1.3. Let $A = [a_1 \cdots a_n]$ be a row vector, and let $B = \begin{bmatrix} b_1 \\ \vdots \\ b_n \end{bmatrix}$ be a column vector. Compute the products AB and BA.

1.4. Verify the associative law for the matrix product $\begin{bmatrix} 1 & 2 \\ 0 & 1 \end{bmatrix}\begin{bmatrix} 0 & 1 & 2 \\ 1 & 1 & 3 \end{bmatrix}\begin{bmatrix} 1 \\ 4 \\ 3 \end{bmatrix}$.

Note: This is a self-checking problem. It won't come out unless you multiply correctly. If you need to practice matrix multiplication, use this problem as a model.

1.5. [3]Let A, B, and C be matrices of sizes $\ell \times m$, $m \times n$, and $n \times p$. How many multiplications are required to compute the product AB? In which order should the triple product ABC be computed, so as to minimize the number of multiplications required?

1.6. Compute $\begin{bmatrix} 1 & a \\ & 1 \end{bmatrix}\begin{bmatrix} 1 & b \\ & 1 \end{bmatrix}$ and $\begin{bmatrix} 1 & a \\ & 1 \end{bmatrix}^n$.

1.7. Find a formula for $\begin{bmatrix} 1 & 1 & 1 \\ & 1 & 1 \\ & & 1 \end{bmatrix}^n$, and prove it by induction.

---

[3]Suggested by Gilbert Strang.

1.8. Compute the following products by block multiplication:

$$\left[\begin{array}{cc|cc} 1 & 1 & 1 & 5 \\ 0 & 1 & 0 & 1 \\ \hline 1 & 0 & 0 & 1 \\ 0 & 1 & 1 & 0 \end{array}\right]\left[\begin{array}{cc|cc} 1 & 2 & 1 & 0 \\ 0 & 1 & 0 & 1 \\ \hline 1 & 0 & 0 & 1 \\ 0 & 1 & 1 & 3 \end{array}\right], \quad \left[\begin{array}{c|cc} 0 & 1 & 2 \\ \hline 0 & 1 & 0 \\ 3 & 0 & 1 \end{array}\right]\left[\begin{array}{c|cc} 1 & 2 & 3 \\ \hline 4 & 2 & 3 \\ 5 & 0 & 4 \end{array}\right].$$

1.9. Let A, B be square matrices.

(a) When is $(A + B)(A - B) = A^2 - B^2$? (b) Expand $(A + B)^3$.

1.10. Let D be the diagonal matrix with diagonal entries $d_1, \ldots, d_n$, and let $A = (a_{ij})$ be an arbitrary $n \times n$ matrix. Compute the products DA and AD.

1.11. Prove that the product of upper triangular matrices is upper triangular.

1.12. In each case, find all 2×2 matrices that commute with the given matrix.

(a) $\begin{bmatrix} 1 & 0 \\ 0 & 0 \end{bmatrix}$, (b) $\begin{bmatrix} 0 & 1 \\ 0 & 0 \end{bmatrix}$, (c) $\begin{bmatrix} 2 & 0 \\ 0 & 6 \end{bmatrix}$, (d) $\begin{bmatrix} 1 & 3 \\ 0 & 1 \end{bmatrix}$, (e) $\begin{bmatrix} 2 & 3 \\ 0 & 6 \end{bmatrix}$.

1.13. A square matrix A is *nilpotent* if $A^k = 0$ for some $k > 0$. Prove that if A is nilpotent, then $I + A$ is invertible. Do this by finding the inverse.

1.14. Find infinitely many matrices B such that $BA = I_2$ when

$$A = \begin{bmatrix} 2 & 3 \\ 1 & 2 \\ 1 & 1 \end{bmatrix},$$

and prove that there is no matrix C such that $AC = I_3$.

1.15. With A arbitrary, determine the products $e_{ij}A$, Ae_{ij}, e_jAe_k, $e_{ii}Ae_{jj}$, and $e_{ij}Ae_{k\ell}$.

Section 2   Row Reduction

2.1. For the reduction of the matrix M (1.2.8) given in the text, determine the elementary matrices corresponding to each operation. Compute the product P of these elementary matrices and verify that PM is indeed the end result.

2.2. Find all solutions of the system of equations $AX = B$ when

$$A = \begin{bmatrix} 1 & 2 & 1 & 1 \\ 3 & 0 & 0 & 4 \\ 1 & -4 & -2 & 2 \end{bmatrix} \quad \text{and} \quad B = \textbf{(a)} \begin{bmatrix} 0 \\ 0 \\ 0 \end{bmatrix}, \quad \textbf{(b)} \begin{bmatrix} 1 \\ 1 \\ 0 \end{bmatrix}, \quad \textbf{(c)} \begin{bmatrix} 0 \\ 2 \\ 2 \end{bmatrix}.$$

2.3. Find all solutions of the equation $x_1 + x_2 + 2x_3 - x_4 = 3$.

2.4. Determine the elementary matrices used in the row reduction in Example (1.2.18), and verify that their product is A^{-1}.

2.5. Find inverses of the following matrices:

$$\begin{bmatrix} & 1 \\ 1 & \end{bmatrix}, \quad \begin{bmatrix} 3 & 5 \\ 1 & 2 \end{bmatrix}, \quad \begin{bmatrix} 1 & 1 \\ & 1 \end{bmatrix}\begin{bmatrix} & 1 \\ 1 & \end{bmatrix}\begin{bmatrix} 3 & 5 \\ 1 & 2 \end{bmatrix}.$$

2.6. The matrix below is based on the Pascal triangle. Find its inverse.

$$\begin{bmatrix} 1 & & & & \\ 1 & 1 & & & \\ 1 & 2 & 1 & & \\ 1 & 3 & 3 & 1 & \\ 1 & 4 & 6 & 4 & 1 \end{bmatrix}.$$

2.7. Make a sketch showing the effect of multiplication by the matrix $A = \begin{bmatrix} 2 & -1 \\ 2 & 3 \end{bmatrix}$ on the plane $\mathbb{R}^2$.

2.8. Prove that if a product AB of $n \times n$ matrices is invertible, so are the factors A and B.

2.9. Consider an arbitrary system of linear equations $AX = B$, where A and B are real matrices.

(a) Prove that if the system of equations $AX = B$ has more than one solution then it has infinitely many.

(b) Prove that if there is a solution in the complex numbers then there is also a real solution.

2.10. Let A be a square matrix. Show that if the system $AX = B$ has a unique solution for some particular column vector B, then it has a unique solution for all B.

Section 3   The Matrix Transpose

3.1. A matrix B is *symmetric* if $B = B^t$. Prove that for any square matrices B, BB^t and $B + B^t$ are symmetric, and that if A is invertible, then $(A^{-1})^t = (A^t)^{-1}$.

3.2. Let A and B be symmetric $n \times n$ matrices. Prove that the product AB is symmetric if and only if $AB = BA$.

3.3. Suppose we make first a row operation, and then a column operation, on a matrix A. Explain what happens if we switch the order of these operations, making the column operation first, followed by the row operation.

3.4. How much can a matrix be simplified if both row and column operations are allowed?

Section 4   Determinants

4.1. Evaluate the following determinants:

(a) $\begin{bmatrix} 1 & i \\ 2-i & 3 \end{bmatrix}$, **(b)** $\begin{bmatrix} 1 & 1 \\ 1 & -1 \end{bmatrix}$, **(c)** $\begin{bmatrix} 2 & 0 & 1 \\ 0 & 1 & 0 \\ 1 & 0 & 2 \end{bmatrix}$, **(d)** $\begin{bmatrix} 1 & 0 & 0 & 0 \\ 5 & 2 & 0 & 0 \\ 8 & 6 & 3 & 0 \\ 0 & 9 & 7 & 4 \end{bmatrix}$.

4.2. (self-checking) Verify the rule $\det AB = (\det A)(\det B)$ for the matrices

$$A = \begin{bmatrix} 2 & 3 \\ 1 & 4 \end{bmatrix}, \text{ and } B = \begin{bmatrix} 1 & 1 \\ 5 & -2 \end{bmatrix}.$$

4.3. Compute the determinant of the following $n \times n$ matrix using induction on n:

$$\begin{bmatrix} 2 & -1 \\ -1 & 2 & -1 \\ & -1 & 2 & -1 \\ & & -1 & \cdot \\ & & & & \cdot \\ & & & & & 2 & -1 \\ & & & & & -1 & 2 \end{bmatrix}.$$

4.4. Let A be an $n \times n$ matrix. Determine $\det(-A)$ in terms of $\det A$.

4.5. Use row reduction to prove that $\det A^t = \det A$.

4.6. Prove that $\det \begin{bmatrix} A & B \\ 0 & D \end{bmatrix} = (\det A)(\det D)$, if A and D are square blocks.

Section 5   Permutation Matrices

5.1. Write the following permutations as products of disjoint cycles:

$(\mathbf{12})(\mathbf{13})(\mathbf{14})(\mathbf{15}), \ (\mathbf{123})(\mathbf{234})(\mathbf{345}), \ (\mathbf{1234})(\mathbf{2345}), \ (\mathbf{12})(\mathbf{23})(\mathbf{34})(\mathbf{45})(\mathbf{51}),$

5.2. Let p be the permutation $(\mathbf{1342})$ of four indices.

(a) Find the associated permutation matrix P.

(b) Write p as a product of transpositions and evaluate the corresponding matrix product.

(c) Determine the sign of p.

5.3. Prove that the inverse of a permutation matrix P is its transpose.

5.4. What is the permutation matrix associated to the permutation of n indices defined by $p(\mathbf{i}) = \mathbf{n} - \mathbf{i} + \mathbf{1}$? What is the cycle decomposition of p? What is its sign?

5.5. In the text, the products qp and pq of the permutations (1.5.2) and (1.5.5) were seen to be different. However, both products turned out to be 3-cycles. Is this an accident?

Section 6   Other Formulas for the Determinant

6.1. (a) Compute the determinants of the following matrices by expansion on the bottom row:

$$\begin{bmatrix} 1 & 2 \\ 3 & 4 \end{bmatrix}, \quad \begin{bmatrix} 1 & 1 & 2 \\ 2 & 4 & 2 \\ 0 & 2 & 1 \end{bmatrix}, \quad \begin{bmatrix} 4 & -1 & 1 \\ 1 & 1 & -2 \\ 1 & -1 & 1 \end{bmatrix}, \quad \begin{bmatrix} a & b & c \\ 1 & 0 & 1 \\ 1 & 1 & 1 \end{bmatrix}.$$

(b) Compute the determinants of these matrices using the complete expansion.

(c) Compute the cofactor matrices of these matrices, and verify Theorem 1.6.9 for them.

6.2. Let A be an $n \times n$ matrix with integer entries a_{ij}. Prove that A is invertible, and that its inverse A^{-1} has integer entries, if and only if $\det A = \pm 1$.

Miscellaneous Problems

*M.1. Let a $2n \times 2n$ matrix be given in the form $M = \begin{bmatrix} A & B \\ C & D \end{bmatrix}$, where each block is an $n \times n$ matrix. Suppose that A is invertible and that $AC = CA$. Use block multiplication to prove that $\det M = \det(AD - CB)$. Give an example to show that this formula need not hold if $AC \neq CA$.

M.2. Let A be an $m \times n$ matrix with $m < n$. Prove that A has no left inverse by comparing A to the square $n \times n$ matrix obtained by adding $(n - m)$ rows of zeros at the bottom.

M.3. The *trace* of a square matrix is the sum of its diagonal entries:

$$\text{trace } A = a_{11} + a_{22} + \cdots + a_{nn},$$

Show that $\text{trace }(A + B) = \text{trace } A + \text{trace } B$, that $\text{trace } AB = \text{trace } BA$, and that if B is invertible, then $\text{trace } A = \text{trace } BAB^{-1}$.

M.4. Show that the equation $AB - BA = I$ has no solution in real $n \times n$ matrices A and B.

M.5. Write the matrix $\begin{bmatrix} 1 & 2 \\ 3 & 4 \end{bmatrix}$ as a product of elementary matrices, using as few as you can, and prove that your expression is as short as possible.

M.6. Determine the smallest integer n such that every invertible 2×2 matrix can be written as a product of at most n elementary matrices.

M.7. (*Vandermonde determinant*)

(a) Prove that $\det \begin{bmatrix} 1 & 1 & 1 \\ a & b & c \\ a^2 & b^2 & c^2 \end{bmatrix} = (a - b)(b - c)(c - a)$.

(b) Prove an analogous formula for $n \times n$ matrices, using appropriate row operations to clear out the first column.

(c) Use the Vandermonde determinant to prove that there is a unique polynomial $p(t)$ of degree n that takes arbitrary prescribed values at $n + 1$ points $t_0, \ldots, t_n$.

*M.8. (*an exercise in logic*) Consider a general system $AX = B$ of m linear equations in n unknowns, where m and n are not necessarily equal. The coefficient matrix A may have a left inverse L, a matrix such that $LA = I_n$. If so, we may try to solve the system as we learn to do in school:

$$AX = B, \quad LAX = LB, \quad X = LB.$$

But when we try to check our work by running the solution backward, we run into trouble: If $X = LB$, then $AX = ALB$. We seem to want L to be a right inverse, which isn't what was given.

(a) Work some examples to convince yourself that there is a problem here.

(b) Exactly what does the sequence of steps made above show? What would the existence of a right inverse show? Explain clearly.

M.9. Let A be a real 2×2 matrix, and let A_1, A_2 be the columns of A. Let P be the parallelogram whose vertices are $0, A_1, A_2, A_1 + A_2$. Determine the effect of elementary row operations on the area of P, and use this to prove that the absolute value $|\det A|$ of the determinant of A is equal to the area of P.

*M.10. Let A, B be $m \times n$ and $n \times m$ matrices. Prove that $I_m - AB$ is invertible if and only if $I_n - BA$ is invertible.

Hint: Perhaps the only approach available to you at this time is to find an explicit expression for one inverse in terms of the other. As a heuristic tool, you could try substituting into the power series expansion for $(1 - x)^{-1}$. The substitution will make no sense unless some series converge, and this needn't be the case. But any way to guess a formula is permissible, provided that you check your guess afterward.

M.11. [4](*discrete Dirichlet problem*) A function $f(u, v)$ is harmonic if it satisfies the Laplace equation $\frac{\partial^2 f}{\partial u^2} + \frac{\partial^2 f}{\partial v^2} = 0$. The Dirichlet problem asks for a harmonic function on a plane region R with prescribed values on the boundary. This exercise solves the discrete version of the Dirichlet problem.

Let f be a real valued function whose domain of definition is the set of integers $\mathbb{Z}$. To avoid asymmetry, the discrete derivative is defined on the shifted integers $\mathbb{Z} + \frac{1}{2}$, as the first difference $f'(n + \frac{1}{2}) = f(n + 1) - f(n)$. The discrete second derivative is back on the integers: $f''(n) = f'(n + \frac{1}{2}) - f'(n - \frac{1}{2}) = f(n + 1) - 2f(n) + f(n - 1)$.

Let $f(u, v)$ be a function whose domain is the lattice of points in the plane with integer coordinates. The formula for the discrete second derivative shows that the discrete version of the Laplace equation for f is

$$f(u + 1, v) + f(u - 1, v) + f(u, v + 1) + f(u, v - 1) - 4f(u, v) = 0.$$

So f is harmonic if its value at a point (u, v) is the average of the values at its four neighbors.

A *discrete region* R in the plane is a finite set of integer lattice points. Its *boundary* ∂R is the set of lattice points that are not in R, but which are at a distance 1 from some point of R. We'll call R the *interior* of the region $\overline{R} = R \cup \partial R$. Suppose that a function β is given on the boundary ∂R. The discrete Dirichlet problem asks for a function f defined on $\overline{R}$, that is equal to β on the boundary, and that satisfies the discrete Laplace equation at all points in the interior. This problem leads to a system of linear equations that we abbreviate as $LX = B$. To set the system up, we write β_{uv} for the given value of the function β at a boundary point. So $f(u, v) = \beta_{uv}$ at a boundary point (u, v). Let x_{uv} denote the unknown value of the function $f(u, v)$ at a point (u, v) of R. We order the points of R arbitrarily and assemble the unknowns x_{uv} into a column vector X. The coefficient matrix L expresses the discrete Laplace equation, except that when a point of R has some neighbors on the boundary, the corresponding terms will be the given boundary values. These terms are moved to the other side of the equation to form the vector B.

(a) When R is the set of five points $(0, 0)$, $(0, \pm 1)$, $(\pm 1, 0)$, there are eight boundary points. Write down the system of linear equations in this case, and solve the Dirichlet problem when β is the function on ∂R defined by $\beta_{uv} = 0$ if $v \le 0$ and $\beta_{uv} = 1$ if $v > 0$.

(b) The *maximum principle* states that a harmonic function takes on its maximal value on the boundary. Prove the maximum principle for discrete harmonic functions.

(c) Prove that the discrete Dirichlet problem has a unique solution for every region R and every boundary function β.

---

[4]I learned this problem from Peter Lax, who told me that he had learned it from my father, Emil Artin.

CHAPTER 2

Groups

*Il est peu de notions en mathématiques qui soient plus primitives
que celle de loi de composition.*

—Nicolas Bourbaki

2.1 LAWS OF COMPOSITION

A *law of composition* on a set S is any rule for combining pairs a, b of elements of S to get
another element, say p, of S. Some models for this concept are addition and multiplication
of real numbers. Matrix multiplication on the set of $n \times n$ matrices is another example.

Formally, a law of composition is a function of two variables, or a map

$$S \times S \to S.$$

Here $S \times S$ denotes, as always, the *product set*, whose elements are pairs a, b of elements
of S.

The element obtained by applying the law to a pair a, b is usually written using a
notation resembling one used for multiplication or addition:

$$p = ab, \ a \times b, \ a \circ b, \ a + b,$$

or whatever, a choice being made for the particular law in question. The element p may be
called the product or the sum of a and b, depending on the notation chosen.

We will use the product notation ab most of the time. Anything done with product
notation can be rewritten using another notation such as addition, and it will continue to be
valid. The rewriting is just a change of notation.

It is important to note right away that ab stands for a certain element of S, namely for
the element obtained by applying the given law to the elements denoted by a and b. Thus
if the law is matrix multiplication and if $a = \begin{bmatrix} 1 & 3 \\ 0 & 2 \end{bmatrix}$ and $b = \begin{bmatrix} 1 & 0 \\ 2 & 1 \end{bmatrix}$, then ab denotes
the matrix $\begin{bmatrix} 7 & 3 \\ 4 & 2 \end{bmatrix}$. Once the product ab has been evaluated, the elements a and b cannot
be recovered from it.

With multiplicative notation, a law of composition is *associative* if the rule

(2.1.1) $(ab)c = a(bc)$ (associative law)

holds for all a, b, c in S, where $(ab)c$ means first multiply (apply the law to) a and b, then multiply the result ab by c. A law of composition is *commutative* if

(2.1.2) $$ab = ba \quad \textit{(commutative law)}$$

holds for all a and b in S. Matrix multiplication is associative, but not commutative.

It is customary to reserve additive notation $a + b$ for commutative laws – laws such that $a + b = b + a$ for all a and b. Multiplicative notation carries no implication either way concerning commutativity.

The associative law is more fundamental than the commutative law, and one reason for this is that composition of functions is associative. Let T be a set, and let g and f be maps (or functions) from T to T. Let $g \circ f$ denote the composed map $t \rightsquigarrow g(f(t))$: first apply f, then g. The rule

$$g, f \rightsquigarrow g \circ f$$

is a law of composition on the set of maps $T \to T$. This law is associative. If f, g, and h are three maps from T to T, then $(h \circ g) \circ f = h \circ (g \circ f)$:

$$T \xrightarrow{\ f\ } T \xrightarrow{\ g\ } T \xrightarrow{\ h\ } T.$$

Both of the composed maps send an element t to $h(g(f(t)))$.

When T contains two elements, say $T = \{a, b\}$, there are four maps $T \to T$:

i: the *identity* map, defined by $i(a) = a$, $i(b) = b$;

τ: the *transposition*, defined by $\tau(a) = b$, $\tau(b) = a$;

α: the constant function $\alpha(a) = \alpha(b) = a$;

β: the constant function $\beta(a) = \beta(b) = b$.

The law of composition on the set $\{i, \tau, \alpha, \beta\}$ of maps $T \to T$ can be exhibited in a *multiplication table*:

(2.1.3)

	i	τ	α	β
i	i	τ	α	β
τ	τ	i	β	α
α	α	α	α	α
β	β	β	β	β

which is to be read in this way:

	f
	$\vdots$
g	$\cdots \quad g \circ f$

Thus $\tau \circ \alpha = \beta$, while $\alpha \circ \tau = \alpha$. Composition of functions is not a commutative law.

Going back to a general law of composition, suppose we want to define the product of a string of n elements of a set: $a_1 a_2 \cdots a_n = ?$ There are various ways to do this using the given law, which tells us how to multiply two elements. For instance, we could first use the law to find the product $a_1 a_2$, then multiply this element by a_3, and so on:

$$((a_1 a_2) a_3) a_4 \cdots .$$

There are several other ways to form a product with the elements in the given order, but if the law is *associative*, then all of them yield the same element of S. This allows us to speak of the product of an arbitrary string of elements.

Proposition 2.1.4 Let an associative law of composition be given on a set S. There is a unique way to define, for every integer n, a product of n elements $a_1, \ldots, a_n$ of S, denoted temporarily by $[a_1 \cdots a_n]$, with the following properties:

(i) The product $[a_1]$ of one element is the element itself.

(ii) The product $[a_1 a_2]$ of two elements is given by the law of composition.

(iii) For any integer i in the range $1 \le i < n$, $[a_1 \cdots a_n] = [a_1 \cdots a_i][a_{i+1} \ldots a_n]$.

The right side of equation (iii) means that the two products $[a_1 \ldots a_i]$ and $[a_{i+1} \ldots a_n]$ are formed first, and the results are then multiplied using the law of composition.

Proof. We use induction on n. The product is defined by (i) and (ii) for $n \le 2$, and it does satisfy (iii) when $n = 2$. Suppose that we have defined the product of r elements when $r \le n - 1$, and that it is the unique product satisfying (iii). We then define the product of n elements by the rule

$$[a_1 \cdots a_n] = [a_1 \cdots a_{n-1}][a_n],$$

where the terms on the right side are those already defined. If a product satisfying (iii) exists, then this formula gives the product because it is (iii) when $i = n - 1$. So if the product of n elements exists, it is unique. We must now check (iii) for $i < n - 1$:

$$
\begin{aligned}
[a_1 \cdots a_n] &= [a_1 \cdots a_{n-1}][a_n] && \text{(our definition)} \\
&= ([a_1 \cdots a_i][a_{i+1} \cdots a_{n-1}])[a_n] && \text{(induction hypothesis)} \\
&= [a_1 \cdots a_i]([a_{i+1} \cdots a_{n-1}][a_n]) && \text{(associative law)} \\
&= [a_1 \cdots a_i][a_{i+1} \cdots a_n] && \text{(induction hypothesis)}.
\end{aligned}
$$

This completes the proof. We will drop the brackets from now on and denote the product by $a_1 \cdots a_n$. $\square$

An *identity* for a law of composition is an element e of S such that

(2.1.5) $ea = a$ and $ae = a$, for all a in S.

There can be at most one identity, for if e and e' are two such elements, then since e is an identity, $ee' = e'$, and since e' is an identity, $e = ee'$. Thus $e = ee' = e'$.

Both matrix multiplication and composition of functions have an identity. For $n \times n$ matrices it is the identity matrix I, and for the set of maps $T \to T$ it is the identity map – the map that carries each element of T to itself.

• The identity element will often be denoted by 1 if the law of composition is written multiplicatively, and by 0 if the law is written additively. These elements do not need to be related to the *numbers* 1 and 0, but they share the property of being identity elements for their laws of composition.

Suppose that a law of composition on a set S, written multiplicatively, is associative and has an identity 1. An element a of S is *invertible* if there is another element b such that

$$ab = 1 \quad \text{and} \quad ba = 1,$$

and if so, then b is called the *inverse* of a. The inverse of an element is usually denoted by a^{-1}, or when additive notation is being used, by $-a$.

We list without proof some elementary properties of inverses. All but the last have already been discussed for matrices. For an example that illustrates the last statement, see Exercise 1.3.

• If an element a has both a left inverse ℓ and a right inverse r, i.e., if $\ell a = 1$ and $ar = 1$, then $\ell = r$, a is invertible, r is its inverse.

• If a is invertible, its inverse is unique.

• Inverses multiply in the opposite order: If a and b are invertible, so is the product ab, and $(ab)^{-1} = b^{-1}a^{-1}$.

• An element a may have a left inverse or a right inverse, though it is not invertible.

Power notation may be used for an associative law: With $n > 0$, $a^n = a \cdots a$ (n factors), $a^{-n} = a^{-1} \cdots a^{-1}$, and $a^0 = 1$. The usual rules for manipulation of powers hold: $a^r a^s = a^{r+s}$ and $(a^r)^s = a^{rs}$. When additive notation is used for the law of composition, the power notation a^n is replaced by the notation $na = a + \cdots + a$.

Fraction notation $\frac{b}{a}$ is not advisable unless the law of composition is commutative, because it isn't clear from the notation whether the fraction stands for ba^{-1} or for $a^{-1}b$, and these two elements may be different.

2.2 GROUPS AND SUBGROUPS

A *group* is a set G together with a law of composition that has the following properties:

• The law of composition is associative: $(ab)c = a(bc)$ for all a, b, c in G.

• G contains an identity element 1, such that $1a = a$ and $a1 = a$ for all a in G.

• Every element a of G has an inverse, an element b such that $ab = 1$ and $ba = 1$.

An *abelian group* is a group whose law of composition is commutative.

For example, the set of nonzero real numbers forms an abelian group under multiplication, and the set of all real numbers forms an abelian group under addition. The set of invertible $n \times n$ matrices, the general linear group, is a very important group in which the law of composition is matrix multiplication. It is not abelian unless $n = 1$.

When the law of composition is evident, it is customary to denote a group and the set of its elements by the same symbol.

The *order* of a group G is the number of elements that it contains. We will often denote the order by $|G|$:

(2.2.1) $|G| =$ number of elements, the order, of G.

If the order is finite, G is said to be a *finite group*. If not, G is an *infinite group*. The same terminology is used for any set. The order $|S|$ of a set S is the number of its elements.

Here is our notation for some familiar infinite abelian groups:

(2.2.2) $\mathbb{Z}^+$:   the set of integers, with addition as its law of composition – the additive group of integers,

$\mathbb{R}^+$:   the set of real numbers, with addition as its law of composition – the additive group of real numbers;

$\mathbb{R}^\times$:   the set of nonzero real numbers, with multiplication as its law of composition – the multiplicative group,

$\mathbb{C}^+, \mathbb{C}^\times$:   the analogous groups, where the set $\mathbb{C}$ of complex numbers replaces the set $\mathbb{R}$ of real numbers.

Warning: Others might use the symbol $\mathbb{R}^+$ to denote the set of *positive* real numbers. To be unambiguous, it might be better to denote the additive group of reals by $(\mathbb{R}, +)$, thus displaying its law of composition explicitly. However, our notation is more compact. Also, the symbol $\mathbb{R}^\times$ denotes the multiplicative group of *nonzero* real numbers. The set of all real numbers is not a group under multiplication because 0 isn't invertible.   □

Proposition 2.2.3 Cancellation Law. Let a, b, c be elements of a group G whose law of composition is written multiplicatively. If $ab = ac$ or if $ba = ca$, then $b = c$. If $ab = a$ or if $ba = a$, then $b = 1$.

Proof. Multiply both sides of $ab = ac$ on the left by a^{-1} to obtain $b = c$. The other proofs are analogous.   □

Multiplication by a^{-1} is essential for this proof. The Cancellation Law needn't hold when the element a is not invertible. For instance,

$$\begin{bmatrix} 1 & 1 \\ & \end{bmatrix}\begin{bmatrix} 1 & 1 \\ 2 & \end{bmatrix} = \begin{bmatrix} 1 & 1 \\ & \end{bmatrix}\begin{bmatrix} 3 & \\ & 1 \end{bmatrix}.$$

Two basic examples of groups are obtained from laws of composition that we have considered – multiplication of matrices and composition of functions – by leaving out the elements that are not invertible.

• The $n \times n$ *general linear group* is the group of all invertible $n \times n$ matrices. It is denoted by

(2.2.4) $GL_n = \{n \times n \text{ invertible matrices } A\}.$

If we want to indicate that we are working with real or with complex matrices, we write $GL_n(\mathbb{R})$ or $GL_n(\mathbb{C})$, according to the case.

Let M be the set of maps from a set T to itself. A map $f : T \rightarrow T$ has an inverse function if and only if it is bijective, in which case we say f is a *permutation* of T. The permutations of T form a group, the law being composition of maps. As in section 1.5, we use multiplicative notation for the composition of permutations, writing qp for $q \circ p$.

• The group of permutations of the set of indices $\{\mathbf{1, 2, \ldots, n}\}$ is called the *symmetric group*, and is denoted by S_n:

(2.2.5) S_n is the group of permutations of the indices $\mathbf{1, 2, \ldots, n}$.

There are $n!$ ('n factorial' $= 1 \cdot 2 \cdot 3 \cdots n$) permutations of a set of n elements, so the symmetric group S_n is a finite group of order $n!$.

The permutations of a set $\{a, b\}$ of two elements are the identity i and the transposition τ (see 2.1.3). They form a group of order two. If we replace a by $\mathbf{1}$ and b by $\mathbf{2}$, we see that this is the same group as the symmetric group S_2. There is essentially only one group G of order two. To see this, we note that one of its elements must be the identity 1; let the other element be g. The multiplication table for the group contains the four products $11, 1g, g1$, and gg. All except gg are determined by the fact that 1 is the identity element. Moreover, the Cancellation Law shows that $gg \neq g$. The only possibility is $gg = 1$. So the multiplication table is completely determined. There is just one group law.

We describe the symmetric group S_3 next. This group, which has order six, serves as a convenient example because it is the smallest group whose law of composition isn't commutative. We will refer to it often. To describe it, we pick two particular permutations in terms of which we can write all others. We take the cyclic permutation $(\mathbf{1\,2\,3})$, and the transposition $(\mathbf{1\,2})$, and label them as x and y, respectively. The rules

$$(2.2.6) \qquad\qquad x^3 = 1, \quad y^2 = 1, \quad yx = x^2 y$$

are easy to verify. Using the cancellation law, one sees that the six elements $1, x, x^2, y, xy, x^2 y$ are distinct. So they are the six elements of the group:

$$(2.2.7) \qquad\qquad S_3 = \{1, x, x^2; y, xy, x^2 y\}.$$

In the future, we will refer to (2.2.6) and (2.2.7) as our "usual presentation" of the symmetric group S_3. Note that S_3 is not a commutative group, because $yx \neq xy$.

The rules (2.2.6) suffice for computation. Any product of the elements x and y and of their inverses can be shown to be equal to one of the products (2.2.7) by applying the rules repeatedly. To do so, we move all occurrences of y to the right side using the last rule, and we use the first two rules to keep the exponents small. For instance,

$$(2.2.8) \qquad x^{-1} y^3 x^2 y = x^2 y x^2 y = x^2 (yx)xy = x^2 (x^2 y)xy = xyxy = x(x^2 y)y = 1.$$

One can write out a multiplication table for S_3 with the aid of the rules (2.2.6), and because of this, those rules are called *defining relations* for the group. We study defining relations in Chapter 7.

We stop here. The structure of S_n becomes complicated very rapidly as n increases.

One reason that the general linear groups and the symmetric groups are important is that many other groups are contained in them as subgroups. A subset H of a group G is a *subgroup* if it has the following properties:

(2.2.9)
- *Closure*: If a and b are in H, then ab is in H.
- *Identity*: 1 is in H.
- *Inverses*: If a is in H, then a^{-1} is in H.

These conditions are explained as follows: The first one tells us that the law of composition on the group G defines a law of composition on H, called the *induced law*. The second and third conditions say that H is a group with respect to this induced law. Notice that (2.2.9) mentions all parts of the definition of a group except for the associative law. We don't need to mention associativity. It carries over automatically from G to the subset H.

Notes: (i) In mathematics, it is essential to learn the definition of each term. An intuitive feeling will not suffice. For example, the set T of invertible real (upper) triangular 2×2 matrices is a subgroup of the general linear group GL_2, and there is only one way to verify this, namely to go back to the definition. It is true that T is a subset of GL_2. One must verify that the product of invertible triangular matrices is triangular, that the identity is triangular, and that the inverse of an invertible triangular matrix is triangular. Of course these points are very easy to check.

(ii) Closure is sometimes mentioned as one of the axioms for a group, to indicate that the product ab of elements of G is again an element of G. We include closure as a part of what is meant by a law of composition. Then it doesn't need to be mentioned separately in the definition of a group.   □

Examples 2.2.10

(a) The set of complex numbers of absolute value 1, the set of points on the unit circle in the complex plane, is a subgroup of the multiplicative group $\mathbb{C}^\times$ called the *circle group*.

(b) The group of real $n \times n$ matrices with determinant 1 is a subgroup of the general linear group GL_n, called the *special linear group*. It is denoted by SL_n:

(2.2.11) $SL_n(\mathbb{R})$ is the set of real $n \times n$ matrices A with determinant equal to 1.

The defining properties (2.2.9) are often very easy to verify for a particular subgroup, and we may not carry the verification out.

• Every group G has two obvious subgroups: the group G itself, and the *trivial subgroup* that consists of the identity element alone. A subgroup is a *proper subgroup* if it is not one of those two.

2.3   SUBGROUPS OF THE ADDITIVE GROUP OF INTEGERS

We review some elementary number theory here, in terms of subgroups of the additive group $\mathbb{Z}^+$ of integers. To begin, we list the axioms for a subgroup when additive notation is used in the group: A subset S of a group G with law of composition written additively is a subgroup if it has these properties:

(2.3.1)

• *Closure*: If a and b are in S, then $a + b$ is in S.
• *Identity*: 0 is in S.
• *Inverses*: If a is in S then $-a$ is in S.

Let a be an integer different from 0. We denote the subset of $\mathbb{Z}$ that consists of all multiples of a by $\mathbb{Z}a$:

(2.3.2) $\mathbb{Z}a = \{n \in \mathbb{Z} \mid n = ka \text{ for some } k \text{ in } \mathbb{Z}\}.$

This is a subgroup of $\mathbb{Z}^+$. Its elements can also be described as the integers divisible by a.

Theorem 2.3.3 Let S be a subgroup of the additive group $\mathbb{Z}^+$. Either S is the trivial subgroup $\{0\}$, or else it has the form $\mathbb{Z}a$, where a is the smallest positive integer in S.

Proof. Let S be a subgroup of $\mathbb{Z}^+$. Then 0 is in S, and if 0 is the only element of S then S is the trivial subgroup. So that case is settled. Otherwise, S contains an integer n different from 0, and either n or $-n$ is positive. The third property of a subgroup tells us that $-n$ is in S, so in either case, S contains a positive integer. We must show that S is equal to $\mathbb{Z}a$, when a is the smallest positive integer in S.

We first show that $\mathbb{Z}a$ is a subset of S, in other words, that ka is in S for every integer k. If k is a positive integer, then $ka = a + a + \cdots + a$ (k terms). Since a is in S, closure and induction show that ka is in S. Since inverses are in S, $-ka$ is in S. Finally, $0 = 0a$ is in S.

Next we show that S is a subset of $\mathbb{Z}a$, that is, every element n of S is an integer multiple of a. We use division with remainder to write $n = qa + r$, where q and r are integers and where the remainder r is in the range $0 \le r < a$. Since $\mathbb{Z}a$ is contained in S, qa is in S, and of course n is in S. Since S is a subgroup, $r = n - qa$ is in S too. Now by our choice, a is the smallest positive integer in S, while the remainder r is in the range $0 \le r < a$. The only remainder that can be in S is 0. So $r = 0$ and n is the integer multiple qa of a. $\square$

There is a striking application of Theorem 2.3.3 to subgroups that contain *two* integers a and b. The set of all integer combinations $ra + sb$ of a and b,

$$(2.3.4) \qquad S = \mathbb{Z}a + \mathbb{Z}b = \{n \in \mathbb{Z} \mid n = ra + sb \text{ for some integers } r, s\}$$

is a subgroup of $\mathbb{Z}^+$. It is called the subgroup *generated by* a and b because it is the smallest subgroup that contains both a and b. Let's assume that a and b aren't both zero, so that S is not the trivial subgroup $\{0\}$. Theorem 2.3.3 tells us that this subgroup S has the form $\mathbb{Z}d$ for some positive integer d; it is the set of integers divisible by d. The generator d is called the *greatest common divisor* of a and b, for reasons that are explained in parts **(a)** and **(b)** of the next proposition. The greatest common divisor of a and b is sometimes denoted by $\gcd(a, b)$.

Proposition 2.3.5 Let a and b be integers, not both zero, and let d be their greatest common divisor, the positive integer that generates the subgroup $S = \mathbb{Z}a + \mathbb{Z}b$. So $\mathbb{Z}d = \mathbb{Z}a + \mathbb{Z}b$. Then

(a) d divides a and b.

(b) If an integer e divides both a and b, it also divides d.

(c) There are integers r and s such that $d = ra + sb$.

Proof. Part **(c)** restates the fact that d is an element of S. Next, a and b are elements of S and $S = \mathbb{Z}d$, so d divides a and b. Finally, if an integer e divides both a and b, then e divides the integer combination $ra + sb = d$. $\square$

Note: If e divides a and b, then e divides any integer of the form $ma + nb$. So **(c)** implies **(b)**. But **(b)** does not imply **(c)**. As we shall see, property **(c)** is a powerful tool. $\square$

One can compute a greatest common divisor easily by repeated division with remainder: For example, if $a = 314$ and $b = 136$, then

$$314 = 2 \cdot 136 + 42 \,, \quad 136 = 3 \cdot 42 + 10 \,, \quad 42 = 4 \cdot 10 + 2.$$

Using the first of these equations, one can show that any integer combination of 314 and 136 can also be written as an integer combination of 136 and the remainder 42, and vice versa. So $\mathbb{Z}(314) + \mathbb{Z}(136) = \mathbb{Z}(136) + \mathbb{Z}(42)$, and therefore $\gcd(314, 136) = \gcd(136, 42)$. Similarly, $\gcd(136, 42) = \gcd(42, 10) = \gcd(10, 2) = 2$. So the greatest common divisor of 314 and 136 is 2. This iterative method of finding the greatest common divisor of two integers is called the *Euclidean Algorithm*.

If integers a and b are given, a second way to find their greatest common divisor is to factor each of them into prime integers and then to collect the common prime factors. Properties **(a)** and **(b)** of Proposition 2.3.5 are easy to verify using this method. But without Theorem 2.3.3, property **(c)**, that the integer determined by this method is an integer combination of a and b wouldn't be clear at all. Let's not discuss this point further here. We come back to it in Chapter 12.

Two nonzero integers a and b are said to be *relatively prime* if the only positive integer that divides both of them is 1. Then their greatest common divisor is 1: $\mathbb{Z}a + \mathbb{Z}b = \mathbb{Z}$.

Corollary 2.3.6 A pair a, b of integers is relatively prime if and only if there are integers r and s such that $ra + sb = 1$. □

Corollary 2.3.7 Let p be a prime integer. If p divides a product ab of integers, then p divides a or p divides b.

Proof. Suppose that the prime p divides ab but does not divide a. The only positive divisors of p are 1 and p. Since p does not divide a, $\gcd(a, p) = 1$. Therefore there are integers r and s such that $ra + sp = 1$. We multiply by b: $rab + spb = b$, and we note that p divides both rab and spb. So p divides b. □

There is another subgroup of $\mathbb{Z}^+$ associated to a pair a, b of integers, namely the intersection $\mathbb{Z}a \cap \mathbb{Z}b$, the set of integers contained both in $\mathbb{Z}a$ and in $\mathbb{Z}b$. We assume now that neither a nor b is zero. Then $\mathbb{Z}a \cap \mathbb{Z}b$ is a subgroup. It is not the trivial subgroup $\{0\}$ because it contains the product ab, which isn't zero. So $\mathbb{Z}a \cap \mathbb{Z}b$ has the form $\mathbb{Z}m$ for some positive integer m. This integer m is called the *least common multiple* of a and b, sometimes denoted by $\text{lcm}(a, b)$, for reasons that are explained in the next proposition.

Proposition 2.3.8 Let a and b be integers different from zero, and let m be their least common multiple – the positive integer that generates the subgroup $S = \mathbb{Z}a \cap \mathbb{Z}b$. So $\mathbb{Z}m = \mathbb{Z}a \cap \mathbb{Z}b$. Then

(a) m is divisible by both a and b.

(b) If an integer n is divisible by a and by b, then it is divisible by m.

Proof. Both statements follow from the fact that an integer is divisible by a and by b if and only if it is contained in $\mathbb{Z}m = \mathbb{Z}a \cap \mathbb{Z}b$. □

Corollary 2.3.9 Let $d = \gcd(a, b)$ and $m = \text{lcm}(a, b)$ be the greatest common divisor and least common multiple of a pair a, b of positive integers, respectively. Then $ab = dm$.

Proof. Since b/d is an integer, a divides ab/d. Similarly, b divides ab/d. So m divides ab/d, and dm divides ab. Next, we write $d = ra + sb$. Then $dm = ram + sbm$. Both terms on the right are divisible by ab, so ab divides dm. Since ab and dm are positive and each one divides the other, $ab = dm$. $\square$

2.4 CYCLIC GROUPS

We come now to an important abstract example of a subgroup, the *cyclic subgroup* generated by an arbitrary element x of a group G. We use multiplicative notation. The cyclic subgroup H generated by x is the set of all elements that are powers of x:

$$(2.4.1) \qquad\qquad H = \{\ldots, x^{-2}, x^{-1}, 1, x, x^2, \ldots\}.$$

This is the smallest subgroup of G that contains x, and it is often denoted by $\langle x \rangle$. But to interpret (2.4.1) correctly, we must remember that the notation x^n represents an element of the group that is obtained in a particular way. Different powers may represent the same element. For example, if G is the multiplicative group $\mathbb{R}^\times$ and $x = -1$, then all elements in the list are equal to 1 or to -1, and H is the set $\{1, -1\}$.

There are two possibilities: Either the powers x^n represent distinct elements, or they do not. We analyze the case that the powers of x are not distinct.

Proposition 2.4.2 Let $\langle x \rangle$ be the cyclic subgroup of a group G generated by an element x, and let S denote the set of integers k such that $x^k = 1$.

(a) The set S is a subgroup of the additive group $\mathbb{Z}^+$.

(b) Two powers $x^r = x^s$, with $r \geq s$, are equal if and only if $x^{r-s} = 1$, i.e., if and only if $r - s$ is in S.

(c) Suppose that S is not the trivial subgroup. Then $S = \mathbb{Z}n$ for some positive integer n. The powers $1, x, x^2, \ldots, x^{n-1}$ are the distinct elements of the subgroup $\langle x \rangle$, and the order of $\langle x \rangle$ is n.

Proof. **(a)** If $x^k = 1$ and $x^\ell = 1$, then $x^{k+\ell} = x^k x^\ell = 1$. This shows that if k and ℓ are in S, then $k + \ell$ is in S. So the first property (2.3.1) for a subgroup is verified. Also, $x^0 = 1$, so 0 is in S. Finally, if k is in S, i.e., $x^k = 1$, then $x^{-k} = (x^k)^{-1} = 1$ too, so $-k$ is in S.

(b) This follows from the Cancellation Law 2.2.3.

(c) Suppose that $S \neq \{0\}$. Theorem 2.3.3 shows that $S = \mathbb{Z}n$, where n is the smallest positive integer in S. If x^k is an arbitrary power, we divide k by n, writing $k = qn + r$ with r in the range $0 \leq r < n$. Then $x^{qn} = 1^q = 1$, and $x^k = x^{qn} x^r = x^r$. Therefore x^k is equal to one of the powers $1, x, \ldots, x^{n-1}$. It follows from **(b)** that these powers are distinct, because x^n is the smallest positive power equal to 1. $\square$

The group $\langle x \rangle = \{1, x, \ldots, x^{n-1}\}$ described by part **(c)** of this proposition is called a *cyclic group of order n*. It is called cyclic because repeated multiplication by x cycles through the n elements.

An element x of a group has *order n* if n is the smallest positive integer with the property $x^n = 1$, which is the same thing as saying that the cyclic subgroup $\langle x \rangle$ generated by x has order n.

With the usual presentation of the symmetric group S_3, the element x has order 3, and y has order 2. In any group, the identity element is the only element of order 1.

If $x^n \neq 1$ for all $n > 0$, one says that x has *infinite order*. The matrix $\begin{bmatrix} 1 & 1 \\ 0 & 1 \end{bmatrix}$ has infinite order in $GL_2(\mathbb{R})$, while $\begin{bmatrix} 1 & 1 \\ -1 & 0 \end{bmatrix}$ has order 6.

When x has infinite order, the group $\langle x \rangle$ is said to be *infinite cyclic*. We won't have much to say about that case.

Proposition 2.4.3 Let x be an element of finite order n in a group, and let k be an integer that is written as $k = nq + r$ where q and r are integers and r is in the range $0 \leq r < n$.

- $x^k = x^r$.
- $x^k = 1$ if and only if $r = 0$.
- Let d be the greatest common divisor of k and n. The order of x^k is equal to n/d. $\square$

One may also speak of the subgroup of a group G generated by a subset U. This is the smallest subgroup of G that contains U, and it consists of all elements of G that can be expressed as a product of a string of elements of U and of their inverses. A subset U of G is said to *generate* G if every element of G is such a product. For example, we saw in (2.2.7) that the set $U = \{x, y\}$ generates the symmetric group S_3. The elementary matrices generate GL_n (1.2.16). In both of these examples, inverses aren't needed. That isn't always true. An infinite cyclic group $\langle x \rangle$ is generated by the element x, but negative powers are needed to fill out the group.

The *Klein four group* V, the group consisting of the four matrices

(2.4.4)
$$\begin{bmatrix} \pm 1 & \\ & \pm 1 \end{bmatrix},$$

is the simplest group that is not cyclic. Any two of its elements different from the identity generate V. The *quaternion group* H is another example of a small group. It consists of the eight matrices

(2.4.5)
$$H = \{\pm 1, \pm i, \pm j, \pm k\},$$

where

$$\mathbf{1} = \begin{bmatrix} 1 & 0 \\ 0 & 1 \end{bmatrix}, \quad \mathbf{i} = \begin{bmatrix} i & 0 \\ 0 & -i \end{bmatrix}, \quad \mathbf{j} = \begin{bmatrix} 0 & 1 \\ -1 & 0 \end{bmatrix}, \quad \mathbf{k} = \begin{bmatrix} 0 & i \\ i & 0 \end{bmatrix}.$$

These matrices can be obtained from the *Pauli matrices* of physics by multiplying by i. The two elements $\mathbf{i}$ and $\mathbf{j}$ generate H. Computation leads to the formulas

(2.4.6) $\mathbf{i}^2 = \mathbf{j}^2 = \mathbf{k}^2 = -\mathbf{1}$, $\mathbf{ij} = -\mathbf{ji} = \mathbf{k}$, $\mathbf{jk} = -\mathbf{kj} = \mathbf{i}$, $\mathbf{ki} = -\mathbf{ik} = \mathbf{j}$.

2.5 HOMOMORPHISMS

Let G and G' be groups, written with multiplicative notation. A *homomorphism* $\varphi : G \to G'$ is a map from G to G' such that for all a and b in G,

(2.5.1)
$$\varphi(ab) = \varphi(a)\varphi(b).$$

The left side of this equation means

> *first multiply a and b in G, then send the product to G' using the map φ,*

while the right side means

> *first send a and b individually to G' using the map φ, then multiply their images in G'.*

Intuitively, a homomorphism is a map that is compatible with the laws of composition in the two groups, and it provides a way to relate different groups.

Examples 2.5.2   The following maps are homomorphisms:

(a) the determinant function $\det: GL_n(\mathbb{R}) \to \mathbb{R}^\times$ (1.4.10),
(b) the sign homomorphism $\sigma: S_n \to \{\pm 1\}$ that sends a permutation to its sign (1.5.11),
(c) the exponential map $\exp: \mathbb{R}^+ \to \mathbb{R}^\times$ defined by $x \rightsquigarrow e^x$,
(d) the map $\varphi: \mathbb{Z}^+ \to G$ defined by $\varphi(n) = a^n$, where a is a given element of G,
(e) the absolute value map $|\ |: \mathbb{C}^\times \to \mathbb{R}^\times$.

In examples **(c)** and **(d)**, the law of composition is written additively in the domain and multiplicatively in the range. The condition (2.5.1) for a homomorphism must be rewritten to take this into account. It becomes

$$\varphi(a + b) = \varphi(a)\varphi(b).$$

The formula showing that the exponential map is a homomorphism is $e^{a+b} = e^a e^b$.

The following homomorphisms need to be mentioned, though they are less interesting. The *trivial homomorphism* $\varphi: G \to G'$ between any two groups maps every element of G to the identity in G'. If H is a subgroup of G, the *inclusion map* $i: H \to G$ defined by $i(x) = x$ for x in H is a homomorphism.

Proposition 2.5.3   Let $\varphi: G \to G'$ be a group homomorphism.

(a) If $a_1, \ldots, a_k$ are elements of G, then $\varphi(a_1 \cdots a_k) = \varphi(a_1) \cdots \varphi(a_k)$.
(b) φ maps the identity to the identity: $\varphi(1_G) = 1_{G'}$.
(c) φ maps inverses to inverses: $\varphi(a^{-1}) = \varphi(a)^{-1}$.

Proof. The first assertion follows by induction from the definition. Next, since $1 \cdot 1 = 1$ and since φ is a homomorphism, $\varphi(1)\varphi(1) = \varphi(1 \cdot 1) = \varphi(1)$. We cancel $\varphi(1)$ from both sides (2.2.3) to obtain $\varphi(1) = 1$. Finally, $\varphi(a^{-1})\varphi(a) = \varphi(a^{-1}a) = \varphi(1) = 1$. Hence $\varphi(a^{-1})$ is the inverse of $\varphi(a)$. □

A group homomorphism determines two important subgroups: its image and its kernel.

• The *image* of a homomorphism $\varphi: G \to G'$, often denoted by $\operatorname{im} \varphi$, is simply the image of φ as a map of sets:

(2.5.4) $\operatorname{im} \varphi = \{x \in G' \mid x = \varphi(a) \text{ for some } a \text{ in } G\},$

Another notation for the image would be $\varphi(G)$.

The image of the map $\mathbb{Z}^+ \to G$ that sends $n \rightsquigarrow a^n$ is the cyclic subgroup $\langle a \rangle$ generated by a.

The image of a homomorphism is a subgroup of the range. We will verify closure and omit the other verifications. Let x and y be elements of the image. This means that there are elements a and b in G such that $x = \varphi(a)$ and $y = \varphi(b)$. Since φ is a homomorphism, $xy = \varphi(a)\varphi(b) = \varphi(ab)$. So xy is equal to $\varphi(\textit{something})$. It is in the image too.

- The *kernel* of a homomorphism is more subtle and also more important. The kernel of φ, often denoted by $\ker \varphi$, is the set of elements of G that are mapped to the identity in G':

(2.5.5) $$\ker \varphi = \{a \in G \mid \varphi(a) = 1\}.$$

The kernel is a subgroup of G because, if a and b are in the kernel, then $\varphi(ab) = \varphi(a)\varphi(b) = 1 \cdot 1 = 1$, so ab is in the kernel, and so on.

The kernel of the determinant homomorphism $GL_n(\mathbb{R}) \to \mathbb{R}^\times$ is the special linear group $SL_n(\mathbb{R})$ (2.2.11). The kernel of the sign homomorphism $S_n \to \{\pm 1\}$ is called the *alternating group*. It consists of the even permutations, and is denoted by A_n:

(2.5.6)   The alternating group A_n is the group of even permutations.

The kernel is important because it controls the entire homomorphism. It tells us not only which elements of G are mapped to the identity in G', but also which pairs of elements have the same image in G'.

- If H is a subgroup of a group G and a is an element of G, the notation aH will stand for the set of all products ah with h in H:

(2.5.7) $$aH = \{g \in G \mid g = ah \text{ for some } h \text{ in } H\}.$$

This set is called a *left coset* of H in G, the word "left" referring to the fact that the element a appears on the left.

Proposition 2.5.8 Let $\varphi: G \to G'$ be a homomorphism of groups, and let a and b be elements of G. Let K be the kernel of φ. The following conditions are equivalent:

- $\varphi(a) = \varphi(b)$,
- $a^{-1}b$ is in K,
- b is in the coset aK,
- The cosets bK and aK are equal.

Proof. Suppose that $\varphi(a) = \varphi(b)$. Then $\varphi(a^{-1}b) = \varphi(a^{-1})\varphi(b) = \varphi(a)^{-1}\varphi(b) = 1$. Therefore $a^{-1}b$ is in the kernel K. To prove the converse, we turn this argument around. If $a^{-1}b$ is in K, then $1 = \varphi(a^{-1}b) = \varphi(a)^{-1}\varphi(b)$, so $\varphi(a) = \varphi(b)$. This shows that the first two bullets are equivalent. Their equivalence with the other bullets follows. $\qquad \square$

Corollary 2.5.9 A homomorphism $\varphi: G \to G'$ is injective if and only if its kernel K is the trivial subgroup $\{1\}$ of G.

Proof. If $K = \{1\}$, Proposition 2.5.8 shows that $\varphi(a) = \varphi(b)$ only when $a^{-1}b = 1$, i.e., $a = b$. Conversely, if φ is injective, then the identity is the only element of G such that $\varphi(a) = 1$, so $K = \{1\}$. $\qquad\square$

The kernel of a homomorphism has another important property that is explained in the next proposition. If a and g are elements of a group G, the element gag^{-1} is called the *conjugate* of a by g.

Definition 2.5.10 A subgroup N of a group G is a *normal subgroup* if for every a in N and every g in G, the conjugate gag^{-1} is in N.

Proposition 2.5.11 The kernel of a homomorphism is a normal subgroup.

Proof. If a is in the kernel of a homomorphism $\varphi: G \to G'$ and if g is any element of G, then $\varphi(gag^{-1}) = \varphi(g)\varphi(a)\varphi(g^{-1}) = \varphi(g)1\varphi(g)^{-1} = 1$. Therefore gag^{-1} is in the kernel too. $\qquad\square$

Thus the special linear group $SL_n(\mathbb{R})$ is a normal subgroup of the general linear group $GL_n(\mathbb{R})$, and the alternating group A_n is a normal subgroup of the symmetric group S_n. Every subgroup of an abelian group is normal, because if G is abelian, then $gag^{-1} = a$ for all a and all g in the group. But subgroups of nonabelian groups needn't be normal. For example, in the symmetric group S_3, with its usual presentation (2.2.7), the cyclic subgroup $\langle y \rangle$ of order two is not normal, because y is in G, but $xyx^{-1} = x^2y$ isn't in $\langle y \rangle$.

• The *center* of a group G, which is often denoted by Z, is the set of elements that commute with every element of G:

$$(2.5.12) \qquad\qquad Z = \{z \in G \mid zx = xz \text{ for all } x \in G\}.$$

It is always a normal subgroup of G. The center of the special linear group $SL_2(\mathbb{R})$ consists of the two matrices $I, -I$. The center of the symmetric group S_n is trivial if $n \geq 3$.

Example 2.5.13 A homomorphism $\varphi: S_4 \to S_3$ between symmetric groups.

There are three ways to partition the set of four indices $\{1, 2, 3, 4\}$ into pairs of subsets of order two, namely

$$(2.5.14) \qquad \Pi_1 : \{1, 2\} \cup \{3, 4\}, \quad \Pi_2 : \{1, 3\} \cup \{2, 4\}, \quad \Pi_3 : \{1, 4\} \cup \{2, 3\}.$$

An element of the symmetric group S_4 permutes the four indices, and by doing so it also permutes these three partitions. This defines the map φ from S_4 to the group of permutations of the set $\{\Pi_1, \Pi_2, \Pi_3\}$, which is the symmetric group S_3. For example, the 4-cycle $p = (1\,2\,3\,4)$ acts on subsets of order two as follows:

$$\{1, 2\} \rightsquigarrow \{2, 3\} \quad \{1, 3\} \rightsquigarrow \{2, 4\} \quad \{1, 4\} \rightsquigarrow \{1, 2\}$$
$$\{2, 3\} \rightsquigarrow \{3, 4\} \quad \{2, 4\} \rightsquigarrow \{1, 3\} \quad \{3, 4\} \rightsquigarrow \{1, 4\}.$$

Looking at this action, one sees that p acts on the set $\{\Pi_1, \Pi_2, \Pi_3\}$ of partitions as the transposition $(\Pi_1\,\Pi_3)$ that fixes Π_2 and interchanges Π_1 and Π_3.

If p and q are elements of S_4, the product pq is the composed permutation $p \circ q$, and the action of pq on the set $\{\Pi_1, \Pi_2, \Pi_3\}$ is the composition of the actions of q and p. Therefore $\varphi(pq) = \varphi(p)\varphi(q)$, and φ is a homomorphism.

The map is surjective, so its image is the whole group S_3. Its kernel can be computed. It is the subgroup of S_4 consisting of the identity and the three products of disjoint transpositions:

(2.5.15) $$K = \{1, \ (\mathbf{12})(\mathbf{34}), \ (\mathbf{13})(\mathbf{24}), \ (\mathbf{14})(\mathbf{23})\}. \qquad \square$$

2.6 ISOMORPHISMS

An *isomorphism* $\varphi: G \to G'$ from a group G to a group G' is a bijective group homomorphism – a bijective map such that $\varphi(ab) = \varphi(a)\varphi(b)$ for all a and b in G.

Examples 2.6.1

- The exponential map e^x is an isomorphism, when it is viewed as a map from the additive group $\mathbb{R}^+$ to its image, the multiplicative group of positive real numbers.
- If a is an element of infinite order in a group G, the map sending $n \rightsquigarrow a^n$ is an isomorphism from the additive group $\mathbb{Z}^+$ to the infinite cyclic subgroup $\langle a \rangle$ of G.
- The set $\mathcal{P}$ of $n \times n$ permutation matrices is a subgroup of GL_n, and the map $S_n \to \mathcal{P}$ that sends a permutation to its associated matrix (1.5.7) is an isomorphism. $\qquad \square$

Corollary 2.5.9 gives us a way to verify that a homomorphism $\varphi: G \to G'$ is an isomorphism. To do so, we check that $\ker \varphi = \{1\}$, which implies that φ is injective, and also that $\operatorname{im} \varphi = G'$, that is, φ is surjective.

Lemma 2.6.2 If $\varphi: G \to G'$ is an isomorphism, the inverse map $\varphi^{-1}: G' \to G$ is also an isomorphism.

Proof. The inverse of a bijective map is bijective. We must show that for all x and y in G', $\varphi^{-1}(x)\varphi^{-1}(y) = \varphi^{-1}(xy)$. We set $a = \varphi^{-1}(x)$, $b = \varphi^{-1}(y)$, and $c = \varphi^{-1}(xy)$. What has to be shown is that $ab = c$, and since φ is bijective, it suffices to show that $\varphi(ab) = \varphi(c)$. Since φ is a homomorphism,

$$\varphi(ab) = \varphi(a)\varphi(b) = xy = \varphi(c). \qquad \square$$

This lemma shows that when $\varphi: G \to G'$ is an isomorphism, we can make a computation in either group, then use φ or φ^{-1} to carry it over to the other. So, for computation with the group law, the two groups have identical properties. To picture this conclusion intuitively, suppose that the elements of one of the groups are put into unlabeled boxes, and that we have an oracle that tells us, when presented with two boxes, which box contains their product. We will have no way to decide whether the elements in the boxes are from G or from G'.

Two groups G and G' are said to be *isomorphic* if there exists an isomorphism φ from G to G'. We sometimes indicate that two groups are isomorphic by the symbol $\approx$

(2.6.3) $$G \approx G' \quad \text{means that} \quad G \text{ is isomorphic to } G'.$$

Since isomorphic groups have identical properties, it is often convenient to identify them with each other when speaking informally. For instance, we often blur the distinction between the symmetric group S_n and the isomorphic group $\mathcal{P}$ of permutation matrices.

- The groups isomorphic to a given group G form what is called the *isomorphism class* of G.

Any two groups in an isomorphism class are isomorphic. When one speaks of *classifying groups*, what is meant is to describe these isomorphism classes. This is too hard to do for all groups, but we will see that every group of prime order p is cyclic. So all groups of order p are isomorphic. There are two isomorphism classes of groups of order 4 (2.11.5) and five isomorphism classes of groups of order 12 (7.8.1).

An interesting and sometimes confusing point about isomorphisms is that there exist isomorphisms $\varphi : G \rightarrow G$ from a group G to itself. Such an isomorphism is called an *automorphism*. The identity map is an automorphism, of course, but there are nearly always others. The most important type of automorphism is conjugation: Let g be a fixed element of a group G. *Conjugation by g* is the map φ from G to itself defined by

(2.6.4)
$$\varphi(x) = gxg^{-1}.$$

This is an automorphism because, first of all, it is a homomorphism:

$$\varphi(xy) = gxyg^{-1} = gxg^{-1}gyg^{-1} = \varphi(x)\varphi(y),$$

and second, it is bijective because it has an inverse function – conjugation by g^{-1}.

If the group is abelian, conjugation by any element g is the identity map: $gxg^{-1} = x$. But any noncommutative group has nontrivial conjugations, and so it has automorphisms different from the identity. For instance, in the symmetric group S_3, presented as usual, conjugation by y interchanges x and x^2.

As was said before, the element gxg^{-1} is the *conjugate* of x by g, and two elements x and x' of a group G are *conjugate* if $x' = gxg^{-1}$ for some g in G. The conjugate gxg^{-1} behaves in much the same way as the element x itself; for example, it has the same order in the group. This follows from the fact that it is the image of x by an automorphism. (See the discussion following Lemma 2.6.2.)

Note: One may sometimes wish to determine whether or not two elements x and y of a group G are conjugate, i.e., whether or not there is an element g in G such that $y = gxg^{-1}$. It is almost always simpler to rewrite the equation to be solved for g as $yg = gx$. □

- The *commutator* $aba^{-1}b^{-1}$ is another element associated to a pair a, b of elements of a group.

The next lemma follows by moving things from one side of an equation to the other.

Lemma 2.6.5 Two elements a and b of a group commute, $ab = ba$, if and only if $aba^{-1} = b$, and this is true if and only if $aba^{-1}b^{-1} = 1$. □

2.7 EQUIVALENCE RELATIONS AND PARTITIONS

A fundamental mathematical construction starts with a set S and forms a new set by equating certain elements of S. For instance, we may divide the set of integers into two classes, the even integers and the odd integers. The new set we obtain consists of two elements that could be called *Even* and *Odd*. Or, it is common to view congruent triangles in the plane as equivalent geometric objects. This very general procedure arises in several ways that we discuss here.

• A *partition* Π of a set S is a subdivision of S into nonoverlapping, nonempty subsets:

(2.7.1) $S = $ union of disjoint nonempty subsets.

The two sets *Even* and *Odd* partition the set of integers. With the usual notation, the sets

(2.7.2) $\{1\}, \{y, xy, x^2y\}, \{x, x^2\}$

form a partition of the symmetric group S_3.

• An *equivalence relation* on a set S is a relation that holds between certain pairs of elements of S. We may write it as $a \sim b$ and speak of it as *equivalence* of a and b. An equivalence relation is required to be:

(2.7.3)

- *transitive*: If $a \sim b$ and $b \sim c$, then $a \sim c$.
- *symmetric*: If $a \sim b$, then $b \sim a$.
- *reflexive*: For all a, $a \sim a$.

Congruence of triangles is an example of an equivalence relation on the set of triangles in the plane. If A, B, and C are triangles, and if A is congruent to B and B is congruent to C, then A is congruent to C, etc.

Conjugacy is an equivalence relation on a group. Two group elements are conjugate, $a \sim b$, if $b = gag^{-1}$ for some group element g. We check transitivity: Suppose that $a \sim b$ and $b \sim c$. This means that $b = g_1ag_1^{-1}$ and $c = g_2bg_2^{-1}$ for some group elements g_1 and g_2. Then $c = g_2(g_1ag_1^{-1})g_2^{-1} = (g_2g_1)a(g_2g_1)^{-1}$, so $a \sim c$.

The concepts of a partition of S and an equivalence relation on S are logically equivalent, though in practice one may be presented with just one of the two.

Proposition 2.7.4 An equivalence relation on a set S determines a partition of S, and conversely.

Proof. Given a partition of S, the corresponding equivalence relation is defined by the rule that $a \sim b$ if a and b lie in the same subset of the partition. The axioms for an equivalence relation are obviously satisfied. Conversely, given an equivalence relation, one defines a partition this way: The subset that contains a is the set of all elements b such that $a \sim b$. This subset is called the *equivalence class* of a. We'll denote it by C_a here:

(2.7.5) $C_a = \{b \in S \mid a \sim b\}.$

The next lemma completes the proof of the proposition.   □

Lemma 2.7.6 Given an equivalence relation on a set S, the subsets of S that are equivalence classes partition S.

Proof. This is an important point, so we will check it carefully. We must remember that the notation C_a stands for a subset defined in a certain way. The partition consists of the subsets, and several notations may describe the same subset.

The reflexive axiom tells us that a is in its equivalence class. Therefore the class C_a is nonempty, and since a can be any element, the union of the equivalence classes is the whole set S. The remaining property of a partition that must be verified is that equivalence classes are disjoint. To show this, we show:

(2.7.7)   If C_a and C_b have an element in common, then $C_a = C_b$.

Since we can interchange the roles of a and b, it will suffice to show that if C_a and C_b have an element, say d, in common, then $C_b \subset C_a$, i.e., any element x of C_b is also in C_a. If x is in C_b, then $b \sim x$. Since d is in both sets, $a \sim d$ and $b \sim d$, and the symmetry property tells us that $d \sim b$. So we have $a \sim d$, $d \sim b$, and $b \sim x$. Two applications of transitivity show that $a \sim x$, and therefore that x is in C_a. $\square$

For example, the relation on a group defined by $a \sim b$ if a and b are elements of the same order is an equivalence relation. The corresponding partition is exhibited in (2.7.2) for the symmetric group S_3.

If a partition of a set S is given, we may construct a new set $\overline{S}$ whose elements are the subsets. We imagine putting the subsets into separate piles, and we regard the piles as the elements of our new set $\overline{S}$. It seems advisable to have a notation to distinguish a subset from the element of the set $\overline{S}$ (the pile) that it represents. If U is a subset, we will denote by $[U]$ the corresponding element of $\overline{S}$. Thus if S is the set of integers and if *Even* and *Odd* denote the subsets of even and odd integers, respectively, then $\overline{S}$ contains the two elements $[Even]$ and $[Odd]$.

We will use this notation more generally. When we want to regard a subset U of S as an element of a set of subsets of S, we denote it by $[U]$.

When an equivalence relation on S is given, the equivalence classes form a partition, and we obtain a new set $\overline{S}$ whose elements are the equivalence classes $[C_a]$. We can think of the elements of this new set in another way, as the set obtained by changing what we mean by equality among elements. If a and b are in S, we interpret $a \sim b$ to mean that a and b become equal in $\overline{S}$, because $C_a = C_b$. With this way of looking at it, the difference between the two sets S and $\overline{S}$ is that in $\overline{S}$ more elements have been declared "equal," i.e., equivalent. It seems to me that we often treat congruent triangles this way in school.

For any equivalence relation, there is a natural surjective map

(2.7.8) $\pi : S \to \overline{S}$

that maps an element a of S to its equivalence class: $\pi(a) = [C_a]$. When we want to regard $\overline{S}$ as the set obtained from S by changing the notion of equality, it will be convenient to denote the element $[C_a]$ of $\overline{S}$ by the symbol $\overline{a}$. Then the map π becomes

$$\pi(a) = \overline{a}$$

We can work in $\overline{S}$ with the symbols used for elements of S, but with bars over them to remind us of the new rule:

(2.7.9)   If a and b are in S, then $\overline{a} = \overline{b}$ means $a \sim b$.

A disadvantage of this bar notation is that many symbols represent the same element of $\overline{S}$. Sometimes this disadvantage can be overcome by choosing a particular element, a *representative element*, in each equivalence class. For example, the even and the odd integers are often represented by $\overline{0}$ and $\overline{1}$:

(2.7.10) $\{[Even], [Odd]\} = \{\overline{0}, \overline{1}\}$.

Though the pile picture may be easier to grasp at first, the second way of viewing $\overline{S}$ is often better because the bar notation is easier to manipulate algebraically.

The Equivalence Relation Defined by a Map

Any map of sets $f : S \to T$ gives us an equivalence relation on its domain S. It is defined by the rule $a \sim b$ if $f(a) = f(b)$.

• The *inverse image* of an element t of T is the subset of S consisting of all elements s such that $f(s) = t$. It is denoted symbolically as

(2.7.11) $f^{-1}(t) = \{s \in S \mid f(s) = t\}$.

This is symbolic notation. Please remember that unless f is bijective, f^{-1} will not be a map. The inverse images are also called the *fibres* of the map f, and the fibres that are not empty are the equivalence classes for the relation defined above.

Here the set $\overline{S}$ of equivalence classes has another incarnation, as the image of the map. The elements of the image correspond bijectively to the nonempty fibres, which are the equivalence classes.

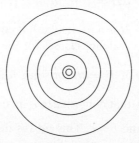

(2.7.12)   Some Fibres of the Absolute Value Map $\mathbb{C}^\times \to \mathbb{R}^\times$.

Example 2.7.13 If G is a finite group, we can define a map $f : G \to \mathbb{N}$ to the set $\{1, 2, 3, \ldots\}$ of natural numbers, letting $f(a)$ be the order of the element a of G. The fibres of this map are the sets of elements with the same order (see (2.7.2), for example).   □

We go back to a group homomorphism $\varphi : G \to G'$. The equivalence relation on G defined by φ is usually denoted by $\equiv$, rather than by $\sim$, and is referred to as *congruence*:

$$(2.7.14) \qquad\qquad a \equiv b \text{ if } \varphi(a) = \varphi(b).$$

We have seen that elements a and b of G are congruent, i.e., $\varphi(a) = \varphi(b)$, if and only if b is in the coset aK of the kernel K (2.5.8).

Proposition 2.7.15 Let K be the kernel of a homomorphism $\varphi : G \to G'$. The fibre of φ that contains an element a of G is the coset aK of K. These cosets partition the group G, and they correspond to elements of the image of φ. $\qquad\qquad\square$

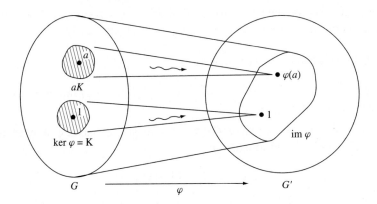

$$(2.7.16) \qquad\qquad \text{A Schematic Diagram of a Group Homomorphism.}$$

2.8 COSETS

As before, if H is a subgroup of G and if a is an element of G, the subset

$$(2.8.1) \qquad\qquad aH = \{ah \mid h \text{ in } H\}.$$

is called a *left coset*. The subgroup H is a particular left coset because $H = 1H$.

The cosets of H in G are equivalence classes for the congruence relation

$$(2.8.2) \qquad\qquad a \equiv b \text{ if } b = ah \text{ for some } h \text{ in } H.$$

This is very simple, but let's verify that congruence is an equivalence relation.

Transitivity: Suppose that $a \equiv b$ and $b \equiv c$. This means that $b = ah$ and $c = bh'$ for some elements h and h' of H. Therefore $c = ahh'$. Since H is a subgroup, hh' is in H, and thus $a \equiv c$.
Symmetry: Suppose $a \equiv b$, so that $b = ah$. Then $a = bh^{-1}$ and h^{-1} is in H, so $b \equiv a$.
Reflexivity: $a = a1$ and 1 is in H, so $a \equiv a$.

Notice that we have made use of all the defining properties of a subgroup here: closure, inverses, and identity.

Corollary 2.8.3 The left cosets of a subgroup H of a group G partition the group.

Proof. The left cosets are the equivalence classes for the congruence relation (2.8.2).   □

Keep in mind that the notation aH defines a certain subset of G. As with any equivalence relation, several notations may define the same subset. For example, in the symmetric group S_3, with the usual presentation (2.2.6), the element y generates a cyclic subgroup $H = \langle y \rangle$ of order 2. There are three left cosets of H in G:

$$(2.8.4) \qquad H = \{1, y\} = yH, \quad xH = \{x, xy\} = xyH, \quad x^2H = \{x^2, x^2y\} = x^2yH.$$

These sets do partition the group.

Recapitulating, let H be a subgroup of a group G and let a and b be elements of G. The following are equivalent:

(2.8.5)

- $b = ah$ for some h in H, or, $a^{-1}b$ is an element of H,
- b is an element of the left coset aH,
- the left cosets aH and bH are equal.

The number of left cosets of a subgroup is called the *index* of H in G. The index is denoted by

$$(2.8.6) \qquad\qquad\qquad [G:H].$$

Thus the index of the subgroup $\langle y \rangle$ of S_3 is 3. When G is infinite, the index may be infinite too.

Lemma 2.8.7 All left cosets aH of a subgroup H of a group G have the same order.

Proof. Multiplication by a defines a map $H \rightarrow aH$ that sends $h \rightsquigarrow ah$. This map is bijective because its inverse is multiplication by a^{-1}.   □

Since the cosets all have the same order, and since they partition the group, we obtain the important *Counting Formula*

$$(2.8.8) \qquad\qquad\qquad |G| = |H|\,[G:H]$$
$$(order\ of\ G) = (order\ of\ H)(number\ of\ cosets),$$

where, as always, $|G|$ denotes the order of the group. The equality has the obvious meaning if some terms are infinite. For the subgroup $\langle y \rangle$ of S_3, the formula reads $6 = 2 \cdot 3$.

It follows from the counting formula that the terms on the right side of (2.8.8) divide the left side. One of these facts is called Lagrange's Theorem:

Theorem 2.8.9 Lagrange's Theorem. Let H be a subgroup of a finite group G. The order of H divides the order of G.   □

Corollary 2.8.10 The order of an element of a finite group divides the order of the group.

Proof. The order of an element a of a group G is equal to the order of the cyclic subgroup $\langle a \rangle$ generated by a (Proposition 2.4.2). □

Corollary 2.8.11 Suppose that a group G has prime order p. Let a be any element of G other than the identity. Then G is the cyclic group $\langle a \rangle$ generated by a.

Proof. The order of an element $a \neq 1$ is greater than 1 and it divides the order of G, which is the prime integer p. So the order of a is equal to p. This is also the order of the cyclic subgroup $\langle a \rangle$ generated by a. Since G has order p, $\langle a \rangle = G$. □

This corollary classifies groups of prime order p. They form one isomorphism class, the class of the cyclic groups of order p.

The counting formula can also be applied when a homomorphism $\varphi : G \rightarrow G'$ is given. As we have seen (2.7.15), the left cosets of the kernel $\ker \varphi$ are the nonempty fibres of the map φ. They are in bijective correspondence with the elements of the image.

(2.8.12)
$$[G : \ker \varphi] = |\operatorname{im} \varphi|.$$

Corollary 2.8.13 Let $\varphi : G \rightarrow G'$ be a homomorphism of finite groups. Then

- $|G| = |\ker \varphi| \cdot |\operatorname{im} \varphi|$,
- $|\ker \varphi|$ divides $|G|$, and
- $|\operatorname{im} \varphi|$ divides both $|G|$ and $|G'|$.

Proof. The first formula is obtained by combining (2.8.8) and (2.8.12), and it implies that $|\ker \varphi|$ and $|\operatorname{im} \varphi|$ divide $|G|$. Since the image is a subgroup of G', Lagrange's theorem tells us that its order divides $|G'|$ too. □

For example, the sign homomorphism $\sigma : S_n \rightarrow \{\pm 1\}$ (2.5.2)**(b)** is surjective, so its image has order 2. Its kernel, the alternating group A_n, has order $\frac{1}{2} n!$. Half of the elements of S_n are even permutations, and half are odd permutations.

The Counting Formula 2.8.8 has an analogue when a chain of subgroups is given.

Proposition 2.8.14 Multiplicative Property of the Index. Let $G \supset H \supset K$ be subgroups of a group G. Then $[G : K] = [G : H][H : K]$.

Proof. We will assume that the two indices on the right are finite, say $[G : H] = m$ and $[H : K] = n$. The reasoning when one or the other is infinite is similar. We list the m cosets of H in G, choosing representative elements for each coset, say as $g_1 H, \ldots, g_m H$. Then $g_1 H \cup \cdots \cup g_m H$ is a partition of G. Similarly, we choose representative elements for each coset of K in H, obtaining a partition $H = h_1 K \cup \cdots \cup h_n K$. Since multiplication by g_i is an invertible operation, $g_i H = g_i h_1 K \cup \cdots \cup g_i h_n K$ will be a partition of the coset $g_i H$. Putting these partitions together, G is partitioned into the mn cosets $g_i h_j K$. □

Right Cosets

Let us go back to the definition of cosets. We made the decision to work with left cosets aH. One can also define right cosets of a subgroup H and repeat the above discussion for them.

The right cosets of a subgroup H of a group G are the sets

(2.8.15) $$Ha = \{ha \mid h \in H\}.$$

They are equivalence classes for the relation (*right congruence*)

$$a \equiv b \text{ if } b = ha, \text{ for some } h \text{ in } H.$$

Right cosets also partition the group G, but they aren't always the same as left cosets. For instance, the right cosets of the subgroup $\langle y \rangle$ of S_3 are

(2.8.16) $$H = \{1, y\} = Hy, \quad Hx = \{x, x^2y\} = Hx^2y, \quad Hx^2 = \{x^2, xy\} = Hxy.$$

This isn't the same as the partition (2.8.4) into left cosets. However, if a subgroup is normal, its right and left cosets are equal.

Proposition 2.8.17 Let H be a subgroup of a group G. The following conditions are equivalent:

(i) H is a normal subgroup: For all h in H and all g in G, ghg^{-1} is in H.

(ii) For all g in G, $gHg^{-1} = H$.

(iii) For all g in G, the left coset gH is equal to the right coset Hg.

(iv) Every left coset of H in G is a right coset.

Proof. The notation gHg^{-1} stands for the set of all elements ghg^{-1}, with h in H.

Suppose that H is normal. So **(i)** holds, and it implies that $gHg^{-1} \subset H$ for all g in G. Substituting g^{-1} for g shows that $g^{-1}Hg \subset H$ as well. We multiply this inclusion on the left by g and on the right by g^{-1} to conclude that $H \subset gHg^{-1}$. Therefore $gHg^{-1} = H$. This shows that **(i)** implies **(ii)**. It is clear that **(ii)** implies **(i)**. Next, if $gHg^{-1} = H$, we multiply this equation on the right by g to conclude that $gH = Hg$. This shows that **(ii)** implies **(iii)**. One sees similarly that **(iii)** implies **(ii)**. Since **(iii)** implies **(iv)** is obvious, it remains only to check that **(iv)** implies **(iii)**.

We ask: Under what circumstances can a left coset be equal to a right coset? We recall that the right cosets partition the group G, and we note that the left coset gH and the right coset Hg have an element in common, namely $g = g \cdot 1 = 1 \cdot g$. So if the left coset gH is equal to any right coset, that coset must be Hg. $\qquad\square$

Proposition 2.8.18

(a) If H is a subgroup of a group G and g is an element of G, the set gHg^{-1} is also a subgroup.

(b) If a group G has just one subgroup H of order r, then that subgroup is normal.

Proof. **(a)** Conjugation by g is an automorphism of G (see (2.6.4)), and gHg^{-1} is the image of H. **(b)** See (2.8.17): gHg^{-1} is a subgroup of order r. $\qquad\square$

Note: If H is a subgroup of a finite group G, the counting formulas using right cosets or left cosets are the same, so the number of left cosets is equal to the number of right cosets. This is also true when G is infinite, though the proof can't be made by counting (see Exercise M.8). $\qquad\square$

2.9   MODULAR ARITHMETIC

This section contains a brief discussion of one of the most important concepts in number theory, congruence of integers. If you have not run across this concept before, you will want to read more about it. See, for instance, [Stark]. We work with a fixed positive integer n throughout the section.

- Two integers a and b are said to be *congruent modulo n*

$$(2.9.1) \qquad\qquad a \equiv b \text{ modulo } n,$$

if n divides $b - a$, or if $b = a + nk$ for some integer k. For instance, $2 \equiv 17$ modulo 5.

It is easy to check that congruence is an equivalence relation, so we may consider the equivalence classes, called *congruence classes*, that it defines. We use bar notation, and denote the congruence class of an integer a modulo n by the symbol $\bar{a}$. This congruence class is the set of integers

$$(2.9.2) \qquad\qquad \bar{a} = \{ \dots, a - n, a, a + n, a + 2n, \dots \}.$$

If a and b are integers, the equation $\bar{a} = \bar{b}$ means that $a \equiv b$ modulo n, or that n divides $b - a$. The congruence class $\bar{0}$ is the subgroup

$$\bar{0} = \mathbb{Z}n = \{ \dots, -n, 0, n, 2n, \dots \} = \{ kn \mid k \in \mathbb{Z} \}$$

of the additive group $\mathbb{Z}^+$. The other congruence classes are the cosets of this subgroup. Please note that $\mathbb{Z}n$ is not a right coset – it is a subgroup of $\mathbb{Z}^+$. The notation for a coset of a subgroup H analogous to aH, but using additive notation for the law of composition, is $a + H = \{ a + h \mid h \in H \}$. To simplify notation, we denote the subgroup $\mathbb{Z}n$ by H. Then the cosets of H, the congruence classes, are the sets

$$(2.9.3) \qquad\qquad a + H = \{ a + kn \mid k \in \mathbb{Z} \}.$$

The n integers $0, 1, \dots, n - 1$ are representative elements for the n congruence classes.

Proposition 2.9.4   There are n congruence classes modulo n, namely $\bar{0}, \bar{1}, \dots, \overline{n-1}$. The index $[\mathbb{Z} : \mathbb{Z}n]$ of the subgroup $\mathbb{Z}n$ in $\mathbb{Z}$ is n.   □

Let $\bar{a}$ and $\bar{b}$ be congruence classes represented by integers a and b. Their *sum* is defined to be the congruence class of $a + b$, and their *product* is the class of ab. In other words, by definition,

$$(2.9.5) \qquad\qquad \bar{a} + \bar{b} = \overline{a+b} \quad \text{and} \quad \bar{a}\bar{b} = \overline{ab}.$$

This definition needs some justification, because the same congruence class can be represented by many different integers. Any integer a' congruent to a modulo n represents the same class as a does. So it had better be true that if $a' \equiv a$ and $b' \equiv b$, then $a' + b' \equiv a + b$ and $a'b' \equiv ab$. Fortunately, this is so.

Lemma 2.9.6   If $a' \equiv a$ and $b' \equiv b$ modulo n, then $a' + b' \equiv a + b$ and $a'b' \equiv ab$ modulo n.

Proof. Assume that $a' \equiv a$ and $b' \equiv b$, so that $a' = a + rn$ and $b' = b + sn$ for some integers r and s. Then $a' + b' = a + b + (r+s)n$. This shows that $a' + b' \equiv a + b$. Similarly, $a'b' = (a+rn)(b+sn) = ab + (as+rb+rns)n$, so $a'b' \equiv ab$. $\square$

The associative, commutative, and distributive laws hold for addition and multiplication of congruence classes because they hold for addition and multiplication of integers. For example, the distributive law is verified as follows:

$$\overline{a}(\overline{b}+\overline{c}) = \overline{a}(\overline{b+c}) = \overline{a(b+c)} \qquad \textit{(definition of + and} \times \textit{for congruence classes)}$$
$$= \overline{ab+ac} \qquad \textit{(distributive law in the integers)}$$
$$= \overline{ab} + \overline{ac} = \overline{a}\,\overline{b} + \overline{a}\,\overline{c} \qquad \textit{(definition of + and} \times \textit{for congruence classes)}.$$

The verifications of other laws are similar, and we omit them.

The set of congruence classes modulo n may be denoted by any one of the symbols $\mathbb{Z}/\mathbb{Z}n$, $\mathbb{Z}/n\mathbb{Z}$, or $\mathbb{Z}/(n)$. Addition, subtraction, and multiplication in $\mathbb{Z}/\mathbb{Z}n$ can be made explicit by working with integers and taking remainders after division by n. That is what the formulas (2.9.5) mean. They tell us that the map

(2.9.7) $$\mathbb{Z} \to \mathbb{Z}/\mathbb{Z}n$$

that sends an integer a to its congruence class $\overline{a}$ is compatible with addition and multiplication. Therefore computations can be made in the integers and then carried over to $\mathbb{Z}/\mathbb{Z}n$ at the end. However, computations are simpler if the numbers are kept small. This can be done by computing the remainder after some part of a computation has been made.

Thus if $n = 29$, so that $\mathbb{Z}/\mathbb{Z}n = \{\overline{0}, \overline{1}, \overline{2}, \ldots, \overline{28}\}$, then $(\overline{35})(\overline{17}+\overline{7})$ can be computed as $\overline{35} \cdot \overline{24} = \overline{6} \cdot (-\overline{5}) = -\overline{30} = -\overline{1}$.

In the long run, the bars over the numbers become a nuisance. They are often left off. When omitting bars, one just has to remember this rule:

(2.9.8)   To say $a = b$ in $\mathbb{Z}/\mathbb{Z}n$ means that $a \equiv b$ modulo n.

Congruences modulo a prime integer have special properties, which we discuss at the beginning of the next chapter.

2.10 THE CORRESPONDENCE THEOREM

Let $\varphi: G \to \mathcal{G}$ be a group homomorphism, and let H be a subgroup of G. We may *restrict* φ to H, obtaining a homomorphism

(2.10.1) $$\varphi|_H : H \to \mathcal{G}.$$

This means that we take the same map φ but restrict its domain: So by definition, if h is in H, then $[\varphi|_H](h) = \varphi(h)$. (We've added brackets around the symbol $\varphi|_H$ for clarity.) The restriction is a homomorphism because φ is one, and the kernel of $\varphi|_H$ is the intersection of the kernel of φ with H:

(2.10.2) $$\ker(\varphi|_H) = (\ker\varphi) \cap H.$$

This is clear from the definition of the kernel. The image of $\varphi|_H$ is the same as the image $\varphi(H)$ of H under the map φ.

The Counting Formula may help to describe the restriction. According to Corollary (2.8.13), the order of the image divides both $|H|$ and $|\mathcal{G}|$. If $|H|$ and $|\mathcal{G}|$ have no common factor, $\varphi(H) = \{1\}$, so H is contained in the kernel.

Example 2.10.3 The image of the sign homomorphism $\sigma : S_n \rightarrow \{\pm 1\}$ has order 2. If a subgroup H of the symmetric group S_n has odd order, it will be contained in the kernel of σ, the alternating group A_n of even permutations. This will be so when H is the cyclic subgroup generated by a permutation q that is an element of odd order in the group. Every permutation whose order in the group is odd, such as an n-cycle with n odd, is an even permutation. A permutation that has even order in the group may be odd or even. □

Proposition 2.10.4 Let $\varphi : G \rightarrow \mathcal{G}$ be a homomorphism with kernel K and let $\mathcal{H}$ be a subgroup of $\mathcal{G}$. Denote the inverse image $\varphi^{-1}(\mathcal{H})$ by H. Then H is a subgroup of G that contains K. If $\mathcal{H}$ is a normal subgroup of $\mathcal{G}$, then H is a normal subgroup of G. If φ is surjective and if H is a normal subgroup of G, then $\mathcal{H}$ is a normal subgroup of $\mathcal{G}$.

For example, let φ denote the determinant homomorphism $GL_n(\mathbb{R}) \rightarrow \mathbb{R}^\times$. The set of positive real numbers is a subgroup of $\mathbb{R}^\times$; it is normal because $\mathbb{R}^\times$ is abelian. Its inverse image, the set of invertible matrices with positive determinant, is a normal subgroup of $GL_n(\mathbb{R})$.

Proof. This proof is simple, but we must keep in mind that φ^{-1} is not a map. By definition, $\varphi^{-1}(\mathcal{H}) = H$ is the set of elements x of G such that $\varphi(x)$ is in $\mathcal{H}$. First, if x is in the kernel K, then $\varphi(x) = 1$. Since 1 is in $\mathcal{H}$, x is in H. Thus H contains K. We verify the conditions for a subgroup.
Closure: Suppose that x and y are in H. Then $\varphi(x)$ and $\varphi(y)$ are in $\mathcal{H}$. Since $\mathcal{H}$ is a subgroup, $\varphi(x)\varphi(y)$ is in $\mathcal{H}$. Since φ is a homomorphism, $\varphi(x)\varphi(y) = \varphi(xy)$. So $\varphi(xy)$ is in $\mathcal{H}$, and xy is in H.
Identity: 1 is in H because $\varphi(1) = 1$ is in $\mathcal{H}$.
Inverses: Let x be an element of H. Then $\varphi(x)$ is in $\mathcal{H}$, and since $\mathcal{H}$ is a subgroup, $\varphi(x)^{-1}$ is also in $\mathcal{H}$. Since φ is a homomorphism, $\varphi(x)^{-1} = \varphi(x^{-1})$, so $\varphi(x^{-1})$ is in $\mathcal{H}$, and x^{-1} is in H.

Suppose that $\mathcal{H}$ is a normal subgroup. Let x and g be elements of H and G, respectively. Then $\varphi(gxg^{-1}) = \varphi(g)\varphi(x)\varphi(g)^{-1}$ is a conjugate of $\varphi(x)$, and $\varphi(x)$ is in $\mathcal{H}$. Because $\mathcal{H}$ is normal, $\varphi(gxg^{-1})$ is in $\mathcal{H}$, and therefore gxg^{-1} is in H.

Suppose that φ is surjective, and that H is a normal subgroup of G. Let a be in $\mathcal{H}$, and let b be in $\mathcal{G}$. There are elements x of H and y of G such that $\varphi(x) = a$ and $\varphi(y) = b$. Since H is normal, yxy^{-1} is in H, and therefore $\varphi(yxy^{-1}) = bab^{-1}$ is in $\mathcal{H}$. □

Theorem 2.10.5 Correspondence Theorem. Let $\varphi : G \rightarrow \mathcal{G}$ be a *surjective* group homomorphism with kernel K. There is a bijective correspondence between subgroups of $\mathcal{G}$ and subgroups of G that contain K:

$$\{\text{subgroups of } G \text{ that contain } K\} \longleftrightarrow \{\text{subgroups of } \mathcal{G}\}.$$

This correspondence is defined as follows:

a subgroup H of G that contains K $\rightsquigarrow$ its image $\varphi(H)$ in $\mathcal{G}$,

a subgroup $\mathcal{H}$ of $\mathcal{G}$ $\rightsquigarrow$ its inverse image $\varphi^{-1}(\mathcal{H})$ in G.

If H and $\mathcal{H}$ are corresponding subgroups, then H is normal in G if and only if $\mathcal{H}$ is normal in $\mathcal{G}$.

If H and $\mathcal{H}$ are corresponding subgroups, then $|H| = |\mathcal{H}||K|$.

Example 2.10.6 We go back to the homomorphism $\varphi: S_4 \to S_3$ that was defined in Example 2.5.13, and its kernel K (2.5.15).

The group S_3 has six subgroups, four of them proper. With the usual presentation, there is one proper subgroup of order 3, the cyclic group $\langle x \rangle$, and there are three subgroups of order 2, including $\langle y \rangle$. The Correspondence Theorem tells us that there are four proper subgroups of S_4 that contain K. Since $|K| = 4$, there is one subgroup of order 12 and there are three of order 8.

We know a subgroup of order 12, namely the alternating group A_4. That is the subgroup that corresponds to the cyclic group $\langle x \rangle$ of S_3.

The subgroups of order 8 can be explained in terms of symmetries of a square. With vertices of the square labeled as in the figure below, a counterclockwise rotation through the angle $\pi/2$ corresponds to the 4-cycle $(\mathbf{1234})$. Reflection about the diagonal through the vertex 1 corresponds to the transposition $(\mathbf{24})$. These two permutations generate a subgroup of order 8. The other subgroups of order 8 can be obtained by labeling the vertices in other ways.

There are also some subgroups of S_4 that do not contain K. The Correspondence Theorem has nothing to say about those subgroups.   □

Proof of the Correspondence Theorem. Let H be a subgroup of G that contains K, and let $\mathcal{H}$ be a subgroup of $\mathcal{G}$. We must check the following points:

- $\varphi(H)$ is a subgroup of $\mathcal{G}$.
- $\varphi^{-1}(\mathcal{H})$ is a subgroup of G, and it contains K.
- $\mathcal{H}$ is a normal subgroup of $\mathcal{G}$ if and only if $\varphi^{-1}(\mathcal{H})$ is a normal subgroup of G.
- *(bijectivity of the correspondence)* $\varphi(\varphi^{-1}(\mathcal{H})) = \mathcal{H}$ and $\varphi^{-1}(\varphi(H)) = H$.
- $|\varphi^{-1}(\mathcal{H})| = |\mathcal{H}||K|$.

Since $\varphi(H)$ is the image of the homomorphism $\varphi|_H$, it is a subgroup of $\mathcal{G}$. The second and third bullets form Proposition 2.10.4.

Concerning the fourth bullet, the equality $\varphi(\varphi^{-1}(\mathcal{H})) = \mathcal{H}$ is true for any surjective map of sets $\varphi: S \to S'$ and any subset $\mathcal{H}$ of S'. Also, $H \subset \varphi^{-1}(\varphi(H))$ is true for any map

φ of sets and any subset H of S. We omit the verification of these facts. Then the only thing remaining to be verified is that $H \supset \varphi^{-1}(\varphi(H))$. Let x be an element of $\varphi^{-1}(\varphi(H))$. We must show that x is in H. By definition of the inverse image, $\varphi(x)$ is in $\varphi(H)$, say $\varphi(x) = \varphi(a)$, with a in H. Then $a^{-1}x$ is in the kernel K (2.5.8), and since H contains K, $a^{-1}x$ is in H. Since both a and $a^{-1}x$ are in H, x is in H too.

We leave the proof of the last bullet as an exercise. $\qquad\square$

2.11   PRODUCT GROUPS

Let G, G' be two groups. The product set $G \times G'$, the set of pairs of elements (a, a') with a in G and a' in G', can be made into a group by component-wise multiplication – that is, multiplication of pairs is defined by the rule

(2.11.1) $$(a, a') \cdot (b, b') = (ab, a'b').$$

The pair $(1, 1)$ is the identity, and the inverse of (a, a') is (a^{-1}, a'^{-1}). The associative law in $G \times G'$ follows from the fact that it holds in G and in G'.

The group obtained in this way is called the *product* of G and G' and is denoted by $G \times G'$. It is related to the two factors G and G' in a simple way that we can sum up in terms of some homomorphisms

(2.11.2)

They are defined by $i(x) = (x, 1)$, $i'(x') = (1, x')$, $p(x, x') = x$, $p'(x, x') = x'$. The injective homomorphisms i and i' may be used to identify G and G' with their images, the subgroups $G \times 1$ and $1 \times G'$ of $G \times G'$. The maps p and p' are surjective, the kernel of p is $1 \times G'$, and the kernel of p' is $G \times 1$. These are the *projections*.

It is obviously desirable to decompose a given group G as a product, that is, to find groups H and H' such that G is isomorphic to the product $H \times H'$. The groups H and H' will be simpler, and the relation between $H \times H'$ and its factors is easily understood. It is rare that a group is a product, but it does happen occasionally.

For example, it is rather surprising that a cyclic group of order 6 can be decomposed: A cyclic group C_6 of order 6 is isomorphic to the product $C_2 \times C_3$ of cyclic groups of orders 2 and 3. To see this, say that $C_2 = \langle y \rangle$ and $C_3 = \langle z \rangle$, with $y^2 = 1$ and $z^3 = 1$, and let x denote the element (y, z) of the product group $C_2 \times C_3$. The smallest positive integer k such that $x^k = (y^k, z^k)$ is the identity $(1, 1)$ is $k = 6$. So x has order 6. Since $C_2 \times C_3$ also has order 6, it is equal to the cyclic group $\langle x \rangle$. The powers of x, in order, are

$$(1, 1), \ (y, z), \ (1, z^2), \ (y, 1), \ (1, z), \ (y, z^2). \qquad\square$$

There is an analogous statement for a cyclic group of order rs, whenever the two integers r and s have no common factor.

Proposition 2.11.3 Let r and s be relatively prime integers. A cyclic group of order rs is isomorphic to the product of a cyclic group of order r and a cyclic group of order s. $\qquad\square$

On the other hand, a cyclic group of order 4 is *not* isomorphic to a product of two cyclic groups of order 2. Every element of $C_2 \times C_2$ has order 1 or 2, whereas a cyclic group of order 4 contains two elements of order 4.

The next proposition describes product groups.

Proposition 2.11.4 Let H and K be subgroups of a group G, and let $f : H \times K \to G$ be the multiplication map, defined by $f(h, k) = hk$. Its image is the set $HK = \{hk | h \in H, \ k \in K\}$.

(a) f is injective if and only if $H \cap K = \{1\}$.

(b) f is a homomorphism from the product group $H \times K$ to G if and only if elements of K commute with elements of H: $hk = kh$.

(c) If H is a normal subgroup of G, then HK is a subgroup of G.

(d) f is an isomorphism from the product group $H \times K$ to G if and only if $H \cap K = \{1\}$, $HK = G$, and also H and K are normal subgroups of G.

It is important to note that the multiplication map may be bijective though it isn't a group homomorphism. This happens, for instance, when $G = S_3$, and with the usual notation, $H = \langle x \rangle$ and $K = \langle y \rangle$.

Proof. (a) If $H \cap K$ contains an element $x \neq 1$, then x^{-1} is in H, and $f(x^{-1}, x) = 1 = f(1, 1)$, so f is not injective. Suppose that $H \cap K = \{1\}$. Let (h_1, k_1) and (h_2, k_2) be elements of $H \times K$ such that $h_1 k_1 = h_2 k_2$. We multiply both sides of this equation on the left by h_1^{-1} and on the right by k_2^{-1}, obtaining $k_1 k_2^{-1} = h_1^{-1} h_2$. The left side is an element of K and the right side is an element of H. Since $H \cap K = \{1\}$, $k_1 k_2^{-1} = h_1^{-1} h_2 = 1$. Then $k_1 = k_2$, $h_1 = h_2$, and $(h_1, k_1) = (h_2, k_2)$.

(b) Let (h_1, k_1) and (h_2, k_2) be elements of the product group $H \times K$. The product of these elements in the product group $H \times K$ is $(h_1 h_2, k_1 k_2)$, and $f(h_1 h_2, k_1 k_2) = h_1 h_2 k_1 k_2$, while $f(h_1, k_1) f(h_2, k_2) = h_1 k_1 h_2 k_2$. These elements are equal if and only if $h_2 k_1 = k_1 h_2$.

(c) Suppose that H is a normal subgroup. We note that KH is a union of the left cosets kH with k in K, and that HK is a union of the right cosets Hk. Since H is normal, $kH = Hk$, and therefore $HK = KH$. Closure of HK under multiplication follows, because $HKHK = HHKK = HK$. Also, $(hk)^{-1} = k^{-1} h^{-1}$ is in $KH = HK$. This proves closure of HK under inverses.

(d) Suppose that H and K satisfy the conditions given. Then f is both injective and surjective, so it is bijective. According to (b), it is an isomorphism if and only if $hk = kh$ for all h in H and k in K. Consider the commutator $(hkh^{-1})k^{-1} = h(kh^{-1}k^{-1})$. Since K is normal, the left side is in K, and since H is normal, the right side is in H. Since $H \cap K = \{1\}$, $hkh^{-1}k^{-1} = 1$, and $hk = kh$. Conversely, if f is an isomorphism, one may verify the conditions listed in the isomorphic group $H \times K$ instead of in G. $\square$

We use this proposition to classify groups of order 4:

Proposition 2.11.5 There are two isomorphism classes of groups of order 4, the class of the cyclic group C_4 of order 4 and the class of the Klein Four Group, which is isomorphic to the product $C_2 \times C_2$ of two groups of order 2.

Proof. Let G be a group of order 4. The order of any element x of G divides 4, so there are two cases to consider:

Case 1: G contains an element of order 4. Then G is a cyclic group of order 4.

Case 2: Every element of G except the identity has order 2.

In this case, $x = x^{-1}$ for every element x of G. Let x and y be two elements of G. Then xy has order 2, so $xyx^{-1}y^{-1} = (xy)(xy) = 1$. This shows that x and y commute (2.6.5), and since these are arbitrary elements, G is abelian. So every subgroup is normal. We choose distinct elements x and y in G, and we let H and K be the cyclic groups of order 2 that they generate. Proposition 2.11.4(**d**) shows that G is isomorphic to the product group $H \times K$. □

2.12   QUOTIENT GROUPS

In this section we show that a law of composition can be defined on the set of cosets of a *normal* subgroup N of any group G. This law makes the set of cosets of a normal subgroup into a group, called a *quotient group*.

Addition of congruence classes of integers modulo n is an example of the quotient construction. Another familiar example is addition of angles. Every real number represents an angle, and two real numbers represent the same angle if they differ by an integer multiple of 2π. The group N of integer multiples of 2π is a subgroup of the additive group $\mathbb{R}^+$ of real numbers, and angles correspond naturally to (additive) cosets $\theta + N$ of N in G. The group of angles is the quotient group whose elements are the cosets.

The set of cosets of a normal subgroup N of a group G is often denoted by G/N.

(2.12.1) G/N is the set of cosets of N in G.

When we regard a coset C as an element of the set of cosets, the bracket notation $[C]$ may be used. If $C = aN$, we may also use the bar notation to denote the element $[C]$ by $\bar{a}$, and then we would denote the set of cosets by $\overline{G}$:

$$\overline{G} = G/N.$$

Theorem 2.12.2 Let N be a normal subgroup of a group G, and let $\overline{G}$ denote the set of cosets of N in G. There is a law of composition on $\overline{G}$ that makes this set into a group, such that the map $\pi : G \to \overline{G}$ defined by $\pi(a) = \bar{a}$ is a surjective homomorphism whose kernel is N.

• The map π is often referred to as the *canonical map* from G to $\overline{G}$. The word "canonical" indicates that this is the only map that we might reasonably be talking about.

The next corollary is very simple, but it is important enough to single out:

Corollary 2.12.3 Let N be a normal subgroup of a group G, and let $\overline{G}$ denote the set of cosets of N in G. Let $\pi : G \to \overline{G}$ be the canonical homomorphism. Let $a_1, ..., a_k$ be elements of G such that the product $a_1 \cdots a_k$ is in N. Then $\bar{a}_1 \cdots \bar{a}_k = \bar{1}$.

Proof. Let $p = a_1 \cdots a_k$. Then p is in N, so $\pi(p) = \bar{p} = \bar{1}$. Since π is a homomorphism, $\bar{a}_1 \cdots \bar{a}_k = \bar{p}$. □

Proof of Theorem 2.12.2. There are several things to be done. We must

- define a law of composition on $\overline{G}$,
- prove that the law makes $\overline{G}$ into a group,
- prove that π is a surjective homomorphism, and
- prove that the kernel of π is N.

We use the following notation: If A and B are subsets of a group G, then AB denotes the set of products ab:

$$(2.12.4) \qquad AB = \{x \in G \mid x = ab \text{ for some } a \in A \text{ and } b \in B\}.$$

We will call this a *product set*, though in some other contexts the phrase "product set" refers to the set $A \times B$ of pairs of elements.

Lemma 2.12.5 Let N be a normal subgroup of a group G, and let aN and bN be cosets of N. The product set $(aN)(bN)$ is also a coset. It is equal to the coset abN.

We note that the set $(aN)(bN)$ consists of all elements of G that can be written in the form $anbn'$, with n and n' in N.

Proof. Since N is a subgroup, $NN = N$. Since N is normal, left and right cosets are equal: $Nb = bN$ (2.8.17). The lemma is proved by the following formal manipulation:

$$(aN)(bN) = a(Nb)N = a(bN)N = abNN = abN. \qquad \square$$

This lemma allows us to define multiplication on the set $\overline{G} = G/N$. Using the bracket notation (2.7.8), the definition is this: If C_1 and C_2 are cosets, then $[C_1][C_2] = [C_1C_2]$, Where C_1C_2 is the product set. The lemma shows that this product set is another coset. To compute the product $[C_1][C_2]$, take any elements a in C_1 and b in C_2. Then $C_1 = aN$, $C_2 = bN$, and C_1C_2 is the coset abN that contains ab. So we have the very natural formula

$$(2.12.6) \qquad [aN][bN] = [abN] \quad \text{or} \quad \overline{a}\,\overline{b} = \overline{ab}.$$

Then by definition of the map π in (2.12.2),

$$(2.12.7) \qquad \pi(a)\pi(b) = \overline{a}\,\overline{b} = \overline{ab} = \pi(ab).$$

The fact that π is a homomorphism will follow from (2.12.7), once we show that $\overline{G}$ is a group. Since the canonical map π is surjective (2.7.8), the next lemma proves this.

Lemma 2.12.8 Let G be a group, and let Y be a set with a law of composition, both laws written with multiplicative notation. Let $\varphi : G \to Y$ be a surjective map with the homomorphism property, that $\varphi(ab) = \varphi(a)\varphi(b)$ for all a and b in G. Then Y is a group and φ is a homomorphism.

Proof. The group axioms that are true in G are carried over to Y by the surjective map φ. Here is the proof of the associative law: Let y_1, y_2, y_3 be elements of Y. Since φ is surjective, $y_i = \varphi(x_i)$ for some x_i in G. Then

$$(y_1 y_2) y_3 = (\varphi(x_1)\varphi(x_2))\varphi(x_3)=\varphi(x_1 x_2)\varphi(x_3)=\varphi((x_1 x_2)x_3)$$

$$\overset{*}{=} \varphi(x_1(x_2 x_3)) = \varphi(x_1)\varphi(x_2 x_3)=\varphi(x_1)(\varphi(x_2)\varphi(x_3)) = y_1(y_2 y_3).$$

The equality marked with an asterisk is the associative law in G. The other equalities follow from the homomorphism property of φ. The verifications of the other group axioms are similar.   □

The only thing remaining to be verified is that the kernel of the homomorphism π is the subgroup N. Well, $\pi(a) = \pi(1)$ if and only if $\bar{a} = \bar{1}$, or $[aN] = [1N]$, and this is true if and only if a is an element of N.   □

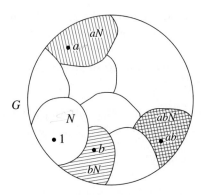

(2.12.9)   A Schematic Diagram of Coset Multiplication.

Note: Our assumption that N be a *normal* subgroup of G is crucial to Lemma 2.12.5. If H is not normal, there will be left cosets C_1 and C_2 of H in G such that the product set $C_1 C_2$ does not lie in a single left coset. Going back once more to the subgroup $H = \langle y \rangle$ of S_3, the product set $(1H)(xH)$ contains four elements: $\{1, y\}\{x, xy\} = \{x, xy, x^2 y, x^2\}$. It is not a coset. The subgroup H is not normal.   □

The next theorem relates the quotient group construction to a general group homomorphism, and it provides a fundamental method of identifying quotient groups.

Theorem 2.12.10 First Isomorphism Theorem. Let $\varphi : G \to G'$ be a surjective group homomorphism with kernel N. The quotient group $\overline{G} = G/N$ is isomorphic to the image G'. To be precise, let $\pi : G \to \overline{G}$ be the canonical map. There is a unique isomorphism $\overline{\varphi} : \overline{G} \to G'$ such that $\varphi = \overline{\varphi} \circ \pi$.

Proof. The elements of $\overline{G}$ are the cosets of N, and they are also the fibres of the map φ (2.7.15). The map $\overline{\varphi}$ referred to in the theorem is the one that sends a nonempty fibre to its image: $\overline{\varphi}(\overline{x}) = \varphi(x)$. For any surjective map of sets $\varphi: G \to G'$, one can form the set $\overline{G}$ of fibres, and then one obtains a diagram as above, in which $\overline{\varphi}$ is the bijective map that sends a fibre to its image. When φ is a group homomorphism, $\overline{\varphi}$ is an isomorphism because $\overline{\varphi}(\overline{a}\overline{b}) = \varphi(ab) = \varphi(a)\varphi(b) = \overline{\varphi}(\overline{a})\overline{\varphi}(\overline{b})$. $\qquad\square$

Corollary 2.12.11 Let $\varphi: G \to G'$ be a group homomorphism with kernel N and image H'. The quotient group $\overline{G} = G/N$ is isomorphic to the image H'. $\qquad\square$

Two quick examples: The image of the absolute value map $\mathbb{C}^\times \to \mathbb{R}^\times$ is the group of positive real numbers, and its kernel is the unit circle U. The theorem asserts that the quotient group $\mathbb{C}^\times/U$ is isomorphic to the multiplicative group of positive real numbers. The determinant is a surjective homomorphism $GL_n(\mathbb{R}) \to \mathbb{R}^\times$, whose kernel is the special linear group $SL_n(\mathbb{R})$. So the quotient $GL_n(\mathbb{R})/SL_n(\mathbb{R})$ is isomorphic to $\mathbb{R}^\times$.

There are also theorems called the Second and the Third Isomorphism Theorems, though they are less important.

> Es giebt alfo fehr viel verfchiedene Arten von Größen,
> welche fich nicht wohl herzehlen laßen;
> und daher entftehen die derfchiedene Theile der Mathematic,
> deren eine jegliche mit einer befondern Art von Größen befchäftiget ift.
>
> —Leonhard Euler

EXERCISES

Section 1   Laws of Composition

1.1. Let S be a set. Prove that the law of composition defined by $ab = a$ for all a and b in S is associative. For which sets does this law have an identity?

1.2. Prove the properties of inverses that are listed near the end of the section.

1.3. Let $\mathbb{N}$ denote the set $\{1, 2, 3, \ldots, \}$ of natural numbers, and let $s: \mathbb{N} \to \mathbb{N}$ be the *shift* map, defined by $s(n) = n + 1$. Prove that s has no right inverse, but that it has infinitely many left inverses.

Section 2   Groups and Subgroups

2.1. Make a multiplication table for the symmetric group S_3.

2.2. Let S be a set with an associative law of composition and with an identity element. Prove that the subset consisting of the invertible elements in S is a group.

2.3. Let x, y, z, and w be elements of a group G.

   (a) Solve for y, given that $xyz^{-1}w = 1$.

   (b) Suppose that $xyz = 1$. Does it follow that $yzx = 1$? Does it follow that $yxz = 1$?

2.4. In which of the following cases is H a subgroup of G?

    (a) $G = GL_n(\mathbb{C})$ and $H = GL_n(\mathbb{R})$.

    (b) $G = \mathbb{R}^\times$ and $H = \{1, -1\}$.

    (c) $G = \mathbb{Z}^+$ and H is the set of positive integers.

    (d) $G = \mathbb{R}^\times$ and H is the set of positive reals.

    (e) $G = GL_2(\mathbb{R})$ and H is the set of matrices $\begin{bmatrix} a & 0 \\ 0 & 0 \end{bmatrix}$, with $a \neq 0$.

2.5. In the definition of a subgroup, the identity element in H is required to be the identity of G. One might require only that H have an identity element, not that it need be the same as the identity in G. Show that if H has an identity at all, then it is the identity in G. Show that the analogous statement is true for inverses.

2.6. Let G be a group. Define an *opposite group* G° with law of composition $a * b$ as follows: The underlying set is the same as G, but the law of composition is $a * b = ba$. Prove that G° is a group.

Section 3   Subgroups of the Additive Group of Integers

3.1. Let $a = 123$ and $b = 321$. Compute $d = \gcd(a, b)$, and express d as an integer combination $ra + bs$.

3.2. Prove that if a and b are positive integers whose sum is a prime p, their greatest common divisor is 1.

3.3. **(a)** Define the greatest common divisor of a set $\{a_1, \ldots, a_n\}$ of n integers. Prove that it exists, and that it is an integer combination of $a_1, \ldots, a_n$.

    (b) Prove that if the greatest common divisor of $\{a_1, \ldots, a_n\}$ is d, then the greatest common divisor of $\{a_1/d, \ldots, a_n/d\}$ is 1.

Section 4   Cyclic Groups

4.1. Let a and b be elements of a group G. Assume that a has order 7 and that $a^3 b = ba^3$. Prove that $ab = ba$.

4.2. An nth root of unity is a complex number z such that $z^n = 1$.

    (a) Prove that the nth roots of unity form a cyclic subgroup of $\mathbb{C}^\times$ of order n.

    (b) Determine the product of all the nth roots of unity.

4.3. Let a and b be elements of a group G. Prove that ab and ba have the same order.

4.4. Describe all groups G that contain no proper subgroup.

4.5. Prove that every subgroup of a cyclic group is cyclic. Do this by working with exponents, and use the description of the subgroups of $\mathbb{Z}^+$.

4.6. **(a)** Let G be a cyclic group of order 6. How many of its elements generate G? Answer the same question for cyclic groups of orders 5 and 8.

    (b) Describe the number of elements that generate a cyclic group of arbitrary order n.

4.7. Let x and y be elements of a group G. Assume that each of the elements x, y, and xy has order 2. Prove that the set $H = \{1, x, y, xy\}$ is a subgroup of G, and that it has order 4.

4.8. (a) Prove that the elementary matrices of the first and third types (1.2.4) generate $GL_n(\mathbb{R})$.

(b) Prove that the elementary matrices of the first type generate $SL_n(\mathbb{R})$. Do the 2×2 case first.

4.9. How many elements of order 2 does the symmetric group S_4 contain?

4.10. Show by example that the product of elements of finite order in a group need not have finite order. What if the group is abelian?

4.11. (a) Adapt the method of row reduction to prove that the transpositions generate the symmetric group S_n.

(b) Prove that, for $n \geq 3$, the three-cycles generate the alternating group A_n.

Section 5   Homomorphisms

5.1. Let $\varphi: G \to G'$ be a surjective homomorphism. Prove that if G is cyclic, then G' is cyclic, and if G is abelian, then G' is abelian.

5.2. Prove that the intersection $K \cap H$ of subgroups of a group G is a subgroup of H, and that if K is a normal subgroup of G, then $K \cap H$ is a normal subgroup of H.

5.3. Let U denote the group of invertible upper triangular 2×2 matrices $A = \begin{bmatrix} a & b \\ 0 & d \end{bmatrix}$, and let $\varphi: U \to \mathbb{R}^\times$ be the map that sends $A \rightsquigarrow a^2$. Prove that φ is a homomorphism, and determine its kernel and image.

5.4. Let $f: \mathbb{R}^+ \to \mathbb{C}^\times$ be the map $f(x) = e^{ix}$. Prove that f is a homomorphism, and determine its kernel and image.

5.5. Prove that the $n \times n$ matrices that have the block form $M = \begin{bmatrix} A & B \\ 0 & D \end{bmatrix}$, with A in $GL_r(\mathbb{R})$ and D in $GL_{n-r}(\mathbb{R})$, form a subgroup H of $GL_n(\mathbb{R})$, and that the map $H \to GL_r(\mathbb{R})$ that sends $M \rightsquigarrow A$ is a homomorphism. What is its kernel?

5.6. Determine the center of $GL_n(\mathbb{R})$.

Hint: You are asked to determine the invertible matrices A that commute with every invertible matrix B. Do not test with a general matrix B. Test with elementary matrices.

Section 6   Isomorphisms

6.1. Let G' be the group of real matrices of the form $\begin{bmatrix} 1 & x \\ & 1 \end{bmatrix}$. Is the map $\mathbb{R}^+ \to G'$ that sends x to this matrix an isomorphism?

6.2. Describe all homomorphisms $\varphi: \mathbb{Z}^+ \to \mathbb{Z}^+$. Determine which are injective, which are surjective, and which are isomorphisms.

6.3. Show that the functions $f = 1/x, g = (x-1)/x$ generate a group of functions, the law of composition being composition of functions, that is isomorphic to the symmetric group S_3.

6.4. Prove that in a group, the products ab and ba are conjugate elements.

6.5. Decide whether or not the two matrices $A = \begin{bmatrix} 3 & \\ & 2 \end{bmatrix}$ and $B = \begin{bmatrix} 1 & 1 \\ -2 & 4 \end{bmatrix}$ are conjugate elements of the general linear group $GL_2(\mathbb{R})$.

6.6. Are the matrices $\begin{bmatrix} 1 & 1 \\ & 1 \end{bmatrix}, \begin{bmatrix} 1 & \\ 1 & 1 \end{bmatrix}$ conjugate elements of the group $GL_2(\mathbb{R})$? Are they conjugate elements of $SL_2(\mathbb{R})$?

6.7. Let H be a subgroup of G, and let g be a fixed element of G. The *conjugate subgroup* gHg^{-1} is defined to be the set of all conjugates ghg^{-1}, with h in H. Prove that gHg^{-1} is a subgroup of G.

6.8. Prove that the map $A \rightsquigarrow (A^t)^{-1}$ is an automorphism of $GL_n(\mathbb{R})$.

6.9. Prove that a group G and its opposite group G° (Exercise 2.6) are isomorphic.

6.10. Find all automorphisms of
   (a) a cyclic group of order 10,   **(b)** the symmetric group S_3.

6.11. Let a be an element of a group G. Prove that if the set $\{1, a\}$ is a normal subgroup of G, then a is in the center of G.

Section 7   Equivalence Relations and Partitions

7.1. Let G be a group. Prove that the relation $a \sim b$ if $b = gag^{-1}$ for some g in G is an equivalence relation on G.

7.2. An equivalence relation on S is determined by the subset R of the set $S \times S$ consisting of those pairs (a, b) such that $a \sim b$. Write the axioms for an equivalence relation in terms of the subset R.

7.3. With the notation of Exercise 7.2, is the intersection $R \cap R'$ of two equivalence relations R and R' an equivalence relation? Is the union?

7.4. A relation R on the set of real numbers can be thought of as a subset of the (x, y)-plane. With the notation of Exercise 7.2, explain the geometric meaning of the reflexive and symmetric properties.

7.5. With the notation of Exercise 7.2, each of the following subsets R of the (x, y)-plane defines a relation on the set $\mathbb{R}$ of real numbers. Determine which of the axioms (2.7.3) are satisfied: **(a)** the set $\{(s, s) \mid s \in \mathbb{R}\}$,   **(b)** the empty set,   **(c)** the locus $\{xy + 1 = 0\}$, **(d)** the locus $\{x^2 y - xy^2 - x + y = 0\}$.

7.6. How many different equivalence relations can be defined on a set of five elements?

Section 8   Cosets

8.1. Let H be the cyclic subgroup of the alternating group A_4 generated by the permutation $(1\,2\,3)$. Exhibit the left and the right cosets of H explicitly.

8.2. In the additive group $\mathbb{R}^m$ of vectors, let W be the set of solutions of a system of homogeneous linear equations $AX = 0$. Show that the set of solutions of an inhomogeneous system $AX = B$ is either empty, or else it is an (additive) coset of W.

8.3. Does every group whose order is a power of a prime p contain an element of order p?

8.4. Does a group of order 35 contain an element of order 5? of order 7?

8.5. A finite group contains an element x of order 10 and also an element y of order 6. What can be said about the order of G?

8.6. Let $\varphi : G \to G'$ be a group homomorphism. Suppose that $|G| = 18$, $|G'| = 15$, and that φ is not the trivial homomorphism. What is the order of the kernel?

8.7. A group G of order 22 contains elements x and y, where $x \neq 1$ and y is not a power of x. Prove that the subgroup generated by these elements is the whole group G.

8.8. Let G be a group of order 25. Prove that G has at least one subgroup of order 5, and that if it contains only one subgroup of order 5, then it is a cyclic group.

8.9. Let G be a finite group. Under what circumstances is the map $\varphi : G \to G$ defined by $\varphi(x) = x^2$ an automorphism of G?

8.10. Prove that every subgroup of index 2 is a normal subgroup, and show by example that a subgroup of index 3 need not be normal.

8.11. Let G and H be the following subgroups of $GL_2(\mathbb{R})$:

$$G = \left\{ \begin{bmatrix} x & y \\ 0 & 1 \end{bmatrix} \right\}, \; H = \left\{ \begin{bmatrix} x & 0 \\ 0 & 1 \end{bmatrix} \right\},$$

with x and y real and $x > 0$. An element of G can be represented by a point in the right half plane. Make sketches showing the partitions of the half plane into left cosets and into right cosets of H.

8.12. Let S be a subset of a group G that contains the identity element 1, and such that the left cosets aS, with a in G, partition G. Prove that S is a subgroup of G.

8.13. Let S be a set with a law of composition. A partition $\Pi_1 \cup \Pi_2 \cup \cdots$ of S is *compatible* with the law of composition if for all i and j, the product set

$$\Pi_i \Pi_j = \{xy \mid x \in \Pi_i, y \in \Pi_j\}$$

is contained in a single subset Π_k of the partition.

(a) The set $\mathbb{Z}$ of integers can be partitioned into the three sets [Pos], [Neg], [{0}]. Discuss the extent to which the laws of composition $+$ and $\times$ are compatible with this partition.

(b) Describe all partitions of the integers that are compatible with the operation $+$.

Section 9 Modular Arithmetic

9.1. For which integers n does 2 have a multiplicative inverse in $\mathbb{Z}/\mathbb{Z}n$?

9.2. What are the possible values of a^2 modulo 4? modulo 8?

9.3. Prove that every integer a is congruent to the sum of its decimal digits modulo 9.

9.4. Solve the congruence $2x \equiv 5$ modulo 9 and modulo 6.

9.5. Determine the integers n for which the pair of congruences $2x - y \equiv 1$ and $4x + 3y \equiv 2$ modulo n has a solution.

9.6. Prove the *Chinese Remainder Theorem*: Let a, b, u, v be integers, and assume that the greatest common divisor of a and b is 1. Then there is an integer x such that $x \equiv u$ modulo a and $x \equiv v$ modulo b.

Hint: Do the case $u = 0$ and $v = 1$ first.

9.7. Determine the order of each of the matrices $A = \begin{bmatrix} 1 & 1 \\ 0 & 1 \end{bmatrix}$ and $B = \begin{bmatrix} 1 & 1 \\ 1 & 0 \end{bmatrix}$ when the matrix entries are interpreted modulo 3.

Section 10 The Correspondence Theorem

10.1. Describe how to tell from the cycle decomposition whether a permutation is odd or even.

10.2. Let H and K be subgroups of a group G.

    (a) Prove that the intersection $xH \cap yK$ of two cosets of H and K is either empty or else is a coset of the subgroup $H \cap K$.

    (b) Prove that if H and K have finite index in G then $H \cap K$ also has finite index in G.

10.3. Let G and G' be cyclic groups of orders 12 and 6, generated by elements x and y, respectively, and let $\varphi : G \to G'$ be the map defined by $\varphi(x^i) = y^i$. Exhibit the correspondence referred to in the Correspondence Theorem explicitly.

10.4. With the notation of the Correspondence Theorem, let H and H' be corresponding subgroups. Prove that $[G:H] = [G':H']$.

10.5. With reference to the homomorphism $S_4 \to S_3$ described in Example 2.5.13, determine the six subgroups of S_4 that contain K.

Section 11 Product Groups

11.1. Let x be an element of order r of a group G, and let y be an element of G' of order s. What is the order of (x, y) in the product group $G \times G'$?

11.2. What does Proposition 2.11.4 tell us when, with the usual notation for the symmetric group S_3, K and H are the subgroups $\langle y \rangle$ and $\langle x \rangle$?

11.3. Prove that the product of two infinite cyclic groups is not infinite cyclic.

11.4. In each of the following cases, determine whether or not G is isomorphic to the product group $H \times K$.

    (a) $G = \mathbb{R}^\times$, $H = \{\pm 1\}$, $K = \{\text{positive real numbers}\}$.

    (b) $G = \{\text{invertible upper triangular } 2 \times 2 \text{ matrices}\}$, $H = \{\text{invertible diagonal matrices}\}$, $K = \{\text{upper triangular matrices with diagonal entries 1}\}$.

    (c) $G = \mathbb{C}^\times$, $H = \{\text{unit circle}\}$, $K = \{\text{positive real numbers}\}$.

11.5. Let G_1 and G_2 be groups, and let Z_i be the center of G_i. Prove that the center of the product group $G_1 \times G_2$ is $Z_1 \times Z_2$.

11.6. Let G be a group that contains normal subgroups of orders 3 and 5, respectively. Prove that G contains an element of order 15.

11.7. Let H be a subgroup of a group G, let $\varphi : G \to H$ be a homomorphism whose restriction to H is the identity map, and let N be its kernel. What can one say about the product map $H \times N \to G$?

11.8. Let G, G', and H be groups. Establish a bijective correspondence between homomorphisms $\Phi : H \to G \times G'$ from H to the product group and pairs (φ, φ') consisting of a homomorphism $\varphi : H \to G$ and a homomorphism $\varphi' : H \to G'$.

11.9. Let H and K be subgroups of a group G. Prove that the product set HK is a subgroup of G if and only if $HK = KH$.

Section 12 Quotient Groups

12.1. Show that if a subgroup H of a group G is not normal, there are left cosets aH and bH whose product is not a coset.

12.2. In the general linear group $GL_3(\mathbb{R})$, consider the subsets

$$H = \begin{bmatrix} 1 & * & * \\ 0 & 1 & * \\ 0 & 0 & 1 \end{bmatrix}, \quad \text{and} \quad K = \begin{bmatrix} 1 & 0 & * \\ 0 & 1 & 0 \\ 0 & 0 & 1 \end{bmatrix},$$

where $*$ represents an arbitrary real number. Show that H is a subgroup of GL_3, that K is a normal subgroup of H, and identify the quotient group H/K. Determine the center of H.

12.3. Let P be a partition of a group G with the property that for any pair of elements A, B of the partition, the product set AB is contained entirely within another element C of the partition. Let N be the element of P that contains 1. Prove that N is a normal subgroup of G and that P is the set of its cosets.

12.4. Let $H = \{\pm 1, \pm i\}$ be the subgroup of $G = \mathbb{C}^\times$ of fourth roots of unity. Describe the cosets of H in G explicitly. Is G/H isomorphic to G?

12.5. Let G be the group of upper triangular real matrices $\begin{bmatrix} a & b \\ 0 & d \end{bmatrix}$, with a and d different from zero. For each of the following subsets, determine whether or not S is a subgroup, and whether or not S is a normal subgroup. If S is a normal subgroup, identify the quotient group G/S.

   (i) S is the subset defined by $b = 0$.

  (ii) S is the subset defined by $d = 1$.

 (iii) S is the subset defined by $a = d$.

Miscellaneous Problems

M.1. Describe the column vectors $(a, c)^t$ that occur as the first column of an integer matrix A whose inverse is also an integer matrix.

M.2. **(a)** Prove that every group of even order contains an element of order 2.

   (b) Prove that every group of order 21 contains an element of order 3.

M.3. Classify groups of order 6 by analyzing the following three cases:

   (i) G contains an element of order 6.

  (ii) G contains an element of order 3 but none of order 6.

 (iii) All elements of G have order 1 or 2.

M.4. A *semigroup* S is a set with an associative law of composition and with an identity. Elements are not required to have inverses, and the Cancellation Law need not hold. A semigroup S is said to be generated by an element s if the set $\{1, s, s^2, \ldots\}$ of nonnegative powers of s is equal to S. Classify semigroups that are generated by one element.

M.5. Let S be a finite semigroup (see Exercise M.4) in which the Cancellation Law 2.2.3 holds. Prove that S is a group.

*M.6. Let $a = (a_1, \ldots, a_k)$ and $b = (b_1, \ldots, b_k)$ be points in k-dimensional space $\mathbb{R}^k$. A *path* from a to b is a continuous function on the unit interval $[0, 1]$ with values in $\mathbb{R}^k$, a function $X:[0, 1] \to \mathbb{R}^k$, sending $t \rightsquigarrow X(t) = (x_1(t), \ldots, x_k(t))$, such that $X(0) = a$ and $X(1) = b$. If S is a subset of $\mathbb{R}^k$ and if a and b are in S, define $a \sim b$ if a and b can be joined by a path lying entirely in S.

(a) Show that $\sim$ is an equivalence relation on S. Be careful to check that any paths you construct stay within the set S.

(b) A subset S is *path connected* if $a \sim b$ for any two points a and b in S. Show that every subset S is partitioned into path-connected subsets with the property that two points in different subsets cannot be connected by a path in S.

(c) Which of the following loci in $\mathbb{R}^2$ are path-connected: $\{x^2 + y^2 = 1\}$, $\{xy = 0\}$, $\{xy = 1\}$?

*M.7. The set of $n \times n$ matrices can be identified with the space $\mathbb{R}^{n \times n}$. Let G be a subgroup of $GL_n(\mathbb{R})$. With the notation of Exercise M.6, prove:

(a) If A, B, C, D are in G, and if there are paths in G from A to B and from C to D, then there is a path in G from AC to BD.

(b) The set of matrices that can be joined to the identity I forms a normal subgroup of G. (It is called the *connected component* of G.)

*M.8. (a) The group $SL_n(\mathbb{R})$ is generated by elementary matrices of the first type (see Exercise 4.8). Use this fact to prove that $SL_n(\mathbb{R})$ is path-connected.

(b) Show that $GL_n(\mathbb{R})$ is a union of two path-connected subsets, and describe them.

M.9. (*double cosets*) Let H and K be subgroups of a group G, and let g be an element of G. The set $HgK = \{x \in G \mid x = hgk \text{ for some } h \in H, k \in K\}$ is called a *double coset*. Do the double cosets partition G?

M.10. Let H be a subgroup of a group G. Show that the double cosets (see Exercise M.9)

$$HgH = \{h_1 g h_2 \mid h_1, h_2 \in H\}$$

are the left cosets gH if and only if H is normal.

*M.11. Most invertible matrices can be written as a product $A = LU$ of a lower triangular matrix L and an upper triangular matrix U, where in addition all diagonal entries of U are 1.

(a) Explain how to compute L and U when the matrix A is given.

(b) Prove uniqueness, that there is at most one way to write A as such a product.

(c) Show that every invertible matrix can be written as a product LPU, where L, U are as above and P is a permutation matrix.

(d) Describe the double cosets LgU (see Exercise M.9).

M.12. (*postage stamp problem*) Let a and b be positive, relatively prime integers.

(a) Prove that every sufficiently large positive integer n can be obtained as $ra + sb$, where r and s are positive integers.

(b) Determine the largest integer that is not of this form.

M.13. (*a game*) The starting position is the point $(1, 1)$, and a permissible "move" replaces a point (a, b) by one of the points $(a + b, b)$ or $(a, a + b)$. So the position after the first move will be either $(2, 1)$ or $(1, 2)$. Determine the points that can be reached.

M.14. (*generating $SL_2(\mathbb{Z})$*) Prove that the two matrices

$$E = \begin{bmatrix} 1 & 1 \\ 0 & 1 \end{bmatrix}, \quad E' = \begin{bmatrix} 1 & 0 \\ 1 & 1 \end{bmatrix}$$

generate the group $SL_2(\mathbb{Z})$ of all *integer* matrices with determinant 1. Remember that the subgroup they generate consists of all elements that can be expressed as products using the four elements E, E', E^{-1}, E'^{-1}.

Hint: Do not try to write a matrix directly as a product of the generators. Use row reduction.

M.15. (*the semigroup generated by elementary matrices*) Determine the semigroup S (see Exercise M.4) of matrices A that can be written as a product, of arbitrary length, each of whose terms is one of the two matrices

$$\begin{bmatrix} 1 & 1 \\ 0 & 1 \end{bmatrix} , \quad \text{or} \quad \begin{bmatrix} 1 & 0 \\ 1 & 1 \end{bmatrix} .$$

Show that every element of S can be expressed as such a product in exactly one way.

M.16. [1](*the homophonic group: a mathematical diversion*) By definition, English words have the same pronunciation if their phonetic spellings in the dictionary are the same. The homophonic group $\mathcal{H}$ is generated by the letters of the alphabet, subject to the following relations: English words with the same pronunciation represent equal elements of the group. Thus $be = bee$, and since $\mathcal{H}$ is a group, we can cancel be to conclude that $e = 1$. Try to determine the group $\mathcal{H}$.

---

[1]I learned this problem from a paper by Mestre, Schoof, Washington and Zagier.

C H A P T E R   3

Vector Spaces

Immer mit den einfachsten Beispielen anfangen.

—David Hilbert

3.1 SUBSPACES OF $\mathbb{R}^n$

Our basic models of vector spaces, the topic of this chapter, are subspaces of the space $\mathbb{R}^n$ of n-dimensional real vectors. We discuss them in this section. The definition of a vector space is given in Section 3.3.

Though row vectors take up less space, the definition of matrix multiplication makes column vectors more convenient, so we usually work with them. To save space, we sometimes use the matrix transpose to write a column vector in the form $(a_1, \ldots, a_n)^{\mathrm{t}}$. As mentioned in Chapter 1, we don't distinguish a column vector from the point of $\mathbb{R}^n$ with the same coordinates. Column vectors will often be denoted by lowercase letters such as v or w, and if v is equal to $(a_1, \ldots, a_n)^{\mathrm{t}}$, we call $(a_1, \ldots, a_n)^{\mathrm{t}}$ the *coordinate vector* of v.

We consider two operations on vectors:

(3.1.1)

$$\textit{vector addition:} \quad \begin{bmatrix} a_1 \\ \vdots \\ a_n \end{bmatrix} + \begin{bmatrix} b_1 \\ \vdots \\ b_n \end{bmatrix} = \begin{bmatrix} a_1 + b_1 \\ \vdots \\ a_n + b_n \end{bmatrix}, \text{ and}$$

$$\textit{scalar multiplication:} \quad c \begin{bmatrix} a_1 \\ \vdots \\ a_n \end{bmatrix} = \begin{bmatrix} ca_1 \\ \vdots \\ ca_n \end{bmatrix}$$

These operations make $\mathbb{R}^n$ into a vector space.

A subset W of $\mathbb{R}^n$ (3.1.1) is a *subspace* if it has these properties:

(3.1.2)

(a) If w and w' are in W, then $w + w'$ is in W.

(b) If w is in W and c is in $\mathbb{R}$, then cw is in W.

(c) The zero vector is in W.

There is another way to state the conditions for a subspace:

(3.1.3) W is not empty, and if $w_1, \ldots, w_n$ are elements of W and $c_1, \ldots, c_n$ are scalars, the *linear combination* $c_1 w_1 + \cdots + c_n w_n$ is also in W.

Systems of homogeneous linear equations provide examples. Given an $m \times n$ matrix A with coefficients in $\mathbb{R}$, the set of vectors in $\mathbb{R}^n$ whose coordinate vectors solve the homogeneous equation $AX = 0$ is a subspace, called the *nullspace* of A. Though this is very simple, we'll check the conditions for a subspace:

- $AX = 0$ and $AY = 0$ imply $A(X + Y) = 0$: If X and Y are solutions, so is $X + Y$.
- $AX = 0$ implies $AcX = 0$: If X is a solution, so is cX.
- $A0 = 0$: The zero vector is a solution.

The zero space $W = \{0\}$ and the whole space $W = \mathbb{R}^n$ are subspaces. A subspace is *proper* if it is not one of these two. The next proposition describes the proper subspaces of $\mathbb{R}^2$.

Proposition 3.1.4 Let W be a proper subspace of the space $\mathbb{R}^2$, and let w be a nonzero vector in W. Then W consists of the scalar multiples cw of w. Distinct proper subspaces have only the zero vector in common.

The subspace consisting of the scalar multiples cw of a given nonzero vector w is called the subspace *spanned* by w. Geometrically, it is a line through the origin in the plane $\mathbb{R}^2$.

Proof of the proposition. We note first that a subspace W that is spanned by a nonzero vector w is also spanned by any other nonzero vector w' that it contains. This is true because if $w' = cw$ with $c \neq 0$, then any multiple aw can also be written in the form $ac^{-1}w'$. Consequently, if two subspaces W_1 and W_2 that are spanned by vectors w_1 and w_2 have a nonzero element v in common, then they are equal.

Next, a subspace W of $\mathbb{R}^2$, not the zero space, contains a nonzero element w_1. Since W is a subspace, it contains the space W_1 spanned by w_1, and if $W_1 = W$, then W consists of the scalar multiples of one nonzero vector. We show that if W is not equal to W_1, then it is the whole space $\mathbb{R}^2$. Let w_2 be an element of W not in W_1, and let W_2 be the subspace spanned by w_2. Since $W_1 \neq W_2$, these subspaces intersect only in 0. So neither of the two vectors w_1 and w_2 is a multiple of the other. Then the coordinate vectors, call them A_i, of w_i aren't proportional, and the 2×2 block matrix $A = [A_1 | A_2]$ with these vectors as columns has a nonzero determinant. In that case we can solve the equation $AX = B$ for the coordinate vector B of an arbitrary vector v, obtaining the linear combination $v = w_1 x_1 + w_2 x_2$. This shows that W is the whole space $\mathbb{R}^2$. $\square$

It can also be seen geometrically from the parallelogram law for vector addition that every vector is a linear combination $c_1 w_1 + c_2 w_2$.

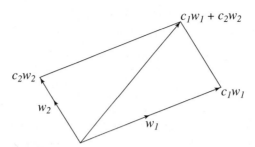

The description of subspaces of $\mathbb{R}^2$ that we have given is clarified in Section 3.4 by the concept of dimension.

3.2 FIELDS

As mentioned at the beginning of Chapter 1, essentially all that was said about matrix operations is true for complex matrices as well as for real ones. Many other number systems serve equally well. To describe these number systems, we list the properties of the "scalars" that are needed, and are led to the concept of a field. We introduce fields here before turning to vector spaces, the main topic of the chapter.

Subfields of the field $\mathbb{C}$ of complex numbers are the simplest fields to describe. A *subfield* of $\mathbb{C}$ is a subset that is closed under the four operations of addition, subtraction, multiplication, and division, and which contains 1. In other words, F is a subfield of $\mathbb{C}$ if it has these properties:

(3.2.1) $(+, -, \times, \div, 1)$

- If a and b are in F, then $a + b$ is in F.
- If a is in F, then $-a$ is in F.
- If a and b are in F, then ab is in F.
- If a is in F and $a \neq 0$, then a^{-1} is in F.
- 1 is in F.

These axioms imply that $1 - 1 = 0$ is an element of F. Another way to state them is to say that F is a subgroup of the additive group $\mathbb{C}^+$, and that the nonzero elements of F form a subgroup of the multiplicative group $\mathbb{C}^\times$.

Some examples of subfields of $\mathbb{C}$:

(a) the field $\mathbb{R}$ of real numbers,

(b) the field $\mathbb{Q}$ of rational numbers (fractions of integers),

(c) the field $\mathbb{Q}[\sqrt{2}]$ of all complex numbers of the form $a + b\sqrt{2}$, with rational numbers a and b.

The concept of an abstract field is only slightly harder to grasp than that of a subfield, and it contains important new classes of fields, including finite fields.

Definition 3.2.2 A *field* F is a set together with two laws of composition

$$F \times F \xrightarrow{+} F \quad \text{and} \quad F \times F \xrightarrow{\times} F$$

called addition: $a, b \rightsquigarrow a + b$ and multiplication: $a, b \rightsquigarrow ab$, which satisfy these axioms:

(i) Addition makes F into an abelian group F^+; its identity element is denoted by 0.

(ii) Multiplication is commutative, and it makes the set of nonzero elements of F into an abelian group $F^\times$; its identity element is denoted by 1.

(iii) *distributive law*: For all a, b, and c in F, $a(b+c) = ab + ac$.

The first two axioms describe properties of the two laws of composition, addition and multiplication, separately. The third axiom, the distributive law, relates the two laws.

You will be familiar with the fact that the real numbers satisfy these axioms, but the fact that they are the only ones needed for the usual algebraic operations can only be understood after some experience.

The next lemma explains how the zero element multiplies.

Lemma 3.2.3 Let F be a field.

(a) The elements 0 and 1 of F are distinct.

(b) For all a in F, $a0 = 0$ and $0a = 0$.

(c) Multiplication in F is associative, and 1 is an identity element.

Proof. **(a)** Axiom **(ii)** implies that 1 is not equal to 0.

(b) Since 0 is the identity for addition, $0 + 0 = 0$. Then $a0 + a0 = a(0 + 0) = a0$. Since F^+ is a group, we can cancel $a0$ to obtain $a0 = 0$, and then $0a = 0$ as well.

(c) Since $F - \{0\}$ is an abelian group, multiplication is associative when restricted to this subset. We need to show that $a(bc) = (ab)c$ when at least one of the elements is zero. In that case, **(b)** shows that the products in question are equal to zero. Finally, the element 1 is an identity on $F - \{0\}$. Setting $a = 1$ in **(b)** shows that 1 is an identity on all of F. $\qquad\square$

Aside from subfields of the complex numbers, the simplest examples of fields are certain finite fields called prime fields, which we describe next. We saw in the previous chapter that the set $\mathbb{Z}/n\mathbb{Z}$ of congruence classes modulo an integer n has laws of addition and multiplication derived from addition and multiplication of integers. All of the axioms for a field hold for the integers, except for the existence of multiplicative inverses. And as noted in Section 2.9, such axioms carry over to addition and multiplication of congruence classes. But the integers aren't closed under division, so there is no reason to suppose that congruence classes have multiplicative inverses. In fact they needn't. The class of 2, for example, has no multiplicative inverse modulo 6. It is somewhat surprising that when p is a prime integer, all nonzero congruence classes modulo p have inverses, and therefore the set $\mathbb{Z}/p\mathbb{Z}$ is a field. This field is called a *prime field*, and is often denoted by $\mathbb{F}_p$.

Using bar notation and choosing the usual representative elements for the p congruence classes,

(3.2.4) $$\mathbb{F}_p = \{\overline{0}, \overline{1}, \ldots, \overline{p{-}1}\} = \mathbb{Z}/p\mathbb{Z}.$$

Theorem 3.2.5 Let p be a prime integer. Every nonzero congruence class modulo p has a multiplicative inverse, and therefore $\mathbb{F}_p$ is a field of order p.

We discuss the theorem before giving the proof.

If a and b are integers, then $\bar{a} \neq \bar{0}$ means that p does not divide a, and $\overline{ab} = \bar{1}$ means $ab \equiv 1$ modulo p. The theorem can be stated in terms of congruence in this way:

(3.2.6)   Let p be a prime, and let a be an integer not divisible by p.
There is an integer b such that $ab \equiv 1$ modulo p.

Finding the inverse of a congruence class $\bar{a}$ modulo p can be done by trial and error if p is small. A systematic way is to compute the powers of $\bar{a}$. If $p = 13$ and $\bar{a} = \bar{3}$, then $\bar{a}^2 = \bar{9}$ and $\bar{a}^3 = \overline{27} = \bar{1}$. We are lucky: $\bar{a}$ has order 3, and therefore $\bar{3}^{-1} = \bar{3}^2 = \bar{9}$. On the other hand, the powers of $\bar{6}$ run through every nonzero congruence class modulo 13. Computing powers may not be the fastest way to find the inverse of $\bar{6}$. But the theorem tells us that the set $\mathbb{F}_p^{\times}$ of nonzero congruence classes forms a group. So every element $\bar{a}$ of $\mathbb{F}_p^{\times}$ has finite order, and if $\bar{a}$ has order r, its inverse will be $\bar{a}^{(r-1)}$.

To make a proof of the theorem using this reasoning, we need the cancellation law:

Proposition 3.2.7 Cancellation Law.   Let p be a prime integer, and let $\bar{a}, \bar{b}$ and $\bar{c}$ be elements of $\mathbb{F}_p$.

(a) If $\bar{a}\bar{b} = \bar{0}$, then $\bar{a} = \bar{0}$ or $\bar{b} = \bar{0}$.
(b) If $\bar{a} \neq \bar{0}$ and if $\bar{a}\bar{b} = \bar{a}\bar{c}$, then $\bar{b} = \bar{c}$.

Proof. **(a)** We represent the congruence classes $\bar{a}$ and $\bar{b}$ by integers a and b, and we translate into congruence. The assertion to be proved is that if p divides ab then p divides a or p divides b. This is Corollary 2.3.7.

(b) It follows from **(a)** that if $\bar{a} \neq \bar{0}$ and $\bar{a}(\bar{b} - \bar{c}) = \bar{0}$, then $\bar{b} - \bar{c} = \bar{0}$.   □

Proof of Theorem (3.2.5). Let $\bar{a}$ be a nonzero element of $\mathbb{F}_p$. We consider the powers $\bar{1}, \bar{a}, \bar{a}^2, \bar{a}^3, \ldots$ Since there are infinitely many exponents and only finitely many elements in $\mathbb{F}_p$, there must be two powers that are equal, say $\bar{a}^m = \bar{a}^n$, where $m < n$. We cancel $\bar{a}^m$ from both sides: $\bar{1} = \bar{a}^{(n-m)}$. Then $\bar{a}^{(n-m-1)}$ is the inverse of $\bar{a}$.   □

It will be convenient to drop the bars over the letters in what follows, trusting ourselves to remember whether we are working with integers or with congruence classes, and remembering the rule (2.9.8):

If a and b are integers, then $a = b$ in $\mathbb{F}_p$ means $a \equiv b$ modulo p.

As with congruences in general, computation in the field $\mathbb{F}_p$ can be done by working with integers, except that division cannot be carried out in the integers. One can operate with matrices A whose entries are in a field, and the discussion of Chapter 1 can be repeated with no essential change.

Suppose we ask for solutions of a system of n linear equations in n unknowns in the prime field $\mathbb{F}_p$. We represent the system of equations by an integer system, choosing representatives for the congruence classes, say $AX = B$, where A is an $n \times n$ integer matrix and B is an integer column vector. To solve the system in $\mathbb{F}_p$, we invert the matrix A modulo p. The formula $\text{cof}(A)A = \delta I$, where $\delta = \det A$ (Theorem 1.6.9), is valid for integer matrices, so it also holds in $\mathbb{F}_p$ when the matrix entries are replaced by their congruence classes. If the congruence class of δ isn't zero, we can invert the matrix A in $\mathbb{F}_p$ by computing $\delta^{-1} \operatorname{cof}(A)$.

Corollary 3.2.8 Let $AX = B$ be a system of n linear equations in n unknowns, where the entries of A and B are in $\mathbb{F}_p$, and let $\delta = \det A$. If δ is not zero, the system has a unique solution in $\mathbb{F}_p$. $\qquad\square$

Consider, for example, the system $AX = B$, where

$$A = \begin{bmatrix} 8 & 3 \\ 2 & 6 \end{bmatrix} \text{ and } B = \begin{bmatrix} 3 \\ -1 \end{bmatrix}.$$

The coefficients are integers, so $AX = B$ defines a system of equations in $\mathbb{F}_p$ for any prime p. The determinant of A is 42, so the system has a unique solution in $\mathbb{F}_p$ for all p that do not divide 42, i.e., all p different from 2, 3, and 7. For instance, $\det A = 3$ when evaluated modulo 13. Since $3^{-1} = 9$ in $\mathbb{F}_{13}$,

$$A^{-1} = 9 \begin{bmatrix} 6 & -3 \\ -2 & 8 \end{bmatrix} = \begin{bmatrix} 2 & -1 \\ 8 & 7 \end{bmatrix} \text{ and } X = A^{-1}B = \begin{bmatrix} 7 \\ 4 \end{bmatrix}, \text{ modulo 13.}$$

The system has no solution in $\mathbb{F}_2$ or $\mathbb{F}_3$. It happens to have solutions in $\mathbb{F}_7$, though $\det A \equiv 0$ modulo 7.

Invertible matrices with entries in the prime field $\mathbb{F}_p$ provide new examples of finite groups, the general linear groups over finite fields:

$$GL_n(\mathbb{F}_p) = \{n \times n \text{ invertible matrices with entries in } \mathbb{F}_p\}$$
$$SL_n(\mathbb{F}_p) = \{n \times n \text{ matrices with entries in } \mathbb{F}_p \text{ and with determinant } 1\}$$

For example, the group of invertible 2×2 matrices with entries in $\mathbb{F}_2$ contains the six elements

$$(3.2.9) \qquad GL_2(\mathbb{F}_2) = \left\{ \begin{bmatrix} 1 & \\ & 1 \end{bmatrix}, \begin{bmatrix} 1 & 1 \\ & 1 \end{bmatrix}, \begin{bmatrix} & 1 \\ 1 & 1 \end{bmatrix}, \begin{bmatrix} & 1 \\ 1 & \end{bmatrix}, \begin{bmatrix} 1 & 1 \\ & 1 \end{bmatrix}, \begin{bmatrix} 1 & \\ 1 & 1 \end{bmatrix} \right\}.$$

This group is isomorphic to the symmetric group S_3. The matrices have been listed in an order that agrees with our usual list $\{1, x, x^2, y, xy, x^2y\}$ of the elements of S_3.

One property of the prime fields $\mathbb{F}_p$ that distinguishes them from subfields of $\mathbb{C}$ is that adding 1 to itself a certain number of times, in fact p times, gives zero. The *characteristic* of a field F is the order of 1, as an element of the additive group F^+, provided that the order is finite. It is the smallest positive integer m such that the sum $1 + \cdots + 1$ of m copies of 1 evaluates to zero. If the order is infinite, that is, $1 + \cdots + 1$ is never 0 in F, the field is, somewhat perversely, said to have *characteristic zero*. Thus subfields of $\mathbb{C}$ have characteristic zero, while the prime field $\mathbb{F}_p$ has characteristic p.

Lemma 3.2.10 The characteristic of any field F is either zero or a prime number.

Proof. To avoid confusion, we let $\overline{0}$ and $\overline{1}$ denote the additive and the multiplicative identities in the field F, respectively, and if k is a positive integer, we let $\overline{k}$ denote the sum of k copies of $\overline{1}$. Suppose that the characteristic m is not zero. Then $\overline{1}$ generates a cyclic subgroup H of F^+ of order m, and $\overline{m} = \overline{0}$. The distinct elements of the cyclic subgroup H generated by $\overline{1}$ are the elements $\overline{k}$ with $k = 0, 1, \ldots, m\text{-}1$ (Proposition 2.4.2). Suppose that m isn't prime, say $m = rs$, with $1 < r, s < m$. Then $\overline{r}$ and $\overline{s}$ are in the multiplicative group $F^\times = F - \{0\}$, but the product $\overline{r}\,\overline{s}$, which is equal to $\overline{0}$, is not in $F^\times$. This contradicts the fact that $F^\times$ is a group. Therefore m must be prime. $\square$

The prime fields $\mathbb{F}_p$ have another remarkable property:

Theorem 3.2.11 Structure of the Multiplicative Group. Let p be a prime integer. The multiplicative group $\mathbb{F}_p^\times$ of the prime field is a cyclic group of order $p - 1$.

We defer the proof of this theorem to Chapter 15, where we prove that the multiplicative group of every finite field is cyclic (Theorem 15.7.3).

- A generator for the cyclic group $\mathbb{F}_p^\times$ is called a *primitive root* modulo p.

There are two primitive roots modulo 7, namely 3 and 5, and four primitive roots modulo 11. Dropping bars, the powers $3^0, 3^1, 3^2, \ldots$ of the primitive root 3 modulo 7 list the nonzero elements of $\mathbb{F}_7$ in the following order:

(3.2.12) $$\mathbb{F}_7^\times = \{1, 3, 2, 6, 4, 5\} = \{1, 3, 2, -1, -3, -2\}.$$

Thus there are two ways to list the nonzero elements of $\mathbb{F}_p$, additively and multiplicatively. If α is a primitive root modulo p,

(3.2.13) $$\mathbb{F}_p^\times = \{1, 2, 3, \ldots, p\text{-}1\} = \{1, \alpha, \alpha^2, \ldots, \alpha^{p-2}\},$$

3.3 VECTOR SPACES

Having some examples and the concept of a field, we proceed to the definition of a vector space.

Definition 3.3.1 A *vector space* V over a field F is a set together with two laws of composition:

(a) *addition*: $V \times V \to V$, written $v, w \rightsquigarrow v + w$, for v and w in V,

(b) *scalar multiplication* by elements of the field: $F \times V \to V$, written $c, v \rightsquigarrow cv$, for c in F and v in V.

These laws are required to satisfy the following axioms:

- Addition makes V into a commutative group V^+, with identity denoted by 0.
- $1v = v$, for all v in V.
- *associative law*: $(ab)v = a(bv)$, for all a and b in F and all v in V.
- *distributive laws*: $(a + b)v = av + bv$ and $a(v + w) = av + aw$, for all a and b in F and all v and w in V.

The space F^n of column vectors with entries in the field F forms a vector space over F, when addition and scalar multiplication are defined as usual (3.1.1).

Some more examples of real vector spaces (vector spaces over $\mathbb{R}$):

Examples 3.3.2

(a) Let $V = \mathbb{C}$ be the set of complex numbers. Forget about multiplication of two complex numbers. Remember only addition $\alpha + \beta$ and multiplication $r\alpha$ of a complex number α by a real number r. These operations make V into a real vector space.

(b) The set of real polynomials $p(x) = a_n x^n + \cdots + a_0$ is a real vector space, with addition of polynomials and multiplication of polynomials by real numbers as its laws of composition.

(c) The set of continuous real-valued functions on the real line is a real vector space, with addition of functions $f + g$ and multiplication of functions by real numbers as its laws of composition.

(d) The set of solutions of the differential equation $\frac{d^2 y}{dt^2} = -y$ is a real vector space.   □

Each of our examples has more structure than we look at when we view it as a vector space. This is typical. Any particular example is sure to have extra features that distinguish it from others, but this isn't a drawback. On the contrary, the strength of the abstract approach lies in the fact that consequences of the axioms can be applied in many different situations.

Two important concepts, subspace and isomorphism, are analogous to subgroups and isomorphisms of groups. As with subspaces of $\mathbb{R}^n$, a *subspace* W of a vector space V over a field F is a nonempty subset closed under the operations of addition and scalar multiplication. A subspace W is *proper* if it is neither the whole space V nor the zero subspace $\{0\}$. For example, the space of solutions of the differential equation (3.3.2)(d) is a proper subspace of the space of all continuous functions on the real line.

Proposition 3.3.3 Let $V = F^2$ be the vector space of column vectors with entries in a field F. Every proper subspace W of V consists of the scalar multiples $\{cw\}$ of a single nonzero vector w. Distinct proper subspaces have only the zero vector in common.

The proof of Proposition 3.1.4 carries over.   □

Example 3.3.4 Let F be the prime field $\mathbb{F}_p$. The space F^2 contains p^2 vectors, $p^2 - 1$ of which are nonzero. Because there are $p - 1$ nonzero scalars, the subspace $W = \{cw\}$ spanned by a nonzero vector w will contain $p - 1$ nonzero vectors. Therefore F^2 contains $(p^2 - 1)/(p - 1) = p + 1$ proper subspaces.   □

An *isomorphism* φ from a vector space V to a vector space V', both over the same field F, is a bijective map $\varphi : V \to V'$ compatible with the two laws of composition, a bijective map such that

(3.3.5) $$\varphi(v + w) = \varphi(v) + \varphi(w) \quad \text{and} \quad \varphi(cv) = c\varphi(v),$$

for all v and w in V and all c in F.

Examples 3.3.6

(a) Let $F^{n \times n}$ denote the set of $n \times n$ matrices with entries in a field F. This set is a vector space over F, and it is isomorphic to the space of column vectors of length n^2.

(b) If we view the set of complex numbers as a real vector space, as in (3.3.2)**(a)**, the map $\varphi : \mathbb{R}^2 \to \mathbb{C}$ sending $(a, b)^{\mathrm{t}} \rightsquigarrow a + bi$ is an isomorphism. $\square$

3.4   BASES AND DIMENSION

We discuss the terminology used when working with the operations of addition and scalar multiplication in a vector space. The new concepts are *span*, *independence*, and *basis*.

We work with *ordered sets* of vectors here. We put curly brackets around unordered sets, and we enclose ordered sets with round brackets in order to make the distinction clear. Thus the ordered set (v, w) is different from the ordered set (w, v), whereas the unordered sets $\{v, w\}$ and $\{w, v\}$ are equal. Repetitions are allowed in an ordered set. So (v, v, w) is an ordered set, and it is different from (v, w), in contrast to the convention for unordered sets, where $\{v, v, w\}$ and $\{v, w\}$ denote the same sets.

• Let V be a vector space over a field F, and let $S = (v_1, \dots, v_n)$ be an ordered set of elements of V. A *linear combination* of S is a vector of the form

(3.4.1) $w = c_1 v_1 + \cdots + c_n v_n, \quad$ with c_i in F.

It is convenient to allow scalars to appear on either side of a vector. We simply agree that if v is a vector and c is a scalar, then the notations vc and cv stand for the same vector, the one obtained by scalar multiplication. So $v_1 c_1 + \cdots + v_n c_n = c_1 v_1 + \cdots + c_n v_n$.

Matrix notation provides a compact way to write a linear combination, and the way we write ordered sets of vectors is chosen with this in mind. Since its entries are vectors, we call an array $S = (v_1, \dots, v_n)$ a *hypervector*. Multiplication of two elements of a vector space is not defined, but we do have scalar multiplication. This allows us to interpret a product of the hypervector S and a column vector X in F^n, as the matrix product

(3.4.2) $SX = (v_1, \dots, v_n) \begin{bmatrix} x_1 \\ \vdots \\ x_n \end{bmatrix} = v_1 x_1 + \cdots + v_n x_n.$

Evaluating the right side by scalar multiplication and vector addition, we obtain another vector, a linear combination in which the scalar coefficients x_i are on the right.

We carry along the subspace W of $\mathbb{R}^3$ of solutions of the linear equation

(3.4.3) $2x_1 - x_2 - 2x_3 = 0, \quad$ or $AX = 0$, where $A = (2, -1, -2)$

as an example. Two particular solutions w_1 and w_2 are shown below, together with a linear combination $w_1 y_1 + w_2 y_2$.

(3.4.4) $w_1 = \begin{bmatrix} 1 \\ 0 \\ 1 \end{bmatrix}, \quad w_2 = \begin{bmatrix} 1 \\ 2 \\ 0 \end{bmatrix}, \quad w_1 y_1 + w_2 y_2 = \begin{bmatrix} y_1 + y_2 \\ 2 y_2 \\ y_1 \end{bmatrix}.$

If we write $S = (w_1, w_2)$ with w_i as in (3.4.4) and $Y = (y_1, y_2)^t$, then the combination $w_1 y_1 + w_2 y_2$ can be written in matrix form as SY.

- The set of all vectors that are linear combinations of $S = (v_1, \ldots, v_n)$ forms a subspace of V, called the subspace *spanned* by the set.

As in Section 3.1, this span is the smallest subspace of V that contains S, and it will often be denoted by Span S. The span of a single vector (v_1) is the space of scalar multiples cv_1 of v_1.

One can define span also for an infinite set of vectors. We discuss this in Section 3.7. Let's assume for now that the sets are finite.

Lemma 3.4.5 Let S be an ordered set of vectors of V, and let W be a subspace of V. If $S \subset W$, then Span $S \subset W$. $\qquad\qquad\qquad\qquad\qquad\qquad\qquad\qquad\qquad\square$

The *column space* of an $m \times n$ matrix with entries in F is the subspace of F^m spanned by the columns of the matrix. It has an important interpretation:

Proposition 3.4.6 Let A be an $m \times n$ matrix, and let B be a column vector, both with entries in a field F. The system of equations $AX = B$ has a solution for X in F^m if and only if B is in the column space of A.

Proof. Let $A_1, \ldots, A_n$ denote the columns of A. For any column vector $X = (x_1, \ldots, x_n)^t$, the matrix product AX is the column vector $A_1 x_1 + \cdots + A_n x_n$. This is a linear combination of the columns, an element of the column space, and if $AX = B$, then B is this linear combination. $\qquad\qquad\qquad\qquad\qquad\qquad\qquad\qquad\qquad\qquad\qquad\square$

A *linear relation* among vectors $v_1, \ldots, v_n$ is any linear combination that evaluates to zero – any equation of the form

$$(3.4.7) \qquad\qquad v_1 x_1 + v_2 x_2 + \cdots + v_n x_n = 0$$

that holds in V, where the coefficients x_i are in F. A linear relation can be useful because, if x_n is not zero, the equation (3.4.7) can be solved for v_n.

Definition 3.4.8 An ordered set of vectors $S = (v_1, \ldots, v_n)$ is *independent*, or *linearly independent* if there is no linear relation $SX = 0$ except for the trivial one in which $X = 0$, i.e., in which all the coefficients x_i are zero. A set that is not independent is *dependent*.

An independent set S cannot have any repetitions. If two vectors v_i and v_j of S are equal, then $v_i - v_j = 0$ is a linear relation of the form (3.4.7), the other coefficients being zero. Also, no vector v_i in an independent set is zero, because if v_i is zero, then $v_i = 0$ is a linear relation.

Lemma 3.4.9

(a) A set (v_1) of one vector is independent if and only if $v_1 \neq 0$.

(b) A set (v_1, v_2) of two vectors is independent if neither vector is a multiple of the other.

(c) Any reordering of an independent set is independent. $\qquad\qquad\qquad\qquad\qquad\square$

Suppose that V is the space F^m and that we know the coordinate vectors of the vectors in the set $S = (v_1, \ldots, v_n)$. Then the equation $SX = 0$ gives us a system of m homogeneous linear equations in the n unknowns x_i, and we can decide independence by solving this system.

Example 3.4.10 Let $S = (v_1, v_2, v_3, v_4)$ be the set of vectors in $\mathbb{R}^3$ whose coordinate vectors are

$$(3.4.11) \qquad A_1 = \begin{bmatrix} 1 \\ 0 \\ 1 \end{bmatrix}, \quad A_2 = \begin{bmatrix} 1 \\ 2 \\ 0 \end{bmatrix}, \quad A_3 = \begin{bmatrix} 2 \\ 1 \\ 2 \end{bmatrix}, \quad A_4 = \begin{bmatrix} 1 \\ 1 \\ 3 \end{bmatrix}.$$

Let A denote the matrix made up of these column vectors:

$$(3.4.12) \qquad A = \begin{bmatrix} 1 & 1 & 2 & 1 \\ 0 & 2 & 1 & 1 \\ 1 & 0 & 2 & 3 \end{bmatrix}.$$

A linear combination will have the form $SX = v_1x_1 + v_2x_2 + v_3x_3 + v_4x_4$, and its coordinate vector will be $AX = A_1x_1 + A_2x_2 + A_3x_3 + A_4x_4$. The homogeneous equation $AX = 0$ has a nontrivial solution because it is a system of three homogeneous equations in four unknowns. So the set S is dependent. On the other hand, the determinant of the 3×3 matrix A' formed from the first three columns of (3.4.12) is equal to 1, so the equation $A'X = 0$ has only the trivial solution. Therefore (v_1, v_2, v_3) is an independent set.   □

Definition 3.4.13 A *basis* of a vector space V is a set $(v_1, \ldots, v_n)$ of vectors that is independent and also spans V.

We will often use a boldface symbol such as **B** to denote a basis. The set (v_1, v_2, v_3) defined above is a basis of $\mathbb{R}^3$ because the equation $A'X = B$ has a unique solution for all B (see 1.2.21). The set (w_1, w_2) defined in (3.4.4) is a basis of the space of solutions of the equation $2x_1 - x_2 - 2x_3 = 0$, though we haven't verified this.

Proposition 3.4.14 The set $\mathbf{B} = (v_1, \ldots, v_n)$ is a basis of V if and only if every vector w in V can be written *in a unique way* as a combination $w = v_1x_1 + \cdots + v_nx_n = \mathbf{B}X$.

Proof. The definition of independence can be restated by saying that the zero vector can be written as a linear combination in just one way. If every vector can be written uniquely as a combination, then **B** is independent, and spans V, so it is a basis. Conversely, suppose that **B** is a basis. Then every vector w in V can be written as a linear combination. Suppose that w is written as a combination in two ways, say $w = \mathbf{B}X = \mathbf{B}X'$. Let $Y = X - X'$. Then $\mathbf{B}Y = 0$. This is a linear relation among the vectors $v_1, \ldots, v_n$, which are independent. Therefore $X - X' = 0$. The two combinations are the same.   □

Let $V = F^n$ be the space of column vectors. As before, e_i denotes the column vector with 1 in the ith position and zeros elsewhere (see (1.1.24)). The set $\mathbf{E} = (e_1, \ldots, e_n)$ is a basis for F^n called the *standard basis*. If the coordinate vector of a vector v in F^n is

$X = (x_1, \ldots, x_n)^t$, then $v = \mathbf{E}X = e_1 x_1 + \cdots + e_n x_n$ is the unique expression for v in terms of the standard basis.

We now discuss the main facts that relate the three concepts, of span, independence, and basis. The most important one is Theorem 3.4.18.

Proposition 3.4.15 Let $S = (v_1, \ldots, v_n)$ be an ordered set of vectors, let w be any vector in V, and let $S' = (S, w)$ be the set obtained by adding w to S.

(a) Span $S =$ Span S' if and only if w is in Span S.

(b) Suppose that S is independent. Then S' is independent if and only if w is not in Span S.

Proof. This is very elementary, so we omit most of the proof. We show only that if S is independent but S' is not, then w is in the span of S. If S' is dependent, there is some linear relation

$$v_1 x_1 + \cdots + v_n x_n + wy = 0,$$

in which the coefficients $x_1, \ldots, x_n$ and y are not all zero. If the coefficient y were zero, the expression would reduce to $SX = 0$, and since S is assumed to be independent, we could conclude that $X = 0$ too. The relation would be trivial, contrary to our hypothesis. So $y \neq 0$, and then we can solve for w as a linear combination of $v_1, \ldots, v_n$. $\qquad\square$

- A vector space V is *finite-dimensional* if some finite set of vectors spans V. Otherwise, V is *infinite-dimensional*.

For the rest of this section, our vector spaces are finite-dimensional.

Proposition 3.4.16 Let V be a finite-dimensional vector space.

(a) Let S be a finite subset that spans V, and let L be an independent subset of V. One can obtain a basis of V by adding elements of S to L.

(b) Let S be a finite subset that spans V. One can obtain a basis of V by deleting elements from S.

Proof. **(a)** If S is contained in Span L, then L spans V, and so it is a basis (3.4.5). If not, we choose an element v in S, which is not in Span L. By Proposition 3.4.15, $L' = (L, v)$ is independent. We replace L by L'. Since S is finite, we can do this only finitely often. So eventually we will have a basis.

(b) If S is dependent, there is a linear relation $v_1 c_1 + \cdots + v_n c_n = 0$ in which some coefficient, say c_n, is not zero. We can solve this equation for v_n, and this shows that v_n is in the span of the set S_1 of the first $n - 1$ vectors. Proposition 3.4.15(a) shows that Span $S =$ Span S_1. So S_1 spans V. We replace S by S_1. Continuing this way we must eventually obtain a family that is independent but still spans V: a basis.

Note: There is a problem with this reasoning when V is the zero vector space $\{0\}$. Starting with an arbitrary set S of vectors in V, all equal to zero, our procedure will throw them out one at a time until there is only one vector v_1 left. And since v_1 is zero, the set (v_1) is dependent. How can we proceed? The zero space isn't particularly interesting, but it may lurk in a corner, ready to trip us up. We have to allow for the possibility that a vector space that arises in the course of some computation, such as solving a system of homogeneous linear equations, is the zero space, though we aren't aware of this. In order to avoid having to mention this possibility as a special case, we adopt the following definitions:

(3.4.17)
- The empty set is independent.
- The span of the empty set is the zero space $\{0\}$.

Then the empty set is a basis for the zero vector space. These definitions allow us to throw out the last vector v_1, which rescues the proof. $\square$

We come now to the main fact about independence:

Theorem 3.4.18 Let S and L be finite subsets of a vector space V. Assume that S spans V and that L is independent. Then S contains at least as many elements as L does: $|S| \geq |L|$.

As before, $|S|$ denotes the order, the number of elements, of the set S.

Proof. Say that $S = (v_1, \ldots, v_m)$ and that $L = (w_1, \ldots, w_n)$. We assume that $|S| < |L|$, i.e., that $m < n$, and we show that L is dependent. To do this, we show that there is a linear relation $w_1 x_1 + \cdots + w_n x_n = 0$, in which the coefficients x_i aren't all zero. We write this undetermined relation as $LX = 0$.

Because S spans V, each element w_j of L is a linear combination of S, say $w_j = v_1 a_{1j} + \cdots + v_m a_{mj} = SA_j$, where A_j is the column vector of coefficients. We assemble these column vectors into an $m \times n$ matrix

(3.4.19)
$$A = \begin{bmatrix} | & & | \\ A_1 & \cdots & A_n \\ | & & | \end{bmatrix}.$$

Then

(3.4.20) $SA = (SA_1, \ldots, SA_n) = (w_1, \ldots, w_n) = L.$

We substitute SA for L into our undetermined linear combination:

$$LX = (SA)X.$$

The associative law for scalar multiplication implies that $(SA)X = S(AX)$. The proof is the same as for the associative law for multiplication of scalar matrices (which we omitted). If $AX = 0$, then our combination LX will be zero too. Now since A is an $m \times n$ matrix with $m < n$, the homogeneous system $AX = 0$ has a nontrivial solution X. Then $LX = 0$ is the linear relation we are looking for. $\square$

Proposition 3.4.21 Let V be a finite-dimensional vector space.

(a) Any two bases of V have the same order (the same number of elements).

(b) Let **B** be a basis. If a finite set S of vectors spans V, then $|S| \geq |\mathbf{B}|$, and $|S| = |\mathbf{B}|$ if and only if S is a basis.

(c) Let **B** be a basis. If a set L of vectors is independent, then $|L| \leq |\mathbf{B}|$, and $|L| = |\mathbf{B}|$ if and only if L is a basis.

Proof. **(a)** We note here that two finite bases $\mathbf{B}_1$ and $\mathbf{B}_2$ have the same order, and we will show in Corollary 3.7.7 that every basis of a finite-dimensional vector space is finite. Taking $S = \mathbf{B}_1$ and $L = \mathbf{B}_2$ in Theorem 3.4.18 shows that $|\mathbf{B}_1| \geq |\mathbf{B}_2|$, and similarly, $|\mathbf{B}_2| \geq |\mathbf{B}_1|$. Parts **(b)** and **(c)** follow from **(a)** and Proposition 3.4.16.   □

Definition 3.4.22 The *dimension* of a finite-dimensional vector space V is the number of vectors in a basis. This dimension will be denoted by $\dim V$.

The dimension of the space F^n of column vectors is n because the standard basis $\mathbf{E} = (e_1, \ldots, e_n)$ contains n elements.

Proposition 3.4.23 If W is a subspace of a finite-dimensional vector space V, then W is finite-dimensional, and $\dim W \leq \dim V$. Moreover, $\dim W = \dim V$ if and only if $W = V$.

Proof. We start with any independent set L of vectors in W, possibly the empty set. If L doesn't span W, we choose a vector w in W not in the span of L. Then $L' = (L, w)$ will be independent (3.4.15). We replace L by L'.

Now it is obvious that if L is an independent subset of W, then it is also independent when thought of as a subset of V. So Theorem 3.4.18 tells us that $|L| \leq \dim V$. Therefore the process of adding elements to L must come to an end, and when it does, we will have a basis of W. Since L contains at most $\dim V$ elements, $\dim W \leq \dim V$. If $|L| = \dim V$, then Proposition 3.4.21**(c)** shows that L is a basis of V, and therefore $W = V$.   □

3.5   COMPUTING WITH BASES

The purpose of bases is to provide a method of computation, and we learn to use them in this section. We consider two topics: how to express a vector in terms of a basis, and how to relate different bases of the same vector space.

Suppose we are given a basis $\mathbf{B} = (v_1, \ldots, v_n)$ of a vector space V over F. Remember: This means that every vector v in V can be expressed as a combination

$$(3.5.1) \qquad v = v_1 x_1 + \cdots + v_n x_n, \quad \text{with } x_i \text{ in } F,$$

in exactly one way (3.4.14). The scalars x_i are the *coordinates* of v, and the column vector

$$(3.5.2) \qquad X = \begin{bmatrix} x_1 \\ \vdots \\ x_n \end{bmatrix}$$

is the *coordinate vector* of v, with respect to the basis $\mathbf{B}$.

For example, $(\cos t, \sin t)$ is a basis of the space of solutions of the differential equation $y'' = -y$. Every solution of this differential equation is a linear combination of this basis. If we are given another solution $f(t)$, the coordinate vector $(x_1, x_2)^t$ of f is the vector such that $f(t) = (\cos t)x_1 + (\sin t)x_2$. Obviously, we need to know something about f to find X.

Not very much: just enough to determine two coefficients. Most properties of f are implicit in the fact that it solves the differential equation.

What we can always do, given a basis **B** of a vector space of dimension n, is to define an *isomorphism of vector spaces* (see 3.3.5) from the space F^n to V:

(3.5.3) $$\psi: F^n \to V \text{ that sends } X \rightsquigarrow \mathbf{B}X.$$

We will often denote this isomorphism by **B**, because it sends a vector X to $\mathbf{B}X$.

Proposition 3.5.4 Let $S = (v_1, \ldots, v_n)$ be a subset of a vector space V, and let $\psi: F^n \to V$ be the map defined by $\psi(X) = SX$. Then

(a) ψ is injective if and only if S is independent,

(b) ψ is surjective if and only if S spans V, and

(c) ψ is bijective if and only if S is a basis of V.

This follows from the definitions of independence, span, and basis. $\square$

Given a basis, the coordinate vector of a vector v in V is obtained by inverting the map ψ (3.5.3). We won't have a formula for the inverse function unless the basis is given more explicitly, but the existence of the isomorphism is interesting:

Corollary 3.5.5 Every vector space V of dimension n over a field F is isomorphic to the space F^n of column vectors. $\square$

Notice also that F^n is not isomorphic to F^m when $m \neq n$, because F^n has a basis of n elements, and the number of elements in a basis depends only on the vector space. Thus the finite-dimensional vector spaces over a field F are completely classified. The spaces F^n of column vectors are representative elements for the isomorphism classes.

The fact that a vector space of dimension n is isomorphic to F^n will allow us to translate problems on vector spaces to the familiar algebra of column vectors, once a basis is chosen. Unfortunately, the same vector space V will have many bases. Identifying V with the isomorphic space F^n is useful when a natural basis is in hand, but not when a basis is poorly suited to a given problem. In that case, we will need to change coordinates, i.e., to change the basis.

The space of solutions of a homogeneous linear equation $AX = 0$, for instance, almost never has a natural basis. The space W of solutions of the equation $2x_1 - x_2 - 2x_3 = 0$ has dimension 2, and we exhibited a basis before: $\mathbf{B} = (w_1, w_2)$, where $w_1 = (1, 0, 1)^t$ and $w_2 = (1, 2, 0)^t$ (see (3.4.4)). Using this basis, we obtain an isomorphism of vector spaces $\mathbb{R}^2 \to W$ that we may denote by **B**. Since the unknowns in the equation are labeled x_i, we need to choose another symbol for variable elements of $\mathbb{R}^2$ here. We'll use $Y = (y_1, y_2)^t$. The isomorphism **B** sends Y to the coordinate vector of $\mathbf{B}Y = w_1 y_1 + w_2 y_2$ that was displayed in (3.4.4).

However, there is nothing very special about the two particular solutions w_1 and w_2. Most other pairs of solutions would serve just as well. The solutions $w_1' = (0, 2, -1)^t$ and $w_2' = (1, 4, -1)^t$ give us a second basis $\mathbf{B}' = (w_1', w_2')$ of W. Either basis suffices to express the solutions uniquely. A solution can be written in either one of the forms

(3.5.6)
$$\begin{bmatrix} y_1 + y_2 \\ 2y_2 \\ y_1 \end{bmatrix} \quad \text{or} \quad \begin{bmatrix} y'_2 \\ 2y'_1 + 4y'_2 \\ -y'_1 - y'_2 \end{bmatrix}.$$

Change of Basis

Suppose that we are given two bases of the same vector space V, say $\mathbf{B} = (v_1, \ldots, v_n)$ and $\mathbf{B}' = (v'_1, \ldots, v'_n)$. We wish to make two computations. We ask first: How are the two bases related? Second, a vector v in V will have coordinates with respect to each of these bases, but they will be different. So we ask: How are the two coordinate vectors related? These are the basechange computations, and they will be very important in later chapters. They can also drive you nuts if you don't organize the notation carefully.

Let's think of $\mathbf{B}$ as the *old* basis and $\mathbf{B}'$ as a *new* basis. We note that every vector of the new basis $\mathbf{B}'$ is a linear combination of the old basis $\mathbf{B}$. We write this combination as

(3.5.7)
$$v'_j = v_1 p_{1j} + v_2 p_{2j} + \cdots + v_n p_{nj}.$$

The column vector $P_j = (p_{1j}, \ldots, p_{nj})^t$ is the coordinate vector of the new basis vector v'_j, when it is computed using the old basis. We collect these column vectors into a square matrix P, obtaining the matrix equation $\mathbf{B}' = \mathbf{B}P$:

(3.5.8)
$$\mathbf{B}' = (v'_1, \ldots, v'_n) = (v_1, \ldots, v_n) \begin{bmatrix} & & \\ & P & \\ & & \end{bmatrix} = \mathbf{B}P.$$

The jth column of P is the coordinate vector of the new basis vector v'_j with respect to the old basis. This matrix P is the *basechange matrix*. [1]

Proposition 3.5.9

(a) Let $\mathbf{B}$ and $\mathbf{B}'$ be two bases of a vector space V. The basechange matrix P is an invertible matrix that is determined uniquely by the two bases $\mathbf{B}$ and $\mathbf{B}'$.

(b) Let $\mathbf{B} = (v_1, \ldots, v_n)$ be a basis of a vector space V. The other bases are the sets of the form $\mathbf{B}' = \mathbf{B}P$, where P can be any invertible $n \times n$ matrix.

Proof. (a) The equation $\mathbf{B}' = \mathbf{B}P$ expresses the basis vectors v'_i as linear combinations of the basis $\mathbf{B}$. There is just one way to do this (3.4.14), so P is unique. To show that P is an invertible matrix, we interchange the roles of $\mathbf{B}$ and $\mathbf{B}'$. There is a matrix Q such that $\mathbf{B} = \mathbf{B}'Q$. Then

$$\mathbf{B} = \mathbf{B}'Q = \mathbf{B}PQ, \quad \text{or} \quad (v_1, \ldots, v_n) = (v_1, \ldots, v_n) \begin{bmatrix} & & \\ & PQ & \\ & & \end{bmatrix}.$$

This equation expresses each v_i as a combination of the vectors $(v_1, \ldots, v_n)$. The entries of the product matrix PQ are the coefficients. But since $\mathbf{B}$ is a basis, there is just one way to

---

[1] This basechange matrix is the inverse of the one that was used in the first edition.

write v_i as a combination of $(v_1, \ldots, v_n)$, namely $v_i = v_i$, or in matrix notation, $\mathbf{B} = \mathbf{B}I$. So $PQ = I$.

(b) We must show that if $\mathbf{B}$ is a basis and if P is an invertible matrix, then $\mathbf{B}' = \mathbf{B}P$ is also a basis. Since P is invertible, $\mathbf{B} = \mathbf{B}'P^{-1}$. This tells us that the vectors v_i are in the span of $\mathbf{B}'$. Therefore $\mathbf{B}'$ spans V, and since it has the same number of elements as $\mathbf{B}$, it is a basis.   □

Let X and X' be the coordinate vectors of the same arbitrary vector v, computed with respect to the two bases $\mathbf{B}$ and $\mathbf{B}'$, respectively, that is, $v = \mathbf{B}X$ and $v = \mathbf{B}'X'$. Substituting $\mathbf{B} = \mathbf{B}'P^{-1}$ gives us the matrix equation

$$(3.5.10) \qquad\qquad v = \mathbf{B}X = \mathbf{B}'P^{-1}X.$$

This shows that the coordinate vector of v with respect to the new basis $\mathbf{B}'$, which we call X', is $P^{-1}X$. We can also write this as $X = PX'$.

Recapitulating, we have a single matrix P, the basechange matrix, with the dual properties

$$(3.5.11) \qquad\qquad \mathbf{B}' = \mathbf{B}P \quad \text{and} \quad PX' = X,$$

where X and X' denote the coordinate vectors of the same arbitrary vector v, with respect to the two bases. Each of these properties characterizes P. Please take note of the positions of P in the two relations.

Going back once more to the equation $2x_1 - x_2 - 2x_3 = 0$, let $\mathbf{B}$ and $\mathbf{B}'$ be the bases of the space W of solutions described above, in (3.5.6). The basechange matrix solves the equation

$$\begin{bmatrix} 0 & 1 \\ 2 & 4 \\ -1 & -1 \end{bmatrix} = \begin{bmatrix} 1 & 1 \\ 0 & 2 \\ 1 & 0 \end{bmatrix} \begin{bmatrix} p_{11} & p_{12} \\ p_{21} & p_{22} \end{bmatrix}. \quad \text{It is} \quad P = \begin{bmatrix} -1 & -1 \\ 1 & 2 \end{bmatrix}.$$

The coordinate vectors Y and Y' of a given vector v with respect to these two bases, the ones that appear in (3.5.6), are related by the equation

$$PY' = \begin{bmatrix} -1 & -1 \\ 1 & 2 \end{bmatrix} \begin{bmatrix} y'_1 \\ y'_2 \end{bmatrix} = \begin{bmatrix} y_1 \\ y_2 \end{bmatrix} = Y.$$

Another example: Let $\mathbf{B}$ be the basis $(\cos t, \sin t)$ of the space of solutions of the differential equation $\frac{d^2 y}{dt^2} = -y$. If we allow complex valued functions, then the exponential functions $e^{\pm it} = \cos t \pm i \sin t$ are also solutions, and $\mathbf{B}' = (e^{it}, e^{-it})$ is a new basis of the space of solutions. The basechange computation is

$$(3.5.12) \qquad\qquad (e^{it}, e^{-it}) = (\cos t, \sin t) \begin{bmatrix} 1 & 1 \\ i & -i \end{bmatrix}.$$

One case in which the basechange matrix is easy to determine is that V is the space F^n of column vectors, the old basis is the standard basis $\mathbf{E} = (e_1, \ldots, e_n)$, and the new basis, we'll denote it by $\mathbf{B} = (v_1, \ldots, v_n)$ here, is arbitrary. Let the coordinate vector of v_i, with respect to the standard basis, be the column vector B_i. So $v_i = \mathbf{E}B_i$. We assemble these column vectors into an $n \times n$ matrix that we denote by $[\mathbf{B}]$:

(3.5.13)

$$[\mathbf{B}] = \begin{bmatrix} | & & | \\ B_1 & \cdots & B_n \\ | & & | \end{bmatrix}. \text{ Then } (v_1, \ldots, v_n) = (e_1, \ldots, e_n) \begin{bmatrix} | & & | \\ B_1 & \cdots & B_n \\ | & & | \end{bmatrix},$$

i.e., $\mathbf{B} = \mathbf{E}[\mathbf{B}]$. Therefore $[\mathbf{B}]$ is the basechange matrix from the standard basis $\mathbf{E}$ to $\mathbf{B}$.

3.6 DIRECT SUMS

The concepts of independence and span of a set of vectors have analogues for subspaces. If $W_1, \ldots, W_k$ are subspaces of a vector space V, the set of vectors v that can be written as a sum

(3.6.1)
$$v = w_1 + \cdots + w_k,$$

where w_i is in W_i is called the *sum* of the subspaces or their *span*, and is denoted by $W_1 + \cdots + W_k$:

(3.6.2) $W_1 + \cdots + W_k = \{v \in V \mid v = w_1 + \cdots + w_k, \text{ with } w_i \text{ in } W_i\}.$

The sum of the subspaces is the smallest subspace that contains all of the subspaces $W_1, \ldots, W_k$. It is analogous to the span of a set of vectors.

The subspaces $W_1, \ldots, W_k$ are called *independent* if no sum $w_1 + \cdots + w_k$ with w_i in W_i is zero, except for the trivial sum, in which $w_i = 0$ for all i. In other words, the spaces are independent if

(3.6.3) $w_1 + \cdots + w_k = 0, \text{ with } w_i \text{ in } W_i, \text{ implies } w_i = 0 \text{ for all } i.$

Note: Suppose that $v_1, \ldots, v_k$ are elements of V, and let W_i be the span of the vector v_i. Then the subspaces $W_1, \ldots, W_k$ are independent if and only if the set $(v_1, \ldots, v_n)$ is independent. This becomes clear if we compare (3.4.8) and (3.6.3). The statement in terms of subspaces is actually the neater one, because scalar coefficients don't need to be put in front of the vectors w_i in (3.6.3). Since each of the subspaces W_i is closed under scalar multiplication, a scalar multiple cw_i is simply another element of W_i.   □

We omit the proof of the next proposition.

Proposition 3.6.4 Let $W_1, \ldots, W_k$ be subspaces of a finite-dimensional vector space V, and let $\mathbf{B}_i$ be a basis of W_i.

(a) The following conditions are equivalent:
- The subspaces W_i are independent, and the sum $W_1 + \cdots + W_k$ is equal to V.
- The set $\mathbf{B} = (\mathbf{B}_1, \ldots, \mathbf{B}_k)$ obtained by appending the bases $\mathbf{B}_i$ is a basis of V.

(b) $\dim(W_1 + \cdots + W_k) \leq \dim W_1 + \cdots + \dim W_k$, with equality if and only if the spaces are independent.

(c) If W_i' is a subspace of W_i for $i = 1, \ldots, k$, and if the spaces $W_1, \ldots, W_k$ are independent, then so are the $W_1', \ldots, W_k'$.   □

If the conditions of Proposition 3.6.4**(a)** are satisfied, we say that V is the *direct sum* of $W_1, \ldots, W_k$, and we write $V = W_1 \oplus \cdots \oplus W_k$:

(3.6.5)
$$V = W_1 \oplus \cdots \oplus W_k, \quad \text{if } W_1 + \cdots + W_k = V$$
$$\text{and } W_1, \ldots, W_k \text{ are independent.}$$

If V is the direct sum, every vector v in V can be written in the form (3.6.1) in exactly one way.

Proposition 3.6.6 Let W_1 and W_2 be subspaces of a finite-dimensional vector space V.

(a) $\dim W_1 + \dim W_2 = \dim(W_1 \cap W_2) + \dim(W_1 + W_2)$.

(b) W_1 and W_2 are independent if and only if $W_1 \cap W_2 = \{0\}$.

(c) V is the direct sum $W_1 \oplus W_2$ if and only if $W_1 \cap W_2 = \{0\}$ and $W_1 + W_2 = V$.

(d) If $W_1 + W_2 = V$, there is a subspace W_2' of W_2 such that $W_1 \oplus W_2' = V$.

Proof. We prove the key part **(a)**: We choose a basis, $\mathbf{U} = (u_1, \ldots, u_k)$ for $W_1 \cap W_2$, and we extend it to a basis $(\mathbf{U}, \mathbf{V}) = (u_1, \ldots, u_k; v_1, \ldots, v_m)$ of W_1. We also extend $\mathbf{U}$ to a basis $(\mathbf{U}, \mathbf{W}) = (u_1, \ldots, u_k; w_1, \ldots, w_n)$ of W_2. Then $\dim(W_1 \cap W_2) = k$, $\dim W_1 = k + m$, and $\dim W_2 = k + n$. The assertion will follow if we prove that the set of $k + m + n$ elements $(\mathbf{U}, \mathbf{V}, \mathbf{W}) = (u_1, \ldots, u_k; v_1, \ldots, v_m; w_1, \ldots, w_n)$ is a basis of $W_1 + W_2$.

We must show that $(\mathbf{U}, \mathbf{V}, \mathbf{W})$ is independent and spans $W_1 + W_2$. An element v of $W_1 + W_2$ has the form $w' + w''$ where w' is in W_1 and w'' is in W_2. We write w' in terms of our basis $(\mathbf{U}, \mathbf{V})$ for W_1, say $w' = \mathbf{U}X + \mathbf{V}Y = u_1 x_1 + \cdots + u_k x_k + v_1 y_1 + \cdots + v_m y_m$. We also write w'' as a combination $\mathbf{U}X' + \mathbf{W}Z$ of our basis $(\mathbf{U}, \mathbf{W})$ for W_2. Then $\mathbf{V} = w' + w'' = \mathbf{U}(X + X') + \mathbf{V}Y + \mathbf{W}Z$.

Next, suppose we are given a linear relation $\mathbf{U}X + \mathbf{V}Y + \mathbf{W}Z = 0$, among the elements $(\mathbf{U}, \mathbf{V}, \mathbf{W})$. We write this as $\mathbf{U}X + \mathbf{V}Y = -\mathbf{W}Z$. The left side of this equation is in W_1 and the right side is in W_2. Therefore $-\mathbf{W}Z$ is in $W_1 \cap W_2$, and so it is a linear combination $\mathbf{U}X'$ of the basis $\mathbf{U}$. This gives us an equation $\mathbf{U}X' + \mathbf{W}Z = 0$. Since the set $(\mathbf{U}, \mathbf{W})$ is a basis for W_2, it is independent, and therefore X' and Z are zero. The given relation reduces to $\mathbf{U}X + \mathbf{V}Y = 0$. But $(\mathbf{U}, \mathbf{V})$ is also an independent set. So X and Y are zero. The relation was trivial.   □

3.7   INFINITE-DIMENSIONAL SPACES

Vector spaces that are too big to be spanned by any finite set of vectors are called *infinite-dimensional*. We won't need them very often, but they are important in analysis, so we discuss them briefly here.

One of the simplest examples of an infinite-dimensional space is the space $\mathbb{R}^\infty$ of infinite real row vectors

(3.7.1)
$$(a) = (a_1, a_2, a_3, \ldots).$$

An infinite vector can be thought of as a sequence $a_1, a_2, \ldots$ of real numbers.

The space $\mathbb{R}^\infty$ has many infinite-dimensional subspaces. Here are a few; you will be able to make up some more:

Examples 3.7.2

(a) Convergent sequences: $C = \{(a) \in \mathbb{R}^\infty \mid$ the limit $\lim\limits_{n \to \infty} a_n$ exists $\}$.

(b) Absolutely convergent series: $\ell^1 = \{(a) \in \mathbb{R}^\infty \mid \sum\limits_1^\infty |a_n| < \infty\}$.

(c) Sequences with finitely many terms different from zero.

$$Z = \{(a) \in \mathbb{R}^\infty \mid a_n = 0 \text{ for all but finitely many } n\}.$$

Now suppose that V is a vector space, infinite-dimensional or not. What do we mean by the *span* of an infinite set S of vectors? It isn't always possible to assign a value to an infinite combination $c_1 v_1 + c_2 v_2 + \cdots$. If V is the vector space $\mathbb{R}^n$, then a value can be assigned provided that the series $c_1 v_1 + c_2 v_2 + \cdots$ converges. But many series don't converge, and then we don't know what value to assign. In algebra it is customary to speak only of combinations of finitely many vectors. The span of an infinite set S is defined to be the set of the vectors v that are combinations of finitely many elements of S:

(3.7.3) $$v = c_1 v_1 + \cdots + c_r v_r, \quad \text{where } v_1, \ldots, v_r \text{ are in } S.$$

The vectors v_i in S can be arbitrary, and the number r is allowed to depend on the vector v and to be arbitrarily large:

(3.7.4) $$\text{Span } S = \left\{ \begin{array}{c} \text{finite combinations} \\ \text{of elements of } S \end{array} \right\}.$$

For example, let $e_i = (0, \ldots, 0, 1, 0, \ldots)$ be the *row vector* in $\mathbb{R}^\infty$ with 1 in the ith position as its only nonzero coordinate. Let $\mathbf{E} = (e_1, e_2, e_3, \ldots)$ be the set of these vectors. This set does not span $\mathbb{R}^\infty$, because the vector

$$w = (1, 1, 1, \ldots)$$

is not a (finite) combination. The span of the set $\mathbf{E}$ is the subspace Z (3.7.2)**(c)**.

A set S, finite or infinite, is *independent* if there is no finite linear relation

(3.7.5) $$c_1 v_1 + \cdots + c_r v_r = 0, \quad \text{with } v_1, \ldots, v_r \text{ in } S,$$

except for the trivial relation in which $c_1 = \cdots = c_r = 0$. Again, the number r is allowed to be arbitrary, that is, the condition has to hold for arbitrarily large r and arbitrary elements $v_1, \ldots, v_r$ of S. For example, the set $S' = (w; e_1, e_2, e_3, \ldots)$ is independent, if w and e_i are the vectors defined above. With this definition of independence, Proposition 3.4.15 continues to be true.

As with finite sets, a *basis* S of V is an independent set that spans V. The set $S = (e_1, e_2, \ldots)$ is a basis of the space Z. The monomials x^i form a basis for the space of polynomials. It can be shown, using *Zorn's Lemma* or the *Axiom of Choice*, that every vector space V has a basis (see the appendix, Proposition A.3.3). However, a basis for $\mathbb{R}^\infty$ will have uncountably many elements, and cannot be made very explicit.

Let us go back for a moment to the case that our vector space V is finite-dimensional (3.4.16), and ask if there can be an *infinite* basis. We saw in (3.4.21) that any two finite bases have the same number of elements. We complete the picture now, by showing that every basis is finite. This follows from the next lemma.

Lemma 3.7.6 Let V be a finite-dimensional vector space, and let S be any set that spans V. Then S contains a finite subset that spans V.

Proof. By hypothesis, there is a finite set, say $(u_1, \ldots, u_m)$, that spans V. Because S spans V, each of the vectors u_i is a linear combination of finitely many elements of S. The elements of S that we use to write all of these vectors as linear combinations make up a finite subset S' of S. Then the vectors u_i are in Span S', and since $(u_1, \ldots, u_m)$ spans V, so does S'. $\square$

Corollary 3.7.7 Let V be a finite-dimensional vector space.

- Every basis is finite.
- Every set S that spans V contains a basis.
- Every independent set L is finite, and can be extended to a basis. $\square$

I don't need to learn 8 + 7: I'll remember 8 + 8 and subtract 1.

—T. Cuyler Young, Jr.

EXERCISES

Section 1   Fields

1.1. Prove that the numbers of the form $a + b\sqrt{2}$, where a and b are rational numbers, form a subfield of $\mathbb{C}$.

1.2. Find the inverse of 5 modulo p, for $p = 7, 11, 13$, and 17.

1.3. Compute the product polynomial $(x^3 + 3x^2 + 3x + 1)(x^4 + 4x^3 + 6x^2 + 4x + 1)$ when the coefficients are regarded as elements of the field $\mathbb{F}_7$. Explain your answer.

1.4. Consider the system of linear equations $\begin{bmatrix} 6 & -3 \\ 2 & 6 \end{bmatrix} \begin{bmatrix} x_1 \\ x_2 \end{bmatrix} = \begin{bmatrix} 3 \\ 1 \end{bmatrix}$

   (a) Solve the system in $\mathbb{F}_p$ when $p = 5, 11$, and 17.
   (b) Determine the number of solutions when $p = 7$.

1.5. Determine the primes p such that the matrix

$$A = \begin{bmatrix} 1 & 2 & 0 \\ 0 & 3 & -1 \\ -2 & 0 & 2 \end{bmatrix}$$

is invertible, when its entries are considered to be in $\mathbb{F}_p$.

1.6. Solve completely the systems of linear equations $AX = 0$ and $AX = B$, where

$$A = \begin{bmatrix} 1 & 1 & 0 \\ 1 & 0 & 1 \\ 1 & -1 & -1 \end{bmatrix}, \quad \text{and} \quad B = \begin{bmatrix} 1 \\ -1 \\ 1 \end{bmatrix}$$

(a) in $\mathbb{Q}$, (b) in $\mathbb{F}_2$, (c) in $\mathbb{F}_3$, (d) in $\mathbb{F}_7$.

1.7. By finding primitive elements, verify that the multiplicative group $\mathbb{F}_p^{\times}$ is cyclic for all primes $p < 20$.

1.8. Let p be a prime integer.

(a) Prove *Fermat's Theorem*: For every integer a, $a^p \equiv a$ modulo p.
(b) Prove *Wilson's Theorem*: $(p - 1)! \equiv -1 (\text{modulo } p)$.

1.9. Determine the orders of the matrices $\begin{bmatrix} 1 & 1 \\ & 1 \end{bmatrix}$ and $\begin{bmatrix} 2 \\ & 1 \end{bmatrix}$ in the group $GL_2(\mathbb{F}_7)$.

1.10. Interpreting matrix entries in the field $\mathbb{F}_2$, prove that the four matrices

$$\begin{bmatrix} 0 & 0 \\ 0 & 0 \end{bmatrix}, \begin{bmatrix} 1 & 0 \\ 0 & 1 \end{bmatrix}, \begin{bmatrix} 1 & 1 \\ 1 & 0 \end{bmatrix}, \begin{bmatrix} 0 & 1 \\ 1 & 1 \end{bmatrix} \text{ form a field.}$$

Hint: You can cut the work down by using the fact that various laws are known to hold for addition and multiplication of matrices.

1.11. Prove that the set of symbols $\{a + bi \mid a, b \in \mathbb{F}_3\}$ forms a field with nine elements, if the laws of composition are made to mimic addition and multiplication of complex numbers. Will the same method work for $\mathbb{F}_5$? For $\mathbb{F}_7$? Explain.

Section 2   Vector Spaces

2.1. (a) Prove that the scalar product of a vector with the zero element of the field F is the zero vector.

(b) Prove that if w is an element of a subspace W, then $-w$ is in W too.

2.2. Which of the following subsets is a subspace of the vector space $F^{n \times n}$ of $n \times n$ matrices with coefficients in F?

(a) symmetric matrices $(A = A^t)$, (b) invertible matrices, (c) upper triangular matrices.

Section 3   Bases and Dimension

3.1. Find a basis for the space of $n \times n$ symmetric matrices $(A^t = A)$.

3.2. Let $W \subset \mathbb{R}^4$ be the space of solutions of the system of linear equations $AX = 0$, where $A = \begin{bmatrix} 2 & 1 & 2 & 3 \\ 1 & 1 & 3 & 0 \end{bmatrix}$. Find a basis for W.

3.3. Prove that the three functions x^2, $\cos x$, and e^x are linearly independent.

3.4. Let A be an $m \times n$ matrix, and let A' be the result of a sequence of elementary row operations on A. Prove that the rows of A span the same space as the rows of A'.

3.5. Let $V = F^n$ be the space of column vectors. Prove that every subspace W of V is the space of solutions of some system of homogeneous linear equations $AX = 0$.

3.6. Find a basis of the space of solutions in $\mathbb{R}^n$ of the equation

$$x_1 + 2x_2 + 3x_3 + \cdots + nx_n = 0.$$

3.7. Let $(X_1, \ldots, X_m)$ and $(Y_1, \ldots, Y_n)$ be bases for $\mathbb{R}^m$ and $\mathbb{R}^n$, respectively. Do the mn matrices $X_i Y_j^t$ form a basis for the vector space $\mathbb{R}^{m \times n}$ of all $m \times n$ matrices?

3.8. Prove that a set $(v_1, \ldots, v_n)$ of vectors in F^n is a basis if and only if the matrix obtained by assembling the coordinate vectors of v_i is invertible.

Section 4 Computing with Bases

4.1. (a) Prove that the set $\mathbf{B} = ((1, 2, 0)^t, (2, 1, 2)^t, (3, 1, 1)^t)$ is a basis of $\mathbb{R}^3$.

(b) Find the coordinate vector of the vector $v = (1, 2, 3)^t$ with respect to this basis.

(c) Let $\mathbf{B}' = ((0, 1, 0)^t, (1, 0, 1)^t, (2, 1, 0)^t)$. Determine the basechange matrix P from $\mathbf{B}$ to $\mathbf{B}'$.

4.2. (a) Determine the basechange matrix in $\mathbb{R}^2$, when the old basis is the standard basis $\mathbf{E} = (e_1, e_2)$ and the new basis is $\mathbf{B} = (e_1 + e_2, e_1 - e_2)$.

(b) Determine the basechange matrix in $\mathbb{R}^n$, when the old basis is the standard basis $\mathbf{E}$ and the new basis is $\mathbf{B} = (e_n, e_{n-1}, \ldots, e_1)$.

(c) Let $\mathbf{B}$ be the basis of $\mathbb{R}^2$ in which $v_1 = e_1$ and v_2 is a vector of unit length making an angle of $120°$ with v_1. Determine the basechange matrix that relates $\mathbf{E}$ to $\mathbf{B}$.

4.3. Let $\mathbf{B} = (v_1, \ldots, v_n)$ be a basis of a vector space V. Prove that one can get from $\mathbf{B}$ to any other basis $\mathbf{B}'$ by a finite sequence of steps of the following types:

(i) Replace v_i by $v_i + av_j$, $i \neq j$, for some a in F,

(ii) Replace v_i by cv_i for some $c \neq 0$,

(iii) Interchange v_i and v_j.

4.4. Let $\mathbb{F}_p$ be a prime field, and let $V = \mathbb{F}_p^2$. Prove:

(a) The number of bases of V is equal to the order of the general linear group $GL_2(\mathbb{F}_p)$.

(b) The order of the general linear group $GL_2(\mathbb{F}_p)$ is $p(p+1)(p-1)^2$, and the order of the special linear group $SL_2(\mathbb{F}_p)$ is $p(p+1)(p-1)$.

4.5. How many subspaces of each dimension are there in **(a)** $\mathbb{F}_p^3$, **(b)** $\mathbb{F}_p^4$?

Section 5 Direct Sums

5.1. Prove that the space $\mathbb{R}^{n \times n}$ of all $n \times n$ real matrices is the direct sum of the space of symmetric matrices $(A^t = A)$ and the space of skew-symmetric matrices $(A^t = -A)$.

5.2. The trace of a square matrix is the sum of its diagonal entries. Let W_1 be the space of $n \times n$ matrices whose trace is zero. Find a subspace W_2 so that $\mathbb{R}^{n \times n} = W_1 \oplus W_2$.

5.3. Let $W_1, \ldots, W_k$ be subspaces of a vector space V, such that $V = \sum W_i$. Assume that $W_1 \cap W_2 = 0$, $(W_1 + W_2) \cap W_3 = 0$, $\ldots$, $(W_1 + W_2 + \cdots + W_{k-1}) \cap W_k = 0$. Prove that V is the direct sum of the subspaces $W_1, \ldots, W_k$.

Section 6   Infinite-Dimensional Spaces

6.1. Let $\mathbf{E}$ be the set of vectors $(e_1, e_2, \ldots)$ in $\mathbb{R}^\infty$, and let $w = (1, 1, 1, \ldots)$. Describe the span of the set $(w, e_1, e_2, \ldots)$.

6.2. The doubly infinite row vectors $(a) = (\ldots, a_{-1}, a_0, a_1, \ldots)$, with a_i real form a vector space. Prove that this space is isomorphic to $\mathbb{R}^\infty$.

*6.3. For every positive integer, we can define the space ℓ^p to be the space of sequences such that $\sum |a_i|^p < \infty$. Prove that ℓ^p is a proper subspace of ℓ^{p+1}.

*6.4. Let V be a vector space that is spanned by a countably infinite set. Prove that every independent subset of V is finite or countably infinite.

Miscellaneous Problems

M.1. Consider the determinant function $\det : F^{2\times 2} \to F$, where $F = \mathbb{F}_p$ is the prime field of order p and $F^{2\times 2}$ is the space of 2×2 matrices. Show that this map is surjective, that all nonzero values of the determinant are taken on the same number of times, but that there are more matrices with determinant 0 than with determinant 1.

M.2. Let A be a real $n \times n$ matrix. Prove that there is an integer N such that A satisfies a nontrivial polynomial relation $A^N + c_{N-1}A^{N-1} + \cdots + c_1 A + c_0 = 0$.

M.3. (*polynomial paths*) **(a)** Let $x(t)$ and $y(t)$ be quadratic polynomials with real coefficients. Prove that the image of the path $(x(t), y(t))$ is contained in a conic, i.e., that there is a real quadratic polynomial $f(x, y)$ such that $f(x(t), y(t))$ is identically zero.

(b) Let $x(t) = t^2 - 1$ and $y(t) = t^3 - t$. Find a nonzero real polynomial $f(x, y)$ such that $f(x(t), y(t))$ is identically zero. Sketch the locus $\{f(x, y) = 0\}$ and the path $(x(t), y(t))$ in $\mathbb{R}^2$.

(c) Prove that every pair $x(t)$, $y(t)$ of real polynomials satisfies some real polynomial relation $f(x, y) = 0$.

*M.4. Let V be a vector space over an infinite field F. Prove that V is not the union of finitely many proper subspaces.

*M.5. Let α be the real cube root of 2.

(a) Prove that $(1, \alpha, \alpha^2)$ is an independent set over $\mathbb{Q}$, i.e., that there is no relation of the form $a + b\alpha + c\alpha^2 = 0$ with integers a, b, c.
Hint: Divide $x^3 - 2$ by $cx^2 + bx + a$.

(b) Prove that the real numbers $a + b\alpha + c\alpha^2$ with a, b, c in $\mathbb{Q}$ form a field.

M.6. (*Tabasco sauce: a mathematical diversion*) My cousin Phil collects hot sauce. He has about a hundred different bottles on the shelf, and many of them, Tabasco for instance, have only three ingredients other than water: chilis, vinegar, and salt. What is the smallest number of bottles of hot sauce that Phil would need to keep on hand so that he could obtain any recipe that uses only these three ingredients by mixing the ones he had?

CHAPTER 4

Linear Operators

That confusions of thought and errors of reasoning
still darken the beginnings of Algebra,
is the earnest and just complaint of sober and thoughtful men.

—Sir William Rowan Hamilton

4.1 THE DIMENSION FORMULA

A *linear transformation* $T : V \to W$ from one vector space over a field F to another is a map that is compatible with addition and scalar multiplication:

(4.1.1) $$T(v_1 + v_2) = T(v_1) + T(v_2) \quad \text{and} \quad T(cv_1) = cT(v_1),$$

for all v_1 and v_2 in V and all c in F. This is analogous to a homomorphism of groups, and calling it a homomorphism would be appropriate too. A linear transformation is compatible with arbitrary linear combinations:

(4.1.2) $$T\left(\sum_i v_i c_i\right) = \sum_i T(v_i)c_i.$$

Left multiplication by an $m \times n$ matrix A with entries in F, the map

(4.1.3) $$F^n \xrightarrow{A} F^m \quad \text{that sends} \quad X \rightsquigarrow AX$$

is a linear transformation. Indeed, $A(X_1 + X_2) = AX_1 + AX_2$, and $A(cX) = cAX$.

If $\mathbf{B} = (v_1, \dots, v_n)$ is a subset of a vector space V over the field F, the map $F^n \to V$ that sends $X \rightsquigarrow \mathbf{B}X$ is a linear transformation.

Another example: Let P_n be the vector space of real polynomial functions

(4.1.4) $$a_n t^n + a_{n-1} t^{n-1} + \cdots + a_1 t + a_0$$

of degree at most n. The derivative $\frac{d}{dt}$ defines a linear transformation from P_n to P_{n-1}.

There are two important subspaces associated with a linear transformation: its kernel and its image:

(4.1.5) $\begin{aligned} \ker T &= \textit{kernel of } T &&= \{v \in V \,|\, T(v) = 0\}, \\ \operatorname{im} T &= \textit{image of } T &&= \{w \in W \,|\, w = T(v) \text{ for some } v \in V\}. \end{aligned}$

The kernel is often called the *nullspace* of the linear transformation. As one may guess from the analogy with group homomorphisms, the kernel is a subspace of V and the image is a subspace of W.

The main result of this section is the next theorem.

Theorem 4.1.6 Dimension Formula. Let $T: V \to W$ be a linear transformation. Then

$$\dim(\ker T) + \dim(\operatorname{im} T) = \dim V.$$

The *nullity* and the *rank* of a linear transformation T are the dimensions of the kernel and the image, respectively, and the nullity and rank of a matrix A are defined analogously. With this terminology, (4.1.6) becomes

(4.1.7)   nullity + rank = dimension of V.

Proof of Theorem (4.1.6). We'll assume that V is finite-dimensional, say of dimension n. Let k be the dimension of $\ker T$, and let $(u_1, \ldots, u_k)$ be a basis for the kernel. We extend this set to a basis of V:

(4.1.8) $(u_1, \ldots, u_k; v_1, \ldots, v_{n-k})$.

(see (3.4.15)). For $i = 1, \ldots, n - k$, let $w_i = T(v_i)$. If we prove that $\mathbf{C} = (w_1, \ldots, w_{n-k})$ is a basis for the image, it will follow that the image has dimension $n - k$, and this will prove the theorem.

We must show that $\mathbf{C}$ spans the image and that it is an independent set. Let w be an element of the image. Then $w = T(v)$ for some v in V. We write v in terms of the basis:

$$v = a_1 u_1 + \cdots + a_k u_k + b_1 v_1 + \cdots + b_{n-k} v_{n-k}$$

and apply T, noting that $T(u_i) = 0$:

$$w = T(v) = b_1 w_1 + \cdots + b_{n-k} w_{n-k}.$$

Thus w is in the span of $\mathbf{C}$.

Next, we show that $\mathbf{C}$ is independent. Suppose we have a linear relation

(4.1.9) $c_1 w_1 + \cdots + c_{n-k} w_{n-k} = 0.$

Let $v = c_1 v_1 + \cdots + c_{n-k} v_{n-k}$, where v_i are the vectors in (4.1.8). Then

$$T(v) = c_1 w_1 + \cdots + c_{n-k} w_{n-k} = 0,$$

so v is in the nullspace. We write v in terms of the basis $(u_1, \ldots, u_k)$ of the nullspace, say $v = a_1 u_1 + \cdots + a_k u_k$. Then

$$-a_1 u_1 - \cdots - a_k u_k + c_1 v_1 + \cdots + c_{n-k} v_{n-k} = -v + v = 0.$$

But the basis (4.1.8) is independent. So $-a_1 = 0, \ldots, -a_k = 0$, and $c_1 = 0, \ldots, c_{n-k} = 0$. The relation (4.1.9) was trivial. Therefore $\mathbf{C}$ is independent.   □

When T is left multiplication by a matrix A (4.1.3), the kernel of T, the nullspace of A, is the set of solutions of the homogeneous equation $AX = 0$. The image of T is the *column space*, the space spanned by the columns of A, which is also the set of vectors B in F^m such that the linear equation $AX = B$ has a solution (3.4.6).

It is a familiar fact that by adding the solutions of the homogeneous equation $AX = 0$ to a particular solution X_0 of the inhomogeneous equation $AX = B$, one obtains all solutions of the inhomogeneous equation. Another way to say this is that the set of solutions of $AX = B$ is the additive coset $X_0 + N$ of the nullspace N in F^n.

An $n \times n$ matrix A whose determinant isn't zero is invertible, and the system of equations $AX = B$ has a unique solution for every B. In this case, the nullspace is $\{0\}$, and the column space is the whole space F^n. On the other hand, if the determinant is zero, the nullspace N has positive dimension, and the image, the column space, has dimension less than n. Not all equations $AX = B$ have solutions, but those that do have a solution have more than one solution, because the set of solutions is a coset of N.

4.2 THE MATRIX OF A LINEAR TRANSFORMATION

Every linear transformation from one space of column vectors to another is left multiplication by a matrix.

Lemma 4.2.1 Let $T : F^n \to F^m$ be a linear transformation between spaces of column vectors, and let the coordinate vector of $T(e_j)$ be $A_j = (a_{1j}, \dots, a_{mj})^t$. Let A be the $m \times n$ matrix whose columns are $A_1, \dots, A_n$. Then T acts on vectors in F^n as multiplication by A.

Proof. $T(X) = T(\sum_j e_j x_j) = \sum_j T(e_j) x_j = \sum_j A_j x_j = AX.$ $\square$

For example, let $c = \cos\theta$, $s = \sin\theta$. Counterclockwise rotation $\rho : \mathbb{R}^2 \to \mathbb{R}^2$ of the plane through the angle θ about the origin is a linear transformation. Its matrix is

$$(4.2.2) \qquad\qquad R = \begin{bmatrix} c & -s \\ s & c \end{bmatrix}.$$

Let's verify that multiplication by this matrix rotates the plane. We write a vector X in the form $r(\cos\alpha, \sin\alpha)^t$, where r is the length of X. Let $c' = \cos\alpha$ and $s' = \sin\alpha$. The addition formulas for cosine and sine show that

$$RX = r \begin{bmatrix} c & -s \\ s & c \end{bmatrix} \begin{bmatrix} c' \\ s' \end{bmatrix} = r \begin{bmatrix} cc' - ss' \\ sc' + cs' \end{bmatrix} = r \begin{bmatrix} \cos(\theta + \alpha) \\ \sin(\theta + \alpha) \end{bmatrix}.$$

So RX is obtained from X by rotating through the angle θ, as claimed.

One can make a computation analogous to that of Lemma 4.2.1 with any linear transformation $T : V \to W$, once bases of the two spaces are chosen. If $\mathbf{B} = (v_1, \dots, v_n)$ is a basis of V, we use the shorthand notation $T(\mathbf{B})$ to denote the hypervector

$$(4.2.3) \qquad\qquad T(\mathbf{B}) = (T(v_1), \dots, T(v_n)).$$

If $v = \mathbf{B}X = v_1 x_1 + \cdots + v_n x_n$, then

$$(4.2.4) \qquad\qquad T(v) = T(v_1)x_1 + \cdots + T(v_n)x_n = T(\mathbf{B})X.$$

Proposition 4.2.5 Let $T: V \to W$ be a linear transformation, and let $\mathbf{B} = (v_1, \ldots, v_n)$ and $\mathbf{C} = (w_1, \ldots, w_m)$ be bases of V and W, respectively. Let X be the coordinate vector of an arbitrary vector v with respect to the basis $\mathbf{B}$ and let Y be the coordinate vector of its image $T(v)$. So $v = \mathbf{B}X$ and $T(v) = \mathbf{C}Y$. There is an $m \times n$ matrix A with the dual properties

$$(4.2.6) \qquad\qquad T(\mathbf{B}) = \mathbf{C}A \qquad \text{and} \qquad AX = Y.$$

This matrix A is the *matrix of the transformation T* with respect to the two bases. Either of the properties (4.2.6) characterizes the matrix.

Proof. We write $T(v_j)$ as a linear combination of the basis $\mathbf{C}$, say

$$(4.2.7) \qquad\qquad T(v_j) = w_1 a_{1j} + \cdots + w_m a_{mj},$$

and we assemble the coefficients a_{ij} into a column vector $A_j = (a_{1j}, \ldots, a_{mj})^{\mathrm{t}}$, so that $T(v_j) = \mathbf{C}A_j$. Then if A is the matrix whose columns are $A_1, \ldots, A_n$,

$$(4.2.8) \qquad T(\mathbf{B}) = (T(v_1), \ldots, T(v_n)) = (w_1, \ldots, w_m) \begin{bmatrix} & & \\ & A & \\ & & \end{bmatrix} = \mathbf{C}A,$$

as claimed. Next, if $v = \mathbf{B}X$, then

$$T(v) = T(\mathbf{B})X = \mathbf{C}AX.$$

Therefore the coordinate vector of $T(v)$, which we named Y, is equal to AX. $\square$

The isomorphisms $F^n \to V$ and $F^m \to W$ determined by the two bases (3.5.3) help to explain the relationship between T and A. If we use those isomorphisms to identify V and W with F^n and F^m, then T corresponds to multiplication by A, as shown in the diagram below:

$$(4.2.9) \qquad \begin{array}{ccc} F^n & \xrightarrow{A} & F^m \\ \mathbf{B}\downarrow & & \downarrow \mathbf{C} \\ V & \xrightarrow{T} & W \end{array} \qquad \begin{array}{ccc} X & \rightsquigarrow & AX \\ \wr & & \wr \\ \mathbf{B}X & \rightsquigarrow & T(\mathbf{B})X = \mathbf{C}AX \end{array}$$

Going from F^n to W along the two paths gives the same answer. A diagram that has this property is said to be *commutative*. All diagrams in this book are commutative.

Thus any linear transformation between finite-dimensional vector spaces V and W corresponds to matrix multiplication, once bases for the two spaces are chosen. This is a nice result, but if we change bases we can do much better.

Theorem 4.2.10

(a) *Vector space form*: Let $T : V \to W$ be a linear transformation between finite-dimensional vector spaces. There are bases **B** and **C** of V and W, respectively, such that the matrix of T with respect to these bases has the form

$$A' = \begin{array}{|c c|} \hline I_r & \\ \hline & 0 \\ \hline \end{array} \quad ,$$

(4.2.11)

where I_r is the $r \times r$ identity matrix and r is the rank of T.

(b) *Matrix form*: Given an $m \times n$ matrix A, there are invertible matrices Q and P such that $A' = Q^{-1}AP$ has the form shown above.

Proof. (a) Let $(u_1, \ldots, u_k)$ be a basis for the kernel of T. We extend this set to a basis **B** of V, listing the additional vectors first, say $(v_1, \ldots, v_r; u_1, \ldots, u_k)$, where $r + k = n$. Let $w_i = T(v_i)$. Then, as in the proof of (4.1.6), one sees that $(w_1, \ldots, w_r)$ is a basis for the image of T. We extend this set to a basis **C** of W, say $(w_1, \ldots, w_r; z_1, \ldots, z_s)$, listing the additional vectors last. The matrix of T with respect to these bases has the form (4.2.11).

Part (b) of the theorem can be proved using row and column operations. The proof is Exercise 2.4. $\square$

This theorem is a prototype for a number of results that are to come. It shows the advantage of working in vector spaces without fixed bases (or coordinates), because the structure of an arbitrary linear transformation is described by the very simple matrix (4.2.11). But why are (a) and (b) considered two versions of the same theorem? To answer this, we need to analyze the way that the matrix of a linear transformation changes when we make other choices of bases.

Let A be the matrix of T with respect to bases **B** and **C** of V and W, as in (4.2.6), and let $\mathbf{B}' = (v'_1, \ldots, v'_n)$ and $\mathbf{C}' = (w'_1, \ldots, w'_m)$ be new bases for V and W. We can relate the new basis $\mathbf{B}'$ to the old basis **B** by an invertible $n \times n$ matrix P, as in (3.5.11). Similarly, $\mathbf{C}'$ is related to **C** by an invertible $m \times m$ matrix Q. These matrices have the properties

(4.2.12) $\mathbf{B}' = \mathbf{B}P, \quad PX' = X \quad$ and $\quad \mathbf{C}' = \mathbf{C}Q, \quad QY' = Y.$

Proposition 4.2.13 Let A be the matrix of a linear transformation T with respect to given bases **B** and **C**.

(a) Suppose that new bases $\mathbf{B}'$ and $\mathbf{C}'$ are related to the given bases by the matrices P and Q, as above. The matrix of T with respect to the new bases is $A' = Q^{-1}AP$.

(b) The matrices A' that represent T with respect to other bases are those of the form $A' = Q^{-1}AP$, where Q and P can be any invertible matrices of the appropriate sizes.

Proof. (a) We substitute $X = PX'$ and $Y = QY'$ into the equation $Y = AX$ (4.2.6), obtaining $QY' = APX'$. So $Y' = (Q^{-1}AP)X'$. Since A' is the matrix such that $A'X' = Y'$, this shows that $A' = Q^{-1}AP$. Part (b) follows because the basechange matrices can be any invertible matrices (3.5.9). $\square$

It follows from the proposition that the two parts of the theorem amount to the same thing. To derive **(a)** from **(b)**, we suppose given the linear transformation T, and we begin with arbitrary choices of bases for V and W, obtaining a matrix A. Part **(b)** tells us that there are invertible matrices P and Q such that $A' = Q^{-1}AP$ has the form (4.2.11). When we use these matrices to change bases in V and W, the matrix A is changed to A'.

To derive **(b)** from **(a)**, we view an arbitrary matrix A as the matrix of the linear transformation "left multiplication by A" on column vectors. Then A is the matrix of T with respect to the standard bases of F^n and F^m, and **(a)** guarantees the existence of P, Q so that $Q^{-1}AP$ has the form (4.2.11).

We also learn something remarkable about matrix multiplication here, because left multiplication by a matrix is a linear transformation. Left multiplication by an arbitrary matrix A is the same as left multiplication by a matrix of the form (4.2.11), but with reference to different coordinates.

In the future, we will often state a result in two equivalent ways, a vector space form and a matrix form, without stopping to show that the two forms are equivalent. Then we will present whichever proof seems simpler to write down.

We can use Theorem 4.2.10 to derive another interesting property of matrix multiplication. Let N and U denote the nullspace and column space of the transformation $A : F^n \to F^m$. So N is a subspace of F^n and U is a subspace of F^m. Let k and r denote the dimensions of N and U. So k is the nullity of A and r is its rank.

Left multiplication by the transpose matrix A^t defines a transformation $A^t : F^m \to F^n$ in the opposite direction, and therefore two more subspaces, the nullspace N_1 and the column space U_1 of A^t. Here U_1 is a subspace of F^n, and N_1 is a subspace of F^m. Let k_1 and r_1 denote the dimensions of N_1 and U_1, respectively. Theorem 4.1.6 tells us that $k + r = n$, and also that $k_1 + r_1 = m$. Theorem 4.2.14 below gives one more relation among these integers:

Theorem 4.2.14 With the above notation, $r_1 = r$: The rank of a matrix is equal to the rank of its transpose.

Proof. Let P and Q be invertible matrices such that $A' = Q^{-1}AP$ has the form (4.2.11). We begin by noting that the assertion is obvious for the matrix A'. Next, we examine the diagrams

(4.2.15)

$$
\begin{array}{ccc}
F^n & \xrightarrow{\ A\ } & F^m \\
{\scriptstyle P}\uparrow & & \uparrow{\scriptstyle Q} \\
F^n & \xrightarrow{\ A'\ } & F^m
\end{array}
\qquad
\begin{array}{ccc}
F^n & \xleftarrow{\ A^t\ } & F^m \\
{\scriptstyle P^t}\downarrow & & \downarrow{\scriptstyle Q^t} \\
F^n & \xleftarrow{\ A'^t\ } & F^m
\end{array}
$$

The vertical arrows are bijective maps. Therefore, in the left-hand diagram, Q carries the column space of A' (the image of multiplication by A') bijectively to the column space of A. The dimensions of these two column spaces, the ranks of A and A', are equal. Similarly, the ranks of A^t and A'^t are equal. So to prove the theorem, we may replace the matrix A by A'. This reduces the proof to the trivial case of the matrix (4.2.11).   □

We can reinterpret the rank r_1 of the transpose matrix A^t. By definition, it is the dimension of the space spanned by the columns of A^t, and this can equally well be thought of as the dimension of the space of *row vectors* spanned by the *rows* of A. Because of this, people often refer to r_1 as the *row rank* of A, and to r as the *column rank*.

The row rank is the maximal number of independent rows of the matrix, and the column rank is the maximal number of independent columns. Theorem 4.2.14 can be stated this way:

Corollary 4.2.16 The row rank and the column rank of an $m \times n$ matrix A are equal.   □

4.3   LINEAR OPERATORS

In this section, we study linear transformations $T : V \rightarrow V$ that map a vector space to itself. They are called *linear operators*. Left multiplication by a (square) $n \times n$ matrix with entries in a field F defines a linear operator on the space F^n of column vectors.

For example, let $c = \cos \theta$ and $s = \sin \theta$. The rotation matrix (4.2.2)

$$\begin{bmatrix} c & -s \\ s & c \end{bmatrix}$$

is a linear operator on the plane $\mathbb{R}^2$.

The dimension formula $\dim(\ker T) + \dim(\operatorname{im} T) = \dim V$ is valid for linear operators. But here, since the domain and range are equal, we have extra information that can be combined with the formula. Both the kernel and the image of T are subspaces of V.

Proposition 4.3.1 Let K and W denote the kernel and image, respectively, of a linear operator T on a finite-dimensional vector space V.

(a) The following conditions are equivalent:

- T is bijective,
- $K = \{0\}$,
- $W = V$.

(b) The following conditions are equivalent:

- V is the direct sum $K \oplus W$,
- $K \cap W = \{0\}$,
- $K + W = V$.

Proof. **(a)** T is bijective if and only if the kernel K is zero *and* the image W is the whole space V. If the kernel is zero, the dimension formula tells us that $\dim W = \dim V$, and therefore $W = V$. Similarly, if $W = V$, the dimension formula shows that $\dim K = 0$, and therefore $K = 0$. In both cases, T is bijective.

(b) V is the direct sum $K \oplus W$ if and only if both of the conditions $K \cap W = \{0\}$ and $K + W = V$ hold. If $K \cap W = \{0\}$, then K and W are independent, so the sum $U = K + W$ is the direct sum $K \oplus W$, and $\dim U = \dim K + \dim W$ (3.6.6)**(a)**. The dimension formula shows that $\dim U = \dim V$, so $U = V$, and this shows that $K \oplus W = V$. If $K + W = V$, the dimension formula and Proposition 3.6.6**(a)** show that K and W are independent, and again, V is the direct sum.   □

• A linear operator that satisfies the conditions (4.3.1)**(a)** is called an *invertible operator.* Its inverse function is also a linear operator. An operator that is not invertible is a *singular operator.*

The conditions of Proposition 4.3.1**(a)** are not equivalent when the dimension of V is infinite. For example, let $V = \mathbb{R}^\infty$ be the space of infinite row vectors $(a_1, a_2, \ldots)$ (see Section 3.7). The kernel of the *right shift operator* S^+, defined by

(4.3.2) $$S^+(a_1, a_2, \ldots) = (0, a_1, a_2, \ldots),$$

is the zero space, and its image is a proper subspace of V. The kernel of the *left shift operator* S^-, defined by

$$S^-(a_1, a_2, a_3, \ldots) = (a_2, a_3, \ldots),$$

is a proper subspace of V, and its image is the whole space.

The discussion of bases in the previous section must be changed slightly when we are dealing with linear operators. We should pick only one basis **B** for V, and use it in place of both of the bases **B** and **C** in (4.2.6). In other words, to define the matrix A of T with respect to the basis **B**, we should write

(4.3.3) $$T(\mathbf{B}) = \mathbf{B}A, \quad \text{and } AX = Y \text{ as before.}$$

As with any linear transformation (4.2.7), the columns of A are the coordinate vectors of the images $T(v_j)$ of the basis vectors:

(4.3.4) $$T(v_j) = v_1 a_{1j} + \cdots + v_n a_{nj}.$$

A linear operator is invertible if and only if its matrix with respect to an arbitrary basis is an invertible matrix.

When one speaks of the the matrix of a linear operator on the space F^n, it is assumed that the basis is the standard basis **E**, unless a different basis is specified. The operator is then multiplication by that matrix.

A new feature arises when we study the effect of a change of basis. Suppose that **B** is replaced by a new basis **B**′.

Proposition 4.3.5 Let A be the matrix of a linear operator T with respect to a basis **B**.

(a) Suppose that a new basis **B**′ is described by **B**′ = **B**P. The matrix that represents T with respect to this basis is $A' = P^{-1}AP$.

(b) The matrices A' that represent the operator T for different bases are the matrices of the form $A' = P^{-1}AP$, where P can be any invertible matrix.   □

In other words, the matrix changes by conjugation. This is a confusing fact to grasp. So, though it follows from (4.2.13), we will rederive it. Since **B**′ = **B**P and since $T(\mathbf{B}) = \mathbf{B}A$, we have

$$T(\mathbf{B}') = T(\mathbf{B})P = \mathbf{B}AP.$$

We are not done. The formula we have obtained expresses $T(\mathbf{B}')$ in terms of the old basis **B**. To obtain the new matrix, we must write $T(\mathbf{B}')$ in terms of the new basis **B**′. So we substitute $\mathbf{B} = \mathbf{B}'P^{-1}$ into the equation. Doing so gives us $T(\mathbf{B}') = \mathbf{B}'P^{-1}AP$.   □

In general, we say that a square matrix A is *similar* to another matrix A' if $A' = P^{-1}AP$ for some invertible matrix P. Such a matrix A' is obtained from A by conjugating by P^{-1}, and since P can be any invertible matrix, P^{-1} is also arbitrary. It would be correct to use the term *conjugate* in place of similar.

Now if we are given the matrix A, it is natural to look for a similar matrix A' that is particularly simple. One would like to get a result somewhat like Theorem 4.2.10. But here our allowable change is much more restricted, because we have only one basis, and therefore one matrix P, to work with. Having domain and range of a linear transformation equal, which seems at first to be a simplification, actually makes things more difficult.

We can get some insight into the problem by writing the hypothetical basechange matrix as a product of elementary matrices, say $P = E_1 \cdots E_r$. Then

$$P^{-1}AP = E_r^{-1} \cdots E_1^{-1}AE_1 \cdots E_r.$$

In terms of elementary operations, we are allowed to change A by a sequence of steps $A \rightsquigarrow E^{-1}AE$. In other words, we may perform an arbitrary column operation E on A, but we must also make the row operation that corresponds to the inverse matrix E^{-1}. Unfortunately, these row and column operations interact, and analyzing them becomes confusing.

4.4   EIGENVECTORS

The main tools for analyzing a linear operator $T : V \rightarrow V$ are invariant subspaces and eigenvectors.

• A subspace W of V is *invariant*, or more precisely *T-invariant*, if it is carried to itself by the operator:

(4.4.1) $$TW \subset W.$$

In other words, W is invariant if, whenever w is in W, $T(w)$ is also in W. When this is so, T defines a linear operator on W, called its *restriction* to W. We often denote this restriction by $T|_W$.

If W is a T-invariant subspace, we may form a basis $\mathbf{B}$ of V by appending vectors to a basis $(w_1, \ldots, w_k)$ of W, say

(4.4.2) $$\mathbf{B} = (w_1, \ldots, w_k; v_1, \ldots, v_{n-k}).$$

Then the fact that W is invariant is reflected in the matrix of T. The columns of this matrix, we'll call it M, are the coordinate vectors of the image vectors (see (4.3.3)). But $T(w_j)$ is in the subspace W, so it is a linear combination of the basis $(w_1, \ldots, w_k)$. When we write $T(w_j)$ in terms of the basis $\mathbf{B}$, the coefficients of the vectors $v_1, \ldots, v_{n-k}$ will be zero. It follows that M will have the block form

(4.4.3) $$M = \begin{bmatrix} A & B \\ 0 & D \end{bmatrix},$$

where A is a $k \times k$ matrix, the matrix of the restriction of T to W.

If V happens to be the direct sum $W_1 \oplus W_2$ of two T-invariant subspaces, and if we make a basis $\mathbf{B} = (\mathbf{B}_1, \mathbf{B}_2)$ of V by appending bases of W_1 and W_2, the matrix of T will have the block diagonal form

(4.4.4)
$$M = \begin{bmatrix} A_1 & 0 \\ 0 & A_2 \end{bmatrix},$$

where A_i is the matrix of the restriction of T to W_i.

The concept of an eigenvector is closely related to that of an invariant subspace.

• An *eigenvector* v of a linear operator T is a nonzero vector such that

(4.4.5)
$$T(v) = \lambda v$$

for some scalar λ, i.e., some element of F. A nonzero column vector is an eigenvector of a square matrix A if it is an eigenvector for the operation of left multiplication by A.

The scalar λ that appears in (4.4.5) is called the *eigenvalue* associated to the eigenvector v. When we speak of an eigenvalue of a linear operator T or of a matrix A without specifying an eigenvector, we mean a scalar λ that is the eigenvalue associated to *some* eigenvector. An eigenvalue may be any element of F, including zero, but an eigenvector is not allowed to be zero. Eigenvalues are often denoted, as here, by the Greek letter λ (lambda).[1]

An eigenvector with eigenvalue 1 is a *fixed vector*: $T(v) = v$. An eigenvector with eigenvalue zero is in the nullspace: $T(v) = 0$. When $V = \mathbb{R}^n$, a nonzero vector v is an eigenvector if v and $T(v)$ are parallel.

If v is an eigenvector of a linear operator T, with eigenvalue λ, the subspace W spanned by v will be T-invariant, because $T(cv) = c\lambda v$ is in W for all scalars c. Conversely, if the one-dimensional subspace spanned by v is invariant, then v is an eigenvector. So an eigenvector can be described as a basis of a one-dimensional invariant subspace.

It is easy to tell whether or not a given vector X is an eigenvector of a matrix A. We simply check whether or not AX is a multiple of X. And, if A is the matrix of T with respect to a basis $\mathbf{B}$, and if X is the coordinate vector of a vector v, then X is an eigenvector of A if and only if v is an eigenvector for T.

The standard basis vector $e_1 = (1, 0)^t$ is an eigenvector, with eigenvalue 3, of the matrix

$$\begin{bmatrix} 3 & 1 \\ 0 & 2 \end{bmatrix}.$$

The vector $(1, -1)^t$ is another eigenvector, with eigenvalue 2. The vector $(0,1,1)^t$ is an eigenvector, with eigenvalue 2, of the matrix

$$A = \begin{bmatrix} 1 & 1 & -1 \\ 2 & 1 & 1 \\ 3 & 0 & 2 \end{bmatrix}.$$

---

[1] The German word "eigen" means roughly "characteristic." Eigenvectors and eigenvalues are sometimes called *characteristic vectors*.

If $(v_1, \ldots, v_n)$ is a basis of V and if v_1 is an eigenvector of a linear operator T, the matrix of T will have the block form

(4.4.6)
$$\begin{bmatrix} \lambda & B \\ 0 & D \end{bmatrix} = \begin{bmatrix} \lambda & * & \cdots & * \\ 0 & & & \\ \vdots & & * & \\ 0 & & & \end{bmatrix},$$

where λ is the eigenvalue of v_1. This is the block form (4.4.3) in the case of an invariant subspace of dimension 1.

Proposition 4.4.7   Similar matrices $(A' = P^{-1}AP)$ have the same eigenvalues.

This is true because similar matrices represent the same linear transformation.   □

Proposition 4.4.8

(a) Let T be a linear operator on a vector space V. The matrix of T with respect to a basis $\mathbf{B} = (v_1, \ldots, v_n)$ is diagonal if and only if each of the basis vectors v_j is an eigenvector.

(b) An $n \times n$ matrix A is similar to a diagonal matrix if and only if there is a basis of F^n that consists of eigenvectors.

This follows from the definition of the matrix A (see (4.3.4)). If $T(v_j) = \lambda_j v_j$, then

(4.4.9)
$$T(\mathbf{B}) = (v_1 \lambda_1, \ldots v_n \lambda_n) = (v_1, \ldots, v_n) \begin{bmatrix} \lambda_1 & & \\ & \ddots & \\ & & \lambda_n \end{bmatrix}.$$
□

This proposition shows that we can represent a linear operator simply by a diagonal matrix, provided that it has enough eigenvectors. We will see in Section 4.5 that every linear operator on a complex vector space has at least one eigenvector, and in Section 4.6 that in most cases there is a basis of eigenvectors. But a linear operator on a real vector space needn't have any eigenvector. For example, a rotation of the plane through an angle θ doesn't carry any vector to a parallel one unless θ is 0 or π. The rotation matrix (4.2.2) with $\theta \neq 0, \pi$ has no real eigenvector.

• A general example of a real matrix that has at least one real eigenvalue is one all of whose entries are positive. Such matrices, called *positive matrices*, occur often in applications, and one of their most important properties is that they always have an eigenvector whose coordinates are positive (a *positive eigenvector*).

Instead of proving this fact, we'll illustrate it by examining the effect of multiplication by a positive 2×2 matrix A on $\mathbb{R}^2$. Let $w_i = Ae_i$ be the columns of A. The parallelogram law for vector addition shows that A sends the first quadrant S to the sector bounded by the vectors w_1 and w_2. The coordinate vector of w_i is the ith column of A. Since the entries of

A are positive, the vectors w_i lie in the first quadrant. So A carries the first quadrant to itself: $S \supset AS$. Applying A to this inclusion, we find $AS \supset A^2 S$, and so on:

(4.4.10)
$$S \supset AS \supset A^2 S \supset A^3 S \supset \ldots,$$

as is illustrated below for the matrix $A = \begin{bmatrix} 3 & 2 \\ 1 & 4 \end{bmatrix}$.

Now, the intersection of a nested set of sectors is either a sector or a half-line. In our case, the intersection $Z = \bigcap A^r S$ turns out to be a half-line. This is intuitively plausible, and it can be shown in various ways, but we'll omit the proof. We multiply the relation $Z = \bigcap A^r S$ on both sides by A:

$$AZ = A\left(\bigcap_0^\infty A^r S\right) = \bigcap_1^\infty A^r S = Z.$$

Hence $Z = AZ$. Therefore the nonzero vectors in Z are eigenvectors.

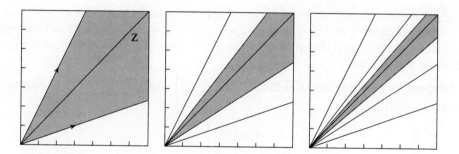

(4.4.11)   Images of the First Quadrant Under Repeated Multiplication by
a Positive Matrix.

4.5   THE CHARACTERISTIC POLYNOMIAL

In this section we determine the eigenvectors of an arbitrary linear operator. We recall that an eigenvector of a linear operator T is a nonzero vector v such that

(4.5.1)
$$T(v) = \lambda v,$$

for some λ in F. If we don't know λ, it can be difficult to find the eigenvector directly when the matrix of the operator is complicated. The trick is to solve a different problem, namely to determine the *eigenvalues* first. Once an eigenvalue λ is determined, equation (4.5.1) becomes linear in the coordinates of v, and solving it presents no problem.

We begin by writing (4.5.1) in the form

(4.5.2)
$$[\lambda I - T](v) = 0,$$

where I stands for the identity operator and $\lambda I - T$ is the linear operator defined by

(4.5.3) $$[\lambda I - T](v) = \lambda v - T(v).$$

It is easy to check that $\lambda I - T$ is indeed a linear operator. We can restate (4.5.2) as follows:

(4.5.4)
> A nonzero vector v is an eigenvector with eigenvalue λ
> if and only if it is in the kernel of $\lambda I - T$.

Corollary 4.5.5 Let T be a linear operator on a finite-dimensional vector space V.

(a) The eigenvalues of T are the scalars λ in F such that the operator $\lambda I - T$ is singular, i.e., its nullspace is not zero.

(b) The following conditions are equivalent:

- T is a singular operator.
- T has an eigenvalue equal to zero.
- If A is the matrix of T with respect to an arbitrary basis, then $\det A = 0$. $\qquad\square$

If A is the matrix of T with respect to some basis, then the matrix of $\lambda I - T$ is $\lambda I - A$. So $\lambda I - T$ is singular if and only if $\det(\lambda I - A) = 0$. This determinant can be computed with indeterminate λ, and doing so provides us, at least in principle, with a method for determining the eigenvalues and eigenvectors.

Suppose for example that A is the matrix $\begin{bmatrix} 3 & 2 \\ 1 & 4 \end{bmatrix}$ whose action on $\mathbb{R}^2$ is illustrated in Figure (4.4.11). Then

$$\lambda I - A = \begin{bmatrix} \lambda\text{-}3 & -2 \\ -1 & \lambda\text{-}4 \end{bmatrix}$$

and

$$\det(\lambda I - A) = \lambda^2 - 7\lambda + 10 = (\lambda - 5)(\lambda - 2).$$

The determinant vanishes when $\lambda = 5$ or 2, so the eigenvalues of A are 5 and 2. To find the eigenvectors, we solve the two systems of equations $[5I - A]X = 0$ and $[2I - A]X = 0$. The solutions are determined up to scalar factor:

(4.5.6) $$v_1 = \begin{bmatrix} 1 \\ 1 \end{bmatrix}, \quad v_2 = \begin{bmatrix} 2 \\ -1 \end{bmatrix}.$$

We now consider the same computation for an indeterminate matrix of arbitrary size. It is customary to replace the symbol λ by a variable t. We form the matrix $tI - A$:

(4.5.7) $$tI - A = \begin{bmatrix} (t\text{-}a_{11}) & -a_{12} & \cdots & -a_{1n} \\ -a_{21} & (t\text{-}a_{22}) & \cdots & -a_{2n} \\ \vdots & & & \vdots \\ -a_{n1} & \cdots & \cdots & (t\text{-}a_{nn}) \end{bmatrix}.$$

The complete expansion of the determinant [Chapter 1 (1.6.4)] shows that $\det(tI - A)$ is a polynomial of degree n in t whose coefficients are scalars, elements of F.

Definition 4.5.8 The *characteristic polynomial* of a linear operator T is the polynomial

$$p(t) = \det(tI - A),$$

where A is the matrix of T with respect to some basis.

The eigenvalues of T are determined by combining (4.5.5) and (4.5.8):

Corollary 4.5.9 The eigenvalues of a linear operator are the roots of its characteristic polynomial.   □

Corollary 4.5.10 Let A be an upper or lower triangular $n \times n$ matrix with diagonal entries $a_{11}, \ldots, a_{nn}$. The characteristic polynomial of A is $(t - a_{11}) \cdots (t - a_{nn})$. The diagonal entries of A are its eigenvalues.

Proof. If A is upper triangular, so is $tI - A$, and the diagonal entries of $tI - A$ are $t - a_{ii}$. The determinant of a triangular matrix is the product of its diagonal entries.   □

Proposition 4.5.11 The characteristic polynomial of an operator T does not depend on the choice of a basis.

Proof. A second basis leads to a matrix $A' = P^{-1}AP$ (4.3.5), and

$$tI - A' = tI - P^{-1}AP = P^{-1}(tI - A)P. \quad \text{Then}$$

$$\det(tI - A') = \det P^{-1} \det(tI - A) \det P = \det(tI - A). \qquad □$$

The characteristic polynomial of the 2×2 matrix $A = \begin{bmatrix} a & b \\ c & d \end{bmatrix}$ is

(4.5.12) $p(t) = \det(tI - A) = \det \begin{bmatrix} t-a & -b \\ -c & t-d \end{bmatrix} = t^2 - (\text{trace } A)t + (\det A),$

where trace $A = a + d$.

An incomplete description of the characteristic polynomial of an $n \times n$ matrix is given by the next proposition, which is proved by computation. It wouldn't be very difficult to determine the remaining coefficients, but explicit formulas for them aren't often used.

Proposition 4.5.13 The characteristic polynomial of an $n \times n$ matrix A has the form

$$p(t) = t^n - (\text{trace } A)t^{n-1} + (\textit{intermediate terms}) + (-1)^n(\det A),$$

where trace A, the *trace* of A, is the sum of its diagonal entries:

$$\text{trace } A = a_{11} + a_{22} + \cdots + a_{nn}. \qquad □$$

Proposition 4.5.11 shows that all coefficients of the characteristic polynomial are independent of the basis. For instance, $\text{trace}(P^{-1}AP) = \text{trace } A$.

Since the characteristic polynomial, the trace, and the determinant are independent of the basis, they depend only on the operator T. So we may define the terms *characteristic polynomial*, *trace*, and *determinant* of a linear operator T. They are the ones obtained using the matrix of T with respect to any basis.

Proposition 4.5.14 Let T be a linear operator on a finite-dimensional vector space V.

(a) If V has dimension n, then T has at most n eigenvalues.

(b) If F is the field of complex numbers and $V \neq \{0\}$, then T has at least one eigenvalue, and hence at least one eigenvector.

Proof. **(a)** The eigenvalues are the roots of the characteristic polynomial, which has degree n. A polynomial of degree n can have at most n roots. This is true for a polynomial with coefficients in any field F (see (12.2.20)).

(b) The Fundamental Theorem of Algebra asserts that every polynomial of positive degree with complex coefficients has at least one complex root. There is a proof of this theorem in Chapter 15 (15.10.1). $\qquad\square$

For example, let R_θ be matrix (4.2.2) that represents the counterclockwise rotation of $\mathbb{R}^2$ through an angle θ. Its characteristic polynomial, $p(t) = t^2 - (2\cos\theta)t + 1$, has no real root provided that $\theta \neq 0, \pi$, so no real eigenvalue. We have observed this before. But the operator on $\mathbb{C}^2$ defined by R_θ does have the complex eigenvalues $e^{\pm i\theta}$.

Note: When we speak of *the* roots of a polynomial $p(t)$ or *the* eigenvalues of a matrix or linear operator, repetitions corresponding to multiple roots are supposed to be included. This terminology is convenient, though imprecise. $\qquad\square$

Corollary 4.5.15 If $\lambda_1, \ldots, \lambda_n$ are the eigenvalues of an $n \times n$ complex matrix A, then $\det A$ is the product $\lambda_1 \cdots \lambda_n$, and $\text{trace } A$ is the sum $\lambda_1 + \ldots + \lambda_n$.

Proof. Let $p(t)$ be the characteristic polynomial of A. Then

$$(t - \lambda_1) \cdots (t - \lambda_n) = p(t) = t^n - (\text{trace } A)t^{n-1} + \cdots \pm (\det A). \qquad\square$$

4.6 TRIANGULAR AND DIAGONAL FORMS

In this section we show that for "most" linear operators on a complex vector space, there is a basis such that the matrix of the operator is diagonal. The key fact, which was noted at the end of Section 4.5, is that every complex polynomial of positive degree has a root. This tells us that every linear operator has at least one eigenvector.

Proposition 4.6.1

(a) *Vector space form*: Let T be a linear operator on a finite-dimensional complex vector space V. There is a basis **B** of V such that the matrix of T with respect to that basis is upper triangular.

(b) *Matrix form*: Every complex $n \times n$ matrix A is similar to an upper triangular matrix: There is a matrix $P \in GL_n(\mathbb{C})$ such that $P^{-1}AP$ is upper triangular.

Proof. The two assertions are equivalent, because of (4.3.5). We will work with the matrix. Let $V = \mathbb{C}^n$. Proposition 4.5.14**(b)** shows that V contains an eigenvector of A, call it v_1. Let λ be its eigenvalue. We extend (v) to a basis $\mathbf{B} = (v_1, \ldots, v_n)$ for V. The new matrix $A' = P^{-1}AP$ has the block form

(4.6.2)
$$A' = \left[\begin{array}{c|c} \lambda & * \\ \hline 0 & D \end{array}\right],$$

where D is an $(n-1)\times(n-1)$ matrix (see (4.4.6)). By induction on n, we may assume that the existence of a matrix $Q \in GL_{n-1}(\mathbb{C})$ such that $Q^{-1}DQ$ is upper triangular will have been proved. Let

$$Q_1 = \left[\begin{array}{c|c} 1 & 0 \\ \hline 0 & Q \end{array}\right]. \quad \text{Then} \quad A'' = Q_1^{-1}A'Q_1 = \left[\begin{array}{c|c} \lambda & * \\ \hline 0 & Q^{-1}DQ \end{array}\right]$$

is upper triangular, and $A'' = (PQ_1)^{-1}A(PQ_1)$.   □

Corollary 4.6.3 Proposition 4.6.1 continues to hold when the phrase "upper triangular" is replaced by "lower triangular."

The lower triangular form is obtained by listing the basis $\mathbf{B}$ of (4.6.1)**(a)** in reverse order.   □

The important point for the proof of Proposition 4.6.1 is that every complex polynomial has a root. The same proof will work for any field F, provided that all the roots of the characteristic polynomial are in the field.

Corollary 4.6.4

(a) *Vector space form*: Let T be a linear operator on a finite-dimensional vector space V over a field F, and suppose that the characteristic polynomial of T is a product of linear factors in the field F. There is a basis $\mathbf{B}$ of V such that the matrix A of T is upper (or lower) triangular.

(b) *Matrix form*: Let A be an $n\times n$ matrix with entries in F, whose characteristic polynomial is a product of linear factors. There is a matrix $P \in GL_n(F)$ such that $P^{-1}AP$ is upper (or lower) triangular.

The proof is the same, except that to make the induction step one has to check that the characteristic polynomial of the matrix D that appears in (4.6.2) is $p(t)/(t-\lambda)$, where $p(t)$ is the characteristic polynomial of A. Then the hypothesis that the characteristic polynomial factors into linear factors carries over from A to D.   □

We now ask which matrices A are similar to *diagonal* matrices. They are called *diagonalizable matrices*. As we saw in (4.4.8) **(b)**, they are the matrices that have bases of eigenvectors. Similarly, a linear operator that has a basis of eigenvectors is called a *diagonalizable* operator. The diagonal entries are determined, except for their order, by the linear operator T. They are the eigenvalues.

Theorem 4.6.6 below gives a partial answer to our question; a more complete answer will be given in the next section.

Proposition 4.6.5 Let $v_1, \ldots, v_r$ be eigenvectors of a linear operator T with distinct eigenvalues $\lambda_1, \ldots, \lambda_r$. The set $(v_1, \ldots, v_r)$ is independent.

Proof. We use induction on r. The assertion is true when $r = 1$, because an eigenvector cannot be zero. Suppose that a dependence relation

$$0 = a_1 v_1 + \cdots + a_r v_r$$

is given. We must show that $a_i = 0$ for all i. We apply the operator T:

$$0 = T(0) = a_1 T(v_1) + \cdots + a_r T(v_r) = a_1 \lambda_1 v_1 + \cdots + a_r \lambda_r v_r.$$

This is a second dependence relation among $(v_1, \ldots, v_r)$. We eliminate v_r from the two relations, multiplying the first relation by λ_r and subtracting the second:

$$0 = a_1(\lambda_r - \lambda_1) v_1 + \cdots + a_{r-1}(\lambda_r - \lambda_{r-1}) v_{r-1}.$$

Applying induction, we may assume that $(v_1, \ldots, v_{r-1})$ is an independent set. This tells us that the coefficients $a_i(\lambda_r - \lambda_i)$, $i < r$, are all zero. Since the λ_i are distinct, $\lambda_r - \lambda_i$ is not zero if $i < r$. Thus $a_1 = \cdots = a_{r-1} = 0$. The original relation reduces to $0 = a_r v_r$. Since an eigenvector cannot be zero, a_r is zero too. $\square$

The next theorem follows by combining (4.4.8) and (4.6.5):

Theorem 4.6.6 Let T be a linear operator on a vector space V of dimension n over a field F. If its characteristic polynomial has n distinct roots in F, there is a basis for V with respect to which the matrix of T is diagonal. $\square$

Note: Diagonalization is a powerful tool. When one is presented with a diagonalizable operator, it should be an automatic response to work with a basis of eigenvectors.

As an example of diagonalization, consider the real matrix

$$(4.6.7) \qquad A = \begin{bmatrix} 3 & 2 \\ 1 & 4 \end{bmatrix}.$$

Its eigenvectors were computed in (4.5.6). These eigenvectors form a basis $\mathbf{B} = (v_1, v_2)$ of $\mathbb{R}^2$. According to (3.5.13), the matrix relating the standard basis $\mathbf{E}$ to this basis $\mathbf{B}$ is

$$(4.6.8) \qquad P = [\mathbf{B}] = \begin{bmatrix} 1 & 2 \\ 1 & -1 \end{bmatrix}, \quad P^{-1} = \frac{1}{3}\begin{bmatrix} 1 & 2 \\ 1 & -1 \end{bmatrix}, \text{ and}$$

$$(4.6.9) \qquad P^{-1}AP = \frac{1}{3}\begin{bmatrix} 1 & 2 \\ 1 & -1 \end{bmatrix}\begin{bmatrix} 3 & 2 \\ 1 & 4 \end{bmatrix}\begin{bmatrix} 1 & 2 \\ 1 & -1 \end{bmatrix} = \begin{bmatrix} 5 & \\ & 2 \end{bmatrix} = \Lambda.$$

The next proposition is a variant of Proposition 4.4.8. We omit the proof.

Proposition 4.6.10 Let F be a field.

(a) Let T be a linear operator on F^n. If $\mathbf{B} = (v_1, \ldots, v_n)$ is a basis of eigenvectors of T, and if $P = [\mathbf{B}]$, then $\Lambda = P^{-1}AP = [\mathbf{B}]^{-1}A[\mathbf{B}]$ is diagonal.

(b) Let $\mathbf{B} = (v_1, \ldots, v_n)$ be a basis of F^n, and let Λ be the diagonal matrix with diagonal entries $\lambda_1, \ldots, \lambda_n$ that are not necessarily distinct. There is a unique matrix A such that, for $i = 1, \ldots, n$, v_i is an eigenvector of A with eigenvalue λ_i, namely the matrix $[\mathbf{B}] \Lambda [\mathbf{B}]^{-1}$. $\square$

A nice way to write the equation $[\mathbf{B}]^{-1}A[\mathbf{B}] = \Lambda$ is

$$(4.6.11) \qquad\qquad A[\mathbf{B}] = [\mathbf{B}]\Lambda.$$

One application of Theorem 4.6.6 is to compute the powers of a diagonalizable matrix. The next lemma needs to be pointed out, though it follows trivially when one expands the left sides of the equations and cancels PP^{-1}.

Lemma 4.6.12 Let A, B, and P be $n \times n$ matrices. If P is invertible, then $(P^{-1}AP)(P^{-1}BP) = P^{-1}(AB)P$, and for all $k \geq 1$, $(P^{-1}AP)^k = P^{-1}A^kP$. $\square$

Thus if A, P, and Λ are as in (4.6.9), then

$$A^k = P\Lambda^k P^{-1} = \frac{1}{3}\begin{bmatrix} 1 & 2 \\ 1 & -1 \end{bmatrix}\begin{bmatrix} 5 & \\ & 2 \end{bmatrix}^k\begin{bmatrix} 1 & 2 \\ 1 & -1 \end{bmatrix} = \frac{1}{3}\begin{bmatrix} 5^k + 2^{k+1} & 2 \cdot 5^k - 2^{k+1} \\ 5^k - 2^k & 2 \cdot 5^k + 2^k \end{bmatrix}.$$

If $f(t) = a_0 + a_1 t + \cdots + a_n t^n$ is a polynomial in t with coefficients in F and if A is an $n \times n$ matrix with entries in F, then $f(A)$ will denote the matrix obtained by substituting A formally for t.

$$(4.6.13) \qquad\qquad f(A) = a_0 I + a_1 A + \cdots + a_n A^n.$$

The constant term a_0 gets replaced by $a_0 I$. Then if $A = P\Lambda P^{-1}$,

$$(4.6.14) \qquad f(A) = f(P\Lambda P^{-1}) = a_0 I + a_1 P\Lambda P^{-1} + \cdots + a_n P\Lambda^n P^{-1} = Pf(\Lambda)P^{-1}.$$

The analogous notation is used for linear operators: If T is a linear operator on a vector space over a field F, the linear operator $f(T)$ on V is defined to be

$$(4.6.15) \qquad\qquad f(T) = a_0 I + a_1 T + \cdots + a_n T^n,$$

where I denotes the identity operator. The operator $f(T)$ acts on a vector by $f(T)v = a_0 v + a_1 Tv + \cdots + a_n T^n v$. (In order to avoid too many parentheses we have omitted some by writing Tv for $T(v)$.)

4.7 JORDAN FORM

Suppose we are given a linear operator T on a finite-dimensional complex vector space V. We have seen that, if the roots of its characteristic polynomial are distinct, there is a basis of eigenvectors, and that the matrix of T with respect to that basis is diagonal. Here we ask what can be done without assuming that the eigenvalues are distinct. When the characteristic polynomial has multiple roots there will most often not be a basis of eigenvectors, but we'll see that, nevertheless, the matrix can be made fairly simple.

An eigenvector with eigenvalue λ of a linear operator T is a nonzero vector v such that $(T - \lambda)v = 0$. (We will write $T - \lambda$ for $T - \lambda I$ here.) Since our operator T may not have enough eigenvectors, we work with generalized eigenvectors.

• A *generalized eigenvector* with eigenvalue λ of a linear operator T is a nonzero vector x such that $(T - \lambda)^k x = 0$ for some $k > 0$. Its *exponent* is the smallest integer d such that $(T - \lambda)^d x = 0$.

Proposition 4.7.1 Let x be a generalized eigenvector of T, with eigenvalue λ and exponent d, and for $j \geq 0$, let $u_j = (T - \lambda)^j x$. Let $\mathbf{B} = (u_0, \dots, u_{d-1})$, and let $X = \text{Span } \mathbf{B}$. Then X is a T-invariant subspace, and $\mathbf{B}$ is a basis of X.

We use the next lemma in the proof.

Lemma 4.7.2 With u_j as above, a linear combination $y = c_j u_j + \cdots + c_{d-1} u_{d-1}$ with $j \leq d - 1$ and $c_j \neq 0$ is a generalized eigenvector, with eigenvalue λ and exponent $d - j$.

Proof. Since the exponent of x is d, $(T - \lambda)^{d-1} x = u_{d-1} \neq 0$. Therefore $(T - \lambda)^{d-j-1} y = c_j u_{d-1}$ isn't zero, but $(T - \lambda)^{d-j} y = 0$. So y is a generalized eigenvector with eigenvalue λ and exponent $d - j$, as claimed.   □

Proof of the Proposition. We note that

$$(4.7.3) \qquad\qquad T u_j = \begin{cases} \lambda u_j + u_{j+1} & \text{if } j < d - 1 \\ \lambda u_j & \text{if } j = d - 1 \\ 0 & \text{if } j > d - 1. \end{cases}$$

Therefore $T u_j$ is in the subspace X for all j. This shows that X is invariant. Next, $\mathbf{B}$ generates X by definition. The lemma shows that every nontrivial linear combination of $\mathbf{B}$ is a generalized eigenvector, so it is not zero. Therefore $\mathbf{B}$ is an independent set.   □

Corollary 4.7.4 Let x be a generalized eigenvector for T, with eigenvalue λ. Then λ is an ordinary eigenvalue – a root of the characteristic polynomial of T.

Proof. If the exponent of x is d, then with notation as above, u_{d-1} is an eigenvector with eigenvalue λ.   □

Formula 4.7.3 determines the matrix that describes the action of T on the basis B of Proposition 4.7.1. It is the $d \times d$ *Jordan block* J_λ. Jordan blocks are shown below for low values of d:

$$(4.7.5) \qquad J_\lambda = [\lambda], \begin{bmatrix} \lambda & \\ 1 & \lambda \end{bmatrix}, \begin{bmatrix} \lambda & & \\ 1 & \lambda & \\ & 1 & \lambda \end{bmatrix}, \begin{bmatrix} \lambda & & & \\ 1 & \lambda & & \\ & 1 & \lambda & \\ & & 1 & \lambda \end{bmatrix}, \cdots$$

The operation of a Jordan block is especially simple when $\lambda = 0$. The $d \times d$ block J_0 operates on the standard basis of $\mathbb{C}^d$ as

$$(4.7.6) \qquad\qquad e_1 \rightsquigarrow e_2 \rightsquigarrow \cdots \rightsquigarrow e_d \rightsquigarrow 0.$$

The 1×1 Jordan block J_0 is zero.

The Jordan Decomposition Theorem below asserts that any complex $n \times n$ matrix is similar to a matrix J made up of diagonal Jordan blocks (4.7.5) – that it has the *Jordan form*

$$(4.7.7) \qquad\qquad J = \begin{bmatrix} J_1 & & & \\ & J_2 & & \\ & & \ddots & \\ & & & J_\ell \end{bmatrix},$$

where $J_i = J_{\lambda_i}$ for some λ_i. The blocks J_i can have various sizes d_i, with $\sum d_i = n$, and the diagonal entries λ_i aren't necessarily distinct. The characteristic polynomial of the matrix J is

$$(4.7.8) \qquad\qquad p(t) = (t - \lambda_1)^{d_1} (t - \lambda_2)^{d_2} \cdots (t - \lambda_\ell)^{d_\ell}.$$

The 2×2 and 3×3 Jordan forms are

$$(4.7.9) \qquad \begin{bmatrix} \lambda_1 & \\ & \lambda_2 \end{bmatrix}, \begin{bmatrix} \lambda_1 & \\ 1 & \lambda_1 \end{bmatrix}; \begin{bmatrix} \lambda_1 & & \\ & \lambda_2 & \\ & & \lambda_3 \end{bmatrix}, \begin{bmatrix} \lambda_1 & & \\ 1 & \lambda_1 & \\ & & \lambda_2 \end{bmatrix}, \begin{bmatrix} \lambda_1 & & \\ 1 & \lambda_1 & \\ & 1 & \lambda_1 \end{bmatrix},$$

where the scalars λ_i may be equal or not, and in the fourth matrix, the blocks may be listed in the other order.

Theorem 4.7.10 Jordan Decomposition.

(a) *Vector space form*: Let T be a linear operator on a finite-dimensional complex vector space V. There is a basis **B** of V such that the matrix of T with respect to **B** has Jordan form (4.7.7).

(b) *Matrix form*: Let A be an $n \times n$ complex matrix. There is an invertible complex matrix P such that $P^{-1} A P$ has Jordan form.

It is also true that the Jordan form of an operator T or a matrix A is unique except for the order of the blocks.

Proof. This proof is due to Filippov [Filippov]. Induction on the dimension of V allows us to assume that the theorem is true for the restriction of T to any proper invariant subspace. So if V is the direct sum of proper T-invariant subspaces, say $V_1 \oplus \cdots \oplus V_r$, with $r > 1$, then the theorem is true for T.

Suppose that we have generalized eigenvectors v_i, for $i = 1, \ldots, r$. Let V_i be the subspace defined as in Proposition 4.7.1, with $x = v_i$. If V is the direct sum $V_1 \oplus \cdots \oplus V_r$, the theorem will be true for V, and we say that $v_1, \ldots, v_r$ are *Jordan generators* for T. We will show that a set of Jordan generators exists.

Step 1: We choose an eigenvalue λ of T, and replace the operator T by $T - \lambda I$. If A is the matrix of T with respect to a basis, the matrix of $T - \lambda I$ with respect to the same basis will be $A - \lambda I$, and if one of the matrices A or $A - \lambda I$ is in Jordan form, so is the other. So replacing T by $T - \lambda I$ is permissible. Having done this, our operator, which we still call T, will have zero as an eigenvalue. This will simplify the notation.

Step 2: We assume that 0 is an eigenvalue of T. Let K_i and U_i denote the kernel and image, respectively, of the ith power T^i. Then $K_1 \subset K_2 \subset \cdots$ and $U_1 \supset U_2 \supset \cdots$. Because V is finite-dimensional, these chains of subspaces become constant for large r, say $K_m = K_{m+1} = \cdots$ and $U_m = U_{m+1} = \cdots$. Let $K = K_m$ and $U = U_m$. We verify that K and U are invariant subspaces, and that V is the direct sum $K \oplus U$.

The subspaces are invariant because $TK_m \subset K_{m-1} \subset K_m$ and $TU_m = U_{m+1} = U_m$. To show that $V = K \oplus U$, it suffices to show that $K \cap U = \{0\}$ (see Proposition 4.3.1(**b**)). Let z be an element of $K \cap U$. Then $T^m z = 0$, and also $z = T^m v$ for some v in V. Therefore $T^{2m} v = 0$, so v is an element of K_{2m}. But $K_{2m} = K_m$, so $T^m v = 0$, i.e., $z = 0$.

Since T has an eigenvalue 0, K is not the zero subspace. Therefore U has smaller dimension than V, and by our induction assumption, the theorem is true for $T|_U$. Unfortunately, we can't use this reasoning on K, because U might be zero. So we must still prove the existence of a Jordan form for $T|_K$. We replace V by K and T by $T|_K$.

- A linear operator T on a vector space V is called *nilpotent* if for some positive integer r, the operator T^r is zero.

We have reduced the proof to the case of a nilpotent operator.

Step 3: We assume that our operator T is nilpotent. Every nonzero vector will be a generalized eigenvector with eigenvalue 0. Let N and W denote the kernel and image of T, respectively. Since T is nilpotent, $N \neq \{0\}$. Therefore the dimension of W is smaller than that of V, and by induction, the theorem is true for the restriction of the operator to W. So there are Jordan generators $w_1, \ldots, w_r$ for $T|_W$. Let e_i denote the exponent of w_i, and let W_i denote the subspace formed as in Proposition 4.7.1, using the generalized eigenvector w_i. So $W = W_1 \oplus \cdots \oplus W_r$.

For each i, we choose an element v_i of V such that $Tv_i = w_i$. The exponent d_i of v_i will be equal to $e_i + 1$. Let V_i denote the subspace formed as in (4.7.1) using the vector v_i. Then $TV_i = W_i$. Let U denote the sum $V_1 + \cdots + V_r$. Since each V_i is an invariant subspace, so is U. We now verify that $v_1, \ldots, v_r$ are Jordan generators for the restriction $T|_U$, i.e., that the subspaces V_i are independent.

We notice two things: First, $TU = W$ because $TV_i = W_i$. Second, $V_i \cap N \subset W_i$. This follows from Lemma 4.7.2, which shows that $V_i \cap N$ is the span of the last basis vector $T^{d_i-1}v_i$. Since $d_i - 1 = e_i$, which is positive, $T^{d_i-1}v_i$ is in the image W_i.

We suppose given a relation $\tilde{v}_1 + \cdots + \tilde{v}_r = 0$, with $\tilde{v}_i$ in V_i. We must show that $\tilde{v}_i = 0$ for all i. Let $\tilde{w}_i = T\tilde{v}_i$. Then $\tilde{w}_1 + \cdots + \tilde{w}_r = 0$, and $\tilde{w}_i$ is in W_i. Since the subspaces W_i are independent, $\tilde{w}_i = 0$ for all i. So $T\tilde{v}_i = 0$, which means that $\tilde{v}_i$ is in $V_i \cap N$. Therefore $\tilde{v}_i$ is in W_i. Using the fact that the subspaces W_i are independent once more, we conclude that, $\tilde{v}_i = 0$ for all i.

Step 4: We show that a set of Jordan generators for T can be obtained by adding some elements of N to the set $\{v_1, \ldots, v_r\}$ of Jordan generators for $T|_U$.

Let v be an arbitrary element of V and let $Tv = w$. Since $TU = W$, there is a vector u in U such that $Tu = w = Tv$. Then $z = v - u$ is in N and $v = u + z$. Therefore $U + N = V$. This being so, we extend a basis of U to a basis of V by adding elements, say $z_1, \ldots, z_\ell$, of N (see Proposition 3.4.16(a)). Let N' be the span of $(z_1, \ldots, z_\ell)$. Then $U \cap N' = \{0\}$ and $U + N' = V$, so V is the direct sum $U \oplus N'$.

The operator T is zero on N', so N' is an invariant subspace, and the matrix of $T|_{N'}$ is the zero matrix, which has Jordan form. Its Jordan blocks are 1×1 zero matrices. Therefore $\{v_1, \ldots, v_r; z_1, \ldots z_\ell\}$ is a set of Jordan generators for T.   □

It isn't difficult to determine the Jordan form for an operator T, provided that the eigenvalues are known, and the analysis also proves uniqueness of the form. However, finding an appropriate basis of V can be painful, and is best avoided.

To determine the Jordan form, one chooses an eigenvalue λ, and replaces T by $T - \lambda I$, to reduce to the case that $\lambda = 0$. Let K_i denote the kernel of T^i, and let k_i be the dimension of K_i. In the case of a single $d \times d$ Jordan block with $\lambda = 0$, these dimensions are:

$$k_i^{block} = \begin{cases} i & \text{if } i \leq d \\ d & \text{if } i \geq d \end{cases} .$$

The dimensions k_i for a general operator T are obtained by adding the numbers k_i^{block} for each block with $\lambda = 0$. So k_1 will be the number of blocks with $\lambda = 0$, $k_2 - k_1$ will be the number of blocks of size $d \geq 2$ with $\lambda = 0$, and so on.

Two simple examples:

$$A = \begin{bmatrix} 0 & 1 & 0 \\ 1 & 0 & 1 \\ 0 & -1 & 0 \end{bmatrix} \quad \text{and} \quad B = \begin{bmatrix} 1 & -1 & 1 \\ 2 & -2 & 2 \\ -1 & 1 & -1 \end{bmatrix}.$$

Here $A^3 = 0$, but $A^2 \neq 0$. If v is a vector such that $A^2 v \neq 0$, for instance $v = e_1$, then (v, Tv, T^2v) will be a basis. The Jordan form consists of a single 3×3 block.

On the other hand, $B^2 = 0$. Taking $v = e_1$ again, the set (v, Tv) is independent, and this gives us a 2×2 block. To obtain the Jordan form, we have to add a vector in N, for example $v' = e_2 + e_3$, which will give a 1×1 block (equal to zero). The required basis is (v, Tv, v').

It is often useful to write the Jordan form as $J = D + N$, where D is the diagonal part of the matrix, and N is the part below the diagonal. For a single Jordan block, we will have $D = \lambda I$ and $N = J_0$, as is illustrated below for a 3×3 block:

$$
J_\lambda = \begin{bmatrix} \lambda & & \\ 1 & \lambda & \\ & 1 & \lambda \end{bmatrix} = \begin{bmatrix} \lambda & & \\ 0 & \lambda & \\ & 0 & \lambda \end{bmatrix} + \begin{bmatrix} 0 & & \\ 1 & 0 & \\ & 1 & 0 \end{bmatrix} = \lambda I + J_0 = D + N.
$$

Writing $J = D + N$ is convenient because D and N commute. The powers of J can be computed by the binomial expansion:

$$(4.7.11) \qquad J^r = (D + N)^r = D^r + \binom{r}{1} D^{r-1} N + \binom{r}{2} D^{r-2} N^2 + \cdots ,$$

When J is an $n \times n$ matrix, $N^n = 0$, and this expansion has at most n terms. In the case of a single block, the formula reads

$$(4.7.12) \qquad J^r = (\lambda I + J_0)^r = \lambda^r I + \binom{r}{1} \lambda^{r-1} J_0 + \binom{r}{2} \lambda^{r-2} J_0^2 + \cdots .$$

Corollary 4.7.13 Let T be a linear operator on a finite-dimensional complex vector space. The following conditions are equivalent:

(a) T is a diagonalizable operator,

(b) every generalized eigenvector is an eigenvector,

(c) all of the blocks in the Jordan form for T are 1×1 blocks.

The analogous statements are true for a square complex matrix A.

Proof. **(a)** $\Rightarrow$ **(b)**: Suppose that T is diagonalizable, say that the matrix of T with respect to the basis $\mathbf{B} = (v_1, \ldots, v_n)$ is the diagonal matrix Λ with diagonal entries $\lambda_1, \ldots, \lambda_n$. Let v be a generalized eigenvector in V, say that $(T - \lambda)^k v = 0$ for some λ and some $k > 0$. We replace T by $T - \lambda$ to reduce to the case that $T^k v = 0$. Let $X = (x_1, \ldots, x_n)^t$ be the coordinate vector of v. The coordinates of $T^k v$ will be $\lambda_i^k x_i$. Since $T^k v = 0$, either $\lambda_i = 0$, or $x_i = 0$, and in either case, $\lambda_i^k x_i = 0$. Therefore $Tv = 0$.

(b) $\Rightarrow$ **(c)**: We prove the contrapositive. If the Jordan form of T has a $k \times k$ Jordan block with $k > 1$, then looking back at the action (4.7.6) of $J_\lambda - \lambda I$, we see that there is a generalized eigenvector that is not an eigenvector. So if **(c)** is false, **(b)** is false too. Finally, it is clear that **(c)** $\Rightarrow$ **(a)**. $\qquad \square$

Here is a nice application of Jordan form.

Theorem 4.7.14 Let T be a linear operator on a finite-dimensional complex vector space V. If some positive power of T is the identity, say $T^r = I$, then T is diagonalizable.

Proof. It suffices to show that every generalized eigenvector is an eigenvector. To do this, we assume that $(T - \lambda I)^2 v = 0$ with $v \neq 0$, and we show that $(T - \lambda)v = 0$. Since λ is an eigenvalue and since $T^r = I$, $\lambda^r = 1$. We divide the polynomial $t^r - 1$ by $t - \lambda$:

$$t^r - 1 = (t^{r-1} + \lambda t^{r-2} + \cdots + \lambda^{r-2} t + \lambda^{r-1})(t - \lambda).$$

We substitute T for t and apply the operators to v. Let $w = (T - \lambda)v$. Since $T^r - I = 0$,

$$
\begin{aligned}
0 = (T^r - I)v &= (T^{r-1} + \lambda T^{r-2} + \cdots + \lambda^{r-2}T + \lambda^{r-1})(T - \lambda)v \\
&= \left(T^{r-1} + \lambda T^{r-2} + \cdots + \lambda^{r-2}T + \lambda^{r-1}\right) w \\
&= r\lambda^{r-1}w.
\end{aligned}
$$

(For the last equality, one uses the fact that $Tw = \lambda w$.) Since $r\lambda^{r-1}w = 0$, $w = 0$.   □

We go back for a moment to the results of this section. Where has the hypothesis that V be a vector space over the complex numbers been used? The answer is that its only use is to ensure that the characteristic polynomial has enough roots.

Corollary 4.7.15 Let V be a finite-dimensional vector space over a field F, and let T be a linear operator on V whose characteristic polynomial factors into linear factors in F. The Jordan Decomposition theorem 4.7.10 is true for T.   □

The proof is identical to the one given for the case that $F = \mathbb{C}$.

Corollary 4.7.16 Let T be a linear operator on a finite-dimensional vector space over a field of characteristic zero. Assume that $T^r = I$ for some $r \geq 1$ and that the polynomial $t^r - 1$ factors into linear factors in F. Then T is diagonalizable.   □

The characteristic zero hypothesis is needed to carry through the last step of the proof of Theorem 4.7.14, where from the relation $r\lambda^{r-1}w = 0$ we want to conclude that $w = 0$. The theorem is false in characteristic different from zero.

—Yvonne Verdier[2]

EXERCISES

Section 1   The Dimension Formula

1.1. Let A be a $\ell \times m$ matrix and let B be an $n \times p$ matrix. Prove that the rule $M \rightsquigarrow AMB$ defines a linear transformation from the space $F^{m \times n}$ of $m \times n$ matrices to the space $F^{\ell \times p}$.

1.2. Let $v_1, \ldots, v_n$ be elements of a vector space V. Prove that the map $\varphi : F^n \to V$ defined by $\varphi(X) = v_1 x_1 + \cdots + v_n x_n$ is a linear transformation.

1.3. Let A be an $m \times n$ matrix. Use the dimension formula to prove that the space of solutions of the linear system $AX = 0$ has dimension at least $n - m$.

1.4. Prove that every $m \times n$ matrix A of rank 1 has the form $A = XY^t$, where X, Y are m- and n-dimensional column vectors. How uniquely determined are these vectors?

---

[2]I've received many emails asking about this rebus. Yvonne, an anthropologist, and her husband Jean-Louis, a mathematician, were close friends who died tragically in 1989. In their memory, I included them among the people quoted. The history of the valentine was one of Yvonne's many interests, and she sent this rebus as a valentine.

1.5. (a) Let U and W be vector spaces over a field F. Show that the operations two $(u, w) + (u', w') = (u + u', w + w')$ and $c(u, w) = (cu, cw)$ on pairs of vectors make the product set $U \times W$ into a vector space. It is called the *product space*.

(b) Let U and W be subspaces of a vector space V. Show that the map $T : U \times W \to V$ defined by $T(u, w) = u + w$ is a linear transformation.

(c) Express the dimension formula for T in terms of the dimensions of subspaces of V.

Section 2   The Matrix of a Linear Transformation

2.1. Let A and B be 2×2 matrices. Determine the matrix of the operator $T : M \rightsquigarrow AMB$ on the space $F^{2 \times 2}$ of 2×2 matrices, with respect to the basis $(e_{11}, e_{12}, e_{21}, e_{22})$ of $F^{2 \times 2}$.

2.2. Let A be an $n \times n$ matrix, and let V denote the space of n-dimensional *row* vectors. What is the matrix of the linear operator "right multiplication by A" with respect to the standard basis of V?

2.3. Find all real 2×2 matrices that carry the line $y = x$ to the line $y = 3x$.

2.4. Prove Theorem 4.2.10**(b)** using row and column operations.

2.5. [3]Let A be an $m \times n$ matrix of rank r, let I be a set of r row indices such that the corresponding rows of A are independent, and let J be a set of r column indices such that the corresponding columns of A are independent. Let M denote the $r \times r$ submatrix of A obtained by taking rows from I and columns from J. Prove that M is invertible.

Section 3   Linear Operators

3.1. Determine the dimensions of the kernel and the image of the linear operator T on the space $\mathbb{R}^n$ defined by $T(x_1, \ldots, x_n)^t = (x_1 + x_n, x_2 + x_{n-1}, \ldots, x_n + x_1)^t$.

3.2. (a) Let $A = \begin{bmatrix} a & b \\ c & d \end{bmatrix}$ be a real matrix, with c not zero. Show that using conjugation by elementary matrices, one can eliminate the "a" entry.

(b) Which matrices with $c = 0$ are similar to a matrix in which the "a" entry is zero?

3.3. Let $T : V \to V$ be a linear operator on a vector space of dimension 2. Assume that T is not multiplication by a scalar. Prove that there is a vector v in V such that $(v, T(v))$ is a basis of V, and describe the matrix of T with respect to that basis.

3.4. Let B be a complex $n \times n$ matrix. Prove or disprove: The linear operator T on the space of all $n \times n$ matrices defined by $T(A) = AB - BA$ is singular.

Section 4   Eigenvectors

4.1. Let T be a linear operator on a vector space V, and let λ be a scalar. The *eigenspace* $V^{(\lambda)}$ is the set of eigenvectors of T with eigenvalue λ, together with 0. Prove that $V^{(\lambda)}$ is a T-invariant subspace.

4.2. (a) Let T be a linear operator on a finite-dimensional vector space V, such that T^2 is the identity operator. Prove that for any vector v in V, $v - Tv$ is either an eigenvector with eigenvalue -1, or the zero vector. With notation as in Exercise 4.1, prove that V is the direct sum of the eigenspaces $V^{(1)}$ and $V^{(-1)}$.

---

[3]Suggested by Robert DeMarco.

(b) Generalize this method to prove that a linear operator T such that $T^4 = I$ decomposes a complex vector space into a sum of four eigenspaces.

4.3. Let T be a linear operator on a vector space V. Prove that if W_1 and W_2 are T-invariant subspaces of V, then $W_1 + W_2$ and $W_1 \cap W_2$ are T-invariant.

4.4. A 2×2 matrix A has an eigenvector $v_1 = (1, 1)^t$ with eigenvalue 2 and also an eigenvector $v_2 = (1, 2)^t$ with eigenvalue 3. Determine A.

4.5. Find all invariant subspaces of the real linear operator whose matrix is

(a) $\begin{bmatrix} 1 & 1 \\ & 1 \end{bmatrix}$,   **(b)** $\begin{bmatrix} 1 & & \\ & 2 & \\ & & 3 \end{bmatrix}$.

4.6. Let P be the real vector space of polynomials $p(x) = a_0 + a_1 + \cdots + a_n x^n$ of degree at most n, and let D denote the derivative $\frac{d}{dx}$, considered as a linear operator on P.

(a) Prove that D is a *nilpotent* operator, meaning that $D^k = 0$ for sufficiently large k.

(b) Find the matrix of D with respect to a convenient basis.

(c) Determine all D-invariant subspaces of P.

4.7. Let $A = \begin{bmatrix} a & b \\ c & d \end{bmatrix}$ be a real 2×2 matrix. The condition that a column vector X be an eigenvector for left multiplication by A is that $AX = Y$ be a scalar multiple of X, which means that the slopes $s = x_2/x_1$ and $s' = y_2/y_1$ are equal.

(a) Find the equation in s that expresses this equality.

(b) Suppose that the entries of A are positive real numbers. Prove that there is an eigenvector in the first quadrant and also one in the second quadrant.

4.8. Let T be a linear operator on a finite-dimensional vector space for which every nonzero vector is an eigenvector. Prove that T is multiplication by a scalar.

Section 5   The Characteristic Polynomial

5.1. Compute the characteristic polynomials and the complex eigenvalues and eigenvectors of

(a) $\begin{bmatrix} -2 & 2 \\ -2 & 3 \end{bmatrix}$,   **(b)** $\begin{bmatrix} 1 & i \\ -i & 1 \end{bmatrix}$,   **(c)** $\begin{bmatrix} \cos\theta & -\sin\theta \\ \sin\theta & \cos\theta \end{bmatrix}$.

5.2. The characteristic polynomial of the matrix below is $t^3 - 4t - 1$. Determine the missing entries.

$$\begin{bmatrix} 0 & 1 & 2 \\ 1 & 1 & 0 \\ 1 & * & * \end{bmatrix}$$

5.3. What complex numbers might be eigenvalues of a linear operator T such that
(a) $T^r = I$,   **(b)** $T^2 - 5T + 6I = 0$?

5.4. Find a recursive relation for the characteristic polynomial of the $k \times k$ matrix

$$\begin{bmatrix} 0 & 1 & & & & \\ 1 & 0 & 1 & & & \\ & 1 & \cdot & \cdot & & \\ & & \cdot & \cdot & \cdot & \\ & & & \cdot & \cdot & 1 \\ & & & & 1 & 0 \end{bmatrix},$$

and compute the polynomial for $k \le 5$.

5.5. Which real 2×2 matrices have real eigenvalues? Prove that the eigenvalues are real if the off-diagonal entries have the same sign.

5.6. Let V be a vector space with basis $(v_0, \ldots, v_n)$ and let $a_0, \ldots, a_n$ be scalars. Define a linear operator T on V by the rules $T(v_i) = v_{i+1}$ if $i < n$ and $T(v_n) = a_0 v_0 + a_1 v_1 + \cdots + a_n v_n$. Determine the matrix of T with respect to the given basis, and the characteristic polynomial of T.

5.7. Do A and A^t have the same eigenvectors? the same eigenvalues?

5.8. Let $A = (a_{ij})$ be a 3×3 matrix. Prove that the coefficient of t in the characteristic polynomial is the sum of the symmetric 2×2 minors

$$\det \begin{bmatrix} a_{11} & a_{12} \\ a_{21} & a_{22} \end{bmatrix} + \det \begin{bmatrix} a_{11} & a_{13} \\ a_{31} & a_{33} \end{bmatrix} + \det \begin{bmatrix} a_{22} & a_{23} \\ a_{32} & a_{33} \end{bmatrix}.$$

5.9. Consider the linear operator of left multiplication by an $m \times m$ matrix A on the space $F^{m \times m}$ of all $m \times m$ matrices. Determine the trace and the determinant of this operator.

5.10. Let A and B be $n \times n$ matrices. Determine the trace and the determinant of the operator on the space $F^{n \times n}$ defined by $M \rightsquigarrow AMB$.

Section 6   Triangular and Diagonal Forms

6.1. Let A be an $n \times n$ matrix whose characteristic polynomial factors into linear factors: $p(t) = (t - \lambda_1) \cdots (t - \lambda_n)$. Prove that trace $A = \lambda_1 + \cdots + \lambda_n$, that $\det A = \lambda_1 \cdots \lambda_n$.

6.2. Suppose that a complex $n \times n$ matrix A has distinct eigenvalues $\lambda_1, \ldots, \lambda_n$, and let $v_1, \ldots, v_n$ be eigenvectors with these eigenvalues.

(a) Show that every eigenvector is a multiple of one of the vectors v_i.

(b) Show how one can recover the matrix from the eigenvalues and eigenvectors.

6.3. Let T be a linear operator that has two linearly independent eigenvectors with the same eigenvalue λ. Prove that λ is a multiple root of the characteristic polynomial of T.

6.4. Let $A = \begin{bmatrix} 2 & 1 \\ 1 & 2 \end{bmatrix}$. Find a matrix P such that $P^{-1}AP$ is diagonal, and find a formula for the matrix A^{30}.

6.5. In each case, find a complex matrix P such that $P^{-1}AP$ is diagonal.

(a) $\begin{bmatrix} 1 & i \\ -i & 1 \end{bmatrix}$, **(b)** $\begin{bmatrix} 0 & 0 & 1 \\ 1 & 0 & 0 \\ 0 & 1 & 0 \end{bmatrix}$, **(c)** $\begin{bmatrix} \cos\theta & -\sin\theta \\ \sin\theta & \cos\theta \end{bmatrix}$.

6.6. Suppose that A is diagonalizable. Can the diagonalization be done with a matrix P in the special linear group?

6.7. Prove that if A and B are $n \times n$ matrices and A is nonsingular, then AB is similar to BA.

6.8. A linear operator T is nilpotent if some positive power T^k is zero. Prove that T is nilpotent if and only if there is a basis of V such that the matrix of T is upper triangular, with diagonal entries zero.

6.9. Find all real 2×2 matrices such that $A^2 = I$, and describe geometrically the way they operate by left multiplication on $\mathbb{R}^2$.

6.10. Let M be a matrix made up of two diagonal blocks: $M = \begin{bmatrix} A & 0 \\ 0 & D \end{bmatrix}$. Prove that M is diagonalizable if and only if A and D are diagonalizable.

6.11. Let $A = \begin{bmatrix} a & b \\ c & d \end{bmatrix}$ be a 2×2 matrix with eigenvalue λ.

(a) Show that unless it is zero, the vector $(b, \lambda - a)^t$ is an eigenvector.

(b) Find a matrix P such that $P^{-1}AP$ is diagonal, assuming that $b \neq 0$ and that A has distinct eigenvalues.

Section 7   Jordan Form

7.1. Determine the Jordan form of the matrix $\begin{bmatrix} 1 & 1 & 0 \\ 0 & 1 & 0 \\ 0 & 1 & 1 \end{bmatrix}$.

7.2. Prove that $A = \begin{bmatrix} 1 & 1 & 1 \\ -1 & -1 & -1 \\ 1 & 1 & 1 \end{bmatrix}$ is an *idempotent* matrix, i.e., that $A^2 = A$, and find its Jordan form.

7.3. Let V be a complex vector space of dimension 5, and let T be a linear operator on V whose characteristic polynomial is $(t - \lambda)^5$. Suppose that the rank of the operator $T - \lambda I$ is 2. What are the possible Jordan forms for T?

7.4. (a) Determine all possible Jordan forms for a matrix whose characteristic polynomial is $(t + 2)^2 (t - 5)^3$.

(b) What are the possible Jordan forms for a matrix whose characteristic polynomial is $(t + 2)^2 (t - 5)^3$, when space of eigenvectors with eigenvalue 2 is one-dimensional, and the space of eigenvectors with eigenvalue 5 is two-dimensional?

7.5. What is the Jordan form of a matrix A all of whose eigenvectors are multiples of a single vector?

7.6. Determine all invariant subspaces of a linear operator whose Jordan form consists of one block.

7.7. Is every complex square matrix A such that $A^2 = A$ diagonalizable?

7.8. Is every complex square matrix A similar to its transpose?

7.9. Find a 2×2 matrix with entries in $\mathbb{F}_p$ that has a power equal to the identity and an eigenvalue in $\mathbb{F}_p$, but is not diagonalizable.

Miscellaneous Problems

M.1. Let $v = (a_1, \ldots, a_n)$ be a real row vector. We may form the $n! \times n$ matrix M whose rows are obtained by permuting the entries of v in all possible ways. The rows can be listed in an arbitrary order. Thus if $n = 3$, M might be

$$\begin{bmatrix} a_1 & a_2 & a_3 \\ a_1 & a_3 & a_2 \\ a_2 & a_3 & a_1 \\ a_2 & a_1 & a_3 \\ a_3 & a_1 & a_2 \\ a_3 & a_2 & a_1 \end{bmatrix}.$$

Determine the possible ranks that such a matrix could have.

M.2. Let A be a complex $n \times n$ matrix with n distinct eigenvalues $\lambda_1, \ldots, \lambda_n$. Assume that λ_1 is the largest eigenvalue, that is, that $|\lambda_1| > |\lambda_i|$ for all $i > 1$.

   (a) Prove that for most vectors X, the sequence $X_k = \lambda_1^{-k} A^k X$ converges to an eigenvector Y with eigenvalue λ_1, and describe precisely what the conditions on X are for this to be true.

   (b) Prove the same thing without assuming that the eigenvalues $\lambda_1, \ldots, \lambda_n$ are distinct.

M.3. Compute the largest eigenvalue of the matrix $\begin{bmatrix} 3 & 1 \\ 3 & 4 \end{bmatrix}$ to three-place accuracy, using a method based on Exercise M.2.

M.4. If $X = (x_1, x_2, \ldots)$ is an infinite real row vector and $A = (a_{ij})$, $0 < i, j < \infty$ is an infinite real matrix, one may or may not able to define the matrix product XA. For which A can one define right multiplication on the space $\mathbb{R}^\infty$ of all infinite row vectors (3.7.1)? on the space Z (3.7.2)?

*__M.5.__ Let $\varphi : F^n \to F^m$ be left multiplication by an $m \times n$ matrix A.

   (a) Prove that the following are equivalent:
- A has a right inverse, a matrix B such that $AB = I$,
- φ is surjective,
- the rank of A is m.

   (b) Prove that the following are equivalent:
- A has a left inverse, a matrix B such that $BA = I$,
- φ is injective,
- the rank of A is n.

M.6. Without using the characteristic polynomial, prove that a linear operator on a vector space of dimension n can have at most n distinct eigenvalues.

*M.7. *(powers of an operator)* Let T be a linear operator on a vector space V. Let K_r and W_r denote the kernel and image, respectively, of T^r.

(a) Show that $K_1 \subset K_2 \subset \cdots$ and that $W_1 \supset W_2 \supset \cdots$.

(b) The following conditions might or might not hold for a particular value of r:

$\quad$ (1) $K_r = K_{r+1}$, $\quad$ (2) $W_r = W_{r+1}$, $\quad$ (3) $W_r \cap K_1 = \{0\}$, $\quad$ (4) $W_1 + K_r = V$.

$\quad$ Find all implications among the conditions (1)–(4) when V is finite dimensional.

(c) Do the same thing when V is infinite dimensional.

M.8. Let T be a linear operator on a finite-dimensional complex vector space V.

(a) Let λ be an eigenvalue of T, and let V_λ be the set of generalized eigenvectors, together with the zero vector. Prove that V_λ is a T-invariant subspace of V. (This subspace is called a *generalized eigenspace*.)

(b) Prove that V is the direct sum of its generalized eigenspaces.

M.9. Let V be a finite-dimensional vector space. A linear operator $T : V \to V$ is called a *projection*: if $T^2 = T$ (not necessarily an "orthogonal projection"). Let K and W be the kernel and image of a linear operator T. Prove

(a) T is a projection onto W if and only if the restriction of T to W is the identity map.

(b) If T is a projection, then V is the direct sum $W \oplus K$.

(c) The trace of a projection T is equal to its rank.

M.10. Let A and B be $m \times n$ and $n \times m$ real matrices.

(a) Prove that if λ is a nonzero eigenvalue of the $m \times m$ matrix AB then it is also an eigenvalue of the $n \times n$ matrix BA. Show by example that this need not be true if $\lambda = 0$.

(b) Prove that $I_m - AB$ is invertible if and only if $I_n - BA$ is invertible.

CHAPTER 5

Applications of Linear Operators

*By relieving the brain from all unnecessary work,
a good notation sets it free to concentrate
on more advanced problems.*

—Alfred North Whitehead

5.1 ORTHOGONAL MATRICES AND ROTATIONS

In this section, the field of scalars is the real number field.

We assume familiarity with the dot product of vectors in $\mathbb{R}^2$. The *dot product* of column vectors $X = (x_1, \ldots, x_n)^t$, $Y = (y_1, \ldots, y_n)^t$ in $\mathbb{R}^n$ is defined to be

$$(5.1.1) \qquad\qquad (X \cdot Y) = x_1 y_1 + \cdots + x_n y_n.$$

It is convenient to write the dot product as the matrix product of a row vector and a column vector:

$$(5.1.2) \qquad\qquad (X \cdot Y) = X^t Y.$$

For vectors in $\mathbb{R}^2$, one has the formula

$$(5.1.3) \qquad\qquad (X \cdot Y) = |X||Y| \cos \theta,$$

where θ is the angle between the vectors. This formula follows from the law of cosines

$$(5.1.4) \qquad\qquad c^2 = a^2 + b^2 - 2ab \cos \theta$$

for the side lengths a, b, c of a triangle, where θ is the angle between the sides a and b. To derive (5.1.3), we apply the law of cosines to the triangle with vertices 0, X, Y. Its side lengths are $|X|$, $|Y|$, and $|X - Y|$, so the law of cosines can be written as

$$((X - Y) \cdot (X - Y)) = (X \cdot X) + (Y \cdot Y) - 2|X||Y| \cos \theta.$$

The left side expands to $(X \cdot X) - 2(X \cdot Y) + (Y \cdot Y)$, and formula (5.1.3) is obtained by comparing this with the right side. The formula is valid for vectors in $\mathbb{R}^n$ too, but it requires understanding the meaning of the angle, and we won't take the time to go into that just now (see (8.5.2)).

The most important points for vectors in $\mathbb{R}^2$ and $\mathbb{R}^3$ are

- the square $|X|^2$ of the length of a vector X is $(X \cdot X) = X^t X$, and
- a vector X is *orthogonal* to another vector Y, written $X \perp Y$, if and only if $X^t Y = 0$.

We take these as the definitions of the length $|X|$ of a vector and of orthogonality of vectors in $\mathbb{R}^n$. Note that the length $|X|$ is positive unless X is the zero vector, because $|X|^2 = X^t X = x_1^2 + \cdots + x_n^2$ is a sum of squares.

Theorem 5.1.5 Pythagoras. If $X \perp Y$ and $Z = X + Y$, then $|Z|^2 = |X|^2 + |Y|^2$.

This is proved by expanding $Z^t Z$. If $X \perp Y$, then $X^t Y = Y^t X = 0$, so

$$Z^t Z = (X + Y)^t (X + Y) = X^t X + X^t Y + Y^t X + Y^t Y = X^t X + Y^t Y. \qquad \square$$

We switch to our lowercase vector notation. If $v_1, \ldots, v_k$ are orthogonal vectors in $\mathbb{R}^n$ and if $w = v_1 + \cdots + v_k$, then Pythagoras's theorem shows by induction that

(5.1.6) $$|w|^2 = |v_1|^2 + \cdots + |v_k|^2.$$

Lemma 5.1.7 Any set $(v_1, \ldots, v_k)$ of orthogonal nonzero vectors in $\mathbb{R}^n$ is independent.

Proof. Let $w = c_1 v_1 + \cdots + c_k v_k$ be a linear combination, where not all c_i are zero, and let $w_i = c_i v_i$. Then w is the sum $w_1 + \cdots + w_k$ of orthogonal vectors, not all of which are zero. By Pythagoras, $|w|^2 = |w_1|^2 + \cdots + |w_k|^2 > 0$, so $w \neq 0$. $\qquad \square$

- An *orthonormal basis* **B** $= (v_1, \ldots, v_n)$ of $\mathbb{R}^n$ is a basis of orthogonal *unit vectors* (vectors of length one). Another way to say this is that **B** is an orthonormal basis if

(5.1.8) $$(v_i \cdot v_j) = \delta_{ij},$$

where δ_{ij}, the *Kronecker delta*, is the i, j-entry of the identity matrix, which is equal to 1 if $i = j$ and to 0 if $i \neq j$.

Definition 5.1.9 A real $n \times n$ matrix A is *orthogonal* if $A^t A = I$, which is to say, A is invertible and its inverse is A^t.

Lemma 5.1.10 An $n \times n$ matrix A is orthogonal if and only if its columns form an orthonormal basis of $\mathbb{R}^n$.

Proof. Let A_i denote the ith column of A. Then A_i^t is the ith row of A^t. The i, j-entry of $A^t A$ is $A_i^t A_j$, so $A^t A = I$ if and only if $A_i^t A_j = \delta_{ij}$ for all i and j. $\qquad \square$

The next properties of orthogonal matrices are easy to verify:

Proposition 5.1.11

(a) The product of orthogonal matrices is orthogonal, and the inverse of an orthogonal matrix, its transpose, is orthogonal. The orthogonal matrices form a subgroup O_n of GL_n, the *orthogonal group*.

(b) The determinant of an orthogonal matrix is ± 1. The orthogonal matrices with determinant 1 form a subgroup SO_n of O_n of index 2, the *special orthogonal group*.   □

Definition 5.1.12 An *orthogonal operator* T on $\mathbb{R}^n$ is a linear operator that preserves the dot product: For every pair X, Y of vectors,

$$(TX \cdot TY) = (X \cdot Y).$$

Proposition 5.1.13 A linear operator T on $\mathbb{R}^n$ is orthogonal if and only if it preserves lengths of vectors, or, if and only if for every vector X, $(TX \cdot TX) = (X \cdot X)$.

Proof. Suppose that lengths are preserved, and let X and Y be arbitrary vectors in $\mathbb{R}^n$. Then
$$(T(X + Y) \cdot T(X + Y)) = ((X + Y) \cdot (X + Y)).$$

The fact that $(TX \cdot TY) = (X \cdot Y)$ follows by expanding the two sides of this equality and cancelling.   □

Proposition 5.1.14 A linear operator T on $\mathbb{R}^n$ is orthogonal if and only if its matrix A with respect to the standard basis is an orthogonal matrix.

Proof. If A is the matrix of T, then
$$(TX \cdot TY) = (AX)^{\mathrm{t}}(AY) = X^{\mathrm{t}}(A^{\mathrm{t}}A)Y.$$

The operator is orthogonal if and only if the right side is equal to $X^{\mathrm{t}}Y$ for all X and Y. We can write this condition as $X^{\mathrm{t}}(A^{\mathrm{t}}A - I)Y = 0$. The next lemma shows that this is true if and only if $A^{\mathrm{t}}A - I = 0$, and therefore A is orthogonal.   □

Lemma 5.1.15 Let M be an $n \times n$ matrix. If $X^{\mathrm{t}}MY = 0$ for all column vectors X and Y, then $M = 0$.

Proof. The product $e_i^{\mathrm{t}} M e_j$ evaluates to the i, j-entry of M. For instance,

$$[0 \quad 1]\begin{bmatrix} m_{11} & m_{12} \\ m_{21} & m_{22} \end{bmatrix}\begin{bmatrix} 1 \\ 0 \end{bmatrix} = m_{21}.$$

If $e_i^{\mathrm{t}} M e_j = 0$ for all i and j, then $M = 0$.   □

We now describe the orthogonal 2×2 matrices.

• A linear operator T on $\mathbb{R}^2$ is a *reflection* if it has orthogonal eigenvectors v_1 and v_2 with eigenvalues 1 and –1, respectively.

Because it fixes v_1 and changes the sign of the orthogonal vector v_2, such an operator reflects the plane about the one-dimensional subspace spanned by v_1. Reflection about the e_1-axis is given by the matrix

(5.1.16)
$$S_0 = \begin{bmatrix} 1 & 0 \\ 0 & -1 \end{bmatrix}.$$

Theorem 5.1.17

(a) The orthogonal 2×2 matrices with determinant 1 are the matrices

(5.1.18)
$$R = \begin{bmatrix} c & -s \\ s & c \end{bmatrix},$$

with $c = \cos\theta$ and $s = \sin\theta$, for some angle θ. The matrix R represents counterclockwise rotation of the plane $\mathbb{R}^2$ about the origin and through the angle θ.

(b) The orthogonal 2×2 matrices A with determinant -1 are the matrices

(5.1.19)
$$S = \begin{bmatrix} c & s \\ s & -c \end{bmatrix} = RS_0$$

with c and s as above. The matrix S reflects the plane about the one-dimensional subspace of $\mathbb{R}^2$ that makes an angle $\frac{1}{2}\theta$ with the e_1-axis.

Proof. Say that

$$A = \begin{bmatrix} c & * \\ s & * \end{bmatrix}$$

is orthogonal. Then its columns are unit vectors (5.1.10), so the point $(c, s)^t$ lies on the unit circle, and $c = \cos\theta$ and $s = \sin\theta$, for some angle θ. We inspect the product $P = R^t A$, where R is the matrix (5.1.18):

(5.1.20)
$$P = R^t A = \begin{bmatrix} 1 & * \\ 0 & * \end{bmatrix}.$$

Since R^t and A are orthogonal, so is P. Lemma 5.1.10 tells us that the second column is a unit vector orthogonal to the first one. So

(5.1.21)
$$P = \begin{bmatrix} 1 & 0 \\ 0 & \pm 1 \end{bmatrix}.$$

Working back, $A = RP$, so $A = R$ if $\det A = 1$ and $A = S = RS_0$ if $\det A = -1$.

We've seen that R represents a rotation (4.2.2), but we must still identify the operator defined by the matrix S. The characteristic polynomial of S is $t^2 - 1$, so its eigenvalues are 1 and -1. Let X_1 and X_2 be unit-length eigenvectors with these eigenvalues. Because S is orthogonal,

$$(X_1 \cdot X_2) = (SX_1 \cdot SX_2) = (X_1 \cdot -X_2) = -(X_1 \cdot X_2).$$

It follows that $(X_1 \cdot X_2) = 0$. The eigenvectors are orthogonal. The span of X_1 will be the line of reflection. To determine this line, we write a unit vector X as $(c', s')^t$, with $c' = \cos\alpha$ and $s' = \sin\alpha$. Then

$$SX = \begin{bmatrix} cc' + ss' \\ sc' - cs' \end{bmatrix} = \begin{bmatrix} \cos(\theta - \alpha) \\ \sin(\theta - \alpha) \end{bmatrix}.$$

When $\alpha = \frac{1}{2}\theta$, X is an eigenvector with eigenvalue 1, a fixed vector.   □

We describe the 3×3 rotation matrices next.

Definition 5.1.22   A *rotation* of $\mathbb{R}^3$ about the origin is a linear operator ρ with these properties:

- ρ fixes a unit vector u, called a *pole* of ρ, and
- ρ rotates the two-dimensional subspace W orthogonal to u.

The *axis of rotation* is the line ℓ spanned by u. We also call the identity operator a rotation, though its axis is indeterminate.

If multiplication by a 3×3 matrix R is a rotation of $\mathbb{R}^3$, R is called a *rotation matrix*.

(5.1.23)   A Rotation of $\mathbb{R}^3$.

The sign of the angle of rotation depends on how the subspace W is oriented. We'll orient W looking at it from the head of the arrow u. The angle θ shown in the figure is positive. (This is the "right hand rule.")

When u is the vector e_1, the set (e_2, e_3) will be a basis for W, and the matrix of ρ will have the form

(5.1.24) $M = \begin{bmatrix} 1 & 0 & 0 \\ 0 & c & -s \\ 0 & s & c \end{bmatrix}$,

where the bottom right 2×2 minor is the rotation matrix (5.1.18).

- A rotation that is not the identity is described by the pair (u, θ), called a *spin*, that consists of a pole u and a nonzero angle of rotation θ.

The rotation with spin (u, θ) may be denoted by $\rho_{(u,\theta)}$. Every rotation ρ different from the identity has two poles, the intersections of the axis of rotation ℓ with the unit sphere in $\mathbb{R}^3$. These are the unit-length eigenvectors of ρ with eigenvalue 1. The choice of a pole

u defines a direction on ℓ, and a change of direction causes a change of sign in the angle of rotation. If (u, θ) is a spin of ρ, so is $(-u, -\theta)$. Thus every rotation has two spins, and $\rho_{(u,\theta)} = \rho_{(-u,-\theta)}$.

Theorem 5.1.25 Euler's Theorem. The 3×3 rotation matrices are the orthogonal 3×3 matrices with determinant 1, the elements of the special orthogonal group SO_3.

Euler's Theorem has a remarkable consequence, which follows from the fact that SO_3 is a group. It is not obvious, either algebraically or geometrically.

Corollary 5.1.26 The composition of rotations about any two axes is a rotation about some other axis.   □

Because their elements represent rotations, the groups SO_2 and SO_3 are called the two- and three-dimensional *rotation groups*. Things become more complicated in dimension greater than 3. The 4×4 matrix

(5.1.27)
$$\begin{bmatrix} \cos \alpha & -\sin \alpha & & \\ \sin \alpha & \cos \alpha & & \\ & & \cos \beta & -\sin \beta \\ & & \sin \beta & \cos \beta \end{bmatrix}$$

is an element of SO_4. Left multiplication by this matrix rotates the two-dimensional subspace spanned by (e_1, e_2) through the angle α, and it rotates the subspace spanned by (e_3, e_4) through the angle β.

Before beginning the proof of Euler's Theorem, we note two more consequences:

Corollary 5.1.28 Let M be the matrix in SO_3 that represents the rotation $\rho_{(u,\alpha)}$ with spin (u, α).

(a) The trace of M is $1 + 2 \cos \alpha$.
(b) Let B be another element of SO_3, and let $u' = Bu$. The conjugate $M' = BMB^t$ represents the rotation $\rho_{(u',\alpha)}$ with spin (u', α).

Proof. **(a)** We choose an orthonormal basis (v_1, v_2, v_3) of $\mathbb{R}^3$ such that $v_1 = u$. The matrix of ρ with respect to this new basis will have the form (5.1.24), and its trace will be $1 + 2 \cos \alpha$. Since the trace doesn't depend on the basis, the trace of M is $1 + 2 \cos \alpha$ too.

(b) Since SO_3 is a group, M' is an element of SO_3. Euler's Theorem tells us that M' is a rotation matrix. Moreover, u' is a pole of this rotation: Since B is orthogonal, $u' = Bu$ has length 1, and

$$M'u' = BMB^{-1}u' = BMu = Bu = u'.$$

Let α' be the angle of rotation of M' about the pole u'. The traces of M and its conjugate M' are equal, so $\cos \alpha = \cos \alpha'$. This implies that $\alpha' = \pm \alpha$. Euler's Theorem tells us that the matrix B also represents a rotation, say with angle β about some pole. Since B and M' depend continuously on β, only one of the two values $\pm\alpha$ for α' can occur. When $\beta = 0$, $B = I$, $M' = M$, and $\alpha' = \alpha$. Therefore $\alpha' = \alpha$ for all β. $\square$

Lemma 5.1.29   A 3×3 orthogonal matrix M with determinant 1 has an eigenvalue equal to 1.

Proof. To show that 1 is an eigenvalue, we show that the determinant of the matrix $M - I$ is zero. If B is an $n\times n$ matrix, $\det(-B) = (-1)^n\det B$. We are dealing with 3×3 matrices, so $\det(M - I) = -\det(I - M)$. Also, $\det(M - I)^t = \det(M - I)$ and $\det M = 1$. Then

$$\det(M - I) = \det(M - I)^t = \det M \det(M - I)^t = \det(M(M^t - I)) = \det(I - M).$$

The relation $\det(M - I) = \det(I - M)$ shows that $\det(M - I) = 0$. $\square$

Proof of Euler's Theorem. Suppose that M represents a rotation ρ with spin (u, α). We form an orthonormal basis **B** of V by appending to u an orthonormal basis of its orthogonal space W. The matrix M' of ρ with respect to this basis will have the form (5.1.24), which is orthogonal and has determinant 1. Moreover, $M = PM'P^{-1}$, where the matrix P is equal to $[\mathbf{B}]$ (3.5.13). Since its columns are orthonormal, $[\mathbf{B}]$ is orthogonal. Therefore M is also orthogonal, and its determinant is equal to 1.

Conversely, let M be an orthogonal matrix with determinant 1, and let T denote left multiplication by M. Let u be a unit-length eigenvector with eigenvalue 1, and let W be the two-dimensional space orthogonal to u. Since T is an orthogonal operator that fixes u, it sends W to itself. So W is a T-invariant subspace, and we can restrict the operator to W.

Since T is orthogonal, it preserves lengths (5.1.13), so its restriction to W is orthogonal too. Now W has dimension 2, and we know the orthogonal operators in dimension 2: they are the rotations and the reflections (5.1.17). The reflections are operators with determinant -1. If an operator T acts on W as a reflection and fixes the orthogonal vector u, its determinant will be -1 too. Since this is not the case, $T|_W$ is a rotation. This verifies the second condition of Definition 5.1.22, and shows that T is a rotation. $\square$

5.2   USING CONTINUITY

Various facts about complex matrices can be deduced by diagonalization, using reasoning based on continuity that we explain here.

A sequence A_k of $n\times n$ matrices converges to an $n\times n$ matrix A if for every i and j, the i, j-entry of A_k converges to the i, j entry of A. Similarly, a sequence $p_k(t), k = 1, 2, \ldots,$ of polynomials of degree n with complex coefficients converges to a polynomial $p(t)$ of degree n if for every j, the coefficient of t^j in p_k converges to the corresponding coefficient of p. We may indicate that a sequence S_k of complex numbers, matrices, or polynomials converges to S by writing $S_k \to S$.

Proposition 5.2.1   Continuity of Roots. Let $p_k(t)$ be a sequence of monic polynomials of degree $\leq n$, and let $p(t)$ be another monic polynomial of degree n. Let $\alpha_{k,1}, \ldots, \alpha_{k,n}$ and $\alpha_1, \ldots \alpha_n$ denote the roots of these polynomials.

(a) If $\alpha_{k,\nu} \to \alpha_\nu$ for $\nu = 1, \ldots, n$, then $p_k \to p$.

(b) Conversely, if $p_k \to p$, the roots $\alpha_{k,\nu}$ of p_k can be numbered in such a way that $\alpha_{k,\nu} \to \alpha_\nu$ for each $\nu = 1, \ldots, n$.

In part **(b)**, the roots of each polynomial p_k must be renumbered individually.

Proof. We note that $p_k(t) = (t - \alpha_{k,1}) \cdots (t - \alpha_{k,n})$ and $p(t) = (t - \alpha_1) \cdots (t - \alpha_n)$. Part **(a)** follows from the fact that the coefficients of $p(t)$ are continuous functions – polynomial functions – of the roots, but **(b)** is less obvious.

Step 1: Let $\alpha_{k,\nu}$ be a root of p_k nearest to α_1, i.e., such that $|\alpha_{k,\nu} - \alpha_1|$ is minimal. We renumber the roots of p_k so that this root becomes $\alpha_{k,1}$. Then

$$|\alpha_1 - \alpha_{k,1}|^n \le |(\alpha_1 - \alpha_{k,1}) \cdots (\alpha_1 - \alpha_{k,n})| = |p_k(\alpha_1)|.$$

The right side converges to $|p(\alpha_1)| = 0$. Therefore the left side does too, and this shows that $\alpha_{k,1} \to \alpha_1$.

Step 2: We divide, writing $p_k(t) = (t - \alpha_{k,1})q_k(t)$ and $p(t) = (t - \alpha_1)q(t)$. Then q_k and q are monic polynomials, and their roots are $\alpha_{k,2}, \ldots, \alpha_{k,n}$ and $\alpha_2, \ldots, \alpha_n$, respectively. If we show that $q_k \to q$, then by induction on the degree n, we will be able to arrange the roots of q_k so that they converge to the roots of q, and we will be done.

To show that $q_k \to q$, we carry the division out explicitly. To simplify notation, we drop the subscript 1 from α_1. Say that $p(t) = t^n + a_{n-1}t^{n-1} + \cdots + a_1 t + a_0$, that $q(t) = t^{n-1} + b_{n-2}t^{n-2} + \cdots + b_1 t + b_0$, and that the notation for p_k and q_k is analogous. The equation $p(t) = (t - \alpha)q(t)$ implies that

$$b_{n-2} = \alpha + a_{n-1},$$
$$b_{n-3} = \alpha^2 + \alpha + a_{n-2},$$
$$\vdots$$
$$b_0 = \alpha^{n-1} + \alpha^{n-2}a_{n-1} + \cdots + \alpha a_2 + a_1.$$

Since $\alpha_{k,1} \to \alpha$ and $a_{k,i} \to a_i$, it is true that $b_{k,i} \to b_i$. $\square$

Proposition 5.2.2 Let A be an $n \times n$ complex matrix.

(a) There is a sequence of matrices A_k that converges to A, and such that for all k the characteristic polynomial $p_k(t)$ of A_k has distinct roots.

(b) If a sequence A_k of matrices converges to A, the sequence $p_k(t)$ of its characteristic polynomials converges to the characteristic polynomial $p(t)$ of A.

(c) Let λ_i be the roots of the characteristic polynomial p. If $A_k \to A$, the roots $\lambda_{k,i}$ of p_k can be numbered so that $\lambda_{k,i} \to \lambda_i$ for each i.

Proof. **(a)** Proposition 4.6.1 tells us that there is an invertible $n \times n$ matrix P such that $A' = P^{-1}AP$ is upper triangular. Its eigenvalues will be the diagonal entries of that matrix. We let A_k' be a sequence of matrices that converges to A', whose off-diagonal entries are the same as those of A', and whose diagonal entries are distinct. Then A'_k is upper triangular, and its characteristic polynomial has distinct roots. Let $A_k = PA'_kP^{-1}$. Since matrix multiplication is continuous, $A_k \to A$. The characteristic polynomial of A_k is the same as that of A'_k, so it has distinct roots.

Part **(b)** follows from **(a)** because the coefficients of the characteristic polynomial depend continuously on the matrix entries, and then **(c)** follows from Proposition 5.2.1.   □

One can use continuity to prove the famous *Cayley-Hamilton Theorem*. We state the theorem in its matrix form.

Theorem 5.2.3 Cayley-Hamilton Theorem. Let $p(t) = t^n + c_{n-1}t^{n-1} + \cdots + c_1 t + c_0$ be the characteristic polynomial of an $n \times n$ complex matrix A. Then $p(A) = A^n + c_{n-1}A^{n-1} + \cdots + c_1 A + c_0 I$ is the zero matrix.

For example, the characteristic polynomial of the 2×2 matrix A, with entries a, b, c, d as usual, is $t^2 - (a+d)t + (ad - bc)$ (4.5.12). The theorem asserts that

$$(5.2.4) \qquad \begin{bmatrix} a & b \\ c & d \end{bmatrix}^2 - (a+d)\begin{bmatrix} a & b \\ c & d \end{bmatrix} + (ad - bc)\begin{bmatrix} 1 & 0 \\ 0 & 1 \end{bmatrix} = \begin{bmatrix} 0 & 0 \\ 0 & 0 \end{bmatrix}.$$

This is easy to verify.

Proof of the Cayley-Hamilton Theorem. *Step 1:* The case that A is a diagonal matrix.
    Let the diagonal entries be $\lambda_1, \ldots, \lambda_n$. The characteristic polynomial is

$$p(t) = (t - \lambda_1) \cdots (t - \lambda_n).$$

Here $p(A)$ is also a diagonal matrix, and its diagonal entries are $p(\lambda_i)$. Since λ_i are the roots of p, $p(\lambda_i) = 0$ and $p(A) = 0$.

Step 2: The case that the eigenvalues of A are distinct.
    In this case, A is diagonalizable; say $A' = P^{-1}AP$ is diagonal. Then the characteristic polynomial of A' is the same as the characteristic polynomial $p(t)$ of A, and moreover,

$$p(A) = Pp(A')P^{-1}$$

(see (4.6.14)). By step 1, $p(A') = 0$, so $p(A) = 0$.

Step 3: The general case.
    We apply proposition 5.2.2. We let A_k be a sequence of matrices with distinct eigenvalues that converges to A. Let p_k be the characteristic polynomial of A_k. Since the sequence p_k converges to the characteristic polynomial p of A, $p_k(A_k) \to p(A)$. Step 2 tells us that $p_k(A_k) = 0$ for all k. Therefore $p(A) = 0$.   □

5.3 SYSTEMS OF DIFFERENTIAL EQUATIONS

We learn in calculus that the solutions of the differential equation

(5.3.1)
$$\frac{dx}{dt} = ax$$

are $x(t) = ce^{at}$, where c is an arbitrary real number. We review the proof because we want to use the argument again. First, ce^{at} does solve the equation. To show that every solution has this form, let $x(t)$ be an arbitrary solution. We differentiate $e^{-at}x(t)$ using the product rule:

(5.3.2)
$$\frac{d}{dt}(e^{-at}x(t)) = (-ae^{-at})x(t) + e^{-at}(ax(t)) = 0.$$

Thus $e^{-at}x(t)$ is a constant c, and $x(t) = ce^{at}$.

To extend this solution to systems of constant coefficient differential equations, we use the following terminology. A *vector-valued function* or *matrix-valued* function is a vector or a matrix whose entries are functions of t:

(5.3.3)
$$X(t) = \begin{bmatrix} x_1(t) \\ \vdots \\ x_n(t) \end{bmatrix}, \quad A(t) = \begin{bmatrix} a_{11}(t) & \cdots & a_{1n}(t) \\ \vdots & & \vdots \\ a_{m1}(t) & \cdots & a_{mn}(t) \end{bmatrix}.$$

The calculus operations of taking limits and differentiating are extended to vector-valued and matrix-valued functions by performing the operations on each entry separately. The derivative of a vector-valued or matrix-valued function is the function obtained by differentiating each entry:

(5.3.4)
$$\frac{dX}{dt} = \begin{bmatrix} x_1'(t) \\ \vdots \\ x_n'(t) \end{bmatrix}, \quad \frac{dA}{dt} = \begin{bmatrix} a_{11}'(t) & \cdots & a_{1n}'(t) \\ \vdots & & \vdots \\ a_{m1}'(t) & \cdots & a_{mn}'(t) \end{bmatrix},$$

where $x_i'(t)$ is the derivative of $x_i(t)$, and so on. So $\frac{dX}{dt}$ is defined if and only if each of the functions $x_i(t)$ is differentiable. The derivative can also be described in matrix notation:

(5.3.5)
$$\frac{dX}{dt} = \lim_{h \to 0} \frac{X(t+h) - X(t)}{h}.$$

Here $X(t+h) - X(t)$ is computed by vector addition and the h in the denominator stands for scalar multiplication by h^{-1}. The limit is obtained by evaluating the limit of each entry separately. So the entries of (5.3.5) are the derivatives $x_i'(t)$. The analogous statement is true for matrix-valued functions.

Many elementary properties of differentiation carry over to matrix-valued functions. The product rule, whose proof is an exercise, is an example:

Lemma 5.3.6 Product Rule.

(a) Let $A(t)$ and $B(t)$ be differentiable matrix-valued functions of t, of suitable sizes so that their product is defined. Then the matrix product $A(t)B(t)$ is differentiable, and its derivative is

$$\frac{d(AB)}{dt} = \frac{dA}{dt}B + A\frac{dB}{dt}.$$

(b) Let $A_1, \ldots, A_k$ be differentiable matrix-valued functions of t, of suitable sizes so that their product is defined. Then the matrix product $A_1 \cdots A_k$ is differentiable, and its derivative is

$$\frac{d}{dt}(A_1 \cdots A_k) = \sum_{i=1}^{k} A_1 \cdots A_{i-1}\left(\frac{dA_i}{dt}\right)A_{i+1} \cdots A_k. \qquad \square$$

A system of homogeneous linear, first-order, constant-coefficient differential equations is a matrix equation of the form

(5.3.7) $$\frac{dX}{dt} = AX,$$

where A is a constant $n \times n$ matrix and $X(t)$ is an n-dimensional vector-valued function. Writing out such a system, we obtain a system of n differential equations

(5.3.8)
$$\frac{dx_1}{dt} = a_{11}x_1(t) + \cdots + a_{1n}x_n(t)$$
$$\vdots \qquad \vdots \qquad \qquad \vdots$$
$$\frac{dx_n}{dt} = a_{n1}x_1(t) + \cdots + a_{nn}x_n(t).$$

The $x_i(t)$ are unknown functions, and the scalars a_{ij} are given. For example, if

(5.3.9) $$A = \begin{bmatrix} 3 & 2 \\ 1 & 4 \end{bmatrix},$$

(5.3.7) becomes a system of two equations in two unknowns:

(5.3.10)
$$\frac{dx_1}{dt} = 3x_1 + 2x_2$$
$$\frac{dx_2}{dt} = x_1 + 4x_2.$$

The simplest systems are those in which A is a diagonal matrix with diagonal entries λ_i. Then equation (5.3.8) reads

(5.3.11) $$\frac{dx_i}{dt} = \lambda_i x_i(t), \quad i = 1, \ldots, n.$$

Here the unknown functions x_i are not mixed up by the equations, so we can solve for each one separately:

$$(5.3.12) \qquad x_i = c_i e^{\lambda_i t},$$

for some arbitrary constants c_i.

The observation that allows us to solve the differential equation (5.3.7) in many cases is this: If V is an eigenvector for A with eigenvalue λ, i.e., if $AV = \lambda V$, then

$$(5.3.13) \qquad X = e^{\lambda t} V$$

is a particular solution of (5.3.7). Here $e^{\lambda t} V$ must be interpreted as the product of the variable scalar $e^{\lambda t}$ and the constant vector V. Differentiation operates on the scalar function, fixing V, while multiplication by A operates on the vector V, fixing the scalar $e^{\lambda t}$. Thus $\frac{d}{dt} e^{\lambda t} V = \lambda e^{\lambda t} V$ and also $A e^{\lambda t} V = \lambda e^{\lambda t} V$. For example,

$$\begin{bmatrix} 1 \\ 1 \end{bmatrix} \quad \text{and} \quad \begin{bmatrix} 2 \\ -1 \end{bmatrix}$$

are eigenvectors of the matrix (5.3.9), with eigenvalue 5 and 2, respectively, and

$$(5.3.14) \qquad \begin{bmatrix} e^{5t} \\ e^{5t} \end{bmatrix} \quad \text{and} \quad \begin{bmatrix} 2e^{2t} \\ -e^{2t} \end{bmatrix}$$

solve the system (5.3.10).

This observation allows us to solve (5.3.7) whenever the matrix A has distinct real eigenvalues. In that case every solution will be a linear combination of the special solutions (5.3.13). To work this out, it is convenient to diagonalize.

Proposition 5.3.15 Let A be an $n \times n$ matrix, and let P be an invertible matrix such that $\Lambda = P^{-1} A P$ is diagonal, with diagonal entries $\lambda_1, \ldots, \lambda_n$. The general solution of the system $\frac{dX}{dt} = AX$ is $X = P\tilde{X}$, where $\tilde{X} = (c_1 e^{\lambda_1 t}, \ldots, c_n e^{\lambda_n t})^{\text{t}}$ solves the equation $\frac{d\tilde{X}}{dt} = \Lambda \tilde{X}$.

The coefficients c_i are arbitrary. They are often determined by assigning *initial conditions* – the value of X at some particular t_0.

Proof. We multiply the equation $\frac{d\tilde{X}}{dt} = \Lambda \tilde{X}$ by P: $P\frac{d\tilde{X}}{dt} = P\Lambda\tilde{X} = AP\tilde{X}$. But since P is constant, $P\frac{d\tilde{X}}{dt} = \frac{d(P\tilde{X})}{dt} = \frac{dX}{dt}$. Thus $\frac{dX}{dt} = AX$. This reasoning can be reversed, so $\tilde{X}$ solves the equation with Λ if and only if X solves the equation with A. $\qquad \square$

The matrix that diagonalizes the matrix (5.3.10) was computed before (4.6.8):

$$(5.3.16) \qquad A = \begin{bmatrix} 3 & 2 \\ 1 & 4 \end{bmatrix}, \quad P = \begin{bmatrix} 1 & 2 \\ 1 & -1 \end{bmatrix}, \quad \text{and} \quad \Lambda = \begin{bmatrix} 5 & \\ & 2 \end{bmatrix}.$$

Thus

(5.3.17) $$X = \begin{bmatrix} x_1 \\ x_2 \end{bmatrix} = P\tilde{X} = \begin{bmatrix} 1 & 2 \\ 1 & -1 \end{bmatrix} \begin{bmatrix} c_1 e^{5t} \\ c_2 e^{2t} \end{bmatrix} = \begin{bmatrix} c_1 e^{5t} + 2c_2 e^{2t} \\ c_1 e^{5t} - c_2 e^{2t} \end{bmatrix}.$$

In other words, every solution is a linear combination of the two basic solutions (5.3.14).

We now consider the case that the coefficient matrix A has distinct eigenvalues, but that they are not all real. To copy the method used above, we first consider differential equations of the form (5.3.1), in which a is a complex number. Properly interpreted, the solutions of such a differential equation still have the form ce^{at}. The only thing to remember is that e^{at} will now be a complex-valued function of the real variable t.

The definition of the derivative of a complex-valued function is the same as for real-valued functions, provided that the limit (5.3.5) exists. There are no new features. We can write any such function $x(t)$ in terms of its real and imaginary parts, which will be real-valued functions, say

(5.3.18) $$x(t) = p(t) + iq(t).$$

Then x is differentiable if and only if p and q are differentiable, and if they are, the derivative of x is $p' + iq'$. This follows directly from the definition. The usual rules for differentiation, such as the product rule, hold for complex-valued functions. These rules can be proved either by applying the corresponding theorem for real functions to p and q, or by copying the proof for real functions.

The exponential of a complex number $a = r + si$ is defined to be

(5.3.19) $$e^a = e^{r+si} = e^r(\cos s + i \sin s).$$

Differentiation of this formula shows that $de^{at}/dt = ae^{at}$. Therefore ce^{at} solves the differential equation (5.3.1), and the proof given at the beginning of the section shows that these are the only solutions.

Having extended the case of one equation to complex coefficients, we can use diagonalization to solve a system of equations (5.3.7) when A is a complex matrix with distinct eigenvalues.

For example, let $A = \begin{bmatrix} 1 & 1 \\ -1 & 1 \end{bmatrix}$. The vectors $v_1 = \begin{bmatrix} 1 \\ i \end{bmatrix}$ and $v_2 = \begin{bmatrix} i \\ 1 \end{bmatrix}$ are eigenvectors, with eigenvalues $1 + i$ and $1 - i$, respectively. Let **B** denote the basis (v_1, v_2). Then A is diagonalized by the matrix $P = [\mathbf{B}]$:

(5.3.20) $$P^{-1}AP = \frac{1}{2}\begin{bmatrix} 1 & -i \\ -i & 1 \end{bmatrix}\begin{bmatrix} 1 & 1 \\ -1 & 1 \end{bmatrix}\begin{bmatrix} 1 & i \\ i & 1 \end{bmatrix} = \begin{bmatrix} 1+i & \\ & 1-i \end{bmatrix} = \Lambda.$$

Then $\tilde{X} = \begin{bmatrix} \tilde{x}_1 \\ \tilde{x}_2 \end{bmatrix} = \begin{bmatrix} c_1 e^{(1+i)t} \\ c_2 e^{(1-i)t} \end{bmatrix}$. The solutions of (5.3.7) are

(5.3.21)
$$\begin{bmatrix} x_1 \\ x_2 \end{bmatrix} = P\tilde{X} = \begin{bmatrix} c_1 e^{(1+i)t} + ic_2 e^{(1-i)t} \\ ic_1 e^{(1+i)t} + c_2 e^{(1-i)t} \end{bmatrix},$$

where c_1, c_2 are arbitrary complex numbers. So every solution is a linear combination of the two basic solutions

(5.3.22)
$$\begin{bmatrix} e^{(1+i)t} \\ ie^{(1+i)t} \end{bmatrix} \quad \text{and} \quad \begin{bmatrix} ie^{(1-i)t} \\ e^{(1-i)t} \end{bmatrix}.$$

However, these solutions aren't very satisfactory, because we began with a system of differential equations with real coefficients, and the answer we obtained is complex. When the equation is real, we will want the real solutions. We note the following lemma:

Lemma 5.3.23 Let A be a real $n \times n$ matrix, and let $X(t)$ be a complex-valued solution of the differential equation $\frac{dX}{dt} = AX$. The real and imaginary parts of $X(t)$ solve the same equation. □

Now *every* solution of the original equation (5.3.7), whether real or complex, has the form (5.3.21) for some complex numbers c_i. So the real solutions are among those we have found. To write them down explicitly, we may take the real and imaginary parts of the complex solutions.

The real and imaginary parts of the basic solutions (5.3.22) are determined using (5.3.19). They are

(5.3.24)
$$\begin{bmatrix} e^t \cos t \\ -e^t \sin t \end{bmatrix} \quad \text{and} \quad \begin{bmatrix} e^t \sin t \\ e^t \cos t \end{bmatrix}.$$

Every real solution is a real linear combination of these particular solutions.

5.4 THE MATRIX EXPONENTIAL

Systems of first-order linear, constant-coefficient differential equations can be solved formally, using the *matrix exponential*.

The exponential of an $n \times n$ real or complex matrix A is the matrix obtained by substituting A for x and I for 1 into the Taylor's series for e^x, which is

(5.4.1)
$$e^x = 1 + \frac{x}{1!} + \frac{x^2}{2!} + \frac{x^3}{3!} + \cdots,$$

Thus by definition,

$$(5.4.2) \qquad e^A = I + \frac{A}{1!} + \frac{A^2}{2!} + \frac{A^3}{3!} + \cdots .$$

We will be interested mainly in the matrix valued function e^{tA} of the variable scalar t, so we substitute tA for A:

$$(5.4.3) \qquad e^{tA} = I + \frac{tA}{1!} + \frac{t^2A^2}{2!} + \frac{t^3A^3}{3!} + \cdots .$$

Theorem 5.4.4

(a) The series (5.4.2) converges absolutely and uniformly on bounded sets of complex matrices.

(b) e^{tA} is a differentiable function of t, and its derivative is the matrix product Ae^{tA}.

(c) Let A and B be complex $n \times n$ matrices that commute: $AB = BA$. Then $e^{A+B} = e^A e^B$.

In order not to break up the discussion, we have moved the proof of this theorem to the end of the section.

The hypothesis that A and B commute is essential for carrying the fundamental property $e^{x+y} = e^x e^y$ over to matrices. Nevertheless, **(c)** is very useful.

Corollary 5.4.5 For any $n \times n$ complex matrix A, the exponential e^A is invertible, and its inverse is e^{-A}.

Proof. Because A and $-A$ commute, $e^A e^{-A} = e^{A-A} = e^0 = I$. □

Since matrix multiplication is relatively complicated, it is often not easy to write down the entries of the matrix e^A. They won't be obtained by exponentiating the entries of A unless A is a diagonal matrix. If A is diagonal, with diagonal entries $\lambda_1, \ldots, \lambda_n$, then inspection of the series shows that e^A is also diagonal, and that its diagonal entries are e^{λ_i}.

The exponential is also fairly easy to compute for a triangular 2×2 matrix. For example, if

$$A = \begin{bmatrix} 1 & 1 \\ & 2 \end{bmatrix},$$

then

$$(5.4.6) \qquad e^A = \begin{bmatrix} 1 & \\ & 1 \end{bmatrix} + \frac{1}{1!}\begin{bmatrix} 1 & 1 \\ & 2 \end{bmatrix} + \frac{1}{2!}\begin{bmatrix} 1 & 3 \\ & 4 \end{bmatrix} + \cdots = \begin{bmatrix} e & * \\ & e^2 \end{bmatrix}.$$

It is a good exercise to calculate the missing entry $*$ directly from the series.

The exponential of e^A can be determined whenever we know a matrix P such that $\Lambda = P^{-1}AP$ is diagonal. Using the rule $P^{-1}A^kP = (P^{-1}AP)^k$ (4.6.12) and the distributive law for matrix multiplication,

(5.4.7) $\qquad P^{-1}e^AP = (P^{-1}IP) + \dfrac{(P^{-1}AP)}{1!} + \dfrac{(P^{-1}AP)^2}{2!} + \cdots = e^{P^{-1}AP} = e^{\Lambda}.$

Suppose that Λ is diagonal, with diagonal entries λ_i. Then e^{Λ} is also diagonal, and its diagonal entries are e^{λ_i}. In this case we can compute e^A explicitly:

(5.4.8) $\qquad\qquad\qquad\qquad e^A = Pe^{\Lambda}P^{-1}.$

For example, if $A = \begin{bmatrix} 1 & 1 \\ & 2 \end{bmatrix}$ and $P = \begin{bmatrix} 1 & 1 \\ & 1 \end{bmatrix}$, then $P^{-1}AP = \Lambda = \begin{bmatrix} 1 & \\ & 2 \end{bmatrix}$. So

$$e^A = Pe^{\Lambda}P^{-1} = \begin{bmatrix} 1 & 1 \\ & 1 \end{bmatrix}\begin{bmatrix} e & \\ & e^2 \end{bmatrix}\begin{bmatrix} 1 & -1 \\ & 1 \end{bmatrix} = \begin{bmatrix} e & e^2-e \\ & e^2 \end{bmatrix}.$$

The next theorem relates the matrix exponential to differential equations:

Theorem 5.4.9 Let A be a real or complex $n \times n$ matrix. The columns of the matrix e^{tA} form a basis for the space of solutions of the differential equation $\frac{dX}{dt} = AX$.

Proof. Theorem 5.4.4**(b)** shows that the columns of e^{tA} solve the differential equation. To show that every solution is a linear combination of the columns, we copy the proof given at the beginning of Section 5.3. Let $X(t)$ be an arbitrary solution. We differentiate the matrix product $e^{-tA}X(t)$ using the product rule (5.3.6):

(5.4.10) $\qquad\qquad \dfrac{d}{dt}\left(e^{-tA}X(t)\right) = \left(-Ae^{-tA}\right)X(t) + e^{-tA}\left(AX(t)\right).$

Fortunately, A and e^{-tA} commute. This follows directly from the definition of the exponential. So the derivative is zero. Therefore $e^{-tA}X(t)$ is a constant column vector, say $C = (c_1, \ldots, c_n)^t$, and $X(t) = e^{tA}C$. This expresses $X(t)$ as a linear combination of the columns of e^{tA}, with coefficients c_i. The expression is unique because e^{tA} is an invertible matrix. $\qquad\qquad\square$

Though the matrix exponential always solves the differential equation (5.3.7), it may not be easy to apply in a concrete situation because computation of the exponential can be difficult. But if A is diagonalizable, the exponential can be computed as in (5.4.8). We can use this method of evaluating e^{tA} to solve equation (5.3.7). Of course we will get the same solutions as we did before. Thus if A, P, and Λ are as in (5.3.16), then

$$e^{tA} = Pe^{t\Lambda}P^{-1} = \begin{bmatrix} 1 & 2 \\ 1 & -1 \end{bmatrix}\begin{bmatrix} e^{5t} & \\ & e^{2t} \end{bmatrix}\left(-\frac{1}{3}\right)\begin{bmatrix} -1 & -2 \\ -1 & 1 \end{bmatrix} = \frac{1}{3}\begin{bmatrix} (e^{5t}+2e^{2t}) & (2e^{5t}-2e^{2t}) \\ (e^{5t}-e^{2t}) & (2e^{5t}+e^{2t}) \end{bmatrix}.$$

The columns of the matrix on the right form a second basis for the space of solutions that was obtained in (5.3.17).

One can also use Jordan form to solve the differential equation. The solutions for an arbitrary $k \times k$ Jordan block J_λ (4.7.5) can be determined by computing the matrix exponential. We write $J_\lambda = \lambda I + N$, as in (4.7.12), where N is the $k \times k$ Jordan block J_0 with $\lambda = 0$. Then $N^k = 0$, so

$$e^{tN} = I + \frac{tN}{1!} + \cdots + \frac{t^{k-1}N^{k-1}}{(k-1)!}$$

Since N and λI commute,

$$e^{tJ} = e^{\lambda t I}e^{tN} = e^{\lambda t}\left(I + \frac{tN}{1!} + \cdots + \frac{t^{k-1}N^{k-1}}{(k-1)!}\right).$$

Thus if J is the 3×3 block

$$J = \begin{bmatrix} 3 & & \\ 1 & 3 & \\ & 1 & 3 \end{bmatrix},$$

then

$$e^{tJ} = \begin{bmatrix} e^{3t} & & \\ & e^{3t} & \\ & & e^{3t} \end{bmatrix}\begin{bmatrix} 1 & & \\ t & 1 & \\ \frac{1}{2!}t^2 & t & 1 \end{bmatrix} = \begin{bmatrix} e^{3t} & & \\ te^{3t} & e^{3t} & \\ \frac{1}{2!}t^2e^{3t} & te^{3t} & e^{3t} \end{bmatrix}.$$

The columns of this matrix form a basis for the space of solutions of the differential equation $\frac{dX}{dt} = JX$.

We now go back to prove Theorem 5.4.4. The main facts about limits of series that we will use are given below, together with references to [Mattuck] and [Rudin]. Those authors consider only real valued functions, but the proofs carry over to complex valued functions because limits and derivatives of complex valued functions can be defined by working on the real and imaginary parts separately.

If r and s are real numbers with $r < s$, the notation $[r, s]$ stands for the interval $r \le t \le s$.

Theorem 5.4.11 ([Mattuck], Theorem 22.2B, [Rudin], Theorem 7.9). Let m_k be a series of positive real numbers such that $\sum m_k$ converges. If $u^{(k)}(t)$ are functions on an interval $[r, s]$, and if $|u^{(k)}(t)| \le m_k$ for all k and all t in the interval, then the series $\sum u^{(k)}(t)$ converges uniformly on the interval.   □

Theorem 5.4.12 ([Mattuck], Theorem 11.5B, [Rudin], Theorem 7.17). Let $u^{(k)}(t)$ be a sequence of functions with continuous derivatives on an interval $[r, s]$. Suppose that the series $\sum u^{(k)}(t)$ converges to a function $f(t)$ and also that the series of derivatives $\sum u'^{(k)}(t)$ converges uniformly to a function $g(t)$, on the interval. Then f is differentiable on the interval, and its derivative is g.   □

Proof of Theorem 5.4.4(a). We denote the i, j-entry of a matrix A by $(A)_{ij}$ here. So $(AB)_{ij}$ stands for the entry of the product matrix AB, and $(A^k)_{ij}$ for the entry of the kth power A^k. With this notation, the i, j-entry of e^A is the sum of the series

$$(5.4.13) \qquad (e^A)_{ij} = (I)_{ij} + \frac{(A)_{ij}}{1!} + \frac{(A^2)_{ij}}{2!} + \frac{(A^3)_{ij}}{3!} + \cdots .$$

To prove that the series for the exponential converges absolutely and uniformly, we need to show that the entries of the powers A^k do not grow too quickly.

We denote by $\|A\|$ the maximum absolute value of the entries of a matrix A, the smallest real number such that

$$(5.4.14) \qquad |(A)_{ij}| \le \|A\| \quad \text{for all } i, j.$$

Its basic property is this:

Lemma 5.4.15 Let A and B be complex $n \times n$ matrices. Then $\|AB\| \le n\|A\| \|B\|$, and for all $k > 0$, $\|A^k\| \le n^{k-1}\|A\|^k$.

Proof. We estimate the size of the i, j-entry of AB:

$$|(AB)_{ij}| = \left| \sum_{v=1}^{n} (A)_{iv}(B)_{vj} \right| \le \sum_{v=1}^{n} |(A)_{iv}||(B)_{vj}| \le n\|A\| \|B\|.$$

The second inequality follows by induction from the first one. $\qquad \square$

We now estimate the exponential series: Let a be a positive real number such that $n\|A\| \le a$. The lemma tells us that $|(A^k)_{ij}| \le a^k$ (with one n to spare). So

$$(5.4.16) \qquad \left|(e^A)_{ij}\right| \le |(I)_{ij}| + |(A)_{ij}| + \frac{1}{2!}\left|(A^2)_{ij}\right| + \frac{1}{3!}\left|(A^3)_{ij}\right| + \cdots$$

$$\le 1 + \frac{a}{1!} + \frac{a^2}{2!} + \frac{a^3}{3!} + \cdots .$$

The ratio test shows that the last series converges (to e^a of course). Theorem 5.4.11 shows that the series for e^A converges absolutely and uniformly for all A with $n\|A\| \le a$. $\qquad \square$

Proof of Theorem 5.4.4(b),(c). We use a trick to shorten the proofs. That is to begin by differentiating the series for e^{tA+B}, assuming that A and B are commuting $n \times n$ matrices. The derivative of $tA + B$ is A, and

$$(5.4.17) \qquad e^{tA+B} = I + \frac{(tA + B)}{1!} + \frac{(tA + B)^2}{2!} + \cdots .$$

Using the product rule (5.3.6), we see that, for $k > 0$, the derivative of the term of degree k of this series is

$$\frac{d}{dt}\left(\frac{(tA + B)^k}{k!}\right) = \left(\frac{1}{k!}\sum_{i=1}^{k}(tA + B)^{i-1} A (tA + B)^{k-i}\right).$$

Since $AB = BA$, we can pull the A in the middle out to the left:

(5.4.18)
$$\frac{d}{dt}\left(\frac{(tA+B)^k}{k!}\right) = kA\frac{(tA+B)^{k-1}}{k!} = A\frac{(tA+B)^{k-1}}{(k-1)!}.$$

This is the product of the matrix A and the term of degree $k-1$ of the exponential series. So term-by-term differentiation of (5.4.17) yields the series for Ae^{tA+B}.

To justify term-by-term differentiation, we apply Theorem 5.4.4(a). The theorem shows that for given A and B, the exponential series e^{tA+B} converges uniformly on any interval $r \le t \le s$. Moreover, the series of derivatives converges uniformly to Ae^{tA+B}. By Theorem 5.4.12, the derivative of e^{tA+B} can be computed term by term, so it is true that

$$\frac{d}{dt}e^{tA+B} = Ae^{tA+B}$$

for any pair A, B of matrices that commute. Taking $B = 0$ proves Theorem 5.4.4(b).

Next, we copy the method used in the proof of Theorem 5.4.9. We differentiate the product $e^{-tA}e^{tA+B}$, again assuming that A and B commute. As in (5.4.10), we find that

$$\frac{d}{dt}\left(e^{-tA}e^{tA+B}\right) = \left(-Ae^{-tA}\right)\left(e^{tA+B}\right) + \left(e^{-tA}\right)\left(Ae^{tA+B}\right) = 0.$$

Therefore $e^{-tA}e^{tA+B} = C$, where C is a constant matrix. Setting $t = 0$ shows that $e^B = C$. Setting $B = 0$ shows that $e^{-tA} = (e^{tA})^{-1}$. Then $(e^{tA})^{-1}e^{tA+B} = e^B$. Setting $t = 1$ shows that $e^{A+B} = e^A e^B$. This proves Theorem 5.4.4(c).   □

We will use the remarkable properties of the matrix exponential again, in Chapter 9.

I have not thought it necessary to undertake the labour of a formal proof of the theorem in the general case.

—Arthur Cayley[1]

EXERCISES

Section 1   Orthogonal Matrices and Rotations

1.1. Determine the matrices that represent the following rotations of $\mathbb{R}^3$:
(a) angle θ, the axis e_2, (b) angle $2\pi/3$, axis contains the vector $(1, 1, 1)^t$, (c) angle $\pi/2$, axis contains the vector $(1, 1, 0)^t$.

1.2. What are the complex eigenvalues of the matrix A that represents a rotation of $\mathbb{R}^3$ through the angle θ about a pole u?

1.3. Is O_n isomorphic to the product group $SO_n \times \{\pm I\}$?

1.4. Describe geometrically the action of an orthogonal 3×3 matrix with determinant -1.

---

[1] Arthur Cayley, one of the mathematicians for whom the Cayley-Hamilton Theorem is named, stated that theorem for $n \times n$ matrices in one of his papers, and then checked the 2×2 case (see (5.2.4)). He closed his discussion of the theorem with the sentence quoted here.

1.5. Let A be a 3×3 orthogonal matrix with $\det A = 1$, whose angle of rotation is different from 0 or π, and let $M = A - A^t$.

(a) Show that M has rank 2, and that a nonzero vector X in the nullspace of M is an eigenvector of A with eigenvalue 1.

(b) Find such an eigenvector explicitly in terms of the entries of the matrix A.

Section 2   Using Continuity

2.1. Use the Cayley-Hamilton Theorem to express A^{-1} in terms of A, $(\det A)^{-1}$, and the coefficients of the characteristic polynomial. Verify your expression in the 2×2 case.

2.2. Let A be $m \times m$ and B be $n \times n$ complex matrices, and consider the linear operator T on the space $\mathbb{C}^{m \times n}$ of all complex matrices defined by $T(M) = AMB$.

(a) Show how to construct an eigenvector for T out of a pair of column vectors X, Y, where X is an eigenvector for A and Y is an eigenvector for B^t.

(b) Determine the eigenvalues of T in terms of those of A and B.

(c) Determine the trace of this operator.

2.3. Let A be an $n \times n$ complex matrix.

(a) Consider the linear operator T defined on the space $\mathbb{C}^{n \times n}$ of all complex $n \times n$ matrices by the rule $T(M) = AM - MA$. Prove that the rank of this operator is at most $n^2 - n$.

(b) Determine the eigenvalues of T in terms of the eigenvalues $\lambda_1, \ldots, \lambda_n$ of A.

2.4. Let A and B be diagonalizable complex matrices. Prove that there is an invertible matrix P such that $P^{-1}AP$ and $P^{-1}BP$ are both diagonal if and only if $AB = BA$.

Section 3   Systems of Differential Equations

3.1. Prove the product rule for differentiation of matrix-valued functions.

3.2. Let $A(t)$ and $B(t)$ be differentiable matrix-valued functions of t. Compute

(a) $\dfrac{d}{dt}(A(t)^3)$,   **(b)** $\dfrac{d}{dt}(A(t)^{-1})$,   **(c)** $\dfrac{d}{dt}(A(t)^{-1}B(t))$.

3.3. Solve the equation $\frac{dX}{dt} = AX$ for the following matrices A:

(a) $\begin{bmatrix} 2 & 1 \\ 1 & 2 \end{bmatrix}$, **(b)** $\begin{bmatrix} 1 & i \\ -i & 1 \end{bmatrix}$, **(c)** $\begin{bmatrix} 1 & 2 & 3 \\ 0 & 0 & 4 \\ 0 & 0 & -1 \end{bmatrix}$, **(d)** $\begin{bmatrix} 0 & 0 & 1 \\ 1 & 0 & 0 \\ 0 & 1 & 0 \end{bmatrix}$.

3.4. Let A and B be constant matrices, with A invertible. Solve the inhomogeneous differential equation $\dfrac{dX}{dt} = AX + B$ in terms of the solutions to the equation $\dfrac{dX}{dt} = AX$.

Section 4   The Matrix Exponential

4.1. Compute e^A for the following matrices A:

(a) $\begin{bmatrix} a & b \\ & \end{bmatrix}$, **(b)** $\begin{bmatrix} 2\pi i & 2\pi i \\ & 2\pi i \end{bmatrix}$, **(c)** $\begin{bmatrix} 0 & -b \\ b & 0 \end{bmatrix}$, **(d)** $\begin{bmatrix} 1 & 0 \\ 1 & 1 \end{bmatrix}$, **(e)** $\begin{bmatrix} 0 & & \\ 1 & 0 & \\ & 1 & 0 \end{bmatrix}$.

4.2. Prove the formula $e^{\text{trace } A} = \det(e^A)$.

4.3. Let X be an eigenvector of an $n \times n$ matrix A, with eigenvalue λ.

(a) Prove that if A is invertible then X is an eigenvector for A^{-1}, with eigenvalue λ^{-1}.

(b) Prove that X is an eigenvector for e^A, with eigenvalue e^λ.

4.4. Let A and B be commuting matrices. To prove that $e^{A+B} = e^A e^B$, one can begin by expanding the two sides into double sums whose terms are multiples of $A^i B^j$. Prove that the two double sums one obtains are the same.

4.5. Solve the differential equation $\dfrac{dX}{dt} = AX$ when A is the given matrix:

(a) $\begin{bmatrix} 2 & \\ 1 & 2 \end{bmatrix}$, **(b)** $\begin{bmatrix} 0 & 0 \\ 1 & 0 \end{bmatrix}$, **(c)** $\begin{bmatrix} 1 & & \\ 1 & 1 & \\ & 1 & 1 \end{bmatrix}$.

4.6. For an $n \times n$ matrix A, define $\sin A$ and $\cos A$ by using the Taylor's series expansions for $\sin x$ and $\cos x$.

(a) Prove that these series converge for all A.

(b) Prove that $\sin(tA)$ is a differentiable function of t and that $\frac{d}{dt} \sin(tA) = A \cos(tA)$.

4.7. Discuss the range of validity of the following identities:

(a) $\cos^2 A + \sin^2 A = I$,

(b) $e^{iA} = \cos A + i \sin A$,

(c) $\sin(A + B) = \sin A \cos B + \cos A \sin B$,

(d) $e^{2\pi i A} = I$,

(e) $\dfrac{d(e^{A(t)})}{dt} = e^{A(t)} \dfrac{dA}{dt}$, when $A(t)$ is a differentiable matrix-valued function of t.

4.8. Let P, B_k, and B be $n \times n$ matrices, with P invertible. Prove that if B_k converges to B, then $P^{-1} B_k P$ converges to $P^{-1} B P$.

Miscellaneous Problems

M.1. Determine the group $O_n(\mathbb{Z})$ of orthogonal matrices with integer entries.

M.2. Prove the Cayley-Hamilton Theorem using Jordan form.

M.3. Let A be an $n \times n$ complex matrix. Prove that if trace $A^k = 0$ for all $k > 0$, then A is nilpotent.

M.4. Let A be a complex $n \times n$ matrix all of whose eigenvalues have absolute value less than 1. Prove that the series $I + A + A^2 + \cdots$ converges to $(I - A)^{-1}$.

M.5. The *Fibonacci numbers* $0, 1, 1, 2, 3, 5, 8, \ldots$, are defined by the recursive relations $f_n = f_{n-1} + f_{n-2}$, with the initial conditions $f_0 = 0$, $f_1 = 1$. This recursive relation can be written in matrix form as $\begin{bmatrix} 0 & 1 \\ 1 & 1 \end{bmatrix} \begin{bmatrix} f_{n-2} \\ f_{n-1} \end{bmatrix} = \begin{bmatrix} f_{n-1} \\ f_n \end{bmatrix}$.

(a) Prove the formula $f_n = \dfrac{1}{\alpha}\left[\left(\dfrac{1+\alpha}{2}\right) - \left(\dfrac{1-\alpha}{2}\right)^2\right]$, where $\alpha = \sqrt{5}$.

(b) Suppose that a sequence a_n is defined by the relation $a_n = \frac{1}{2}(a_{n-1} + a_{n-2})$. Compute the limit of the sequence a_n in terms of a_0, a_1.

M.6. (*an integral operator*) The space $\mathcal{C}$ of continuous functions $f(u)$ on the interval $[0, 1]$ is one of many infinite-dimensional analogues of $\mathbb{R}^n$, and continuous functions $A(u, v)$ on the square $0 \le u, v \le 1$ are infinite-dimensional analogues of matrices. The integral

$$A \cdot f = \int_0^1 A(u, v) f(v) dv$$

is analogous to multiplication of a matrix and a vector. (To visualize this, rotate the unit square in the u, v-plane and the interval $[0, 1]$ by $90°$ in the clockwise direction.) The response of a bridge to a variable load could, with suitable assumptions, be represented by such an integral. For this, f would represent the load along the bridge, and then $A \cdot f$ would compute the vertical deflection of the bridge caused by that load.

This problem treats the integral as a linear operator. For the function $A = u + v$, determine the image of the operator explicitly. Determine its nonzero eigenvalues, and describe its kernel in terms of the vanishing of some integrals. Do the same for the function $A = u^2 + v^2$.

M.7. Let A be a 2×2 complex matrix with distinct eigenvalues, and let X be an indeterminate 2×2 matrix. How many solutions to the matrix equation $X^2 = A$ can there be?

M.8. Find a geometric way to determine the axis of rotation for the composition of two three-dimensional rotations.

CHAPTER 6

Symmetry

*L'algébre n'est qu'une géométrie écrite;
la géométrie n'est qu'une algébre figurée.*

—Sophie Germain

Symmetry provides some of the most appealing applications of groups. Groups were invented to analyze symmetries of certain algebraic structures, field extensions (Chapter 16), and because symmetry is a common phenomenon, it is one of the two main ways in which group theory is applied. The other is through group representations, which are discussed in Chapter 10. The symmetries of plane figures, which we study in the first sections, provide a rich source of examples and a background for the general concept of a group operation that is introduced in Section 6.7.

We allow free use of geometric reasoning. Carrying the arguments back to the axioms of geometry will be left for another occasion.

6.1 SYMMETRY OF PLANE FIGURES

Symmetries of plane figures are usually classified into the types shown below:

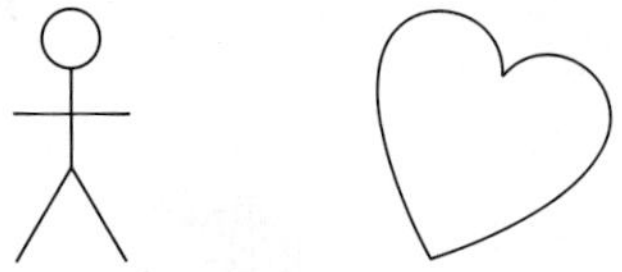

(6.1.1)   Bilateral Symmetry.

(6.1.2)   Rotational Symmetry.

(6.1.3)   Translational Symmetry.

Figures such as these are supposed to extend indefinitely in both directions. There is also a fourth type of symmetry, though its name, *glide symmetry*, may be less familiar:

(6.1.4)   Glide Symmetry.

Figures such as the wallpaper pattern shown below may have two independent translational symmetries,

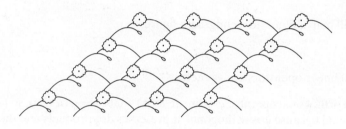

(6.1.5)

and other combinations of symmetries may occur. The star has bilateral as well as rotational symmetry. In the figure below, translational and rotational symmetry are combined:

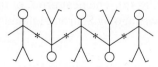

(6.1.6)

Another example:

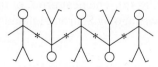

(6.1.7)

A rigid motion of the plane is called an isometry, and if an isometry carries a subset F of the plane to itself, it is called a *symmetry* of F. The set of all symmetries of F forms a subgroup of the group of all isometries of the plane: If m and m' carry F to F, then so does the composed map mm', and so on. This is the *group of symmetries* of F.

Figure 6.1.3 has infinite cyclic groups of symmetry that are generated by the translation t that carries the figure one unit to the left.

$$G = \{\ldots, t^{-2}, t^{-1}, 1, t, t^2, \ldots\}.$$

Figure 6.1.7 has symmetries in addition to translations.

6.2 ISOMETRIES

The distance between points of $\mathbb{R}^n$ is the length $|u - v|$ of the vector $u - v$. An *isometry* of n-dimensional space $\mathbb{R}^n$ is a distance-preserving map f from $\mathbb{R}^n$ to itself, a map such that, for all u and v in $\mathbb{R}^n$,

(6.2.1) $|f(u) - f(v)| = |u - v|.$

An isometry will map a figure to a congruent figure.

Examples 6.2.2

(a) Orthogonal linear operators are isometries.

Because an orthogonal operator φ is linear, $\varphi(u) - \varphi(v) = \varphi(u - v)$, so $|\varphi(u) - \varphi(v)| = |\varphi(u - v)|$, and because φ is orthogonal, it preserves dot products and therefore lengths, so $|\varphi(u - v)| = |u - v|$.

(b) *Translation t_a* by a vector a, the map defined by $t_a(x) = x + a$, is an isometry.

Translations are not linear operators because they don't send 0 to 0, except of course for translation by the zero vector, which is the identity map.

(c) The composition of isometries is an isometry.   □

Theorem 6.2.3 The following conditions on a map $\varphi : \mathbb{R}^n \to \mathbb{R}^n$ are equivalent:

(a) φ is an isometry that fixes the origin: $\varphi(0) = 0$,

(b) φ preserves dot products: $(\varphi(v) \cdot \varphi(w)) = (v \cdot w)$, for all v and w,

(c) φ is an orthogonal linear operator.

We have seen that **(c)** implies **(a)**. The neat proof of the implication **(b)** $\Rightarrow$ **(c)** that we present next was found a few years ago by Sharon Hollander, when she was a student in an MIT algebra class.

Lemma 6.2.4 Let x and y be points of $\mathbb{R}^n$. If the three dot products $(x \cdot x)$, $(x \cdot y)$, and $(y \cdot y)$ are equal, then $x = y$.

Proof. Suppose that $(x \cdot x) = (x \cdot y) = (y \cdot y)$. Then

$$((x - y) \cdot (x - y)) = (x \cdot x) - 2(x \cdot y) + (y \cdot y) = 0.$$

The length of $x - y$ is zero, and therefore $x = y$. $\qquad\square$

Proof of Theorem 6.2.3, (**b**) $\Rightarrow$ (**c**): Let φ be a map that preserves dot product. Then it will be orthogonal, provided that it is a linear operator (5.1.12). To prove that φ is a linear operator, we must show that $\varphi(u + v) = \varphi(u) + \varphi(v)$ and that $\varphi(cv) = c\varphi(v)$, for all u and v and all scalars c.

Given x in $\mathbb{R}^n$, we'll use the symbol x' to stand for $\varphi(x)$. We also introduce the symbol w for the sum, writing $w = u + v$. Then the relation $\varphi(u + v) = \varphi(u) + \varphi(v)$ that is to be shown becomes $w' = u' + v'$.

We substitute $x = w'$ and $y = u' + v'$ into Lemma 6.2.4. To show that $w' = u' + v'$, it suffices to show that the three dot products

$$(w' \cdot w'), \quad (w' \cdot (u' + v')), \quad \text{and} \quad ((u' + v') \cdot (u' + v'))$$

are equal. We expand the second and third dot products. It suffices to show that

$$(w' \cdot w') \;=\; (w' \cdot u') + (w' \cdot v') \;=\; (u' \cdot u') + 2(u' \cdot v') + (v' \cdot v').$$

By hypothesis, φ preserves dot products. So we may drop the primes: $(w' \cdot w') = (w \cdot w)$, etc. Then it suffices to show that

(6.2.5) $\qquad (w \cdot w) \;=\; (w \cdot u) + (w \cdot v) \;=\; (u \cdot u) + 2(u \cdot v) + (v \cdot v).$

Now whereas $w' = u' + v'$ is to be shown, $w = u + v$ is true by definition. So we may substitute $u + v$ for w. Then (6.2.5) becomes true.

To prove that $\varphi(cv) = c\varphi(v)$, we write $u = cv$, and we must show that $u' = cv'$. The proof is analogous to the one we have just given. $\qquad\square$

Proof of Theorem 6.2.3, (**a**) $\Rightarrow$ (**b**): Let φ be an isometry that fixes the origin. With the prime notation, the distance-preserving property of φ reads

(6.2.6) $\qquad ((u' - v') \cdot (u' - v')) = ((u - v) \cdot (u - v)),$

for all u and v in $\mathbb{R}^n$. We substitute $v = 0$. Since $0' = 0$, $(u' \cdot u') = (u \cdot u)$. Similarly, $(v' \cdot v') = (v \cdot v)$. Now (**b**) follows when we expand (6.2.6) and cancel $(u \cdot u)$ and $(v \cdot v)$ from the two sides of the equation. $\qquad\square$

Corollary 6.2.7 Every isometry f of $\mathbb{R}^n$ is the composition of an orthogonal linear operator and a translation. More precisely, if f is an isometry and if $f(0) = a$, then $f = t_a\varphi$, where t_a is a translation and φ is an orthogonal linear operator. This expression for f is unique.

Proof. Let f be an isometry, let $a = f(0)$, and let $\varphi = t_{-a}f$. Then $t_a\varphi = f$. The corollary amounts to the assertion that φ is an orthogonal linear operator. Since φ is the composition of the isometries t_{-a} and f, it is an isometry. Also, $\varphi(0) = t_{-a}f(0) = t_{-a}(a) = 0$, so φ fixes the origin. Theorem 6.2.3 shows that φ is an orthogonal linear operator. The expression $f = t_a\varphi$ is unique because, since $\varphi(0) = 0$, we must have $a = f(0)$, and then $\varphi = t_{-a}f$. □

To work with the expressions $t_a\varphi$ for isometries, we need to determine the product (the composition) of two such expressions. We know that the composition $\varphi\psi$ of orthogonal operators is an orthogonal operator. The other rules are:

$$(6.2.8) \qquad t_a t_b = t_{a+b} \quad \text{and} \quad \varphi t_a = t_{a'}\varphi, \quad \text{where} \ \ a' = \varphi(a).$$

We verify the last relation: $\varphi t_a(x) = \varphi(x + a) = \varphi(x) + \varphi(a) = \varphi(x) + a' = t_{a'}\varphi(x)$.

Corollary 6.2.9 The set of all isometries of $\mathbb{R}^n$ forms a group that we denote by M_n, with composition of functions as its law of composition.

Proof. The composition of isometries is an isometry, and the inverse of an isometry is an isometry too, because orthogonal operators and translations are invertible, and if $f = t_a\varphi$, then $f^{-1} = \varphi^{-1}t_a^{-1} = \varphi^{-1}t_{-a}$. This is a composition of isometries. □

Note: It isn't very easy to verify, directly from the definition, that an isometry is invertible.

The Homomorphism $M_n \to O_n$

There is an important map $\pi : M_n \to O_n$, defined by dropping the translation part of an isometry f. We write f (uniquely) in the form $f = t_a\varphi$, and define $\pi(f) = \varphi$.

Proposition 6.2.10 The map π is a surjective homomorphism. Its kernel is the set $T = \{t_v\}$ of translations, which is a normal subgroup of M_n.

Proof. It is obvious that π is surjective, and once we show that π is a homomorphism, it will be obvious that T is its kernel, hence that T is a normal subgroup. We must show that if f and g are isometries, then $\pi(fg) = \pi(f)\pi(g)$. Say that $f = t_a\varphi$ and $g = t_b\psi$, so that $\pi(f) = \varphi$ and $\pi(g) = \psi$. Then $\varphi t_b = t_{b'}\varphi$, where $b' = \varphi(b)$ and $fg = t_a\varphi t_b\psi = t_{a+b'}\varphi\psi$. So $\pi(fg) = \varphi\psi = \pi(f)\pi(g)$. □

Change of Coordinates

Let P denote an n-dimensional space. The formula $t_a\varphi$ for an isometry depends on our choice of coordinates, so let's ask how the formula changes when coordinates are changed. We will allow changes by orthogonal matrices and also shifts of the origin by translations. In other words, we may change coordinates by any isometry.

To analyze the effect of such a change, we begin with an isometry f, a point p of P, and its image $q = f(p)$, without reference to coordinates. When we introduce our coordinate system, the space P becomes identified with $\mathbb{R}^n$, and the points p and q have coordinates, say $x = (x_1, \ldots, x_n)^t$ and $y = (y_1, \ldots, y_n)^t$. Also, the isometry f will have a formula $t_a\varphi$ in terms of the coordinates; let's call that formula m. The equation $q = f(p)$ translates to

$y = m(x) \ (= t_a\varphi(x))$. We want to determine what happens to the coordinate vectors and to the formula, when we change coordinates. The analogous computation for change of basis in a linear operator gives the clue: m will be changed by conjugation.

Our change in coordinates will be given by some isometry, let's denote it by η (eta). Let the new coordinate vectors of p and q be x' and y'. The new formula m' for f is the one such that $m'(x') = y'$. We also have the formula $\eta(x') = x$ analogous to the change of basis formula $PX' = X$ (3.5.11).

We substitute $\eta(x') = x$ and $\eta(y') = y$ into the equation $m(x) = y$, obtaining $m\eta(x') = \eta(y')$, or $\eta^{-1}m\eta(x') = y'$. The new formula is the conjugate, as expected:

$$(6.2.11) \qquad\qquad m' = \eta^{-1}m\eta.$$

Corollary 6.2.12 The homomorphism $\pi : M_n \rightarrow O_n$ (6.2.10) does not change when the origin is shifted by a translation.

When the origin is shifted by a translation $t_v = \eta$, (6.2.11) reads $m' = t_{-v}mt_v$. Since translations are in the kernel of π and since π is a homomorphism, $\pi(m') = \pi(m)$. □

Orientation

The determinant of an orthogonal operator φ on $\mathbb{R}^n$ is ± 1. The operator is said to be *orientation-preserving* if its determinant is 1 and *orientation-reversing* if its determinant is –1. Similarly, an orientation-preserving (or orientation-reversing) isometry f is one such that, when it is written in the form $f = t_a\varphi$, the operator φ is orientation-preserving (or orientation-reversing). An isometry of the plane is orientation-reversing if it interchanges front and back of the plane, and orientation-preserving if it maps the front to the front.

The map

$$(6.2.13) \qquad\qquad \sigma : M_n \rightarrow \{\pm 1\}$$

that sends an orientation-preserving isometry to 1 and an orientation-reversing isometry to –1 is a group homomorphism.

6.3 ISOMETRIES OF THE PLANE

In this section we describe isometries of the plane, both algebraically and geometrically.

We denote the group of isometries of the plane by M. To compute in this group, we choose some special isometries as generators, and we obtain relations among them. The relations are somewhat analogous to those that define the symmetric group S_3, but because M is infinite, there are more of them.

We choose a coordinate system and use it to identify the plane P with the space $\mathbb{R}^2$. Then we choose as generators the translations, the rotations about the origin, and the reflection about the e_1-axis. We denote the rotation through the angle θ by ρ_θ, and the reflection about the e_1-axis by r. These are linear operators whose matrices R and S_0 were exhibited before (see (5.1.17) and (5.1.16)).

(6.3.1)

1. *translation t_a by a vector a:* $t_a(x) = x + a = \begin{bmatrix} x_1 \\ x_2 \end{bmatrix} + \begin{bmatrix} a_1 \\ a_2 \end{bmatrix}$.

2. *rotation ρ_θ by an angle θ about the origin:* $\rho_\theta(x) = \begin{bmatrix} \cos\theta & -\sin\theta \\ \sin\theta & \cos\theta \end{bmatrix} \begin{bmatrix} x_1 \\ x_2 \end{bmatrix}$.

3. *reflection r about the e_1-axis:* $r(x) = \begin{bmatrix} 1 & 0 \\ 0 & -1 \end{bmatrix} \begin{bmatrix} x_1 \\ x_2 \end{bmatrix}$.

We haven't listed all of the isometries. Rotations about a point other than the origin aren't included, nor are reflections about other lines, or glides. However, every element of M is a product of these isometries, so they generate the group.

Theorem 6.3.2 Let m be an isometry of the plane. Then $m = t_v \rho_\theta$, or else $m = t_v \rho_\theta r$, for a uniquely determined vector v and angle θ, possibly zero.

Proof. Corollary 6.2.7 asserts that any isometry m is written uniquely in the form $m = t_v \varphi$ where φ is an orthogonal operator. And the orthogonal linear operators on $\mathbb{R}^2$ are the rotations ρ_θ about the origin and the reflections about lines through the origin. The reflections have the form $\rho_\theta r$ (see (5.1.17)). $\qquad\square$

An isometry of the form $t_v \rho_\theta$ preserves orientation while $t_v \rho_\theta r$ reverses orientation.

Computation in M can be done with the symbols t_v, ρ_θ, and r, using the following rules for composing them. The rules can be verified using Formulas 6.3.1 (see also (6.2.8)).

$$\rho_\theta t_v = t_{v'} \rho_\theta, \quad \text{where } v' = \rho_\theta(v),$$

(6.3.3) $\quad r t_v = t_{v'} r, \quad \text{where } v' = r(v),$

$$r\rho_\theta = \rho_{-\theta} r.$$

$$t_v t_w = t_{v+w}, \quad \rho_\theta \rho_\eta = \rho_{\theta+\eta}, \quad \text{and} \quad rr = 1.$$

The next theorem describes the isometries of the plane geometrically.

Theorem 6.3.4 Every isometry of the plane has one of the following forms:

(a) *orientation-preserving isometries:*

  (i) *translation*: a map t_v that sends $p \rightsquigarrow p + v$.

  (ii) *rotation*: rotation of the plane through a nonzero angle θ about some point.

(b) *orientation-reversing isometries:*

  (i) *reflection*: a bilateral symmetry about a line ℓ.

  (ii) *glide reflection* (or *glide* for short): reflection about a line ℓ, followed by translation by a nonzero vector parallel to ℓ.

The proof of this remarkable theorem is below. One of its consequences is that the composition of rotations about two different points is a rotation about a third point, unless it is a translation. This isn't obvious, but it follows from the theorem, because the composition preserves orientation.

Some compositions are easier to visualize. The composition of rotations through angles α and β about the same point is a rotation about that point, through the angle $\alpha + \beta$. The composition of translations by the vectors a and b is the translation by their sum $a + b$.

The composition of reflections about nonparallel lines ℓ_1, ℓ_2 is a rotation about the intersection point $p = \ell_1 \cap \ell_2$. This also follows from the theorem, because the composition is orientation-preserving, and it fixes p. The composition of reflections about parallel lines is a translation by a vector orthogonal to the lines.

Proof of Theorem (6.3.4). We consider orientation-preserving isometries first. Let f be an isometry that preserves orientation but is not a translation. We must prove that f is a rotation about some point. We choose coordinates to write the formula for f as $m = t_a\rho_\theta$ as in (6.3.3). Since m is not a translation, $\theta \neq 0$.

Lemma 6.3.5 An isometry f that has the form $m = t_a\rho_\theta$, with $\theta \neq 0$, is a rotation through the angle θ about a point in the plane.

Proof. To simplify notation, we denote ρ_θ by ρ. To show that f represents a rotation with angle θ about some point p, we change coordinates by a translation t_p. We hope to choose p so that the new formula for the isometry f becomes $m' = \rho$. If so, then f will be rotation with angle θ about the point p.

The rule for change of coordinates is $t_p(x') = x$, and therefore the new formula for f is $m' = t_p^{-1}mt_p = t_{-p}t_a\rho t_p$ (6.2.11). We use the rules (6.3.3): $\rho t_p = t_{p'}\rho$, where $p' = \rho(p)$. Then if $b = -p + a + p' = a + \rho(p) - p$, we will have $m' = t_b\rho$. We wish to choose p such that $b = 0$.

Let I denote the identity operator, and let $c = \cos\theta$ and $s = \sin\theta$. The matrix of the linear operator $I - \rho$ is

(6.3.6)
$$\begin{bmatrix} 1\text{-}c & s \\ -s & 1\text{-}c \end{bmatrix}.$$

Its determinant is $2 - 2c = 2 - 2\cos\theta$. The determinant isn't zero unless $\cos\theta = 1$, and this happens only when $\theta = 0$. Since $\theta \neq 0$, the equation $(I - \rho)p = a$ has a unique solution for p. The equation can be solved explicitly when needed. $\square$

The point p is the fixed point of the isometry $t_a\rho_\theta$, and it can be found geometrically, as illustrated below. The line ℓ passes through the origin and is perpendicular to the vector a. The sector with angle θ is situated so as to be bisected by ℓ, and the fixed point p is determined by inserting the vector a into the sector, as shown.

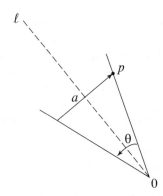

(6.3.7)   The fixed point of the isometry $t_a \rho_\theta$.

To complete the proof of Theorem 6.3.4, we show that an orientation-reversing isometry $m = t_a \rho_\theta r$ is a glide or a reflection. To do this, we change coordinates. The isometry $\rho_\theta r$ is a reflection about a line ℓ_0 through the origin. We may as well rotate coordinates so that ℓ_0 becomes the horizontal axis. In the new coordinate system, the reflection becomes our standard reflection r, and the translation t_a remains a translation, though the coordinates of the vector a will have changed. Let's use the same symbol a for this new vector. In the new coordinate system, the isometry becomes $m = t_a r$. It acts as

$$m \begin{bmatrix} x_1 \\ x_2 \end{bmatrix} = t_a \begin{bmatrix} x_1 \\ -x_2 \end{bmatrix} = \begin{bmatrix} x_1 + a_1 \\ -x_2 + a_2 \end{bmatrix}.$$

This isometry is the glide obtained by reflection about the line $\ell : \{x_2 = \frac{1}{2}a_2\}$, followed by translation by the vector $a_1 e_1$. If $a_1 = 0$, m is a reflection.

This completes the proof of Theorem 6.3.4. □

Corollary 6.3.8 The glide line of the isometry $t_a \rho_\theta r$ is parallel to the line of reflection of $\rho_\theta r$. □

The isometries that fix the origin are the orthogonal linear operators, so when coordinates are chosen, the orthogonal group O_2 becomes a subgroup of the group of isometries M. We may also consider the subgroup of M of isometries that fix a point of the plane other than the origin. The relationship of this group with the orthogonal group is given in the next proposition.

Proposition 6.3.9 Assume that coordinates in the plane have been chosen, so that the orthogonal group O_2 becomes the subgroup of M of isometries that fix the origin. Then the group of isometries that fix a point p of the plane is the conjugate subgroup $t_p O_2 t_p^{-1}$.

Proof. If an isometry m fixes p, then $t_p^{-1} m t_p$ fixes the origin: $t_p^{-1} m t_p o = t_p^{-1} m p = t_p^{-1} p = o$. Conversely, if m fixes o, then $t_p m t_p^{-1}$ fixes p. □

One can visualize the rotation about a point p this way: First translate by t_{-p} to move p to the origin, then rotate about the origin, then translate back to p.

We go back to the homomorphism $\pi : M \rightarrow O_2$ that was defined in (6.2.10). The discussion above shows this:

Proposition 6.3.10 Let p be a point of the plane, and let $\rho_{\theta,p}$ denote rotation through the angle θ about p. Then $\pi(\rho_{\theta,p}) = \rho_\theta$. Similarly, if r_ℓ is reflection about a line ℓ or a glide with glide line ℓ that is parallel to the x-axis, then $\pi(r_\ell) = r$. $\square$

Points and Vectors

In most of this book, there is no convincing reason to distinguish a point p of the plane $P = \mathbb{R}^2$ from the vector that goes from the origin o to p, which is often written as $\overrightarrow{op}$ in calculus books. However, when working with isometries, it is best to maintain the distinction. So we introduce another copy of the plane, we call it V, and we think of its elements as *translation vectors*. Translation by a vector v in V acts on a point p of P as $t_v(p) = p + v$. It shifts every point of the plane by v.

Both V and P are planes. The difference between them becomes apparent only when we change coordinates. Suppose that we shift coordinates in P by a translation: $\eta = t_w$. The rule for changing coordinates is $\eta(p') = p$, or $p' + w = p$. At the same time, an isometry m changes to $m' = \eta^{-1}m\eta = t_{-w}mt_w$ (6.2.11). If we apply this rule with $m = t_v$, then $m' = t_{-w}t_vt_w = t_v$. The points of P get new coordinates, but the translation vectors are unchanged.

On the other hand, if we change coordinates by an orthogonal operator φ, then $\varphi(p') = p$, and if $m = t_v$, then $m' = \varphi^{-1}t_v\varphi = t_{v'}$, where $v' = \varphi^{-1}v$. So $\varphi v' = v$. The effect of change of coordinates by an orthogonal operator is the same on P as on V.

The only difference between P and V is that the origin in P needn't be fixed, whereas the zero vector is picked out as the origin in V.

Orthogonal operators act on V, but they don't act on P unless the origin is chosen.

6.4   FINITE GROUPS OF ORTHOGONAL OPERATORS ON THE PLANE

Theorem 6.4.1 Let G be a finite subgroup of the orthogonal group O_2. There is an integer n such that G is one of the following groups:

(a) C_n: the *cyclic group* of order n generated by the rotation ρ_θ, where $\theta = 2\pi/n$.

(b) D_n: the *dihedral group* of order $2n$ generated by two elements: the rotation ρ_θ, where $\theta = 2\pi/n$, and a reflection r' about a line ℓ through the origin.

We will take a moment to describe the dihedral group D_n before proving the theorem. This group depends on the line of reflection, but if we choose coordinates so that ℓ becomes the horizontal axis, the group will contain our standard reflection r, the one whose matrix is

(6.4.2)
$$\begin{bmatrix} 1 & \\ & -1 \end{bmatrix}.$$

Then if we also write ρ for ρ_θ, the $2n$ elements of the group will be the n powers ρ^i of ρ and the n products $\rho^i r$. The rule for commuting ρ and r is

$$
r\rho = \begin{bmatrix} 1 & \\ & -1 \end{bmatrix}\begin{bmatrix} c & -s \\ s & c \end{bmatrix} = \begin{bmatrix} c & s \\ -s & c \end{bmatrix}\begin{bmatrix} 1 & \\ & -1 \end{bmatrix} = \rho^{-1}r,
$$

where $c = \cos\theta$, $s = \sin\theta$, and $\theta = 2\pi/n$.

To conform with a more customary notation for groups, we denote the rotation $\rho_{2\pi/n}$ by x, and the reflection r by y.

Proposition 6.4.3 The dihedral group D_n has order $2n$. It is generated by two elements x and y that satisfy the relations

$$
x^n = 1, \quad y^2 = 1, \quad yx = x^{-1}y.
$$

The elements of D_n are

$$
1, x, x^2, \ldots, x^{n-1}; \ y, xy, x^2y, \ldots, x^{n-1}y. \qquad \square
$$

Using the first two relations (6.4.3), the third one can be rewritten in various ways. It is equivalent to

(6.4.4) $$ xyxy = 1, \quad \text{and also to} \quad yx = x^{n-1}y. $$

When $n = 3$, the relations are the same as for the symmetric group S_3 (2.2.6).

Corollary 6.4.5 The dihedral group D_3 and the symmetric group S_3 are isomorphic. $\qquad \square$

For $n > 3$, the dihedral and symmetric groups are not isomorphic, because D_n has order $2n$, while S_n has order $n!$.

When $n \geq 3$, the elements of the dihedral group D_n are the orthogonal operators that carry a regular n-sided polygon Δ to itself – the group of symmetries of Δ. This is easy to see, and it follows from the theorem: A regular n-gon is carried to itself by the rotation by $2\pi/n$ about its center, and also by some reflections. Theorem 6.4.1 identifies the group of all symmetries as D_n.

The dihedral groups D_1, D_2 are too small to be symmetry groups of an n-gon in the usual sense. D_1 is the group $\{1, r\}$ of two elements. So it is a cyclic group, as is C_2. But the element r of D_1 is a reflection, while the element different from the identity in C_2 is the rotation with angle π. The group D_2 contains the four elements $\{1, \rho, r, \rho r\}$, where ρ is the rotation with angle π and ρr is the reflection about the vertical axis. This group is isomorphic to the Klein four group.

If we like, we can think of D_1 and D_2 as groups of symmetry of the 1-gon and 2-gon:

1-gon. 2-gon.

We begin the proof of Theorem 6.4.1 now. A subgroup Γ of the additive group $\mathbb{R}^+$ of real numbers is called *discrete* if there is a (small) positive real number ϵ such that every nonzero element c of Γ has absolute value $\geq \epsilon$.

Lemma 6.4.6 Let Γ be a discrete subgroup of $\mathbb{R}^+$. Then either $\Gamma = \{0\}$, or Γ is the set $\mathbb{Z}a$ of integer multiples of a positive real number a.

Proof. This is very similar to the proof of Theorem 2.3.3, that a nonzero subgroup of $\mathbb{Z}^+$ has the form $\mathbb{Z}n$.

If a and b are distinct elements of Γ, then since Γ is a group, $a - b$ is in Γ, and $|a - b| \geq \epsilon$. Distinct elements of Γ are separated by a distance at least ϵ. Since only finitely many elements separated by ϵ can fit into any bounded interval, a bounded interval contains finitely many elements of Γ.

Suppose that $\Gamma \neq \{0\}$. Then Γ contains a nonzero element b, and since it is a group, Γ contains $-b$ as well. So it contains a positive element, say a'. We choose the smallest positive element a in Γ. We can do this because we only need to choose the smallest element of the finite subset of Γ in the interval $0 \leq x \leq a'$.

We show that $\Gamma = \mathbb{Z}a$. Since a is in Γ and Γ is a group, $\mathbb{Z}a \subset \Gamma$. Let b be an element of Γ. Then $b = ra$ for some real number r. We take out the integer part of r, writing $r = m + r_0$ with m an integer and $0 \leq r_0 < 1$. Since Γ is a group, $b' = b - ma$ is in Γ and $b' = r_0 a$. Then $0 \leq b' < a$. Since a is the smallest positive element in Γ, b' must be zero. So $b = ma$, which is in $\mathbb{Z}a$. This shows that $\Gamma \subset \mathbb{Z}a$, and therefore that $\Gamma = \mathbb{Z}a$. $\square$

Proof of Theorem (6.4.1). Let G be a finite subgroup of O_2. We want to show that G is C_n or D_n. We remember that the elements of O_2 are the rotations ρ_θ and the reflections $\rho_\theta r$.

Case 1: All elements of G are rotations.

We must prove that G is cyclic. Let Γ be the set of real numbers α such that ρ_α is in G. Then Γ is a subgroup of the additive group $\mathbb{R}^+$, and it contains 2π. Since G is finite, Γ is discrete. So Γ has the form $\mathbb{Z}\alpha$. Then G consists of the rotations through integer multiples of the angle α. Since 2π is in Γ, it is an integer multiple of α. Therefore $\alpha = 2\pi/n$ for some integer n, and $G = C_n$.

Case 2: G contains a reflection.

We adjust our coordinates so that the standard reflection r is in G. Let H denote the subgroup consisting of the rotations that are elements of G. We apply what has been proved in Case 1 to conclude that H is the cyclic group generated by ρ_θ, for some angle $\theta = 2\pi/n$. Then the $2n$ products ρ_θ^k and $\rho_\theta^k r$, for $0 \leq k < n - 1$, are in G, so G contains the dihedral group D_n. We claim that $G = D_n$, and to show this we take any element g of G. Then g is either a rotation or a reflection. If g is a rotation, then by definition of H, g is in H. The elements of H are also in D_n, so g is in D_n. If g is a reflection, we write it in the form $\rho_\alpha r$ for some rotation ρ_α. Since r is in G, so is the product $gr = \rho_\alpha$. Therefore ρ_α is a power of ρ_θ, and again, g is in D_n. $\square$

Theorem 6.4.7 Fixed Point Theorem. Let G be a finite group of isometries of the plane. There is a point in the plane that is fixed by every element of G, a point p such that $g(p) = p$ for all g in G.

Proof. This is a nice geometric argument. Let s be any point in the plane, and let S be the set of points that are the images of s under the various isometries in G. So each element s' of S has the form $s' = g(s)$ for some g in G. This set is called the *orbit* of s for the action of G. The element s is in the orbit because the identity element 1 is in G, and $s = 1(s)$. A typical orbit for the case that G is the group of symmetries of a regular pentagon is depicted below, together with the fixed point p of the operation.

Any element of G will permute the orbit S. In other words, if s' is in S and h is in G, then $h(s')$ is in S: Say that $s' = g(s)$, with g in G. Since G is a group, hg is in G. Then $hg(s)$ is in S and is equal to $h(s')$.

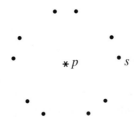

We list the elements of S arbitrarily, writing $S = \{s_1, \ldots, s_n\}$. The fixed point we are looking for is the *centroid*, or *center of gravity* of the orbit, defined as

(6.4.8) $$p = \tfrac{1}{n}(s_1 + \cdots + s_n),$$

where the right side is computed by vector addition, using an arbitrary coordinate system in the plane.

Lemma 6.4.9 Isometries carry centroids to centroids: Let $S = \{s_1, \ldots, s_n\}$ be a finite set of points of the plane, and let p be its centroid, as defined by (6.4.8). Let m be an isometry. Let $m(p) = p'$ and $m(s_i) = s_i'$. Then p' is the centroid of the set $S' = \{s_1', \ldots, s_n'\}$. $\quad\square$

The fact that the centroid of our set S is a fixed point follows. An element g of G permutes the orbit S. It sends S to S and therefore it sends p to p. $\quad\square$

Proof of Lemma 6.4.9 This can be deduced by physical reasoning. It can be shown algebraically too. To do so, it suffices to look separately at the cases $m = t_a$ and $m = \varphi$, where φ is an orthogonal operator. Any isometry is obtained from such isometries by composition.

Case 1: $m = t_a$ is a translation. Then $s_i' = s_i + a$ and $p' = p + a$. It is true that

$$p' = p + a = \tfrac{1}{n}((s_1 + a) + \cdots + (s_n + a)) = \tfrac{1}{n}(s_1' + \cdots + s_n').$$

Case 2: $m = \varphi$ is a linear operator. Then

$$p' = \varphi(p) = \varphi(\tfrac{1}{n}(s_1 + \cdots + s_n)) = \tfrac{1}{n}(\varphi(s_1) + \cdots + \varphi(s_n)) = \tfrac{1}{n}(s_1' + \cdots + s_n'). \quad\square$$

By combining Theorems 6.4.1 and 6.4.7 one obtains a description of the symmetry groups of bounded figures in the plane.

Corollary 6.4.10 Let G be a finite subgroup of the group M of isometries of the plane. If coordinates are chosen suitably, G becomes one of the groups C_n or D_n described in Theorem 6.4.1. □

6.5 DISCRETE GROUPS OF ISOMETRIES

In this section we discuss groups of symmetries of unbounded figures such as the one depicted in Figure 6.1.5. What I call the *kaleidoscope principle* can be used to construct a figure with a given group of symmetries. You have probably looked through a kaleidoscope. One sees a sector at the end of the tube, whose sides are bounded by two mirrors that run the length of the tube and are placed at an angle θ, such as $\theta = \pi/6$. One also sees the reflection of the sector in each mirror, and then one sees the reflection of the reflection, and so on. There are usually some bits of colored glass in the sector, whose reflections form a pattern.

There is a group involved. In the plane at the end of the kaleidoscope tube, let ℓ_1 and ℓ_2 be the lines that bound the sector formed by the mirrors. The group is a dihedral group, generated by the reflections r_i about ℓ_i. The product $r_1 r_2$ of these reflections preserves orientation and fixes the point of intersection of the two lines, so it is a rotation. Its angle of rotation is $\pm 2\theta$.

One can use the same principle with any subgroup G of M. We won't give precise reasoning to show this, but the method can be made precise. We start with a random figure R in the plane. Every element g of our group G will move R to a new position, call it gR. The figure F is the union of all the figures gR. An element h of the group sends gR to hgR, which is also a part of F, so it sends F to itself. If R is sufficiently random, G will be the group of symmetries of F. As we know from the kaleidoscope, the figure F is often very attractive. The result of applying this procedure when G is the group of symmetries of a regular pentagon is shown below.

Of course many different figures have the same group of symmetry. But it is interesting and instructive to describe the groups. We are going to present a rough classification, which will be refined in the exercises.

Some subgroups of M are too wild to have a reasonable geometry. For instance, if the angle θ at which the mirrors in a kaleidoscope are placed were not a rational multiple of 2π, there would be infinitely many distinct reflections of the sector. We need to rule this possibility out.

Definition 6.5.1 A group G of isometries of the plane P is *discrete* if it does not contain arbitrarily small translations or rotations. More precisely, G is discrete if there is a positive real number ϵ so that:

(i) if an element of G is the translation by a nonzero vector a, then the length of a is at least ϵ: $|a| \geq \epsilon$, and

(ii) if an element of G is the rotation through a nonzero angle θ about some point of the plane, then the absolute value of θ is at least ϵ: $|\theta| \geq \epsilon$.

Note: Since the translation vectors and the rotation angles form different sets, it might seem more appropriate to have separate lower bounds for them. However, in this definition we don't care about the *best* bounds for the vectors and the angles, so we choose ϵ small enough to take care of both at the same time. $\square$

The translations and rotations are all of the orientation-preserving isometries (6.3.4), and the conditions apply to all of them. We don't impose a condition on the orientation-reversing isometries. If m is a glide with nonzero glide vector v, then m^2 is the translation t_{2v}. So a lower bound on the translation vectors determines a bound for the glide vectors too.

There are three main tools for analyzing a discrete group G:

(6.5.2) • the translation group L, a subgroup of the group V of translation vectors,

  • the point group $\overline{G}$, a subgroup of the orthogonal group O_2,

  • an operation of $\overline{G}$ on L.

The Translation Group

The *translation group* L of G is the set of vectors v such that the translation t_v is in G.

(6.5.3) $$L = \{v \in V \mid t_v \in G\}.$$

Since $t_v t_w = t_{v+w}$ and $t_v^{-1} = t_{-v}$, L is a subgroup of the additive group V^+ of all translation vectors. The bound ϵ on translations in G bounds the lengths of the vectors in L:

(6.5.4) Every nonzero vector v in L has length $|v| \geq \epsilon$.

• A subgroup L of one of the additive groups V^+ or $\mathbb{R}^{n+}$ that satisfies condition (6.5.4) for some $\epsilon > 0$ is called a *discrete subgroup*. (This is the definition made before for $\mathbb{R}^+$.)

A subgroup L is discrete if and only if the distance between distinct vectors a and b of L is at least ϵ. This is true because the distance is the length of $b - a$, and $b - a$ is in L because L is a group. If (6.5.4) holds, then $|b - a| \geq \epsilon$. $\square$

Theorem 6.5.5 Every discrete subgroup L of V^+ or of $\mathbb{R}^{2+}$ is one of the following:

(a) the zero group: $L = \{0\}$.

(b) the set of integer multiples of a nonzero vector a:

$$L = \mathbb{Z}a = \{ma \mid m \in \mathbb{Z}\}, \quad \text{or}$$

(c) the set of integer combinations of two linearly independent vectors a and b:

$$L = \mathbb{Z}a + \mathbb{Z}b = \{ma + nb \mid m, n \in \mathbb{Z}\}.$$

Groups of the third type listed above are called *lattices*, and the generating set (a, b) is called a *lattice basis*.

(6.5.6)   A Lattice

Lemma 6.5.7 Let L be a discrete subgroup of V^+ or $\mathbb{R}^{2+}$.

(a) A bounded region of the plane contains only finitely many points of L.

(b) If L is not the trivial group, it contains a nonzero vector of minimal length.

Proof. **(a)** Since the elements of L are separated by a distance at least ϵ, a small square can contain at most one point of L. A region of the plane is bounded if it is contained in some large rectangle. We can cover any rectangle by finitely many small squares, each of which contains at most one point of L.

(b) We say that a vector v is a nonzero vector of *minimal length* of L if L contains no shorter nonzero vector. To show that such a vector exists, we use the hypothesis that L is not the trivial group. There is some nonzero vector a in L. Then the disk of radius $|a|$ about the origin is a bounded region that contains a and finitely many other nonzero points of L. Some of those points will have minimal length. $\square$

Given a basis $\mathbf{B} = (u, w)$ of $\mathbb{R}^2$, we let $\Pi(\mathbf{B})$ denote the parallelogram with vertices $0, u, w, u + w$. It consists of the linear combinations $ru + sw$ with $0 \le r \le 1$ and $0 \le s \le 1$. We also denote by $\Pi'(\mathbf{B})$ the region obtained from $\Pi(\mathbf{B})$ by deleting the two edges $[u, u + w]$ and $[w, u + w]$. It consists of the linear combinations $ru + sw$ with $0 \le r < 1$ and $0 \le s < 1$.

Lemma 6.5.8 Let $\mathbf{B} = (u, w)$ be a basis of $\mathbb{R}^2$, and let L be the lattice of integer combinations of $\mathbf{B}$. Every vector v in $\mathbb{R}^2$ can be written uniquely in the form $v = x + v_0$, with x in L and v_0 in $\Pi'(\mathbf{B})$.

Proof. Since **B** is a basis, every vector is a linear combination $ru + sw$, with real coefficieints r and s. We take out their integer parts, writing $r = m + r_0$ and $s = n + s_0$, with m, n integers and $0 \leq r_0, s_0 < 1$. Then $v = x + v_0$, where $x = mu + nv$ is in L and $v_0 = r_0 u + s_0 w$ is in $\Pi'(\mathbf{B})$. There is just one way to do this. $\square$

Proof of Theorem 6.5.5 It is enough to consider a discrete subgroup L of $\mathbb{R}^{2+}$. The case that L is the zero group is included in the list. If $L \neq \{0\}$, there are two possibilities:

Case 1: All vectors in L lie on a line ℓ through the origin.

Then L is a subgroup of the additive group of ℓ^+, which is isomorphic to $\mathbb{R}^+$. Lemma 6.4.6 shows that L has the form $\mathbb{Z}a$.

Case 2: The elements of L do not lie on a line.

In this case, L contains independent vectors a' and b', and then $\mathbf{B}' = (a', b')$ is a basis of $\mathbb{R}^2$. We must show that there is a lattice basis for L.
    We first consider the line ℓ spanned by a'. The subgroup $L \cap \ell$ of ℓ^+ is discrete, and a' isn't zero. So by what has been proved in Case 1, L has the form $\mathbb{Z}a$ for some vector a. We adjust coordinates and rescale so that a becomes the vector $(1, 0)^t$.
    Next, we replace $b' = (b_1', b_2')^t$ by $-b'$ if necessary, so that b_2' becomes positive. We look for a vector $b = (b_1, b_2)^t$ in L with b_2 positive, and otherwise as small as possible. A priori, we have infinitely many elements to inspect. However, since b' is in L, we only need to inspect the elements b such that $0 < b_2 \leq b_2'$. Moreover, we may add a multiple of a to b, so we may also assume that $0 \leq b_1 < 1$. When this is done, b will be in a bounded region that contains finitely many elements of L. We look through this finite set to find the required element b, and we show that $\mathbf{B} = (a, b)$ is a lattice basis for L.
    Let $\tilde{L} = \mathbb{Z}a + \mathbb{Z}b$. Then $\tilde{L} \subset L$. We must show that every element of L is in $\tilde{L}$, and according to Lemma 6.5.8, applied to the lattice $\tilde{L}$, it is enough to show that the only element of L in the region $\Pi'(\mathbf{B})$ is the zero vector. Let $c = (c_1, c_2)^t$ be a point of L in that region, so that $0 \leq c_1 < 1$ and $0 \leq c_2 < b_2$. Since b_2 was chosen minimal, $c_2 = 0$, and c is on the line ℓ. Then c is an integer multiple of a, and since $0 \leq c_1 < 1$, $c = 0$. $\square$

The Point Group

We turn now to the second tool for analyzing a discrete group of isometries. We choose coordinates, and go back to the homomorphism $\pi : M \to O_2$ whose kernel is the group T of translations (6.3.10). When we restrict this homomorphism to a discrete subgroup G, we obtain a homomorphism

(6.5.9) $$\pi|_G : G \to O_2.$$

The *point group* $\overline{G}$ is the image of G in the orthogonal group O_2.
    It is important to make a clear distinction between elements of the group G and those of its point group $\overline{G}$. So to avoid confusion, we will put bars over symbols when they represent elements of $\overline{G}$. For g in G, $\overline{g}$ will be an orthogonal operator.
    By definition, a rotation $\overline{\rho}_\theta$ is in $\overline{G}$ if G contains an element of the form $t_a \rho_\theta$, and this is a rotation through the same angle θ about some point of the plane (6.3.5). The inverse

image in G of an element $\overline{\rho}_\theta$ of $\overline{G}$ consists of the elements of G that are rotations through the angle θ about various points of the plane.

Similarly, let ℓ denote the line of reflection of $\rho_\theta r$. As we have noted before, its angle with the e_1-axis is $\frac{1}{2}\theta$ (5.1.17). The point group $\overline{G}$ contains $\overline{\rho_\theta r}$ if there is an element $t_a \rho_\theta r$ in G, and $t_a \rho_\theta r$ is a reflection or a glide reflection along a line parallel to ℓ (6.3.8). The inverse image of $\overline{\rho_\theta r}$ consists of all of the elements of G that are reflections or glides along lines parallel to ℓ. To sum up:

- The point group $\overline{G}$ records the *angles* of rotation and the *slopes* of the glide lines and the lines of reflection, of elements of G.

Proposition 6.5.10   A discrete subgroup $\overline{G}$ of O_2 is finite, and is therefore either cyclic or dihedral.

Proof. Since $\overline{G}$ contains no small rotations, the set Γ of real numbers θ such that ρ_θ is in $\overline{G}$ is a discrete subgroup of the additive group $\mathbb{R}^+$ that contains 2π. Lemma 6.4.6 tells us that Γ has the form $\mathbb{Z}\theta$, where $\theta = 2\pi/n$ for some integer n. At this point, the proof of Theorem 6.4.1 carries over.   □

The Crystallographic Restriction

If the translation group of a discrete group of isometries G is the trivial group, the restriction of π to G will be injective. In this case G will be isomorphic to its point group $\overline{G}$, and will be cyclic or dihedral. The next proposition is our third tool for analyzing infinite discrete groups. It relates the point group to the translation group.

Unless an origin is chosen, the orthogonal group O_2 doesn't operate on the plane P. But it does operate on the space V of translation vectors.

Proposition 6.5.11   Let G be a discrete subgroup of M. Let a be an element of its translation group L, and let $\overline{g}$ be an element of its point group $\overline{G}$. Then $\overline{g}(a)$ is in L.

We can restate this proposition by saying that the elements of $\overline{G}$ map L to itself. So $\overline{G}$ is contained in the group of symmetries of L, when L is regarded as a figure in the plane V.

Proof of Proposition 6.5.11 Let a and g be elements of L and G, respectively, let $\overline{g}$ be the image of g in $\overline{G}$, and let $a' = \overline{g}(a)$. We will show that $t_{a'}$ is the conjugate $gt_a g^{-1}$. This will show that $t_{a'}$ is in G, and therefore that a' is in L. We write $g = t_b \varphi$. Then φ is in O_2 and $\overline{g} = \overline{\varphi}$. So $a' = \overline{\varphi}(a)$. Using the formulas (6.2.8), we find:

$$gt_a g^{-1} = (t_b \varphi)t_a(\varphi^{-1}t_{-b}) = t_b t_{a'} \varphi \varphi^{-1} t_{-b} = t_{a'}.$$ □

Note: It is important to understand that the group G does *not* operate on its translation group L. Indeed, it makes no sense to ask whether G operates on L, because the elements of G are isometries of the plane P, while L is a subset of V. Unless an origin is fixed, P is not the same as V. If we fix the origin in P, we can identify P with V. Then the question makes sense. We may ask: Is there a point of P so that with that point as the origin, the elements of G carry L to itself? Sometimes yes, sometimes no. That depends on the group.   □

The next theorem describes the point groups that can occur when the translation group L is not trivial.

Theorem 6.5.12 Crystallographic Restriction. Let L be a discrete subgroup of V^+ or $\mathbb{R}^{2+}$, and let $H \subset O_2$ be a subgroup of the group of symmetries of L. Suppose that L is not the trivial group. Then

(a) every rotation in H has order 1, 2, 3, 4, or 6, and

(b) H is one of the groups C_n or D_n, and $n = 1, 2, 3, 4,$ or 6.

In particular, rotations of order 5 are ruled out. There is no wallpaper pattern with five-fold rotational symmetry ("Quasi-periodic" patterns with five-fold symmetry do exist. See, for example, [Senechal].)

Proof of the Crystallographic Restriction We prove **(a)**. Part **(b)** follows from **(a)** and from Theorem 6.4.1. Let ρ be a rotation in H with angle θ, and let a be a nonzero vector in L of minimal length. Since H operates on L, $\rho(a)$ is also in L. Then $b = \rho(a) - a$ is in L too, and since a has a minimal length, $|b| \geq |a|$. Looking at the figure below, one sees that $|b| < |a|$ when $\theta < 2\pi/6$. So we must have $\theta \geq 2\pi/6$. It follows that the group H is discrete, hence finite, and that ρ has order ≤ 6.

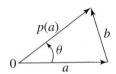

The case that $\theta = 2\pi/5$ can be ruled out too, because for that angle, the element $b' = \rho^2(a) + a$ is shorter than a:

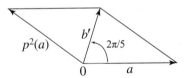

$\square$

6.6   PLANE CRYSTALLOGRAPHIC GROUPS

We go back to our discrete group of isometries $G \subset M$. We have seen that when L is the trivial group, G is cyclic or dihedral. The discrete groups G such that L is infinite cyclic (6.5.5)**(b)** are the symmetry groups of frieze patterns such as those shown in (6.1.3), (6.1.4). We leave the classification of those groups as an exercise.

When L is a lattice, G is called a *two-dimensional crystallographic group*. These crystallographic groups are the symmetry groups of two-dimensional crystals such as graphite. We imagine a crystal to be infinitely large. Then the fact that the molecules are arranged regularly implies that they form an array having two independent translational symmetries. A wallpaper pattern also repeats itself in two different directions – once along the strips of paper because the pattern is printed using a roller, and a second time because strips of paper are glued to the wall side by side. The crystallographic restriction limits the possibilities and allows one to classify crystallographic groups into 17 types. Representative patterns with the various types of symmetry are illustrated in Figure (6.6.2).

The point group $\overline{G}$ and the translation group L do not determine the group G completely. Things are complicated by the fact that a reflection in $\overline{G}$ needn't be the image of a reflection in G. It may be represented in G only by glides, as in the brick pattern that is illustrated below. This pattern (my favorite) is relatively subtle because its group of symmetries doesn't contain a reflection. It has rotational symmetries with angle π about the center of each brick. All of these rotations represent the same element $\overline{\rho}_\pi$ of the point group $\overline{G}$. There are no nontrivial rotational symmetries with angles other than 0 and π. The pattern also has glide symmetry along the dashed line drawn in the figure, so $\overline{G} = D_2 = \{1, \overline{\rho}_\pi, \overline{r}, \overline{\rho}_\pi \overline{r}\}$.

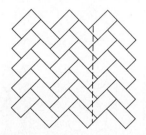

One can determine the point group of a pattern fairly easily, in two steps: One looks first for rotational symmetries. They are usually relatively easy to find. A rotation $\overline{\rho}_\theta$ in the point group $\overline{G}$ is represented by a rotation with the same angle in the group G of symmetries of the pattern. When the rotational symmetries have been found, one will know the integer n such that the point group is C_n or D_n. Then to distinguish D_n from C_n, one looks to see if the pattern has reflection or glide symmetry. If it does, $\overline{G} = D_n$, and if not, $\overline{G} = C_n$.

Plane Crystallographic Groups with a Fourfold Rotation in the Point Group

As an example of the methods used to classify discrete groups of isometries, we analyze groups whose point groups are C_4 or D_4.

Let G be such a group, let $\overline{\rho}$ denote the rotation with angle $\pi/2$ in $\overline{G}$, and let L be the lattice of G, the set of vectors v such that t_v is in G.

Lemma 6.6.2 The lattice L is square.

Proof. We choose a nonzero vector a in L of minimal length. The point group operates on L, so $\overline{\rho}(a) = b$ is in L and is orthogonal to a. We claim that (a, b) is a lattice basis for L.

Suppose not. Then according to Lemma 6.5.8, there will be a point of L in the region Π' consisting of the points $r_1 a + r_2 b$ with $0 \le r_i < 1$. Such a point w will be at a distance less than $|a|$ from one of the four vertices $0, a, b, a + b$ of the square. Call that vertex v. Then $v - w$ is also in L, and $|v - w| < |a|$. This contradicts the choice of a.   □

We choose coordinates and rescale so that a and b become the standard basis vectors e_1 and e_2. Then L becomes the lattice of vectors with integer coordinates, and Π' becomes the set of vectors $(s, t)^t$ with $0 \le s < 1$ and $0 \le t < 1$. This determines coordinates in the plane P up to a translation.

(6.6.2)   Sample Patterns for the 17 Plane Crystallographic Groups.

The orthogonal operators on V that send L to itself form the dihedral group D_4 generated by the rotation $\overline{\rho}$ through the angle $\pi/2$ and the standard reflection $\overline{r}$. Our assumption is that $\overline{\rho}$ is in $\overline{G}$. If $\overline{r}$ is also in $\overline{G}$, then $\overline{G}$ is the dihedral group D_4. If not, $\overline{G}$ is the cyclic group C_4. We describe the group G when $\overline{G}$ is C_4 first. Let g be an element of G whose image in $\overline{G}$ is the rotation $\overline{\rho}$. Then g is a rotation through the angle $\pi/2$ about some point p in the plane. We translate coordinates in the plane P so that the point p becomes the origin. In this coordinate system, G contains the rotation $\rho = \rho_{\pi/2}$ about the origin.

Proposition 6.6.3 Let G be a plane crystallographic group whose point group $\overline{G}$ is the cyclic group C_4. With coordinates chosen so that L is the lattice of points with integer coordinates, and so that $\rho = \rho_{\pi/2}$ is an element of G, the group G consists of the products $t_v \rho^i$, with v in L and $0 \le i < 4$:

$$G = \{t_v \rho^i \,|\, v \in L\}.$$

Proof. Let G' denote the set of elements of the form $t_v \rho^i$ with v in L. We must show that $G' = G$. By definition of L, t_v is in G, and also ρ is in G. So $t_v \rho^i$ is in G, and therefore G' is a subset of G.

To prove the opposite inclusion, let g be any element of G. Since the point group $\overline{G}$ is C_4, every element of G preserves orientation. So g has the form $g = t_u \rho_\alpha$ for some translation vector u and some angle α. The image of this element in the point group is $\overline{\rho}_\alpha$, so α is a multiple of $\pi/2$, and $\rho_\alpha = \rho^i$ for some i. Since ρ is in G, $g\rho^{-i} = t_u$ is in G and u is in L. Therefore g is in G'. $\qquad\square$

We now consider the case that the point group $\overline{G}$ is D_4.

Proposition 6.6.4 Let G be a plane crystallographic group whose point group $\overline{G}$ is the dihedral group D_4. Let coordinates be chosen so that L is the lattice of points with integer coordinates and so that $\rho = \rho_{\pi/2}$ is an element of G. Also, let c denote the vector $(\frac{1}{2}, \frac{1}{2})^t$. There are two possibilities:

(a) The elements of G are the products $t_v \varphi$ where v is in L and φ is in D_4,

$$G = \{t_v \rho^i \,|\, v \in L\} \,\cup\, \{t_v \rho^i r \,|\, v \in L\}, \quad \text{or}$$

(b) the elements of G are products $t_x \varphi$, with φ in D_4. If φ is a rotation, then x is in L, and if φ is a reflection, then x is in the coset $c + L$:

$$G = \{t_v \rho^i \,|\, v \in L\} \,\cup\, \{t_u \rho^i r \,|\, u \in c + L\}.$$

Proof. Let H be the subset of orientation-preserving isometries in G. This is a subgroup of G whose lattice of translations is L, and which contains ρ. So its point group is C_4. Proposition 6.6.3 tells us that H consists of the elements $t_v \rho^i$, with v in L.

The point group also contains the reflection $\overline{r}$. We choose an element g in G such that $\overline{g} = \overline{r}$. It will have the form $g = t_u r$ for some vector u, but we don't know whether or not u is in L. Analyzing this case will require a bit of fiddling. Say that $u = (p, q)^t$.

We can multiply g on the left by a translation t_v in G (i.e., v in L), to move u into the region Π' of points with $0 \le p, q < 1$. Let's suppose this has been done.

We compute with $g = t_u r$, using the formulas (6.3.3):

$$g^2 = t_u r t_u r = t_{u+ru} \text{ and } (g\rho)^2 = (t_u r \rho)^2 = t_{u+r\rho u}.$$

These are elements of G, so $u + ru = (2p, 0)^t$, and $u + r\rho u = (p - q, q - p)^t$ are in the lattice L. They are vectors with integer coordinates. Since $0 \leq p, q < 1$ and $2p$ is an integer, p is either 0 or $\frac{1}{2}$. Since $p - q$ is also an integer, $q = 0$ if $p = 0$ and $q = \frac{1}{2}$ if $p = \frac{1}{2}$. So there are only two possibilities for u: Either $u = (0, 0)^t$, or $u = c = (\frac{1}{2}, \frac{1}{2})^t$. In the first case, $g = r$, so G contains a reflection. This is case **(a)** of the proposition. The second possibility is case **(b)**.   □

6.7   ABSTRACT SYMMETRY: GROUP OPERATIONS

The concept of symmetry can be applied to things other than geometric figures. Complex conjugation $(a+bi) \rightsquigarrow (a-bi)$, for instance, may be thought of as a symmetry of the complex numbers. Since complex conjugation is compatible with addition and multiplication, it is called an *automorphism* of the field $\mathbb{C}$. Geometrically, it is the bilateral symmetry of the complex plane about the real axis, but the statement that it is an automorphism refers to its algebraic structure. The field $F = \mathbb{Q}[\sqrt{2}]$ whose elements are the real numbers of the form $a+b\sqrt{2}$, with a and b rational, also has an automorphism, one that sends $a+b\sqrt{2} \rightsquigarrow a-b\sqrt{2}$. This isn't a geometric symmetry. Another example of abstract "bilateral" symmetry is given by a cyclic group H of order 3. It has an automorphism that interchanges the two elements different from the identity.

The set of automorphisms of an algebraic structure X, such as a group or a field, forms a group, the law of composition being composition of maps. Each automorphism should be thought of as a symmetry of X, in the sense that it is a permutation of the elements of X that is compatible with its algebraic structure. But the structure in this case is algebraic instead of geometric.

So the words "automorphism" and "symmetry" are more or less synonymous, except that "automorphism" is used to describe a permutation of a set that preserves an algebraic structure, while "symmetry" often, though not always, refers to a permutation that preserves a geometric structure.

Both automorphisms and symmetries are special cases of the more general concept of a *group operation*. An operation of a group G on a set S is a rule for combining an element g of G and an element s of S to get another element of S. In other words, it is a map $G \times S \rightarrow S$. For the moment we denote the result of applying this law to elements g and s by $g*s$. An operation is required to satisfy the following axioms:

Example 6.7.1

(a) $1*s = s$ for all s in S. (Here 1 is the identity of G.)

(b) *associative law*: $(gg')*s = g*(g'*s)$, for all g and g' in G and all s in S.

We usually omit the asterisk, and write the operation multiplicatively, as $g, s \rightsquigarrow gs$. With multiplicative notation, the axioms are $1s = s$ and $(gg')s = g(g's)$.

Examples of sets on which a group operates can be found manywhere,[1] and most often, it will be clear that the axioms for an operation hold. The group M of isometries of the plane operates on the set of points of the plane. It also operates on the set of lines in the plane and on the set of triangles in the plane. The symmetric group S_n operates on the set of indices $\{1, 2, \ldots, \mathbf{n}\}$.

The reason that such a law is called an operation is this: If we fix an element g of G but let s vary in S, then *left multiplication by* g (or *operation of* g) defines a map from S to itself. We denote this map, which describes the way the element g operates, by m_g:

(6.7.2) $$m_g : S \to S$$

is the map defined by $m_g(s) = gs$. It is a *permutation* of S, a bijective map, because it has the inverse function $m_{g^{-1}}$: *multiplication by* g^{-1}.

• Given an operation of a group G on a set S, an element s of S will be sent to various other elements by the group operation. We collect together those elements, obtaining a subset called the *orbit* O_s of s:

(6.7.3) $$O_s = \{s' \in S \mid s' = gs \text{ for some } g \text{ in } G\}.$$

When the group M of isometries of the plane operates on the set S of triangles in the plane, the orbit O_Δ of a given triangle Δ is the set of all triangles congruent to Δ. Another orbit was introduced when we proved the existence of a fixed point for the operation of a finite group on the plane (6.4.7).

The orbits for a group action are equivalence classes for the equivalence relation

(6.7.4) $$s \sim s' \text{ if } s' = gs, \text{ for some } g \text{ in } G.$$

So if $s \sim s'$, that is, if $s' = gs$ for some g in G, then the orbits of s and of s' are the same. Since they are are equivalence classes:

(6.7.5)   The orbits partition the set S.

The group operates independently on each orbit. For example, the set of triangles of the plane is partitioned into congruence classes, and an isometry permutes each congruence class separately.

If S consists of just one orbit, the operation of G is called *transitive*. This means that every element of S is carried to every other one by some element of the group. The symmetric group S_n operates transitively on the set of indices $\{1, \ldots, \mathbf{n}\}$. The group M of isometries of the plane operates transitively on the set of points of the plane, and it operates transitively on the set of lines. It does not operate transitively on the set of triangles.

• The *stabilizer* of an element s of S is the set of group elements that leave s fixed. It is a subgroup of G that we often denote by G_s:

(6.7.6) $$G_s = \{g \in G \mid gs = s\}.$$

---

[1] While writing a book, the mathematician Masayoshi Nagata decided that the English language needed this word; then he actually found it in a dictionary.

For instance, in the operation of the group M on the set of points of the plane, the stabilizer of the origin is isomorphic to the group O_2 of orthogonal operators. The stabilizer of the index $\mathbf{n}$ for the operation of the symmetric group S_n is isomorphic to the subgroup S_{n-1} of permutations of $\{1, \dots, \mathbf{n-1}\}$. Or, if S is the set of triangles in the plane, the stabilizer of a particular equilateral triangle $\triangle$ is its group of symmetries, a subgroup of M that is isomorphic to the dihedral group D_3.

Note: It is important to be clear about the following distinction: When we say that an isometry m *stabilizes* a triangle $\triangle$, we don't mean that m fixes the points of $\triangle$. The only isometry that fixes every point of a triangle is the identity. We mean that in permuting the set of triangles, m carries $\triangle$ to itself.   □

Just as the kernel K of a group homomorphism $\varphi : G \to G'$ tells us when two elements x and y of G have the same image, namely, if $x^{-1}y$ is in K, the stabilizer G_s of an element s of S tells us when two elements x and y of G act in the same way on s.

Proposition 6.7.7 Let S be a set on which a group G operates, let s be an element of S, and let H be the stabilizer of s.

(a) If a and b are elements of G, then $as = bs$ if and only if $a^{-1}b$ is in H, and this is true if and only if b is in the coset aH.

(b) Suppose that $as = s'$. The stabilizer H' of s' is a *conjugate subgroup*:

$$H' = aHa^{-1} = \{g \in G \mid g = aha^{-1} \text{ for some } h \text{ in } H\}.$$

Proof. **(a)** $as = bs$ if and only if $s = a^{-1}bs$.

(b) If g is in aHa^{-1}, say $g = aha^{-1}$ with h in H, then $gs' = (aha^{-1})(as) = ahs = as = s'$, so g stabilizes s'. This shows that $aHa^{-1} \subset H'$. Since $s = a^{-1}s'$, we can reverse the roles of s and s', to conclude that $a^{-1}H'a \subset H$, which implies that $H' \subset aHa^{-1}$. Therefore $H' = aHa^{-1}$.   □

Note: Part **(b)** of the proposition explains a phenomenon that we have seen several times before: When $as = s'$, a group element g fixes s if and only if aga^{-1} fixes s'.

6.8 THE OPERATION ON COSETS

Let H be a subgroup of a group G. As we know, the left cosets aH partition G. We often denote the set of left cosets of H in G by G/H, copying this from the notation used for quotient groups when the subgroup is normal (2.12.1), and we use the bracket notation $[C]$ for a coset C, when it is considered as an element of the set G/H.

The set of cosets G/H is not a group unless H is a normal subgroup. However,

- The group G operates on G/H in a natural way.

The operation is quite obvious: If g is an element of the group, and C is a coset, then $g[C]$ is defined to be the coset $[gC]$, where $gC = \{gc \mid c \in C\}$. Thus if $[C] = [aH]$, then $g[C] = [gaH]$. The next proposition is elementary.

Proposition 6.8.1 Let H be a subgroup of a group G.

(a) The operation of G on the set G/H of cosets is transitive.
(b) The stabilizer of the coset $[H]$ is the subgroup H.   □

Note the distinction once more: Multiplication by an element h of H does not act trivially on the elements of the coset H, but it sends the coset $[H]$ to itself.

Please work carefully through the next example. Let G be the symmetric group S_3 with its usual presentation, and let H be the cyclic subgroup $\{1, y\}$. Its left cosets are

(6.8.2) $C_1 = H = \{1, y\}, \quad C_2 = xH = \{x, xy\}, \quad C_3 = x^2H = \{x^2, x^2y\}$

(see (2.8.4)), and G operates on the set of cosets $G/H = \{[C_1], [C_2], [C_3]\}$. The elements x and y operate in the same way as on the set of indices $\{\mathbf{1}, \mathbf{2}, \mathbf{3}\}$:

(6.8.3) $m_x \leftrightarrow (\mathbf{123}) \quad \text{and} \quad m_y \leftrightarrow (\mathbf{23}).$

For instance, $yC_2 = \{yx, yxy\} = \{x^2y, x^2\} = C_3$.

The next proposition, sometimes called the orbit-stabilizer theorem, shows how an arbitrary group operation can be described in terms of operations on cosets.

Proposition 6.8.4 Let S be a set on which a group G operates, and let s be an element of S. Let H and O_s be the stabilizer and orbit of s, respectively. There is a bijective map $\epsilon : G/H \to O_s$ defined by $[aH] \rightsquigarrow as$. This map is compatible with the operations of the group: $\epsilon(g[C]) = g\epsilon([C])$ for every coset C and every element g in G.

For example, the dihedral group D_5 operates on the vertices of a regular pentagon. Let $\mathcal{V}$ denote the set of vertices, and let H be the stabilizer of a particular vertex. There is a bijective map $D_5/H \to \mathcal{V}$. In the operation of the group M of isometries of the plane P, the orbit of a point is the set of all points of P. The stabilizer of the origin is the group O_2 of orthogonal operators, and there is a bijective map $M/O_2 \to P$. Similarly, if H denotes the stabilizer of a line and if $\mathcal{L}$ denotes the set of all lines in the plane, there is a bijective map $M/H \to \mathcal{L}$.

Proof of Proposition (6.8.4). It is clear that the map ϵ defined in the statement of the proposition will be compatible with the operation of the group, if it exists. Symbolically, ϵ simply replaces H by the symbol s. What is not so clear is that the rule $[gH] \rightsquigarrow gs$ defines a map at all. Since many symbols gH represent the same coset, we must show that if a and b are group elements, and if the cosets aH and bH are equal, then as and bs are equal too. Suppose that $aH = bH$. Then $a^{-1}b$ is in H (2.8.5). Since H is the stabilizer of s, $a^{-1}bs = s$, and therefore $as = bs$. Our definition is legitimate, and reading this reasoning backward, we also see that ϵ is an injective map. Since ϵ carries $[gH]$ to gs, which can be an arbitrary element of O_s, ϵ is surjective as well as injective.   □

Note: The reasoning that we made to define the map ϵ occurs frequently. Suppose that a set $\overline{S}$ is presented as the set of equivalence classes of an equivalence relation on a set S, and let

$\pi : S \rightarrow \overline{S}$ be the map that sends an element s to its equivalence class $\overline{s}$. A common way to define a map ϵ from $\overline{S}$ to another set T is this: Given x in $\overline{S}$, one chooses an element s in S such that $x = \overline{s}$, and defines $\epsilon(x)$ in terms of s. Then one must show, as we did above, that the definition doesn't depend on the choice of the element s whose equivalence class is x, but only on x. This process is referred to as showing that the map is *well defined*.   □

6.9   THE COUNTING FORMULA

Let H be a subgroup of a finite group G. As we know, all cosets of H in G have the same number of elements, and with the notation G/H for the set of cosets, the order $|G/H|$ is what is called the index $[G:H]$ of H in G. The Counting Formula 2.8.8 becomes

$$(6.9.1) \qquad\qquad |G| = |H|\,|G/H|.$$

There is a similar formula for an orbit of any group operation:

Proposition 6.9.2 Counting Formula. Let S be a finite set on which a group G operates, and let G_s and O_s be the stabilizer and orbit of an element s of S. Then

$$|G| = |G_s|\,|O_s|, \quad \text{or}$$
$$\text{(order of } G) = (\text{order of stabilizer})\cdot(\text{order of orbit}).$$

This follows from (6.9.1) and Proposition (6.8.4).   □

Thus the order of the orbit is equal to the index of the stabilizer,

$$(6.9.3) \qquad\qquad |O_s| = [G:G_s],$$

and it divides the order of the group. There is one such formula for every element s of S.

Another formula uses the partition of the set S into orbits to count its elements. We number the orbits that make up S arbitrarily, as $O_1, \ldots, O_k$. Then

$$(6.9.4) \qquad\qquad |S| = |O_1| + |O_2| + \cdots + |O_k|.$$

Formulas 6.9.2 and 6.9.4 have many applications.

Examples 6.9.5 **(a)** The group G of rotational symmetries of a regular dodecahedron operates transitively on the set F of its faces. The stabilizer G_f of a particular face f is the group of rotations by multiples of $2\pi/5$ about the center of f; its order is 5. The dodecahedron has 12 faces. Formula 6.9.2 reads $60 = 5 \cdot 12$, so the order of G is 60. Or, G operates transitively on the set V of vertices. The stabilizer G_v of a vertex v is the group of order 3 of rotations by multiples of $2\pi/3$ about that vertex. A dodecahedron has 20 vertices, so $60 = 3 \cdot 20$, which checks. There is a similar computation for edges: G operates transitively on the set of edges, and the stabilizer of an edge e contains the identity and a rotation by π about the center of e. So $|G_e| = 2$. Since $60 = 2 \cdot 30$, a dodecahedron has 30 edges.

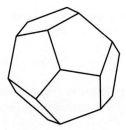

(b) We may also *restrict* an operation of a group G to a subgroup H. By restriction, an operation of G on a set S defines an operation of H on S, and this operation leads to more numerical relations. The H-orbit of an element s will be contained in the G-orbit of s, so a single G-orbit will be partitioned into H-orbits.

For example, let F be the set of 12 faces of the dodecahedron, and let H be the stabilizer of a particular face f, a cyclic group of order 5. The order of any H-orbit is either 1 or 5. So when we partition the set F of 12 faces into H-orbits, we must find two orbits of order 1. We do: H fixes f and it fixes the face opposite to f. The remaining faces make two orbits of order 5. Formula 6.9.4 for the operation of the group H on the set of faces is $12 = 1 + 1 + 5 + 5$. Or, let K denote the stabilizer of a vertex, a cyclic group of order 3. We may also partition the set F into K-orbits. In this case Formula 6.9.4 is $12 = 3 + 3 + 3 + 3$. $\square$

6.10   OPERATIONS ON SUBSETS

Suppose that a group G operates on a set S. If U is a subset of S of order r,

$$(6.10.1) \qquad\qquad gU = \{gu \mid u \in U\}$$

is another subset of order r. This allows us to define an operation of G on the set of subsets of order r of S. The axioms for an operation are verified easily.

For instance, let O be the octahedral group of 24 rotations of a cube, and let F be the set of six faces of the cube. Then O also operates on the subsets of F of order two, that is, on unordered pairs of faces. There are 15 pairs, and they form two orbits: $F = \{pairs\ of\ opposite\ faces\} \cup \{pairs\ of\ adjacent\ faces\}$. These orbits have orders 3 and 12, respectively.

The stabilizer of a subset U is the set of group elements g such that $[gU] = [U]$, which is to say, $gU = U$. The stabilizer of a pair of opposite faces has order 8.

Note this point once more: The stabilizer of U consists of the group elements such that $gU = U$. This means that g permutes the elements within U, that whenever u is in U, gu is also in U.

6.11   PERMUTATION REPRESENTATIONS

In this section we analyze the various ways in which a group G can operate on a set S.

• A *permutation representation* of a group G is a homomorphism from the group to a symmetric group:

(6.11.1) $\varphi : G \to S_n.$

Proposition 6.11.2 Let G be a group. There is a bijective correspondence between operations of G on the set $S = \{\mathbf{1} \ldots, \mathbf{n}\}$ and permutation representations $G \to S_n$:

$$\begin{bmatrix} \text{operations of } G \\ \text{on } S \end{bmatrix} \longleftrightarrow \begin{bmatrix} \text{permutation} \\ \text{representations} \end{bmatrix}.$$

Proof. This is very simple, though it can be confusing when one sees it for the first time. If we are given an operation of G on S, we define a permutation representation φ by setting $\varphi(g) = m_g$, multiplication by g (6.7.2). The associative property $g(h\mathbf{i}) = (gh)\mathbf{i}$ shows that

$$m_g(m_h\mathbf{i}) = g(h\mathbf{i}) = (gh)\mathbf{i} = m_{gh}\mathbf{i}.$$

Hence φ is a homomorphism. Conversely, if φ is a permutation representation, the same formula defines an operation of G on S.   □

For example, the operation of the dihedral group D_n on the vertices $(v_1, \ldots, v_n)$ of a regular n-gon defines a homomorphism $\varphi : D_n \to S_n$.

Proposition 6.11.2 has nothing to do with the fact that it works with a set of indices. If $\text{Perm}(S)$ is the group of permutations of an arbitrary set S, we also call a homomorphism $\varphi : G \to \text{Perm}(S)$ a permutation representation of G.

Corollary 6.11.3 Let $\text{Perm}(S)$ denote the group of permutations of a set S, and let G be a group. There is a bijective correspondence between operations of G on S and permutation representations $\varphi : G \to \text{Perm}(S)$:

$$\begin{bmatrix} \text{operations} \\ \text{of } G \text{ on } S \end{bmatrix} \longleftrightarrow \begin{bmatrix} \text{homomorphisms} \\ G \to \text{Perm}(S) \end{bmatrix}.$$
                                                                                                    □

A permutation representation $G \to \text{Perm}(S)$ needn't be injective. If it happens to be injective, one says that the corresponding operation is *faithful*. To be faithful, an operation must have the property that m_g, multiplication by g, is not the identity map unless $g = 1$:

(6.11.4)   An operation is faithful if it has this property:

The only element g of G such that $gs = s$ for every s in S is the identity.

The operation of the group of isometries M on the set S of equilateral triangles in the plane is faithful, because the only isometry that carries every equilateral triangle to itself is the identity.

Permutation representations $\varphi: G \rightarrow \text{Perm}(S)$ are rarely surjective because the order of $\text{Perm}(S)$ tends to be very large. But one case is given in the next example.

Example 6.11.5 The group $GL_2(\mathbb{F}_2)$ of invertible matrices with mod 2 coefficients is isomorphic to the symmetric group S_3.

We denote the field $\mathbb{F}_2$ by F and the group $GL_2(\mathbb{F}_2)$ by G. The space F^2 of column vectors consists of four vectors:

$$0 = \begin{bmatrix} 0 \\ 0 \end{bmatrix}, \quad e_1 = \begin{bmatrix} 1 \\ 0 \end{bmatrix}, \quad e_2 = \begin{bmatrix} 0 \\ 1 \end{bmatrix}, \quad e_1 + e_2 = \begin{bmatrix} 1 \\ 1 \end{bmatrix}.$$

The group G operates on the set of three nonzero vectors $S = \{e_1, e_2, e_1 + e_2\}$, and this gives us a permutation representation $\varphi: G \rightarrow S_3$. The identity is the only matrix that fixes both e_1 and e_2, so the operation of G on S is faithful, and φ is injective. The columns of an invertible matrix must be an ordered pair of distinct elements of S. There are six such pairs, so $|G| = 6$. Since S_3 also has order six φ is an isomorphism.   □

6.12   FINITE SUBGROUPS OF THE ROTATION GROUP

In this section, we apply the Counting Formula to classify the finite subgroups of SO_3, the group of rotations of $\mathbb{R}^3$. As happens with finite groups of isometries of the plane, all of them are symmetry groups of familiar figures.

Theorem 6.12.1 A finite subgroup of SO_3 is one of the following groups:

C_k: the *cyclic group* of rotations by multiples of $2\pi/k$ about a line, with k arbitrary;

D_k: the *dihedral group* of symmetries of a regular k-gon, with k arbitrary;

T: the *tetrahedral group* of 12 rotational symmetries of a tetrahedron;

O: the *octahedral group* of 24 rotational symmetries of a cube or an octahedron;

I: the *icosahedral group* of 60 rotational symmetries of a dodecahedron or an icosahedron.

Note: The dihedral groups are usually presented as groups of symmetry of a regular polygon in the plane, where reflections reverse orientation. However, a reflection of a plane can be achieved by a rotation through the angle π in three-dimensional space, and in this way the symmetries of a regular polygon can be realized as rotations of $\mathbb{R}^3$. The dihedral group D_n can be generated by a rotation x with angle $2\pi/n$ about the e_1-axis and a rotation y with angle π about the e_2-axis. With $c = \cos 2\pi/n$ and $s = \sin 2\pi/n$, the matrices that represent these rotations are

(6.12.2)
$$x = \begin{bmatrix} 1 & & \\ & c & -s \\ & s & c \end{bmatrix}, \quad \text{and} \quad y = \begin{bmatrix} -1 & & \\ & 1 & \\ & & -1 \end{bmatrix}. \qquad \square$$

Let G be a finite subgroup of SO_3, of order $N > 1$. We'll call a pole of an element $g \neq 1$ of G a *pole* of the group. Any rotation of $\mathbb{R}^3$ except the identity has two poles – the intersections of the axis of rotation with the unit sphere $\mathbb{S}^2$. So a pole of G is a point on the 2-sphere that is fixed by a group element g different from 1.

Example 6.12.3 The group T of rotational symmetries of a tetrahedron Δ has order 12. Its poles are the points of $\mathbb{S}^2$ that lie above the centers of the faces, the vertices, and the centers of the edges. Since Δ has four faces, four vertices, and six edges, there are 14 poles.

$$|poles| = 14 = |faces| + |vertices| + |edges|$$

Each of the 11 elements $g \neq 1$ of T has two spins – two pairs (g, p), where p is a pole of g. So there are 22 spins altogether. The stabilizer of a face has order 3. Its two elements $\neq 1$ share a pole above the center of a face. Similarly, there are two elements with a pole above a vertex, and one element with a pole above the center of an edge.

$$|spins| = 22 = 2\,|faces| + 2\,|vertices| + |edges|$$

$\square$

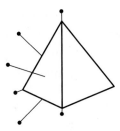

Let $\mathcal{P}$ denote the set of all poles of a finite subgroup G. We will get information about the group by counting these poles. As the example shows, the count can be confusing.

Lemma 6.12.4 The set $\mathcal{P}$ of poles of G is a union of G-orbits. So G operates on $\mathcal{P}$.

Proof. Let p be a pole, say the pole of an element $g \neq 1$ in G, let h be another element of G, and let $q = hp$. We have to show that q is a pole, meaning that q is fixed by some element g' of G other than the identity. The required element is hgh^{-1}. This element is not equal to 1 because $g \neq 1$, and $hgh^{-1}q = hgp = hp = q$. $\square$

The stabilizer G_p of a pole p is the group of all of the rotations about p that are in G. It is a cyclic group, generated by the rotation of smallest positive angle θ. We'll denote its order by r_p. Then $\theta = 2\pi/r_p$.

Since p is a pole, the stabilizer G_p contains an element besides 1, so $r_p > 1$. The set of elements of G with pole p is the stabilizer G_p, with the identity element omitted. So there are $r_p - 1$ group elements that have p as pole. Every group element g except one has two poles. Since $|G| = N$, there are $2N - 2$ spins. This gives us the relation

$$(6.12.5) \qquad\qquad \sum_{p \in P} (r_p - 1) = 2(N - 1).$$

We collect terms to simplify the left side of this equation: Let n_p denote the order of the orbit O_p of p. By the Counting Formula (6.9.2),

$$(6.12.6) \qquad\qquad r_p n_p = N.$$

If two poles p and p' are in the same orbit, their orbits are equal, so $n_p = n_{p'}$, and therefore $r_p = r_{p'}$. We label the various orbits arbitrarily, say as $O_1, O_2, \ldots O_k$, and we let $n_i = n_p$ and $r_i = r_p$ for p in O_i, so that $n_i r_i = N$. Since the orbit O_i contains n_i elements, there are n_i terms equal to $r_i - 1$ on the left side of (6.12.5). We collect those terms together. This gives us the equation

$$\sum_{i=1}^{k} n_i (r_i - 1) = 2N - 2.$$

We divide both sides by N to get a famous formula:

$$(6.12.7) \qquad\qquad \sum_i \left(1 - \frac{1}{r_i}\right) = 2 - \frac{2}{N}.$$

This may not look like a promising tool, but in fact it tells us a great deal. The right side is between 1 and 2, while each term on the left is at least $\frac{1}{2}$. It follows that there can be at most three orbits.

The rest of the classification is made by listing the possibilities:

One orbit: $1 - \frac{1}{r_1} = 2 - \frac{2}{N}$. This is impossible, because $1 - \frac{1}{r_1} < 1$, while $2 - \frac{2}{N} \geq 1$.

Two orbits: $(1 - \frac{1}{r_1}) + (1 - \frac{1}{r_2}) = 2 - \frac{2}{N}$, that is, $\frac{1}{r_1} + \frac{1}{r_2} = \frac{2}{N}$.

Because r_i divides N, this equation holds only when $r_1 = r_2 = N$, and then $n_1 = n_2 = 1$. There are two poles p_1 and p_2, both fixed by every element of the group. So G is the cyclic group C_N of rotations whose axis of rotation is the line ℓ through p_1 and p_2.

Three orbits: $(1 - \frac{1}{r_1}) + (1 - \frac{1}{r_2}) + (1 - \frac{1}{r_3}) = 2 - \frac{2}{N}$.

This is the most interesting case. Since $\frac{2}{N}$ is positive, the formula implies that

$$(6.12.8) \qquad\qquad \frac{1}{r_1} + \frac{1}{r_2} + \frac{1}{r_3} > 1.$$

We arrange the r_i in increasing order. Then $r_1 = 2$: If all r_i were at least 3, the left side would be ≤ 1.

Case 1: $r_1 = r_2 = 2$. The third order $r_3 = k$ can be arbitrary, and $N = 2k$:

$$r_i = 2, 2, k; \quad n_i = k, k, 2; \quad N = 2k.$$

There is one pair of poles $\{p, p'\}$ making the orbit O_3. Half of the elements of G fix p, and the other half interchange p and p'. So the elements of G are rotations about the line ℓ through p and p', or else they are rotations by π about a line perpendicular to ℓ. The group G is the group of rotations fixing a regular k-gon Δ, the dihedral group D_k. The polygon Δ lies in the plane perpendicular to ℓ, and the vertices and the centers of faces of Δ correspond to the remaining poles. The bilateral symmetries of Δ in $\mathbb{R}^2$ have become rotations through the angle π in $\mathbb{R}^3$.

Case 2: $r_1 = 2$ and $2 < r_2 \le r_3$. The equation $1/2 + 1/4 + 1/4 = 1$ rules out the possibility that $r_2 \ge 4$. Therefore $r_2 = 3$. Then the equation $1/2 + 1/3 + 1/6 = 1$ rules out $r_3 \ge 6$. Only three possibilities remain:

(6.12.9)

   (i) $r_i = 2, 3, 3; \quad n_i = 6, 4, 4; \quad N = 12.$

      The poles in the orbit O_3 are the vertices of a regular tetrahedron, and G is the tetrahedral group T of its 12 rotational symmetries.

  (ii) $r_i = 2, 3, 4; \quad n_i = 12, 8, 6; \quad N = 24.$

      The poles in the orbit O_3 are the vertices of a regular octahedron, and G is the octahedral group O of its 24 rotational symmetries.

 (iii) $r_i = 2, 3, 5; \quad n_i = 30, 20, 12; \quad N = 60.$

      The poles in the orbit O_3 are the vertices of a regular icosahedron, and G is the icosahedral group I of its 60 rotational symmetries.

In each case, the integers n_i are the numbers of edges, faces, and vertices, respectively.

    Intuitively, the poles in an orbit should be the vertices of a regular polyhedron because they must be evenly spaced on the sphere. However, this isn't quite correct, because the centers of the edges of a cube, for example, form an orbit, but they do not span a regular polyhedron. The figure they span is called a *truncated polyhedron*.

    We'll verify the assertion of **(iii)**. Let V be the orbit O_3 of order twelve. We want to show that the poles in this orbit are the vertices of a regular icosahedron. Let p be one of the poles in V. Thinking of p as the north pole of the unit sphere gives us an equator and a south pole. Let H be the stabilizer of p. Since $r_3 = 5$, this is a cyclic group, generated by a rotation x about p with angle $2\pi/5$. When we decompose V into H-orbits, we must get at least two H-orbits of order 1. These are the north and south poles. The ten other poles making up V form two H-orbits of order 5. We write them as $\{q_0, \ldots, q_4\}$ and $\{q'_0, \ldots, q'_4\}$, where $q_i = x^i q_0$ and $q'_i = x^i q'_0$. By symmetry between the north and south poles, one of these H-orbits is in the northern hemisphere and one is in the southern hemisphere, or else both are on the equator. Let's say that the orbit $\{q_i\}$ is in the northern hemisphere or on the equator.

Let $|x, y|$ denote the spherical distance between points x and y on the unit sphere. We note that $d = |p, q_i|$ is independent of $i = 0, \ldots, 4$, because there is an element of H that carries $q_0 \rightsquigarrow q_i$, while fixing p. Similarly, $d' = |p, q_i'|$ is independent of i. So as p' ranges over the orbit V the distance $|p, p'|$ takes on only four values $0, d, d'$ and π. The values d and d' are taken on five times each, and 0 and π are taken on once. Since G operates transitively on V, we will obtain the same four values when p is replaced by any other pole in V.

We note that $d \leq \pi/2$ while $d' \geq \pi/2$. Because there are five poles in the orbit $\{q_i\}$, the spherical distance $|q_i, q_{i+1}|$ is less than $\pi/2$, so it is equal to d, and $d < \pi/2$. Therefore that orbit isn't on the equator. The three poles p, q_i, q_{i+1} form an equilateral triangle. There are five congruent equilateral triangles meeting at p, and therefore five congruent triangles meet at each pole. They form the faces of an icosahedron.

Note: There are just five regular polyhedra. This can be proved by counting the number of ways that one can begin to build one by bringing congruent regular polygons together at a vertex. One can assemble three, four, or five equilateral triangles, three squares, or three regular pentagons. (Six triangles, four squares, or three hexagons glue together into flat surfaces.) So there are just five possibilities. But this analysis omits the interesting question of existence. Does an icosahedron exist? Of course, we can build one out of cardboard. But when we do, the triangles never fit together precisely, and we take it on faith that this is due to our imprecision. If we drew the analogous conclusion about the circle of fifths in music, we'd be wrong: the circle of fifths almost closes up, but not quite. The best way to be sure that the icosahedron exists may be to write down the coordinates of its vertices and check the distances. This is Exercise 12.7. □

Our discussion of the isometries of the plane has analogues for the group of isometries of three-space. One can define the notion of a *crystallographic group*, a discrete subgroup whose translation group is a three-dimensional lattice. The crystallographic groups are analogous to two-dimensional lattice groups, and crystals form examples of three-dimensional configurations having such groups as symmetry. It can be shown that there are 230 types of crystallographic groups, analogous to the 17 lattice groups (6.6.2). This is too long a list to be useful, so crystals have been classified more crudely into seven *crystal systems*. For more about this, and for a discussion of the 32 crystallographic point groups, look in a book on crystallography, such as [Schwarzenbach].

> *Un bon héritage vaut mieux que le plus joli problème*
> *de géométrie, parce qu'il tient lieu de méthode*
> *générale, et sert à resoudre bien des problèmes.*
>
> —Gottfried Wilhelm Leibnitz[2]

---

[2]I learned this quote from V.I. Arnold. l'Hôpital had written to Leibniz, apologizing for a long silence, and saying that he had been in the country taking care of an inheritance. In his reply, Leibniz told him not to worry, and continued with the sentence quoted.

EXERCISES

Section 1 Symmetry of Plane Figures

1.1. Determine all symmetries of Figures 6.1.4, 6.1.6, and 6.1.7.

Section 3 Isometries of the Plane

3.1. Verify the rules (6.3.3).

3.2. Let m be an orientation-reversing isometry. Prove algebraically that m^2 is a translation.

3.3. Prove that a linear operator on $\mathbb{R}^2$ is a reflection if and only if its eigenvalues are 1 and –1, and the eigenvectors with these eigenvalues are orthogonal.

3.4. Prove that a conjugate of a glide reflection in M is a glide reflection, and that the glide vectors have the same length.

3.5. Write formulas for the isometries (6.3.1) in terms of a complex variable $z = x + iy$.

3.6. (a) Let s be the rotation of the plane with angle $\pi/2$ about the point $(1, 1)^t$. Write the formula for s as a product $t_a \rho_\theta$.

    (b) Let s denote reflection of the plane about the vertical axis $x = 1$. Find an isometry g such that $grg^{-1} = s$, and write s in the form $t_a \rho_\theta r$.

Section 4 Finite Groups of Orthogonal Operators on the Plane

4.1. Write the product $x^2 y x^{-1} y^{-1} x^3 y^3$ in the form $x^i y^j$ in the dihedral group D_n.

4.2. (a) List all subgroups of the dihedral group D_4, and decide which ones are normal.

    (b) List the proper normal subgroups N of the dihedral group D_{15}, and identify the quotient groups D_{15}/N.

    (c) List the subgroups of D_6 that do not contain x^3.

4.3. (a) Compute the left cosets of the subgroup $H = \{1, x^5\}$ in the dihedral group D_{10}.

    (b) Prove that H is normal and that D_{10}/H is isomorphic to D_5.

    (c) Is D_{10} isomorphic to $D_5 \times H$?

Section 5 Discrete Groups of Isometries

5.1. Let ℓ_1 and ℓ_2 be lines through the origin in $\mathbb{R}^2$ that intersect in an angle π/n, and let r_i be the reflection about ℓ_i. Prove that r_1 and r_2 generate a dihedral group D_n.

5.2. What is the crystallographic restriction for a discrete group of isometries whose translation group L has the form $\mathbb{Z}a$ with $a \neq 0$?

5.3. How many sublattices of index 3 are contained in a lattice L in $\mathbb{R}^2$?

5.4. Let (a, b) be a lattice basis of a lattice L in $\mathbb{R}^2$. Prove that every other lattice basis has the form $(a', b') = (a, b)P$, where P is a 2×2 integer matrix with determinant ± 1.

5.5. Prove that the group of symmetries of the frieze pattern ◁◁◁◁◁◁ is isomorphic to the direct product $C_2 \times C_\infty$ of a cyclic group of order 2 and an infinite cyclic group.

5.6. Let G be the group of symmetries of the frieze pattern ﹁ᒋ﹁ᒋ﹁ᒋ﹁ᒋ . Determine the point group $\overline{G}$ of G, and the index in G of its subgroup of translations.

5.7. Let N denote the group of isometries of a line $\mathbb{R}^1$. Classify discrete subgroups of N, identifying those that differ in the choice of origin and unit length on the line.

*5.8. Let N' be the group of isometries of an infinite ribbon

$$R = \{(x, y) \,|\, -1 \le y \le 1\}.$$

It can be viewed as a subgroup of the group M. The following elements are in N':

$$t_a: (x, y) \rightarrow (x + a, y)$$
$$s: (x, y) \rightarrow (-x, y)$$
$$r: (x, y) \rightarrow (x, -y)$$
$$\rho: (x, y) \rightarrow (-x, -y).$$

(a) State and prove analogues of (6.3.3) for these isometries.

(b) A frieze pattern is a pattern on the ribbon that is periodic and whose group of symmetries is discrete. Classify the corresponding symmetry groups, identifying those that differ in the choice of origin and unit length on the ribbon. Begin by making some patterns with different symmetries. Make a careful case analysis when proving your results.

5.9. Let G be a discrete subgroup of M whose translation group is not trivial. Prove that there is a point p_0 in the plane that is not fixed by any element of G except the identity.

5.10. Let f and g be rotations of the plane about distinct points, with arbitrary nonzero angles of rotation θ and ϕ. Prove that the group generated by f and g contains a translation.

5.11. If S and S' are subsets of $\mathbb{R}^n$ with $S \subset S'$, then S is *dense* in S' if for every element s' of S', there are elements of S arbitrarily near to s'.

(a) Prove that a subgroup Γ of $\mathbb{R}^+$ is either dense in $\mathbb{R}$, or else discrete.

(b) Prove that the subgroup of $\mathbb{R}^+$ generated by 1 and $\sqrt{2}$ is dense in $\mathbb{R}^+$.

(c) Let H be a subgroup of the group G of angles. Prove that H is either a cyclic subgroup of G or else it is dense in G.

5.12. Classify discrete subgroups of the additive group $\mathbb{R}^{3+}$.

Section 6   Plane Crystallographic Groups

6.1. (a) Determine the point group $\overline{G}$ for each of the patterns depicted in Figure (6.6.2).

(b) For which of the patterns can coordinates be chosen so that the group G operates on the lattice L?

6.2. Let G be the group of symmetries of an equilateral triangular lattice L. Determine the index in G of the subgroup of translations in G.

6.3. With each of the patterns shown, determine the point group and find a pattern with the same type of symmetry in Table 6.6.2.

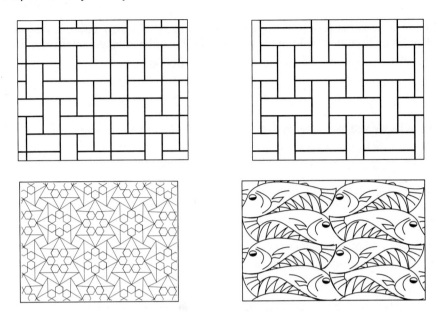

*6.4. Classify plane crystallographic groups with point group $D_1 = \{\bar{1}, \bar{r}\}$.

6.5. (a) Prove that if the point group of a two-dimensional crystallographic group G is C_6 or D_6, the translation group L is an equilateral triangular lattice.

   (b) Classify those groups.

*6.6. Prove that symmetry groups of the figures in Figure 6.6.2 exhaust the possibilities.

Section 7   Abstract Symmetry: Group Operations

7.1. Let $G = D_4$ be the dihedral group of symmetries of the square.

   (a) What is the stabilizer of a vertex? of an edge?

   (b) G operates on the set of two elements consisting of the diagonal lines. What is the stabilizer of a diagonal?

7.2. The group M of isometries of the plane operates on the set of lines in the plane. Determine the stabilizer of a line.

7.3. The symmetric group S_3 operates on two sets U and V of order 3. Decompose the product set $U \times V$ into orbits for the "diagonal action" $g(u, v) = (gu, gv)$, when

   (a) the operations on U and V are transitive,

   (b) the operation on U is transitive, the orbits for the operation on V are $\{v_1\}$ and $\{v_2, v_3\}$.

7.4. In each of the figures in Exercise 6.3, find the points that have nontrivial stabilizers, and identify the stabilizers.

7.5. Let G be the group of symmetries of a cube, including the orientation-reversing symmetries. Describe the elements of G geometrically.

7.6. Let G be the group of symmetries of an equilateral triangular prism P, including the orientation-reversing symmetries. Determine the stabilizer of one of the rectangular faces of P and the order of the group.

• **7.7.** Let $G = GL_n(\mathbb{R})$ operate on the set $V = \mathbb{R}^n$ by left multiplication.

   (a) Describe the decomposition of V into orbits for this operation.

   (b) What is the stabilizer of e_1?

7.8. Decompose the set $\mathbb{C}^{2\times 2}$ of 2×2 complex matrices into orbits for the following operations of $GL_2(\mathbb{C})$:   **(a)** left multiplication,   **(b)** conjugation.

• **7.9. (a)** Let S be the set $\mathbb{R}^{m\times n}$ of real $m\times n$ matrices, and let $G = GL_m(\mathbb{R})\times GL_n(\mathbb{R})$. Prove that the rule $(P, Q) * A = PAQ^{-1}$ define an operation of G on S.

   (b) Describe the decomposition of S into G-orbits.

   (c) Assume that $m \le n$. What is the stabilizer of the matrix $[I \,|\, 0]$?

7.10. (a) Describe the orbit and the stabilizer of the matrix $\begin{bmatrix} 1 & 0 \\ 0 & 2 \end{bmatrix}$ under conjugation in the general linear group $GL_n(\mathbb{R})$.

   (b) Interpreting the matrix in $GL_2(\mathbb{F}_5)$, find the order of the orbit.

7.11. Prove that the only subgroup of order 12 of the symmetric group S_4 is the alternating group A_4.

Section 8   The Operation on Cosets

8.1. Does the rule $P * A = PAP^t$ define an operation of GL_n on the set of $n\times n$ matrices?

8.2. What is the stabilizer of the coset $[aH]$ for the operation of G on G/H?

8.3. Exhibit the bijective map (6.8.4) explicitly, when G is the dihedral group D_4 and S is the set of vertices of a square.

8.4. Let H be the stabilizer of the index 1 for the operation of the symmetric group $G = S_n$ on the set of indices $\{1, \dots, n\}$. Describe the left cosets of H in G and the map (6.8.4) in this case.

Section 9   The Counting Formula

9.1. Use the counting formula to determine the orders of the groups of rotational symmetries of a cube and of a tetrahedron.

9.2. Let G be the group of rotational symmetries of a cube, let G_v, G_e, G_f be the stabilizers of a vertex v, an edge e, and a face f of the cube, and let V, E, F be the sets of vertices, edges, and faces, respectively. Determine the formulas that represent the decomposition of each of the three sets into orbits for each of the subgroups.

9.3. Determine the order of the group of symmetries of a dodecahedron, when orientation-reversing symmetries such as reflections in planes are allowed.

9.4. Identify the group T' of all symmetries of a regular tetrahedron, including orientation-reversing symmetries.

9.5. Let F be a section of an I-beam, which one can think of as the product set of the letter I and the unit interval. Identify its group of symmetries, orientation-reversing symmetries included.

9.6. Identify the group of symmetries of a baseball, taking the seam (but not the stitching) into account and allowing orientation-reversing symmetries.

Section 10   Operations on Subsets

10.1. Determine the orders of the orbits for left multiplication on the set of subsets of order 3 of D_3.

10.2. Let S be a finite set on which a group G operates transitively, and let U be a subset of S. Prove that the subsets gU cover S evenly, that is, that every element of S is in the same number of sets gU.

10.3. Consider the operation of left multiplication by G on the set of its subsets. Let U be a subset such that the sets gU partition G. Let H be the unique subset in this orbit that contains 1. Prove that H is a subgroup of G.

Section 11   Permutation Representations

11.1. Describe all ways in which S_3 can operate on a set of four elements.

11.2. Describe all ways in which the tetrahedral group T can operate on a set of two elements.

11.3. Let S be a set on which a group G operates, and let H be the subset of elements g such that $gs = s$ for all s in S. Prove that H is a normal subgroup of G.

11.4. Let G be the dihedral group D_4 of symmetries of a square. Is the action of G on the vertices a faithful action? on the diagonals?

11.5. A group G operates faithfully on a set S of five elements, and there are two orbits, one of order 3 and one of order 2. What are the possible groups?

Hint: Map G to a product of symmetric groups.

11.6. Let $F = \mathbb{F}_3$. There are four one-dimensional subspaces of the space of column vectors F^2. List them. Left multiplication by an invertible matrix permutes these subspaces. Prove that this operation defines a homomorphism $\varphi: GL_2(F) \to S_4$. Determine the kernel and image of this homomorphism.

11.7. For each of the following groups, find the smallest integer n such that the group has a faithful operation on a set of order n: **(a)** D_4, **(b)** D_6, **(c)** the quaternion group H.

11.8. Find a bijective correspondence between the multiplicative group $\mathbb{F}_p^\times$ and the set of automorphisms of a cyclic group of order p.

11.9. Three sheets of rectangular paper S_1, S_2, S_3 are made into a stack. Let G be the group of all symmetries of this configuration, including symmetries of the individual sheets as well as permutations of the set of sheets. Determine the order of G, and the kernel of the map $G \to S_3$ defined by the permutations of the set $\{S_1, S_2, S_3\}$.

Section 12   Finite Subgroups of the Rotation Group

12.1. Explain why the groups of symmetries of the dodecahedron and the icosahedron are isomorphic.

12.2. Describe the orbits of poles for the group of rotations of an octahedron.

12.3. Let O be the group of rotations of a cube, and let S be the set of four diagonal lines connecting opposite vertices. Determine the stabilizer of one of the diagonals.

12.4. Let $G = O$ be the group of rotations of a cube, and let H be the subgroup carrying one of the two inscribed tetrahedra to itself (see Exercise 3.4). Prove that $H = T$.

12.5. Prove that the icosahedral group has a subgroup of order 10.

12.6. Determine all subgroups of **(a)** the tetrahedral group, **(b)** the icosahedral group.

12.7. The 12 points $(\pm 1, \pm \alpha, 0)^t$, $(0, \pm 1, \pm \alpha)^t$, $(\pm \alpha, 0, \pm 1)^t$ form the vertices of a regular icosahedron if $\alpha > 1$ is chosen suitably. Verify this, and determine α.

*12.8. Prove the crystallographic restriction for three-dimensional crystallographic groups: A rotational symmetry of a crystal has order 2, 3, 4, or 6.

Miscellaneous Problems

*M.1. Let G be a two-dimensional crystallographic group such that no element $g \neq 1$ fixes any point of the plane. Prove that G is generated by two translations, or else by one translation and one glide.

M.2. (a) Prove that the set Aut G of automorphisms of a group G forms a group, the law of composition being composition of functions.

    (b) Prove that the map $\varphi : G \to$ Aut G defined by $g \rightsquigarrow$ (conjugation by g) is a homomorphism, and determine its kernel.

    (c) The automorphisms that are obtained as conjugation by a group element are called *inner automorphisms*. Prove that the set of inner automorphisms, the image of φ, is a normal subgroup of the group Aut G.

M.3. Determine the groups of automorphisms (see Exercise M.2) of the group
    (a) C_4, **(b)** C_6, **(c)** $C_2 \times C_2$, **(d)** D_4, **(e)** the quaternion group H.

*M.4. With coordinates $x_1, \ldots, x_n$ in $\mathbb{R}^n$ as usual, the set of points defined by the inequalities $-1 \le x_i \le +1$, for $i = 1, \ldots, n$, is an n-dimensional *hypercube* C_n. The 1-dimensional hypercube is a line segment and the 2-dimensional hypercube is a square. The 4-dimensional hypercube has eight *face cubes*, the 3-dimensional cubes defined by $\{x_i = 1\}$ and by $\{x_i = -1\}$, for $i = 1, 2, 3, 4$, and it has 16 *vertices* $(\pm 1, \pm 1, \pm 1, \pm 1)$.

    Let G_n denote the subgroup of the orthogonal group O_n of elements that send the hypercube to itself, the group of symmetries of C_n, including the orientation-reversing symmetries. Permutations of the coordinates and sign changes are among the elements of G_n.

    (a) Use the counting formula and induction to determine the order of the group G_n.

    (b) Describe G_n explicitly, and identify the stabilizer of the vertex $(1, \ldots, 1)$. Check your answer by showing that G_2 is isomorphic to the dihedral group D_4.

*M.5. (a) Find a way to determine the area of one of the hippo heads that make up the first pattern in Figure 6.6.2. Do the same for one of the fleurs-de-lys in the pattern at the bottom of the figure.

    (b) A *fundamental domain* D for a plane crystallographic group is a bounded region of the plane such that the images gD, g in G, cover the plane exactly once, without overlap.

Find two noncongruent fundamental domains for group of symmetries of the hippo pattern. Do the same for the fleur-de-lys pattern.

(c) Prove that if D and D' are fundamental domains for the same pattern, then D can be cut into finitely many pieces and reassembled to form D'.

(d) Find a formula relating the area of a fundamental domain to the order of the point group of the pattern.

*M.6. Let G be a discrete subgroup of M. Choose a point p in the plane whose stablilizer in G is trivial, and let S be the orbit of p. For every point q of S other than p, let ℓ_q be the line that is the perpendicular bisector of the line segment $[p, q]$, and let H_q be the half plane that is bounded by ℓ_q and that contains p. Prove that $D = \bigcap H_q$ is a fundamental domain for G (see Exercise M.5).

*M.7. Let G be a finite group operating on a finite set S. For each element g of G, let S^g denote the subset of elements of S fixed by $g : S^g = \{s \in S \mid gs = s\}$, and let G_s be the stabilizer of s.

(a) We may imagine a true–false table for the assertion that $gs = s$, say with rows indexed by elements of G and columns indexed by elements of S. Construct such a table for the action of the dihedral group D_3 on the vertices of a triangle.

(b) Prove the formula $\sum_{s \in S} |G_s| = \sum_{g \in G} |S^g|$.

(c) Prove *Burnside's Formula*: $|G| \cdot (\text{number of orbits}) = \sum_{g \in G} |S^g|$.

M.8. There are $70 = \binom{8}{4}$ ways to color the edges of an octagon, with four black and four white. The group D_8 operates on this set of 70, and the orbits represent equivalent colorings. Use Burnside's Formula (see Exercise M.7) to count the number of equivalence classes.

CHAPTER 7

More Group Theory

The more to do or to prove, the easier the doing or the proof.

—James Joseph Sylvester

We discuss three topics in this chapter: conjugation, the most important group operation, the Sylow Theorems, which describe subgroups of prime power order in a finite group, and generators and relations for a group.

7.1 CAYLEY'S THEOREM

Every group G operates on itself in several ways, *left multiplication* being one of them:

$$
\begin{aligned}
G \times G &\to G \\
g, x &\rightsquigarrow gx.
\end{aligned}
$$
(7.1.1)

This is a transitive operation – there is just one orbit. The stabilizer of any element is the trivial subgroup $\langle 1 \rangle$, so the operation is faithful, and the permutation representation

$$
\begin{aligned}
G &\to \mathrm{Perm}(G) \\
g &\rightsquigarrow m_g - \text{left multiplication by } g
\end{aligned}
$$
(7.1.2)

defined by this operation is injective (see Section 6.11).

Theorem 7.1.3 Cayley's Theorem. Every finite group is isomorphic to a subgroup of a permutation group. A group of order n is isomorphic to a subgroup of the symmetric group S_n.

Proof. Since the operation by left multiplication is faithful, G is isomorphic to its image in $\mathrm{Perm}(G)$. If G has order n, $\mathrm{Perm}(G)$ is isomorphic to S_n. $\qquad\square$

Cayley's Theorem is interesting, but it is difficult to use because the order of S_n is usually too large in comparison with n.

7.2 THE CLASS EQUATION

Conjugation, the operation of G on itself defined by

$$
(g, x) \rightsquigarrow gxg^{-1}.
$$
(7.2.1)

is more subtle and more important than left multiplication. Obviously, we shouldn't use multiplicative notation for this operation. We'll verify the associative law (6.7.1) for the operation, using $g * x$ as a temporary notation for the conjugate gxg^{-1}:

$$(gh) * x = (gh)x(gh)^{-1} = ghxh^{-1}g^{-1} = g(h * x)g^{-1} = g * (h * x).$$

Having checked the axiom, we return to the usual notation gxg^{-1}.

• The stabilizer of an element x of G for the operation of conjugation is called the *centralizer* of x. It is often denoted by $Z(x)$:

(7.2.2) $$Z(x) = \{g \in G \mid gxg^{-1} = x\} = \{g \in G \mid gx = xg\}.$$

The centralizer of x is the set of elements that commute with x.

• The orbit of x for conjugation is called the *conjugacy class* of x, and is often denoted by $C(x)$. It consists of all of the conjugates gxg^{-1}:

(7.2.3) $$C(x) = \{x' \in G \mid x' = gxg^{-1} \text{ for some } g \text{ in } G\}.$$

The counting formula (6.9.2) tells us that

(7.2.4) $$|G| = |Z(x)| \cdot |C(x)|$$
$$|G| = |centralizer| \cdot |conj. \ class|$$

The *center* Z of a group G was defined in Chapter 2. It is the set of elements that commute with every element of the group: $Z = \{z \in G \mid zy = yz \text{ for all } y \text{ in } G\}$.

Proposition 7.2.5

(a) The centralizer $Z(x)$ of an element x of G contains x, and it contains the center Z.

(b) An element x of G is in the center if and only if its centralizer $Z(x)$ is the whole group G, and this happens if and only if the conjugacy class $C(x)$ consists of the element x alone. □

Since the conjugacy classes are orbits for a group operation, they partition the group. This fact gives us the *class equation* of a finite group:

(7.2.6) $$|G| = \sum_{\substack{\text{conjugacy} \\ \text{classes } C}} |C|.$$

If we number the conjugacy classes, writing them as $C_1, \dots, C_k$, the class equation reads

(7.2.7) $$|G| = |C_1| + \cdots + |C_k|.$$

The conjugacy class of the identity element 1 consists of that element alone. It seems natural to list that class first, so that $|C_1| = 1$. The other occurences of 1 on the right side of the class equation correspond to the elements of the center Z of G. Note also that each term on the right side divides the left side, because it is the order of an orbit.

(7.2.8) The numbers on the right side of the class equation divide the order of the group, and at least one of them is equal to 1.

This is a strong restriction on the combinations of integers that may occur in such an equation.

The symmetric group S_3 has order 6. With our usual notation, the element x has order 3. Its centralizer $Z(x)$ contains x, so its order is 3 or 6. Since $yx = x^2y$, x is not in the center of the group, and $|Z(x)| = 3$. It follows that $Z(x) = \langle x \rangle$, and the counting formula (7.2.4) shows that the conjugacy class $C(x)$ has order 2. Similar reasoning shows that the conjugacy class $C(y)$ of the element y has order 3. The class equation of the symmetric group S_3 is

(7.2.9)
$$6 = 1 + 2 + 3.$$

As we see, the counting formula helps to determine the class equation. One can determine the order of a conjugacy class directly, or one can compute the order of its centralizer. The centralizer, being a subgroup, has more structure, and computing its order is often the better way. We will see a case in which it is easier to determine the conjugacy classes in the next section, but let's look at another case in which one should use the centralizer.

Let G be the special linear group $SL_2(\mathbb{F}_3)$ of matrices of determinant 1 with entries in the field $\mathbb{F}_3$. The order of this group is 24 (see Exercise 4.4). To start computing the class equation by listing the elements of G would be incredibly boring. It is better to begin by computing the centralizers of a few matrices A. This is done by solving the equation $PA = AP$, for the matrix P. It is easier to use this equation, rather than $PAP^{-1} = A$. For instance, let

$$A = \begin{bmatrix} & -1 \\ 1 & \end{bmatrix} \quad \text{and} \quad P = \begin{bmatrix} a & b \\ c & d \end{bmatrix}.$$

The equation $PA = AP$ imposes the conditions $b = -c$ and $a = d$, and then the equation $\det P = 1$ becomes $a^2 + c^2 = 1$. This equation has four solutions in $\mathbb{F}_3$: $a = \pm 1, c = 0$ and $a = 0, c = \pm 1$. So $|Z(A)| = 4$ and $|C(A)| = 6$. This gives us a start for the class equation: $24 = 1 + 6 + \cdots$. To finish the computation, one needs to compute centralizers of a few more matrices. Since conjugate elements have the same characteristic polynomial, one can begin by choosing elements with different characteristic polynomials.

The class equation of $SL_2(\mathbb{F}_3)$ is

(7.2.10)
$$24 = 1 + 1 + 4 + 4 + 4 + 4 + 6.$$

7.3   *p*-GROUPS

The class equation has several applications to groups whose orders are positive powers of a prime p. They are called *p-group*s.

Proposition 7.3.1 The center of a *p*-group is not the trivial group.

Proof. Say that $|G| = p^e$ with $e \geq 1$. Every term on the right side of the class equation divides p^e, so it is a power of p too, possibly $p^0 = 1$. The positive powers of p are divisible by p. If the class C_1 of the identity made the only contribution of 1 to the right side, the equation would read

$$p^e = 1 + \sum (\textit{multiples of } p).$$

This is impossible, so there must be more 1's on the right. The center is not trivial.   □

A similar argument can be used to prove the following theorem for operations of p-groups. We'll leave its proof as an exercise.

Theorem 7.3.2 Fixed Point Theorem. Let G be a p-group, and let S be a finite set on which G operates. If the order of S is not divisible by p, there is a fixed point for the operation of G on S – an element s whose stabilizer is the whole group.   □

Proposition 7.3.3   Every group of order p^2 is abelian.

Proof. Let G be a group of order p^2. According to the previous proposition, its center Z is not the trivial group. So the order of Z must be p or p^2. If the order of Z is p^2, then $Z = G$, and G is abelian as the proposition asserts. Suppose that the order of Z is p, and let x be an element of G that is not in Z. The centralizer $Z(x)$ contains x as well as Z, so it is strictly larger than Z. Since $|Z(x)|$ divides $|G|$, it must be equal to p^2, and therefore $Z(x) = G$. This means that x commutes with every element of G, so it is in the center after all, which is a contradiction. Therefore the center cannot have order p.   □

Corollary 7.3.4   A group of order p^2 is either cyclic, or the product of two cyclic groups of order p.

Proof. Let G be a group of order p^2. If G contains an element of order p^2, it is cyclic. If not, every element of G different from 1 has order p. We choose elements x and y of order p such that y is not in the subgroup $\langle x \rangle$. Proposition 2.11.4 shows that G is isomorphic to the product $\langle x \rangle \times \langle y \rangle$.   □

The number of isomorphism classes of groups of order p^e increases rapidly with e. There are five isomorphism classes of groups of order eight, 14 isomorphism classes of groups of order 16, and 51 isomorphism classes of groups of order 32.

7.4   THE CLASS EQUATION OF THE ICOSAHEDRAL GROUP

In this section we use the conjugacy classes in the icosahedral group I – the group of rotational symmetries of a dodecahedron, to study this interesting group. You may want to refer to a model of a dodecahedron or to an illustration while thinking about this.

Let $\theta = 2\pi/3$. The icosahedral group contains the rotation by θ about a vertex v. This rotation has spin (v, θ), so we denote it by $\rho_{(v,\theta)}$. The 20 vertices form an I-orbit orbit, and if v' is another vertex, then $\rho_{(v,\theta)}$ and $\rho_{(v',\theta)}$ are conjugate elements of I. This follows from Corollary 5.1.28(b). The vertices form an orbit of order 20, so all of the rotations $\rho_{(v,\theta)}$ are conjugate. They are distinct, because the only spin that defines the same rotation as (v, θ) is $(-v, -\theta)$ and $-\theta \neq \theta$. So these rotations form a conjugacy class of order 20.

Next, I contains rotations with angle $2\pi/5$ about the center of a face, and the 12 faces form an orbit. Reasoning as above, we find a conjugacy class of order 12. Similarly, the rotations with angle $4\pi/5$ form a conjugacy class of order 12.

Finally, I contains a rotation with angle π about the center of an edge. There are 30 edges, which gives us 30 spins (e, π). But $\pi = -\pi$. If e is the center of an edge, so is $-e$, and the spins (e, π) and $(-e, -\pi)$ represent the same rotation. This conjugacy class contains only 15 distinct rotations.

The class equation of the icosahedral group is

(7.4.1) $$60 = 1 + 20 + 12 + 12 + 15.$$

Note: Calling (v, θ) and (e, π) spins isn't accurate, because v and e can't both have unit length. But this is obviously not an important point.

Simple Groups

A group G is *simple* if it is not the trivial group and if it contains no proper normal subgroup – no normal subgroup other than $\langle 1 \rangle$ and G. (This use of the word *simple* does not mean "uncomplicated." Its meaning here is roughly "not compound.") Cyclic groups of prime order contain no proper subgroup at all; they are therefore simple groups. All other groups except the trivial group contain proper subgroups, though not necessarily proper normal subgroups.

The proof of the following lemma is straightforward.

Lemma 7.4.2 Let N be a normal subgroup of a group G.

(a) If N contains an element x, then it contains the conjugacy class $C(x)$ of x.
(b) N is a union of conjugacy classes.
(c) The order of N is the sum of the orders of the conjugacy classes that it contains.   □

We now use the class equation to prove the following theorem.

Theorem 7.4.3 The icosahedral group I is a simple group.

Proof. The order of a proper normal subgroup of the icosahedral group is a proper divisor of 60, and according to the lemma, it is also the sum of some of the terms on the right side of the class equation (7.4.1), including the term 1, which is the order of the conjugacy class of the identity element. There is no integer that satisfies both of those requirements, and this proves the theorem.   □

The property of being simple can be useful because one may run across normal subgroups, as the next theorem illustrates.

Theorem 7.4.4 The icosahedral group is isomorphic to the alternating group A_5. Therefore A_5 is a simple group.

Proof. To describe this isomorphism, we need to find a set S of five elements on which I operates. This is rather subtle, but the five cubes that can be inscribed into a dodecahedron, one of which is shown below, form such a set.

The icosahedral group operates on this set of five cubes, and this operation defines a homomorphism $\varphi : I \to S_5$, the associated permutation representation. We show that φ defines an isomorphism from I to the alternating group A_5. To do this, we use the fact that I is a simple group, but the only information that we need about the operation is that it isn't trivial.

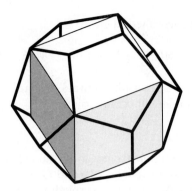

(7.4.5)   One of the Cubes Inscribed in a Dodecahedron.

The kernel of φ is a normal subgroup of I. Since I is a simple group, the kernel is either the trivial group $\langle 1 \rangle$ or the whole group I. If the kernel were the whole group, the operation of I on the set of five cubes would be the trivial operation, which it is not. Therefore $\ker \varphi = \langle 1 \rangle$. This shows that φ is injective. It defines an isomorphism from I to its image in S_5.

Next, we compose the homomorphism φ with the sign homomorphism $\sigma : S_5 \to \{\pm 1\}$, obtaining a homomorphism $\sigma\varphi : I \to \{\pm 1\}$. If this homomorphism were surjective, its kernel would be a proper normal subgroup of I. This is not the case because I is simple. Therefore the restriction is the trivial homomorphism, which means that the image of φ is contained in the kernel of σ, the alternating group A_5. Both I and A_5 both have order 60, and φ is injective. So the image of φ, which is isomorphic to I, is A_5. $\square$

7.5   CONJUGATION IN THE SYMMETRIC GROUP

The least confusing way to describe conjugation in the symmetric group is to think of relabeling the indices. If the given indices are $\mathbf{1, 2, 3, 4, 5}$, and if we relabel them as $\mathbf{a, b, c, d, e}$, respectively, the permutation $p = (\mathbf{134})(\mathbf{25})$ is changed to $(\mathbf{a\,c\,d})(\mathbf{b\,e})$.

To write a formula for this procedure, we let $\varphi : I \to L$ denote the relabeling map that goes from the set I of indices to the set L of letters: $\varphi(\mathbf{1}) = \mathbf{a}$, $\varphi(\mathbf{2}) = \mathbf{b}$, etc. Then the relabeled permutation is $\varphi \circ p \circ \varphi^{-1}$. This is explained as follows:

First map letters to indices using φ^{-1}.

Next, permute the indices by p.

Finally, map indices back to letters using φ.

We can use a permutation q of the indices to relabel in the same way. The result, the conjugate $p' = qpq^{-1}$, will be a new permutation of the same set of indices. For example, if we use $q = (\mathbf{1452})$ to relabel, we will get

$$qpq^{-1} = (\mathbf{1452}) \circ (\mathbf{134})(\mathbf{25}) \circ (\mathbf{2541}) = (\mathbf{435})(\mathbf{12}) = p'.$$

There are two things to notice. First, the relabeling will produce a permutation whose cycles have the same lengths as the original one. Second, by choosing the permutation q

suitably, we can obtain any other permutation that has cycles of those same lengths. If we write one permutation above the other, ordered so that the cycles correspond, we can use the result as a table to define q. For example, to obtain $p' = (435)(12)$ as a conjugate of the permutation $p = (134)(25)$, as we did above, we could write

$$\frac{(134)\,(25)}{(435)\,(12)}.$$

The relabeling permutation q is obtained by reading this table down: $1 \rightsquigarrow 4$, etc.

Because a cycle can start from any of its indices, there will most often be several permutations q that yield the same conjugate.

The next proposition sums up the discussion.

Proposition 7.5.1 Two permutations p and p' are conjugate elements of the symmetric group if and only if their cycle decompositions have the same orders. $\square$

We use Proposition 7.5.1 to determine the class equation of the symmetric group S_4. The cycle decomposition of a permutation gives us a partition of the set $\{1, 2, 3, 4\}$. The orders of the subsets making a partition of four can be

$$1, 1, 1, 1; \quad 2, 1, 1; \quad 2, 2; \quad 3, 1; \quad \text{or } 4.$$

The permutations with cycles of these orders are the identity, the transpositions, the products of (disjoint) transpositions, the 3-cycles, and the 4-cycles, respectively.

There are six transpositions, three products of transpositions, eight 3-cycles, and six 4-cycles. The proposition tells us that each of these sets forms one conjugacy class, so the class equation of S_4 is

(7.5.2) $$24 = 1 + 3 + 6 + 6 + 8.$$

A similar computation shows that the class equation of the symmetric group S_5 is

(7.5.3) $$120 = 1 + 10 + 15 + 20 + 20 + 30 + 24.$$

We saw in the previous section (7.4.4) that the alternating group A_5 is a simple group because it is isomorphic to the icosahedral group I, which is simple. We now prove that most alternating groups are simple.

Theorem 7.5.4 For every $n \geq 5$, the alternating group A_n is a simple group.

To complete the picture we note that A_2 is the trivial group, A_3 is cyclic of order three, and that A_4 is not simple. The group of order four that consists of the identity and the three products of transpositions $(12)(34)$, $(13)(24)$, $(14)(23)$ is a normal subgroup of S_4 and of A_4 (see (2.5.13)**(b)**).

Lemma 7.5.5

(a) For $n \geq 3$, the alternating group A_n is generated by 3-cycles.
(b) For $n \geq 5$, the 3-cycles form a single conjugacy class in the alternating group A_n.

Proof. **(a)** This is analogous to the method of row reduction. Say that an even permutation p, not the identity, fixes m of the indices. We show that if we multiply p on the left by a suitable 3-cycle q, the product qp will fix at least $m+1$ indices. Induction on m will complete the proof.

If p is not the identity, it will contain either a k-cycle with $k \geq 3$, or a product of two 2-cycles. It does not matter how we number the indices, so we may suppose that $p = (123\cdots\mathbf{k})\cdots$ or $p = (12)(34)\cdots$. Let $q = (321)$. The product qp fixes the index $\mathbf{1}$ as well as all indices fixed by p.

(b) Suppose that $n \geq 5$, and let $q = (123)$. According to Proposition 7.5.1, the 3-cycles are conjugate in the symmetric group S_n. So if q' is another 3-cycle, there is a permutation p such that $pqp^{-1} = q'$. If p is an even permutation, then q and q' are conjugate in A_n. Suppose that p is odd. The transposition $\tau = (45)$ is in S_n because $n \geq 5$, and $\tau q \tau^{-1} = q$. Then $p\tau$ is even, and $(p\tau)q(p\tau)^{-1} = q'$. $\qquad\square$

Proof. We now proceed to the proof of the Theorem. Let N be a nontrivial normal subgroup of the alternating group A_n with $n \geq 5$. We must show that N is the whole group A_n. It suffices to show that N contains a 3-cycle. If so, then (7.5.5)**(b)** will show that N contains every three-cycle, and (7.5.5)**(a)** will show that $N = A_n$.

We are given that N is a normal subgroup and that it contains a permutation x different from the identity. Three operations are allowed: We may multiply, invert, and conjugate. For example, if g is any element of A_n, then gxg^{-1} and x^{-1} are in N too. So is their product, the commutator $gxg^{-1}x^{-1}$. And since g can be arbitrary, these commutators give us many elements that must be in N.

Our first step is to note that a suitable power of x will have prime order, say order ℓ. We may replace x by this power, so we may assume that x has order ℓ. Then the cycle decomposition of x will consist of ℓ-cycles and 1-cycles.

Unfortunately, the rest of the proof requires looking separately at several cases. In each of the cases, we compute a commutator $gxg^{-1}x^{-1}$, hoping to be led to a 3-cycle. Appropriate elements can be found by experiment.

Case 1: x has order $\ell \geq 5$.

How the indices are numbered is irrelevant, so we may suppose that x contains the ℓ-cycle $(12345\cdots\ell)$, say $x = (12345\cdots\ell)y$, where y is a permutation of the remaining indices. Let $g = (432)$. Then

first do this

$$gxg^{-1}x^{-1} = [(432)] \circ [(12345\cdots\ell)y] \circ [(234)] \circ [y^{-1}(\ell\cdots54321)] = (245).$$

The commutator is a 3-cycle.

Case 2: x has order 3.

There is nothing to prove if x is a 3-cycle. If not, then x contains at least two 3-cycles, say $x = (123)(456)y$. Let $g = (432)$. Then $gxg^{-1}x^{-1} = (15243)$. The commutator has order 5. We go back to Case 1.

Case 3a: x has order 2 and it contains a 1-cycle.

Since it is an even permutation, x must contain at least two 2-cycles, say $x = (1\,2)(3\,4)(5)\,y$. Let $g = (5\,3\,1)$. Then $gxg^{-1}x^{-1} = (1\,5\,2\,4\,3)$. The commutator has order 5, and we go back to Case 1 again.

Case 3b: x has order $\ell = 2$, and contains no 1-cycles.

Since $n \geq 5$, x contains more than two 2-cycles. Say $x = (1\,2)(3\,4)(5\,6)\,y$. Let $g = (5\,3\,1)$. Then $gxg^{-1}x^{-1} = (1\,5\,3)(2\,4\,6)$. The commutator has order 3 and we go back to Case 2.

These are the possibilities for an even permutation of prime order, so the proof of the theorem is complete. $\qquad\square$

7.6 NORMALIZERS

We consider the orbit of a subgroup H of a group G for the operation of conjugation by G. The orbit of $[H]$ is the set of *conjugate subgroups* $[gHg^{-1}]$, with g in G. The stabilizer of $[H]$ for this operation is called the *normalizer* of H, and is denoted by $N(H)$:

$$(7.6.1) \qquad\qquad N(H) = \{g \in G \mid gHg^{-1} = H\}.$$

The Counting Formula reads

$$(7.6.2) \qquad\qquad |G| = |N(H)| \cdot (\textit{number of conjugate subgroups}).$$

The number of conjugate subgroups is equal to the index $[G:N(H)]$.

Proposition 7.6.3 Let H be a subgroup of a group G, and let N be the normalizer of H.

(a) H is a normal subgroup of N.
(b) H is a normal subgroup of G if and only if $N = G$.
(c) $|H|$ divides $|N|$ and $|N|$ divides $|G|$. $\qquad\qquad\square$

For example, let H be the cyclic subgroup of order two of the symmetric group S_5 that is generated by the element $p = (1\,2)(3\,4)$. The conjugacy class $C(p)$ contains the 15 pairs of disjoint transpositions, each of which generates a conjugate subgroup of H. The counting formula shows that the normalizer $N(H)$ has order eight: $120 = 8 \cdot 15$.

7.7 THE SYLOW THEOREMS

The Sylow Theorems describe the subgroups of prime power order of an arbitrary finite group. They are named after the Norwegian mathematician Ludwig Sylow, who discovered them in the 19th century.

Let G be a group of order n, and let p be a prime integer that divides n. Let p^e denote the largest power of p that divides n, so that

$$(7.7.1) \qquad\qquad n = p^e m,$$

where m is an integer not divisible by p. Subgroups H of G of order p^e are called *Sylow p-subgroups* of G. A Sylow p-subgroup is a p-group whose index in the group isn't divisible by p.

Theorem 7.7.2 First Sylow Theorem. A finite group whose order is divisible by a prime p contains a Sylow p-subgroup.

Proofs of the Sylow Theorems are at the end of the section.

Corollary 7.7.3 A finite group whose order is divisible by a prime p contains an element of order p.

Proof. Let G be such a group, and let H be a Sylow p-subgroup of G. Then H contains an element x different from 1. The order of x divides the order of H, so it is a positive power of p, say p^k. Then $x^{p^{k-1}}$ has order p.   □

This corollary isn't obvious. We already know that the order of any element divides the order of the group, but we might imagine a group of order 6, for example, made up of the identity 1 and five elements of order 2. No such group exists. A group of order 6 must contain an element of order 3 and an element of order 2.

The remaining Sylow Theorems give additional information about the Sylow subgroups.

Theorem 7.7.4 Second Sylow Theorem. Let G be a finite group whose order is divisible by a prime p.

(a) The Sylow p-subgroups of G are conjugate subgroups.

(b) Every subgroup of G that is a p-group is contained in a Sylow p-subgroup.

A conjugate subgroup of a Sylow p-subgroup will be a Sylow p-subgroup too.

Corollary 7.7.5 A group G has just one Sylow p-subgroup H if and only if that subgroup is normal.   □

Theorem 7.7.6 Third Sylow Theorem. Let G be a finite group whose order n is divisible by a prime p. Say that $n = p^e m$, where p does not divide m, and let s denote the number of Sylow p-subgroups. Then s divides m and s is congruent to 1 modulo p: $s = kp + 1$ for some integer $k \geq 0$.

Before proving the Sylow theorems, we will use them to classify groups of orders 6, 15, and 21. These examples show the power of the theorems, but the classification of groups of order n is not easy when n has many factors. There are just too many possibilities.

Proposition 7.7.7

(a) Every group of order 15 is cyclic.

(b) There are two isomorphism classes of groups of order 6, the class of the cyclic group C_6 and the class of the symmetric group S_3.

(c) There are two isomorphism classes of groups of order 21: the class of the cyclic group C_{21}, and the class of a group G generated by two elements x and y that satisfy the relations $x^7 = 1$, $y^3 = 1$, $yx = x^2 y$.

Proof. **(a)** Let G be a group of order 15. According to the Third Sylow Theorem, the number of its Sylow 3-subgroups divides 5 and is congruent 1 modulo 3. The only such integer is 1. Therefore there is one Sylow 3-subgroup, say H, and it is a normal subgroup. For similar reasons, there is just one Sylow 5-subgroup, say K, and it is normal. The subgroup H is cyclic of order 3, and K is cyclic of order 5. The intersection $H \cap K$ is the trivial group. Proposition 2.11.4**(d)** tells us that G is isomorphic to the product group $H \times K$. So all groups of order 15 are isomorphic to the product $C_3 \times C_5$ of cyclic groups and to each other. The cyclic group C_{15} is one such group, so all groups of order 15 are cyclic.

(b) Let G be a group of order 6. The First Sylow Theorem tells us that G contains a Sylow 3-subgroup H, a cyclic group of order 3, and a Sylow 2-subgroup K, cyclic of order 2. The Third Sylow Theorem tells us that the number of Sylow 3-subgroups divides 2 and is congruent 1 modulo 3. The only such integer is 1. So there is one Sylow 3-subgroup H, and it is a normal subgroup. The same theorem also tells us that the number of Sylow two-subgroups divides 3 and is congruent 1 modulo 2. That number is either 1 or 3.

Case 1: Both H and K are normal subgroups.

As in the previous example, G is isomorphic to the product group $H \times K$, which is abelian. All abelian groups of order 6 are cyclic.

Case 2: G contains 3 Sylow 2-subgroups, say K_1, K_2, K_3.

The group G operates by conjugation on the set $S = \{[K_1], [K_2], [K_3]\}$ of order three, and this gives us a homomorphism $\varphi : G \to S_3$ from G to the symmetric group, the associated permutation representation (6.11.2). The Second Sylow Theorem tells us that the operation on S is transitive, so the stabilizer in G of the element $[K_i]$, which is the normalizer $N(K_i)$, has order 2. It is equal to K_i. Since $K_1 \cap K_2 = \{1\}$, the identity is the only element of G that fixes all elements of S. The operation is faithful, and the permutation representation φ is injective. Since G and S_3 have the same order, φ is an isomorphism.

(c) Let G be a group of order 21. The Third Sylow Theorem shows that the Sylow 7-subgroup K must be normal, and that the number of Sylow 3-subgroups is 1 or 7. Let x be a generator for K, and let y be a generator for one of the Sylow 3-subgroups H. Then $x^7 = 1$ and $y^3 = 1$, so $H \cap K = \{1\}$, and therefore the product map $H \times K \to G$ is injective (2.11.4)**(a)**. Since G has order 21, the product map is bijective. The elements of G are the products $x^i y^j$ with $0 \le i < 7$ and $0 \le j < 3$.

Since K is a normal subgroup, yxy^{-1} is an element of K, a power of x, say x^i, with i in the range $1 \le i < 7$. So the elements x and y satisfy the relations

(7.7.8) $$x^7 = 1, \quad y^3 = 1, \quad yx = x^i y.$$

These relations are enough to determine the multiplication table for the group. However, the relation $y^3 = 1$ restricts the possible exponents i, because it implies that $y^3 x y^{-3} = x$:

$$x = y^3 x y^{-3} = y^2 x^i y^{-2} = yx^{i^2} y^{-1} = x^{i^3}.$$

Therefore $i^3 \equiv 1$ modulo 7. This tells us that i must be 1, 2, or 4.

The exponent $i = 3$, for instance, would imply $x = x^{3^3} = x^6 = x^{-1}$. Then $x^2 = 1$ and also $x^7 = 1$, from which it follows that $x = 1$. The group defined by the relations (7.7.8) with $i = 3$ is a cyclic group of order 3, generated by y.

Case 1: $yxy^{-1} = x$. Then x commutes with y. Both H and K are normal subgroups. As before, G is isomorphic to a direct product of cyclic groups of orders 3 and 7, and is a cyclic group.

Case 2: $yxy^{-1} = x^2$. As noted above, the multiplication table is determined. But we still have to show that this group actually exists. This comes down to showing that the relations don't cause the group to collapse, as happens when $i = 3$. We'll learn a systematic method for doing this, the Todd-Coxeter Algorithm, in Section 7.11. Another way is to exhibit the group explicitly, for example as a group of matrices. Some experimentation is required to do this.

Since the group we are looking for is supposed to contain an element of order 7, it is natural to try to find suitable matrices with entries modulo 7. At least we can write down a 2×2 matrix with entries in $\mathbb{F}_7$ that has order 7, namely the matrix x below. Then y can be found by trial and error. The matrices

$$x = \begin{bmatrix} 1 & 1 \\ & 1 \end{bmatrix}, \quad \text{and} \quad y = \begin{bmatrix} 2 & \\ & 1 \end{bmatrix}$$

with entries in $\mathbb{F}_7$ satisfy the relations $x^7 = 1$, $y^3 = 1$, $yx = x^2y$, and they generate a group of order 21.

Case 3: $yxy^{-1} = x^4$. Then $y^2xy^{-2} = x^2$. We note that y^2 is also an element of order 3. So we may replace y by y^2, which is another generator for H. The result is that the exponent 4 is replaced by 2, which puts us back in the previous case.

Thus there are two isomorphism classes of groups of order 21, as claimed.   □

We use two lemmas in the proof of the first Sylow Theorem.

Lemma 7.7.9 Let U be a subset of a group G. The order of the stabilizer $\text{Stab}([U])$ of $[U]$ for the operation of left multiplication by G on the set of its subsets divides both of the orders $|U|$ and $|G|$.

Proof. If H is a subgroup of G, the H-orbit of an element u of G for left multiplication by H is the right coset Hu. Let H be the stabilizer of $[U]$. Then multiplication by H permutes the elements of U, so U is partitioned into H-orbits, which are right cosets. Each coset has order $|H|$, so $|H|$ divides $|U|$. Because H is a subgroup, $|H|$ divides $|G|$.   □

Lemma 7.7.10 Let n be an integer of the form $p^e m$, where $e > 0$ and p does not divide m. The number N of subsets of order p^e in a set of order n is not divisible by p.

Proof. The number N is the binomial coefficient

$$\binom{n}{p^e} = \frac{n \ (n-1) \cdots \ (n-k) \cdots (n - p^e + 1)}{p^e (p^e - 1) \cdots (p^e - k) \ \cdots \ 1}.$$

The reason that $N \not\equiv 0$ modulo p is that every time p divides a term $(n - k)$ in the numerator of N, it also divides the term $(p^e - k)$ of the denominator the same number of times: If we write k in the form $k = p^i \ell$, where p does not divide ℓ, then $i < e$. Therefore $(m - k) = (p^e - k)$ and $(n - k) = (p^e m - k)$ are both divisible by p^i but not by p^{i+1}. $\square$

Proof of the First Sylow Theorem. Let S be the set of all subsets of G of order p^e. One of the subsets is a Sylow subgroup, but instead of finding it directly we look at the operation of left multiplication by G on S. We will show that one of the subsets $[U]$ of order p^e has a stabilizer of order p^e. That stabilizer will be the subgroup we are looking for.

We decompose S into orbits for the operation of left multiplication, obtaining an equation of the form

$$N = |\mathcal{S}| = \sum_{\text{orbits } O} |O|.$$

According to Lemma 7.7.10, p doesn't divide N. So at least one orbit has an order that isn't divisible by p, say the orbit $O_{[U]}$ of the subset $[U]$. Let H be the stabilizer of $[U]$. Lemma 7.7.9 tells us that the order of H divides the order of U, which is p^e. So $|H|$ is a power of p. We have $|H| \cdot |O_{[U]}| = |G| = p^e m$, and $|O_{[U]}|$ isn't divisible by p. Therefore $|O_{[U]}| = m$ and $|H| = p^e$. So H is a Sylow p-subgroup. $\square$

Proof of the Second Sylow Theorem. Suppose that we are given a p-subgroup K and a Sylow p-subgroup H. We will show that some conjugate subgroup H' of H contains K, which will prove **(b)**. If K is also a Sylow p-subgroup, it will be equal to the conjugate subgroup H', so **(a)** will be proved as well.

We choose a set $\mathcal{C}$ on which the group G operates, with these properties: p does not divide the order $|\mathcal{C}|$, the operation is transitive, and $\mathcal{C}$ contains an element c whose stabilizer is H. The set of left cosets of H in G has these properties, so such a set exists. (We prefer not to clutter up the notation by explicit reference to cosets.)

We restrict the operation of G on $\mathcal{C}$ to the p-group K. Since p doesn't divide $|\mathcal{C}|$, there is a fixed point c' for the operation of K. This is the Fixed Point Theorem 7.3.2. Since the operation of G is transitive, $c' = gc$ for some g in G. The stabilizer of c' is the conjugate subgroup gHg^{-1} of H (6.7.7), and since K fixes c', the stabilizer contains K. $\square$

Proof of the Third Sylow Theorem. We write $|G| = p^e m$ as before. Let s denote the number of Sylow p-subgroups. The Second Sylow Theorem tells us that the operation of G on the set S of Sylow p-subgroups is transitive. The stabilizer of a particular Sylow p-subgroup $[H]$ is the normalizer $N = N(H)$ of H. The counting formula tells us that the order of S, which is s, is equal to the index $[G:N]$. Since N contains H (7.6.3) and since $[G:H]$ is equal to m, s divides m.

Next, we decompose the set S into orbits for the operation of conjugation by H. The H-orbit of $[H]$ has order 1. Since H is a p-group, the order of any H-orbit is a power of p. To show that $s \equiv 1$ modulo p, we show that no element of S except $[H]$ is fixed by H.

Suppose that H' is a p-Sylow subgroup and that conjugation by H fixes $[H']$. Then H is contained in the normalizer N' of H', so both H and H' are Sylow p-subgroups of N'. The second Sylow theorem tells us that the p-Sylow subgroups of N' are conjugate subgroups of N'. But H' is a normal subgroup of N' (7.6.3)**(a)**. Therefore $H' = H$. $\square$

7.8   GROUPS OF ORDER 12

We use the Sylow Theorems to classify groups of order 12. This theorem serves to illustrate the fact that classifying groups becomes complicated when the order has several factors.

Theorem 7.8.1   There are five isomorphism classes of groups of order 12. They are represented by:

- the product of cyclic groups $C_4 \times C_3$,
- the product of cyclic groups $C_2 \times C_2 \times C_3$,
- the alternating group A_4,
- the dihedral group D_6,
- the group generated by elements x and y, with relations $x^4 = 1$, $y^3 = 1$, $xy = y^2 x$.

All but the last of these groups should be familiar. The product group $C_4 \times C_3$ is isomorphic to C_{12}, and $C_2 \times C_2 \times C_3$ is isomorphic to $C_2 \times C_6$ (see Proposition 2.11.3).

Proof. Let G be a group of order 12, let H be a Sylow 2-subgroup of G, which has order 4, and let K be a Sylow 3-subgroup of order 3. It follows from the Third Sylow Theorem that the number of Sylow 2-subgroups is either 1 or 3, and that the number of Sylow 3-subgroups is 1 or 4. Also, H is a group of order 4 and is therefore either a cyclic group C_4 or the Klein four group $C_2 \times C_2$ (Proposition 2.11.5). Of course K is cyclic.

Though this is not necessary for the proof, begin by showing that at least one of the two subgroups, H or K, is normal. If K is not normal, there will be four Sylow 3-subgroups conjugate to K, say $K_1, \ldots, K_4$, with $K_1 = K$. These groups have prime order, so the intersection of any two of them is the trivial group $\langle 1 \rangle$. Then there are only three elements of G that are not in any of the groups K_i. This fact is shown schematically below.

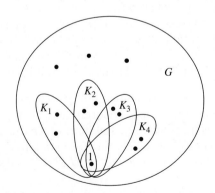

A Sylow 2-subgroup H has order 4, and $H \cap K_i = \langle 1 \rangle$. Therefore H consists of the three elements not in any of the groups K_i, together with 1. This describes H for us and shows that there is only one Sylow 2-subgroup. Thus H is normal.

Next, we note that $H \cap K = \langle 1 \rangle$, so the product map $H \times K \to G$ is a bijective map of sets (2.11.4). Every element of G has a unique expression as a product hk, with h in H and k in K.

Case 1: H and K are both normal.

Then G is isomorphic to the product group $H \times K$ (2.11.4). Since there are two possibilities for H and one for K, there are two possibilities for G:

$$G \approx C_4 \times C_3 \quad \text{or} \quad G \approx C_2 \times C_2 \times C_3.$$

These are the abelian groups of order 12.

Case 2: K is not normal.

There are four conjugate Sylow 3-subgroups, $K_1, \ldots, K_4$, and G operates by conjugation on this set of four. This operation determines a permutation representation, a homomorphism $\varphi: G \to S_4$ to the symmetric group. We'll show that φ maps G isomorphically to the alternating group A_4.

The normalizer N_i of K_i contains K_i, and the counting formula shows that $|N_i| = 3$. Therefore $N_i = K_i$. Since the only element in common to the subgroups K_i is the identity, only the identity stabilizes all of these subgroups. Thus the operation of G is faithful, φ is injective, and G is isomorphic to its image in S_4.

Since G has four subgroups of order 3, it contains eight elements of order 3. Their images are the 3-cycles in S_4, which generate A_4 (7.5.5). So the image of G contains A_4. Since G and A_4 have the same order, the image is equal to A_4.

Case 3: K is normal, but H is not.

Then H operates by conjugation on $K = \{1, y, y^2\}$. Since H is not normal, it contains an element x that doesn't commute with y, and then $xyx^{-1} = y^2$.

Case 3a: K is normal, H is not normal, and H is a cyclic group.

The element x generates H, so G is generated by elements x and y, with the relations

(7.8.2) $$x^4 = 1, \ y^3 = 1, \ xy = y^2 x.$$

These relations determine the multiplication table of G, so there is at most one isomorphism class of such groups. But we must show that these relations don't collapse the group further, and as with groups of order 21 (see 7.7.8), it is simplest to represent the group by matrices. We'll use complex matrices here. Let ω be the complex cube root of unity $e^{2\pi i/3}$. The complex matrices

(7.8.3) $$x = \begin{bmatrix} & -1 \\ 1 & \end{bmatrix}, \quad y = \begin{bmatrix} \omega & \\ & \omega^2 \end{bmatrix}$$

satisfy the three relations, and they generate a group of order 12.

Case 3b: K is normal, H is not normal, and $H \approx C_2 \times C_2$.

The stabilizer of y for the operation of H by conjugation on the set $\{y, y^2\}$ has order 2. So H contains an element $z \neq 1$ such that $zy = yz$ and also an element x such that $xy = y^2 x$. Since H is abelian, $xz = zx$. Then G is generated by three elements x, y, z, with relations

$$x^2 = 1, \ y^3 = 1, \ z^2 = 1, \ yz = zy, \ xz = zx, \ xy = y^2 x.$$

These relations determine the multiplication table of the group, so there is at most one isomorphism class of such groups. The dihedral group D_6 isn't one of the four groups described before, so it must be this one. Therefore G is isomorphic to D_6. $\square$

7.9 THE FREE GROUP

We have seen that one can compute in the symmetric group S_3 using the usual generators x and y, together with the relations $x^3 = 1$, $y^2 = 1$, and $yx = x^2 y$. In the rest of the chapter, we study generators and relations in other groups.

We first consider groups with generators that satisfy *no* relations other than ones (such as the associative law) that are implied by the group axioms. A set of group elements that satisfy no relations except those implied by the axioms is called *free*, and a group that has a free set of generators is called a *free group*.

To describe free groups, we start with an arbitrary set, say $S = \{a, b, c, \ldots\}$. We call its elements "symbols," and we define a *word* to be a finite string of symbols, in which repetition is allowed. For instance a, aa, ba, and $aaba$ are words. Two words can be composed by juxtaposition, that is, placing them side by side:

$$aa, ba \rightsquigarrow aaba.$$

This is an associative law of composition on the set W of words. We include the "empty word" in W as an identity element, and we use the symbol 1 to denote it. Then the set W becomes what is called the *free semigroup* on the set S. It isn't a group because it lacks inverses, and adding inverses complicates things a little.

Let S' be the set that consists of symbols a and a^{-1} for every a in S:

(7.9.1)
$$S' = \{a, a^{-1}, b, b^{-1}, c, c^{-1}, \ldots\},$$

and let W' be the semigroup of words made using the symbols in S'. If a word looks like

$$\cdots x x^{-1} \cdots \quad \text{or} \quad \cdots x^{-1} x \cdots$$

for some x in S, we may agree to *cancel* the two symbols x and x^{-1} to reduce the length of the word. A word is called *reduced* if no such cancellation can be made. Starting with any word w in W', we can perform a finite sequence of cancellations and must eventually get a reduced word w_0, possibly the empty word 1. We call w_0 a *reduced form* of w.

There may be more than one way to proceed with cancellation. For instance, starting with $w = abb^{-1} c^{-1} cb$, we can proceed in two ways:

$$
\begin{array}{cc}
\underline{a}\cancel{b}\cancel{b}^{-1}c^{-1}c\underline{b} & \underline{abb}^{-1}\cancel{c}^{-1}\cancel{c}b \\
\downarrow & \downarrow \\
\underline{a}\cancel{c}^{-1}\cancel{c}\underline{b} & \underline{ab}\cancel{b}^{-1}\cancel{b} \\
\downarrow & \downarrow \\
ab & ab
\end{array}
$$

The same reduced word is obtained at the end, though the symbols come from different places in the original word. (The ones that remain at the end have been underlined.) This is always true.

Proposition 7.9.2 There is only one reduced form of a given word w.

Proof. We use induction on the length of w. If w is reduced, there is nothing to show. If not, there must be some pair of symbols that can be cancelled, say the underlined pair

$$w = \cdots \underline{xx^{-1}} \cdots .$$

(Let's allow x to denote any element of S', with the understanding that if $x = a^{-1}$ then $x^{-1} = a$.) If we show that we can obtain every reduced form of w by cancelling the pair $\underline{xx}^{-1}$ first, the proposition will follow by induction, because the word $\cdots \cancel{x}\cancel{x}^{-1} \cdots$ is shorter.

Let w_0 be a reduced form of w. It is obtained from w by some sequence of cancellations. The first case is that our pair $\underline{xx}^{-1}$ is cancelled at some step in this sequence. If so, we may as well cancel $\underline{xx}^{-1}$ first. So this case is settled. On the other hand, since w_0 is reduced, the pair $\underline{xx}^{-1}$ cannot remain in w_0. At least one of the two symbols must be cancelled at some time. If the pair itself is not cancelled, the first cancellation involving the pair must look like

$$\cdots \cancel{x}^{-1}\underline{\cancel{x}x^{-1}} \cdots \quad \text{or} \quad \cdots \underline{x\cancel{x}^{-1}}\cancel{x} \cdots .$$

Notice that the word obtained by this cancellation is the same as the one obtained by cancelling the pair $\underline{xx}^{-1}$. So at this stage we may cancel the original pair instead. Then we are back in the first case, so the proposition is proved. $\square$

We call two words w and w' in W' *equivalent*, and we write $w \sim w'$, if they have the same reduced form. This is an equivalence relation.

Proposition 7.9.3 Products of equivalent words are equivalent: If $w \sim w'$ and $v \sim v'$, then $wv \sim w'v'$.

Proof. To obtain the reduced word equivalent to the product wv, we may first cancel as much as possible in w and in v, to reduce w to w_0 and v to v_0. Then wv is reduced to w_0v_0. Now we continue, cancelling in w_0v_0 until the word is reduced. If $w \sim w'$ and $v \sim v'$, the same process, when applied to $w'v'$, passes through w_0v_0 too, so it leads to the same reduced word. $\square$

It follows from this proposition that equivalence classes of words can be multiplied:

Proposition 7.9.4 The set $\mathcal{F}$ of equivalence classes of words in W' is a group, with the law of composition induced from multiplication (juxtaposition) in W'.

Proof. The facts that multiplication is associative and that the class of the empty word 1 is an identity follow from the corresponding facts in W' (see Lemma 2.12.8). We must check that all elements of $\mathcal{F}$ are invertible. But clearly, if w is the product $xy \cdots z$ of elements of S', then the class of $z^{-1} \cdots y^{-1}x^{-1}$ inverts the class of w. $\square$

The group $\mathcal{F}$ of equivalence classes of words in S' is called the *free group* on the set S. An element of $\mathcal{F}$ corresponds to exactly one reduced word in W'. To multiply reduced words, combine and cancel: $(abc^{-1})(cb) \rightsquigarrow abc^{-1}cb = abb$.

Power notation may be used: $aaab^{-1}b^{-1} = a^3b^{-2}$.

Note: The free group on a set $S = \{a\}$ of one element is simply an infinite cyclic group. In contrast, the free group on a set of two or more elements is quite complicated.

7.10   GENERATORS AND RELATIONS

Having described free groups, we now consider the more common case, that a set of generators of a group is not free – that there are some nontrivial relations among them.

Definition 7.10.1 A *relation* R among elements $x_1, \ldots, x_n$ of a group G is a word r in the free group on the set $\{x_1, \ldots, x_n\}$ that evaluates to 1 in G. We will write such a relation either as r, or for emphasis, as $r = 1$.

For example, the dihedral group D_n of symmetries of a regular n-sided polygon is generated by the rotation x with angle $2\pi/n$ and a reflection y, and these generators satisfy relations that were listed in (6.4.3):

(7.10.2) $$x^n = 1, \; y^2 = 1, \; xyxy = 1.$$

(The last relation is often written as $yx = x^{-1}y$, but it is best to write every relation in the form $r = 1$ here.)

One can use these relations to write the elements of D_n in the form $x^i y^j$ with $0 \le i < n$ and $0 \le j < 2$, and then one can compute the multiplication table for the group. So the relations determine the group. They are therefore called *defining relations*. When the relations are more complicated, it can be difficult to determine the elements of the group and the multiplication table explicitly, but, using the free group and the next lemma, we will define the concept of a group generated by a given set of elements, with a given set of relations.

Lemma 7.10.3 Let R be a subset of a group G. There exists a unique smallest normal subgroup N of G that contains R, called the *normal subgroup generated by R*. If a normal subgroup of G contains R, it contains N. The elements of N can be described in either of the following ways:

(a) An element of G is in N if it can be obtained from the elements of R using a finite sequence of the operations of multiplication, inversion, and conjugation.

(b) Let R' be the set consisting of elements r and r^{-1} with r in R. An element of G is in N if it can be written as a product $y_1 \cdots y_r$ of some arbitrary length, where each y_ν is a conjugate of an element of R'.

Proof. Let N denote the set of elements obtained by a sequence of the operations mentioned in **(a)**. A nonempty subset is a normal subgroup if and only if it is closed under those operations. Since N is closed under those operations, it is a normal subgroup. Moreover, any normal subgroup that contains R must contain N. So the smallest normal subgroup containing R exists, and is equal to N. Similar reasoning identifies N as the subset described in **(b)**. $\square$

As usual, we must take care of the empty set. We say that the empty set generates the trivial subgroup $\{1\}$.

Definition 7.10.4 Let $\mathcal{F}$ be the free group on a set $S = \{x_1, \ldots, x_n\}$, and let $R = \{r_1, \ldots, r_k\}$ be a set of elements of $\mathcal{F}$. The group *generated by S, with relations $r_1 = 1, \ldots, r_k = 1$*, is the quotient group $\mathcal{G} = \mathcal{F}/\mathcal{R}$, where $\mathcal{R}$ is the normal subgroup of $\mathcal{F}$ generated by R.

The group $\mathcal{G}$ will often be denoted by

(7.10.5) $$\langle x_1, \ldots, x_n \,|\, r_1, \ldots, r_k \rangle.$$

Thus the dihedral group D_n is isomorphic to the group

(7.10.6) $$\langle x, y \,|\, x^n, y^2, xyxy \rangle.$$

Example 7.10.7 In the tetrahedral group T of rotational symmetries of a regular tetrahedron, let x and y denote rotations by $2\pi/3$ about the center of a face and about a vertex, and let z denote rotation by π about the center of an edge, as shown below. With vertices numbered as in the figure, x acts on the vertices as the permutation $(\mathbf{234})$, y acts as $(\mathbf{123})$, and z acts as $(\mathbf{13})(\mathbf{24})$. Computing the product of these permutations shows that xyz acts trivially on the vertices. Since the only isometry that fixes all vertices is the identity, $xyz = 1$.

(7.10.8)

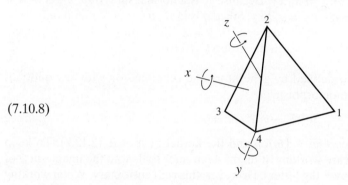

So the following relations hold in the tetrahedral group:

(7.10.9) $$x^3 = 1 \,, \ y^3 = 1 \,, \ z^2 = 1 \,, \ xyz = 1. \qquad \square$$

Two questions arise:

1. Is this a set of defining relations for T? In other words, is the group

(7.10.10) $$\langle x, y, z \,|\, x^3, y^3, z^2, xyz \rangle$$

isomorphic to T?

It is easy to verify that the rotations x, y, z generate T, but it isn't particularly easy to work with the relations. It is confusing enough to list the 12 elements of the group as products of the generators without repetition. We show in the next section that the answer to our question is yes, but we don't do that by writing the elements of the group explicitly.

2. How can one compute in a group $\mathcal{G} = \langle x_1, \ldots, x_n \,|\, r_1, \ldots, r_k \rangle$ that is presented by generators and relations?

Because computation in the free group $\mathcal{F}$ is easy, the only problem is to decide when an element w of the free group represents the identity element of $\mathcal{G}$, i.e., when w is an element of the subgroup $\mathcal{R}$. This is the *word problem* for $\mathcal{G}$. If we can solve the word problem, then because the relation $w_1 = w_2$ is equivalent to $w_1^{-1} w_2 = 1$, we will be able to decide when two elements of the free group represent equal elements of $\mathcal{G}$. This will enable us to compute.

The word problem can be solved in any finite group, but not in every group. However, we won't discuss this point, because some work is required to give a precise meaning to the statement that the word problem can or cannot be solved. If you are interested, see [Stillwell].

The next example shows that computation in $\mathcal{R}$ can become complicated, even in a relatively simple case.

Example 7.10.11 The element $w = yxyx$ is equal to 1 in the group T. Let's verify that w is in the normal subgroup $\mathcal{R}$ generated by the four relations (7.10.9). We use what you will recognize as a standard method: reducing w to the identity by the allowed operations.

The relations that we will use are z^2 and xyz, and we'll denote them by p and q, respectively. First, let $w_1 = y^{-1} w y = xyxy$. Because $\mathcal{R}$ is a normal subgroup, w_1 is in $\mathcal{R}$ if and only if w is. Next, let $w_2 = q^{-1} w_1 = z^{-1} xy$. Since q is in $\mathcal{R}$, w_2 is in $\mathcal{R}$ if and only if w_1 is. Continuing, $w_3 = zw_2 z^{-1} = xyz^{-1}$, $w_4 = q^{-1} w_3 = z^{-1} z^{-1}$, $pw_4 = 1$. Solving back, $w = yqz^{-1} q p^{-1} zy^{-1}$ is in $\mathcal{R}$. Thus $w = 1$ in the group (7.10.10). $\qquad \square$

We return to the group $\mathcal{G}$ defined by generators and relations. As with any quotient group, we have a canonical homomorphism

$$\pi : \mathcal{F} \longrightarrow \mathcal{F} / \mathcal{R} = \mathcal{G}$$

that sends a word w to the coset $\overline{w} = [w\mathcal{R}]$, and the kernel of π is $\mathcal{R}$ (2.12.2). To keep track of the group in which we are working, it might seem safer to denote the images in $\mathcal{G}$ of elements of $\mathcal{F}$ by putting bars over the letters. However, this isn't customary. When working in $\mathcal{G}$, one simply remembers that elements w_1 and w_2 of the free group are *equal in $\mathcal{G}$* if the cosets $w_1 \mathcal{R}$ and $w_2 \mathcal{R}$ are equal, or if $w_1^{-1} w_2$ is in $\mathcal{R}$.

Since the defining relations r_i are in $\mathcal{R}$, $r_i = 1$ is true in $\mathcal{G}$. If we write r_i out as words, then because π is a homomorphism, the corresponding product in $\mathcal{G}$ will be equal to 1 (see Corollary 2.12.3). For instance, $xyz = 1$ is true in the group $\langle x, y, z \mid x^3, y^3, z^2, xyz \rangle$.

We go back once more to the example of the tetrahedral group and to the first question. How is the group $\langle x, y, z \mid x^3, y^3, z^2, xyz \rangle$ related to T? A partial explanation is based on the mapping properties of free groups and of quotient groups. Both of these properties are intuitive. Their proofs are simple enough that we leave them as exercises.

Proposition 7.10.12 Mapping Property of the Free Group. Let $\mathcal{F}$ be the free group on a set $S = \{a, b, \dots\}$, and let G be a group. Any map of sets $f : S \to G$ extends in a unique way to a group homomorphism $\varphi : \mathcal{F} \to G$. If we denote the image $f(x)$ of an element x of S by $\underline{x}$, then φ sends a word in $S' = \{a, a^{-1}, b, b^{-1}, \dots\}$ to the corresponding product of the elements $\{\underline{a}, \underline{a}^{-1}, \underline{b}, \underline{b}^{-1} \dots\}$ in G. $\qquad \square$

This property reflects the fact that the elements of S satisfy no relations in $\mathcal{F}$ except those implied by the group axioms. It is the reason for the adjective "free."

Proposition 7.10.13 Mapping Property of Quotient Groups. Let $\varphi : G' \to G$ be a group homomorphism with kernel K, and let N be a normal subgroup of G' that is contained in K.

Let $\overline{G}' = G'/N$, and let $\pi: G' \to \overline{G}'$ be the canonical map $a \rightsquigarrow \overline{a}$. The rule $\overline{\varphi}(\overline{a}) = \varphi(a)$ defines a homomorphism $\overline{\varphi}: \overline{G}' \to G$, and $\overline{\varphi} \circ \pi = \varphi$.

This mapping property generalizes the First Isomorphism Theorem. The hypothesis that N be contained in the kernel K is, of course, essential.

The next corollary uses notation introduced previously: $S = \{x_1, \ldots, x_n\}$ is a subset of a group G, $R = \{r_1, \ldots, r_k\}$ is a set of relations among the elements of S of G, $\mathcal{F}$ is the free group on S, and $\mathcal{R}$ is the normal subgroup of $\mathcal{F}$ generated by R. Finally, $\mathcal{G} = \langle x_1, \ldots, x_n | r_1 \ldots, r_k \rangle = \mathcal{F}/\mathcal{R}$.

Corollary 7.10.14

   (i)   There is a canonical homomorphism $\psi: \mathcal{G} \to G$ that sends $x_i \rightsquigarrow x_i$.

  (ii) ψ is surjective if and only if the set S generates G.

 (iii) ψ is injective if and only if every relation among the elements of S is in $\mathcal{R}$.

Proof. We will prove **(i)**, and omit the verification of **(ii)** and **(iii)**. The mapping property of the free group gives us a homomorphism $\varphi: \mathcal{F} \to G$ with $\varphi(x_i) = x_i$. Since the relations r_i evaluate to 1 in G, R is contained in the kernel K of φ. Since the kernel is a normal subgroup, $\mathcal{R}$ is also contained in K. Then the mapping property of quotient groups gives us a map $\overline{\varphi}: \mathcal{G} \to G$. This is the map ψ:

If the map ψ described in the corollary is bijective, one says that R forms a *complete set of relations* among the generators S. To decide whether this is true requires knowing more about G. Going back to the tetrahedral group, the corollary gives us a homomorphism $\psi: \mathcal{G} \to T$, where $\mathcal{G} = \langle x, y, z | x^3, y^3, z^2, xyz \rangle$. It is surjective because x, y, z generate T. And we saw in Example 7.10.11 that the relation $yxyx$, which holds among the elements of T, is in the normal subgroup $\mathcal{R}$ generated by the set $\{x^3, y^3, z^2, xyz\}$. Is every relation among x, y, z in $\mathcal{R}$? If not, we'd want to add some more relations to our list. It may seem disappointing not to have the answer to this question yet, but we will see in the next section that ψ is indeed bijective.

Recapitulating, when we speak of a group defined by generators S and relations R, we mean the quotient group $\mathcal{G} = \mathcal{F}/\mathcal{R}$, where $\mathcal{F}$ is the free group on S and $\mathcal{R}$ is the normal subgroup of $\mathcal{F}$ generated by R. *Any* set of relations will define a group. The larger R is, the larger $\mathcal{R}$ becomes, and the more collapsing takes place in the homomorphism $\pi: \mathcal{F} \to \mathcal{G}$. The extreme case is $\mathcal{R} = \mathcal{F}$, in which case $\mathcal{G}$ is the trivial group. All relations become true in the trivial group. Problems arise because computation in $\mathcal{F}/\mathcal{R}$ may be difficult. But because generators and relations allow efficient computation in many cases, they are a useful tool.

7.11   THE TODD-COXETER ALGORITHM

The Todd-Coxeter Algorithm, which is described in this section, is an amazing method for determining the operation of a finite group G on the set of cosets of a subgroup H.

In order to compute, both G and H must be given explicitly. So we consider a group

$$(7.11.1) \qquad G = \langle x_1, \ldots, x_m \mid r_1, \ldots, r_k \rangle$$

presented by generators and relations, as in the previous section.

We also assume that the subgroup H of G is given explicitly, by a set of words

$$(7.11.2) \qquad \{h_1, \ldots, h_s\}$$

in the free group $\mathcal{F}$, whose images in G generate H.

The algorithm proceeds by constructing some tables that become easier to read when one works with *right cosets* Hg. The group G operates by *right multiplication* on the set of right cosets, and this changes the order of composition of operations. A product gh acts by right multiplication as "first multiply by g, then by h". Similarly, when we want permutations to operate on the right, we must read a product this way:

$$\underset{\text{first do this}}{(\mathbf{234})} \circ \underset{\text{then this}}{(\mathbf{123})} = (\mathbf{12})(\mathbf{34}).$$

The following rules suffice to determine the operation of G on the right cosets:

Rules 7.11.3

1. The operation of each generator is a permutation.
2. The relations operate trivially: they fix every coset.
3. The generators of H fix the coset $[H]$.
4. The operation is transitive.

The first rule follows from the fact that group elements are invertible, and the second one reflects the fact that the relations represent the identity element of G. Rules 3 and 4 are special properties of the operation on cosets.

When applying these rules, the cosets are usually denoted by indices $\mathbf{1, 2, 3}, \ldots$, with $\mathbf{1}$ standing for the coset $[H]$. At the start, one doesn't know how many indices will be needed; new ones are added as necessary.

We begin with a simple example, in which we replace y^3 by y^2 in the relations (7.10.9).

Example 7.11.4 Let G be the group $\langle x, y, z \mid x^3, y^2, z^2, xyz \rangle$, and let H be the cyclic subgroup $\langle z \rangle$ generated by z. First, Rule 3 tells us that z sends $\mathbf{1}$ to itself, $\mathbf{1} \overset{z}{\to} \mathbf{1}$. This exhausts the information in Rule 3, so Rules 1 and 2 take over. Rule 4 will only appear implicitly.

Nothing we have done up to now tells us what x does to the index $\mathbf{1}$. In such a case, the procedure is simply to assign a new index, $\mathbf{1} \overset{x}{\to} \mathbf{2}$. (Since $\mathbf{1}$ stands for the coset $[H]$, the index **2** stands for $[Hx]$, but it is best to ignore this.) Continuing, we don't know where x sends the index **2**, so we assign a third index, $2 \xrightarrow{x} 3$. Then $1 \xrightarrow{x^2} 3$.

What we have so far is a *partial operation*, meaning that the operations of some generators on some indices have been assigned. It is helpful to keep track of the partial operation as one goes along. The partial operation that we have so far is

$$z = (1) \cdots \quad \text{and} \quad x = (123 \cdots .$$

There is no closing parenthesis for the partial operation of x because we haven't determined the index to which x sends **3**.

Rule 2 now comes into play. It tells us that because x^3 is a relation, it fixes every index. Since x^2 sends **1** to **3**, x must send **3** back to **1**. It is customary to sum this information up in a table that exhibits the operation of x on the indices:

$$\frac{x \quad x \quad x}{1 \quad 2 \quad 3 \quad 1}$$

The relation xxx appears on top, and Rule 2 is reflected in the fact that the same index **1** appears at both ends. We have now determined the partial operation

$$x = (123) \cdots ,$$

except that we don't yet know whether or not the indices **1, 2, 3** represent distinct cosets.

Next, we ask for the operation of y on the index **1**. Again, we don't know it, so we assign a new index: $1 \xrightarrow{y} 4$. Rule 2 applies again. Since y^2 is a relation, y must send **4** back to **1**. This is exhibited in the table

$$\frac{y \quad y}{1 \quad 4 \quad 1}, \quad \text{so} \quad y = (14) \cdots .$$

For review, we have now determined the entries in the table below. The four defining relations appear on top.

x	x	x		y	y		z	z		x	y	z
1	**2**	**3**	**1**	**1**	**4**	**1**	**1**	**1**	**1**	**1**	**2**		**1**

The missing entry in the table for xyz is **1**. This follows from the fact that z acts as a permutation that fixes the index **1**. Entering **1** into the table, we see that $2 \xrightarrow{y} 1$. But we also have $4 \xrightarrow{y} 1$. Therefore $4 = 2$. We replace **4** by **2** and continue constructing a table.

The entries below have been determined:

x	x	x		y	y		z	z		x	y	z	
1	**2**	**3**	**1**	**1**	**2**	**1**	**1**	**1**	**1**	**1**	**2**	**1**	**1**
2	**3**	**1**	**2**	**2**	**1**	**2**	**2**		**2**	**2**	**3**		**2**
3	**1**	**2**	**3**	**3**		**3**	**3**		**3**	**3**	**1**	**2**	**3**

The third row of the table for xyz shows that $2 \xrightarrow{z} 3$, and this determines the rest of the table. There are three indices, and the complete operation is

$$x = (123), \quad y = (12), \quad z = (23).$$

At the end of the section, we will show that this is indeed the permutation representation defined by the operation of G on the cosets of H. □

What such a table tells us depends on the particular case. It will always tell us the number of cosets, the index $[G:H]$, which will be equal to the number of distinct indices: 3 in our example. It may also tell us something about the order of the generators. In our example, we are given the relation $z^2 = 1$, so the order of z must be 1 or 2. But z acts on indices as the transposition $(\mathbf{23})$, and this tells us that we can't have $z = 1$. So the order of z is 2, and $|H| = 2$. The counting formula $|G| = |H|[G:H]$ shows that G has order $2 \cdot 3 = 6$. The three permutations shown above generate the symmetric group S_3, so the permutation representation $G \rightarrow S_3$ defined by this operation is an isomorphism.

If one takes for H the trivial subgroup $\{1\}$, the cosets correspond bijectively to the group elements, and the permutation representation determines G completely. The cost of doing this is that there will be many indices. In other cases, the permutation representation may not suffice to determine the order of G.

We'll compute two more examples.

Example 7.11.5 We show that the relations (7.10.9) form a complete set of relations for the tetrahedral group. The verification is simplified a little if one uses the relation $xyz = 1$ to eliminate the generator z. Since $z^2 = 1$, that relation implies that $xy = z^{-1} = z$. The remaining elements x, y suffice to generate T. So we substitute $z = xy$ into z^2, and replace the relation z^2 by $xyxy$. The relations become

$$(7.11.6) \qquad\qquad x^3 = 1,\ y^3 = 1,\ xyxy = 1.$$

These relations among x and y are equivalent to the relations (7.10.9) among x, y, and z, so they hold in T.

Let G denote the group $\langle x, y \,|\, x^3, y^3, xyxy \rangle$. Corollary (7.10.14) gives us a homomorphism $\psi: G \rightarrow T$. To show that (7.11.6) are defining relations for T, we show that ψ is bijective. Since x and y generate T, ψ is surjective. So it suffices to show that the order of G is equal to the order of T, which is 12.

We choose the subgroup $H = \langle x \rangle$. This subgroup has order 1 or 3 because x^3 is one of the relations. If we show that H has order 3 and that the index of H in G is 4, it will follow that G has order 12, and we will be done. Here is the resulting table. To fill it in, work from both ends of the relations.

x	x	x		y	y	y		x	y	x	y	
1	1	1	1	1	2	3	1	1	1	2	3	1
2	3	4	2	2	3	1	2	2	3	1	1	2
3	4	2	3	3	1	2	3	3	4	4	2	3
4	2	3	4	4	4	4	4	4	2	3	4	4

The permutation representation is

$$(7.11.7) \qquad\qquad x = (\mathbf{234}),\ y = (\mathbf{123}).$$

Since there are four indices, the index of H is 4. Also, x does have order 3, not 1, because the permutation associated to x has order 3. The order of G is 12, as predicted.

Incidentally, we see that T is isomorphic to the alternating group A_4, because the permutations (7.11.7) generate that group. □

Example 7.11.8 We modify the relations (7.10.9) slightly, to illustrate how "bad" relations may collapse the group. Let G be the group $\langle x, y \mid x^3, y^3, yxyxy \rangle$, and let H be the subgroup $\langle y \rangle$. Here is a start for a table:

x	x	x		y	y	y		y	x	y	x	y
1	2	3	1	1	1	1	1	1	1	2	3	1	1
2			2	2			2	2	3	1	1	2	2

In the table for $yxyxy$, the first three entries in the first row are determined by working from the left, and the last three by working from the right. That row shows that $2 \overset{y}{\to} 3$. The second row is determined by working from the left, and it shows that $2 \overset{y}{\to} 2$. So $2 = 3$. Looking at the table for xxx, we see that then $2 = 1$. There is just one index left, so one coset, and consequently $H = G$. The group G is generated by y. It is a cyclic group of order 3. □

Warning: Care is essential when constructing such a table. Any mistake will cause the operation to collapse.

In our examples, we took for H the subgroup generated by one of the generators of G. If H is generated by a word h, one can introduce a new generator u and the new relation $u^{-1}h = 1$ (i.e., $u = h$). Then G (7.11.1) is isomorphic to the group

$$\langle x_1, \ldots, x_m, u \mid r_1, \ldots, r_k, u^{-1}h \rangle,$$

and H becomes the subgroup generated by u. If H has several generators, we do this for each of them.

We now address the question of why the procedure we have described determines the operation on cosets. A formal proof of this fact is not possible without first defining the algorithm formally, and we have not done this. We will discuss the question informally. (See [Todd-Coxeter] for a more complete discussion.) We describe the procedure this way: At a given stage of the computation, we will have some set **I** of indices, and a *partial operation* on **I**, the operation of some generators on some indices, will have been determined. A partial operation need not be consistent with Rules 1, 2, and 3, but it should be transitive; that is, all indices should be in the "partial orbit" of **1**. This is where Rule 4 comes in. It tells us not to introduce any indices that we don't need. In the starting position, **I** is the set {**1**} of one element, and no operations have been assigned.

At any stage there are two possible steps:

(7.11.9) (i) We may equate two indices **i** and **j** if the the rules tell us that they are equal, or (ii) we may choose a generator x and an index **i** such that **i**x has not been determined, and define **i**$x = $ **j**, where **j** is a new index.

We never equate indices unless their equality is implied by the rules.

We stop the process when an operation has been determined that is consistent with the rules. There are two questions to ask: First, will this procedure terminate? Second, if it terminates, is the operation the right one? The answer to both questions is yes. It can be shown that the process does terminate, provided that the group G is *finite*, and that preference is given to steps of type (i). We will not prove this. More important for applications is the fact that, if the process terminates, the resulting permutation representation is the right one.

Theorem 7.11.10 Suppose that a finite number of repetitions of steps (i) and (ii) yields a consistent table compatible with the rules (7.11.3). Then the table defines a permutation representation that, by suitable numbering, is the representation on the right cosets of H in G.

Proof. Say that the group is $G = \langle x_1, \ldots, x_n | r_1, \ldots, r_k \rangle$, and let $\mathbf{I}^*$ denote the final set of indices. For each generator x_i, the table determines a permutation of the indices, and the relations operate trivially. Corollary 7.10.14 gives us a homomorphism from G to the group of permutations of $\mathbf{I}^*$, and therefore an operation, on the right, of G on $\mathbf{I}^*$ (see Proposition 6.11.2). Provided that we have followed the rules, the table will show that the operation of G is transitive, and that the subgroup H fixes the index **1**.

Let C denote the set of right cosets of H. We prove the proposition by defining a bijective map $\varphi^* : \mathbf{I}^* \to C$ from $\mathbf{I}^*$ to C that is compatible with the operations of the group on the two sets. We define φ^* inductively, by defining at each stage a map $\varphi : \mathbf{I} \to C$ from the set of indices determined at that stage to C, compatible with the partial operation on $\mathbf{I}$ that has been determined. To start, $\varphi_0 : \{\mathbf{1}\} \to C$ sends $\mathbf{1} \rightsquigarrow [H]$. Suppose that $\varphi : \mathbf{I} \to C$ has been defined, and let $\mathbf{I}'$ be the result of applying one of the steps (7.11.9) to $\mathbf{I}$.

In case of step (ii), there is no difficulty in extending φ to a map $\varphi' : \mathbf{I}' \to C$. Say that $\varphi(\mathbf{i})$ is the coset $[Hg]$, and that the operation of a generator x on $\mathbf{i}$ has been defined to be a new index, say $\mathbf{i}x = \mathbf{j}$. Then we define $\varphi'(\mathbf{j}) = [Hgx]$, and we define $\varphi'(\mathbf{k}) = \varphi(\mathbf{k})$ for all other indices.

Next, suppose that we use step (i) to equate the indices $\mathbf{i}$ and $\mathbf{j}$, so that $\mathbf{I}$ is collapsed to form the new index set $\mathbf{I}'$. The next lemma allows us to define the map $\varphi' : \mathbf{I}' \to C$.

Lemma 7.11.11 Suppose that a map $\varphi : \mathbf{I} \to C$ is given, compatible with a partial operation on $\mathbf{I}$. Let $\mathbf{i}$ and $\mathbf{j}$ be indices in $\mathbf{I}$, and suppose that one of the rules forces $\mathbf{i} = \mathbf{j}$. Then $\varphi(\mathbf{i}) = \varphi(\mathbf{j})$.

Proof. This is true because, as we have remarked before, the operation on cosets does satisfy the rules.   □

The surjectivity of the map φ follows from the fact that the operation of the group on the set C of right cosets is transitive. As we now verify, the injectivity follows from the facts that the stabilizer of the coset $[H]$ is the subgroup H, and that the stabilizer of the index **1** contains H. Let $\mathbf{i}$ and $\mathbf{j}$ be indices. Since the operation on $\mathbf{I}^*$ is transitive, $\mathbf{i} = \mathbf{1}a$ for some group element a, and then $\varphi(\mathbf{i}) = \varphi(\mathbf{1})a = [Ha]$. Similarly, if $\mathbf{j} = \mathbf{1}b$, then $\varphi(\mathbf{j}) = [Hb]$. Suppose that $\varphi(\mathbf{i}) = \varphi(\mathbf{j})$, i.e., that $Ha = Hb$. Then $H = Hba^{-1}$, so ba^{-1} is an element of H. Since H stabilizes the index **1**, $\mathbf{1} = \mathbf{1}ba^{-1}$ and $\mathbf{i} = \mathbf{1}a = \mathbf{1}b = \mathbf{j}$.   □

The method of postulating what we want has many advantages;
they are the same as the advantages of theft over honest toil.

—Bertrand Russell

EXERCISES

Section 1   Cayley's Theorem

1.1. Does the rule $g * x = xg^{-1}$ define an operation of G on G?

1.2. Let H be a subgroup of a group G. Describe the orbits for the operation of H on G by left multiplication.

Section 2   The Class Equation

2.1. Determine the centralizer and the order of the conjugacy class of

(a) the matrix $\begin{bmatrix} 1 & 1 \\ & 1 \end{bmatrix}$ in $GL_2(\mathbb{F}_3)$,   (b) the matrix $\begin{bmatrix} 1 & \\ & 2 \end{bmatrix}$ in $GL_2(\mathbb{F}_5)$.

2.2. A group of order 21 contains a conjugacy class $C(x)$ of order 3. What is the order of x in the group?

2.3. A group G of order 12 contains a conjugacy class of order 4. Prove that the center of G is trivial.

2.4. Let G be a group, and let φ be the nth power map: $\varphi(x) = x^n$. What can be said about how φ acts on conjugacy classes?

2.5. Let G be the group of matrices of the form $\begin{bmatrix} x & y \\ & 1 \end{bmatrix}$, where $x, y \in \mathbb{R}$ and $x > 0$. Determine the conjugacy classes in G, and sketch them in the (x, y)-plane.

2.6. Determine the conjugacy classes in the group M of isometries of the plane.

2.7. Rule out as many as you can, as class equations for a group of order 10:
$$1 + 1 + 1 + 2 + 5, \quad 1 + 2 + 2 + 5, \quad 1 + 2 + 3 + 4, \quad 1 + 1 + 2 + 2 + 2 + 2.$$

2.8. Determine the possible class equations of nonabelian groups of order   (a) 8,   (b) 21.

2.9. Determine the class equation for the following groups: (a) the quaternion group, (b) D_4, (c) D_5, (d) the subgroup of $GL_2(\mathbb{F}_3)$ of invertible upper triangular matrices.

2.10. (a) Let A be an element of SO_3 that represents a rotation with angle π. Describe the centralizer of A geometrically.

(b) Determine the centralizer of the reflection r about the e_1-axis in the group M of isometries of the plane.

2.11. Determine the centralizer in $GL_3(\mathbb{R})$ of each matrix:

$$\begin{bmatrix} 1 & & \\ & 2 & \\ & & 3 \end{bmatrix}, \begin{bmatrix} 1 & & \\ & 1 & \\ & & 2 \end{bmatrix}, \begin{bmatrix} 1 & 1 & \\ & 1 & \\ & & 1 \end{bmatrix}, \begin{bmatrix} 1 & 1 & \\ & 1 & 1 \\ & & 1 \end{bmatrix}, \begin{bmatrix} & 1 & \\ & & 1 \\ 1 & & \end{bmatrix}.$$

*2.12. Determine all finite groups that contain at most three conjugacy classes.

2.13. Let N be a normal subgroup of a group G. Suppose that $|N| = 5$ and that $|G|$ is an odd integer. Prove that N is contained in the center of G.

2.14. The class equation of a group G is $1 + 4 + 5 + 5 + 5$.

(a) Does G have a subgroup of order 5? If so, is it a normal subgroup?

(b) Does G have a subgroup of order 4? If so, is it a normal subgroup?

2.15. Verify the class equation (7.2.10) of $SL_2(\mathbb{F}_3)$.

2.16. Let $\varphi: G \to G'$ be a surjective group homomorphism, let C denote the conjugacy class of an element x of G, and let C' denote the conjugacy class in G' of its image $\varphi(x)$. Prove that φ maps C surjectively to C', and that $|C'|$ divides $|C|$.

2.17. Use the class equation to show that a group of order pq, with p and q prime, contains an element of order p.

2.18. Which pairs of matrices $\begin{bmatrix} 0 & 1 \\ -1 & d \end{bmatrix}$, $\begin{bmatrix} 0 & -1 \\ 1 & d \end{bmatrix}$ are conjugate elements of **(a)** $GL_n(\mathbb{R})$, **(b)** $SL_n(\mathbb{R})$?

Section 3   *p*-Groups

3.1. Prove the Fixed Point Theorem (7.3.2).

3.2. Let Z be the center of a group G. Prove that if G/Z is a cyclic group, then G is abelian, and therefore $G = Z$.

3.3. A nonabelian group G has order p^3, where p is prime.

  (a) What are the possible orders of the center Z?
  (b) Let x be an element of G that isn't in Z. What is the order of its centralizer $Z(x)$?
  (c) What are the possible class equations for G?

3.4. Classify groups of order 8.

Section 4   The Class Equation of the Icosahedral Group

4.1. The icosahedral group operates on the set of five inscribed cubes in the dodecahedron. Determine the stabilizer of one of the cubes.

4.2. Is A_5 the only proper normal subgroup of S_5?

4.3. What is the centralizer of an element of order 2 of the icosahedral group I?

4.4. (a) Determine the class equation of the tetrahedral group T.
  (b) Prove that T has a normal subgroup of order 4, and no subgroup of order 6.

4.5. (a) Determine the class equation of the octahedral group O.
  (b) This group contains two proper normal subgroups. Find them, show that they are normal, and show that there are no others.

4.6. (a) Prove that the tetrahedral group T is isomorphic to the alternating group A_4, and that the octahedral group O is isomorphic to the symmetric group S_4.
    Hint: Find sets of four elements on which the groups operate.
  (b) Two tetrahedra can be inscribed into a cube C, each one using half the vertices. Relate this to the inclusion $A_4 \subset S_4$.

4.7. Let G be a group of order n that operates nontrivially on a set of order r. Prove that if $n > r!$, then G has a proper normal subgroup.

4.8. (a) Suppose that the centralizer $Z(x)$ of a group element x has order 4. What can be said about the center of the group?
  (b) Suppose that the conjugacy class $C(y)$ of an element y has order 4. What can be said about the center of the group?

4.9. Let x be an element of a group G, not the identity, whose centralizer $Z(x)$ has order pq, where p and q are primes. Prove that $Z(x)$ is abelian.

Section 5   Conjugation in the Symmetric Group

5.1. (a) Prove that the transpositions $(12), (23), \ldots, (n-1, n)$ generate the symmetric group S_n.

(b) How many transpositions are needed to write the cycle $(123 \cdots n)$?

(c) Prove that the cycles $(12 \cdots n)$ and (12) generate the symmetric group S_n.

5.2. What is the centralizer of the element (12) in S_5?

5.3. Determine the orders of the elements of the symmetric group S_7.

5.4. Describe the centralizer $Z(\sigma)$ of the permutation $\sigma = (153)(246)$ in the symmetric group S_7, and compute the orders of $Z(\sigma)$ and of $C(\sigma)$.

5.5. Let p and q be permutations. Prove that the products pq and qp have cycles of equal sizes.

5.6. Find all subgroups of S_4 of order 4, and decide which ones are normal.

5.7. Prove that A_n is the only subgroup of S_n of index 2.

5.8. [1]Determine the integers n such that there is a surjective homomorphism from the symmetric group S_n to S_{n-1}.

5.9. Let q be a 3-cycle in S_n. How many even permutations p are there such that $pqp^{-1} = q$?

5.10. Verify formulas (7.5.2) and (7.5.3) for the class equations of S_4 and S_5, and determine the centralizer of a representative element in each conjugacy class.

5.11. (a) Let C be the conjugacy class of an even permutation p in S_n. Show that C is either a conjugacy class in A_n, or else the union of two conjugacy classes in A_n of equal order. Explain how to decide which case occurs in terms of the centralizer of p.

(b) Determine the class equations of A_4 and A_5.

(c) One may also decompose the conjugacy classes of permutations of odd order into A_n-orbits. Describe this decomposition.

5.12. Determine the class equations of S_6 and A_6.

Section 6   Normalizers

6.1. Prove that the subgroup B of invertible upper triangular matrices in $GL_n(\mathbb{R})$ is conjugate to the subgroup L of invertible lower triangular matrices.

6.2. Let B be the subgroup of $G = GL_n(\mathbb{C})$ of invertible upper triangular matrices, and let $U \subset B$ be the set of upper triangular matrices with diagonal entries 1. Prove that $B = N(U)$ and that $B = N(B)$.

*6.3. Let P denote the subgroup of $GL_n(\mathbb{R})$ consisting of the permutation matrices. Determine the normalizer $N(P)$.

6.4. Let H be a normal subgroup of prime order p in a finite group G. Suppose that p is the smallest prime that divides the order of G. Prove that H is in the center $Z(G)$.

---

[1]Suggested by Ivan Borsenko.

6.5. Let p be a prime integer and let G be a p-group. Let H be a proper subgroup of G. Prove that the normalizer $N(H)$ of H is strictly larger than H, and that H is contained in a normal subgroup of index p.

*6.6. Let H be a proper subgroup of a finite group G. Prove:

   (a) The group G is not the union of the conjugate subgroups of H.

   (b) There is a conjugacy class C that is disjoint from H.

Section 7   The Sylow Theorems

7.1. Let $n = p^e m$, as in (4.5.1), and let N be the number of subsets of order p^e in a set of order n. Determine the congruence class of N modulo p.

7.2. Let $G_1 \subset G_2$ be groups whose orders are divisible by p, and let H_1 be a Sylow p-subgroup of G_1. Prove that there is a Sylow p-subgroup H_2 of G_2 such that $H_1 = H_2 \cap G_1$.

7.3. How many elements of order 5 might be contained in a group of order 20?

7.4. (a) Prove that no simple group has order pq, where p and q are prime.

   (b) Prove that no simple group has order $p^2 q$, where p and q are prime.

7.5. Find Sylow 2-subgroups of the following groups: **(a)** D_{10}, **(b)** T, **(c)** O, **(d)** I.

7.6. Exhibit a subgroup of the symmetric group S_7 that is a nonabelian group of order 21.

7.7. Let $n = pm$ be an integer that is divisible exactly once by p, and let G be a group of order n. Let H be a Sylow p-subgroup of G, and let S be the set of all Sylow p-subgroups. Explain how S decomposes into H-orbits.

*7.8. Compute the order of $GL_n(\mathbb{F}_p)$. Find a Sylow p-subgroup of $GL_n(\mathbb{F}_p)$, and determine the number of Sylow p-subgroups.

7.9. Classify groups of order **(a)** 33, **(b)** 18, **(c)** 20, **(d)** 30.

7.10. Prove that the only simple groups of order <60 are the groups of prime order.

Section 8   The Groups of Order 12

8.1. Which of the groups of order 12 described in Theorem 7.8.1 is isomorphic to $S_3 \times C_2$?

8.2. (a) Determine the smallest integer n such that the symmetric group S_n contains a subgroup isomorphic to the group (7.8.2).

   (b) Find a subgroup of $SL_2(\mathbb{F}_5)$ that is isomorphic to that group.

8.3. Determine the class equations of the groups of order 12.

8.4. Prove that a group of order $n = 2p$, where p is prime, is either cyclic or dihedral.

8.5. Let G be a nonabelian group of order 28 whose sylow 2 subgroups are cyclic.

   (a) Determine the numbers of sylow 2 - subgroups and of sylow 7 - subgroups.

   (b) Prove that there is at most one isomorphism class of such groups.

   (c) Determine the numbers of elements of each order, and the class equation of G.

8.6. Let G be a group of order 55.

    (a) Prove that G is generated by two elements x and y, with the relations $x^{11} = 1$, $y^5 = 1$, $yxy^{-1} = x^r$, for some r, $1 \leq r < 11$.

    (b) Decide which values of r are possible.

    (c) Prove that there are two isomorphism classes of groups of order 55.

Section 9   The Free Group

9.1. Let F be the free group on $\{x, y\}$. Prove that the three elements $u = x^2$, $v = y^2$, and $z = xy$ generate a subgroup isomorphic to the free group on u, v, and z.

9.2. We may define a *closed word* in S' to be the oriented loop obtained by joining the ends of a word. Reading counterclockwise,

$$
\begin{array}{ccc}
 & c\ a^{-1} & \\
b & & b \\
a & & b \\
a & & c \\
 & b\ b\ d & \\
\end{array}
$$

is a closed word. Establish a bijective correspondence between reduced closed words and conjugacy classes in the free group.

Section 10   Generators and Relations

10.1. Prove the mapping properties of free groups and of quotient groups.

10.2. Let $\varphi : G \to G'$ be a surjective group homomorphism. Let S be a subset of G whose image $\varphi(S)$ generates G', and let T be a set of generators of $\ker \varphi$. Prove that $S \cup T$ generates G.

10.3. Can every finite group G be presented by a finite set of generators and a finite set of relations?

10.4. The group $G = \langle x, y; xyx^{-1}y^{-1} \rangle$ is called a *free abelian group*. Prove a mapping property of this group: If u and v are elements of an abelian group A, there is a unique homomorphism $\varphi : G \to A$ such that $\varphi(x) = u$, $\varphi(y) = v$.

10.5. Prove that the group generated by x, y, z with the single relation $yxyz^{-2} = 1$ is actually a free group.

10.6. A subgroup H of a group G is *characteristic* if it is carried to itself by all automorphisms of G.

    (a) Prove that every characteristic subgroup is normal, and that the center Z is a characteristic subgroup.

    (b) Determine the normal subgroups and the characteristic subgroups of the quaternion group.

10.7. The *commutator subgroup* C of a group G is the smallest subgroup that contains all commutators. Prove that the commutator subgroup is a characteristic subgroup (see Exercise 10.6), and that G/C is an abelian group.

10.8. Determine the commutator subgroups (Exercise 10.7) of the following groups:
(a) SO_2, (b) O_2, (c) the group M of isometries of the plane, (d) S_n, (d) SO_3.

10.9. Let G denote the group of 3×3 upper triangular matrices with diagonal entries equal to 1 and with entries in the field $\mathbb{F}_p$. For each prime p, determine the center, the commutator subgroup (Exercise 10.6), and the orders of the elements of G.

10.10. Let $\mathcal{F}$ be the free group on x, y and let $\mathcal{R}$ be the smallest normal subgroup containing the commutator $xyx^{-1}y^{-1}$.

(a) Show that $x^2y^2x^{-2}y^{-2}$ is in $\mathcal{R}$.

(b) Prove that $\mathcal{R}$ is the commutator subgroup (Exercise 10.7) of $\mathcal{F}$.

Section 11   The Todd-Coxeter Algorithm

11.1. Complete the proof that the group given in Example 7.11.8 is cyclic of order 3.

11.2. Use the Todd-Coxeter algorithm to show that the group defined by the relations (7.8.2) has order 12 and that the group defined by the relations (7.7.8) has order 21.

11.3. Use the Todd-Coxeter Algorithm to analyze the group generated by two elements x, y, with the following relations. Determine the order of the group and identify the group if you can:
(a) $x^2 = y^2 = 1$, $xyx = yxy$,   (b) $x^3 = y^3 = 1$, $xyx = yxy$,
(c) $x^4 = y^2 = 1$, $xyx = yxy$,   (d) $x^4 = y^4 = x^2y^2 = 1$,
(e) $x^3 = 1$, $y^2 = 1$, $yxyxy = 1$,   (f) $x^3 = y^3 = yxyxy = 1$,
(g) $x^4 = 1$, $y^3 = 1$, $xy = y^2x$,   (h) $x^7 = 1$, $y^3 = 1$, $yx = x^2y$,
(i) $x^{-1}yx = y^{-1}$, $y^{-1}xy = x^{-1}$,   (j) $y^3 = 1$, $x^2yxy = 1$.

11.4. How is normality of a subgroup H of G reflected in the table that displays the operation on cosets?

11.5. Let G be the group generated by elements x, y, with relations $x^4 = 1$, $y^3 = 1$, $x^2 = yxy$. Prove that this group is trivial in two ways: using the Todd-Coxeter Algorithm, and working directly with the relations.

11.6. A *triangle group* G^{pqr} is a group $\langle x, y, z \mid x^p, y^q, z^r, xyz \rangle$, where $p \le q \le r$ are positive integers. In each case, prove that the triangle group is isomorphic to the group listed.

(a) the dihedral group D_n, when $p, q, r = 2, 2, n$,

(b) the octahedral group, when $p, q, r = 2, 3, 4$,

(c) the icosahedral group, when $p, q, r = 2, 3, 5$.

11.7. Let Δ denote an equilateral triangle, and let a, b, c denote the reflections of the plane about the three sides of Δ. Let $x = ab$, $y = bc$, $z = ca$. Prove that x, y, z generate a triangle group (Exercise 11.6).

11.8. (a) Prove that the group G generated by elements x, y, z with relations $x^2 = y^3 = z^5 = 1$, $xyz = 1$ has order 60.

(b) Let H be the subgroup generated by x and zyz^{-1}. Determine the permutation representation of G on G/H, and identify H.

(c) Prove that G is isomorphic to the alternating group A_5.

(d) Let K be the subgroup of G generated by x and yxz. Determine the permutation representation of G on G/K, and identify K.

Miscellaneous Problems

M.1. Classify groups that are generated by two elements x and y of order 2.

Hint: It will be convenient to make use of the element $z = xy$.

M.2. With the presentation (6.4.3), determine the double cosets (see Exercise M.9) HgH of the subgroup $H = \{1, y\}$ in the dihedral group D_n. Show that each double coset has either two or four elements.

M.3. **(a)** Suppose that a group G operates transitively on a set S, and that H is the stabilizer of an element s_0 of S. Consider the operation of G on $S \times S$ defined by $g(s_1, s_2) = (gs_1, gs_2)$. Establish a bijective correspondence between double cosets of H in G and G-orbits in $S \times S$.

(b) Work out the correspondence explicitly for the case that G is the dihedral group D_5 and S is the set of vertices of a pentagon.

(c) Work it out for the case that $G = T$ and that S is the set of edges of a tetrahedron.

M.4. Let H and K be subgroups of a group G, with $H \subset K$. Suppose that H is normal in K, and that K is normal in G. Is H normal in G?

M.5. Let H and N be subgroups of a group G, and assume that N is a normal subgroup.

(a) Determine the kernels of the restrictions of the canonical homomorphism $\pi : G \to G/N$ to the subgroups H and HN.

(b) Applying First Isomorphism Theorem to these restrictions, prove the *Second Isomorphism Theorem*: $H/(H \cap N)$ is isomorphic to $(HN)/N$.

M.6. Let H and N be normal subgroups of a group G such that $H \supset N$. Let $\overline{H} = H/N$ and $\overline{G} = G/N$.

(a) Prove that $\overline{H}$ is a normal subgroup of $\overline{G}$.

(b) Use the composed homomorphism $G \to \overline{G} \to \overline{G}/\overline{H}$ to prove the

Third Isomorphism Theorem: G/H is isomorphic to $\overline{G}/\overline{H}$.

M.7. [2]Let p_1, p_2 be permutations of the set $S = \{1, 2, ..., n\}$, and let U_i be the subset of S of indices that are *not* fixed by p_i. Prove:

(a) If $U_1 \cap U_2 = \emptyset$, the commutator $p_1 p_2 p_1^{-1} p_2^{-1}$ is the identity.

(b) If $U_1 \cap U_2$ contains exactly one element, the commutator $p_1 p_2 p_1^{-1} p_2^{-1}$ is a three-cycle.

M.8. Let H be a subgroup of a group G. Prove that the number of left cosets is equal to the number of right cosets also when G is an infinite group.

M.9. Let x be an element, not the identity, of a group of odd order. Prove that the elements x and x^{-1} are not conjugate.

---

[2]Suggested by Benedict Gross.

M.10. Let G be a finite group that operates transitively on a set S of order ≥ 2. Show that G contains an element g that doesn't fix any element of S.

M.11. Determine the conjugacy classes of elements order 2 in $GL_2(\mathbb{Z})$.

*M.12. *(class equation of SL_2)* Many, though not all, conjugacy classes in $SL_2(F)$ contain matrices of the form $A = \begin{bmatrix} & -1 \\ 1 & a \end{bmatrix}$.

   (a) Determine the centralizers in $SL_2(\mathbb{F}_5)$ of the matrices A, for $a = 0, 1, 2, 3, 4$.

   (b) Determine the class equation of $SL_2(\mathbb{F}_5)$.

   (c) How many solutions of an equation of the form $x^2 + axy + y^2 = 1$ in $\mathbb{F}_p$ might there be? To analyze this, one can begin by setting $y = \lambda x + 1$. For most values of λ there will be two solutions, one of which is $x = 0$, $y = 1$.

   (d) Determine the class equation of $SL_2(\mathbb{F}_p)$.

CHAPTER 8

Bilinear Forms

8.1 BILINEAR FORMS

The dot product $(X \cdot Y) = X^t Y = x_1 y_1 + \cdots + x_n y_n$ on $\mathbb{R}^n$ was discussed in Chapter 5. It is *symmetric*: $(Y \cdot X) = (X \cdot Y)$, and *positive definite*: $(X \cdot X) > 0$ for every $X \neq 0$. We examine several analogues of dot product in this chapter. The most important ones are symmetric forms and Hermitian forms. All vector spaces in this chapter are assumed to be finite-dimensional.

Let V be a real vector space. A *bilinear form* on V is a real-valued function of two vector variables – a map $V \times V \to \mathbb{R}$. Given a pair v, w of vectors, the form returns a real number that will usually be denoted by $\langle v, w \rangle$. A bilinear form is required to be linear in each variable:

(8.1.1) $\qquad \langle rv_1, w_1 \rangle = r \langle v_1, w_1 \rangle \quad \text{and} \quad \langle v_1 + v_2, w_1 \rangle = \langle v_1, w_1 \rangle + \langle v_2, w_1 \rangle$

$\qquad\qquad\quad \langle v_1, rw_1 \rangle = r \langle v_1, w_1 \rangle \quad \text{and} \quad \langle v_1, w_1 + w_2 \rangle = \langle v_1, w_1 \rangle + \langle v_1, w_2 \rangle$

for all v_i and w_i in V and all real numbers r. Another way to say this is that the form is compatible with linear combinations in each variable:

(8.1.2) $\qquad\qquad\qquad \langle \sum x_i v_i, w \rangle = \sum x_i \langle v_i, w \rangle$

$\qquad\qquad\qquad\qquad \langle v, \sum w_j y_j \rangle = \sum \langle v, w_j \rangle y_j$

for all vectors v_i and w_i and all real numbers x_i and y_i. (It is often convenient to bring scalars in the second variable out to the right side.)

The form on $\mathbb{R}^n$ defined by

(8.1.3) $\qquad\qquad\qquad\qquad \langle X, Y \rangle = X^t A Y,$

where A is an $n \times n$ matrix, is an example of a bilinear form. The dot product is the case $A = I$, and when one is working with real column vectors, one always assumes that the form is dot product unless a different form has been specified.

If a basis $\mathbf{B} = (v_1, \ldots, v_n)$ of V is given, a bilinear form $\langle \, , \, \rangle$ can be related to a form of the type (8.1.3) by the *matrix of the form*. This matrix is simply $A = (a_{ij})$, where

$$(8.1.4) \qquad\qquad a_{ij} = \langle v_i, v_j \rangle.$$

Proposition 8.1.5 Let $\langle \, , \, \rangle$ be a bilinear form on a vector space V, let $\mathbf{B} = (v_1, \ldots, v_n)$ be a basis of V, and let A be the matrix of the form with respect to that basis. If X and Y are the coordinate vectors of the vectors v and w, respectively, then

$$\langle v, w \rangle = X^{\mathrm{t}} A Y.$$

Proof. If $v = \mathbf{B}X$ and $w = \mathbf{B}Y$, then

$$\langle v, w \rangle = \left\langle \sum_i v_i x_i, \sum_j v_j y_j \right\rangle = \sum_{i,j} x_i \langle v_i, v_j \rangle y_j = \sum_{i,j} x_i a_{ij} y_j = X^{\mathrm{t}} A Y. \qquad \square$$

A bilinear form is *symmetric* if $\langle v, w \rangle = \langle w, v \rangle$ for all v and w in V, and *skew-symmetric* if $\langle v, w \rangle = -\langle w, v \rangle$ for all v and w in V. When we refer to a symmetric form, we mean a bilinear symmetric form, and similarly, reference to a skew-symmetric form implies bilinearity.

Lemma 8.1.6

(a) Let A be an $n \times n$ matrix. The form $X^{\mathrm{t}} A Y$ is symmetric: $X^{\mathrm{t}} A Y = Y^{\mathrm{t}} A X$ for all X and Y, if and only if the matrix A is symmetric: $A^{\mathrm{t}} = A$.

(b) A bilinear form $\langle \, , \, \rangle$ is symmetric if and only if its matrix with respect to an arbitrary basis is a symmetric matrix.

The analogous statements are true when the word *symmetric* is replaced by *skew-symmetric*.

Proof. (a) Assume that $A = (a_{ij})$ is a symmetric matrix. Thinking of $X^{\mathrm{t}} A Y$ as a 1×1 matrix, it is equal to its transpose. Then $X^{\mathrm{t}} A Y = (X^{\mathrm{t}} A Y)^{\mathrm{t}} = Y^{\mathrm{t}} A^{\mathrm{t}} X = Y^{\mathrm{t}} A X$. Thus the form is symmetric. To derive the other implication, we note that $e_i{}^{\mathrm{t}} A e_j = a_{ij}$, while $e_j^{\mathrm{t}} A e_i = a_{ji}$. In order for the form to be symmetric, we must have $a_{ij} = a_{ji}$.

(b) This follows from (a) because $\langle v, w \rangle = X^{\mathrm{t}} A Y$. $\qquad \square$

The effect of a change of basis on the matrix of a form is determined in the usual way.

Proposition 8.1.7 Let $\langle \, , \, \rangle$ be a bilinear form on a real vector space V, and let A and A' be the matrices of the form with respect to two bases $\mathbf{B}$ and $\mathbf{B}'$. If P is the matrix of change of basis, so that $\mathbf{B}' = \mathbf{B}P$, then

$$A' = P^{\mathrm{t}} A P.$$

Proof. Let X and X' be the coordinate vectors of a vector v with respect to the bases $\mathbf{B}$ and $\mathbf{B}'$. Then $v = \mathbf{B}X = \mathbf{B}'X'$, and $PX' = X$. With analogous notation, $w = \mathbf{B}Y = \mathbf{B}'Y'$,

$$\langle v, w \rangle = X^{\mathrm{t}} A Y = (PX')^{\mathrm{t}} A (PY') = X'^{\mathrm{t}} (P^{\mathrm{t}} A P) Y'.$$

This identifies $P^{\mathrm{t}} A P$ as the matrix of the form with respect to the basis $\mathbf{B}'$. $\qquad \square$

Corollary 8.1.8 Let A be the matrix of a bilinear form with respect to a basis. The matrices that represent the same form with respect to different bases are the matrices $P^t A P$, where P can be any invertible matrix. $\qquad\square$

Note: There is an important observation to be made here. When a basis is given, both linear operators and bilinear forms are described by matrices. It may be tempting to think that the theories of linear operators and of bilinear forms are equivalent in some way. They are not equivalent. When one makes a change of basis, the matrix of the bilinear form $X^t A Y$ changes to $P^t A P$, while the matrix of the linear operator $Y = AX$ changes to $P^{-1} A P$. The matrices obtained with respect to the new basis will most often be different. $\qquad\square$

8.2 SYMMETRIC FORMS

Let V be a real vector space. A symmetric form on V is *positive definite* if $\langle v, v \rangle > 0$ for all nonzero vectors v, and *positive semi-definite* if $\langle v, v \rangle \geq 0$ for all nonzero vectors v. *Negative definite* and *negative semidefinite* forms are defined analogously. Dot product is a symmetric, positive definite form on $\mathbb{R}^n$.

A symmetric form that is not positive definite is called *indefinite*. The *Lorentz form*

$$(8.2.1) \qquad \langle X, Y \rangle = x_1 y_1 + x_2 y_2 + x_3 y_3 - x_4 y_4$$

is an indefinite symmetric form on "space–time" $\mathbb{R}^4$, where x_4 is the "time" coordinate, and the speed of light is normalized to 1. Its matrix with respect to the standard basis of $\mathbb{R}^4$ is

$$(8.2.2) \qquad \begin{bmatrix} 1 & & & \\ & 1 & & \\ & & 1 & \\ & & & -1 \end{bmatrix}.$$

As an introduction to the study of symmetric forms, we ask what happens to dot product when we change coordinates. The effect of the change of basis from the standard basis $\mathbf{E}$ to a new basis $\mathbf{B}'$ is given by Proposition 8.1.7. If $\mathbf{B}' = \mathbf{E}P$, the matrix I of dot product changes to $A' = P^t I P = P^t P$, or in terms of the form, if $PX' = X$ and $PY' = Y$, then

$$(8.2.3) \qquad X^t Y = X'^t A' Y', \quad \text{where} \quad A' = P^t P.$$

If the change of basis is orthogonal, then $P^t P$ is the identity matrix, and $(X \cdot Y) = (X' \cdot Y')$. But under a general change of basis, the formula for dot product changes as indicated.

This raises a question: Which of the bilinear forms $X^t A Y$ are equivalent to dot product, in the sense that they represent dot product with respect to some basis of $\mathbb{R}^n$? Formula (8.2.3) gives a theoretical answer:

Corollary 8.2.4 The matrices A that represent a form $\langle X, Y \rangle = X^t A Y$ equivalent to dot product are those that can be written as a product $P^t P$, for some invertible matrix P. $\qquad\square$

This answer won't be satisfactory until we can decide which matrices A can be written as such a product. One condition that A must satisfy is very simple: It must be symmetric, because $P^t P$ is always a symmetric matrix. Another condition comes from the fact that dot product is positive definite.

In analogy with the terminology for symmetric forms, a symmetric real matrix A is called *positive definite* if $X^t A X > 0$ for all nonzero column vectors X. If the form $X^t A Y$ is equivalent to dot product, the matrix A will be positive definite.

The two conditions, symmetry and positive definiteness, characterize matrices that represent dot product.

Theorem 8.2.5   The following properties of a real $n \times n$ matrix A are equivalent:

(i) The form $X^t A Y$ represents dot product, with respect to some basis of $\mathbb{R}^n$.

(ii) There is an invertible matrix P such that $A = P^t P$.

(iii) The matrix A is symmetric and positive definite.

We have seen that (i) and (ii) are equivalent (Corollary 8.2.4) and that (i) implies (iii). We will prove that (iii) implies (i) in Section 8.4 (see (8.4.18)).

8.3   HERMITIAN FORMS

The most useful way to extend the concept of symmetric forms to complex vector spaces is to Hermitian forms. A *Hermitian form* on a complex vector space V is a map $V \times V \to \mathbb{C}$, denoted by $\langle v, w \rangle$, that is conjugate linear in the first variable, linear in the second variable, and Hermitian symmetric:

$$(8.3.1) \qquad \langle c v_1, w_1 \rangle = \overline{c} \langle v_1, w_1 \rangle \quad \text{and} \quad \langle v_1 + v_2, w_1 \rangle = \langle v_1, w_1 \rangle + \langle v_2, w_1 \rangle$$
$$\langle v_1, c w_1 \rangle = c \langle v_1, w_1 \rangle \quad \text{and} \quad \langle v_1, w_1 + w_2 \rangle = \langle v_1, w_1 \rangle + \langle v_1, w_2 \rangle$$
$$\langle w_1, v_1 \rangle \quad = \quad \overline{\langle v_1, w_1 \rangle}$$

for all v_i and w_i in V, and all complex numbers c, where the overline denotes complex conjugation. As with bilinear forms (8.1.2), this condition can be expressed in terms of linear combinations in the variables:

$$(8.3.2) \qquad \left\langle \sum x_i v_i, w \right\rangle = \sum \overline{x_i} \langle v_i, w \rangle$$
$$\left\langle v, \sum w_j y_j \right\rangle = \sum \langle v, w_j \rangle y_j$$

for any vectors v_i and w_j and any complex numbers x_i and y_j. Because of Hermitian symmetry, $\langle v, v \rangle = \overline{\langle v, v \rangle}$, and therefore $\langle v, v \rangle$ is a real number, for all vectors v.

The *standard Hermitian form* on $\mathbb{C}^n$ is the form

$$(8.3.3) \qquad \langle X, Y \rangle = X^* Y = \overline{x}_1 y_1 + \cdots + \overline{x}_n y_n,$$

where the notation X^* stands for the conjugate transpose $(\overline{x}_1, \ldots, \overline{x}_n)$ of $X = (x_1, \ldots, x_n)^t$. When working with $\mathbb{C}^n$, one always assumes that the form is the standard Hermitian form, unless another form has been specified.

The reason that the complication caused by complex conjugation is introduced is that $\langle X, X \rangle$ becomes a positive real number for every nonzero complex vector X. If we use the bijective correspondence of complex n-dimensional vectors with real $2n$-dimensional vectors, by

(8.3.4) $(x_1, \ldots, x_n)^t \longleftrightarrow (a_1, b_1, \ldots, a_n, b_n)^t,$

where $x_\nu = a_\nu + b_\nu i$, then $\bar{x}_\nu = a_\nu - b_\nu i$ and

$$\langle X, X \rangle = \bar{x}_1 x_1 + \cdots + \bar{x}_n x_n = a_1^2 + b_1^2 + \cdots + a_n^2 + b_n^2.$$

Thus $\langle X, X \rangle$ is the square length of the corresponding real vector, a positive real number.

For arbitrary vectors X and Y, the symmetry property of dot product is replaced by *Hermitian symmetry*: $\langle Y, X \rangle = \overline{\langle X, Y \rangle}$. Bear in mind that when $X \neq Y$, $\langle X, Y \rangle$ is likely to be a complex number, whereas dot product of the corresponding real vectors would be real. Though elements of $\mathbb{C}^n$ correspond bijectively to elements of $\mathbb{R}^{2n}$, as above, these two vector spaces aren't equivalent, because scalar multiplication by a complex number isn't defined on $\mathbb{R}^{2n}$.

The *adjoint* A^* of a complex matrix $A = (a_{ij})$ is the complex conjugate of the transpose matrix A^t, a notation that was used above for column vectors. So the i, j entry of A^* is $\bar{a}_{ji}$. For example, $\begin{bmatrix} 1 & 1+i \\ 2 & i \end{bmatrix}^* = \begin{bmatrix} 1 & 2 \\ 1-i & -i \end{bmatrix}$.

Here are some rules for computing with adjoint matrices:

(8.3.5) $(cA)^* = \bar{c}A^*, \quad (A+B)^* = A^* + B^*, \quad (AB)^* = B^*A^*, \quad A^{**} = A.$

A square matrix A is *Hermitian* (or *self-adjoint*) if

(8.3.6) $A^* = A.$

The entries of a Hermitian matrix A satisfy the relation $a_{ji} = \bar{a}_{ij}$. Its diagonal entries are real and the entries below the diagonal are the complex conjugates of those above it:

(8.3.7) $A = \begin{bmatrix} r_1 & & a_{ij} \\ & \ddots & \\ \bar{a}_{ij} & & r_n \end{bmatrix}, \quad r_i \in \mathbb{R}, \quad a_{ij} \in \mathbb{C}.$

For example, $\begin{bmatrix} 2 & i \\ -i & 3 \end{bmatrix}$ is a Hermitian matrix. A real matrix is Hermitian if and only if it is symmetric.

The matrix of a Hermitian form with respect to a basis $\mathbf{B} = (v_1, \ldots, v_n)$ is defined as for bilinear forms. It is $A = (a_{ij})$, where $a_{ij} = \langle v_i, v_j \rangle$. The matrix of the standard Hermitian form on $\mathbb{C}^n$ is the identity matrix.

Proposition 8.3.8 Let A be the matrix of a Hermitian form $\langle \, , \, \rangle$ on a complex vector space V, with respect to a basis $\mathbf{B}$. If X and Y are the coordinate vectors of the vectors v and w, respectively, then $\langle v, w \rangle = X^*AY$ and A is a Hermitian matrix. Conversely, if A is a Hermitian matrix, then the form on $\mathbb{C}^n$ defined by $\langle X, Y \rangle = X^*AY$ is a Hermitian form.

The proof is analogous to that of Proposition 8.1.5. $\square$

Recall that if the form is Hermitian, $\langle v, v \rangle$ is a real number. A Hermitian form is *positive definite* if $\langle v, v \rangle$ is positive for every nonzero vector v, and a Hermitian matrix is positive definite if X^*AX is positive for every nonzero complex column vector X. A Hermitian form is positive definite if and only if its matrix with respect to an arbitrary basis is positive definite.

The rule for a change of basis $\mathbf{B}' = \mathbf{B}P$ in the matrix of a Hermitian form is determined, as usual, by substituting $PX' = X$ and $PY' = Y$:

$$X^*AY = (PX')^*A(PY') = X'^*(P^*AP)Y'.$$

The matrix of the form with respect to the new basis is

(8.3.9) $$A' = P^*AP.$$

Corollary 8.3.10

(a) Let A be the matrix of a Hermitian form with respect to a basis. The matrices that represent the same form with respect to different bases are those of the form $A' = P^*AP$, where P can be any invertible complex matrix.

(b) A change of basis $\mathbf{B}' = \mathbf{E}P$ in $\mathbb{C}^n$ changes the standard Hermitian form X^*Y to $X'^*A'Y'$, where $A' = P^*P$.   □

The next theorem gives the first of the many special properties of Hermitian matrices.

Theorem 8.3.11 The eigenvalues, the trace, and the determinant of a Hermitian matrix A are real numbers.

Proof. Since the trace and determinant can be expressed in terms of the eigenvalues, it suffices to show that the eigenvalues of a Hermitian matrix A are real. Let X be an eigenvector of A with eigenvalue λ. Then

$$X^*AX = X^*(AX) = X^*(\lambda X) = \lambda X^*X.$$

We note that $(\lambda X)^* = \overline{\lambda} X^*$. Since $A^* = A$,

$$X^*AX = (X^*A)X = (X^*A^*)X = (AX)^*X = (\lambda X)^*X = \overline{\lambda}X^*X.$$

So $\lambda X^*X = \overline{\lambda}X^*X$. Since X^*X is a positive real number, it is not zero. Therefore $\lambda = \overline{\lambda}$, which means that λ is real.   □

Please go over this proof carefully. It is simple, but so tricky that it seems hard to trust. Here is a startling corollary:

Corollary 8.3.12 The eigenvalues of a real symmetric matrix are real numbers.

Proof. When a real symmetric matrix is regarded as a complex matrix, it is Hermitian, so the corollary follows from the theorem.   □

This corollary would be difficult to prove without going over to complex matrices, though it can be checked directly for a real symmetric 2×2 matrix.

A matrix P such that

(8.3.13) $$P^*P = I, \quad \left(\text{or} \quad P^* = P^{-1} \right)$$

is called a *unitary* matrix. A matrix P is unitary if and only if its columns $P_1, \ldots, P_n$ are orthonormal with respect to the standard Hermitian form, i.e., if and only if $P_i^* P_i = 1$ and $P_i^* P_j = 0$ when $i \neq j$. For example, the matrix $\frac{1}{\sqrt{2}} \begin{bmatrix} 1 & -i \\ 1 & i \end{bmatrix}$ is unitary.

The unitary matrices form a subgroup of the complex general linear group called the *unitary group*. It is denoted by U_n:

(8.3.14) $$U_n = \{P \mid P^*P = I\}.$$

We have seen that a change of basis in $\mathbb{R}^n$ preserves dot product if and only if the change of basis matrix is orthogonal 5.1.14. Similarly, a change of basis in $\mathbb{C}^n$ preserves the standard Hermitian form X^*Y if and only if the change of basis matrix is unitary. (see (8.3.10)(b)).

8.4   ORTHOGONALITY

In this section we describe, at the same time, symmetric (bilinear) forms on a real vector space and Hermitian forms on a complex vector space. Throughout the section, we assume that we are given either a finite-dimensional real vector space V with a symmetric form, or a finite-dimensional complex vector space V with a Hermitian form. We won't assume that the given form is positive definite. Reference to a symmetric form indicates that V is a real vector space, while reference to a Hermitian form indicates that V is a complex vector space. Though everything we do applies to both cases, it may be best for you to think of a symmetric form on a real vector space when reading this for the first time.

In order to include Hermitian forms, bars will have to be put over some symbols. Since complex conjugation is the identity operation on the real numbers, we can ignore bars when considering symmetric forms. Also, the adjoint of a real matrix is equal to its transpose. When a matrix A is real, A^* is the transpose of A.

We assume given a symmetric or Hermitian form on a finite-dimensional vector space V. The basic concept used to study the form is orthogonality.

- Two vectors v and w are *orthogonal* (written $v \perp w$) if

$$\langle v, w \rangle = 0.$$

This extends the definition given before when the form is dot product. Note that $v \perp w$ if and only if $w \perp v$.

What orthogonality of real vectors means geometrically depends on the form and also on a basis. One peculiar thing is that, when the form is indefinite, a nonzero vector v may be self-orthogonal: $\langle v, v \rangle = 0$. Rather than trying to understand the geometric meaning of orthogonality for each symmetric form, it is best to work algebraically with the definition of orthogonality, $\langle v, w \rangle = 0$, and let it go at that.

If W is a subspace of V, we may *restrict* the form on V to W, which means simply that we take the same form but look at it only when the vectors are in W. It is obvious that if the form on V is symmetric, Hermitian, or positive definite, then its restriction to W will have the same property.

• The *orthogonal space* to a subspace W of V, often denoted by $W^\perp$, is the subspace of vectors v that are orthogonal to every vector in W, or symbolically, such that $v \perp W$:

$$(8.4.1) \qquad W^\perp = \{v \in V \mid \langle v, w \rangle = 0 \text{ for all } w \text{ in } W\}.$$

• An *orthogonal basis* $\mathbf{B} = (v_1, \ldots, v_n)$ of V is a basis whose vectors are mutually orthogonal: $\langle v_i, v_j \rangle = 0$ for all indices i and j with $i \neq j$. The matrix of the form with respect to an orthogonal basis will be a diagonal matrix, and the form will be nondegenerate (see below) if and only if the diagonal entries $\langle v_i, v_i \rangle$ of the matrix are nonzero (see (8.4.4)(b)).

• A *null vector* v in V is a vector orthogonal to every vector in V, and the *nullspace N* of the form is the set of null vectors. The nullspace can be described as the orthogonal space to the whole space V:

$$N = \{v \mid v \perp V\} = V^\perp.$$

• The form on V is *nondegenerate* if its nullspace is the zero space $\{0\}$. This means that for every nonzero vector v, there is a vector v' such that $\langle v, v' \rangle \neq 0$. A form that isn't nondegenerate is *degenerate*. The most interesting forms are nondegenerate.

• The form on V is *nondegenerate on a subspace W* if its restriction to W is a nondegenerate form, which means that for every nonzero vector w in W, there is a vector w', *also in W*, such that $\langle w, w' \rangle \neq 0$. A form may be degenerate on a subspace, though it is nondegenerate on the whole space, and vice versa.

Lemma 8.4.2 The form is nondegenerate on W if and only if $W \cap W^\perp = \{0\}$. $\square$

There is an important criterion for equality of vectors in terms of a nondegenerate form.

Proposition 8.4.3 Let $\langle\,,\,\rangle$ be a nondegenerate symmetric or Hermitian form on V, and let v and v' be vectors in V. If $\langle v, w \rangle = \langle v', w \rangle$ for all vectors w in V, then $v = v'$.

Proof. If $\langle v, w \rangle = \langle v', w \rangle$, then $v - v'$ is orthogonal to w. If this is true for all w in V, then $v - v'$ is a null vector, and because the form is nondegenerate, $v - v' = 0$. $\square$

Proposition 8.4.4 Let $\langle\,,\,\rangle$ be a symmetric form on a real vector space or a Hermitian form on a complex vector space, and let A be its matrix with respect to a basis.

(a) A vector v is a null vector if and only if its coordinate vector Y solves the homogeneous equation $AY = 0$.

(b) The form is nondegenerate if and only if the matrix A is invertible.

Proof. Via the basis, the form corresponds to the form X^*AY, so we may as well work with that form. If Y is a vector such that $AY = 0$, then $X^*AY = 0$ for all X, which means that Y

is orthogonal to every vector, i.e., it is a null vector. Conversely, if $AY \neq 0$, then AY has a nonzero coordinate. The matrix product $e_i^* AY$ picks out the ith coordinate of AY. So one of those products is not zero, and therefore Y is not a null vector. This proves **(a)**. Because A is invertible if and only if the equation $AY = 0$ has no nontrivial solution, **(b)** follows.   □

Theorem 8.4.5 Let $\langle\ ,\ \rangle$ be a symmetric form on a real vector space V or a Hermitian form on a complex vector space V, and let W be a subspace of V.

(a) The form is nondegenerate on W if and only if V is the direct sum $W \oplus W^{\perp}$.

(b) If the form is nondegenerate on V and on W, then it is nondegenerate on $W^{\perp}$.

When a vector space V is a direct sum $W_1 \oplus \cdots \oplus W_k$ and W_i is orthogonal to W_j for $i \neq j$, V is said to be the *orthogonal sum* of the subspaces. The theorem asserts that if the form is nondegenerate on W, then V is the orthogonal sum of W and $W^{\perp}$.

Proof of Theorem 8.4.5. **(a)** The conditions for a direct sum are $W \cap W^{\perp} = \{0\}$ and $V = W + W^{\perp}$ (3.6.6)**(c)**. The first condition simply restates the hypothesis that the form be nondegenerate on the subspace. So if V is the direct sum, the form is nondegenerate. We must show that if the form is nondegenerate on W, then every vector v in V can be expressed as a sum $v = w + u$, with w in W and u in $W^{\perp}$.

We extend a basis $(w_1, \ldots, w_k)$ of W to a basis $\mathbf{B} = (w_1, \ldots, w_k; v_1, \ldots, v_{n-k})$ of V, and we write the matrix of the form with respect to this basis in block form

$$(8.4.6) \qquad\qquad M = \begin{bmatrix} A & B \\ C & D \end{bmatrix},$$

where A is the upper left $k \times k$ submatrix.

The entries of the block A are $\langle w_i, w_j \rangle$ for $i, j = 1, \ldots, k$, so A is the matrix of the form restricted to W. Since the form is nondegenerate on W, A is invertible. The entries of the block B are $\langle w_i, v_j \rangle$ for $i = 1, \ldots, k$ and $j = 1, \ldots, n - k$. If we can choose the vectors $v_1, \ldots, v_{n-k}$ so that B becomes zero, those vectors will be orthogonal to the basis of W, so they will be in the orthogonal space $W^{\perp}$. Then since $\mathbf{B}$ is a basis of V, it will follow that $V = W + W^{\perp}$, which is what we want to show.

To achieve $B = 0$, we change basis using a matrix with a block form

$$(8.4.7) \qquad\qquad P = \begin{bmatrix} I & Q \\ 0 & I \end{bmatrix},$$

where the block Q remains to be determined. The new basis $\mathbf{B}' = \mathbf{B}P$ will have the form $(w_1, \ldots, w_k; v'_1, \ldots, v'_{n-k})$. The basis of W will not change. The matrix of the form with respect to the new basis will be

$$(8.4.8) \qquad M' = P^* M P = \begin{bmatrix} I & 0 \\ Q^* & I \end{bmatrix} \begin{bmatrix} A & B \\ C & D \end{bmatrix} \begin{bmatrix} I & Q \\ 0 & I \end{bmatrix} = \begin{bmatrix} A & AQ + B \\ \cdot & \cdot \end{bmatrix}.$$

We don't need to compute the other entries. When we set $Q = -A^{-1}B$, the upper right block of M' becomes zero, as desired.

(b) Suppose that the form is nondegenerate on V and on W. **(a)** shows that $V = W \oplus W^\perp$. If we choose a basis for V by appending bases for W and $W^\perp$, the matrix of the form on V will be a diagonal block matrix, where the blocks are the matrices of the form restricted to W and to $W^\perp$. The matrix of the form on V is invertible (8.4.4), so the blocks are invertible. It follows that the form is nondegenerate on $W^\perp$. $\qquad\square$

Lemma 8.4.9 If a symmetric or Hermitian form is not identically zero, there is a vector v in V such that $\langle v, v \rangle \neq 0$.

Proof. If the form is not identically zero, there will be vectors x and y such that $\langle x, y \rangle$ is not zero. If the form is Hermitian, we replace y by cy where c is a nonzero complex number, to make $\langle x, y \rangle$ real and still not zero. Then $\langle y, x \rangle = \langle x, y \rangle$. We expand:

$$\langle x + y, x + y \rangle = \langle x, x \rangle + 2\langle x, y \rangle + \langle y, y \rangle.$$

Since the term $2\langle x, y \rangle$ isn't zero, at least one of the three other terms in the equation isn't zero. $\qquad\square$

Theorem 8.4.10 Let $\langle \, , \, \rangle$ be a symmetric form on a real vector space V or a Hermitian form on a complex vector space V. There exists an orthogonal basis for V.

Proof. Case 1: The form is identically zero. Then every basis is orthogonal.

Case 2: The form is not identically zero. By induction on dimension, we may assume that there is an orthogonal basis for the restriction of the form to any proper subspace of V. We apply Lemma 8.4.9 and choose a vector v_1 with $\langle v_1, v_1 \rangle \neq 0$ as the first vector in our basis. Let W be the span of (v_1). The matrix of the form restricted to W is the 1×1 matrix whose entry is $\langle v_1, v_1 \rangle$. It is an invertible matrix, so the form is nondegenerate on W. By Theorem 8.4.5, $V = W \oplus W^\perp$. By our induction assumption, $W^\perp$ has an orthogonal basis, say $(v_2, \ldots, v_n)$. Then $(v_1, v_2, \ldots, v_n)$ will be an orthogonal basis of V. $\qquad\square$

Orthogonal Projection

Suppose that our given form is nondegenerate on a subspace W. Theorem 8.4.5 tells us that V is the direct sum $W \oplus W^\perp$. Every vector v in V can be written uniquely in the form $v = w + u$, with w in W and u in $W^\perp$. The *orthogonal projection* from V to W is the map $\pi : V \to W$ defined by $\pi(v) = w$. The decomposition $v = w + u$ is compatible with sums of vectors and with scalar multiplication, so π is a linear transformation.

The orthogonal projection is the unique linear transformation from V to W such that $\pi(w) = w$ if w is in W and $\pi(u) = 0$ if u is in $W^\perp$.

Note: If the form is degenerate on a subspace W, the orthogonal projection to W doesn't exist. The reason is that $W \cap W^\perp$ will contain a nonzero element x, and it will be impossible to have both $\pi(x) = x$ and $\pi(x) = 0$. $\qquad\square$

The next theorem provides a very important formula for orthogonal projection.

Theorem 8.4.11 Projection Formula. Let $\langle\,,\,\rangle$ be a symmetric form on a real vector space V or a Hermitian form on a complex vector space V, and let W be a subspace of V on which the form is nondegenerate. If $(w_1, \ldots, w_k)$ is an orthogonal basis for W, the orthogonal projection $\pi : V \to W$ is given by the formula $\pi(v) = w_1 c_1 + \cdots + w_k c_k$, where

$$c_i = \frac{\langle w_i, v \rangle}{\langle w_i, w_i \rangle}.$$

Proof. Because the form is nondegenerate on W and its matrix with respect to an orthogonal basis is diagonal, $\langle w_i, w_i \rangle \neq 0$. The formula makes sense. Given a vector v, let w denote the vector $w_1 c_1 + \cdots + w_k c_k$, with c_i as above. This is an element of W, so if we show that $v - w = u$ is in $W^\perp$, it will follow that $\pi(v) = w$, as the theorem asserts. To show that u is in $W^\perp$, we show that $\langle w_i, u \rangle = 0$ for $i = 1, \ldots, k$. We remember that $\langle w_i, w_j \rangle = 0$ if $i \neq j$. Then

$$\langle w_i, u \rangle = \langle w_i, v \rangle - \langle w_i, w \rangle = \langle w_i, v \rangle - \big(\langle w_i, w_1 \rangle c_1 + \cdots + \langle w_i, w_k \rangle c_k\big)$$
$$= \langle w_i, v \rangle - \langle w_i, w_i \rangle c_i = 0. \qquad \square$$

Warning: This projection formula is not correct unless the basis is orthogonal.

Example 8.4.12 Let V be the space $\mathbb{R}^3$ of column vectors, and let $\langle v, w \rangle$ denote the dot product form. Let W be the subspace spanned by the vector w_1 whose coordinate vector is $(1, 1, 1)^t$. Let $(x_1, x_2, x_3)^t$ be the coordinate vector of a vector v. Then $\langle w_1, v \rangle = x_1 + x_2 + x_3$. The projection formula reads $\pi(v) = w_1 c$, where $c = (x_1 + x_2 + x_3)/3$. $\qquad \square$

If a form is nondegenerate on the whole space V, the orthogonal projection from V to V will be the identity map. The projection formula is interesting in this case too, because it can be used to compute the coordinates of a vector v with respect to an orthogonal basis.

Corollary 8.4.13 Let $\langle\,,\,\rangle$ be a nondegenerate symmetric form on a real vector space V or a nondegenerate Hermitian form on a complex vector space V, let $(v_1, \ldots, v_n)$ be an orthogonal basis for V, and let v be any vector. Then $v = v_1 c_1 + \cdots + v_n c_n$, where

$$c_i = \frac{\langle v_i, v \rangle}{\langle v_i, v_i \rangle}. \qquad \square$$

Example 8.4.14 Let $\mathbf{B} = (v_1, v_2, v_3)$ be the orthogonal basis of $\mathbb{R}^3$ whose coordinate vectors are

$$\begin{bmatrix} 1 \\ 1 \\ 1 \end{bmatrix}, \begin{bmatrix} 1 \\ -1 \\ 0 \end{bmatrix}, \begin{bmatrix} 1 \\ 1 \\ -2 \end{bmatrix}.$$

Let v be a vector with coordinate vector $(x_1, x_2, x_3)^t$. Then $v = v_1 c_1 + v_2 c_2 + v_3 c_3$ and

$$c_1 = (x_1 + x_2 + x_3)/3, \quad c_2 = (x_1 - x_2)/2, \quad c_3 = (x_1 + x_2 - 2x_3)/6. \qquad \square$$

Next, we consider scaling of the vectors that make up an orthogonal basis.

Corollary 8.4.15 Let $\langle\ ,\ \rangle$ be a symmetric form on a real vector space V or a Hermitian form on a complex vector space V.

(a) There is an orthogonal basis $\mathbf{B} = (v_1, \ldots, v_n)$ for V with the property that for each i, $\langle v_i, v_i \rangle$ is equal to 1, –1, or 0.

(b) *Matrix form*: If A is a real symmetric $n \times n$ matrix, there is an invertible real matrix P such that $P^t A P$ is a diagonal matrix, each of whose diagonal entries is 1, –1, or 0. If A is a complex Hermitian $n \times n$ matrix, there is an invertible complex matrix P such that $P^* A P$ is a diagonal matrix, each of whose diagonal entries is 1, –1, or 0.

Proof. **(a)** Let $(v_1, \ldots, v_n)$ be an orthogonal basis. If v is a vector, then for any nonzero *real* number c, $\langle cv, cv \rangle = c^2 \langle v, v \rangle$, and c^2 can be any positive real number. So if we multiply v_i by a scalar, we can adjust the real number $\langle v_i, v_i \rangle$ by an arbitrary positive real number. This proves **(a)**. Part **(b)** follows in the usual way, by applying **(a)** to the form $X^* A Y$. $\quad\square$

If we arrange an orthogonal basis that has been scaled suitably, the matrix of the form will have a block decomposition

$$(8.4.16) \qquad A = \begin{bmatrix} I_p & & \\ & -I_m & \\ & & 0_z \end{bmatrix},$$

where p, m, and z are the numbers of 1's, –1's, and 0's on the diagonal, and $p + m + z = n$. The form is nondegenerate if and only if $z = 0$.

If the form is nondegenerate, the pair of integers (p, m) is called the *signature* of the form. Sylvester's Law (see Exercise 4.21) asserts that the signature does not depend on the choice of the orthogonal basis.

The notation $I_{p,m}$ is often used to denote the diagonal matrix

$$(8.4.17) \qquad I_{p,m} = \begin{bmatrix} I_p & \\ & -I_m \end{bmatrix}.$$

With this notation, the matrix (8.2.2) that represents the Lorentz form is $I_{3,1}$.

The form is positive definite if and only if m and z are both zero. Then the normalized basis has the property that $\langle v_i, v_i \rangle = 1$ for each i, and $\langle v_i, v_j \rangle = 0$ when $i \neq j$. This is called an *orthonormal basis*, in agreement with the terminology introduced before, for bases of $\mathbb{R}^n$ (5.1.8). An orthonormal basis $\mathbf{B}$ refers the form back to dot product on $\mathbb{R}^n$ or to the standard Hermitian form on $\mathbb{C}^n$. That is, if $v = \mathbf{B}X$ and $w = \mathbf{B}Y$, then $\langle v, w \rangle = X^* Y$. An orthonormal basis exists if and only if the form is positive definite.

Note: If $\mathbf{B}$ is an orthonormal basis for a subspace W of V, the projection from V to W is given by the formula $\pi(v) = w_1 c_1 + \cdots w_k c_k$, where $c_i = \langle w_i, v \rangle$. The projection formula is simpler because the denominators $\langle w_i, w_i \rangle$ in (8.4.11) are equal to 1. However, normalizing the vectors requires extracting a square root, and because of this, it is sometimes preferable to work with an orthogonal basis without normalizing. $\quad\square$

The proof of the remaining implication (iii) $\Rightarrow$ (i) of Theorem 8.2.5 follows from this discussion:

Corollary 8.4.18   If a real matrix A is symmetric and positive definite, then the form X^tAY represents dot product with respect to some basis of $\mathbb{R}^n$.

When a positive definite symmetric or Hermitian form is given, the projection formula provides an inductive method, called the *Gram-Schmidt procedure*, to produce an orthonormal basis, starting with an arbitrary basis $(v_1, \ldots, v_n)$. The procedure is as follows: Let V_k denote the space spanned by the basis vectors $(v_1, \ldots, v_k)$. Suppose that, for some $k \leq n$, we have found an orthonormal basis $(w_1, \ldots, w_{k-1})$ for V_{k-1}. Let π denote the orthogonal projection from V to V_{k-1}. Then $\pi(v_k) = w_1 c_1 + \cdots + w_{k-1} c_{k-1}$, where $c_i = \langle w_i, v_k \rangle$, and $w_k = v_k - \pi(v_k)$ is orthogonal to V_{k-1}. When we normalize $\langle w_k, w_k \rangle$ to 1, the set $(w_1, \ldots, w_k)$ will be an orthonormal basis for V_k.   □

The last topic of this section is a criterion for a symmetric form to be positive definite in terms of its matrix with respect to an arbitrary basis. Let $A = (a_{ij})$ be the matrix of a symmetric form with respect to a basis $\mathbf{B} = (v_1, \ldots, v_n)$ of V, and let A_k denote the $k \times k$ minor made up of the matrix entries a_{ij} with $i, j \leq k$:

$$A_1 = [a_{11}], \quad A_2 = \begin{bmatrix} a_{11} & a_{12} \\ a_{21} & a_{22} \end{bmatrix}, \quad \ldots, \quad A_n = A.$$

Theorem 8.4.19   The form and the matrix are positive definite if and only if $\det A_k > 0$ for $k = 1, \ldots, n$.
We leave the proof as an exercise.   □

For example, the matrix $A = \begin{bmatrix} 2 & 1 \\ 1 & 1 \end{bmatrix}$ is positive definite, because $\det[2]$ and $\det A$ are both positive.

8.5   EUCLIDEAN SPACES AND HERMITIAN SPACES

When we work in $\mathbb{R}^n$, we may wish to change the basis. But if our problem involves dot products – if length or orthogonality of vectors is involved – a change to an arbitrary new basis may be undesirable, because it will not preserve length and orthogonality. It is best to restrict oneself to orthonormal bases, so that dot products are preserved. The concept of a Euclidean space provides us with a framework in which to do this. A real vector space together with a positive definite symmetric form is called a *Euclidean space*, and a complex vector space together with a positive definite Hermitian form is called a *Hermitian space*.

The space $\mathbb{R}^n$, with dot product, is the *standard Euclidean space*. An orthonormal basis for any Euclidean space will refer the space back to the standard Euclidean space. Similarly, the standard Hermitian form $\langle X, Y \rangle = X^*Y$ makes $\mathbb{C}^n$ into the *standard Hermitian space*, and an orthonormal basis for any Hermitian space will refer the form back to the standard Hermitian space. The only significant difference between an arbitrary Euclidean or Hermitian space and the standard Euclidean or Hermitian space is that no orthonormal basis is preferred. Nevertheless, when working in such spaces we always use orthonormal bases, though none have been picked out for us. A change of orthonormal bases will be given by a matrix that is orthogonal or unitary, according to the case.

Corollary 8.5.1 Let V be a Euclidean or a Hermitian space, with positive definite form $\langle\,,\,\rangle$, and let W be a subspace of V. The form is nondegenerate on W, and therefore $V = W \oplus W^{\perp}$.

Proof. If w is a nonzero vector in W, then $\langle w, w \rangle$ is a positive real number. It is not zero, and therefore w is not a null vector in V or in W. The nullspaces are zero. □

What we have learned about symmetric forms allows us to interpret the length of a vector and the angle between two vectors v and w in a Euclidean space V. Let's set aside the special case that these vectors are dependent, and assume that they span a two-dimensional subspace W. When we restrict the form, W becomes a Euclidean space of dimension 2. So W has an orthonormal basis (w_1, w_2), and via this basis, the vectors v and w will have coordinate vectors in $\mathbb{R}^2$. We'll denote these two-dimensional coordinate vectors by lowercase letters x and y. They aren't the coordinate vectors that we would obtain using an orthonormal basis for the whole space V, but we will have $\langle v, w \rangle = x^t y$, and this allows us to interpret geometric properties of the form in terms of dot product in $\mathbb{R}^2$.

The *length* $|v|$ of a vector v is defined by the formula $|v|^2 = \langle v, v \rangle$. If x is the coordinate vector of v in $\mathbb{R}^2$, then $|v|^2 = x^t x$. The *law of cosines* $(x \cdot y) = |x||y| \cos\theta$ in $\mathbb{R}^2$ becomes

(8.5.2) $$\langle v, w \rangle = |v||w| \cos\theta,$$

where θ is the angle between x and y. Since this formula expresses $\cos\theta$ in terms of the form, it defines the unoriented *angle* θ between vectors v and w. But the ambiguity of sign in the angle that arises because $\cos\theta = \cos(-\theta)$ can't be eliminated. When one views a plane in $\mathbb{R}^3$ from its front and its back, the angles one sees differ by sign.

8.6 THE SPECTRAL THEOREM

In this section, we analyze certain linear operators on a Hermitian space.

Let $T: V \to V$ be a linear operator on a Hermitian space V, and let A be the matrix of T with respect to an orthonormal basis **B**. The *adjoint operator* $T^*: V \to V$ is the operator whose matrix with respect to the same basis is the adjoint matrix A^*.

If we change to a new orthonormal basis **B**$'$, the basechange matrix P will be unitary, and the new matrix of T will have the form $A' = P^*AP = P^{-1}AP$. Its adjoint will be $A'^* = P^*A^*P$. This is the matrix of T^* with respect to the new basis. So the definition of T^* makes sense: It is independent of the orthonormal basis.

The rules (8.3.5) for computing with adjoint matrices carry over to adjoint operators:

(8.6.1) $$(T + U)^* = T^* + U^*, \quad (TU)^* = U^*T^*, \quad T^{**} = T.$$

A *normal* matrix is a complex matrix A that commutes with its adjoint: $A^*A = AA^*$. In itself, this isn't a particularly important class of matrices, but is the natural class for which to state the Spectral Theorem that we prove in this section, and it includes two important classes: Hermitian matrices ($A^* = A$) and unitary matrices ($A^* = A^{-1}$).

Lemma 8.6.2 Let A be a complex $n \times n$ matrix and let P be an $n \times n$ unitary matrix. If A is normal, Hermitian, or unitary, so is P^*AP. □

A linear operator T on a Hermitian space is called *normal*, *Hermitian*, or *unitary* if its matrix with respect to an orthonormal basis has the same property. So T is normal if $T^*T = TT^*$, Hermitian if $T^* = T$, and unitary if $T^*T = I$. A Hermitian operator is sometimes called a *self-adjoint operator*, but we won't use that terminology.

The next proposition interprets these conditions in terms of the form.

Proposition 8.6.3 Let T be a linear operator on a Hermitian space V, and let T^* be the adjoint operator.

(a) For all v and w in V, $\langle Tv, w \rangle = \langle v, T^*w \rangle$ and $\langle v, Tw \rangle = \langle T^*v, w \rangle$

(b) T is normal if and only if, for all v and w in V, $\langle Tv, Tw \rangle = \langle T^*v, T^*w \rangle$.

(c) T is Hermitian if and only if, for all v and w in V, $\langle Tv, w \rangle = \langle v, Tw \rangle$.

(d) T is unitary if and only if, for all v and w in V, $\langle Tv, Tw \rangle = \langle v, w \rangle$.

Proof. **(a)** Let A be the matrix of the operator T with respect to an orthonormal basis **B**. With $v = \mathbf{B}X$ and $w = \mathbf{B}Y$ as usual, $\langle Tv, w \rangle = (AX)^*Y = X^*A^*Y$ and $\langle v, T^*w \rangle = X^*A^*Y$. Therefore $\langle Tv, w \rangle = \langle v, T^*w \rangle$. The proof of the other formula of **(a)** is similar.

(b) We substitute T^*v for v into the first equation of **(a)**: $\langle TT^*v, w \rangle = \langle T^*v, T^*w \rangle$. Similarly, substituting Tv for v into the second equation of **(a)**: $\langle Tv, Tw \rangle = \langle T^*Tv, w \rangle$. So if T is normal, then $\langle Tv, Tw \rangle = \langle T^*v, T^*w \rangle$. The converse follows by applying Proposition 8.4.3 to the two vectors T^*Tv and TT^*v. The proofs of **(c)** and **(d)** are similar. $\qquad\square$

Let T be a linear operator on a Hermitian space V. As before, a subspace W of V is *T-invariant* if $TW \subset W$. A linear operator T will restrict to a linear operator on a T-invariant subspace, and if T is normal, Hermitian, or unitary, the restricted operator will have the same property. This follows from Proposition 8.6.3.

Proposition 8.6.4 Let T be a linear operator on a Hermitian space V and let W be a subspace of V. If W is T-invariant, then the orthogonal space $W^\perp$ is T^*-invariant. If W is T^*-invariant then $W^\perp$ is T-invariant.

Proof. Suppose that W is T-invariant. To show that $W^\perp$ is T^*-invariant, we must show that if u is in $W^\perp$, then T^*u is also in $W^\perp$, which by definition of $W^\perp$ means that $\langle w, T^*u \rangle = 0$ for all w in W. By Proposition 8.6.3, $\langle w, T^*u \rangle = \langle Tw, u \rangle$. Since W is T-invariant, Tw is in W. Then since u is in $W^\perp$, $\langle Tw, u \rangle = 0$. So $\langle w, T^*u \rangle = 0$, as required. Since $T^{**} = T$, one obtains the second assertion by interchanging the roles of T and T^*. $\qquad\square$

The next theorem is the main place that we use the hypothesis that the form given on V be positive definite.

Theorem 8.6.5 Let T be a normal operator on a Hermitian space V, and let v be an eigenvector of T with eigenvalue λ. Then v is also an eigenvector of T^*, with eigenvalue $\overline{\lambda}$.

Proof. Case 1: $\lambda = 0$. Then $Tv = 0$, and we must show that $T^*v = 0$. Since the form is positive definite, it suffices to show that $\langle T^*v, T^*v \rangle = 0$. By Proposition 8.6.3, $\langle T^*v, T^*v \rangle = \langle Tv, Tv \rangle = \langle 0, 0 \rangle = 0$.

Case 2: λ is arbitrary. Let S denote the linear operator $T - \lambda I$. Then v is an eigenvector for S with eigenvalue zero: $Sv = 0$. Moreover, $S^* = T^* - \overline{\lambda}I$. You can check that S is a normal operator. By Case 1, v is an eigenvector for S^* with eigenvalue 0: $S^*v = T^*v - \overline{\lambda}v = 0$. This shows that v is an eigenvector of T^* with eigenvalue $\overline{\lambda}$. $\qquad\square$

Theorem 8.6.6 Spectral Theorem for Normal Operators

(a) Let T be a normal operator on a Hermitian space V. There is an orthonormal basis of V consisting of eigenvectors for T.

(b) *Matrix form:* Let A be a normal matrix. There is a unitary matrix P such that P^*AP is diagonal.

Proof. (a) We choose an eigenvector v_1 for T, and normalize its length to 1. Theorem 8.6.5 tells us that v_1 is also an eigenvector for T^*. Therefore the one-dimensional subspace W spanned by v_1 is T^*-invariant. By Proposition 8.6.4, $W^\perp$ is T-invariant. We also know that $V = W \oplus W^\perp$. The restriction of T to any invariant subspace, including $W^\perp$, is a normal operator. By induction on dimension, we may assume that $W^\perp$ has an orthonormal basis of eigenvectors, say $(v_2, \ldots, v_n)$. Adding v_1 to this set yields an orthonormal basis of V of eigenvectors for T.

(b) This is proved from (a) in the usual way. We regard A as the matrix of the normal operator of multiplication by A on $\mathbb{C}^n$. By (a) there is an orthonormal basis **B** consisting of eigenvectors. The matrix P of change of basis from **E** to **B** is unitary, and the matrix of the operator with respect to the new basis, which is P^*AP, is diagonal. $\qquad\square$

The next corollaries are obtained by applying the Spectral Theorem to the two most important types of normal matrices.

Corollary 8.6.7 Spectral Theorem for Hermitian Operators.

(a) Let T be a Hermitian operator on a Hermitian space V.

    (i) There is an orthonormal basis of V consisting of eigenvectors of T.

    (ii) The eigenvalues of T are real numbers.

(b) *Matrix form:* Let A be a Hermitian matrix.

    (i) There is a unitary matrix P such that P^*AP is a real diagonal matrix.

    (ii) The eigenvalues of A are real numbers.

Proof. Part (b)(ii) has been proved before (Theorem 8.3.11) and (a)(i) follows from the Spectral Theorem for normal operators. The other assertions are variants. $\qquad\square$

Corollary 8.6.8 Spectral Theorem for Unitary Matrices.

(a) Let A be a unitary matrix. There is a unitary matrix P such that P^*AP is diagonal.

(b) Every conjugacy class in the unitary group U_n contains a diagonal matrix. $\qquad\square$

To diagonalize a Hermitian matrix M, one can proceed by determining its eigenvectors. If the eigenvalues are distinct, the corresponding eigenvectors will be orthogonal, and one can normalize their lengths to 1. This follows from the Spectral Theorem. For example, $v_1' = \begin{bmatrix} 1 \\ -i \end{bmatrix}$ and $v_2' = \begin{bmatrix} 1 \\ i \end{bmatrix}$ are eigenvectors of the Hermitian matrix $M = \begin{bmatrix} 2 & i \\ -i & 2 \end{bmatrix}$, with eigenvalues 3 and 1, respectively. We normalize their lengths to 1 by the factor $1/\sqrt{2}$, obtaining the unitary matrix $P = \frac{1}{\sqrt{2}}\begin{bmatrix} 1 & 1 \\ -i & i \end{bmatrix}$. Then $P^*MP = \begin{bmatrix} 3 & \\ & 1 \end{bmatrix}$.

However, the Spectral Theorem asserts that a Hermitian matrix can be diagonalized even when its eigenvalues aren't distinct. For instance, the only 2×2 Hermitian matrix whose characteristic polynomial has a double root λ is λI.

What we have proved for Hermitian matrices has analogues for real symmetric matrices. A *symmetric operator* T on a Euclidean space V is a linear operator whose matrix with respect to an orthonormal basis is symmetric. Similarly, an *orthogonal operator* T on a Euclidean space V is a linear operator whose matrix with respect to an orthonormal basis is orthogonal.

Proposition 8.6.9 Let T be a linear operator on a Euclidean space V.

(a) T is symmetric if and only if, for all v and w in V, $\langle Tv, w \rangle = \langle v, Tw \rangle$.
(b) T is orthogonal if and only if, for all v and w in V, $\langle Tv, Tw \rangle = \langle v, w \rangle$. □

Theorem 8.6.10 Spectral Theorem for Symmetric Operators.

(a) Let T be a symmetric operator on a Euclidean space V.

   (i) There is an orthonormal basis of V consisting of eigenvectors of T.
   (ii) The eigenvalues of T are real numbers.

(b) *Matrix form:* Let A be a real symmetric matrix.

   (i) There is an orthogonal matrix P such that $P^t A P$ is a real diagonal matrix.
   (ii) The eigenvalues of A are real numbers.

Proof. We have noted **(b)**(ii) before (Corollary 8.3.12), and **(a)**(ii) follows. Knowing this, the proof of **(a)**(i) follows the pattern of the proof of Theorem 8.6.6. □

The Spectral Theorem is a powerful tool. When faced with a Hermitian operator or a Hermitian matrix, it should be an automatic response to apply that theorem.

8.7 CONICS AND QUADRICS

Ellipses, hyperbolas, and parabolas are called *conics*. They are loci in $\mathbb{R}^2$ defined by quadratic equations $f = 0$, where

(8.7.1) $f(x_1, x_2) = a_{11}x_1^2 + 2a_{12}x_1x_2 + a_{22}x_2^2 + b_1x_1 + b_2x_2 + c,$

and the coefficients a_{ij}, b_i, and c are real numbers. (The reason that the coefficient of x_1x_2 is written as $2a_{12}$ will be explained presently.) If the locus $f = 0$ of a quadratic equation is not a conic, we call it a *degenerate conic*. A degenerate conic can be a pair of lines, a single line, a point, or empty, depending on the equation. To emphasize that a particular locus is not degenerate, we may sometimes refer to it as a *nondegenerate conic*. The term *quadric* is used to designate an analogous locus in three or more dimensions.

We propose to describe the orbits of the conics under the action of the group of isometries of the plane. Two nondegenerate conics are in the same orbit if and only if they are congruent geometric figures.

The quadratic part of the polynomial $f(x_1, x_2)$ is called a *quadratic form*:

$$(8.7.2) \qquad q(x_1, x_2) = a_{11}x_1^2 + 2a_{12}x_1x_2 + a_{22}x_2^2.$$

A quadratic form in any number of variables is a polynomial, each of whose terms has degree 2 in the variables. It is convenient to express the quadratic form q in matrix notation. To do this, we introduce the symmetric matrix

$$(8.7.3) \qquad A = \begin{bmatrix} a_{11} & a_{12} \\ a_{12} & a_{22} \end{bmatrix}$$

Then if $X = (x_1, x_2)^t$, the quadratic form can be written as $q(x_1, x_2) = X^t A X$. We put the coefficient 2 into Formulas 8.7.1 and 8.7.2 in order to avoid some coefficients $\frac{1}{2}$ in this matrix. If we also introduce the 1×2 matrix $B = [b_1\, b_2]$, the equation $f = 0$ can be written compactly in matrix notation as

$$(8.7.4) \qquad X^t A X + B X + c = 0.$$

Theorem 8.7.5 Every nondegenerate conic is congruent to one of the following loci, where the coefficients a_{11} and a_{22} are positive:

$$
\begin{array}{llll}
\textit{Ellipse:} & a_{11}x_1^2 + a_{22}x_2^2 & -1 & = 0, \\
\textit{Hyperbola:} & a_{11}x_1^2 - a_{22}x_2^2 & -1 & = 0, \\
\textit{Parabola:} & a_{11}x_1^2 & -x_2 & = 0.
\end{array}
$$

The coefficients a_{11} and a_{22} are determined by the congruence class of the conic, except that they can be interchanged in the equation of an ellipse.

Proof. We simplify the equation (8.7.4) in two steps, first applying an orthogonal transformation to diagonalize the matrix A and then applying a translation to eliminate the linear terms and the constant term when possible.

The Spectral Theorem for symmetric operators (8.6.10) asserts that there is a 2×2 orthogonal matrix P such that $P^t A P$ is diagonal. We make the change of variable $PX' = X$, and substitute into (8.7.4):

$$(8.7.6) \qquad X'^t A' X' + B' X' + c = 0$$

where $A' = P^t A P$ and $B' = BP$. With this orthogonal change of variable, the quadratic form becomes diagonal, that is, the coefficient of $x_1' x_2'$ is zero. We drop the primes. When the quadratic form is diagonal, f has the form

$$f(x_1, x_2) = a_{11}x_1^2 + a_{22}x_2^2 + b_1x_1 + b_2x_2 + c.$$

To continue, we eliminate b_i by "completing squares," with the substitutions

(8.7.7)
$$x_i = \left(x_i' - \frac{b_i}{2a_{ii}} \right).$$

This substitution corresponds to a translation of coordinates. Dropping primes again, f becomes

(8.7.8)
$$f(x_1, x_2) = a_{11}x_1^2 + a_{22}x_2^2 + c = 0,$$

where the constant term c has changed. The new constant can be computed when needed. When it is zero, the locus is degenerate. Assuming that $c \neq 0$, we can multiply f by a scalar to change c to -1. If a_{ii} are both negative, the locus is empty, hence degenerate. So at least one of the coefficients is positive, and we may assume that $a_{11} > 0$. Then we are left with the equations of the ellipses and the hyperbolas in the statement of the theorem.

The parabola arises because the substitution made to eliminate the linear coefficient b_i requires a_{ii} to be nonzero. Since the equation f is supposed to be quadratic, these coefficients aren't both zero, and we may assume $a_{11} \neq 0$. If $a_{22} = 0$ but $b_2 \neq 0$, we eliminate b_1 and use the substitution

(8.7.9)
$$x_2 = x_2' - c/b_2$$

to eliminate the constant term. Adjusting f by a scalar factor and eliminating degenerate cases leaves us with the equation of the parabola.   □

Example 8.7.10 Let f be the quadratic polynomial $x_1^2 + 2x_1x_2 - x_2^2 + 2x_1 + 2x_2 - 1$. Then

$$A = \begin{bmatrix} 1 & 1 \\ 1 & -1 \end{bmatrix}, \quad B = [2 \ \ 2], \quad \text{and} \quad c = -1.$$

The eigenvalues of A are $\pm \sqrt{2}$. Setting $a = \sqrt{2} - 1$ and $b = \sqrt{2} + 1$, the vectors

$$v_1 = \begin{bmatrix} 1 \\ a \end{bmatrix}, \quad v_2 = \begin{bmatrix} -1 \\ b \end{bmatrix}$$

are eigenvectors with eigenvalues $\sqrt{2}$ and $-\sqrt{2}$, respectively. They are orthogonal, and when we normalize their lengths to 1, they will form an orthonormal basis **B** such that $[\mathbf{B}]^{-1}A[\mathbf{B}]$ is diagonal. Unfortunately, the square length of v_1 is $4 - 2\sqrt{2}$. To normalize its length to 1, we must divide by $\sqrt{4 - 2\sqrt{2}}$. It is unpleasant to continue this computation by hand.

If a quadratic equation $f(x_1, x_2) = 0$ is given, we can determine the type of conic that it represents most simply by allowing arbitrary changes of basis, not necessarily orthogonal ones. A nonorthogonal change will distort an ellipse but it will not change an ellipse into a hyperbola, a parabola, or a degenerate conic. If we wish only to identify the type of conic, arbitrary changes of basis are permissible.

We proceed as in (8.7.6), but with a nonorthogonal change of basis:

$$P = \begin{bmatrix} 1 & -1 \\ & 1 \end{bmatrix}, \quad P^tAP = \begin{bmatrix} 1 & \\ -1 & 1 \end{bmatrix}\begin{bmatrix} 1 & 1 \\ 1 & -1 \end{bmatrix}\begin{bmatrix} 1 & -1 \\ & 1 \end{bmatrix} = \begin{bmatrix} 1 & \\ & -2 \end{bmatrix}, \quad BP = [2 \ \ 0].$$

Dropping primes, the new equation becomes $x_1^2 - 2x_2^2 + 2x_1 - 1 = 0$, and completing the square yields $x_1^2 - 2x_2^2 - 2 = 0$, a hyperbola. So the original locus is a hyperbola too.

By the way, the matrix A is positive or negative definite in the equation of an ellipse and indefinite in the equation of a hyperbola. The matrix A shown above is indefinite. We could have seen right away that the locus we have just inspected was either a hyperbola or a degenerate conic. $\square$

The method used to describe conics can be applied to classify quadrics in any dimension. The general quadratic equation has the form $f = 0$, where

$$(8.7.11) \qquad f(x_1, \ldots, x_n) = \sum_i a_{ii}x_i^2 + \sum_{i<j} 2a_{ij}x_ix_j + \sum_i b_ix_i + c.$$

Let matrices A and B be defined by

$$A = \begin{bmatrix} a_{11} & \cdots & a_{1n} \\ \vdots & & \vdots \\ a_{1n} & \cdots & a_{nn} \end{bmatrix}, \quad B = [\, b_1 \quad \cdots \quad b_n \,].$$

Then

$$(8.7.12) \qquad f(x_1, \ldots, x_n) = X^tAX + BX + c.$$

The associated quadratic form is

$$(8.7.13) \qquad q(x_1, \ldots, x_n) = X^tAX.$$

According to the Spectral Theorem for symmetric operators, the matrix A can be diagonalized by an orthogonal transformation P. When A is diagonal, the linear terms and the constant term may be eliminated, so far as possible, as above. Here is the classification in three variables:

Theorem 8.7.14 The congruence classes of nondegenerate quadrics in $\mathbb{R}^3$ are represented by the following loci, in which a_{ii} are positive real numbers:

Ellipsoids:	$a_{11}x_1^2 + a_{22}x_2^2 + a_{33}x_3^2 - 1 = 0,$
One-sheeted hyperboloids:	$a_{11}x_1^2 + a_{22}x_2^2 - a_{33}x_3^2 - 1 = 0,$
Two-sheeted hyperboloids:	$a_{11}x_1^2 - a_{22}x_2^2 - a_{33}x_3^2 - 1 = 0,$
Elliptic paraboloids:	$a_{11}x_1^2 + a_{22}x_2^2 - x_3 = 0,$
Hyperbolic paraboloids:	$a_{11}x_1^2 - a_{22}x_2^2 - x_3 = 0.$ $\quad\square$

A word is in order about the case that B and c are zero in the quadratic polynomial $f(x_1, x_2, x_3)$ (8.7.12), i.e, that f is equal to its quadratic form q (8.7.13). The locus $\{q = 0\}$

is considered degenerate, but is interesting. Let's call it Q. Since all of the terms $a_{ij}x_ix_j$ that appear in q have degree 2,

(8.7.15)
$$q(\lambda x_1, \lambda x_2, \lambda x_3) = \lambda^2 q(x_1, x_2, x_3).$$

for any real number λ. Consequently, if a point $X \neq 0$ lies on Q, i.e., if $q(X) = 0$, then $q(\lambda X) = 0$ too, so λX lies on Q for every real number λ. Therefore Q is a union of lines through the origin, a double cone.

For example, suppose that q is the diagonal quadratic form

$$a_{11}x_1^2 + a_{22}x_2^2 - x_3^2,$$

where a_{ii} are positive. When we intersect the locus Q with the plane $x_3 = 1$, we obtain an ellipse $a_{11}x_1^2 + a_{22}x_2^2 = 1$ in the remaining variables. In this case Q is the union of lines through the origin and the points of this ellipse.

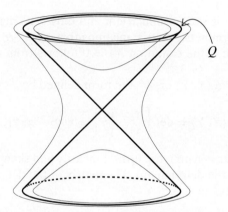

(8.7.16)   Hyperboloids Near to a Cone.

Notice that $q(x)$ is positive in the exterior of the double cone, and negative in its interior. (The value of $q(x)$ changes sign only when one crosses Q.) So for any $r > 0$, the locus $a_{11}x_1^2 + a_{22}x_2^2 - x_3^2 - r = 0$ lies in the exterior of the double cone. It is a one-sheeted hyperboloid, while the locus $a_{11}x_1^2 + a_{22}x_2^2 - x_3^2 + r = 0$ lies in the interior, and is a two-sheeted hyperboloid.

Similar reasoning can be applied to any homogeneous polynomial $g(x_1, \ldots, x_n)$, any polynomial in which all of the terms have the same degree d. If g is homogeneous of degree d, $g(\lambda x) = \lambda^d g(x)$, and because of this, the locus $\{g = 0\}$ will also be a union of lines through the origin.

8.8   SKEW-SYMMETRIC FORMS

The description of skew-symmetric bilinear forms is the same for any field of scalars, so in this section we allow vector spaces over an arbitrary field F. However, as usual, it may be best to think of real vector spaces when going through this for the first time.

A bilinear form $\langle\,,\,\rangle$ on a vector space V is *skew-symmetric* if it has either one of the following equivalent properties:

(8.8.1) $\langle v, v \rangle = 0$ for all v in V,   or

(8.8.2) $\langle u, v \rangle = -\langle v, u \rangle$ for all u and v in V.

To be more precise, these conditions are equivalent whenever the field of scalars has characteristic different from 2. If F has characteristic 2, the first condition (8.8.1) is the correct one. The fact that (8.8.1) implies (8.8.2) is proved by expanding $\langle u + v, u + v \rangle$:

$$\langle u + v, u + v \rangle = \langle u, u \rangle + \langle u, v \rangle + \langle v, u \rangle + \langle v, v \rangle,$$

and using the fact that $\langle u, u \rangle = \langle v, v \rangle = \langle u + v, u + v \rangle = 0$. Conversely, if the second condition holds, then setting $u = v$ gives us $\langle v, v \rangle = -\langle v, v \rangle$, hence $2\langle v, v \rangle = 0$, and it follows that $\langle v, v \rangle = 0$, unless $2 = 0$.

A bilinear form $\langle\,,\,\rangle$ is skew-symmetric if and only if its matrix A with respect to an arbitrary basis is a skew-symmetric matrix, meaning that $a_{ji} = -a_{ij}$ and $a_{ii} = 0$, for all i and j. Except in characteristic 2, the condition $a_{ii} = 0$ follows from $a_{ji} = -a_{ij}$ when one sets $i = j$.

The determinant form $\langle X, Y \rangle$ on $\mathbb{R}^2$, the form defined by

(8.8.3) $\langle X, Y \rangle = \det \begin{bmatrix} x_1 & y_1 \\ x_2 & y_2 \end{bmatrix} = x_1 y_2 - x_2 y_1,$

is a simple example of a skew-symmetric form. Linearity and skew symmetry in the columns are familiar properties of the determinant. The matrix of the determinant form (8.8.3) with respect to the standard basis of $\mathbb{R}^2$ is

(8.8.4) $\Sigma = \begin{bmatrix} & 1 \\ -1 & \end{bmatrix}.$

We will see in Theorem 8.8.7 below that every nondegenerate skew-symmetric form looks very much like this one.

Skew-symmetric forms also come up when one counts intersections of paths on a surface. To obtain a count that doesn't change when the paths are deformed, one can adopt the rule used for traffic flow: A vehicle that enters an intersection from the right has the right of way. If two paths X and Y on the surface intersect at a point p, we define the intersection number $\langle X, Y \rangle_p$ at p as follows: If X enters the intersection to the right of Y, then $\langle X, Y \rangle_p = 1$, and if X enters to the left of Y, then $\langle X, Y \rangle_p = -1$. Then in either case, $\langle X, Y \rangle_p = -\langle Y, X \rangle_p$. The total intersection number $\langle X, Y \rangle$ is obtained by adding these contributions for all intersection points. In this way the contributions arising when X crosses Y and then turns back to cross again cancel. This is how topologists define a product in "homology."

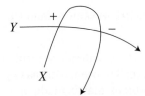

(8.8.5)   Oriented Intersections $\langle X, Y \rangle$.

Many of the definitions given in Section 8.4 can be used also with skew-symmetric forms. In particular, two vectors v and w are orthogonal if $\langle v, w \rangle = 0$. It is true once more that $v \perp w$ if and only if $w \perp v$, but there is a difference: When the form is skew-symmetric, every vector v is self-orthogonal: $v \perp v$. And since all vectors are self-orthogonal, there can be no orthogonal bases.

As is true for symmetric forms, a skew-symmetric form is nondegenerate if and only if its matrix with respect to an arbitrary basis is nonsingular. The proof of the next theorem is the same as for Theorem 8.4.5.

Theorem 8.8.6 Let $\langle \, , \, \rangle$ be a skew-symmetric form on a vector space V, and let W be a subspace of V on which the form is nondegenerate. Then V is the orthogonal sum $W \oplus W^{\perp}$. If the form is nondegenerate on V and on W, it is nondegenerate on $W^{\perp}$ too.   □

Theorem 8.8.7

(a) Let V be a vector space of positive dimension m over a field F, and let $\langle \, , \, \rangle$ be a nondegenerate skew-symmetric form on V. The dimension of V is even, and V has a basis **B** such that the matrix S_0 of the form with respect to that basis is made up of diagonal blocks, where all blocks are equal to the 2×2 matrix S shown above (8.8.4):

$$S_0 = \begin{bmatrix} \Sigma & & \\ & \cdot & \\ & & \cdot \\ & & & \Sigma \end{bmatrix}$$

(b) *Matrix form*: Let A be an invertible skew-symmetric $m \times m$ matrix. There is an invertible matrix P such that $P^t A P = S_0$ is as above.

Proof. **(a)** Since the form is nondegenerate, we may choose nonzero vectors v_1 and v_2 such that $\langle v_1, v_2 \rangle = c$ is not zero. We adjust v_2 by a scalar factor to make $c = 1$. Since $\langle v_1, v_2 \rangle \neq 0$ but $\langle v_1, v_1 \rangle = 0$, these vectors are independent. Let W be the two-dimensional subspace with basis (v_1, v_2). The matrix of the form restricted to W is Σ. Since this matrix is invertible, the form is nondegenerate on W, so V is the direct sum $W \oplus W^{\perp}$, and the form is nondegenerate on $W^{\perp}$. By induction, we may assume that there is a basis $(v_3, \ldots, v_n)$ for $W^{\perp}$ such that the matrix of the form on this subspace has the form (8.8.7). Then $(v_1, v_2, v_3, \ldots, v_n)$ is the required basis for V.   □

Corollary 8.8.8 If A is an invertible $m \times m$ skew-symmetric matrix, then m is an even integer. $\qquad \square$

Let $\langle \, , \, \rangle$ be a nondegenerate skew-symmetric form on a vector space of dimension $2n$. We rearrange the basis referred to in Theorem 8.8.7 as $(v_1, v_3, \ldots, v_{2n-1}; v_2, v_4, \cdots, v_{2n})$. The matrix will be changed into a block matrix made up of $n \times n$ blocks

$$(8.8.9) \qquad\qquad S = \begin{bmatrix} 0 & I \\ -I & 0 \end{bmatrix}.$$

8.9 SUMMARY

We collect some of the terms that we have used together here. They are used for a symmetric or a skew-symmetric form on a real vector space and also for a Hermitian form on a complex vector space.

orthogonal vectors: Two vectors v and w are orthogonal (written $v \perp w$) if $\langle v, w \rangle = 0$.

orthogonal space to a subspace: The orthogonal space $W^\perp$ to a subspace W of V is the set of vectors v that are orthogonal to every vector in W:

$$W^\perp = \Big\{ v \in V \mid \langle v, W \rangle = 0 \Big\}.$$

null vector: A null vector is a vector that is orthogonal to every vector in V.

nullspace: The nullspace N of the given form is the set of null vectors:

$$N = \Big\{ v \mid \langle v, V \rangle = 0 \Big\}.$$

nondegenerate form: The form is nondegenerate if its nullspace is the zero space $\{0\}$. This means that for every nonzero vector v, there is a vector v' such that $\langle v, v' \rangle \neq 0$.

nondegeneracy on a subspace: The form is nondegenerate on a subspace W if its restriction to W is a nondegenerate form, or if $W \cap W^\perp = \{0\}$. If the form is nondegenerate on a subspace W, then $V = W \oplus W^\perp$.

orthogonal basis: A basis $\mathbf{B} = (v_1, \ldots, v_n)$ of V is orthogonal if the vectors are mutually orthogonal, that is, if $\langle v_i, v_j \rangle = 0$ for all indices i and j with $i \neq j$. The matrix of the form with respect to an orthogonal basis is a diagonal matrix. Orthogonal bases exist for any symmetric or Hermitian form, but not for a skew-symmetric form.

orthonormal basis: A basis $\mathbf{B} = (v_1, \ldots, v_n)$ is orthonormal if $\langle v_i, v_j \rangle = 0$ for $i \neq j$ and $\langle v_i, v_i \rangle = 1$. An orthonormal basis for a symmetric or Hermitian form exists if and only if the form is positive definite.

orthogonal projection: If a symmetric or Hermitian form is nondegenerate on a subspace W, the orthogonal projection to W is the unique linear transformation $\pi: V \rightarrow W$ such that: $\pi(v) = v$ if v is in W, and $\pi(v) = 0$ if v is in the orthogonal space $W^\perp$.

If the form is nondegenerate on a subspace W and if $(w_1, \ldots, w_k)$ is an orthogonal basis for W, the orthogonal projection is given by the formula $\pi(v) = w_1 c_1 + \cdots w_k c_k$, where

$$c_i = \frac{\langle w_i, v \rangle}{\langle w_i, w_i \rangle}.$$

Spectral Theorem:

- If A is normal, there is a unitary matrix P such that $P^* A P$ is diagonal.
- If A is Hermitian, there is a unitary matrix P such that $P^* A P$ is a real diagonal matrix.
- In the unitary group U_n, every matrix is conjugate to a diagonal matrix.
- If A is a real symmetric matrix, there is an orthogonal matrix P such that $P^t A P$ is diagonal.

The table below compares various concepts used for real and for complex vector spaces.

Real Vector Spaces	**Complex Vector Spaces**
forms	
symmetric	Hermitian
$\langle v, w \rangle = \langle w, v \rangle$	$\langle v, w \rangle = \overline{\langle w, v \rangle}$
matrices	
symmetric	Hermitian
$A^t = A$	$A^* = A$
orthogonal	unitary
$A^t A = I$	$A^* A = I$
	normal
	$A^* A = A A^*$
operators	
symmetric	Hermitian
$\langle Tv, w \rangle = \langle v, Tw \rangle$	$\langle Tv, w \rangle = \langle v, Tw \rangle$
orthogonal	unitary
$\langle v, w \rangle = \langle Tv, Tw \rangle$	$\langle v, w \rangle = \langle Tv, Tw \rangle$
	normal
	$\langle Tv, Tw \rangle = \langle T^*v, T^*w \rangle$
	arbitrary
	$\langle v, Tw \rangle = \langle T^*v, w \rangle$

In helping geometry, modern algebra is helping itself above all.

—Oscar Zariski

EXERCISES

Section 1   Real Bilinear Forms

1.1. Show that a bilinear form $\langle \ , \ \rangle$ on a real vector space V is a sum of a symmetric form and a skew-symmetric form.

Section 2   Symmetric Forms

2.1. Prove that the maximal entries of a positive definite, symmetric, real matrix are on the diagonal.

2.2. Let A and A' be symmetric matrices related by $A' = P^t A P$, where P is invertible. Is it true that the ranks of A and of A' are equal?

Section 3   Hermitian Forms

3.1. Is a complex $n \times n$ matrix A such that $X^* A X$ is real for all X Hermitian?

3.2. Let $\langle \ , \ \rangle$ be a positive definite Hermitian form on a complex vector space V, and let $\{ \ , \ \}$ and $[\ , \ ]$ be its real and imaginary parts, the real-valued forms defined by

$$\langle v, w \rangle = \{v, w\} + [v, w]i.$$

Prove that when V is made into a real vector space by restricting scalars to $\mathbb{R}$, $\{ \ , \ \}$ is a positive definite symmetric form, and $[\ , \ ]$ is a skew-symmetric form.

3.3. The set of $n \times n$ Hermitian matrices forms a *real* vector space. Find a basis for this space.

3.4. Prove that if A is an invertible matrix, then $A^* A$ is Hermitian and positive definite.

3.5. Let A and B be positive definite Hermitian matrices. Decide which of the following matrices are necessarily positive definite Hermitian: A^2, A^{-1}, AB, $A + B$.

3.6. Use the characteristic polynomial to prove that the eigenvalues of a 2×2 Hermitian matrix A are real.

Section 4   Orthogonality

4.1. What is the inverse of a matrix whose columns are orthogonal?

4.2. Let $\langle \ , \ \rangle$ be a bilinear form on a real vector space V, and let v be a vector such that $\langle v, v \rangle \neq 0$. What is the formula for orthogonal projection to the space $W = v^\perp$ orthogonal to v?

4.3. Let A be a real $m \times n$ matrix. Prove that $B = A^t A$ is positive semidefinite, i.e., that $X^t B X \geq 0$ for all X, and that A and B have the same rank.

4.4. Make a sketch showing the positions of some orthogonal vectors in $\mathbb{R}^2$, when the form is $\langle X, Y \rangle = x_1 y_1 - x_2 y_2$.

4.5. Find an orthogonal basis for the form on $\mathbb{R}^n$ whose matrix is

(a) $\begin{bmatrix} 1 & 1 \\ 1 & 1 \end{bmatrix}$,   (b) $\begin{bmatrix} 1 & 0 & 1 \\ 0 & 2 & 1 \\ 1 & 1 & 1 \end{bmatrix}$.

4.6. Extend the vector $X_1 = \frac{1}{2}(1, -1, 1, 1)^t$ to an orthonormal basis for $\mathbb{R}^4$.

4.7. Apply the Gram–Schmidt procedure to the basis $(1, 1, 0)^t$, $(1, 0, 1)^t$, $(0, 1, 1)^t$ of $\mathbb{R}^3$.

4.8. Let $A = \begin{bmatrix} 2 & 1 \\ 1 & 2 \end{bmatrix}$. Find an orthonormal basis for $\mathbb{R}^2$ with respect to the form $X^t A Y$.

4.9. Find an orthonormal basis for the vector space P of all real polynomials of degree at most 2, with the symmetric form defined by

$$\langle f, g \rangle = \int_{-1}^{1} f(x) g(x) dx.$$

4.10. Let V denote the vector space of real $n \times n$ matrices. Prove that $\langle A, B \rangle = \text{trace}(A^t B)$ defines a positive definite bilinear form on V, and find an orthonormal basis for this form.

4.11. Let W_1, W_2 be subspaces of a vector space V with a symmetric bilinear form. Prove
  (a) $(W_1 + W_2)^{\perp} = W_1^{\perp} \cap W_2^{\perp}$, **(b)** $W \subset W^{\perp\perp}$, **(c)** If $W_1 \subset W_2$, then $W_1^{\perp} \supset W_2^{\perp}$.

4.12. Let $V = \mathbb{R}^{2 \times 2}$ be the vector space of real 2×2 matrices.

  (a) Determine the matrix of the bilinear form $\langle A, B \rangle = \text{trace}(AB)$ on V with respect to the standard basis $\{e_{ij}\}$.

  (b) Determine the signature of this form.

  (c) Find an orthogonal basis for this form.

  (d) Determine the signature of the form trace AB on the space $\mathbb{R}^{n \times n}$ of real $n \times n$ matrices.

*4.13. (a) Decide whether or not the rule $\langle A, B \rangle = \text{trace}(A^* B)$ defines a Hermitian form on the space $\mathbb{C}^{n \times n}$ of complex matrices, and if so, determine its signature.

  (b) Answer the same question for the form defined by $\langle A, B \rangle = \text{trace}(\overline{A} B)$.

4.14. The matrix form of Theorem 8.4.10 asserts that if A is a real symmetric matrix, there exists an invertible matrix P such that $P^t A P$ is diagonal. Prove this by row and column operations.

4.15. Let W be the subspace of $\mathbb{R}^3$ spanned by the vectors $(1, 1, 0)^t$ and $(0, 1, 1)^t$. Determine the orthogonal projection of the vector $(1, 0, 0)^t$ to W.

4.16. Let V be the real vector space of 3×3 matrices with the bilinear form $\langle A, B \rangle = \text{trace } A^t B$, and let W be the subspace of skew-symmetric matrices. Compute the orthogonal projection to W with respect to this form, of the matrix

$$\begin{bmatrix} 1 & 2 & 0 \\ 0 & 0 & 1 \\ 1 & 3 & 0 \end{bmatrix}.$$

4.17. Use the method of (3.5.13) to compute the coordinate vector of the vector $(x_1, x_2, x_3)^t$ with respect to the basis **B** described in Example 8.4.14, and compare your answer with the projection formula.

4.18. Find the matrix of a projection $\pi : \mathbb{R}^3 \to \mathbb{R}^2$ such that the image of the standard bases of $\mathbb{R}^3$ forms an equilateral triangle and $\pi(e_1)$ points in the direction of the x-axis.

4.19. Let W be a two-dimensional subspace of $\mathbb{R}^3$, and consider the orthogonal projection π of $\mathbb{R}^3$ onto W. Let $(a_i, b_i)^t$ be the coordinate vector of $\pi(e_i)$, with respect to a chosen orthonormal basis of W. Prove that (a_1, a_2, a_3) and (b_1, b_2, b_3) are orthogonal unit vectors.

4.20. Prove the criterion for positive definiteness given in Theorem 8.4.19. Does the criterion carry over to Hermitian matrices?

4.21. Prove Sylvester's Law (see 8.4.17).

Hint: Begin by showing that if W_1 and W_2 are subspaces of V and if the form is positive definite on W_1 and negative semi-definite on W_2, then W_1 and W_2 are independent.

Section 5   Euclidean Spaces and Hermitian Spaces

5.1. Let V be a Euclidean space.

(a) Prove the *Schwarz inequality* $|\langle v, w \rangle| \le |v||w|$.

(b) Prove the *parallelogram law* $|v + w|^2 + |v - w|^2 = 2|v|^2 + 2|w|^2$.

(c) Prove that if $|v| = |w|$, then $(v + w) \perp (v - w)$.

5.2. Let W be a subspace of a Euclidean space V. Prove that $W = W^{\perp\perp}$.

*5.3. Let $w \in \mathbb{R}^n$ be a vector of length 1, and let U denote the orthogonal space $w^\perp$. The *reflection* r_w about U is defined as follows: We write a vector v in the form $v = cw + u$, where $u \in U$. Then $r_w(v) = -cw + u$.

(a) Prove that the matrix $P = I - 2ww^t$ is orthogonal.

(b) Prove that multiplication by P is a reflection about the orthogonal space U.

(c) Let u, v be vectors of equal length in $\mathbb{R}^n$. Determine a vector w such that $Pu = v$.

5.4. Let T be a linear operator on $V = \mathbb{R}^n$ whose matrix A is a real symmetric matrix.

(a) Prove that V is the orthogonal sum $V = (\ker T) \oplus (\operatorname{im} T)$.

(b) Prove that T is an orthogonal projection onto $\operatorname{im} T$ if and only if, in addition to being symmetric, $A^2 = A$.

5.5. Let P be a unitary matrix, and let X_1 and X_2 be eigenvectors for P, with distinct eigenvalues λ_1 and λ_2. Prove that X_1 and X_2 are orthogonal with respect to the standard Hermitian form on $\mathbb{C}^n$.

5.6. What complex numbers might occur as eigenvalues of a unitary matrix?

Section 6   The Spectral Theorem

6.1. Prove Proposition 8.6.3**(c), (d)**.

6.2. Let T be a symmetric operator on a Euclidean space. Using Proposition 8.6.9, prove that if v is a vector and if $T^2v = 0$, then $Tv = 0$.

6.3. What does the Spectral Theorem tell us about a real 3×3 matrix that is both symmetric and orthogonal?

6.4. What can be said about a matrix A such that A^*A is diagonal?

6.5. Prove that if A is a real skew-symmetric matrix, then iA is a Hermitian matrix. What does the Spectral Theorem tell us about a real skew-symmetric matrix?

6.6. Prove that an invertible matrix A is normal if and only if A^*A^{-1} is unitary.

6.7. Let P be a real matrix that is normal and has real eigenvalues. Prove that P is symmetric.

6.8. Let V be the space of differentiable complex-valued functions on the unit circle in the complex plane, and for $f, g \in V$, define

$$\langle f, g \rangle = \int_0^{2\pi} \overline{f(\theta)} g(\theta) d\theta.$$

(a) Show that this form is Hermitian and positive definite.

(b) Let W be the subspace of V of functions $f(e^{i\theta})$, where f is a polynomial of degree $\leq n$. Find an orthonormal basis for W.

(c) Show that $T = i\frac{d}{d\theta}$ is a Hermitian operator on V, and determine its eigenvalues on W.

6.9. Determine the signature of the form on $\mathbb{R}^2$ whose matrix is $\begin{bmatrix} & 1 \\ 1 & \end{bmatrix}$, and determine an orthogonal matrix P such that $P^t A P$ is diagonal.

6.10. Prove that if T is a Hermitian operator on a Hermitian space V, the rule $\{v, w\} = \langle v, Tw \rangle$ defines a second Hermitian form on V.

6.11. Prove that eigenvectors associated to distinct eigenvalues of a Hermitian matrix A are orthogonal.

6.12. Find a unitary matrix P so that $P^* A P$ is diagonal, when $A = \begin{bmatrix} 1 & i \\ -i & 1 \end{bmatrix}$.

6.13. 5. Find a real orthogonal matrix P so that $P^t A P$ is diagonal, when A is the matrix

(a) $\begin{bmatrix} 1 & 2 \\ 2 & 1 \end{bmatrix}$, (b) $\begin{bmatrix} 1 & 1 & 1 \\ 1 & 1 & 1 \\ 1 & 1 & 1 \end{bmatrix}$, (c) $\begin{bmatrix} 1 & 0 & 1 \\ 0 & 1 & 0 \\ 1 & 0 & 0 \end{bmatrix}$.

6.14. Prove that a real symmetric matrix A is positive definite if and only if its eigenvalues are positive.

6.15. Prove that for any square matrix A, $\ker A = (\operatorname{im} A^*)^\perp$, and that if A is normal, $\ker A = (\operatorname{im} A)^\perp$.

*6.16. Let $\zeta = e^{2\pi i/n}$, and let A be the $n \times n$ matrix whose entries are $a_{jk} = \zeta^{jk}/\sqrt{n}$. Prove that A is unitary.

*6.17. Let A, B be Hermitian matrices that commute. Prove that there is a unitary matrix P such that $P^* A P$ and $P^* B P$ are both diagonal.

6.18. Use the Spectral Theorem to prove that a positive definite real symmetric $n \times n$ matrix A has the form $A = P^t P$ for some P.

6.19. Prove that the cyclic shift operator

$$\begin{bmatrix} 0 & 1 & & & \\ & 0 & 1 & & \\ & & \cdot & \cdot & \\ & & & \cdot & \cdot \\ & & & & \cdot & 1 \\ 1 & & & & & 0 \end{bmatrix}$$

is unitary, and determine its diagonalization.

6.20. Prove that the *circulant*, the matrix below, is normal.

$$\begin{bmatrix} c_0 & c_1 & \cdots & & c_n \\ c_n & c_0 & \cdots & & c_{n-1} \\ \vdots & & & & \vdots \\ c_1 & c_2 & \cdots & & c_0 \end{bmatrix}$$

6.21. What conditions on the eigenvalues of a normal matrix A imply that A is Hermitian? That A is unitary?

6.22. Prove the Spectral Theorem for symmetric operators.

Section 7   Conics and Quadrics

7.1. Determine the type of the quadric $x^2 + 4xy + 2xz + z^2 + 3x + z - 6 = 0$.

7.2. Suppose that the quadratic equation (8.7.1) represents an ellipse. Instead of diagonalizing the form and then making a translation to reduce to the standard type, we could make the translation first. How can one determine the required translation?

7.3. Give a necessary and sufficient condition, in terms of the coefficients of its equation, for a conic to be a circle.

7.4. Describe the degenerate quadrics geometrically.

Section 8   Skew-Symmetric Forms

8.1. Let A be an invertible, real, skew-symmetric matrix. Prove that A^2 is symmetric and negative definite.

8.2. Let W be a subspace on which a real skew-symmetric form is nondegenerate. Find a formula for the orthogonal projection $\pi : V \to W$.

8.3. Let S be a real skew-symmetric matrix. Prove that $I + S$ is invertible, and that $(I - S)(I + S)^{-1}$ is orthogonal.

*8.4. Let A be a real skew-symmetric matrix.

   (a) Prove that $\det A \geq 0$.

   (b) Prove that if A has integer entries, then $\det A$ is the square of an integer.

Miscellaneous Problems

M.1. According to Sylvester's Law, every 2×2 real symmetric matrix is congruent to exactly one of six standard types. List them. If we consider the operation of GL_2 on 2×2 matrices by $P * A = PAP^t$, then Sylvester's Law asserts that the symmetric matrices form six orbits. We may view the symmetric matrices as points in $\mathbb{R}^3$, letting (x, y, z) correspond to the matrix $\begin{bmatrix} x & y \\ y & z \end{bmatrix}$. Describe the decomposition of $\mathbb{R}^3$ into orbits geometrically, and make a clear drawing depicting it.

   Hint: If you don't get a beautiful result, you haven't understood the configuration.

M.2. Describe the symmetry of the matrices $AB + BA$ and $AB - BA$ in the following cases.
   (a) A, B symmetric,   **(b)** A, B Hermitian,   **(c)** A, B skew-symmetric,
   (d) A symmetric, B skew-symmetric.

M.3. With each of the following types of matrices, describe the possible determinants and eigenvalues.

(a) real orthogonal, **(b)** unitary, **(c)** Hermitian, **(d)** real symmetric, negative definite, **(e)** real skew-symmetric.

M.4. Let E be an $m \times n$ complex matrix. Prove that the matrix $\begin{bmatrix} I & E^* \\ -E & I \end{bmatrix}$ is invertible.

M.5. The vector cross product is $x \times y = (x_2 y_3 - x_3 y_2, \, x_3 y_1 - x_1 y_3, \, x_1 y_2 - x_2 y_1)^t$. Let v be a fixed vector in $\mathbb{R}^3$, and let T be the linear operator $T(x) = (x \times v) \times v$.

(a) Show that this operator is symmetric. You may use general properties of the scalar triple product $\det[x|y|z] = (x \times y) \cdot z$, but not the matrix of the operator.

(b) Compute the matrix.

M.6. (a) What is wrong with the following argument? Let P be a real orthogonal matrix. Let X be a (possibly complex) eigenvector of P, with eigenvalue λ. Then $X^t P^t X = (PX)^t X = \lambda X^t X$. On the other hand, $X^t P^t X = X^t (P^{-1} X) = \lambda^{-1} X^t X$. Therefore $\lambda = \lambda^{-1}$, and so $\lambda = \pm 1$.

(b) State and prove a correct theorem based on the error in this argument.

*M.7. Let A be a real $m \times n$ matrix. Prove that there are orthogonal matrices P in O_m, and Q in O_n such that PAQ is diagonal, with non-negative diagonal entries.

M.8. (a) Show that if A is a nonsingular complex matrix, there is a positive definite Hermitian matrix B such that $B^2 = A^*A$, and that B is uniquely determined by A.

(b) Let A be a nonsingular matrix, and let B be a positive definite Hermitian matrix such that $B^2 = A^*A$. Show that AB^{-1} is unitary.

(c) Prove the *Polar decomposition*: Every nonsingular matrix A is a product $A = UP$, where P is positive definite Hermitian and U is unitary.

(d) Prove that the Polar decomposition is unique.

(e) What does this say about the operation of left multiplication by the unitary group U_n on the group GL_n?

*M.9. Let V be a Euclidean space of dimension n, and let $S = (v_1, \ldots, v_k)$ be a set of vectors in V. A *positive combination* of S is a linear combination $p_1 v_1 + \cdots + p_k v_k$ in which all coefficients p_i are positive. The subspace $U = \{v | \langle v, w \rangle = 0\}$ of V of vectors orthogonal to a vector w is called a *hyperplane*. A hyperplane divides the space V into two *half spaces* $\{v | \langle v, w \rangle \geq 0\}$ and $\{v | \langle v, w \rangle \leq 0\}$.

(a) Prove that the following are equivalent:
- S is not contained in any half space.
- For every nonzero vector w in V, $\langle v_i, w \rangle < 0$ for some $i = 1, \ldots, k$.

(b) Let S' be the set obtained by deleting v_k from S. Prove that if S is not contained in a half space, then S' spans V.

(c) Prove that the following conditions are equivalent:

(i) S is not contained in a half space.

(ii) Every vector in V is a positive combination of S.

(iii) S spans V and 0 is a positive combination of S.

Hint: To show that (i) implies (ii) or (iii), I recommend projecting to the space U orthogonal to v_k. That will allow you to use induction.

M.10. The row and column indices in the $n \times n$ *Fourier matrix* A run from 0 to $n - 1$, and the i, j entry is ζ^{ij}, with $\zeta = e^{2\pi i/n}$. This matrix solves the following interpolation problem: Given complex numbers $b_0, \ldots, b_{n-1}$, find a complex polynomial $f(t) = c_0 + c_1 t + \cdots + c_{n-1} t^{n-1}$ such that $f(\zeta^v) = b_v$.

  (a) Explain how the matrix solves the problem.

  (b) Prove that A is symmetric and normal, and compute A^2.

  ∗**(c)** Determine the eigenvalues of A.

M.11. Let A be a real $n \times n$ matrix. Prove that A defines an orthogonal projection to its image W if and only if $A^2 = A = A^t A$.

M.12. Let A be a real $n \times n$ orthogonal matrix.

  (a) Let X be a complex eigenvector of A with complex eigenvalue λ. Prove that $X^t X = 0$. Write the eigenvector as $X = R + Si$ where R and S are real vectors. Show that the space W spanned by R and S is A-invariant, and describe the restriction of the operator A to W.

  (b) Prove that there is a real orthogonal matrix P such that $P^t A P$ is a block diagonal matrix made up of 1×1 and 2×2 blocks, and describe those blocks.

M.13. Let $V = \mathbb{R}^n$, and let $\langle X, Y \rangle = X^t A Y$, where A is a symmetric matrix. Let W be the subspace of V spanned by the columns of an $n \times r$ matrix M of rank r, and let $\pi : V \to W$ denote the orthogonal projection of V to W with respect to the form $\langle\ ,\ \rangle$. One can compute π in the form $\pi(X) = MY$ by setting up and solving a suitable system of linear equations for Y. Determine the matrix of π explicitly in terms of A and M. Check your result in the case that $r = 1$ and $\langle\ ,\ \rangle$ is dot product. What hypotheses on A and M are necessary?

M.14. What is the maximal number of vectors v_i in $\mathbb{R}^n$ such that $(v_i \cdot v_j) < 0$ for all $i \neq j$?

M.15. [1]This problem is about the space V of real polynomials in the variables x and y. If f is a polynomial, ∂_f will denote the operator $f(\frac{\partial}{\partial x}, \frac{\partial}{\partial y})$, and $\partial_f(g)$ will denote the result of applying this operator to a polynomial g.

  (a) The rule $\langle f, g \rangle = \partial_f(g)_0$ defines a bilinear form on V, the subscript 0 denoting evaluation of a polynomial at the origin. Prove that this form is symmetric and positive definite, and that the monomials $x^i y^j$ form an orthogonal basis of V (not an orthonormal basis).

  (b) We also have the operator of multiplication by f, which we write as m_f. So $m_f(g) = fg$. Prove that ∂_f and m_f are adjoint operators.

  (c) When $f = x^2 + y^2$, the operator ∂_f is the Laplacian, which is often written as Δ. A polynomial h is harmonic if $\Delta h = 0$. Let H denote the space of harmonic polynomials. Identify the space $H^\perp$ orthogonal to H with respect to the given form.

---

[1]Suggested by Serge Lang

CHAPTER 9

Linear Groups

*In these days the angel of topology and the devil of abstract algebra
fight for the soul of every individual discipline of mathematics.*

—Hermann Weyl[1]

9.1 THE CLASSICAL GROUPS

Subgroups of the general linear group GL_n are called *linear groups*, or *matrix groups*. The most important ones are the special linear, orthogonal, unitary, and symplectic groups – the classical groups. Some of them will be familiar, but let's review the definitions.

The real *special linear group* SL_n is the group of real matrices with determinant 1:

(9.1.1) $$SL_n = \{P \in GL_n(\mathbb{R}) \mid \det P = 1\}.$$

The *orthogonal group* O_n is the group of real matrices P such that $P^t = P^{-1}$:

(9.1.2) $$O_n = \{P \in GL_n(\mathbb{R}) \mid P^t P = I\}.$$

A change of basis by an orthogonal matrix preserves the dot product $X^t Y$ on $\mathbb{R}^n$.

The *unitary group* U_n is the group of complex matrices P such that $P^* = P^{-1}$:

(9.1.3) $$U_n = \{P \in GL_n(\mathbb{C}) \mid P^* P = I\}.$$

A change of basis by a unitary matrix preserves the standard Hermitian product $X^* Y$ on $\mathbb{C}^n$.

The *symplectic group* is the group of real matrices that preserve the skew-symmetric form $X^t S Y$ on $\mathbb{R}^{2n}$, where

$$S = \begin{bmatrix} 0 & I \\ -I & 0 \end{bmatrix},$$

(9.1.4) $$SP_{2n} = \{P \in GL_{2n}(\mathbb{R}) \mid P^t S P = S\}.$$

---

[1]This quote is taken from Morris Kline's book *Mathematical Thought from Ancient to Modern Times*.

There are analogues of the orthogonal group for indefinite forms. The *Lorentz group* is the group of real matrices that preserve the Lorentz form (8.2.2)

$$(9.1.5) \qquad\qquad O_{3,1} = \{P \in GL_n \mid P^t I_{3,1} P = I_{3,1}\}.$$

The linear operators represented by these matrices are called *Lorentz transformations*. An analogous group $O_{p,m}$ can be defined for any signature p, m.

The word *special* is added to indicate the subgroup of matrices with determinant 1:

> *Special orthogonal group* SO_n: real orthogonal matrices with determinant 1,
>
> *Special unitary group* SU_n: unitary matrices with determinant 1.

Though this is not obvious from the definition, symplectic matrices have determinant 1, so the two uses of the letter S do not conflict.

Many of these groups have complex analogues, defined by the same relations. But except in Section 9.8, GL_n, SL_n, O_n, and SP_{2n} stand for the real groups in this chapter. Note that the complex orthogonal group is not the same as the unitary group. The defining properties of these two groups are $P^t P = I$ and $P^* P = I$, respectively.

We plan to describe geometric properties of the classical groups, viewing them as subsets of the spaces of matrices. The word "homeomorphism" from topology will come up. A *homeomorphism* $\varphi: X \to Y$ is a continuous bijective map whose inverse function is also continuous [Munkres, p. 105]. Homeomorphic sets are topologically equivalent. It is important not to confuse the words "homomorphism" and "homeomorphism," though, unfortunately, their only difference is that "homeomorphism" has one more letter.

The geometry of a few linear groups will be familiar. The unit circle,

$$x_0^2 + x_1^2 = 1,$$

for instance, has several incarnations as a group, all isomorphic. Writing $(x_0, x_1) = (\cos\theta, \sin\theta)$ identifies the circle as the additive group of angles. Or, thinking of it as the unit circle in the complex plane by $e^{i\theta}$, it becomes a multiplicative group, the group of unitary 1×1 matrices:

$$(9.1.6) \qquad\qquad U_1 = \{p \in \mathbb{C}^\times \mid \overline{p}p = 1\}.$$

The unit circle can also be embedded into $\mathbb{R}^{2\times2}$ by the map

$$(9.1.7) \qquad\qquad (\cos\theta, \sin\theta) \rightsquigarrow \begin{bmatrix} \cos\theta & -\sin\theta \\ \sin\theta & \cos\theta \end{bmatrix}.$$

It is isomorphic to the special orthogonal group SO_2, the group of rotations of the plane. These are three descriptions of what is essentially the same group, the *circle group*.

The *dimension* of a linear group G is, roughly speaking, the number of degrees of freedom of a matrix in G. The circle group has dimension 1. The group SL_2 has dimension 3, because the equation $\det P = 1$ eliminates one degree of freedom from the four matrix entries. We discuss dimension more carefully in Section 9.7, but we want to describe some of the low-dimensional groups first. The smallest dimension in which really interesting nonabelian groups appear is 3, and the most important ones are SU_2, SO_3, and SL_2. We examine the special unitary group SU_2 and the rotation group SO_3 in Sections 9.3 and 9.4.

9.2 INTERLUDE: SPHERES

By analogy with the unit sphere in $\mathbb{R}^3$, the locus

$$\{x_0^2 + x_1^2 + \cdots + x_n^2 = 1\}$$

in $\mathbb{R}^{n+1}$ is called the *n-dimensional unit sphere*, or the *n-sphere*, for short. We'll denote it by $\mathbb{S}^n$. Thus the unit sphere in $\mathbb{R}^3$ is the 2-sphere $\mathbb{S}^2$, and the unit circle in $\mathbb{R}^2$ is the 1-sphere $\mathbb{S}^1$. A space that is homeomorphic to a sphere may sometimes be called a sphere too.

We review stereographic projection from the 2-sphere to the plane, because it can be used to give topological descriptions of the sphere that have analogues in other dimensions. We think of the x_0-axis as the vertical axis in (x_0, x_1, x_2)-space $\mathbb{R}^3$. The *north pole* on the sphere is the point $p = (1, 0, 0)$. We also identify the locus $\{x_0 = 0\}$ with a plane that we call $\mathbb{V}$, and we label the coordinates in $\mathbb{V}$ as v_1, v_2. The point (v_1, v_2) of $\mathbb{V}$ corresponds to $(0, v_1, v_2)$ in $\mathbb{R}^3$.

Stereographic projection $\pi : \mathbb{S}^2 \to \mathbb{V}$ is defined as follows: To obtain the image $\pi(x)$ of a point x on the sphere, one constructs the line ℓ that passes through p and x. The projection $\pi(x)$ is the intersection of ℓ with $\mathbb{V}$. The projection is bijective at all points of $\mathbb{S}^2$ except the north pole, which is "sent to infinity."

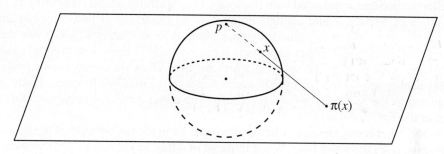

(9.2.1)   Stereographic Projection.

One way to construct the sphere topologically is as the union of the plane $\mathbb{V}$ and a single point, the north pole. The inverse function to π does this. It shrinks the plane a lot near infinity, because a small circle about p on the sphere corresponds to a large circle in the plane.

Stereographic projection is the identity map on the equator. It maps the southern hemisphere bijectively to the unit disk $\{v_1^2 + v_2^2 \le 1\}$ in $\mathbb{V}$, and the northern hemisphere to the exterior $\{v_1^2 + v_2^2 \ge 1\}$ of the disk, except that the north pole is missing from the exterior. On the other hand, stereographic projection from the south pole would map the northern hemisphere to the disk. Both hemispheres correspond bijectively to disks. This provides a second way to build the sphere topologically, as the union of two unit disks glued together along their boundaries. The disks need to be stretched, like blowing up a balloon, to make the actual sphere.

To determine the formula for stereographic projection, we write the line through p and x in the parametric form $q(t) = p + t(x - p) = (1 + t(x_0 - 1), tx_1, tx_2)$. The point $q(t)$ is in the plane $\mathbb{V}$ when $t = \frac{1}{1 - x_0}$. So

(9.2.2) $$\pi(x) = (v_1, v_2) = \left(\frac{x_1}{1 - x_0}, \frac{x_2}{1 - x_0} \right).$$

Stereographic projection π from the n-sphere to n-space is defined in exactly the same way. The *north pole* on the n-sphere is the point $p = (1, 0, \ldots, 0)$, and we identify the locus $\{x_0 = 0\}$ in $\mathbb{R}^{n+1}$ with an n-space $\mathbb{V}$. A point $(v_1, \ldots, v_n)$ of $\mathbb{V}$ corresponds to $(0, v_1, \ldots, v_n)$ in $\mathbb{R}^{n+1}$. The image $\pi(x)$ of a point x on the sphere is the intersection of the line ℓ through the north pole p and x with $\mathbb{V}$. As before, the north pole p is sent to infinity, and π is bijective at all points of $\mathbb{S}^n$ except p. The formula for π is

$$(9.2.3) \qquad \pi(x) = \left(\frac{x_1}{1 - x_0}, \ldots, \frac{x_n}{1 - x_0} \right).$$

This projection maps the lower hemisphere $\{x_0 \leq 0\}$ bijectively to the n-dimensional *unit ball* in $\mathbb{V}$, the locus $\{v_1^2 + \cdots + v_n^2 \leq 1\}$, while projection from the south pole maps the upper hemisphere $\{x_0 \geq 0\}$ to the unit ball. So, as is true for the 2-sphere, the n-sphere can be constructed topologically in two ways: as the union of an n-space $\mathbb{V}$ and a single point p, or as the union of two copies of the n-dimensional unit ball, glued together along their boundaries, which are $(n - 1)$-spheres, and stretched appropriately.

We are particularly interested in the three-dimensional sphere $\mathbb{S}^3$, and it is worth making some effort to become acquainted with this locus. Topologically, $\mathbb{S}^3$ can be constructed either as the union of 3-space $\mathbb{V}$ and a single point p, or as the union of two copies of the unit ball $\{v_1^2 + v_2^2 + v_3^2 \leq 1\}$ in $\mathbb{R}^3$, glued together along their boundaries (which are ordinary 2-spheres) and stretched. Neither construction can be made in three-dimensional space.

We can think of $\mathbb{V}$ as the space in which we live. Then via stereographic projection, the lower hemisphere of the 3-sphere $\mathbb{S}^3$ corresponds to the unit ball in space. Traditionally, it is depicted as the *terrestrial sphere*, the Earth. The upper hemisphere corresponds to the exterior of the Earth, the sky.

On the other hand, the upper hemisphere can be made to correspond to the unit ball via projection from the south pole. When thinking of it this way, it is depicted traditionally as the *celestial sphere*. (The phrases "terrestial ball" and "celestial ball" would fit mathematical terminology better, but they wouldn't be traditional.)

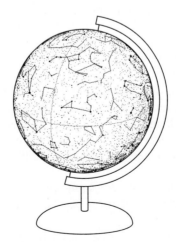

(9.2.4)   A Model of the Celestial Sphere.

To understand this requires some thought. When the upper hemisphere is represented as the celestial sphere, the center of the ball corresponds to the north pole of $\mathbb{S}^3$, and to infinity in our space $\mathbb{V}$. While looking at a celestial globe from its exterior, you must imagine that you are standing on the Earth, looking out at the sky. It is a common mistake to think of the Earth as the center of the celestial sphere.

Latitudes and Longitudes on the 3-Sphere

The curves of constant latitude on the globe, the 2-sphere $\{x_0^2 + x_1^2 + x_2^2 = 1\}$, are the horizontal circles $x_0 = c$, with $-1 < c < 1$, and the curves of constant longitude are the vertical great circles through the poles. The longitude curves can be described as intersections of the 2-sphere with the two-dimensional subspaces of $\mathbb{R}^3$ that contain the pole $(1, 0, 0)$.

When we go to the 3-sphere $\{x_0^2 + x_1^2 + x_2^2 + x_3^2 = 1\}$, the dimension increases, and one has to make some decisions about what the analogues should be. We use analogues that will have algebraic significance for the group SU_2 that we study in the next section.

As analogues of latitude curves on the 3-sphere, we take the "horizontal" surfaces, the surfaces on which the x_0-coordinate is constant. We call these loci *latitudes*. They are two-dimensional spheres, embedded into $\mathbb{R}^4$ by

(9.2.5) $\qquad x_0 = c, \quad x_1^2 + x_2^2 + x_3^2 = (1 - c^2), \quad \text{with } -1 < c < 1.$

The particular latitude defined by $x_0 = 0$ is the intersection of the 3-sphere with the horizontal space $\mathbb{V}$. It is the unit 2-sphere $\{v_1^2 + v_2^2 + v_3^2 = 1\}$ in $\mathbb{V}$. We call this latitude the *equator*, and we denote it by $\mathbb{E}$.

Next, as analogues of the longitude curves, we take the great circles through the north pole $(1, 0, 0, 0)$. They are the intersections of the 3-sphere with two-dimensional subspaces W of $\mathbb{R}^4$ that contain the pole. The intersection $L = W \cap \mathbb{S}^3$ will be the unit circle in W, and we call L a *longitude*. If we choose an orthonormal basis (p, v) for the space W, the first vector being the north pole, the longitude will have the parametrization

(9.2.6) $\qquad\qquad\qquad L : \ell(\theta) = \cos\theta\, p + \sin\theta\, v.$

This is elementary, but we verify it below.

Thus, while the latitudes on $\mathbb{S}^3$ are 2-spheres, the longitudes are 1-spheres.

Lemma 9.2.7 Let (p, v) be an orthonormal basis for a subspace W of $\mathbb{R}^4$, the first vector being the north pole p, and let L be the longitude of unit vectors in W.

(a) L meets the equator $\mathbb{E}$ in two points. If v is one of those points, the other one is $-v$.

(b) L has the parametrization (9.2.6). If q is a point of L, then replacing v by $-v$ if necessary, one can express q in the form $\ell(\theta)$ with θ in the interval $0 \le \theta \le \pi$, and then this representation of a point of L is unique for all $\theta \ne 0, \pi$.

(c) Except for the two poles, every point of the sphere $\mathbb{S}^3$ lies on a unique longitude.

Proof. We omit the proof of **(a)**.

(b) This is seen by computing the length of a vector $ap + bv$ of W:

$$|ap + bv|^2 = a^2(p \cdot p) + 2ab(p \cdot v) + b^2(v \cdot v) = a^2 + b^2.$$

So $ap + bv$ is a unit vector if and only if the point (a, b) lies on the unit circle, in which case $a = \cos\theta$ and $b = \sin\theta$ for some θ.

(c) Let x be a unit vector in $\mathbb{R}^4$, not on the vertical axis. Then the set (p, x) is independent, and therefore spans a two-dimensional subspace W containing p. So x lies in just one such subspace, and in just one longitude. $\square$

9.3   THE SPECIAL UNITARY GROUP SU_2

The elements of SU_2 are complex 2×2 matrices of the form

$$(9.3.1) \qquad\qquad P = \begin{bmatrix} a & b \\ -\bar{b} & \bar{a} \end{bmatrix}, \quad \text{with } \bar{a}a + \bar{b}b = 1.$$

Let's verify this. Let $P = \begin{bmatrix} a & b \\ u & v \end{bmatrix}$ be an element of SU_2, with a, b, u, v in $\mathbb{C}$. The equations that define SU_2 are $P^* = P^{-1}$ and $\det P = 1$. When $\det P = 1$, the equation $P^* = P^{-1}$ becomes

$$\begin{bmatrix} \bar{a} & \bar{u} \\ \bar{b} & \bar{v} \end{bmatrix} = P^* = P^{-1} = \begin{bmatrix} v & -b \\ -u & a \end{bmatrix}.$$

Therefore $v = \bar{a}, u = -\bar{b}$, and then $\det P = \bar{a}a + \bar{b}b = 1$. $\square$

Writing $a = x_0 + x_1 i$ and $b = x_2 + x_3 i$ defines a bijective correspondence of SU_2 with the unit 3-sphere $\{x_0^2 + x_1^2 + x_2^2 + x_3^2 = 1\}$ in $\mathbb{R}^4$.

$$\begin{array}{ccc} SU_2 & \longleftrightarrow & \mathbb{S}^3 \end{array}$$
$$(9.3.2) \qquad P = \begin{bmatrix} x_0 + x_1 i & x_2 + x_3 i \\ -x_2 + x_3 i & x_0 - x_1 i \end{bmatrix} \longleftrightarrow (x_0, x_1, x_2, x_3)$$

This gives us two notations for an element of SU_2. We use the matrix notation as much as possible, because it is best for computation in the group, but length and orthogonality refer to dot product in $\mathbb{R}^4$.

Note: The fact that the 3-sphere has a group structure is remarkable. There is no way to make the 2-sphere into a group. A famous theorem of topology asserts that the only spheres on which one can define continuous group laws are the 1-sphere and the 3-sphere. $\square$

In matrix notation, the north pole $e_0 = (1, 0, 0, 0)$ on the sphere is the identity matrix I. The other standard basis vectors are the matrices that define the quaternion group (2.4.5). We list them again for reference:

$$(9.3.3) \qquad \mathbf{i} = \begin{bmatrix} i & 0 \\ 0 & -i \end{bmatrix}, \mathbf{j} = \begin{bmatrix} 0 & 1 \\ -1 & 0 \end{bmatrix}, \mathbf{k} = \begin{bmatrix} 0 & i \\ i & 0 \end{bmatrix} \longleftrightarrow e_1, e_2, e_3.$$

These matrices satisfy relations such as $\mathbf{ij} = \mathbf{k}$ that were displayed in (2.4.6). The real vector space with basis $(I, \mathbf{i}, \mathbf{j}, \mathbf{k})$ is called the *quaternion algebra*. So SU_2 can be thought of as the set of unit vectors in the quaternion algebra.

Lemma 9.3.4 Except for the two special matrices $\pm I$, the eigenvalues of P (9.3.2) are complex conjugate numbers of absolute value 1.

Proof. The characteristic polynomial of P is $t^2 - 2x_0 t + 1$, and its discriminant D is $4x_0^2 - 4$. When (x_0, x_1, x_2, x_3) is on the unit sphere, x_0 is in the interval $-1 \le x_0 \le 1$, and $D \le 0$. (In fact, the eigenvalues of any unitary matrix have absolute value 1.) $\square$

We now describe the algebraic structures on SU_2 that correspond to the latitudes and longitudes on $\mathbb{S}^3$ that were defined in the previous section.

Proposition 9.3.5 The latitudes in SU_2 are conjugacy classes. For a given c in the interval $-1 < c < 1$, the latitude $\{x_0 = c\}$ consists of the matrices P in SU_2 such that trace $P = 2c$. The remaining conjugacy classes are $\{I\}$ and $\{-I\}$. They make up the center of SU_2.

The proposition follows from the next lemma.

Lemma 9.3.6 Let P be an element of SU_2 with eigenvalues λ and $\bar{\lambda}$. There is an element Q in SU_2 such that Q^*PQ is the diagonal matrix Λ with diagonal entries λ and $\bar{\lambda}$. Therefore all elements of SU_2 with the same eigenvalues, or with the same trace, are conjugate.

Proof. One can base a proof of the lemma on the Spectral Theorem for unitary operators, or verify it directly as follows: Let $X = (u, v)^t$ be an eigenvector of P of length 1, with eigenvalue λ, and let $Y = (-\bar{v}, \bar{u})^t$. You will be able to check that Y is an eigenvector of P with eigenvalue $\bar{\lambda}$, that the matrix $Q = \begin{bmatrix} u & -\bar{v} \\ v & \bar{u} \end{bmatrix}$ is in SU_2, and that $PQ = Q\Lambda$. $\square$

The *equator* $\mathbb{E}$ of SU_2 is the latitude defined by the equation trace $P = 0$ (or $x_0 = 0$). A point on the equator has the form

$$(9.3.7) \qquad A = \begin{bmatrix} x_1 i & x_2 + x_3 i \\ -x_2 + x_3 i & -x_1 i \end{bmatrix} = x_1 \mathbf{i} + x_2 \mathbf{j} + x_3 \mathbf{k}.$$

Notice that the matrix A is *skew-Hermitian*: $A^* = -A$, and that its trace is zero. We haven't run across skew-Hermitian matrices before, but they are closely related to Hermitian matrices: a matrix A is skew-Hermitian if and only if iA is Hermitian.

The 2×2 skew-Hermitian matrices with trace zero form a *real* vector space of dimension 3 that we denote by $\mathbb{V}$, in agreement with the notation used in the previous section. The space $\mathbb{V}$ is the orthogonal space to I. It has the basis $(\mathbf{i}, \mathbf{j}, \mathbf{k})$, and $\mathbb{E}$ is the unit 2-sphere in $\mathbb{V}$.

Proposition 9.3.8 The following conditions on an element A of SU_2 are equivalent:

- A is on the equator, i.e., trace $A = 0$,
- the eigenvalues of A are i and $-i$,
- $A^2 = -I$.

Proof. The equivalence of the first two statements follows by inspection of the characteristic polynomial $t^2 - (\text{trace } A)t + 1$. For the third statement, we note that $-I$ is the only matrix in SU_2 with an eigenvalue -1. If λ is an eigenvalue of A, then λ^2 is an eigenvalue of A^2. So $\lambda = \pm i$ if and only if A^2 has eigenvalues -1, in which case $A^2 = -I$. $\qquad\square$

Next, we consider the longitudes of SU_2, the intersections of SU_2 with two-dimensional subspaces of $\mathbb{R}^4$ that contain the pole I. We use matrix notation.

Proposition 9.3.9 Let W be a two-dimensional subspace of $\mathbb{R}^4$ that contains I, and let L be the longitude of unit vectors in W.

(a) L meets the equator $\mathbb{E}$ in two points. If A is one of them, the other one is $-A$. Moreover, (I, A) is an orthonormal basis of W.

(b) The elements of L can be written in the form $P_\theta = (\cos\theta)I + (\sin\theta)A$, with A on $\mathbb{E}$ and $0 \le \theta \le 2\pi$. When $P \ne \pm I$, A and θ can be chosen with $0 < \theta < \pi$, and then the expression for P is unique.

(c) Every element of SU_2 except $\pm I$ lies on a unique longitude. The elements $\pm I$ lie on every longitude.

(d) The longitudes are conjugate subgroups of SU_2.

Proof. When one translates to matrix notation, the first three assertions become Lemma 9.2.7. To prove (d), we first verify that a longitude L is a subgroup. Let c, s and c', s' denote the cosine and sine of the angles α and α', respectively, and let $\beta = \alpha + \alpha'$. Then because $A^2 = -I$, the addition formulas for cosine and sine show that

$$(cI + sA)(c'I + s'A) = (cc' - ss')I + (cs' + sc')A = (\cos\beta)I + (\sin\beta)A.$$

So L is closed under multiplication. It is also closed under inversion.

Finally, we verify that the longitudes are conjugate. Say that L is the longitude $P_\theta = cI + sA$, as above. Proposition 9.3.5 tells us that A is conjugate to $\mathbf{i}$, say $\mathbf{i} = QAQ^*$. Then $QP_\theta Q^* = cQIQ^* + sQAQ^* = cI + s\mathbf{i}$. So L is conjugate to the longitude $cI + s\mathbf{i}$. $\qquad\square$

Examples 9.3.10

- The longitude $cI + s\mathbf{i}$, with $c = \cos\theta$ and $s = \sin\theta$, is the group of diagonal matrices in SU_2. We denote this longitude by T. Its elements have the form

$$c\begin{bmatrix} 1 & \\ & 1 \end{bmatrix} + s\begin{bmatrix} i & \\ & -i \end{bmatrix} = \begin{bmatrix} e^{i\theta} & \\ & e^{-i\theta} \end{bmatrix}.$$

- The longitude $cI + s\mathbf{j}$ is the group of real matrices in SU_2, the rotation group SO_2. The matrix $cI + s\mathbf{i}$ represents rotation of the plane through the angle $-\theta$.

$$c\begin{bmatrix} 1 & \\ & 1 \end{bmatrix} + s\begin{bmatrix} & 1 \\ -1 & \end{bmatrix} = \begin{bmatrix} c & s \\ -s & c \end{bmatrix}.$$

We haven't run across the the longitude $cI + s\mathbf{k}$ before. $\qquad\square$

The figure below was made by Bill Schelter. It shows a projection of the 3-sphere SU_2 onto the unit disc in the plane. The elliptical disc shown is the image of the equator. Just as the orthogonal projection of a circle from $\mathbb{R}^3$ to $\mathbb{R}^2$ is an ellipse, the projection of the

2-sphere $\mathbb{E}$ from $\mathbb{R}^4$ to $\mathbb{R}^3$ is an ellipsoid, and the further projection of this ellipsoid to the plane maps it onto an elliptical disc. Every point in the interior of the disc is the image of two points of $\mathbb{E}$.

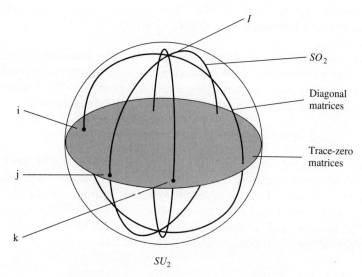

(9.3.11)   Some Latitudes and Longitudes in SU_2.

9.4   THE ROTATION GROUP SO_3

Since the equator $\mathbb{E}$ of SU_2 is a conjugacy class, the group operates on it by conjugation. We will show that conjugation by an element P of SU_2, an operation that we denote by γ_P, *rotates* this sphere. This will allow us to describe the three-dimensional rotation group SO_3 in terms of the special unitary group SU_2.

The poles of a nontrivial rotation of $\mathbb{E}$ are its fixed points, the intersections of $\mathbb{E}$ with the axis of rotation (5.1.22). If A is on $\mathbb{E}$, (A, α) will denote the spin that rotates $\mathbb{E}$ with angle α about the pole A. The two spins (A, α) and $(-A, -\alpha)$ represent the same rotation.

Theorem 9.4.1

(a) The rule $P \rightsquigarrow \gamma_P$ defines a surjective homomorphism $\gamma : SU_2 \rightarrow SO_3$, the *spin homomorphism*. Its kernel is the center $\{\pm I\}$ of SU_2.

(b) Suppose that $P = \cos\theta I + \sin\theta A$, with $0 < \theta < \pi$ and with A on $\mathbb{E}$. Then γ_P rotates $\mathbb{E}$ about the pole A, through the angle 2θ. So γ_P is represented by the spin $(A, 2\theta)$.

The homomorphism γ described by this theorem is called the *orthogonal representation* of SU_2. It sends a matrix P in SU_2, a complex 2×2 matrix, to a mysterious real 3×3 rotation matrix, the matrix of γ_P. The theorem tells us that every element of SU_2 except $\pm I$ can be described as a nontrivial rotation together with a choice of spin. Because of this, SU_2 is often called the *spin group*.

We discuss the geometry of the map γ before proving the theorem. If P is a point of SU_2, the point $-P$ is its *antipodal point*. Since γ is surjective and since its kernel is the center

$Z = \{\pm I\}$, SO_3 is isomorphic to the quotient group SU_2/Z, whose elements are pairs of antipodal points, the cosets $\{\pm P\}$ of Z. Because γ is two-to-one, SU_2 is called a *double covering* of SO_3.

The homomorphism $\mu : SO_2 \rightarrow SO_2$ of the 1-sphere to itself defined by $\rho_\theta \rightsquigarrow \rho_{2\theta}$ is another, closely related, example of a double covering. Every fibre of μ consists of two rotations, ρ_θ and $\rho_{\theta+\pi}$.

The orthogonal representation helps to describe the topological structure of the rotation group. Since elements of SO_3 correspond to pairs of antipodal points of SU_2, we can obtain SO_3 topologically by identifying antipodal points on the 3-sphere. The space obtained in this way is called *(real) projective 3-space*, and is denoted by $\mathbb{P}^3$.

(9.4.2) SO_3 is homeomorphic to projective 3-space $\mathbb{P}^3$.

Points of $\mathbb{P}^3$ are in bijective correspondence with one-dimensional subspaces of $\mathbb{R}^4$. Every one-dimensional subspace meets the unit 3-sphere in a pair of antipodal points.

The projective space $\mathbb{P}^3$ is much harder to visualize than the sphere $\mathbb{S}^3$. However, it is easy to describe projective 1-space $\mathbb{P}^1$, the set obtained by identifying antipodal points of the unit circle $\mathbb{S}^1$. If we wrap $\mathbb{S}^1$ around so that it becomes the lefthand figure of (9.4.3), the figure on the right will be $\mathbb{P}^1$. Topologically, $\mathbb{P}^1$ is a circle too.

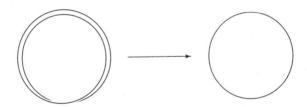

(9.4.3)   A Double Covering of the 1-Sphere.

We'll describe $\mathbb{P}^1$ again, in a way that one can attempt to extend to higher dimensional projective spaces. Except for the two points on the horizontal axis, every pair of antipodal points of the unit circle contains just one point in the lower semicircle. So to obtain $\mathbb{P}^1$, we simply identify a point pair with a single point in the lower semicircle. But the endpoints of the semicircle, the two points on the horizontal axis, must still be identified. So we glue the endpoints together, obtaining a circle as before.

In principle, the same method can be used to describe $\mathbb{P}^2$. Except for points on the equator of the 2-sphere, a pair of antipodal points contains just one point in the lower hemisphere. So we can form $\mathbb{P}^2$ from the lower hemisphere by identifying opposite points of the equator. Let's imagine that we start making this identification by gluing a short segment of the equator to the opposite segment. Unfortunately, when we orient the equator to keep track, we see that the opposite segment gets the opposite orientation. So when we glue the two segments together, we have to insert a twist. This gives us, topologically, a Möbius band, and $\mathbb{P}^2$ contains this Möbius band. It is not an orientable surface.

Then to visualize $\mathbb{P}^3$, we would take the lower hemisphere in $\mathbb{S}^3$ and identify antipodal points of its equator $\mathbb{E}$. Or, we could take the terrestial ball and identify antipodal points of its boundary, the surface of the Earth. This is quite confusing.   □

We begin the proof of Theorem 9.4.1 now. We recall that the equator $\mathbb{E}$ is the unit 2-sphere in the three-dimensional space $\mathbb{V}$ of trace zero, skew-Hermitian matrices (9.3.7). Conjugation by an element P of SU_2 preserves both the trace and the skew-Hermitian property, so this conjugation, which we are denoting by γ_P, operates on the whole space $\mathbb{V}$. The main point is to show that γ_P is a rotation. This is done in Lemma 9.4.5 below.

Let $\langle U, V \rangle$ denote the form on $\mathbb{V}$ that is carried over from dot product on $\mathbb{R}^3$. The basis of $\mathbb{V}$ that corresponds to the standard basis of $\mathbb{R}^3$ is $(\mathbf{i}, \mathbf{j}, \mathbf{k})$ (9.3.3). We write $U = u_1\mathbf{i} + u_2\mathbf{j} + u_3\mathbf{k}$ and use analogous notation for V. Then

$$\langle U, V \rangle = u_1v_1 + u_2v_2 + u_3v_3.$$

Lemma 9.4.4 With notation as above, $\langle U, V \rangle = -\frac{1}{2}\text{trace}(UV)$.

Proof. We compute the product UV using the quaternion relations (2.4.6):

$$UV = (u_1\mathbf{i} + u_2\mathbf{j} + u_3\mathbf{k})(v_1\mathbf{i} + v_2\mathbf{j} + v_3\mathbf{k})$$
$$= -(u_1v_1 + u_2v_2 + u_3v_3)I + U \times V,$$

where $U \times V$ is the vector cross product

$$U \times V = (u_2v_3 - u_3v_2)\mathbf{i} + (u_3v_1 - u_1v_3)\mathbf{j} + (u_1v_2 - u_2v_1)\mathbf{k}.$$

Then because trace $I = 2$, and because $\mathbf{i}, \mathbf{j}, \mathbf{k}$ have trace zero,

$$\text{trace}(UV) = -2(u_1v_1 + u_2v_2 + u_3v_3) = -2\langle U, V \rangle. \qquad \square$$

Lemma 9.4.5 The operator γ_P is a rotation of $\mathbb{E}$ and of $\mathbb{V}$.

Proof. For review, γ_P is the operator defined by $\gamma_P U = PUP^*$. The safest way to prove that this operator is a rotation may be to compute its matrix. But the matrix is too complicated to give much insight. It is nicer to describe γ indirectly. We will show that γ_P is an orthogonal linear operator with determinant 1. Euler's Theorem 5.1.25 will tell us that it is a rotation.

To show that γ_P is a linear operator, we must show that for all U and V in $\mathbb{V}$ and all real numbers r, $\gamma_P(U + V) = \gamma_P U + \gamma_P V$ and $\gamma_P(rU) = r(\gamma_P U)$. We omit this routine verification. To prove that γ_P is orthogonal, we verify the criterion (8.6.9) for orthogonality, which is

(9.4.6) $$\langle \gamma_P U, \gamma_P V \rangle = \langle U, V \rangle.$$

This follows from the previous lemma, because trace is preserved by conjugation.

$$\langle \gamma_P U, \gamma_P V \rangle = -\frac{1}{2}\,\text{trace}((\gamma_P U)(\gamma_P V)) = -\frac{1}{2}\,\text{trace}(PUP^*PVP^*)$$
$$= -\frac{1}{2}\,\text{trace}(PUVP^*) = -\frac{1}{2}\,\text{trace}(UV) = \langle U, V \rangle.$$

Finally, to show that the determinant of γ_P is 1, we recall that the determinant of any orthogonal matrix is ± 1. Since SU_2 is a sphere, it is path connected, and since the determinant is a continuous function, only one of the two values ± 1 can be taken on by $\det \gamma_P$. When $P = I$, γ_P is the identity operator, which has determinant 1. So $\det \gamma_P = 1$ for every P. $\qquad \square$

We now prove part **(a)** of the theorem. Because γ_P is a rotation, γ maps SU_2 to SO_3. The verification that γ is a homomorphism is simple: $\gamma_P\gamma_Q = \gamma_{PQ}$ because

$$\gamma_P(\gamma_Q U) = P(QUQ^*)P^* = (PQ)U(PQ)^* = \gamma_{PQ}U.$$

We show next that the kernel of γ is $\pm I$. If P is in the kernel, conjugation by P fixes every element of $\mathbb{E}$, which means that P commutes with every such element. Any element of SU_2 can be written in the form $Q = cI + sB$ with B in $\mathbb{E}$. Then P commutes with Q too. So P is in the center $\{\pm I\}$ of SU_2. The fact that γ is surjective will follow, once we identify 2θ as the angle of rotation, because every angle α has the form 2θ, with $0 \le \theta \le \pi$.

Let P be an element of SU_2, written in the form $P = \cos\theta I + \sin\theta A$ with A in $\mathbb{E}$. It is true that $\gamma_P A = A$, so A is a pole of γ_P. Let α denote the angle of rotation of γ_P about the pole A. To identify this angle, we show first that it is enough to identify the angle for a single matrix P in a conjugacy class.

Say that $P' = QPQ^*(= \gamma_Q P)$ is a conjugate, where Q is another element of SU_2. Then $P' = \cos\theta I + \sin\theta A'$, where $A' = \gamma_Q A = QAQ^*$. The angle θ has not changed.

Next, we apply Corollary 5.1.28, which asserts that if M and N are elements of SO_3, and if M is a rotation with angle α about the pole X, then the conjugate $M' = NMN^{-1}$ is a rotation with the same angle α about the pole NX. Since γ is a homomorphism, $\gamma_{P'} = \gamma_Q\gamma_P\gamma_Q^{-1}$. Since γ_P is a rotation with angle α about A, $\gamma_{P'}$ is a rotation with angle α about $A' = \gamma_Q A$. The angle α hasn't changed either.

This being so, we make the computation for the matrix $P = \cos\theta I + \sin\theta\mathbf{i}$, which is the diagonal matrix with diagonal entries $e^{i\theta}$ and $e^{-i\theta}$. We apply γ_P to $\mathbf{j}$:

(9.4.7)
$$\gamma_P\mathbf{j} = P\mathbf{j}P^* = \begin{bmatrix} e^{i\theta} & \\ & e^{-i\theta} \end{bmatrix}\begin{bmatrix} & 1 \\ -1 & \end{bmatrix}\begin{bmatrix} e^{-i\theta} & \\ & e^{i\theta} \end{bmatrix} = \begin{bmatrix} & e^{2i\theta} \\ -e^{-2i\theta} & \end{bmatrix}$$
$$= \cos 2\theta\,\mathbf{j} + \sin 2\theta\,\mathbf{k}.$$

The set $(\mathbf{j}, \mathbf{k})$ is an orthonormal basis of the orthogonal space W to $\mathbf{i}$, and the equation above shows that γ_P rotates the vector $\mathbf{j}$ through the angle 2θ in W. The angle of rotation is 2θ, as predicted. This completes the proof of Theorem (9.4.1). $\qquad\square$

9.5   ONE-PARAMETER GROUPS

In Chapter 5, we used the matrix-valued function

(9.5.1)
$$e^{tA} = I + \frac{tA}{1!} + \frac{t^2A^2}{2!} + \frac{t^3A^3}{3!} + \cdots$$

to describe solutions of the differential equation $\frac{dX}{dt} = AX$. The same function describes the *one-parameter groups* in the general linear group – the differentiable homomorphisms from the additive group $\mathbb{R}^+$ of real numbers to GL_n.

Theorem 9.5.2

(a) Let A be an arbitrary real or complex matrix, and let GL_n denote $GL_n(\mathbb{R})$ or $GL_n(\mathbb{C})$. The map $\varphi:\mathbb{R}^+ \to GL_n$ defined by $\varphi(t) = e^{tA}$ is a group homomorphism.

(b) Conversely, let $\varphi:\mathbb{R}^+ \to GL_n$ be a differentiable map that is a homomorphism, and let A denote its derivative $\varphi'(0)$ at the origin. Then $\varphi(t) = e^{tA}$ for all t.

Proof. For any real numbers r and s, the matrices rA and sA commute. So (see (5.4.4))

$$(9.5.3) \qquad\qquad e^{(r+s)A} = e^{rA}e^{sA}.$$

This shows that e^{tA} is a homomorphism. Conversely, let $\varphi:\mathbb{R}^+ \to GL_n$ be a differentiable homomorphism. Then $\varphi(\Delta t + t) = \varphi(\Delta t)\varphi(t)$ and $\varphi(t) = \varphi(0)\varphi(t)$, so we can factor $\varphi(t)$ out of the difference quotient:

$$(9.5.4) \qquad\qquad \frac{\varphi(\Delta t + t) - \varphi(t)}{\Delta t} = \frac{\varphi(\Delta t) - \varphi(0)}{\Delta t}\varphi(t).$$

Taking the limit as $\Delta t \to 0$, we see that $\varphi'(t) = \varphi'(0)\varphi(t) = A\varphi(t)$. Therefore $\varphi(t)$ is a matrix-valued function that solves the differential equation

$$(9.5.5) \qquad\qquad \frac{d\varphi}{dt} = A\varphi.$$

The function e^{tA} is another solution, and when $t = 0$, both solutions take the value I. Therefore $\varphi(t) = e^{tA}$ (see (5.4.9)). ☐

Examples 9.5.6

(a) Let A be the 2×2 matrix unit e_{12}. Then $A^2 = 0$. All but two terms of the series expansion for the exponential are zero, and $e^{tA} = I + e_{12}t$.

If $A = \begin{bmatrix} 0 & 1 \\ 0 & 0 \end{bmatrix}$, then $e^{tA} = \begin{bmatrix} 1 & t \\ & 1 \end{bmatrix}$.

(b) The usual parametrization of SO_2 is a one-parameter group.

If $A = \begin{bmatrix} 0 & -1 \\ 1 & 0 \end{bmatrix}$, then $e^{tA} = \begin{bmatrix} \cos t & -\sin t \\ \sin t & \cos t \end{bmatrix}$.

(c) The usual parametrization of the unit circle in the complex plane is a one-parameter group in U_1.

If a is a nonzero real number and $\alpha = ai$, then $e^{t\alpha} = [\cos at + i \sin at]$. ☐

If α is a nonreal complex number of absolute value $\neq 1$, the image of $e^{t\alpha}$ in $\mathbb{C}^\times$ will be a logarithmic spiral. If a is a nonzero real number, the image of e^{ta} is the positive real axis, and if $a = 0$ the image consists of the point 1 alone.

If we are given a subgroup H of GL_n, we may also ask for one-parameter groups in H, meaning one-parameter groups whose images are in H, or differentiable homomorphisms $\varphi : \mathbb{R}^+ \to H$. It turns out that linear groups of positive dimension always have one-parameter groups, and they are usually not hard to determine for a particular group.

Since the one-parameter groups are in bijective correspondence with $n \times n$ matrices, we are asking for the matrices A such that e^{tA} is in H for all t. We will determine the one-parameter groups in the orthogonal, unitary, and special linear groups.

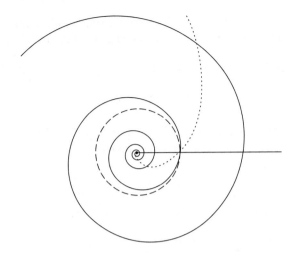

(9.5.7)   Images of Some One-Parameter Groups in $\mathbb{C}^\times = GL_1(\mathbb{C})$.

Proposition 9.5.8

(a) If A is a real skew-symmetric matrix ($A^t = -A$), then e^A is orthogonal. If A is a complex skew-Hermitian matrix ($A^* = -A$), then e^A is unitary.

(b) The one-parameter groups in the orthogonal group O_n are the homomorphisms $t \rightsquigarrow e^{tA}$, where A is a real skew-symmetric matrix.

(c) The one-parameter groups in the unitary group U_n are the homomorphisms $t \rightsquigarrow e^{tA}$, where A is a complex skew-Hermitian matrix.

Proof. We discuss the complex case.

The relation $(e^A)^* = e^{(A^*)}$ follows from the definition of the exponential, and we know that $(e^A)^{-1} = e^{-A}$ (5.4.5). So if A is skew-Hermitian, i.e., $A^* = -A$, then $(e^A)^* = (e^A)^{-1}$, and e^A is unitary. This proves (a) for complex matrices.

Next, if A is skew-Hermitian, so is tA, and by what was shown above, e^{tA} is unitary for all t, so it is a one-parameter group in the unitary group. Conversely, suppose that e^{tA} is unitary for all t. We write this as $e^{tA^*} = e^{-tA}$. Then the derivatives of the two sides of this equation, evaluated at $t = 0$, must be equal, so $A^* = -A$, and A is skew-Hermitian.

The proof for the orthogonal group is the same, when we interpret A^* as A^t. □

We consider the special linear group SL_n next.

Lemma 9.5.9 For any square matrix A, $e^{\operatorname{trace} A} = \det e^A$.

Proof. An eigenvector X of A with eigenvalue λ is also an eigenvector of e^A with eigenvalue e^λ. So, if $\lambda_1, \ldots, \lambda_n$ are the eigenvalues of A, then the eigenvalues of e^A are e^{λ_i}. The trace of A is the sum $\lambda_1 + \cdots + \lambda_n$, and the determinant of e^A is the product $e^{\lambda_1} \cdots e^{\lambda_n}$ (4.5.15). Therefore $e^{\operatorname{trace} A} = e^{\lambda_1 + \cdots + \lambda_n} = e^{\lambda_1} \cdots e^{\lambda_n} = \det e^A$. □

Proposition 9.5.10 The one-parameter groups in the special linear group SL_n are the homomorphisms $t \rightsquigarrow e^{tA}$, where A is a real $n \times n$ matrix whose trace is zero.

Proof. Lemma 9.5.9 shows that if trace $A = 0$, then $\det e^{tA} = e^{t \operatorname{trace} A} = e^0 = 1$ for all t, so e^{tA} is a one-parameter group in SL_n. Conversely, if $\det e^{tA} = 1$ for all t, the derivative of $e^{t \operatorname{trace} A}$, evaluated at $t = 0$, is zero. The derivative is trace A. $\square$

The simplest one-parameter group in SL_2 is the one in Example 9.5.6(a). The one-parameter groups in SU_2 are the longitudes described in (9.3.9).

9.6   THE LIE ALGEBRA

The space of tangent vectors to a matrix group G at the identity is called the *Lie algebra* of the group. We denote it by $\operatorname{Lie}(G)$. It is called an algebra because it has a law of composition, the bracket operation that is defined below.

For instance, when we represent the circle group as the unit circle in the complex plane, the Lie algebra is the space of real multiples of i.

The observation from which the definition of tangent vector is derived is something we learn in calculus: If $\varphi(t) = (\varphi_1(t), \dots, \varphi_k(t))$ is a differentiable path in $\mathbb{R}^k$, the velocity vector $v = \varphi'(0)$ is tangent to the path at the point $x = \varphi(0)$. A vector v is said to be *tangent* to a subset S of $\mathbb{R}^k$ at a point x if there is a differentiable path $\varphi(t)$, defined for sufficiently small t and lying entirely in S, such that $\varphi(0) = x$ and $\varphi'(0) = v$.

The elements of a linear group G are matrices, so a path $\varphi(t)$ in G will be a matrix-valued function. Its derivative $\varphi'(0)$ at $t = 0$ will be represented naturally as a matrix, and if $\varphi(0) = I$, the matrix $\varphi'(0)$ will be an element of $\operatorname{Lie}(G)$. For example, the usual parametrization (9.5.6)(b) of the group SO_2 shows that the matrix $\begin{bmatrix} 0 & -1 \\ 1 & 0 \end{bmatrix}$ is in $\operatorname{Lie}(SO_2)$.

We already know a few paths in the orthogonal group O_n: the one-parameter groups $\varphi(t) = e^{At}$, where A is a skew-symmetric matrix (9.5.8). Since $(e^{At})_{t=0} = I$ and $\left(\frac{d}{dt} e^{At}\right)_{t=0} = A$, every skew-symmetric matrix A is a tangent vector to O_n at the identity – an element of its Lie algebra. We show now that the Lie algebra consists precisely of those matrices. Since one-parameter groups are very special, this isn't completely obvious. There are many other paths.

Proposition 9.6.1 The Lie algebra of the orthogonal group O_n consists of the skew-symmetric matrices.

Proof. We denote transpose by $*$. If φ is a path in O_n with $\varphi(0) = I$ and $\varphi'(0) = A$, then $\varphi(t)^* \varphi(t) = I$ identically, and so $\frac{d}{dt}(\varphi(t)^* \varphi(t)) = 0$. Then

$$\frac{d}{dt}(\varphi^* \varphi)_{t=0} = \left(\frac{d\varphi^*}{dt} \varphi + \varphi^* \frac{d\varphi}{dt} \right)_{t=0} = A^* + A = 0.$$
$\square$

Next, we consider the special linear group SL_n. The one-parameter groups in SL_n have the form $\varphi(t) = e^{At}$, where A is a trace-zero matrix (9.5.10). Since $(e^{At})_{t=0} = I$ and $\left(\frac{d}{dt} e^{At}\right)_{t=0} = A$, every trace-zero matrix A is a tangent vector to SL_n at the identity – an element of its Lie algebra.

Lemma 9.6.2 Let φ be a path in GL_n with $\varphi(0) = I$ and $\varphi'(0) = A$. Then $\left(\frac{d}{dt}(\det \varphi)\right)_{t=0} =$ trace A.

Proof. We write the matrix entries of φ as φ_{ij}, and we compute $\frac{d}{dt} \det \varphi$ using the complete expansion (1.6.4) of the determinant:

$$\det \varphi = \sum_{p \in S_n} (\text{sign } p)\, \varphi_{1,p1} \cdots \varphi_{n,pn}.$$

By the product rule,

$$(9.6.3) \qquad \frac{d}{dt}(\varphi_{1,p1} \cdots \varphi_{n,pn}) = \sum_{i=1}^{n} \varphi_{1,p1} \cdots \varphi'_{i,pi} \cdots \varphi_{n,pn}.$$

We evaluate at $t = 0$. Since $\varphi(0) = I$, $\varphi_{ij}(0) = 0$ if $i \neq j$ and $\varphi_{ii}(0) = 1$. So in the sum (9.6.3), the term $\varphi_{1,p1} \cdots \varphi'_{i,pi} \cdots \varphi_{n,pn}$ evaluates to zero unless $pj = j$ for all $j \neq i$, and if $pj = j$ for all $j \neq i$, then since p is a permutation, $pi = i$ too, and therefore p is the identity. So (9.6.3) evaluates to zero except when $p = 1$, and when $p = 1$, it becomes $\sum_i \varphi'_{ii}(0) = \text{trace } A$. This is the derivative of $\det \varphi$. $\qquad \square$

Proposition 9.6.4 The Lie algebra of the special linear group SL_n consists of the trace-zero matrices. $\qquad \square$

Proof. If φ is a path in the special linear group with $\varphi(0) = I$ and $\varphi'(0) = A$, then $\det(\varphi(t)) = 1$ identically, and therefore $\frac{d}{dt} \det(\varphi(t)) = 0$. Evaluating at $t = 0$, we obtain trace $A = 0$. $\qquad \square$

Similar methods are used to describe the Lie algebras of other classical groups. Note also that the Lie algebras of O_n and SL_n are real vector spaces, subspaces of the space of matrices. It is usually easy to verify for other groups that $\text{Lie}(G)$ is a real vector space.

The Lie Bracket

The Lie algebra has an additional structure, an operation called the *bracket*, the law of composition defined by the rule

$$(9.6.5) \qquad\qquad [A, B] = AB - BA.$$

The bracket is a version of the commutator: It is zero if and only if A and B commute. It isn't an associative law, but it satisfies an identity called the *Jacobi identity*:

$$(9.6.6) \qquad\qquad [A, [B, C]] + [B, [C, A]] + [C, [A, B]] = 0.$$

To show that the bracket is defined on the Lie algebra, we must check that if A and B are in $\text{Lie}(G)$, then $[A, B]$ is also in $\text{Lie}(G)$. This can be done easily for any particular group. For the special linear group, the required verification is that if A and B have trace zero, then $AB - BA$ also has trace zero, which is true because trace $AB = $ trace BA. The Lie algebra of the orthogonal group is the space of skew-symmetric matrices. For that group, we must verify that if A and B are skew-symmetric, then $[A, B]$ is skew-symmetric:

$$[A, B]^{\mathrm{t}} = (AB)^{\mathrm{t}} - (BA)^{\mathrm{t}} = B^{\mathrm{t}}A^{\mathrm{t}} - A^{\mathrm{t}}B^{\mathrm{t}} = (-B)(-A) - (-A)(-B) = -[A, B].$$

The definition of an abstract Lie algebra includes a bracket operation.

Definition 9.6.7 A *Lie algebra* V is a real vector space together with a law of composition $V \times V \to V$ denoted by $v, w \rightsquigarrow [v, w]$ and called the *bracket*, which satisfies these axioms for all u, v, w in V and all c in $\mathbb{R}$:

$$\begin{aligned} \text{bilinearity:} \quad & [v_1 + v_2, w] = [v_1, w] + [v_2, w] \quad \text{and} \quad [cv, w] = c[v, w], \\ & [v, w_1 + w_2] = [v, w_1] + [v, w_2] \quad \text{and} \quad [v, cw] = c[v, w], \\ \text{skew symmetry:} \quad & [v, w] = -[w, v], \quad \text{or} \quad [v, v] = 0, \\ \text{Jacobi identity:} \quad & [u, [v, w]] + [v, [w, u]] + [w, [u, v]] = 0. \end{aligned}$$

Lie algebras are useful because, being vector spaces, they are easier to work with than linear groups. And, though this is not easy to prove, many linear groups, including the classical groups, are nearly determined by their Lie algebras.

9.7   TRANSLATION IN A GROUP

Let P be an element of a matrix group G. Left multiplication by P is a bijective map from G to itself:

(9.7.1)
$$G \xrightarrow{m_P} G$$
$$X \rightsquigarrow PX.$$

Its inverse function is left multiplication by P^{-1}. The maps m_P and $m_{P^{-1}}$ are continuous because matrix multiplication is continuous. Thus m_P is a homeomorphism from G to G (not a homomorphism). It is also called *left translation* by P, in analogy with translation in the plane, which is left translation in the additive group $\mathbb{R}^{2+}$.

The important property of a group that is implied by the existence of these maps is *homogeneity*. Multiplication by P is a homeomorphism that carries the identity element I to P. Intuitively, the group looks the same at P as it does at I, and since P is arbitrary, it looks the same at any two points. This is analogous to the fact that the plane looks the same everywhere.

Left multiplication in the circle group SO_2 rotates the circle, and left multiplication in SU_2 is also a rigid motion of the 3-sphere. But homogeneity is weaker in other matrix groups. For example, let G be the group of real invertible diagonal 2×2 matrices. If we identify the elements of G with the points (a, d) in the plane and not on the coordinate axes, multiplication by the matrix

(9.7.2)
$$P = \begin{bmatrix} 2 & 0 \\ 0 & 1 \end{bmatrix}$$

distorts the group G, but it does this continuously.

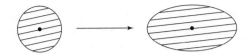

(9.7.3)   Left Multiplication in a Group.

Now the only geometrically reasonable subsets of $\mathbb{R}^k$ that have such a homogeneity property are manifolds. A *manifold M* of dimension d is a set in which every point has a neighborhood that is homeomorphic to an open set in $\mathbb{R}^d$ (see [Munkres], p. 155). It isn't surprising that the classical groups are manifolds, though there are subgroups of GL_n that aren't. The group $GL_n(\mathbb{Q})$ of invertible matrices with rational coefficients is an interesting group, but it is a countable dense subset of the space of matrices.

The following theorem gives a satisfactory answer to the question of which linear groups are manifolds:

Theorem 9.7.4 A subgroup of GL_n that is a closed subset of GL_n is a manifold.

Proving this theorem here would take us too far afield, but we illustrate it by showing that the orthogonal groups are manifolds. Proofs for the other classical groups are similar.

Lemma 9.7.5 The matrix exponential $A \rightsquigarrow e^A$ maps a small neighborhood U of 0 in $\mathbb{R}^{n \times n}$ homeomorphically to a neighborhood V of I in $GL_n(\mathbb{R})$.

The fact that the exponential series converges uniformly on bounded sets of matrices implies that it is a continuous function ([Rudin] Thm 7.12). To prove the lemma, one needs to show that it has a continuous inverse function for matrices sufficiently near to I. This can be proved using the inverse function theorem, or the series for $\log(1+x)$:

$$(9.7.6) \qquad \log(1+x) = x - \tfrac{1}{2}x^2 + \tfrac{1}{3}x^3 - \cdots .$$

The series $\log(I + B)$ converges for small matrices B, and it inverts the exponential. $\square$

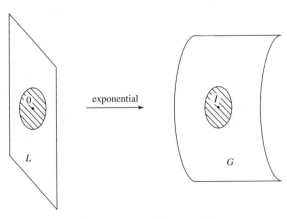

(9.7.7)   The Matrix Exponential.

Proposition 9.7.8 The orthogonal group O_n is a manifold of dimension $\frac{1}{2}n(n-1)$.

Proof. We denote the group O_n by G, and its Lie algebra, the space of skew-symmetric matrices, by L. If A is skew-symmetric, then e^A is orthogonal (9.5.8). So the exponential maps L to G. Conversely, suppose that A is near 0. Then, denoting transpose by $*$, A^* and $-A$ are also near zero, and e^{A^*} and e^{-A} are near to I. If e^A is orthogonal, i.e., if $e^{A^*} = e^{-A}$, Lemma (9.7.5) tells us that $A^* = -A$, so A is skew-symmetric. Therefore a matrix A near 0 is in L if and only if e^A is in G. This shows that the exponential defines a homeomorphism from a neighborhood V of 0 in L to a neighborhood U of I in G. Since L is a vector space, it is a manifold. The condition for a manifold is satisfied by the orthogonal group at the identity. Homogeneity implies that it is satisfied at all points. Therefore G is a manifold, and its dimension is the same as that of L, namely $\frac{1}{2}n(n-1)$. $\qquad\square$

Here is another application of the principle of homogeneity.

Proposition 9.7.9 Let G be a path-connected matrix group, and let H be a subgroup of G that contains a nonempty open subset U of G. Then $H = G$.

Proof. A subset of $\mathbb{R}^n$ is *path connected* if any two points of S can be joined by a continuous path lying entirely in S (see [Munkres, p. 155] or Chapter 2, Exercise M.6).

Since left multiplication by an element g is a homeomorphism from G to G, the set gU is also open, and it is contained in a single coset of H, namely in gH. Since the translates of U cover G, the ones contained in a coset C cover that coset. So each coset is a union of open subsets of G, and therefore is open itself. Then G is partitioned into open subsets, the cosets of H. A path-connected set is not a disjoint union of proper open subsets (see [Munkres, p. 155]). Thus there can be only one coset, and $H = G$. $\qquad\square$

We use this proposition to determine the normal subgroups of SU_2.

Theorem 9.7.10

(a) The only proper normal subgroup of SU_2 is its center $\{\pm I\}$.
(b) The rotation group SO_3 is a simple group.

Proof. **(a)** Let N be a normal subgroup of SU_2 that contains an element $P \neq \pm I$. We must show that N is equal to SU_2. Since N is normal, it contains the conjugacy class C of P, which is a latitude, a 2-sphere.

We choose a continuous map $P(t)$ from the unit interval $[0, 1]$ to C such that $P(0) = P$ and $P(1) \neq P$, and we form the path $Q(t) = P(t)P^{-1}$. Then $Q(0) = I$, and $Q(1) \neq I$, so this path leads out from the identity I, as in the figure below. Since N is a group that contains P and $P(t)$, it also contains $Q(t)$ for every t in the interval $[0, 1]$. We don't need to know anything else about the path $Q(t)$.

We note that trace $Q \leq 2$ for any Q in SU_2, and that I is the only matrix with trace equal to 2. Therefore trace $Q(0) = 2$ and trace $Q(1) = \tau < 2$. By continuity, all values between τ and 2 are taken on by trace $Q(t)$. Since N is normal, it contains the conjugacy class of $Q(t)$ for every t. Therefore N contains all elements of SU_2 whose traces are sufficiently near to 2, and this includes all matrices near to the identity. So N contains an open neighborhood of the identity in SU_2. Since SU_2 is path-connected, Proposition 9.7.9 shows that $N = SU_2$.

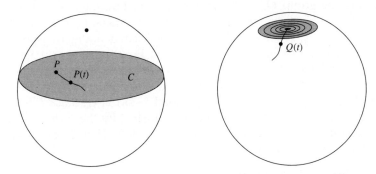

(b) There is a surjective map $\varphi : SU_2 \to SO_3$ whose kernel is $\{\pm I\}$ (9.4.1). By the Correspondence Theorem 2.10.5, the inverse image of a normal subgroup in SO_3 is a normal subgroup of SU_2 that contains $\{\pm I\}$. Part **(a)** tells us that there are no proper subgroups of SU_2 except $\{\pm I\}$, so SO_3 contains no proper normal subgroup at all.   □

One can apply translation in a group G to tangent vectors too. If A is a tangent vector at the identity and if P is an element of G, the vector PA is tangent to G at P, and if A isn't zero, neither is PA. As P ranges over the group, the family of these vectors forms what is called a *tangent vector field*. Now just the existence of a continuous tangent vector field that is nowhere zero puts strong restrictions on the space G. It is a theorem of topology, sometimes called the "Hairy Ball Theorem," that any tangent vector field on the 2-sphere must vanish at some point (see [Milnor]). This is one reason that the 2-sphere has no group structure. But since the 3-sphere is a group, it has tangent vector fields that are nowhere zero.

9.8   NORMAL SUBGROUPS OF SL_2

Let F be a field. The center of the group $SL_2(F)$ is $\{\pm I\}$. (This is Exercise 8.5.) The quotient group $SL_2(F)/\{\pm I\}$ is called the *projective group*, and is denoted by $PSL_2(F)$. Its elements are the cosets $\{\pm P\}$.

Theorem 9.8.1   Let F be a field of order at least four.

(a)   The only proper normal subgroup of $SL_2(F)$ is its center $Z = \{\pm I\}$.
(b)   The projective group $PSL_2(F)$ is a simple group.

Part **(b)** of the theorem follows from **(a)** and the Correspondence Theorem 2.10.5, and it identifies an interesting class of finite simple groups: the projective groups $PSL_2(F)$ when F is a finite field. The other finite, nonabelian simple groups that we have seen are the alternating groups (7.5.4).

We will show in Chapter 15 that the order of a finite field is always a power of a prime, that for every prime power $q = p^e$, there is a field $\mathbb{F}_q$ of order q, and that $\mathbb{F}_q$ has characteristic p (Theorem 15.7.3). Finite fields of order 2^e have characteristic 2. In those fields, $1 = -1$ and $I = -I$. Then the center of $SL_2(\mathbb{F}_q)$ is the trivial group. Let's assume these facts for now.

We omit the proof of the next lemma. (See Chapter 3, Exercise 4.4 for the case that q is a prime.)

Lemma 9.8.2 Let q be a power of a prime. The order of $SL_2(\mathbb{F}_q)$ is $q^3 - q$. If q is not a power of 2, the order of $PSL_2(\mathbb{F}_q)$ is $\frac{1}{2}(q^3 - q)$. If q is a power of 2, then $PSL_2(\mathbb{F}_q) \approx SL_2(\mathbb{F}_q)$, and the order of $PSL_2(\mathbb{F}_q)$ is $q^3 - q$. $\square$

The orders of PSL_2 for small q are listed below, along with the orders of the first three simple alternating groups.

$|F|$	4	5	7	8	9	11	13	16	17	19
$|PSL_2|$	60	60	168	504	360	660	1092	4080	2448	3420

n	5	6	7
$|A_n|$	60	360	2520

The orders of the ten smallest nonabelian simple groups appear in this list. The next smallest would be $PSL_3(\mathbb{F}_3)$, which has order 5616.

The projective group is not simple when $|F| = 2$ or 3. $PSL_2(\mathbb{F}_2)$ is isomorphic to the symmetric group S_3 and $PSL_2(\mathbb{F}_3)$ is isomorphic to the alternating group A_4.

As shown in these tables, $PSL_2(\mathbb{F}_4)$, $PSL_2(\mathbb{F}_5)$, and A_5 have order 60. These three groups happen to be isomorphic. (This is Exercise 8.3.) The other coincidences among orders are the groups $PSL_2(\mathbb{F}_9)$ and A_6, which have order 360. They are isomorphic too. $\square$

For the proof, we will leave the cases $|F| = 4$ and 5 aside, so that we can use the next lemma.

Lemma 9.8.3 A field F of order greater than 5 contains an element r whose square is not 0, 1, or –1.

Proof. The only element with square 0 is 0, and the elements with square 1 are ± 1. There are at most two elements whose squares are –1: If $a^2 = b^2 = -1$, then $(a - b)(a + b) = 0$, so $b = \pm a$. $\square$

Proof of Theorem 9.8.1. We assume given the field F, we let SL_2 and PSL_2 stand for $SL_2(F)$ and $PSL_2(F)$, respectively, and we denote the space F^2 by V. We choose a nonzero element r of F whose square s is not ± 1.

Let N be a normal subgroup of SL_2 that contains an element $A \neq \pm I$. We must show that N is the whole group SL_2. Since A is arbitrary, it is hard to work with directly. The strategy is to begin by showing that N contains a matrix that has eigenvalue s.

Step 1: There is a matrix P in SL_2 such that the commutator $B = APA^{-1}P^{-1}$ is in N, and has eigenvalues s and s^{-1}.

This is a nice trick. We choose a vector v_1 in V that is *not* an eigenvector of A and we let $v_2 = Av_1$. Then v_1 and v_2 are independent, so $\mathbf{B} = (v_1, v_2)$ is a basis of V. (It is easy to check that the only matrices in SL_2 for which every vector is an eigenvector are I and $-I$.)

Let R be the diagonal matrix with diagonal entries r and r^{-1}. The matrix $P = [\mathbf{B}]R[\mathbf{B}]^{-1}$ has determinant 1, and v_1 and v_2 are eigenvectors, with eigenvalues r and r^{-1}, respectively

(4.6.10). Because N is a normal subgroup, the commutator $B = APA^{-1}P^{-1}$ is an element of N (see (7.5.4)). Then

$$Bv_2 = APA^{-1}P^{-1}v_2 = APA^{-1}(rv_2) = rAPv_1 = r^2Av_1 = sv_2.$$

Therefore s is an eigenvalue of B. Because $\det B = 1$, the other eigenvalue is s^{-1}.

Step 2: The matrices having eigenvalues s and s^{-1} form a single conjugacy class C in SL_2, and this conjugacy class is contained in N.

The elements s and s^{-1} are distinct because $s \neq \pm 1$. Let S be a diagonal matrix with diagonal entries s and s^{-1}. Every matrix Q with eigenvalues s and s^{-1} is a conjugate of S in $GL_2(F)$ (4.4.8)**(b)**, say $Q = LSL^{-1}$. Since S is diagonal, it commutes with any other diagonal matrix. We can multiply L on the right by a suitable diagonal matrix, to make $\det L = 1$, while preserving the equation $Q = LSL^{-1}$. So Q is a conjugate of S in SL_2. This shows that the matrices with eigenvalues s and s^{-1} form a single conjugacy class. By Step 1, the normal subgroup N contains one such matrix. So $C \subset N$.

Step 3: The elementary matrices $E = \begin{bmatrix} 1 & x \\ 0 & 1 \end{bmatrix}$ and $E^t = \begin{bmatrix} 1 & 0 \\ x & 1 \end{bmatrix}$, with x in F, are in N.

For any element x of F, the terms on the left side of the equation

$$\begin{bmatrix} s^{-1} & 0 \\ 0 & s \end{bmatrix} \begin{bmatrix} s & sx \\ 0 & s^{-1} \end{bmatrix} = \begin{bmatrix} 1 & x \\ 0 & 1 \end{bmatrix} = E$$

are in C and in N, so E is in N. One sees similarly that E^t is in N.

Step 4: The matrices E and E^t, with x in F, generate SL_2. Therefore $N = SL_2$.

The proof of this is Exercise 4.8 of Chapter 2 . $\square$

As is shown by the alternating groups and the projective groups, simple groups arise frequently, and this is one of the reasons that they have been studied intensively. On the other hand, simplicity is a very strong restriction on a group. There couldn't be too many of them. A famous theorem of Cartan is one manifestation of this.

A *complex algebraic group* is a subgroup of the complex general linear group $GL_n(\mathbb{C})$ which is the locus of complex solutions of a finite system of complex polynomial equations in the matrix entries. Cartan's theorem lists the simple complex algebraic groups. In the statement of the theorem, we use the symbol Z to denote the center of a group.

Theorem 9.8.4

(a) The centers of the groups $SL_n(\mathbb{C})$, $SO_n(\mathbb{C})$, and $SP_{2n}(\mathbb{C})$ are finite cyclic groups.

(b) For $n \geq 1$, the groups $SL_n(\mathbb{C})/Z$, $SO_n(\mathbb{C})/Z$, and $SP_{2n}(\mathbb{C})/Z$ are path-connected complex algebraic groups. Except for $SO_2(\mathbb{C})/Z$ and $SO_4(\mathbb{C})/Z$, they are simple.

(c) In addition to the isomorphism classes of these groups, there are exactly five isomorphism classes of simple, path-connected complex algebraic groups, called the *exceptional groups*.

Theorem 9.8.4 is based on a classification of the corresponding Lie algebras. It is too hard to prove here.

A large project, the classification of the *finite* simple groups, was completed in 1980. The finite simple groups we have seen are the groups of prime order, the alternating groups A_n with $n \geq 5$, and the groups $PSL_2(F)$ when F is a finite field of order at least 4. Matrix groups play a dominant role in the classification of the finite simple groups too. Each of the forms (9.8.4) leads to a whole series of finite simple groups when finite fields are substituted for the complex field. There are also some finite simple groups analogous to the unitary groups. All of these finite linear groups are said to be of *Lie type*. In addition to the groups of prime order, the alternating groups, and the groups of Lie type, there are 26 finite simple groups called the *sporadic groups*. The smallest sporadic group is the *Mathieu group* M_{11}, whose order is 7920. The largest, the *Monster*, has order roughly 10^{53}.

> *It seems unfair to crow about the successes of a theory*
> *and to sweep all its failures under the rug.*
>
> —Richard Brauer

EXERCISES

Section 1   The Classical Linear Groups

1.1. (a) Is $GL_n(\mathbb{C})$ isomorphic to a subgroup of $GL_{2n}(\mathbb{R})$?

(b) Is $SO_2(\mathbb{C})$ a bounded subset of $\mathbb{C}^{2 \times 2}$?

1.2. A matrix P is orthogonal if and only if its columns form an orthonormal basis. Describe the properties of the columns of a matrix in the Lorentz group $O_{3,1}$.

1.3. Prove that there is no continuous isomorphism from the orthogonal group O_4 to the Lorentz group $O_{3,1}$.

1.4. Describe by equations the group $O_{1,1}$ and show that it has four path-connected components.

1.5. Prove that $SP_2 = SL_2$, but that $SP_4 \neq SL_4$.

1.6. Prove that the following matrices are symplectic, if the blocks are $n \times n$:
$$\begin{bmatrix} & -I \\ I & \end{bmatrix}, \begin{bmatrix} A^t & \\ & A^{-1} \end{bmatrix}, \begin{bmatrix} I & B \\ & I \end{bmatrix}, \text{ where } B = B^t \text{ and } A \text{ is invertible.}$$

*1.7. Prove that

(a) the symplectic group SP_{2n} operates transitively on $\mathbb{R}^{2n}$,

(b) SP_{2n} is path-connected,   **(c)** symplectic matrices have determinant 1.

Section 2   Interlude: Spheres

2.1. Compute the formula for the inverse of the stereographic projection $\pi:\mathbb{S}^3 \to \mathbb{R}^3$.

2.2. One can parametrize proper subspaces of $\mathbb{R}^2$ by a circle in two ways. First, if a subspace W intersects the horizontal axis with angle θ, one can use the double angle $\alpha = 2\theta$. The double angle eliminates the ambiguity between θ and $\theta + \pi$. Or, one can choose a nonzero vector (y_1, y_2) in W, and use the inverse of stereographic projection to map the slope $\lambda = y_2/y_1$ to a point of $\mathbb{S}^1$. Compare these two parametrizations.

2.3. *(unit vectors and subspaces in $\mathbb{C}^2$)* A proper subspace W of the vector space $\mathbb{C}^2$ has dimension 1. Its slope is defined to be $\lambda = y_2/y_1$, where (y_1, y_2) is a nonzero vector in W. The slope can be any complex number, or when $y_1 = 0$, $\lambda = \infty$.

    (a) Let $z = v_1 + v_2 i$. Write the formula for sterographic projection π (9.2.2) and its inverse function σ in terms of z.

    (b) The function that sends a unit vector (y_1, y_2) to $\sigma(y_2/y_1)$ defines a map form the unit sphere $\mathbb{S}^3$ in $\mathbb{C}^2$ to the two-sphere $\mathbb{S}^2$. This map can be used to parametrize subspaces by points of $\mathbb{S}^2$. Compute the function $\sigma(y_2/y_1)$ on unit vectors (y_1, y_2).

    (c) What pairs of points of $\mathbb{S}^2$ correspond to pairs of subspaces W and W' that are orthogonal with respect to the standard Hermitian form on $\mathbb{C}^2$?

Section 3   The Special Unitary Group SU_2

3.1. Let P and Q be elements of SU_2, represented by the real vectors (x_0, x_1, x_2, x_3) and (y_0, y_1, y_2, y_3), respectively. Compute the real vector that corresponds to the product PQ.

3.2. Prove that U_2 is homeomorphic to the product $\mathbb{S}^3 \times \mathbb{S}^1$.

3.3. Prove that every great circle in SU_2 (circle of radius 1) is a coset of one of the longitudes.

3.4. Determine the centralizer of $\mathbf{j}$ in SU_2.

Section 4   The Rotation Group SO_3

4.1. Let W be the space of real skew-symmetric 3×3 matrices. Describe the orbits for the operation $P * A = PAP^t$ of SO_3 on W.

4.2. The rotation group SO_3 may be mapped to a 2-sphere by sending a rotation matrix to its first column. Describe the fibres of this map.

4.3. Extend the orthogonal representation $\varphi:SU_2 \to SO_3$ to a homomorphism
$\Phi:U_2 \to SO_3$, and describe the kernel of Φ.

4.4. (a) With notation as in (9.4.1), compute the matrix of the rotation γ_P, and show that its trace is $1 + 2\cos 2\theta$.

    (b) Prove directly that the matrix is orthogonal.

4.5. Prove that conjugation by an element of SU_2 rotates every latitude.

4.6. Describe the conjugacy classes in SO_3 in two ways:

    (a) Its elements operate on $\mathbb{R}^3$ as rotations. Which rotations make up a conjugacy class?

(b) The spin homomorphism $SU_2 \rightarrow SO_3$ can be used to relate the conjugacy classes in the two groups. Do this.

(c) The conjugacy classes in SU_2 are spheres. Describe the conjugacy classes in SO_3 geometrically. Be careful.

4.7. (a) Calculate left multiplication by a fixed matrix P in SU_2 explicitly, in terms of the coordinate vector (x_0, x_1, x_2, x_3). Prove that it is given as multiplication by a 4×4 orthogonal matrix Q.

(b) Prove that Q is orthogonal by a method similar to that used in describing the orthogonal representation: Express dot product of the vectors (x_0, x_1, x_2, x_3) and (x_0', x_1', x_2', x_3') that correspond to matrices P and P' in SU_2, in matrix terms.

4.8. Let W be the real vector space of Hermitian 2×2 matrices.

(a) Prove that the rule $P \cdot A = PAP^*$ defines an operation of $SL_2(\mathbb{C})$ on W.

(b) Prove that the function $\langle A, A' \rangle = \det(A + A') - \det A - \det A'$ is a bilinear form on W, and that its signature is $(3, 1)$.

(c) Use **(a)** and **(b)** to define a homomorphism $\varphi : SL_2(\mathbb{C}) \rightarrow O_{3,1}$, whose kernel is $\{\pm I\}$.

***4.9. (a)** Let H_i be the subgroup of SO_3 of rotations about the x_i-axis, $i = 1, 2, 3$. Prove that every element of SO_3 can be written as a product ABA', where A and A' are in H_1 and B is in H_2. Prove that this representation is unique unless $B = I$.

(b) Describe the double cosets $H_1 Q H_1$ geometrically (see Chapter 2, Exercise M.9).

Section 5   One-Parameter Groups

5.1. Can the image of a one-parameter group in GL_n cross itself?

5.2. Determine the one-parameter groups in U_2.

5.3. Describe by equations the images of the one-parameter groups in the group of real, invertible, 2×2 diagonal matrices, and make a drawing showing some of them in the plane.

5.4. Find the conditions on a matrix A so that e^{tA} is a one-parameter group in

(a) the special unitary group SU_n, **(b)** the Lorentz group $O_{3,1}$.

5.5. Let G be the group of real matrices of the form $\begin{bmatrix} x & y \\ & 1 \end{bmatrix}$, with $x > 0$.

(a) Determine the matrices A such that e^{tA} is a one-parameter group in G.

(b) Compute e^{tA} explicitly for the matrices in **(a)**.

(c) Make a drawing showing some one-parameter groups in the (x, y)-plane.

5.6. Let G be the subgroup of GL_2 of matrices $\begin{bmatrix} x & y \\ & x^{-1} \end{bmatrix}$ with $x > 0$ and y arbitrary. Determine the conjugacy classes in G, and the matrices A such that e^{tA} is a one-parameter group in G.

5.7. Determine the one-parameter groups in the group of invertible $n \times n$ upper triangular matrices.

5.8. Let $\varphi(t) = e^{tA}$ be a one-parameter group in a subgroup G of GL_n. Prove that the cosets of its image are matrix solutions of the differential equation $dX/dt = AX$.

5.9. Let $\varphi: \mathbb{R}^+ \to GL_n$ be a one-parameter group. Prove that $\ker\varphi$ is either trivial, or an infinite cyclic group, or the whole group.

5.10. Determine the differentiable homomorphisms from the circle group SO_2 to GL_n.

Section 6 The Lie Algebra

6.1. Verify the Jacobi identity for the bracket operation $[A, B] = AB - BA$.

6.2. Let V be a real vector space of dimension 2, with a law of composition $[v, w]$ that is bilinear and skew-symmetric (see (9.6.7)). Prove that the Jacobi identity holds.

6.3. The group SL_2 operates by conjugation on the space of trace-zero matrices. Decompose this space into orbits.

6.4. Let G be the group of invertible real matrices of the form $\begin{bmatrix} a & b \\ & a^2 \end{bmatrix}$. Determine the Lie algebra L of G, and compute the bracket on L.

6.5. Show that the set defined by $xy = 1$ is a subgroup of the group of invertible diagonal 2×2 matrices, and compute its Lie algebra.

6.6. **(a)** Show that O_2 operates by conjugation on its Lie algebra.

(b) Show that this operation is compatible with the bilinear form $\langle A, B \rangle = \frac{1}{2}$ trace AB.

(c) Use the operation to define a homomorphism $O_2 \to O_2$, and describe this homomorphism explicitly.

6.7. Determine the Lie algebras of the following groups.

(a) U_n, **(b)** SU_n, **(c)** $O_{3,1}$, **(d)** $SO_n(\mathbb{C})$.

6.8. Determine the Lie algebra of SP_{2n}, using block form $M = \begin{bmatrix} A & B \\ \hline C & D \end{bmatrix}$.

6.9. **(a)** Show that the vector cross product makes $\mathbb{R}^3$ into a Lie algebra L_1.

(b) Let $L_2 = \text{Lie}(SU_2)$, and let $L_3 = \text{Lie}(SO_3)$. Prove that the three Lie algebras L_1, L_2 and L_3 are isomorphic.

6.10. Classify complex Lie algebras of dimension ≤ 3.

6.11. Let B be a real $n \times n$ matrix, and let $\langle \ , \ \rangle$ be the bilinear form $X^t BY$. The orthogonal group G of this form is defined to be the group of matrices P such that $P^t BP = B$. Determine the one-parameter groups in G, and the Lie algebra of G.

Section 7 Translation in a Group

7.1. Prove that the unitary group U_n is path connected.

7.2. Determine the dimensions of the following groups:

(a) U_n, **(b)** SU_n, **(c)** $SO_n(\mathbb{C})$, **(d)** $O_{3,1}$, **(e)** SP_{2n}.

7.3. Using the exponential, find all solutions near I of the equation $P^2 = I$.

7.4. Find a path-connected, nonabelian subgroup of GL_2 of dimension 2.

*7.5. (a) Prove that the exponential map defines a bijection between the set of all Hermitian matrices and the set of positive definite Hermitian matrices.

(b) Describe the topological structure of $GL_2(\mathbb{C})$ using the Polar decomposition (Chapter 8, Exercise M.8) and **(a)**.

7.6. Sketch the tangent vector field PA to the group $\mathbb{C}^\times$, when $A = 1 + i$.

7.7. Let H be a finite normal subgroup of a path connected group G. Prove that H is contained in the center of G.

Section 8 Normal Subgroups of SL_2

8.1. Prove Theorem 9.8.1 for the cases $F = \mathbb{F}_4$ and $\mathbb{F}_5$.

8.2. Describe isomorphisms $PSL_2(\mathbb{F}_2) \approx S_3$ and $PSL_2(\mathbb{F}_3) \approx A_4$.

8.3. (a) Determine the numbers of Sylow p-subgroups of $PSL_2(\mathbb{F}_5)$, for $p = 2, 3, 5$.

(b) Prove that the three groups A_5, $PSL_2(\mathbb{F}_4)$, and $PSL_2(\mathbb{F}_5)$ are isomorphic.

8.4. (a) Write the polynomial equations that define the symplectic group.

(b) Show that the unitary group U_n can be defined by real polynomial equations in the real and imaginary parts of the matrix entries.

8.5. Determine the centers of the groups $SL_n(\mathbb{R})$ and $SL_n(\mathbb{C})$.

8.6. Determine all normal subgroups of $GL_2(\mathbb{R})$ that contain its center.

8.7. With Z denoting the center of a group, is $PSL_n(\mathbb{C})$ isomorphic to $GL_n(\mathbb{C})/Z$? Is $PSL_n(\mathbb{R})$ isomorphic to $GL_n(\mathbb{R})/Z$?

8.8. (a) Let P be a matrix in the center of SO_n, and let A be a skew-symmetric matrix. Prove that $PA = AP$.

(b) Prove that the center of SO_n is trivial if n is odd and is $\{\pm I\}$ if n is even and $n \geq 4$.

8.9. Compute the orders of the groups

(a) $SO_2(\mathbb{F}_3)$, **(b)** $SO_3(\mathbb{F}_3)$, **(c)** $SO_2(\mathbb{F}_5)$, **(d)** $SO_3(\mathbb{F}_5)$.

*8.10. (a) Let V be the space V of complex 2×2 matrices, with the basis $(e_{11}, e_{12}, e_{21}, e_{22})$. Write the matrix of conjugation by $A = \begin{bmatrix} a & b \\ c & d \end{bmatrix}$ on V in block form.

(b) Prove that conjugation defines a homomorphism $\varphi: SL_2(\mathbb{C}) \to GL_4(\mathbb{C})$, and that the image of φ is isomorphic to $PSL_2(\mathbb{C})$.

(c) Prove that $PSL_2(\mathbb{C})$ is a complex algebraic group by finding polynomial equations in the entries y_{ij} of a 4×4 matrix whose solutions are the matrices in the image of φ.

Miscellaneous Exercises

M.1. Let $G = SL_2(\mathbb{R})$, let $A = \begin{bmatrix} x & y \\ z & w \end{bmatrix}$ be a matrix in G, and let t be its trace. Substituting $t - x$ for w, the condition $\det A = 1$ becomes $x(t - x) - yz = 1$. For fixed trace t, the locus of solutions of this equation is a quadric in x, y, z-space. Describe the quadrics that arise this way, and decompose them into conjugacy classes.

*M.2. Which elements of $SL_2(\mathbb{R})$ lie on a one-parameter group?

M.3. Are the conjugacy classes in a path connected group G path connected?

M.4. *Quaternions* are expressions of the form $\alpha = a + b\mathbf{i} + c\mathbf{j} + d\mathbf{k}$, where a, b, c, d are real numbers (see (9.3.3)).

   (a) Let $\bar{\alpha} = a - b\mathbf{i} - c\mathbf{j} - d\mathbf{k}$. Compute $\bar{\alpha}\alpha$.

   (b) Prove that every $\alpha \neq 0$ has a multiplicative inverse.

   (c) Prove that the set of quaternions α such that $a^2 + b^2 + c^2 + d^2 = 1$ forms a group under multiplication that is isomorphic to SU_2.

M.5. The *affine group* A_n is the group of transformations of $\mathbb{R}^n$ generated by GL_n and the group T_n of translations: $t_a(x) = x + a$. Prove that T_n is a normal subgroup of A_n and that A_n/T_n is isomorphic to GL_n.

M.6. (*Cayley transform*) Let U denote the set of matrices A such that $I + A$ is invertible, and define $A' = (I - A)(I + A)^{-1}$.

   (a) Prove that if A is in U, then so is A', and that $(A')' = A$.

   (b) Let V denote the vector space of real skew-symmetric $n \times n$ matrices. Prove that the rule $A \rightsquigarrow (I - A)(I + A)^{-1}$ defines a homeomorphism from a neighborhood of 0 in V to a neighborhood of I in SO_n .

   (c) Is there an analogous statement for the unitary group?

   (d) Let $S = \begin{bmatrix} 0 & I \\ -I & 0 \end{bmatrix}$. Show that a matrix A in U is symplectic if and only if $(A')^t S = -SA'$.

M.7. Let $G = SL_2$. A *ray* in $\mathbb{R}^2$ is a half line leading from the origin to infinity. The rays are in bijective correspondence with the points on the unit 1-sphere in $\mathbb{R}^2$.

   (a) Determine the stabilizer H of the ray $\{r e_1 | r \geq 0\}$.

   (b) Prove that the map $f : H \times SO_2 \to G$ defined by $f(P, B) = PB$ is a homeomorphism (not a homomorphism).

   (c) Use **(b)** to identify the topological structure of SL_2.

M.8. Two-dimensional space-time is the space of real three-dimensional column vectors, with the Lorentz form $\langle Y, Y' \rangle = Y^t I_{2,1} Y' = y_1 y_1' + y_2 y_2' - y_3 y_3'$.
The space W of real trace-zero 2×2 matrices has a basis $\mathbf{B} = (w_1, w_2, w_3)$, where

$$w_1 = \begin{bmatrix} 1 & \\ & -1 \end{bmatrix}, \quad w_2 = \begin{bmatrix} & 1 \\ 1 & \end{bmatrix}, \quad w_3 = \begin{bmatrix} & 1 \\ -1 & \end{bmatrix}.$$

   (a) Show that if $A = \mathbf{B}Y$ and $A' = \mathbf{B}Y'$ are trace-zero matrices, the Lorentz form carries over to $\langle A, A' \rangle = y_1 y_1' + y_2 y_2' - y_3 y_3' = \frac{1}{2} \text{trace}(AA')$.

   (b) The group SL_2 operates by conjugation on the space W. Use this operation to define a homomorphism $\varphi : SL_2 \to O_{2,1}$ whose kernel is $\{\pm I\}$.

   *(c) Prove that the Lorentz group $O_{2,1}$ has four connected components and that the image of φ is the component that contains the identity.

M.9. The icosahedral group is a subgroup of index 2 in the group G_1 of all symmetries of a dodecahedron, including orientation-reversing symmetries. The alternating group A_5 is a subgroup of index 2 of the symmetric group $G_2 = S_5$. Finally, consider the spin homomorphism $\varphi : SU_2 \to SO_3$. Let G_3 be the inverse image of the icosahedral group in SU_2. Are any of the groups G_i isomorphic?

*M.10. Let P be the matrix (9.3.1) in SU_2, and let T denote the subgroup of SU_2 of diagonal matrices. Prove that if the entries a, b of P are not zero, then the double coset TPT is homeomorphic to a torus, and describe the remaining double cosets (see Chapter 2, Exercise M.9).

*M.11. The *adjoint representation* of a linear group G is the representation by conjugation on its Lie algebra: $G \times L \to L$ defined by $P, A \rightsquigarrow PAP^{-1}$. The form $\langle A, A' \rangle = \text{trace}(AA')$ on L is called the *Killing form*. For the following groups, verify that if P is in G and A is in L, then PAP^{-1} is in L. Prove that the Killing form is symmetric and bilinear and that the operation is compatible with the form, i.e., that $\langle A, A \rangle = \langle PAP^{-1}, PA'P^{-1} \rangle$.

   (a) U_n, **(b)** $O_{3,1}$, **(c)** $SO_n(\mathbb{C})$, **(d)** SP_{2n}.

*M.12. Determine the signature of the Killing form (Exercise M.11) on the Lie algebra of

   (a) SU_n, **(b)** SO_n, **(c)** SL_n.

*M.13. Use the adjoint representation of $SL_2(\mathbb{C})$ (Exercise M.11) to define an isomorphism $SL_2(\mathbb{C})/\{\pm I\} \approx SO_3(\mathbb{C})$.

C H A P T E R 10

Group Representations

A tremendous effort has been made by mathematicians
for more than a century to clear up the chaos in group theory.
Still, we cannot answer some of the simplest questions.

—Richard Brauer

Group representations arise in mathematics and in other sciences when a structure with symmetry is being studied. If one makes all possible measurements of some sort (in chemistry, it might be vibrations of a molecule) and assembles the results into a "state vector," a symmetry of the molecule will transform that vector. This produces an operation of the symmetry group on the space of vectors, a representation of the group, that can help to analyze the structure.

10.1 DEFINITIONS

In this chapter, GL_n denotes the complex general linear group $GL_n(\mathbb{C})$.

A *matrix representation* of a group G is a homomorphism

$$(10.1.1) \qquad\qquad R: G \to GL_n,$$

from G to one of the complex general linear groups. The number n is the *dimension* of the representation.

We use the notation R_g instead of $R(g)$ for the image of a group element g. Each R_g is an invertible matrix, and the statement that R is a homomorphism reads

$$(10.1.2) \qquad\qquad R_{gh} = R_g R_h.$$

If a group is given by generators and relations, say $\langle x_1, \ldots, x_n \mid r_1, \ldots, r_k \rangle$, a matrix representation can be defined by assigning matrices $R_{x_1}, \ldots, R_{x_n}$ that satisfy the relations. For example, the symmetric group S_3 can be presented as $\langle x, y \mid x^3, y^2, xyxy \rangle$, so a representation of S_3 is defined by matrices R_x and R_y such that $R_x^3 = I$, $R_y^2 = I$, and $R_x R_y R_x R_y = I$. Some relations in addition to these required ones may hold.

Because S_3 is isomorphic to the dihedral group D_3, it has a two-dimensional matrix representation that we denote by A. We place an equilateral triangle with its center at the origin, and so that one vertex is on the e_1-axis. Then its group of symmetries will be generated by the rotation A_x with angle $2\pi/3$ and the reflection A_y about the e_1-axis. With $c = \cos 2\pi/3$ and $s = \sin 2\pi/3$,

(10.1.3)
$$A_x = \begin{bmatrix} c & -s \\ s & c \end{bmatrix}, \quad A_y = \begin{bmatrix} 1 & 0 \\ 0 & -1 \end{bmatrix}.$$

We call this the *standard representation* of the dihedral group D_3 and of S_3.

• A representation R is *faithful* if the homomorphism $R : G \to GL_n$ is injective, and there-fore maps G isomorphically to its image, a subgroup of GL_n. The standard representation of S_3 is faithful.

Our second representation of S_3 is the one-dimensional *sign representation* Σ. Its value on a group element is the 1×1 matrix whose entry is the sign of the permutation:

(10.1.4)
$$\Sigma_x = [1], \quad \Sigma_y = [-1].$$

This is not a faithful representation.

Finally, every group has the *trivial representation*, the one-dimensional representation that takes the value 1 identically:

(10.1.5)
$$T_x = [1], \quad T_y = [1].$$

There are other representations of S_3, including the representation by permutation matrices and the representation as a group of rotations of $\mathbb{R}^3$. But we shall see that every representation of this group can be built up out of the three representations A, Σ, and T.

Because they involve several matrices, each of which may have many entries, repre-sentations are notationally complicated. The secret to understanding them is to throw out most of the information that the matrices contain, keeping only one essential part, its trace, or character.

• The *character* χ_R of a matrix representation R is the complex-valued function whose domain is the group G, defined by $\chi_R(g) = \text{trace } R_g$.

Characters are usually denoted by χ ('chi'). The characters of the three representations of the symmetric group that we have defined are displayed below in tabular form, with the group elements listed in their usual order.

(10.1.6)

	1	x	x^2	y	xy	x^2y
χ_T	1	1	1	1	1	1
χ_Σ	1	1	1	-1	-1	-1
χ_A	2	-1	-1	0	0	0

Several interesting phenomena can be observed in this table:

• The rows form orthogonal vectors of length equal to six, which is also the order of S_3. The columns are orthogonal too.

These astonishing facts illustrate the beautiful Main Theorem 10.4.6 on characters.

Two other phenomena are more elementary:

- $\chi_R(1)$ is the dimension of the representation, also called the *dimension* of the character.

Since a representation is a homomorphism, it sends the identity in the group to the identity matrix. So $\chi_R(1)$ is the trace of the identity matrix.

- The characters are constant on conjugacy classes.

(The conjugacy classes in S_3 are the sets $\{1\}$, $\{x, x^2\}$, and $\{y, xy, x^2 y\}$.)

This phenomenon is explained as follows: Let g and g' be conjugate elements of a group G, say $g' = hgh^{-1}$. Because a representation R is a homomorphism, $R_{g'} = R_h R_g R_h^{-1}$. So $R_{g'}$ and R_g are conjugate matrices. Conjugate matrices have the same trace.

It is essential to work as much as possible without fixing a basis, and to facilitate this, we introduce the concept of a representation of a group on a vector space V. We denote by

(10.1.7) $GL(V)$

the group of invertible linear operators on V, the law of composition being composition of operators. We always assume that V is a finite-dimensional complex vector space, and not the zero space.

- A *representation* of a group G on a complex vector space V is a homomorphism

(10.1.8) $\rho : G \to GL(V)$.

So a representation assigns a linear operator to every group element. A matrix representation can be thought of as a representation of G on the space of column vectors.

The elements of a finite rotation group (6.12) are rotations of a three-dimensional Euclidean space V without reference to a basis, and these orthogonal operators give us what we call the *standard representation* of the group. (We use this term in spite of the fact that, for D_3, it conflicts with (10.1.3).) We also use the symbol ρ for other representations, and this will *not* imply that the operators ρ_g are rotations.

If ρ is a representation, we denote the image of an element g in $GL(V)$ by ρ_g rather than by $\rho(g)$, to keep the symbol g out of the way. The result of applying ρ_g to a vector v will be written as

$$\rho_g(v) \quad \text{or as} \quad \rho_g v.$$

Since ρ is a homomorphism,

(10.1.9) $\rho_{gh} = \rho_g \rho_h$.

The choice of a basis $\mathbf{B} = (v_1, \ldots, v_n)$ for a vector space V defines an isomorphism from $GL(V)$ to the general linear group GL_n:

(10.1.10)
$$
\begin{aligned}
GL(V) &\to\ GL_n \\
T &\ \rightsquigarrow\ \text{matrix of } T,
\end{aligned}
$$

and a representation ρ defines a matrix representation R, by the rule

(10.1.11) $\rho_g \rightsquigarrow$ its matrix $= R_g$.

Thus every representation of G on a finite-dimensional vector space can be made into a matrix representation, if we are willing to choose a basis. We may want to choose a basis in order to make explicit calculations, but we must determine which properties are independent of the basis, and which bases are the good ones.

A change of basis in V by a matrix P changes the matrix representation R associated to ρ to a *conjugate representation* $R' = P^{-1}RP$, i.e.,

$$(10.1.12) \qquad\qquad R'_g = P^{-1}R_gP,$$

with the same P for every g in G. This follows from Rule 4.3.5 for a change of basis.

• An *operation* of a group G *by linear operators* on a vector space V is an operation on the underlying set:

$$(10.1.13) \qquad\qquad 1v = v \quad\text{and}\quad (gh)v = g(hv),$$

and in addition every group element acts as a linear operator. Writing out what this means, we obtain the rules

$$(10.1.14) \qquad\qquad g(v + v') = gv + gv' \quad\text{and}\quad g(cv) = cgv,$$

which, when added to (10.1.13), give a complete list of axioms for such an operation. We can speak of orbits and stabilizers as before.

The two concepts "operation by linear operators on V" and "representation on V" are equivalent. Given a representation ρ of G on V, we can define an operation of G on V by

$$(10.1.15) \qquad\qquad gv = \rho_g(v).$$

Conversely, given an operation, the same formula can be used to define the operator ρ_g.

We now have two notations (10.1.15) for the operation of g on v, and we use them interchangeably. The notation gv is more compact, so we use it when possible, though it is ambiguous because it doesn't specify ρ.

• An *isomorphism* from one representation $\rho : G \rightarrow GL(V)$ of a group G to another representation $\rho' : G \rightarrow GL(V')$ is an isomorphism of vector spaces $T : V \rightarrow V'$, an invertible linear transformation, that is compatible with the operations of G:

$$(10.1.16) \qquad\qquad T(gv) = gT(v)$$

for all v in V and all g in G. If $T : V \rightarrow V'$ is an isomorphism, and if **B** and **B**$'$ are corresponding bases of V and V', the associated matrix representations R_g and R'_g will be *equal*.

The main topic of the chapter is the determination of the isomorphism classes of *complex representations* of a group G, representations on finite-dimensional, nonzero complex vector spaces. Any real matrix representation, such as one of the representations of S_3 described above, can be used to define a complex representation, simply by interpreting the real matrices as complex matrices. We will do this without further comment. And except in the last section, our groups will be finite.

10.2 IRREDUCIBLE REPRESENTATIONS

Let ρ be a representation of a finite group G on the (nonzero, finite-dimensional) complex vector space V. A vector v is G-*invariant* if the operation of every group element fixes the vector:

$$(10.2.1) \qquad gv = v \text{ or } \rho_g(v) = v, \text{ for all } g \text{ in } G.$$

Most vectors aren't G-invariant. However, starting with any vector v, one can produce a G-invariant vector by *averaging over the group*. Averaging is an important procedure that will be used often. We used it once before, in Chapter 6, to find a fixed point of a finite group operation on the plane. The G-invariant averaged vector is

$$(10.2.2) \qquad \tilde{v} = \frac{1}{|G|} \sum_{g \in G} gv.$$

The reason for the normalization factor $\frac{1}{|G|}$ is that, if v happens to be G-invariant itself, then $\tilde{v} = v$.

We verify that $\tilde{v}$ is G-invariant: Since the symbol g is used in the summation (10.2.2), we write the condition for G-invariance as $h\tilde{v} = \tilde{v}$ for all h in G. The proof is based on the fact that left multiplication by h defines a bijective map from G to itself. We make the substitution $g' = hg$. Then as g runs through the elements of the group G, g' does too, though in a different order, and

$$(10.2.3) \qquad h\tilde{v} = h\frac{1}{|G|} \sum_{g \in G} gv = \frac{1}{|G|} \sum_{g \in G} g'v = \frac{1}{|G|} \sum_{g \in G} gv = \tilde{v}.$$

This reasoning can be confusing when one sees it for the first time, so we illustrate it by an example, with $G = S_3$. We list the elements of the group as usual: $g = 1, x, x^2, y, xy, x^2y$. Let $h = y$. Then $g' = hg$ lists the group in the order $g' = y, x^2y, xy, 1, x^2, x$. So

$$\sum_{g \in G} g'v = yv + x^2yv + xyv + 1v + x^2v + xv = \sum_{g \in G} gv$$

The fact that multiplication by h is bijective implies that g' will always run over the group in some order. Please study this reindexing trick.

The averaging process may fail to yield an interesting vector. It is possible that $\tilde{v} = 0$.

Next, we turn to G-invariant subspaces.

• Let ρ be a representation of G on V. A subspace W of V is called G-*invariant* if gw is in W for all w in W and g in G. So the operation by a group element must carry W to itself: For all g,

$$(10.2.4) \qquad gW \subset W, \quad \text{or} \quad \rho_g W \subset W.$$

This is an extension of the concept of T-invariant subspace that was introduced in Section 4.4. Here we ask that W be an invariant subspace for each of the operators ρ_g.

When W is G-invariant, we can restrict the operation of G to obtain a representation of G on W.

Lemma 10.2.5 If W is an invariant subspace of V, then $gW = W$ for all g in G.

Proof. Since group elements are invertible, their operations on V are invertible. So gW and W have the same dimension. If $gW \subset W$, then $gW = W$. $\square$

• If V is the direct sum of G-invariant subspaces, say $V = W_1 \oplus W_2$, the representation ρ on V is called the *direct sum* of its restrictions to W_1 and W_2, and we write

$$(10.2.6) \qquad\qquad \rho = \alpha \oplus \beta,$$

where α and β denote the restrictions of ρ to W_1 and W_2, respectively. Suppose that this is the case, and let $\mathbf{B} = (\mathbf{B}_1, \mathbf{B}_2)$ be a basis of V obtained by listing bases of W_1 and W_2 in succession. Then the matrix of ρ_g will have the block form

$$(10.2.7) \qquad\qquad R_g = \begin{bmatrix} A_g & 0 \\ 0 & B_g \end{bmatrix},$$

where A_g is the matrix of α and B_g is the matrix of β, with respect to the chosen bases. The zeros below the block A_g reflect the fact that the operation of g does not spill vectors out of the subspace W_1, and the zeros above the block B_g reflect the analogous fact for W_2.

Conversely, if R is a matrix representation and if all of the matrices R_g have a block form (10.2.7), with A_g and B_g square, we say that the matrix representation R is the direct sum $A \oplus B$.

For example, since the symmetric group S_3 is isomorphic to the dihedral group D_3, it is a rotation group, a subgroup of SO_3. We choose coordinates so that x acts on $\mathbb{R}^3$ as a rotation with angle $2\pi/3$ about the e_3-axis, and y acts as a rotation by π about the e_1-axis. This gives us a three-dimensional matrix representation M:

$$(10.2.8) \qquad\qquad M_x = \begin{bmatrix} c & -s & \\ s & c & \\ & & 1 \end{bmatrix}, \quad M_y = \begin{bmatrix} 1 & & \\ & -1 & \\ & & -1 \end{bmatrix},$$

with $c = \cos 2\pi/3$ and $s = \sin 2\pi/3$. We see that M has a block decomposition, and that it is the direct sum $A \oplus \Sigma$ of the standard representation and the sign representation.

Even when a representation ρ is a direct sum, the matrix representation obtained using a basis will not have a block form unless the basis is compatible with the direct sum decomposition. Until we have made a further analysis, it may be difficult to tell that a representation is a direct sum, when it is presented using the wrong basis. But if we find such a decomposition of our representation ρ, we may try to decompose the summands α and β further, and we may continue until no further decomposition is possible.

• If ρ is a representation of a group G on V and if V has no proper G-invariant subspace, ρ is called an *irreducible* representation. If V has a proper G-invariant subspace, ρ is *reducible*.

The standard representation of S_3 is irreducible.

Suppose that our representation ρ is reducible, and let W be a proper G-invariant subspace of V. Let α be the restriction of ρ to W. We extend a basis of W to a basis of V, say $\mathbf{B} = (w_1, \ldots, w_k; v_{k+1}, \ldots v_d)$. The matrix of ρ_g will have the block form

$$(10.2.9) \qquad\qquad R_g = \begin{bmatrix} A_g & * \\ 0 & B_g \end{bmatrix},$$

where A is the matrix of α and B_g is some other matrix representation of G. I think of the block indicated by $*$ as "junk." Maschke's theorem, which is below, tells us that we can get rid of that junk. But to do so we must choose the basis more carefully.

Theorem 10.2.10 Maschke's Theorem.   Every representation of a finite group G on a nonzero, finite-dimensional complex vector space is a direct sum of irreducible representations.

This theorem will be proved in the next section. We'll illustrate it here by one more example in which G is the symmetric group S_3. We consider the representation of S_3 by the permutation matrices that correspond to the permutations $x = (\mathbf{123})$ and $y = (\mathbf{12})$. Let's denote this representation by N:

$$(10.2.11) \qquad N_x = \begin{bmatrix} 0 & 0 & 1 \\ 1 & 0 & 0 \\ 0 & 1 & 0 \end{bmatrix}, \quad N_y = \begin{bmatrix} 0 & 1 & 0 \\ 1 & 0 & 0 \\ 0 & 0 & 1 \end{bmatrix}.$$

There is no block decomposition for this pair of matrices. However, the vector $w_1 = (1, 1, 1)^{\mathrm{t}}$ is fixed by both matrices, so it is G-invariant, and the one-dimensional subspace W spanned by w_1 is also G-invariant. The restriction of N to this subspace is the trivial representation T. Let's change the standard basis of $\mathbb{C}^3$ to the basis $\mathbf{B} = (w_1, e_2, e_3)$. With respect to this new basis, the representation N is changed as follows:

$$P = [\mathbf{B}] = \begin{bmatrix} 1 & 0 & 0 \\ 1 & 1 & 0 \\ 1 & 0 & 1 \end{bmatrix} : \quad P^{-1}N_xP = \begin{bmatrix} 1 & 0 & 1 \\ 0 & 0 & -1 \\ 0 & 1 & -1 \end{bmatrix}, \quad P^{-1}N_yP = \begin{bmatrix} 1 & 1 & 0 \\ 0 & -1 & 0 \\ 0 & -1 & 1 \end{bmatrix}.$$

The upper right blocks aren't zero, so we don't have a decomposition of the representation as a direct sum.

There is a better approach: The matrices N_x and N_y are unitary, so N_g is unitary for all g in G. (They are orthogonal, but we are considering complex representations.) Unitary matrices preserve orthogonality. Since W is G-invariant, the orthogonal space $W^{\perp}$ is G-invariant too (see (10.3.4)). If we form a basis by choosing vectors w_2 and w_3 from $W^{\perp}$, the junk disappears. The permutation representation N is isomorphic to the direct sum $T \oplus A$. We'll soon have techniques that make verifying this extremely simple, so we won't bother doing so here.

This decomposition of the representation using orthogonal spaces illustrates a general method that we investigate next.

10.3   UNITARY REPRESENTATIONS

Let V be a Hermitian space – a complex vector space together with a positive definite Hermitian form $\langle\ ,\ \rangle$. A unitary operator T on V is a linear operator with the property

$$(10.3.1) \qquad \langle Tv, Tw \rangle = \langle v, w \rangle$$

for all v and w in V (8.6.3). If A is the matrix of a linear operator T with respect to an orthonormal basis, then T is unitary if and only if A is a unitary matrix: $A^* = A^{-1}$.

• A representation $\rho: G \to GL(V)$ on a Hermitian space V is called *unitary* if ρ_g is a unitary operator for every g. We can write this condition as

(10.3.2) $$\langle gv, gw \rangle = \langle v, w \rangle \quad \text{or} \quad \langle \rho_g v, \rho_g w \rangle = \langle v, w \rangle,$$

for all v and w in V and all g in G. Similarly, a matrix representation $R: G \to GL_n$ is *unitary* if R_g is a unitary matrix for every g in G. A unitary matrix representation is a homomorphism from G to the unitary group:

(10.3.3) $$R: G \to U_n.$$

A representation ρ on a Hermitian space will be unitary if and only if the matrix representation obtained using an orthonormal basis is unitary.

Lemma 10.3.4 Let ρ be a unitary representation of G on a Hermitian space V, and let W be a G-invariant subspace. The orthogonal complement $W^\perp$ is also G-invariant, and ρ is the direct sum of its restrictions to the Hermitian spaces W and $W^\perp$. These restrictions are also unitary representations.

Proof. It is true that $V = W \oplus W^\perp$ (8.5.1). Since ρ is unitary, it preserves orthogonality: If W is invariant and $u \perp W$, then $gu \perp gW = W$. This means that if $u \in W^\perp$, then $gu \in W^\perp$. $\square$

The next corollary follows from the lemma by induction.

Corollary 10.3.5 Every unitary representation $\rho: G \to GL(V)$ on a Hermitian vector space V is an orthogonal sum of irreducible representations. $\square$

The trick now is to turn the condition (10.3.2) for a unitary representation around, and think of it as a condition on the form instead of on the representation. Suppose we are given a representation $\rho: G \to GL(V)$ on a vector space V, and let $\langle \, , \, \rangle$ be a positive definite Hermitian form on V. We say that the form is *G-invariant* if (10.3.2) holds. This is exactly the same as saying that the representation is unitary, when we use the form to make V into a Hermitian space. But if only the representation ρ is given, we are free to choose the form.

Theorem 10.3.6 Let $\rho: G \to GL(V)$ be a representation of a finite group on a vector space V. There exists a G-invariant, positive definite Hermitian form on V.

Proof. We begin with an arbitrary positive definite Hermitian form on V that we denote by $\{ , \}$. For example, we may choose a basis for V and use it to transfer the standard Hermitian form X^*Y on $\mathbb{C}^n$ over to V. Then we use the averaging process to construct another form. The averaged form is defined by

(10.3.7) $$\langle v, w \rangle = \frac{1}{|G|} \sum_{g \in G} \{gv, gw\}.$$

We claim that this form is Hermitian, positive definite, and G-invariant. The verifications of these properties are easy. We omit the first two, but we will verify G-invariance. The proof is almost identical to the one used to show that averaging produces an G-invariant vector

(10.2.3), except that it is based here on the fact that *right* multiplication by an element h of G defines a bijective map $G \to G$.

Let h be an element of G. We must show that $\langle hv, hw \rangle = \langle v, w \rangle$ for all v and w in V (10.3.2). We make the substitution $g' = gh$. As g runs over the group, so does g'. Then

$$\langle hv, hw \rangle = \tfrac{1}{|G|} \sum_g \{ghv, ghw\} = \tfrac{1}{|G|} \sum_g \{g'v, g'w\} = \tfrac{1}{|G|} \sum_g \{gv, gw\} = \langle v, w \rangle. \qquad \square$$

Theorem 10.3.6 has remarkable consequences:

Corollary 10.3.8

(a) (*Maschke's Theorem*): Every representation of a finite group G is a direct sum of irreducible representations.

(b) Let $\rho : G \to GL(V)$ be a representation of a finite group G on a vector space V. There exists a basis **B** of V such that the matrix representation R obtained from ρ using this basis is unitary.

(c) Let $R : G \to GL_n$ be a matrix representation of a finite group G. There is an invertible matrix P such that $R'_g = P^{-1} R_g P$ is unitary for all g, i.e., such that R' is a homomorphism from G to the unitary group U_n.

(d) Every finite subgroup of GL_n is conjugate to a subgroup of the unitary group U_n.

Proof. (a) This follows from Theorem 10.3.6 and Corollary 10.3.5.

(b) Given ρ, we choose a G-invariant positive definite Hermitian form on V, and we take for **B** an orthonormal basis with respect to this form. The associated matrix representation will be unitary.

(c) This is the matrix form of (b), and it is derived in the usual way, by viewing R as a representation on the space $\mathbb{C}^n$ and then changing basis.

(d) This is obtained from (c) by viewing the inclusion of a subgroup H into GL_n as a matrix representation of H. $\qquad \square$

This corollary provides another proof of Theorem 4.7.14:

Corollary 10.3.9 Every matrix A of finite order in $GL_n(\mathbb{C})$ is diagonalizable.

Proof. The matrix A generates a finite cyclic subgroup of GL_n. By Theorem 10.3.8(d), this subgroup is conjugate to a subgroup of the unitary group. Hence A is conjugate to a unitary matrix. The Spectral Theorem 8.6.8 tells us that a unitary matrix is diagonalizable. Therefore A is diagonalizable. $\qquad \square$

10.4 CHARACTERS

As mentioned in the first section, one works almost exclusively with characters, one reason being that representations are complicated. The *character* χ of a representation ρ is the complex-valued function whose domain is the group G, defined by

(10.4.1) $$\chi(g) = \text{trace } \rho_g.$$

If R is the matrix representation obtained from ρ by a choice of basis, then χ is also the character of R. The dimension of the vector space V is called the *dimension* of the representation ρ, and also the *dimension* of its character χ. The character of an irreducible representation is called an *irreducible character*.

Here are some basic properties of the character.

Proposition 10.4.2 Let χ be the character of a representation ρ of a finite group G.

(a) $\chi(1)$ is the dimension of χ.

(b) The character is constant on conjugacy classes: If $g' = hgh^{-1}$, then $\chi(g') = \chi(g)$.

(c) Let g be an element of G of order k. The roots of the characteristic polynomial of ρ_g are powers of the k-th root of unity $\zeta = e^{2\pi i/k}$. If ρ has dimension d, then $\chi(g)$ is a sum of d such powers.

(d) $\chi(g^{-1})$ is the complex conjugate $\overline{\chi(g)}$ of $\chi(g)$.

(e) The character of a direct sum $\rho \oplus \rho'$ of representations is the sum $\chi + \chi'$ of their characters.

(f) Isomorphic representations have the same character.

Proof. Parts **(a)** and **(b)** were discussed before, for matrix representations (see (10.1.6)).

(c) The trace of ρ_g is the sum of its eigenvalues. If λ is an eigenvalue of ρ, then λ^k is an eigenvalue of ρ_g^k, and if $g^k = 1$, then $\rho_g^k = I$ and $\lambda^k = 1$. So λ is a power of ζ.

(d) The eigenvalues $\lambda_1, \ldots, \lambda_d$ of R_g have absolute value 1 because they are roots of unity. For any complex number λ of absolute value 1, $\lambda^{-1} = \bar{\lambda}$. Therefore $\chi(g^{-1}) = \lambda_1^{-1} + \cdots + \lambda_d^{-1} = \bar{\lambda}_1 + \cdots + \bar{\lambda}_d = \overline{\chi(g)}$.

Parts **(e)** and **(f)** are obvious. $\qquad\qquad\square$

Two things simplify the computation of a character χ. First, since χ is constant on conjugacy classes, we need only determine the value of χ on one element in each class – a representative element. Second, since trace is independent of a basis, we may select a convenient basis for each individual group element to compute it. We don't need to use the same basis for all elements.

There is a *Hermitian product* on characters, defined by

(10.4.3) $$\langle \chi, \chi' \rangle = \tfrac{1}{|G|} \sum_g \overline{\chi(g)} \chi'(g).$$

When χ and χ' are viewed as vectors, as in Table 10.1.6, this is the standard Hermitian product (8.3.3), scaled by the factor $\tfrac{1}{|G|}$.

It is convenient to rewrite this formula by grouping the terms for each conjugacy class. This is permissible because the characters are constant on them. We number the conjugacy classes arbitrarily, as $C_1, \ldots, C_r$, and we let c_i denote the order of the class C_i. We also choose a representative element g_i in the class C_i. Then

(10.4.4)
$$\langle \chi, \chi' \rangle = \tfrac{1}{|G|} \sum_{i=1}^{r} c_i \overline{\chi(g_i)} \chi'(g_i).$$

We go back to our usual example: Let G be the symmetric group S_3. Its class equation is $6 = 1 + 2 + 3$, and the elements $1, x, y$ represent the conjugacy classes of orders $1, 2, 3$, respectively. Then

$$\langle \chi, \chi' \rangle = \tfrac{1}{6} \left(\overline{\chi(1)} \chi'(1) + 2\overline{\chi(x)} \chi'(x) + 3\overline{\chi(y)} \chi'(y) \right).$$

Looking at Table 10.1.6, we find

(10.4.5) $\langle \chi_A, \chi_A \rangle = \tfrac{1}{6}(4 + 2 + 0) = 1$ and $\langle \chi_A, \chi_\Sigma \rangle = \tfrac{1}{6}(2 + -2 + 0) = 0.$

The characters $\chi_T, \chi_\Sigma, \chi_A$ are orthonormal with respect to the Hermitian product $\langle\ ,\ \rangle$.

These computations illustrate the Main Theorem on characters. It is one of the most beautiful theorems of algebra, both because it is so elegant, and because it simplifies the problem of classifying representations so much.

Theorem 10.4.6 Main Theorem. Let G be a finite group.

(a) *(orthogonality relations)*   The irreducible characters of G are orthonormal: If χ_i is the character of an irreducible representation ρ_i, then $\langle \chi_i, \chi_i \rangle = 1$. If χ_i and χ_j are the characters of nonisomorphic irreducible representations ρ_i and ρ_j, then $\langle \chi_i, \chi_j \rangle = 0$.

(b) There are finitely many isomorphism classes of irreducible representations, the same number as the number of conjugacy classes in the group.

(c) Let $\rho_1, \ldots, \rho_r$ represent the isomorphism classes of irreducible representations of G, and let $\chi_1, \ldots, \chi_r$ be their characters. The dimension d_i of ρ_i (or of χ_i) divides the order $|G|$ of the group, and $|G| = d_1^2 + \cdots + d_r^2.$

This theorem is proved in Section 10.8, except we won't prove that d_i divides $|G|$.

One should compare (c) with the class equation. Let the conjugacy classes be $C_1, \ldots, C_r$ and let $c_i = |C_i|$. Then c_i divides $|G|$, and $|G| = c_1 + \cdots + c_r$.

The Main Theorem allows us to decompose any character as a linear combination of the irreducible characters, using the formula for orthogonal projection (8.4.11). Maschke's Theorem tells us that every representation ρ is isomorphic to a direct sum of the irreducible representations $\rho_1, \ldots, \rho_r$. We write this symbolically as

(10.4.7)
$$\rho \approx n_1 \rho_1 \oplus \cdots \oplus n_r \rho_r,$$

where n_i are non-negative integers, and $n_i \rho_i$ stands for the direct sum of n_i copies of ρ_i.

Corollary 10.4.8 Let $\rho_1, \ldots, \rho_r$ represent the isomorphism classes of irreducible representations of a finite group G, and let ρ be any representation of G. Let χ_i and χ be the characters of ρ_i and ρ, respectively, and let $n_i = \langle \chi, \chi_i \rangle$. Then

(a) $\chi = n_1 \chi_1 + \cdots + n_r \chi_r$, and
(b) ρ is isomorphic to $n_1 \rho_1 \oplus \cdots \oplus n_r \rho_r$.
(c) Two representations ρ and ρ' of a finite group G are isomorphic if and only if their characters are equal.

Proof. Any representation ρ is isomorphic to an integer combination $m_1 \rho_1 \oplus \cdots \oplus m_r \rho_r$ of the representations ρ_i, and then $\chi = m_1 \chi_1 + \cdots + m_r \chi_r$ (Lemma 10.4.2). Since the characters χ_i are orthonormal, the projection formula shows that $m_i = n_i$. This proves **(a)** and **(b)**, and **(c)** follows. $\qquad\square$

Corollary 10.4.9 For any characters χ and χ', $\langle \chi, \chi' \rangle$ is an integer. $\qquad\square$

Note also that, with χ as in (10.4.8)**(a)**,

(10.4.10)
$$\langle \chi, \chi \rangle = n_1^2 + \cdots + n_r^2.$$

Some consequences of this formula are:

$\langle \chi, \chi \rangle = 1 \iff \chi$ is an irreducible character,
$\langle \chi, \chi \rangle = 2 \iff \chi$ is the sum of two distinct irreducible characters,
$\langle \chi, \chi \rangle = 3 \iff \chi$ is the sum of three distinct irreducible characters,
$\langle \chi, \chi \rangle = 4 \iff \chi$ is either the sum of four distinct irreducible characters, or
$\chi = 2\chi_i$ for some irreducible character χ_i. $\qquad\square$

A complex-valued function on the group, such as a character, that is constant on each conjugacy class, is called a *class function*. A class function φ can be given by assigning arbitrary values to each conjugacy class. So the complex vector space $\mathcal{H}$ of class functions has dimension equal to the number of conjugacy classes. We use the same product as (10.4.3) to make $\mathcal{H}$ into a Hermitian space:

$$\langle \varphi, \psi \rangle = \tfrac{1}{|G|} \sum_g \overline{\varphi(g)}\psi(g).$$

Corollary 10.4.11 The irreducible characters form on orthonormal basis of the space $\mathcal{H}$ of class functions.

This follows from parts **(a)** and **(b)** of the Main Theorem. The characters are independent because they are orthonormal, and they span $\mathcal{H}$ because the dimension of $\mathcal{H}$ is equal to the number of conjugacy classes. $\qquad\square$

Using the Main Theorem, it becomes easy to see that T, Σ, and A represent all of the isomorphism classes of irreducible representations of the group S_3 (see Section 10.1). Since there are three conjugacy classes, there are three irreducible representations. We verified above (10.4.5) that $\langle \chi_A, \chi_A \rangle = 1$, so A is an irreducible representation. The representations T and Σ are obviously irreducible because they are one-dimensional. And, these three representations are not isomorphic because their characters are distinct.

The irreducible characters of a group can be assembled into a table, the *character table* of the group. It is customary to list the values of the character on a conjugacy class just once. Table 10.1.6, showing the irreducible characters of S_3, gets compressed into three columns. In the table below, the three conjugacy classes in S_3 are described by the representative elements 1, x, y, and for reference, the orders of the conjugacy classes are given above them in parentheses. We have assigned indices to the irreducible characters: $\chi_T = \chi_1$, $\chi_\Sigma = \chi_2$, and $\chi_A = \chi_3$.

		conjugacy class			
		(1)	(2)	(3)	order of the class
		1	x	y	representative element
irreducible	χ_1	1	1	1	
character	χ_2	1	1	-1	value of the character
	χ_3	2	-1	0	

(10.4.12)   Character table of the symmetric group S_3

In such a table, we put the trivial character, the character of the trivial representation, into the top row. It consists entirely of 1's. The first column lists the dimensions of the representations (10.4.2)(a).

We determine the character table of the tetrahedral group T of 12 rotational symmetries of a tetrahedron next. Let x denote rotation by $2\pi/3$ about a face, and let z denote rotation by π about the center of an edge, as in Figure 7.10.8. The conjugacy classes are $C(1)$, $C(x)$, $C(x^2)$, and $C(z)$, and their orders are 1, 4, 4, and 3, respectively. So there are four irreducible characters; let their dimensions be d_i. Then $12 = d_1^2 + \cdots + d_4^2$. The only solution of this equation is $12 = 1^2 + 1^2 + 1^2 + 3^2$, so the dimensions of the irreducible representations are 1, 1, 1, 3. We write the table first with undetermined entries:

	(1)	(4)	(4)	(3)
	1	x	x^2	z
χ_1	1	1	1	1
χ_2	1	a	b	c
χ_3	1	a'	b'	c'
χ_4	3	*	*	*

and we evaluate the form (10.4.4) on the orthogonal characters χ_1 and χ_2.

(10.4.13) $\langle \chi_1, \chi_2 \rangle = \frac{1}{12}(1 + 4a + 4b + 3c) = 0.$

Since χ_2 is a one-dimensional character, $\chi_2(z) = c$ is the trace of a 1×1 matrix. It is the unique entry in that matrix, and since $z^2 = 1$, its square is 1. So c is equal to 1 or -1. Similarly, since $x^3 = 1$, $\chi_2(x) = a$ will be a power of $\omega = e^{2\pi i/3}$. So a is equal to 1, ω, or ω^2. Moreover, $b = a^2$. Looking at (10.4.13), one sees that $a = 1$ is impossible. The possible values are

$a = \omega$ or ω^2, and then $c = 1$. The same reasoning applies to the character χ_3. Since χ_2 and χ_3 are distinct, and since we can interchange them, we may assume that $a = \omega$ and $a' = \omega^2$. It is natural to guess that the irreducible three-dimensional character χ_4 might be the character of the standard representation of T by rotations, and it is easy to verify this by computing that character and checking that $\langle \chi, \chi \rangle = 1$. Since we know the other characters, χ_4 is also determined by the fact that the characters are orthonormal. The character table is

	(1)	(4)	(4)	(3)
	1	x	x^2	z
χ_1	1	1	1	1
χ_2	1	ω	ω^2	1
χ_3	1	ω^2	ω	1
χ_4	3	0	0	-1

(10.4.14)   Character table of the tetrahedral group

The columns in these tables are orthogonal. This is a general phenomenon, whose proof we leave as Exercise 4.6.

10.5   ONE-DIMENSIONAL CHARACTERS

A one-dimensional character is the character of a representation of G on a one-dimensional vector space. If ρ is a one-dimensional representation, then ρ_g is represented by a 1×1 matrix R_g, and $\chi(g)$ is the unique entry in that matrix. Speaking loosely,

(10.5.1) $$\chi(g) = \rho_g = R_g.$$

A one-dimensional character χ is a homomorphism from G to $GL_1 = \mathbb{C}^\times$, because

$$\chi(gh) = \rho_{gh} = \rho_g \rho_h = \chi(g)\chi(h).$$

If χ is one-dimensional and if g is an element of G of order k, then $\chi(g)$ is a power of the primitive root of unity $\zeta = e^{2\pi i/k}$. And since $\mathbb{C}^\times$ is abelian, any commutator is in the kernel of such a character.

Normal subgroups are among the many things that can be determined by looking at a character table. The kernel of a one-dimensional character χ is the union of the conjugacy classes $C(g)$ such that $\chi(g) = 1$. For instance, the kernel of the character χ_2 in the character table of the tetrahedral group T is the union of the two conjugacy classes $C(1) \cup C(y)$. It is a normal subgroup of order four that we have seen before.

Warning: A character of dimension greater than 1 is *not* a homomorphism. The values taken on by such a character are *sums* of roots of unity.

Theorem 10.5.2 Let G be a finite abelian group.

(a) Every irreducible character of G is one-dimensional. The number of irreducible characters is equal to the order of the group.

(b) Every matrix representation R of G is diagonalizable: There is an invertible matrix P such that $P^{-1}R_g P$ is diagonal for all g.

Proof. In an abelian group of order N, there will be N conjugacy classes, each containing a single element. Then according to the main theorem, the number of irreducible representations is also equal N. The formula $N = d_1^2 + \cdots + d_N^2$ shows that $d_i = 1$ for all i. □

A simple example: The cyclic group $C_3 = \{1, x, x^2\}$ of order 3 has three irreducible characters of dimension 1. If χ is a one of them, then $\chi(x)$ will be a power of $\omega = e^{2\pi i/3}$, and $\chi(x^2) = \chi(x)^2$. Since there are three distinct powers of ω and three irreducible characters, $\chi_i(x)$ must take on all three values. The character table of C_3 is therefore

	(1)	(1)	(1)
	1	x	x^2
χ_1	1	1	1
χ_2	1	ω	ω^2
χ_3	1	ω^2	ω

(10.5.3)   Character table of the cyclic group C_3

10.6   THE REGULAR REPRESENTATION

Let $S = (s_1, \ldots, s_n)$ be a finite ordered set on which a group G operates, and let R_g denote the permutation matrix that describes the operation of a group element g on S. If g operates on S as the permutation p, i.e., if $gs_i = s_{pi}$, that matrix is (see (1.5.7))

$$(10.6.1) \qquad R_g = \sum_i e_{pi,i},$$

and $R_g e_i = e_{pi}$. The map $g \rightsquigarrow R_g$ defines a matrix representation R of G that we call a *permutation representation*, though that phrase had a different meaning in Section 6.11. The representation (10.2.11) of S_3 is an example of a permutation representation.

The ordering of S is used only so that we can assemble R_g into a matrix. It is nicer to describe a permutation representation without reference to an ordering. To do this we introduce a vector space V_S that has the unordered basis $\{e_s\}$ indexed by elements of S. Elements of V_S are linear combinations $\sum_g c_g e_g$, with complex coefficients c_g. If we are given an operation of G on the set S, the associated *permutation representation* ρ of G on V_S is defined by

$$(10.6.2) \qquad \rho_g(e_s) = e_{gs}.$$

When we choose an ordering of S, the basis $\{e_s\}$ becomes an ordered basis, and the matrix of ρ_g has the form described above.

The character of a permutation representation is especially easy to compute:

Lemma 10.6.3   Let ρ be the permutation representation associated to an operation of a group G on a nonempty finite set S. For all g in G, $\chi(g)$ is equal to the number of elements of S that are fixed by g.

Proof. We order the set S arbitrarily. Then every element s that is fixed by g, there is a 1 on the diagonal of the matrix R_g (10.6.1), and for every element that is not fixed, there is a 0. □

When we decompose a set on which G operates into orbits, we will obtain a decomposition of the permutation representation ρ or R as a direct sum. This is easy to see. But there is an important new feature: The fact that linear combinations are available allows us to decompose the representation further. Even when the operation of G on S is transitive, ρ will not be irreducible unless S is a set of one element.

Lemma 10.6.4 Let R be the permutation representation associated to an operation of G on a finite nonempty ordered set S. When its character χ is written as an integer combination of the irreducible characters, the trivial character χ_1 appears.

Proof. The vector $\sum_g e_g$ of V_S, which corresponds to $(1, 1, \ldots, 1)^t$ in $\mathbb{C}^n$, is fixed by every permutation of S, so it spans a G-invariant subspace of dimension 1 on which the group operates trivially. $\square$

Example 10.6.5 Let G be the tetrahedral group T, and let S be the set $(v_1 \ldots, v_4)$ of vertices of the tetrahedron. The operation of G on S defines a four-dimensional representation of G. Let x denote the rotation by $2\pi/3$ about a face and z the rotation by π about an edge, as before (see 7.10.8). Then x acts as the 3-cycle $(\mathbf{234})$ and z acts as $(\mathbf{13})(\mathbf{24})$. The associated permutation representation is

$$(10.6.6) \qquad R_x = \begin{bmatrix} 1 & 0 & 0 & 0 \\ 0 & 0 & 0 & 1 \\ 0 & 1 & 0 & 0 \\ 0 & 0 & 1 & 0 \end{bmatrix}, \quad R_z = \begin{bmatrix} 0 & 0 & 1 & 0 \\ 0 & 0 & 0 & 1 \\ 1 & 0 & 0 & 0 \\ 0 & 1 & 0 & 0 \end{bmatrix}.$$

Its character is

$$(10.6.7) \qquad \begin{array}{c|cccc} & 1 & x & x^2 & z \\ \hline \chi^{vert} & 4 & 1 & 1 & 0 \end{array}.$$

The character table (10.4.14) shows that $\chi^{vert} = \chi_1 + \chi_4$. By the way, another way to determine the character χ_4 in the character table is to check that $\langle \chi^{vert}, \chi^{vert} \rangle = 2$. Then χ^{vert} is a sum of two irreducible characters. Lemma 10.6.4 shows that one of them is the trivial character χ_1. So $\chi^{vert} - \chi_1$ is an irreducible character. It must be χ_4. $\square$

• The *regular representation* ρ^{reg} of a group G is the representation associated to the operation of G on itself by left multiplication. It is a representation on the vector space V_G that has a basis $\{e_g\}$ indexed by elements of G. If h is an element of G, then

$$(10.6.8) \qquad \rho_g^{reg}(e_h) = e_{gh}.$$

This operation of G on itself by left multiplication isn't particularly interesting, but the associated permutation representation ρ^{reg} is very interesting. Its character χ^{reg} is simple:

$$(10.6.9) \qquad \chi^{reg}(1) = |G|, \quad \text{and} \quad \chi^{reg}(g) = 0, \text{ if } g \neq 1.$$

This is true because the dimension of χ^{reg} is the order of the group, and because multiplication by g doesn't fix any element of G unless $g = 1$.

This simple formula makes it easy to compute $\langle \chi^{reg}, \chi \rangle$ for any character χ:

(10.6.10) $\langle \chi^{reg}, \chi \rangle = \frac{1}{|G|} \sum_g \overline{\chi^{reg}(g)} \chi(g) = \frac{1}{|G|} \overline{\chi^{reg}(1)} \chi(1) = \chi(1) = \dim \chi.$

Corollary 10.6.11 Let $\chi_1, \ldots, \chi_r$ be the irreducible characters of a finite group G, let ρ_i be a representation with character χ_i, and let $d_i = \dim \chi_i$. Then $\chi^{reg} = d_1 \chi_1 + \cdots + d_r \chi_r$, and ρ^{reg} is isomorphic to $d_1 \rho_1 \oplus \cdots \oplus d_r \rho_r$.

This follows from (10.6.10) and the projection formula. Isn't it nice? Counting dimensions,

(10.6.12) $$|G| = \dim \chi^{reg} = \sum_{i=1}^r d_i \dim \chi_i = \sum_{i=1}^r d_i^2.$$

This is the formula in **(c)** of the Main Theorem. So that formula follows from the orthogonality relations (10.4.6)**(a)**.

For instance, the character of the regular representation of the symmetric group S_3 is

	1	x	y
χ^{reg}	6	0	0

Looking at the character table (10.4.12) for S_3, one sees that $\chi^{reg} = \chi_1 + \chi_2 + 2\chi_3$, as expected.

Still one more way to determine the last character χ_4 of the tetrahedral group (see (10.4.14) is to use the relation $\chi^{reg} = \chi_1 + \chi_2 + \chi_3 + 3\chi_4$.

We determine the character table of the icosahedral group I next. As we know, I is isomorphic to the alternating group A_5 (7.4.4). The conjugacy classes have been determined before (7.4.1). They are listed below, with representative elements taken from A_5:

	class	representative
	$C_1 = \{1\}$	(1)
	$C_2 = 15$ edge rotations, angle π	$(12)(34)$
(10.6.13)	$C_3 = 20$ vertex rotations, angles $\pm 2\pi/3$	(123)
	$C_4 = 12$ face rotations, angles $\pm 2\pi/5$	(12345)
	$C_5 = 12$ face rotations, angles $\pm 4\pi/5$	(13524)

Since there are five conjugacy classes, there are five irreducible characters. The character table is

	(1)	(15)	(20)	(12)	(12)	
	0	π	$2\pi/3$	$2\pi/5$	$4\pi/5$	angle
χ_1	1	1	1	1	1	
χ_2	3	-1	0	α	β	
χ_3	3	-1	0	β	α	
χ_4	4	0	1	-1	-1	
χ_5	5	1	-1	0	0	

(10.6.14)   Character table of the icosahedral group I

The entries α and β are explained below. One way to find the irreducible characters is to decompose some permutation representations. The alternating group A_5 operates on the set of five indices. This gives us a five-dimensional permutation representation; we'll call it ρ'. Its character χ' is

	0	π	$2\pi/3$	$2\pi/5$	$4\pi/5$	angle
χ'	5	1	2	0	0	

Then $\langle \chi', \chi' \rangle = \frac{1}{60}\left(1 \cdot 5^2 + 15 \cdot 1^2 + 20 \cdot 2^2\right) = 2$. Therefore χ' is the sum of two distinct irreducible characters. Since the trivial representation is a summand, $\chi' - \chi_1$ is an irreducible character, the one labeled χ_4 in the table.

Next, the icosahedral group I operates on the set of six pairs of opposite faces of the dodecahedron; let the corresponding six-dimensional character be χ''. A similar computation shows that $\chi'' - \chi_1$ is the irreducible character χ_5.

We also have the representation of dimension 3 of I as a rotation group. Its character is χ_2. To compute that character, we remember that the trace of a rotation of $\mathbb{R}^3$ with angle θ is $1 + 2\cos\theta$, which is also equal to $1 + e^{i\theta} + e^{-i\theta}$ (5.1.28). The second and third entries for χ_2 are $1 + 2\cos\pi = -1$ and $1 + 2\cos 2\pi/3 = 0$. The last two entries are labeled

$$\alpha = 1 + 2\cos(2\pi/5) = 1 + \zeta + \zeta^4 \quad \text{and} \quad \beta = 1 + 2\cos(4\pi/5) = 1 + \zeta^2 + \zeta^3,$$

where $\zeta = e^{2\pi i/5}$. The remaining character χ_3 can be determined by orthogonality, or by using the relation

$$\chi^{reg} = \chi_1 + 3\chi_2 + 3\chi_3 + 4\chi_4 + 5\chi_5.$$

10.7 SCHUR'S LEMMA

Let ρ and ρ' be representations of a group G on vector spaces V and V'. A linear transformation $T: V' \to V$ is called G-*invariant* if it is compatible with the operation of G, meaning that for all g in G,

(10.7.1) $$T(gv') = gT(v'), \quad \text{or} \quad T \circ \rho'_g = \rho_g \circ T,$$

as indicated by the diagram

(10.7.2)
$$
\begin{array}{ccc}
V' & \xrightarrow{\ T\ } & V \\
{\scriptstyle \rho'_g}\downarrow & & \downarrow{\scriptstyle \rho_g} \\
V' & \xrightarrow{\ T\ } & V
\end{array}
$$

A bijective G-invariant linear transformation is an *isomorphism* of representations (10.1.16).
    It is useful to rewrite the condition for G-invariance in the form

$$T(v') = g^{-1}T(gv'), \quad \text{or} \quad \rho_g^{-1}T\rho'_g = T.$$

This definition of a G-invariant linear transformation T makes sense only when the representations ρ and ρ' are given. It is important to keep this in mind when the ambiguous group operation notation $T(gv') = gT(v')$ is used.

If bases **B** and **B'** for V and V' are given, and if R_g, R'_g, and M denote the matrices of ρ_g, ρ'_g, and T with respect to these bases, the condition (10.7.1) becomes

$$(10.7.3) \qquad\qquad\qquad MR'_g = R_g M \quad \text{or} \quad R_g^{-1} M R'_g = M$$

for all g in G. A matrix M is called G-*invariant* if it satisfies this condition.

Lemma 10.7.4 The kernel and the image of a G-invariant linear transformation $T : V' \to V$ are G-invariant subspaces of V' and V, respectively.

Proof. The kernel and image of any linear transformation are subspaces. To show that the kernel is G-invariant, we must show that if x is in $\ker T$, then gx is in $\ker T$, i.e., that if $T(x) = 0$, then $T(gx) = 0$. This is true: $T(gx) = gT(x) = g0 = 0$. If y is in the image of T, i.e., $y = T(x)$ for some x in V', then $gy = gT(x) = T(gx)$, so gy is in the image too. $\qquad\square$

Similarly, if ρ is a representation of G on V, a linear operator on V is G-*invariant* if

$$(10.7.5) \qquad\quad T(gv) = gT(v), \quad \text{or} \quad \rho_g \circ T = T \circ \rho_g, \quad \text{for all } g \text{ in } G,$$

which means that T commutes with each of the operators ρ_g. The matrix form of this condition is

$$R_g M = M R_g \quad \text{or} \quad M = R_g^{-1} M R_g, \quad \text{for all } g \text{ in } G.$$

Because a G-invariant linear operator T must commute with all of the operators ρ_g, invariance is a strong condition. Schur's Lemma shows this.

Theorem 10.7.6 Schur's Lemma.

(a) Let ρ and ρ' be irreducible representations of G on vector spaces V and V', respectively, and let $T : V' \to V$ be a G-invariant transformation. Either T is an isomorphism, or else $T = 0$.

(b) Let ρ be an irreducible representation of G on a vector space V, and let $T : V \to V$ be a G-invariant linear operator. Then T is multiplication by a scalar: $T = cI$.

Proof. **(a)** Suppose that T is not the zero map. Since ρ' is irreducible and since $\ker T$ is a G-invariant subspace, $\ker T$ is either V' or $\{0\}$. It is not V' because $T \neq 0$. Therefore $\ker T = \{0\}$, and T is injective. Since ρ is irreducible and $\mathrm{im}\, T$ is G-invariant, $\mathrm{im}\, T$ is either $\{0\}$ or V. It is not $\{0\}$ because $T \neq 0$. Therefore $\mathrm{im}\, T = V$ and T is surjective.

(b) Suppose that T is a G-invariant linear operator on V. We choose an eigenvalue λ of T. The linear operator $S = T - \lambda I$ is also G-invariant. The kernel of S isn't zero because it contains an eigenvector of T. Therefore S is not an isomorphism. By **(a)**, $S = 0$ and $T = \lambda I$. $\qquad\square$

Suppose that we are given representations ρ and ρ' on spaces V and V'. Though G-invariant linear tranformations are rare, the averaging process can be used to create a

G-invariant transformation from any linear transformation $T: V' \to V$. The average is the linear transformation $\tilde{T}$ defined by

(10.7.7) $$\tilde{T}(v') = \tfrac{1}{|G|} \sum_{g \in G} g^{-1}(T(gv')), \quad \text{or} \quad \tilde{T} = \tfrac{1}{|G|} \sum_{g \in G} \rho_g^{-1} T \rho_g'.$$

Similarly, if we are given matrix representations R and R', of G of dimensions n and m, and if M is any $m \times n$ matrix, then the averaged matrix is

(10.7.8) $$\tilde{M} = \tfrac{1}{|G|} \sum_{g \in G} R_g^{-1} M R_g'.$$

Lemma 10.7.9 With the above notation, $\tilde{T}$ is a G-invariant linear transformation, and $\tilde{M}$ is a G-invariant matrix. If T is G-invariant, then $\tilde{T} = T$, and if M is G-invariant, then $\tilde{M} = M$.

Proof. Since compositions and sums of linear transformations are linear, $\tilde{T}$ is a linear transformation, and it is easy to see that $\tilde{T} = T$ if T is invariant. To show that $\tilde{T}$ is invariant, we let h be an element of G and we show that $\tilde{T} = h^{-1}\tilde{T}h$. We make the substitution $g_1 = gh$. Reindexing as in (10.2.3),

$$h^{-1}\tilde{T}h = h^{-1}\left(\tfrac{1}{|G|} \sum_g g^{-1}Tg\right) = \tfrac{1}{|G|} \sum_g (gh)^{-1}T(gh)$$

$$= \tfrac{1}{|G|} \sum_g g_1^{-1}Tg_1 = \tfrac{1}{|G|} \sum_g g^{-1}Tg = \tilde{T}.$$

The proof that $\tilde{M}$ is invariant is analogous.   □

The averaging process may yield $\tilde{T} = 0$, the trivial transformation, though T was not zero. Schur's Lemma tells us that this *must* happen if ρ and ρ' are irreducible and not isomorphic. This fact is the basis of the proof given in the next section that distinct irreducible characters are orthogonal. For linear operators, the average is often not zero, because trace is preserved by the averaging process.

Proposition 10.7.10 Let ρ be an irreducible representation of G on a vector space V. Let $T: V \to V$ be a linear operator, and let $\tilde{T}$ be as in (10.7.7), with $\rho' = \rho$. Then trace $\tilde{T} =$ trace T. If trace $T \neq 0$, then $\tilde{T} \neq 0$.   □

10.8 PROOF OF THE ORTHOGONALITY RELATIONS

We will now prove **(a)** of the Main Theorem. We use matrix notation. Let $\mathcal{M}$ denote the space $\mathbb{C}^{m \times n}$ of $m \times n$ matrices.

Lemma 10.8.1 Let A and B be $m \times m$ and $n \times n$ matrices respectively, and let F be the linear operator on $\mathcal{M}$ defined by $F(M) = AMB$. The trace of F is the product (trace A)(trace B).

Proof. The trace of an operator is the sum of its eigenvalues. Let $\alpha_1, \ldots \alpha_m$ and $\beta_1, \ldots, \beta_n$ be the eigenvalues of A and B^t respectively. If X_i is an eigenvector of A with eigenvalue α_i, and Y_j is an eigenvector of B^t with eigenvalue β_j, the $m \times n$ matrix $M = X_i Y_j^t$ is an eigenvector for the operator F, with eigenvalue $\alpha_i \beta_j$. Since the dimension of $\mathcal{M}$ is mn, the mn complex numbers $\alpha_i \beta_j$ are all of the eigenvalues, provided that they are distinct. If so, then

$$\text{trace } F = \sum_{i,j} \alpha_i \beta_j = \left(\sum_i \alpha_i \right) \left(\sum_j \beta_j \right) = (\text{trace } A)(\text{trace } B).$$

In general, there will be matrices A' and B' arbitrarily close to A and B such that the products of their eigenvalues are distinct, and the lemma follows by continuity (see Section 5.2). $\square$

Let ρ' and ρ be representations of dimensions m and n, with characters χ' and χ respectively, and let R' and R be the matrix representations obtained from ρ' and ρ using some arbitrary bases. We define a linear operator Φ on the space $\mathcal{M}$ by

$$(10.8.2) \qquad\qquad \Phi(M) = \tfrac{1}{|G|} \sum_g R_g^{-1} M R'_g = \tilde{M}.$$

In the last section, we saw that $\tilde{M}$ is a G-invariant matrix, and that $\tilde{M} = M$ if M is invariant. Therefore the image of Φ is the space of G-invariant matrices. We denote that space by $\tilde{\mathcal{M}}$.

Parts **(a)** and **(b)** of the next lemma compute the trace of the operator Φ in two ways. The orthogonality relations are part **(c)**.

Lemma 10.8.3 With the above notation,

(a) trace $\Phi = \langle \chi, \chi' \rangle$.

(b) trace $\Phi = \dim \tilde{\mathcal{M}}$.

(c) If ρ is an irreducible representation, $\langle \chi, \chi \rangle = 1$, and if ρ and ρ' are non-isomorphic irreducible representations, $\langle \chi, \chi' \rangle = 0$.

Proof. **(a)** We recall that $\chi(g^{-1}) = \overline{\chi(g)}$ (10.4.2)**(d)**. Let F_g denote the linear operator on $\mathcal{M}$ defined by $F_g(M) = R_g^{-1} M R'_g$.

Since trace is linear, Lemma 10.8.1 shows that

$$(10.8.4) \qquad \begin{aligned} \text{trace } \Phi &= \tfrac{1}{|G|} \sum_g \text{trace } F_g = \tfrac{1}{|G|} \sum_g (\text{trace } R_g^{-1})(\text{trace } R'_g) \\ &= \tfrac{1}{|G|} \sum_g \chi(g^{-1}) \chi'(g) = \tfrac{1}{|G|} \sum_g \overline{\chi(g)} \chi'(g) = \langle \chi, \chi' \rangle. \end{aligned}$$

(b) Let $\mathcal{N}$ be the kernel of Φ. If M is in the intersection $\tilde{\mathcal{M}} \cap \mathcal{N}$, then $\Phi(M) = M$ and also $\Phi(M) = 0$, so $M = 0$. The intersection is the zero space. Therefore $\mathcal{M}$ is the direct sum $\tilde{\mathcal{M}} \oplus \mathcal{N}$ (4.3.1)**(b)**. We choose a basis for $\mathcal{M}$ by appending bases of $\tilde{\mathcal{M}}$ and $\mathcal{N}$. Since $\tilde{M} = M$ if M is invariant, Φ is the identity on $\tilde{\mathcal{M}}$. So the matrix of Φ will have the block form

$$\begin{bmatrix} I & \\ & 0 \end{bmatrix},$$

where I is the identity matrix of size $\dim \tilde{\mathcal{M}}$. Its trace is equal to the dimension of $\tilde{\mathcal{M}}$.

(c) We apply **(a)** and **(b)**: $\langle \chi, \chi' \rangle = \dim \widetilde{\mathcal{M}}$. If ρ' and ρ are irreducible and not isomorphic, Schur's Lemma tells us that the only G-invariant operator is zero, and so the only G-invariant matrix is the zero matrix. Therefore $\widetilde{\mathcal{M}} = \{0\}$ and $\langle \chi, \chi' \rangle = 0$. If $\rho' = \rho$, Schur's Lemma says that the G-invariant matrices have the form cI. Then $\widetilde{\mathcal{M}}$ has dimension 1, and $\langle \chi, \chi' \rangle = 1$.   □

We go over to operator notation for the proof of Theorem 10.4.6**(b)**, that the number of irreducible characters is equal to the number of conjugacy classes in the group. As before, $\mathcal{H}$ denotes the space of class functions. Its dimension is equal to the number of conjugacy classes (see (10.4.11)). Let $\mathcal{C}$ denote the subspace of $\mathcal{H}$ spanned by the characters. We show that $\mathcal{C} = \mathcal{H}$ by showing that the orthogonal space to $\mathcal{C}$ in $\mathcal{H}$ is zero. The next lemma does this.

Lemma 10.8.5

(a) Let φ be a class function on G that is orthogonal to every character. For any representation ρ of G, $\frac{1}{|G|} \sum_g \overline{\varphi(g)} \rho_g$ is the zero operator.

(b) Let ρ^{reg} be the regular representation of G. The operators ρ_g^{reg} with g in G are linearly independent.

(c) The only class function φ that is orthogonal to every character is the zero function.

Proof. **(a)** Since any representation is a direct sum of irreducible representations, we may assume that ρ is irreducible. Let $T = \frac{1}{|G|} \sum_g \overline{\varphi(g)} \rho_g$. We first show that T is a G-invariant operator, i.e., that $T = \rho_h^{-1} T \rho_h$ for every h in G. Let $g'' = h^{-1} g h$. Then as g runs over the group G, so does g''. Since ρ is a homomorphism, $\rho_h^{-1} \rho_g \rho_h = \rho_{g''}$, and because φ is a class function, $\varphi(g) = \varphi(g'')$. Therefore

$$\rho_h^{-1} T \rho_h = \frac{1}{|G|} \sum_g \overline{\varphi(g)} \rho_{g''} = \frac{1}{|G|} \sum_g \overline{\varphi(g'')} \rho_{g''} = \frac{1}{|G|} \sum_g \overline{\varphi(g)} \rho_g = T.$$

Let χ be the character of ρ. The trace of T is $\frac{1}{|G|} \sum_g \overline{\varphi(g)} \chi(g) = \langle \varphi, \chi \rangle$. The trace is zero because φ is orthogonal to χ. Since ρ is irreducible, Schur's lemma tells us that T is multiplication by a scalar, and since its trace is zero, $T = 0$.

(b) We apply Formula 10.6.8 to the basis element e_1 of V_G: $\rho_g^{reg}(e_1) = e_g$. Then since the vectors e_g are independent elements of V_G, the operators ρ_g^{reg} are independent too.

(c) Let φ be a class function orthogonal to every character. **(a)** tells us that $\sum_g \overline{\varphi(g)} \rho_g^{reg} = 0$ is a linear relation among the operators ρ_g^{reg}, which are independent by **(b)**. Therefore all of the coefficients $\overline{\varphi(g)}$ are zero, and φ is the zero function.   □

10.9   REPRESENTATIONS OF SU_2

Remarkably, the orthogonality relations carry over to *compact groups*, matrix groups that are compact subsets of spaces of matrices, when summation over the group is replaced by an integral. In this section, we verify this for some representations of the special unitary group SU_2.

We begin by defining the representations that we will analyze. Let H_n denote the complex vector space of homogeneous polynomials of degree n in the variables u, v, of the form

(10.9.1) $$f(u, v) = c_0 u^n + c_{n-1} u^{n-1} v + \cdots + c_{n-1} uv^{n-1} + c_n v^n.$$

We define a representation

(10.9.2) $$\rho_n : SU_2 \to GL(H_n)$$

as follows: The result of operating by an element P of SU_2 on a polynomial f in H_n will be another polynomial that we denote by $[Pf]$. The definition is

(10.9.3) $$[Pf](u, v) = f(ua + vb, -u\bar{b} + v\bar{a}), \quad \text{where} \quad P = \begin{bmatrix} a & -\bar{b} \\ b & \bar{a} \end{bmatrix}.$$

In words, P operates by substituting $(u, v)P$ for the variables (u, v). Thus

$$[Pu^i v^j] = (ua + vb)^i (-u\bar{b} + v\bar{a})^j.$$

It is easy to compute the matrix of this operator when P is diagonal. Let $\alpha = e^{i\theta}$, and let

(10.9.4) $$A_\theta = \begin{bmatrix} e^{i\theta} & \\ & e^{-i\theta} \end{bmatrix} = \begin{bmatrix} \alpha & \\ & \bar{\alpha} \end{bmatrix} = \begin{bmatrix} \alpha & \\ & \alpha^{-1} \end{bmatrix}.$$

Then $[A_\theta u^i v^j] = (u\alpha)^i (v\bar{\alpha})^j = u^i v^j \alpha^{i-j}$. So A_θ acts on the basis $(u^n, u^{n-1}v, \ldots, uv^{n-1}, v^n)$ of the space H_n as the diagonal matrix

$$\begin{bmatrix} \alpha^n & & & \\ & \alpha^{n-2} & & \\ & & \ddots & \\ & & & \alpha^{-n} \end{bmatrix}.$$

The character χ_n of the representation ρ_n is defined as before: $\chi_n(g) = \text{trace } \rho_{n,g}$. It is constant on the conjugacy classes, which are the latitudes on the sphere SU_2. Because of this, it is enough to compute the characters χ_n on one matrix in each latitude, and we use A_θ. To simplify notation, we write $\chi_n(\theta)$ for $\chi_n(A_\theta)$. The character is

$$\chi_0(\theta) = 1$$
$$\chi_1(\theta) = \alpha + \alpha^{-1}$$
$$\chi_2(\theta) = \alpha^2 + 1 + \alpha^{-2}$$
$$\cdots$$

(10.9.5) $$\chi_n(\theta) = \alpha^n + \alpha^{n-2} + \cdots + \alpha^{-n} = \frac{\alpha^{n+1} - \alpha^{-(n+1)}}{\alpha - \alpha^{-1}}.$$

The Hermitian product that replaces (10.4.3) is

$$(10.9.6) \qquad \langle \chi_m, \chi_n \rangle = \tfrac{1}{|G|} \int_G \overline{\chi_m(g)} \chi_n(g) \, dV.$$

In this formula G stands for the group SU_2, the unit 3-sphere, $|G|$ is the three-dimensional volume of the unit sphere, and dV stands for the integral with respect to three-dimensional volume. The characters happen to be real-valued functions, so the complex conjugation that appears in the formula is irrelevant.

Theorem 10.9.7 The characters of SU_2 that are defined above are orthonormal: $\langle \chi_m, \chi_n \rangle = 0$ if $m \neq n$, and $\langle \chi_n, \chi_n \rangle = 1$.

Proof. Since the characters are constant on the latitudes, we can evaluate the integral (10.9.6) by slicing, as we learn to do in calculus. We use the unit circle $x_0 = \cos\theta$, $x_1 = \sin\theta$, and $x_2 = \cdots = x_n = 0$ to parametrize the slices of the unit n-sphere $\mathbb{S}^n : \{x_0^2 + x_1^2 + \cdots + x_n^2 = 1\}$. So $\theta = 0$ is the north pole, and $\theta = \pi$ is the south pole (see Section 9.2). For $0 < \theta < \pi$, the slice of the unit n-sphere is an $(n-1)$-sphere of radius $\sin\theta$.

To compute an integral by slicing, we integrate with respect to arc length on the unit circle. Let $\mathrm{vol}_n(r)$ denote the n-dimensional volume of the n-sphere of radius r. So $\mathrm{vol}_1(r)$ is the arc length of the circle of radius r, and $\mathrm{vol}_2(r)$ is the surface area of the 2-sphere of radius r. If f is a function on the unit n-sphere $\mathbb{S}^n$ that is constant on the slices $\theta = c$, its integral will be

$$(10.9.8) \qquad \int_{\mathbb{S}^n} f \, dV_n = \int_0^\pi f(\theta) \, \mathrm{vol}_{n-1}(\sin\theta) \, d\theta,$$

where dV_n denotes integration with respect to n-dimensional volume, and $f(\theta)$ denotes the value of f on the slice.

Integration by slicing provides a recursive formula for the volumes of the spheres:

$$(10.9.9) \qquad \mathrm{vol}_n(1) = \int_{\mathbb{S}^n} 1 \, dV_n = \int_0^\pi \mathrm{vol}_{n-1}(\sin\theta) \, d\theta,$$

and $\mathrm{vol}_n(r) = r^n \, \mathrm{vol}_n(1)$. The zero-sphere $x_0^2 = r^2$ consists of two points. Its zero-dimensional volume is 2. So

$$
\begin{aligned}
\mathrm{vol}_1(r) &= r \int_0^\pi \mathrm{vol}_0(\sin\theta) d\theta = r \int_0^\pi 2 \, d\theta = 2\pi r, \\
(10.9.10) \qquad \mathrm{vol}_2(r) &= r^2 \int_0^\pi \mathrm{vol}_1(\sin\theta) d\theta = r^2 \int_0^\pi 2\pi \sin\theta \, d\theta = 4\pi r^2, \\
\mathrm{vol}_3(r) &= r^3 \int_0^\pi \mathrm{vol}_2(\sin\theta) d\theta = r^3 \int_0^\pi 4\pi \sin^2\theta \, d\theta = 2\pi^2 r^3.
\end{aligned}
$$

To evaluate the last integral, it is convenient to use the formula $\sin\theta = -i(\alpha - \alpha^{-1})/2$.

$$(10.9.11) \qquad \mathrm{vol}_2(\sin\theta) = 4\pi \sin^2\theta = -\pi(\alpha - \alpha^{-1})^2.$$

Expanding, $\text{vol}_2(\sin\theta) = \pi(2 - (\alpha + \alpha^{-1}))$. The integral of $\alpha^2 + \alpha^{-2}$ is zero:

(10.9.12)
$$\int_0^\pi (\alpha^k + \alpha^{-k})\,d\theta = \int_0^{2\pi} \alpha^k\,d\theta = \begin{cases} 0 & \text{if } k > 0 \\ 2\pi & \text{if } k = 0. \end{cases}$$

We now compute the integral (10.9.6). The volume of the group SU_2 is

(10.9.13)
$$\text{vol}_3(1) = 2\pi^2.$$

The latitude sphere that contains A_θ has radius $\sin\theta$. Since the characters are real, integration by slicing gives

(10.9.14)
$$\langle \chi_m, \chi_n \rangle = \frac{1}{2\pi^2} \int_0^\pi \chi_m(\theta)\chi_n(\theta)\,\text{vol}_2(\sin\theta)\,d\theta$$

$$= \frac{1}{2\pi^2} \int_0^\pi \left(\frac{\alpha^{m+1} - \alpha^{-(m+1)}}{\alpha - \alpha^{-1}} \right) \left(\frac{\alpha^{n+1} - \alpha^{-(n+1)}}{\alpha - \alpha^{-1}} \right) \left(-\pi(\alpha - \alpha^{-1})^2 \right) d\theta$$

$$= -\frac{1}{2\pi} \int_0^\pi \left(\alpha^{m+n+2} + \alpha^{-(m+n+2)} \right) d\theta + \frac{1}{2\pi} \int_0^\pi \left(\alpha^{m-n} + \alpha^{n-m} \right) d\theta$$

This evaluates to 1 if $m = n$ and to zero otherwise (see (10.9.12)). The characters χ_n are orthonormal.   □

We won't prove the next theorem, though the proof follows the case of finite groups fairly closely. If you are interested, see [Sepanski].

Theorem 10.9.15   Every continuous representation of SU_2 is isomorphic to a direct sum of the representations ρ_n (10.9.2).

We leave the obvious generalizations to the reader.

—Israel Herstein

EXERCISES

Section 1   Definitions

   1.1. Show that the image of a representation of dimension 1 of a finite group is a cyclic group.

   1.2. (a) Choose a suitable basis for $\mathbb{R}^3$ and write the standard representation of the octahedral group O explicitly.   **(b)** Do the same for the dihedral group D_n.

Section 2   Irreducible Representations

   2.1. Prove that the standard three-dimensional representation of the tetrahedral group T is irreducible as a complex representation.

2.2. Consider the standard two-dimensional representation of the dihedral group D_n. For which n is this an irreducible complex representation?

2.3. Suppose given a representation of the symmetric group S_3 on a vector space V. Let x and y denote the usual generators for S_3.

(a) Let u be a nonzero vector in V. Let $v = u + xu + x^2u$ and $w = u + yu$. By analyzing the G-orbits of v, w, show that V contains a nonzero invariant subspace of dimension at most 2.

(b) Prove that all irreducible two-dimensional representations of G are isomorphic, and determine all irreducible representations of G.

Section 3 Unitary Representations

3.1. Let G be a cyclic group of order 3. The matrix $A = \begin{bmatrix} -1 & -1 \\ 1 & 0 \end{bmatrix}$ has order 3, so it defines a matrix representation of G. Use the averaging process to produce a G-invariant form from the standard Hermitian product X^*Y on $\mathbb{C}^2$.

3.2. Let $\rho: G \to GL(V)$ be a representation of a finite group on a real vector space V. Prove the following:

(a) There exists a G-invariant, positive definite symmetric form $\langle\,,\,\rangle$ on V.

(b) ρ is a direct sum of irreducible representations.

(c) Every finite subgroup of $GL_n(\mathbb{R})$ is conjugate to a subgroup of O_n.

3.3. (a) Let $R: G \to SL_2(\mathbb{R})$ be a faithful representation of a finite group by real 2×2 matrices with determinant 1. Use the results of Exercise 3.2 to prove that G is a cyclic group.

(b) Determine the finite groups that have faithful real two-dimensional representations.

(c) Determine the finite groups that have faithful real three-dimensional representations with determinant 1.

3.4. Let $\langle\,,\,\rangle$ be a nondegenerate skew-symmetric form on a vector space V, and let ρ be a representation of a finite group G on V. Prove that the averaging process (10.3.7) produces a G-invariant skew-symmetric form on V, and show by example that the form obtained in this way needn't be nondegenerate.

3.5. Let x be a generator of a cyclic group G of order p. Sending $x \rightsquigarrow \begin{bmatrix} 1 & 1 \\ & 1 \end{bmatrix}$ defines a matrix representation $G \to GL_2(\mathbb{F}_p)$. Prove that this representation is not the direct sum of irreducible representations.

Section 4 Characters

4.1. Find the dimensions of the irreducible representations of the octahedral group, the quaternion group, and the dihedral groups D_4, D_5, and D_6.

4.2. A nonabelian group G has order 55. Determine its class equation and the dimensions of its irreducible characters.

4.3. Determine the character tables for

(a) the Klein four group,

(b) the quaternion group,

(c) the dihedral group D_4,

(d) the dihedral group D_6,

(e) a nonabelian group of order 21 (see Proposition 7.7.7).

4.4. Let G be the dihedral group D_5, presented with generators x, y and relations $x^5 = 1$, $y^2 = 1$, $yxy^{-1} = x^{-1}$, and let χ be an arbitrary two-dimensional character of G.

(a) What does the relation $x^5 = 1$ tell us about $\chi(x)$?

(b) What does the fact that x and x^{-1} are conjugate tell us about $\chi(x)$?

(c) Determine the character table of G.

(d) Decompose the restriction of each irreducible character of D_5 into irreducible characters of C_5.

4.5. Let $G = \langle x, y \,|\, x^5, y^4, yxy^{-1}x^{-2}\rangle$. Determine the character table of G.

4.6. Explain how to adjust the entries of a character table to produce a unitary matrix, and prove that the columns of a character table are orthogonal.

4.7. Let $\pi: G \to G' = G/N$ be the canonical map from a finite group to a quotient group, and let ρ' be an irreducible representation of G'. Prove that the representation $\rho = \rho' \circ \pi$ of G is irreducible in two ways: directly, and using Theorem 10.4.6.

4.8. Find the missing rows in the character table below:

	(1)	(3)	(6)	(6)	(8)
χ_1	1	1	1	1	1
χ_2	1	1	-1	-1	1
χ_3	3	-1	1	-1	0
χ_4	3	-1	-1	1	0

*4.9. Below is a partial character table. One conjugacy class is missing.

	(1)	(1)	(2)	(2)	(3)
	1	u	v	w	x
χ_1	1	1	1	1	1
χ_2	1	1	1	1	-1
χ_3	1	-1	1	-1	i
χ_4	1	-1	1	-1	$-i$
χ_5	2	2	-1	-1	0

(a) Complete the table.

(b) Determine the orders of representative elements in each conjugacy class.

(c) Determine the normal subgroups.

(d) Describe the group.

4.10. (a) Find the missing rows in the character table below.

(b) Determine the orders of the elements a, b, c, d.

(c) Show that the group G with this character table has a subgroup H of order 10, and describe this subgroup as a union of conjugacy classes.

(d) Decide whether H is C_{10} or D_5.

(e) Determine all normal subgroups of G.

	(1)	(4)	(5)	(5)	(5)
	1	a	b	c	d
χ_1	1	1	1	1	1
χ_2	1	1	-1	-1	1
χ_3	1	1	$-i$	i	-1
χ_4	1	1	i	$-i$	-1

*4.11. In the character table below, $\omega = e^{2\pi i/3}$.

	(1)	(6)	(7)	(7)	(7)	(7)	(7)
	1	a	b	c	d	e	f
χ_1	1	1	1	1	1	1	1
χ_2	1	1	1	ω	$\overline{\omega}$	ω	$\overline{\omega}$
χ_3	1	1	1	$\overline{\omega}$	ω	$\overline{\omega}$	ω
χ_4	1	1	-1	$-\omega$	$-\overline{\omega}$	ω	$\overline{\omega}$
χ_5	1	1	-1	$-\overline{\omega}$	$-\omega$	$\overline{\omega}$	ω
χ_6	1	1	-1	-1	-1	1	1
χ_7	6	-1	0	0	0	0	0

(a) Show that G has a normal subgroup N isomorphic to D_7.

(b) Decompose the restrictions of each character to N into irreducible N-characters.

(c) Determine the numbers of Sylow p-subgroups, for $p = 2, 3$, and 7.

(d) Determine the orders of the representative elements c, d, e, f.

(e) Determine all normal subgroups of G.

4.12. Let H be a subgroup of index 2 of a group G, and let $\sigma\colon H \to GL(V)$ be a representation. Let a be an element of G not in H. Define a *conjugate* representation $\sigma'\colon H \to GL(V)$ by the rule $\sigma'(h) = \sigma(a^{-1}ha)$. Prove that

(a) σ' is a representation of H.

(b) If σ is the restriction to H of a representation of G, then σ' is isomorphic to σ.

(c) If b is another element of G not in H, then the representation $\sigma''(h) = \sigma(b^{-1}hb)$ is isomorphic to σ'.

Section 5 One-Dimensional Characters

5.1. Decompose the standard two-dimensional representation of the cyclic group C_n by rotations into irreducible (complex) representations.

5.2. Prove that the sign representation $p \rightsquigarrow$ sign p and the trivial representation are the only one-dimensional representations of the symmetric group S_n.

5.3. Suppose that a group G has exactly two irreducible characters of dimension 1, and let χ denote the nontrivial one-dimensional character. Prove that for all g in G, $\chi(g) = \pm 1$.

5.4. Let χ be the character of a representation ρ of dimension d. Prove that $|\chi(g)| \leq d$ for all g in G, and that if $|\chi(g)| = d$, then $\rho(g) = \zeta I$, for some root of unity ζ. Moreover, if $\chi(g) = d$, then ρ_g is the identity operator.

5.5. Prove that the one-dimensional characters of a group G form a group under multiplication of functions. This group is called the *character group* of G, and is often denoted by $\hat{G}$. Prove that if G is abelian, then $|\hat{G}| = |G|$ and $\hat{G} \approx G$.

5.6. Let G be a cyclic group of order n, generated by an element x, and let $\zeta = e^{2\pi i/n}$.

   (a) Prove that the irreducible representations are $\rho_0, \ldots, \rho_{n-1}$, where $\rho_k : G \rightsquigarrow \mathbb{C}^\times$ is defined by $\rho_k(x) = \zeta^k$.

   (b) Identify the character group of G (see Exercise 5.5).

5.7. (a) Let $\varphi : G \to G'$ be a homomorphism of abelian groups. Define an induced homomorphism $\hat{\varphi} : \hat{G}' \to \hat{G}$ between their character groups (see Exercise 5.5).

   (b) Prove that if φ is injective, then $\hat{\varphi}$ is surjective, and conversely.

Section 6   The Regular Representation

6.1. Let R^{reg} denote the regular matrix representation of a group G. Determine $\sum_g R_g^{reg}$.

6.2. Let ρ be the permutation representation associated to the operation of D_3 on itself by conjugation. Decompose the character of ρ into irreducible characters.

6.3. Let χ^e denote the character of the representation of the tetrahedral group T on the six edges of the tetrahedron. Decompose this character into irreducible characters.

6.4. (a) Identify the five conjugacy classes in the octahedral group O, and find the orders of its irreducible representations.

   (b) The group O operates on these sets:

   - six faces of the cube,
   - three pairs of opposite faces,
   - eight vertices,
   - four pairs of opposite vertices,
   - six pairs of opposite edges,
   - two inscribed tetrahedra.

   Decompose the corresponding characters into irreducible characters.

   (c) Compute the character table for O.

6.5. The symmetric group S_n operates on $\mathbb{C}^n$ by permuting the coordinates. Decompose this representation explicitly into irreducible representations.

   Hint: I recommend against using the orthogonality relations. This problem is closely related to Exercise M.1 from Chapter 4.

6.6. Decompose the characters of the representations of the icosahedral group on the sets of faces, edges, and vertices into irreducible characters.

6.7. The group S_5 operates by conjugation on its normal subgroup A_5. How does this action operate on the isomorphism classes of irreducible representations of A_5?

6.8. The stabilizer in the icosahedral group of one of the cubes inscribed in a dodecahedron is the tetrahedral group T. Decompose the restrictions to T of the irreducible characters of I.

6.9. (a) Explain how one can prove that a group is simple by looking at its character table.

(b) Use the character table of the icosahedral group to prove that it is a simple group.

6.10. Determine the character tables for the nonabelian groups of order 12 (see (7.8.1)).

6.11. The character table for the group $G = PSL_2(\mathbb{F}_7)$ is below, with $\gamma = \frac{1}{2}(-1 + \sqrt{7}i)$, $\gamma' = \frac{1}{2}(-1 - \sqrt{7}i)$.

	(1)	(21)	(24)	(24)	(42)	(56)
	1	a	b	c	d	e
χ_1	1	1	1	1	1	1
χ_2	3	-1	γ	γ'	1	0
χ_3	3	-1	γ'	γ	1	0
χ_4	6	2	-1	-1	0	0
χ_5	7	-1	0	0	-1	1
χ_6	8	0	1	1	0	-1

(a) Use it to give two proofs that this group is simple.

(b) Identify, so far as possible, columns that corresponds to the conjugacy classes of the elements

$$\begin{bmatrix} 1 & 1 \\ & 1 \end{bmatrix}, \quad \begin{bmatrix} 2 & \\ & 4 \end{bmatrix},$$

and find matrices that represent the remaining conjugacy classes.

(c) G operates on the set of eight one-dimensional subspaces of $\mathbb{F}_7^2$. Decompose the associated character into irreducible characters.

Section 7   Schur's Lemma

7.1. Prove a converse to Schur's Lemma: If ρ is a representation, and if the only G-invariant linear operators on V are multiplications by scalars, then ρ is irreducible.

7.2. Let A be the standard representation (10.1.3) of the symmetric group S_3, and let $B = \begin{bmatrix} 1 & 1 \\ & 1 \end{bmatrix}$. Use the averaging process to produce a G-invariant linear operator from left multiplication by B.

7.3. The matrices $R_x = \begin{bmatrix} 1 & 1 & -1 \\ & 1 & 1 \\ 1 & & -1 \end{bmatrix}$, $R_y = \begin{bmatrix} -1 & & -1 \\ -1 & & 1 \\ & & -1 \end{bmatrix}$ define a representation R of the group S_3. Let φ be the linear transformation $\mathbb{C}^1 \rightarrow \mathbb{C}^3$ whose matrix is $(1, 0, 0)^t$. Use the averaging method to produce a G-invariant linear transformation from φ, using the sign representation Σ of (10.1.4) on $\mathbb{C}^1$ and the representation R on $\mathbb{C}^3$.

7.4. Let ρ be a representation of G and let C be a conjugacy class in G. Show that the linear operator $T = \sum_{g \in C} \rho_g$ is G-invariant.

7.5. Let ρ be a representation of a group G on V, and let χ be a character of G, not necessarily the character of ρ. Prove that the linear operator $T = \sum_g \chi(g)\rho_g$ on V is G-invariant.

7.6. Compute the matrix of the operator F of Lemma 10.8.1, and use the matrix to verify the formula for its trace.

Section 8   Representations of SU_2

8.1. Calculate the four-dimensional volume of the 4-ball $\mathbb{B}^4$ of radius r in $\mathbb{R}^4$, the locus $x_0^2 + \cdots + x_3^2 \le r^2$, by slicing with three-dimensional slices. Check your answer by differentiating.

8.2. Verify the associative law $[Q[Pf]] = [(QP)f]$ for the operation (10.9.3).

8.3. Prove that the orthogonal representation (9.4.1) $SU_2 \to SO_3$ is irreducible.

8.4. Left multiplication defines a representation of SU_2 on the space $\mathbb{R}^4$ with coordinates $x_0, \ldots, x_3$, as in Section 9.3. Decompose the associated complex representation into irreducible representations.

8.5. Use Theorem 10.9.14 to determine the irreducible representations of the rotation group SO_3.

8.6. *(representations of the circle group)* All representations here are assumed to be differentiable functions of θ. Let G be the circle group $\{e^{i\theta}\}$.

  (a) Let ρ be a representation of G on a vector space V. Show that there exists a positive definite G-invariant Hermitian form on V.

  (b) Prove Maschke's Theorem for G.

  (c) Describe the representations of G in terms of one-parameter groups, and use that description to prove that the irreducible representations are one-dimensional.

  (d) Verify the orthogonality relations, using an analogue of the Hermitian product (10.9.6).

8.7. Using the results of Exercise 8.6, determine the irreducible representations of the orthogonal group O_2.

Miscellaneous Problems

M.1. The representations in this problem are *real*. A molecule M in 'Flatland' (a two-dimensional world) consists of three like atoms a_1, a_2, a_3 forming a triangle. The triangle is equilateral at time t_0, its center is at the origin, and a_1 is on the positive x-axis. The group G of symmetries of M at time t_0 is the dihedral group D_3. We list the velocities of the individual atoms at t_0 and call the resulting six-dimensional vector $\mathbf{v} = (v_1, v_2, v_3)^t$ the *state* of M. The operation of G on the space V of state vectors defines a six-dimensional matrix representation S. For example, the rotation ρ by $2\pi/3$ about the origin permutes the atoms cyclically, and at the same time it rotates them.

  (a) Let r be the reflection about the x-axis. Determine the matrices S_ρ and S_r.

  (b) Determine the space W of vectors fixed by S_ρ, and show that W is G-invariant.

  (c) Decompose W and V explicitly into direct sums of irreducible G-invariant subspaces.

  (d) Explain the subspaces found in (c) in terms of motions and vibrations of the molecule.

M.2. What can be said about a group that has exactly three irreducible characters, of dimensions 1, 2, and 3, respectively?

M.3. Let ρ be a representation of a group G. In each of the following cases, decide whether or not ρ' is a representation, and whether or not it is necessarily isomorphic to ρ.

    (a) x is a fixed element of G, and $\rho'_g = \rho_{xgx^{-1}}$

    (b) φ is an automorphism of G, and $\rho'_g = \rho_{\varphi(g)}$.

    (c) σ is a one-dimensional representation of G, and $\rho'_g = \sigma_g \rho_g$.

M.4. Prove that an element z of a group G is in the center of G if and only if for all irreducible representations ρ, $\rho(z)$ is multiplication by a scalar.

M.5. Let A, B be commuting matrices such that some positive power of each matrix is the identity. Prove that there is an invertible matrix P such that PAP^{-1} and PBP^{-1} are both diagonal.

M.6. Let ρ be an irreducible representation of a finite group G. How unique is the positive definite G-invariant Hermitian form?

M.7. Describe the commutator subgroup of a group G in terms of the character table.

M.8. Prove that a finite simple group that is not of prime order has no nontrivial representation of dimension 2.

*M.9. Let H be a subgroup of index 2 of a finite group G. Let a be an element of G that is not in H, so that H and aH are the two cosets of H.

    (a) Given a matrix representation $S : H \rightarrow GL_n$ of the subgroup H, the *induced representation* $ind\, S : G \rightarrow GL_{2n}$ of the group G is defined by

$$(ind\,S)_h = \begin{bmatrix} S_h & 0 \\ 0 & S_{a^{-1}ha} \end{bmatrix}, \quad (ind\,S)_g = \begin{bmatrix} 0 & S_{ga} \\ S_{a^{-1}g} & 0 \end{bmatrix}$$

    for h in H and g in aH. Prove that $ind\, S$ is a representation of G, and describe its character.

    Note: The element $a^{-1}ha$ will be in H, but because a is not in H, it needn't be a conjugate of h in H.

    (b) If $R: G \rightarrow GL_n$ is a matrix representation of G, we may restrict it to H. We denote the restriction by $res\, R: H \rightarrow GL_n$. Prove that $res(ind\, S) \approx S \oplus S'$, where S' is the *conjugate representation* defined by $S'_h = S_{a^{-1}ha}$.

    (c) Prove *Frobenius reciprocity:* $\langle \chi_{ind\,S}, \chi_R \rangle = \langle \chi_S, \chi_{res\,R} \rangle$.

    (d) Let S be an irreducible representation of H. Use Frobenius reciprocity to prove that if S not isomorphic to the conjugate representation S', then the induced representation $ind\, S$ is irreducible, and on the other hand, if S and S' are isomorphic, then $ind\, S$ is a sum of two non-isomorphic representations of G.

*M.10. Let H be a subgroup of index 2 of a group G, and let R be a matrix representation of G. Let R' denote the representation defined by $R'_g = R_g$ if $g \in H$, and $R'_g = -R_g$ otherwise.

    (a) Show that R' is isomorphic to R if and only if the character of R is identically zero on the coset gH not equal to H.

    (b) Use Frobenius reciprocity (Exercise M.9) to show that $ind(res\, R) \approx R \oplus R'$.

   (c) Suppose that R is irreducible. Show that if R is not isomorphic to R', then $res\, R$ is irreducible, and if these two representations are isomorphic, then $res\, R$ is a sum of two irreducible representations of H.

*__M.11.__ Derive the character table of S_n using induced representations from A_n, when
   (a) $n = 3$,   **(b)** $n = 4$,   **(c)** $n = 5$.

*__M.12.__ Derive the character table of the dihedral group D_n, using induced representations from C_n.

__M.13.__ Let G be a finite subgroup of $GL_n(\mathbb{C})$. Prove that if $\sum_g \text{trace}\, g = 0$, then $\sum_g g = 0$.

__M.14.__ Let $\rho: G \rightarrow GL(V)$ be a two-dimensional representation of a finite group G, and assume that 1 is an eigenvalue of ρ_g for every g in G. Prove that ρ is a sum of two one-dimensional representations.

__M.15.__ Let $\rho: G \rightarrow GL_n(\mathbb{C})$ be an irreducible representation of a finite group G. Given a representation $\sigma: GL_n \rightarrow GL(V)$ of GL_n, we can consider the composition $\sigma \circ \rho$ as a representation of G.

   (a) Determine the character of the representation obtained in this way when σ is left multiplication of GL_n on the space V of $n \times n$ matrices. Decompose $\sigma \circ \rho$ into irreducible representations in this case.

   (b) Determine the character of $\sigma \circ \rho$ when σ is the operation of conjugation on $\mathbb{C}^{n \times n}$.

CHAPTER 11

Rings

11.1 DEFINITION OF A RING

Rings are algebraic structures closed under addition, subtraction, and multiplication, but not under division. The integers form our basic model for this concept.

Before going to the definition of a ring, we look at a few examples, subrings of the complex numbers. A *subring* of $\mathbb{C}$ is a subset which is closed under addition, subtraction and multiplication, and which contains 1.

• The *Gauss integers* , the complex numbers of the form $a + bi$, where a and b are integers, form a subring of $\mathbb{C}$ that we denote by $\mathbb{Z}[i]$:

$$(11.1.1) \qquad \mathbb{Z}[i] = \{a + bi \mid a, b \in \mathbb{Z}\}.$$

Its elements are the points of a square lattice in the complex plane.

We can form a subring $\mathbb{Z}[\alpha]$ analogous to the ring of Gauss integers, starting with any complex number α: the subring *generated by* α. This is the smallest subring of $\mathbb{C}$ that contains α, and it can be described in a general way. If a ring contains α, then it contains all positive powers of α because it is closed under multiplication. It also contains sums and differences of such powers, and it contains 1. Therefore it contains every complex number β that can be expressed as an integer combination of powers of α, or, saying this another way, can be obtained by evaluating a polynomial with integer coefficients at α:

$$(11.1.2) \qquad \beta = a_n\alpha^n + \cdots + a_1\alpha + a_0, \quad \text{where} \quad a_i \text{ are in } \mathbb{Z}.$$

On the other hand, the set of all such numbers is closed under the operations $+$, $-$, and $\times$, and it contains 1. So it is the subring generated by α.

In most cases, $\mathbb{Z}[\alpha]$ will not be represented as a lattice in the complex plane. For example, the ring $\mathbb{Z}\left[\frac{1}{2}\right]$ consists of the rational numbers that can be expressed as a polynomial in $\frac{1}{2}$ with integer coefficients. These rational numbers can be described simply as those whose denominators are powers of 2. They form a dense subset of the real line.

323

• A complex number α is *algebraic* if it is a root of a (nonzero) polynomial with integer coefficients – that is, if some expression of the form (11.1.2) evaluates to zero. If there is no polynomial with integer coefficients having α as a root, α is *transcendental*. The numbers e and π are transcendental, though it isn't very easy to prove this.

When α is transcendental, two distinct polynomial expressions (11.1.2) represent distinct complex numbers. Then the elements of the ring $\mathbb{Z}[\alpha]$ correspond bijectively to polynomials $p(x)$ with integer coefficients, by the rule $p(x) \rightsquigarrow p(\alpha)$. When α is algebraic there will be many polynomial expressions that represent the same complex number. Some examples of algebraic numbers are: $i + 3$, $1/7$, $7 + \sqrt[3]{2}$, and $\sqrt{3} + \sqrt{-5}$.

The definition of a ring is similar to that of field (3.2.2). The only difference is that multiplicative inverses aren't required:

Definition 11.1.3 $(+, -, \times, 1)$ A *ring* R is a set with two laws of composition $+$ and $\times$, called addition and multiplication, that satisfy these axioms:

(a) With the law of composition $+$, R is an abelian group that we denote by R^+; its identity is denoted by 0.

(b) Multiplication is commutative and associative, and has an identity denoted by 1.

(c) *distributive law*: For all a, b, and c in R, $(a + b)c = ac + bc$.

A *subring* of a ring is a subset that is closed under the operations of addition, subtraction, and multiplication and that contains the element 1.

Note: There is a related concept, of a *noncommutative ring* – a structure that satisfies all axioms of (11.1.3) except the commutative law for multiplication. The set of all real $n \times n$ matrices is one example. Since we won't be studying noncommutative rings, we use the word "ring" to mean "commutative ring." □

Aside from subrings of $\mathbb{C}$, the most important rings are polynomial rings. A polynomial in x with coefficients in a ring R is an expression of the form

$$(11.1.4) \qquad a_n x^n + \cdots + a_1 x + a_0,$$

with a_i in R. The set of these polynomials forms a ring that we discuss in the next section.

Another example: The set $\mathcal{R}$ of continuous real-valued functions of a real variable x forms a ring, with addition and multiplication of functions: $[f + g](x) = f(x) + g(x)$ and $[fg](x) = f(x)g(x)$.

There is a ring that contains just one element, 0; it is called the *zero ring*. In the definition of a field (3.2.2), the set $F^\times$ obtained by deleting 0 is a group that contains the multiplicative identity 1. So 1 is not equal to 0 in a field. The relation $1 = 0$ hasn't been ruled out in a ring, but it occurs only once:

Proposition 11.1.5   A ring R in which the elements 1 and 0 are equal is the zero ring.

Proof. We first note that $0a = 0$ for every element a of a ring R. The proof is the same as for vector spaces: $0 = 0a - 0a = (0 - 0)a = 0a$. Assume that $1 = 0$ in R, and let a be any element. Then $a = 1a = 0a = 0$. The only element of R is 0. □

Though elements of a ring aren't required to have multiplicative inverses, a particular element may have an inverse, and the inverse is unique if it exists.

• A *unit* of a ring is an element that has a multiplicative inverse.

The units in the ring of integers are 1 and –1, and the units in the ring of Gauss integers are ± 1 and $\pm i$. The units in the ring $\mathbb{R}[x]$ of real polynomials are the nonzero constant polynomials. Fields are rings in which $0 \neq 1$ and in which every nonzero element is a unit.

The identity element 1 of a ring is always a unit, and any reference to "the" unit element in R refers to the identity element. The ambiguous term "unit" is poorly chosen, but it is too late to change it.

11.2 POLYNOMIAL RINGS

• A *polynomial* with coefficients in a ring R is a (finite) linear combination of powers of the variable:

(11.2.1) $$f(x) = a_n x^n + a_{n-1} x^{n-1} + \cdots + a_1 x + a_0,$$

where the *coefficients* a_i are elements of R. Such an expression is sometimes called a *formal polynomial*, to distinguish it from a polynomial function. Every formal polynomial with real coefficients determines a polynomial function on the real numbers. But we use the word *polynomial* to mean formal polynomial.

The set of polynomials with coefficients in a ring R will be denoted by $R[x]$. Thus $\mathbb{Z}[x]$ denotes the set of polynomials with integer coefficients – the set of *integer polynomials*.

The *monomials* x^i are considered independent. So if

(11.2.2) $$g(x) = b_m x^m + b_{m-1} x^{m-1} + \cdots + b_1 x + b_0$$

is another polynomial with coefficients in R, then $f(x)$ and $g(x)$ are equal if and only if $a_i = b_i$ for all $i = 0, 1, 2, \ldots$.

• The *degree* of a nonzero polynomial, which may be denoted by $\deg f$, is the largest integer n such that the coefficient a_n of x_n is not zero. A polynomial of degree zero is called a *constant* polynomial. The zero polynomial is also called a constant polynomial, but its degree will not be defined.

The nonzero coefficient of highest degree of a polynomial is its *leading coefficient*, and a *monic* polynomial is one whose leading coefficient is 1.

The possibility that some coefficients of a polynomial may be zero creates a nuisance. We have to disregard terms with zero coefficient, so the polynomial $f(x)$ can be written in more than one way. This is irritating because it isn't an interesting point. One way to avoid ambiguity is to imagine listing the coefficients of all monomials, whether zero or not. This allows efficient verification of the ring axioms. So for the purpose of defining the ring operations, we write a polynomial as

(11.2.3) $$f(x) = a_0 + a_1 x + a_1 x^2 + \cdots,$$

where the coefficients a_i are all in the ring R and only finitely many of them are different from zero. This polynomial is determined by its vector (or sequence) of coefficients a_i:

$$(11.2.4) \qquad\qquad a = (a_0, a_1, \ldots),$$

where a_i are elements of R, all but a finite number zero. Every such vector corresponds to a polynomial.

When R is a field, these infinite vectors form the vector space Z with the infinite basis e_i that was defined in (3.7.2). The vector e_i corresponds to the monomial x^i, and the monomials form a basis of the space of all polynomials.

The definitions of addition and multiplication of polynomials mimic the familiar operations on polynomial functions. If $f(x)$ and $g(x)$ are polynomials, then with notation as above, their sum is

$$(11.2.5) \qquad f(x) + g(x) \;=\; (a_0 + b_0) + (a_1 + b_1)x + \cdots \;=\; \sum_k (a_k + b_k)x^k,$$

where the notation $(a_i + b_i)$ refers to addition in R. So if we think of a polynomial as a vector, addition is vector addition: $a + b = (a_0 + b_0, \, a_1 + b_1, \ldots)$.

The product of polynomials f and g is computed by expanding the product:

$$(11.2.6) \qquad f(x)g(x) \;=\; (a_0 + a_1 x + \cdots)(b_0 + b_1 x + \cdots) \;=\; \sum_{i,j} a_i b_j \, x^{i+j},$$

where the products $a_i b_j$ are to be evaluated in the ring R. There will be finitely many nonzero coefficients $a_i b_j$. This is a correct formula, but the right side is not in the standard form (11.2.3), because the same monomial x^n appears several times – once for each pair i, j of indices such that $i + j = n$. So terms have to be collected on the right side. This leads to the definition

$$(11.2.7) \qquad\qquad f(x)g(x) = p_0 + p_1 x + p_2 x^2 + \cdots,$$

with

$$p_k = \sum_{i+j=k} a_i b_j,$$

$$p_0 = a_0 b_0, \quad p_1 = a_0 b_1 + a_1 b_0, \quad p_2 = a_0 b_2 + a_1 b_1 + a_2 b_0, \ \ldots$$

Each p_k is evaluated using the laws of composition in the ring. However, when making computations, it may be desirable to defer the collection of terms temporarily.

Proposition 11.2.8 There is a unique commutative ring structure on the set of polynomials $R[x]$ having these properties:

- Addition of polynomials is defined by (11.2.5).
- Multiplication of polynomials is defined by (11.2.7).
- The ring R becomes a subring of $R[x]$ when the elements of R are identified with the constant polynomials.

Since polynomial algebra is familiar and since the proof of this proposition has no interesting features, we omit it.   □

Division with remainder is an important operation on polynomials.

Proposition 11.2.9 Division with Remainder. Let R be a ring, let f be a monic polynomial and let g be any polynomial, both with coefficients in R. There are uniquely determined polynomials q and r in $R[x]$ such that

$$g(x) = f(x)q(x) + r(x),$$

and such that the remainder r, if it is not zero, has degree less than the degree of f. Moreover, f divides g in $R[x]$ if and only if the remainder r is zero.

The proof of this proposition follows the algorithm for division of polynomials that one learns in school.   □

Corollary 11.2.10 Division with remainder can be done whenever the leading coefficient of f is a unit. In particular, it can be done whenever the coefficient ring is a field and $f \neq 0$.

If the leading coefficient is a unit u, we can factor it out of f.   □

However, one cannot divide $x^2 + 1$ by $2x + 1$ in the ring $\mathbb{Z}[x]$ of integer polynomials.

Corollary 11.2.11 Let $g(x)$ be a polynomial in $R[x]$, and let α be an element of R. The remainder of division of $g(x)$ by $x - \alpha$ is $g(\alpha)$. Thus $x - \alpha$ divides g in $R[x]$ if and only if $g(\alpha) = 0$.

This corollary is proved by substituting $x = \alpha$ into the equation $g(x) = (x - \alpha)q(x) + r$ and noting that r is a constant.   □

Polynomials are fundamental to the theory of rings, and we will also want to use polynomials in several variables. There is no major change in the definitions.

- A *monomial* is a formal product of some variables $x_1, \ldots, x_n$ of the form

$$x_1^{i_1} x_2^{i_2} \cdots x_n^{i_n},$$

where the exponents i_ν are non-negative integers. The *degree* of a monomial, sometimes called the *total degree*, is the sum $i_1 + \cdots + i_n$.

An n-tuple $(i_1, \ldots, i_n)$ is called a *multi-index*, and vector notation $i = (i_1, \ldots, i_n)$ for multi-indices is convenient. Using multi-index notation, we may write a monomial symbolically as x^i:

(11.2.12) $$x^i = x_1^{i_1} x_2^{i_2} \cdots x_n^{i_n}.$$

The monomial x^0, with $0 = (0, \ldots, 0)$, is denoted by 1. A *polynomial* in the variables $x_1, \ldots, x_n$, with coefficients in a ring R, is a linear combination of finitely many monomials, with coefficients in R. With multi-index notation, a polynomial $f(x) = f(x_1, \ldots, x_n)$ can be written in exactly one way in the form

$$(11.2.13) \qquad\qquad f(x) = \sum_i a_i x^i,$$

where i runs through all multi-indices $(i_1, \ldots, i_n)$, the coefficients a_i are in R, and only finitely many of these coefficients are different from zero.

A polynomial in which all monomials with nonzero coefficients have (total) degree d is called a *homogeneous* polynomial.

Using multi-index notation, formulas (11.2.5) and (11.2.7) define addition and multiplication of polynomials in several variables, and the analogue of Proposition 11.2.8 is true. However, division with remainder requires more thought. We will come back to it below (see Corollary 11.3.9).

The ring of polynomials with coefficients in R is usually denoted by one of the symbols

$$(11.2.14) \qquad\qquad R[x_1, \ldots, x_n] \quad \text{or} \quad R[x],$$

where the symbol x is understood to refer to the set of variables $\{x_1, \ldots, x_n\}$. When no set of variables has been introduced, $R[x]$ denotes the polynomial ring in one variable.

11.3 HOMOMORPHISMS AND IDEALS

• A *ring homomorphism* $\varphi : R \to R'$ is a map from one ring to another which is compatible with the laws of composition and which carries the unit element 1 of R to the unit element 1 in R' – a map such that, for all a and b in R,

$$(11.3.1) \qquad \varphi(a + b) = \varphi(a) + \varphi(b), \quad \varphi(ab) = \varphi(a)\varphi(b), \quad \text{and} \quad \varphi(1) = 1.$$

The map

$$(11.3.2) \qquad\qquad \varphi : \mathbb{Z} \to \mathbb{F}_p$$

that sends an integer to its congruence class modulo p is a ring homomorphism.

An *isomorphism* of rings is a bijective homomorphism, and if there is an isomorphism from R to R', the two rings are said to be *isomorphic*. We often use the notation $R \approx R'$ to indicate that two rings R and R' are isomorphic.

A word about the third condition of (11.3.1): The assumption that a homomorphism φ is compatible with addition implies that it is a homomorphism from the additive group R^+ of R to the additive group R'^+. A group homomorphism carries the identity to the identity, so $\varphi(0) = 0$. But we can't conclude that $\varphi(1) = 1$ from compatibility with multiplication, so that condition must be listed separately. (R is not a group with respect to $\times$.) For example, the *zero map* $R \to R'$ that sends all elements of R to zero is compatible with $+$ and $\times$, but it doesn't send 1 to 1 unless $1 = 0$ in R'. The zero map is not called a ring homomorphism unless R' is the zero ring (see (11.1.5)).

The most important ring homomorphisms are obtained by evaluating polynomials. Evaluation of real polynomials at a real number a defines a homomorphism

$$(11.3.3) \qquad\qquad \mathbb{R}[x] \to \mathbb{R}, \quad \text{that sends} \quad p(x) \rightsquigarrow p(a).$$

One can also evaluate real polynomials at a complex number such as i, to obtain a homomorphism $\mathbb{R}[x] \to \mathbb{C}$ that sends $p(x) \rightsquigarrow p(i)$.

The general formulation of the principle of evaluation of polynomials is this:

Proposition 11.3.4 Substitution Principle. Let $\varphi : R \to R'$ be a ring homomorphism, and let $R[x]$ be the ring of polynomials with coefficients in R.

(a) Let α be an element of R'. There is a unique homomorphism $\Phi : R[x] \to R'$ that agrees with the map φ on constant polynomials, and that sends $x \rightsquigarrow \alpha$.

(b) More generally, given elements $\alpha_1, \ldots, \alpha_n$ of R', there is a unique homomorphism $\Phi : R[x_1, \ldots, x_n] \to R'$, from the polynomial ring in n variables to R', that agrees with φ on constant polynomials and that sends $x_\nu \rightsquigarrow \alpha_\nu$, for $\nu = 1, \ldots, n$.

Proof. **(a)** Let us denote the image $\varphi(a)$ of an element a of R by a'. Using the fact that Φ is a homomorphism that restricts to φ on R and sends x to α, we see that it acts on a polynomial $f(x) = \sum a_i x^i$ by sending

(11.3.5) $$\Phi\left(\sum a_i x^i\right) = \sum \Phi(a_i)\Phi(x)^i = \sum a_i'\alpha^i.$$

In words, Φ acts on the coefficients of a polynomial as φ, and it substitutes α for x. Since this formula describes Φ, we have proved the uniqueness of the substitution homomorphism. To prove its existence, we take this formula as the definition of Φ, and we show that Φ is a homomorphism $R[x] \to R'$. It is clear that 1 is sent to 1, and it is easy to verify compatibility with addition of polynomials. Compatibility with multiplication is checked using formula (11.2.6):

$$\Phi(fg) = \Phi\left(\sum a_i b_j x^{i+j}\right) = \sum \Phi(a_i b_j x^{i+j}) = \sum_{i,j} a_i' b_j' \alpha^{i+j}$$

$$= \left(\sum_i a_i'\alpha^i\right)\left(\sum_j b_j'\alpha^i\right) = \Phi(f)\Phi(g).$$

With multi-index notation, the proof of **(b)** becomes the same as that of **(a)**.   □

Here is a simple example of the substitution principle in which the coefficient ring R changes. Let $\psi : R \to S$ be a ring homomorphism. Composing ψ with the inclusion of S as a subring of the polynomial ring $S[x]$, we obtain a homomorphism $\varphi : R \to S[x]$. The substitution principle asserts that there is a unique extension of φ to a homomorphism $\Phi : R[x] \to S[x]$ that sends $x \rightsquigarrow x$. This map operates on the coefficients of a polynomial, while leaving the variable x fixed. If we denote $\psi(a)$ by a', then it sends a polynomial $a_n x^n + \cdots + a_1 x + a_0$ to $a_n' x^n + \cdots + a_1' x + a_0'$.

A particularly interesting case is that φ is the homomorphism $\mathbb{Z} \to \mathbb{F}_p$ that sends an integer a to its residue $\bar{a}$ modulo p. This map extends to a homomorphism $\Phi : \mathbb{Z}[x] \to \mathbb{F}_p[x]$, defined by

(11.3.6) $$f(x) = a_n x^n + \cdots + a_0 \;\rightsquigarrow\; \bar{a}_n x^n + \cdots + \bar{a}_0 = \bar{f}(x),$$

where $\bar{a}_i$ is the residue class of a_i modulo p. It is natural to call the polynomial $\overline{f}(x)$ the *residue* of $f(x)$ modulo p.

Another example: Let R be any ring, and let P denote the polynomial ring $R[x]$. One can use the substitution principle to construct an isomorphism

$$(11.3.7) \qquad\qquad R[x, y] \rightarrow P[y] = (R[x])[y].$$

This is stated and proved below in Proposition 11.3.8. The domain is the ring of polynomials in two variables x and y, and the range is the ring of polynomials in y whose coefficients are polynomials in x. The statement that these rings are isomorphic is a formalization of the procedure of collecting terms of like degree in y in a polynomial $f(x, y)$. For example,

$$x^4y + x^3 - 3x^2y + y^2 + 2 \;=\; y^2 + (x^4 - 3x^2)y + (x^3 + 2).$$

This procedure can be useful. For one thing, one may end up with a polynomial that is monic in the variable y, as happens in the example above. If so, one can do division with remainder (see Corollary 11.3.9 below).

Proposition 11.3.8 Let $x = (x_1, \ldots, x_m)$ and $y = (y_1, \ldots, y_n)$ denote sets of variables. There is a unique isomorphism $R[x, y] \rightarrow R[x][y]$, which is the identity on R and which sends the variables to themselves.

This is very elementary, but it would be boring to verify compatibility of multiplication in the two rings directly.

Proof. We note that since R is a subring of $R[x]$ and $R[x]$ is a subring of $R[x][y]$, R is also a subring of $R[x][y]$. Let φ be the inclusion of R into $R[x][y]$. The substitution principle tells us that there is a unique homomorphism $\Phi: R[x, y] \rightarrow R[x][y]$, which extends φ and sends the variables x_μ and y_ν wherever we want. So we can send the variables to themselves. The map Φ thus constructed is the required isomorphism. It isn't difficult to see that Φ is bijective. One way to show this would be to use the substitution principle again, to define the inverse map. $\qquad\square$

Corollary 11.3.9 Let $f(x, y)$ and $g(x, y)$ be polynomials in two variables, elements of $R[x, y]$. Suppose that, when regarded as a polynomial in y, f is a monic polynomial of degree m. There are uniquely determined polynomials $q(x, y)$ and $r(x, y)$ such that $g = fq + r$, and such that if $r(x, y)$ is not zero, its degree in the variable y is less than m.

This follows from Propositions 11.2.9 and 11.3.8. $\qquad\square$

Another case in which one can describe homomorphisms easily is when the domain is the ring of integers.

Proposition 11.3.10 Let R be a ring. There is exactly one homomorphism $\varphi: \mathbb{Z} \rightarrow R$ from the ring of integers to R. It is the map defined, for $n \geq 0$, by $\varphi(n) = 1 + \cdots + 1$ (n terms) and $\varphi(-n) = -\varphi(n)$.

Sketch of Proof. Let $\varphi: \mathbb{Z} \rightarrow R$ be a homomorphism. By definition of a homomorphism, $\varphi(1) = 1$ and $\varphi(n + 1) = \varphi(n) + \varphi(1)$. This recursive definition describes φ on the natural numbers, and together with $\varphi(-n) = -\varphi(n)$ if $n > 0$ and $\varphi(0) = 0$, it determines φ uniquely. So it is the only map $\mathbb{Z} \to R$ that could be a homomorphism, and it isn't hard to convince oneself that it is one. To prove this formally, one would go back to the definitions of addition and multiplication of integers (see Appendix).   □

Proposition (11.3.10) allows us to identify the image of an integer in an arbitrary ring R. We interpet the symbol 3, for example, as the element $1 + 1 + 1$ of R.

• Let $\varphi : R \to R'$ be a ring homomorphism. The *kernel* of φ is the set of elements of R that map to zero:

(11.3.11) $\ker \varphi = \{s \in R \mid \varphi(s) = 0\}.$

This is the same as the kernel obtained when one regards φ as a homomorphism of additive groups $R^+ \to R'^+$. So what we have learned about kernels of group homomorphisms applies. For instance, φ is injective if and only if $\ker \varphi = \{0\}$.

As you will recall, the kernel of a group homomorphism is not only a subgroup, it is a normal subgroup. Similarly, the kernel of a ring homomorphism is closed under the operation of addition, and it has a property that is stronger than closure under multiplication:

(11.3.12)   If s is in $\ker \varphi$, then for every element r of R, rs is in $\ker \varphi$.

For, if $\varphi(s) = 0$, then $\varphi(rs) = \varphi(r)\varphi(s) = \varphi(r)0 = 0$.

This property is abstracted in the concept of an ideal.

Definition 11.3.13   An *ideal* I of a ring R is a nonempty subset of R with these properties:

- I is closed under addition, and
- If s is in I and r is in R, then rs is in I.

The kernel of a ring homomorphism is an ideal.

The peculiar term "ideal" is an abbreviation of the phrase "ideal element" that was formerly used in number theory. We will see in Chapter 13 how it arose. A good way, probably a better way, to think of the definition of an ideal is this equivalent formulation:

(11.3.14) I is not empty, and a linear combination $r_1 s_1 + \cdots + r_k s_k$
                   of elements s_i of I with coefficients r_i in R is in I.

• In any ring R, the multiples of a particular element a form an ideal called the *principal ideal* generated by a. An element b of R is in this ideal if and only if b is a multiple of a, which is to say, if and only if a divides b in R.

There are several notations for this principal ideal:

(11.3.15) $(a) = aR = Ra = \{ra \mid r \in R\}.$

The ring R itself is the principal ideal (1), and because of this it is called the *unit ideal*. It is the only ideal that contains a unit of the ring. The set consisting of zero alone is the principal ideal (0), and is called the *zero ideal*. An ideal I is *proper* if it is neither the zero ideal nor the unit ideal.

Every ideal I satisfies the requirements for a subring, except that the unit element 1 of R will not be in I unless I is the whole ring. Unless I is equal to R, it will not be what we call a subring.

Examples 11.3.16

(a) Let φ be the homomorphism $\mathbb{R}[x] \to \mathbb{R}$ defined by substituting the real number 2 for x. Its kernel, the set of polynomials that have 2 as a root, can be described as the set of polynomials divisible by $x - 2$. This is a principal ideal that might be denoted by $(x - 2)$.

(b) Let $\Phi : \mathbb{R}[x, y] \to \mathbb{R}[t]$ be the homomorphism that is the identity on the real numbers, and that sends $x \rightsquigarrow t^2$, $y \rightsquigarrow t^3$. Then it sends $g(x, y) \rightsquigarrow g(t^2, t^3)$. The polynomial $f(x, y) = y^2 - x^3$ is in the kernel of Φ. We'll show that the kernel is the principal ideal (f) generated by f, i.e., that if $g(x, y)$ is a polynomial and if $g(t^2, t^3) = 0$, then f divides g. To show this, we regard f as a polynomial in y whose coefficients are polynomials in x (see (11.3.8)). It is a monic polynomial in y, so we can do division with remainder: $g = fq + r$, where q and r are polynomials, and where the remainder r, if not zero, has degree at most 1 in y. We write the remainder as a polynomial in $y : r(x, y) = r_1(x)y + r_0(x)$. If $g(t^2, t^3) = 0$, then both g and fq are in the kernel of Φ, so r is too: $r(t^2, t^3) = r_1(t^2)t^3 + r_0(t^2) = 0$. The monomials that appear in $r_0(t^2)$ have even degree, while those in $r_1(t^2)t^3$ have odd degree. Therefore, in order for $r(t^2, t^3)$ to be zero, $r_0(x)$ and $r_1(x)$ must both be zero. Since the remainder is zero, f divides g. $\square$

The notation (a) for a principal ideal is convenient, but it is ambiguous because the ring isn't mentioned. For instance, $(x - 2)$ could stand for an ideal of $\mathbb{R}[x]$ or of $\mathbb{Z}[x]$, depending on the circumstances. When several rings are being discussed, a different notation may be preferable.

• The ideal I *generated by a set of elements* $\{a_1, \ldots, a_n\}$ of a ring R is the smallest ideal that contains those elements. It can be described as the set of all linear combinations

(11.3.17) $$r_1 a_1 + \cdots + r_n a_n$$

with coefficients r_i in the ring. This ideal is often denoted by $(a_1, \ldots, a_n)$:

(11.3.18) $$(a_1, \ldots, a_n) = \{r_1 a_1 + \cdots + r_n a_n \mid r_i \in R\}.$$

For instance, the kernel K of the homomorphism $\varphi : \mathbb{Z}[x] \to \mathbb{F}_p$ that sends $f(x)$ to the residue of $f(0)$ modulo p is the ideal (p, x) of $\mathbb{Z}[x]$ generated by p and x. Let's check this. First, p and x are in the kernel, so $(p, x) \subset K$. To show that $K \subset (p, x)$, we let $f(x) = a_n x^n + \cdots + a_1 x + a_0$ be an integer polynomial. Then $f(0) = a_0$. If $a_0 \equiv 0$ modulo p, say $a_0 = bp$, then f is the linear combination $bp + (a_n x^{n-1} + \cdots + a_1)x$ of p and x. So f is in the ideal (p, x).

The number of elements required to generate an ideal can be arbitrarily large. The ideal $(x^3, x^2 y, xy^2, y^3)$ of the polynomial ring $\mathbb{C}[x, y]$ consists of the polynomials in which every term has degree at least 3. It cannot be generated by fewer than four elements.

In the rest of this section, we describe ideals in some simple cases.

Proposition 11.3.19

(a) The only ideals of a field are the zero ideal and the unit ideal.

(b) A ring that has exactly two ideals is a field.

Proof. If an ideal I of a field F contains a nonzero element a, that element is invertible. Then I contains $a^{-1}a = 1$, and is the unit ideal. The only ideals of F are (0) and (1).

Assume that R has exactly two ideals. The properties that distinguish fields among rings are that $1 \neq 0$ and that every nonzero element a of R has a multiplicative inverse. We have seen that $1 = 0$ happens only in the zero ring. The zero ring has only one ideal, the zero ideal. Since our ring has two ideals, $1 \neq 0$ in R. The two ideals (1) and (0) are different, so they are the only two ideals of R.

To show that every nonzero element a of R has an inverse, we consider the principal ideal (a). It is not the zero ideal because it contains the element a. Therefore it is the unit ideal. The elements of (a) are the multiples of a, so 1 is a multiple of a, and therefore a is invertible. $\qquad\square$

Corollary 11.3.20 Every homomorphism $\varphi: F \to R$ from a field F to a nonzero ring R is injective.

Proof. The kernel of φ is an ideal of F. So according to Proposition 11.3.19, the kernel is either (0) or (1). If $\ker \varphi$ were the unit ideal (1), φ would be the zero map. But the zero map isn't a homomorphism when R isn't the zero ring. Therefore $\ker \varphi = (0)$, and φ is injective. $\qquad\square$

Proposition 11.3.21 The ideals in the ring of integers are the subgroups of $\mathbb{Z}^+$, and they are principal ideals.

An ideal of the ring $\mathbb{Z}$ of integers will be a subgroup of the additive group $\mathbb{Z}^+$. It was proved before (2.3.3) that every subgroup of $\mathbb{Z}^+$ has the form $\mathbb{Z}n$. $\qquad\square$

The proof that subgroups of $\mathbb{Z}^+$ have the form $\mathbb{Z}n$ can be adapted to the polynomial ring $F[x]$.

Proposition 11.3.22 Every ideal in the ring $F[x]$ of polynomials in one variable x over a field F is a principal ideal. A nonzero ideal I in $F[x]$ is generated by the unique monic polynomial of lowest degree that it contains.

Proof. Let I be an ideal of $F[x]$. The zero ideal is principal, so we may assume that I is not the zero ideal. The first step in finding a generator for a nonzero subgroup of $\mathbb{Z}$ is to choose its smallest positive element. The substitute here is to choose a nonzero polynomial f in I of minimal degree. Since F is a field, we may choose f to be monic. We claim that I is the principal ideal (f) of polynomial multiples of f. Since f is in I, every multiple of f is in I, so $(f) \subset I$. To prove that $I \subset (f)$, we choose an element g of I, and we use division with remainder to write $g = fq + r$, where r, if not zero, has lower degree than f. Since g and f are in I, $g - fq = r$ is in I too. Since f has minimal degree among nonzero elements of I, the only possibility is that $r = 0$. Therefore f divides g, and g is in (f).

If f_1 and f_2 are two monic polynomials of lowest degree in I, their difference is in I and has lower degree than n, so it must be zero. Therefore the monic polynomial of lowest degree is unique. $\square$

Example 11.3.23 Let $\gamma = \sqrt[3]{2}$ be the real cube root of 2, and let $\Phi : \mathbb{Q}[x] \to \mathbb{C}$ be the substitution map that sends $x \leadsto \gamma$. The kernel of this map is a principal ideal, generated by the monic polynomial of lowest degree in $\mathbb{Q}[x]$ that has γ as a root (11.3.22). The polynomial $x^3 - 2$ is in the kernel, and because $\sqrt[3]{2}$ is not a rational number, it is not the product $f = gh$ of two nonconstant polynomials with rational coefficients. So it is the lowest degree polynomial in the kernel, and therefore it generates the kernel.

We restrict the map Φ to the integer polynomial ring $\mathbb{Z}[x]$, obtaining a homomorphism $\Phi' : \mathbb{Z}[x] \to \mathbb{C}$. The next lemma shows that the kernel of Φ' is the principal ideal of $\mathbb{Z}[x]$ generated by the same polynomial f.

Lemma 11.3.24 Let f be a monic integer polynomial, and let g be another integer polynomial. If f divides g in $\mathbb{Q}[x]$, then f divides g in $\mathbb{Z}[x]$.

Proof. Since f is monic, we can do division with remainder in $\mathbb{Z}[x]$: $g = fq + r$. This equation remains true in the ring $\mathbb{Q}[x]$, and division with remainder in $\mathbb{Q}[x]$ gives the same result. In $\mathbb{Q}[x]$, f divides g. Therefore $r = 0$, and f divides g in $\mathbb{Z}[x]$. $\square$

The proof of the following corollary is similar to the proof of existence of the greatest common divisor in the ring of integers ((2.3.5), see also (12.2.8)).

Corollary 11.3.25 Let R denote the polynomial ring $F[x]$ in one variable over a field F, and let f and g be elements of R, not both zero. Their *greatest common divisor* $d(x)$ is the unique monic polynomial that generates the ideal (f, g). It has these properties:

(a) $Rd = Rf + Rg$.

(b) d divides f and g.

(c) If a polynomial $e = e(x)$ divides both f and g, it also divides d.

(d) There are polynomials p and q such that $d = pf + qg$. $\square$

The definition of the *characteristic* of a ring R is the same as for a field. It is the non-negative integer n that generates the kernel of the homomorphism $\varphi : \mathbb{Z} \to R$ (11.3.10). If $n = 0$, the characteristic is zero, and this means that no positive multiple of 1 in R is equal to zero. Otherwise n is the smallest positive integer such that "n times 1" is zero in R. The characteristic of a ring can be any non-negative integer.

11.4 QUOTIENT RINGS

Let I be an ideal of a ring R. The cosets of the additive subgroup I^+ of R^+ are the subsets $a + I$. It follows from what has been proved for groups that the set of cosets $\overline{R} = R/I$ is a group under addition. It is also a ring:

Theorem 11.4.1 Let I be an ideal of a ring R. There is a unique ring structure on the set $\overline{R}$ of additive cosets of I such that the map $\pi: R \to \overline{R}$ that sends $a \rightsquigarrow \overline{a} = [a + I]$ is a ring homomorphism. The kernel of π is the ideal I.

As with quotient groups, the map π is referred to as the *canonical map*, and $\overline{R}$ is called the *quotient ring*. The image $\overline{a}$ of an element a is called the *residue* of the element.

Proof. This proof has already been carried out for the ring of integers (Section 2.9). We want to put a ring structure on $\overline{R}$, and if we forget about multiplication and consider only the addition law, I becomes a normal subgroup of R^+, for which the proof has been given (2.12.2). What is left to do is to define multiplication, to verify the ring axioms, and to prove that π is a homomorphism. Let $\overline{a} = [a + I]$ and $\overline{b} = [b + I]$ be elements of $\overline{R}$. We would like to define the product by the setting $\overline{a}\overline{b} = [ab + I]$. The set of products

$$P = (a + I)(b + I) = \{rs \mid r \in a + I, \, s \in b + I\}$$

isn't always a coset of I. However, as in the case of the ring of integers, P is always contained in the coset $ab + I$. If we write $r = a + u$ and $s = b + v$ with u and v in I, then

$$(a + u)(b + v) = ab + (av + bu + uv).$$

Since I is an ideal that contains u and v, it contains $av + bu + uv$. This is all that is needed to define the product coset: It is the coset that contains the set of products. That coset is unique because the cosets partition R.

The proofs of the remaining assertions follow the patterns set in Section 2.9.   □

As with groups, one often drops the bars over the letters that represent elements of a quotient ring $\overline{R}$, remembering that "$a = b$ in $\overline{R}$" means $\overline{a} = \overline{b}$.

The next theorems are analogous to ones that we have seen for groups:

Theorem 11.4.2 Mapping Property of Quotient Rings. Let $f: R \to R'$ be a ring homomorphism with kernel K and let I be another ideal. Let $\pi: R \to \overline{R}$ be the canonical map from R to $\overline{R} = R/I$.
(a) If $I \subset K$, there is a unique homomorphism $\overline{f}: \overline{R} \to R'$ such that $\overline{f}\pi = f$:

(b) *(First Isomorphism Theorem)* If f is surjective and $I = K$, $\overline{f}$ is an isomorphism.   □

The First Isomorphism Theorem is our fundamental method of identifying quotient rings. However, it doesn't apply very often. Quotient rings will be new rings in most cases, and this is one reason that the quotient construction is important. The ring $\mathbb{C}[x, y]/(y^2 - x^3 + 1)$, for example, is completely different from any ring we have seen up to now. Its elements are functions on an elliptic curve (see [Silverman]).

The Correspondence Theorem for rings describes the fundamental relationship between ideals in a ring and a quotient ring.

Theorem 11.4.3 Correspondence Theorem. Let $\varphi: R \to \mathcal{R}$ be a *surjective* ring homomorphism with kernel K. There is a bijective correspondence between the set of all ideals of $\mathcal{R}$ and the set of ideals of R that contain K:

$$\{\text{ideals of } R \text{ that contain } K\} \longleftrightarrow \{\text{ideals of } \mathcal{R}\}.$$

This correspondence is defined as follows:

- If I is a ideal of R and if $K \subset I$, the corresponding ideal of $\mathcal{R}$ is $\varphi(I)$.
- If $\mathcal{I}$ is a ideal of $\mathcal{R}$, the corresponding ideal of R is $\varphi^{-1}(\mathcal{I})$.

If the ideal I of R corresponds to the ideal $\mathcal{I}$ of $\mathcal{R}$, the quotient rings R/I and $\mathcal{R}/\mathcal{I}$ are naturally isomorphic.

Note that the inclusion $K \subset I$ is the reverse of the one in the mapping property.

Proof of the Correspondence Theorem. We let $\mathcal{I}$ be an ideal of $\mathcal{R}$ and we let I be an ideal of R that contains K. We must check the following points:

- $\varphi(I)$ is an ideal of $\mathcal{R}$.
- $\varphi^{-1}(\mathcal{I})$ is an ideal of R, and it contains K.
- $\varphi(\varphi^{-1}(\mathcal{I})) = \mathcal{I}$, and $\varphi^{-1}(\varphi(I)) = I$.
- If $\varphi(I) = \mathcal{I}$, then $R/I \approx \mathcal{R}/\mathcal{I}$.

We go through these points in order, referring to the proof of the Correspondence Theorem 2.10.5 for groups when it applies. We have seen before that the image of a subgroup is a subgroup. So to show that $\varphi(I)$ is an ideal of $\mathcal{R}$, we need only prove that it is closed under multiplication by elements of $\mathcal{R}$. Let $\tilde{r}$ be in $\mathcal{R}$ and let $\tilde{x}$ be in $\varphi(I)$. Then $\tilde{x} = \varphi(x)$ for some x in I, and because φ is surjective, $\tilde{r} = \varphi(r)$ for some r in R. Since I is an ideal, rx is in I, and $\tilde{r}\tilde{x} = \varphi(rx)$, so $\tilde{r}\tilde{x}$ is in $\varphi(I)$.

Next, we verify that $\varphi^{-1}(\mathcal{I})$ is an ideal of R that contains K. This is true whether or not φ is surjective. Let's write $\varphi(a) = \tilde{a}$. By definition of the inverse image, a is in $\varphi^{-1}(\mathcal{I})$ if and only if $\tilde{a}$ is in $\mathcal{I}$. If a is in $\varphi^{-1}(\mathcal{I})$ and r is in R, then $\varphi(ra) = \tilde{r}\tilde{a}$ is in $\mathcal{I}$ because $\mathcal{I}$ is an ideal, and hence ra is in $\varphi^{-1}(\mathcal{I})$. The facts that $\varphi^{-1}(\mathcal{I})$ is closed under sums and that it contains K were shown in (2.10.4).

The third assertion, the bijectivity of the correspondence, follows from the case of a group homomorphism.

Finally, suppose that an ideal I of R that contains K corresponds to an ideal $\mathcal{I}$ of $\mathcal{R}$, that is, $\mathcal{I} = \varphi(I)$ and $I = \varphi^{-1}(\mathcal{I})$. Let $\tilde{\pi}: \mathcal{R} \to \mathcal{R}/\mathcal{I}$ be the canonical map, and let f denote the composed map $\tilde{\pi}\varphi: R \to \mathcal{R} \to \mathcal{R}/\mathcal{I}$. The kernel of f is the set of elements x in R such that $\tilde{\pi}\varphi(x) = 0$, which translates to $\varphi(x) \in \mathcal{I}$, or to $x \in \varphi^{-1}(\mathcal{I}) = I$. The kernel of f is I. The mapping property, applied to the map f, gives us a homomorphism $\overline{f}: R/I \to \mathcal{R}/\mathcal{I}$, and the First Isomorphism Theorem asserts that $\overline{f}$ is an isomorphism. $\square$

To apply the Correspondence Theorem, it helps to know the ideals of one of the rings. The next examples illustrate this in very simple situations, in which one of the two rings is $\mathbb{C}[t]$. We will be able to use the fact that every ideal of $\mathbb{C}[t]$ is principal (11.3.22).

Example 11.4.4 **(a)** Let $\varphi:\mathbb{C}[x, y] \to \mathbb{C}[t]$ be the homomorphism that sends $x \rightsquigarrow t$ and $y \rightsquigarrow t^2$. This is a surjective map, and its kernel K is the principal ideal of $\mathbb{C}[x, y]$ generated by $y - x^2$. (The proof of this is similar to the one given in Example 11.3.16.)

The Correspondence Theorem relates ideals I of $\mathbb{C}[x, y]$ that contain $y - x^2$ to ideals J of $\mathbb{C}[t]$, by $J = \varphi(I)$ and $I = \varphi^{-1}(J)$. Here J will be a principal ideal, generated by a polynomial $p(t)$. Let I_1 denote the ideal of $\mathbb{C}[x, y]$ generated by $y - x^2$ and $p(x)$. Then I_1 contains K, and its image is equal to J. The Correspondence Theorem asserts that $I_1 = I$. Every ideal of the polynomial ring $\mathbb{C}[x, y]$ that contains $y - x^2$ has the form $I = (y - x^2, p(x))$, for some polynomial $p(x)$.

(b) We identify the ideals of the quotient ring $R' = \mathbb{C}[t]/(t^2 - 1)$ using the canonical homomorphism $\pi : \mathbb{C}[t] \to R'$. The kernel of π is the principal ideal $(t^2 - 1)$. Let I be an ideal of $\mathbb{C}[t]$ that contains $t^2 - 1$. Then I is principal, generated by a monic polynomial f, and the fact that $t^2 - 1$ is in I means that f divides $t^2 - 1$. The monic divisors of $t^2 - 1$ are: $1, t - 1, t + 1$ and $t^2 - 1$. Therefore the ring R' contains exactly four ideals. They are the principal ideals generated by the residues of the divisors of $t^2 - 1$. $\square$

Adding Relations

We reinterpret the quotient ring construction when the ideal I is principal, say $I = (a)$. In this situation, we think of $\overline{R} = R/I$ as the ring obtained by imposing the relation $a = 0$ on R, or of killing the element a. For instance, the field $\mathbb{F}_7$ will be thought of as the ring obtained by killing 7 in the ring $\mathbb{Z}$ of integers.

Let's examine the collapsing that takes place in the map $\pi : R \to \overline{R}$. Its kernel is the ideal I, so a is in the kernel: $\pi(a) = 0$. If b is any element of R, the elements that have the same image in $\overline{R}$ as b are those in the coset $b + I$, and since $I = (a)$ those elements have the form $b + ra$. We see that imposing the relation $a = 0$ in the ring R forces us also to set $b = b + ra$ for all b and r in R, and that these are the only consequences of killing a.

Any number of relations $a_1 = 0, \ldots, a_n = 0$ can be introduced, by working modulo the ideal I generated by $a_1, \ldots, a_n$, the set of linear combinations $r_1 a_1 + \cdots + r_n a_n$, with coefficients r_i in R. The quotient ring $\overline{R} = R/I$ is viewed as the ring obtained by killing the n elements. Two elements b and b' of R have the same image in $\overline{R}$ if and only if b' has the form $b + r_1 a_1 + \cdots + r_n a_n$ for some r_i in R.

The more relations we add, the more collapsing takes place in the map π. If we add relations carelessly, the worst that can happen is that we may end up with $I = R$ and $\overline{R} = 0$. All relations $a = 0$ become true when we collapse R to the zero ring.

Here the Correspondence Theorem asserts something that is intuitively clear: Introducing relations one at a time or all together leads to isomorphic results. To spell this out, let a and b be elements of a ring R, and let $\overline{R} = R/(a)$ be the result of killing a in R. Let $\overline{b}$ be the residue of b in $\overline{R}$. The Correspondence Theorem tells us that the principal ideal $(\overline{b})$ of $\overline{R}$ corresponds to the ideal (a, b) of R, and that $R/(a, b)$ is isomorphic to $\overline{R}/(\overline{b})$. Killing a and b in R at the same time gives the same result as killing $\overline{b}$ in the ring $\overline{R}$ that is obtained by killing a first.

Example 11.4.5 We ask to identify the quotient ring $\overline{R} = \mathbb{Z}[i]/(i-2)$, the ring obtained from the Gauss integers by introducing the relation $i - 2 = 0$. Instead of analyzing this directly, we note that the kernel of the map $\mathbb{Z}[x] \to \mathbb{Z}[i]$ sending $x \rightsquigarrow i$ is the principal ideal of $\mathbb{Z}[x]$ generated by $f = x^2 + 1$. The First Isomorphism Theorem tells us that $\mathbb{Z}[x]/(f) \approx \mathbb{Z}[i]$. The image of $g = x - 2$ is $i - 2$, so $\overline{R}$ can also be obtained by introducing the two relations $f = 0$ and $g = 0$ into the integer polynomial ring. Let $I = (f, g)$ be the ideal of $\mathbb{Z}[x]$ generated by the two polynomials f and g. Then $\overline{R} \approx \mathbb{Z}[x]/I$.

To form $\overline{R}$, we may introduce the two relations in the opposite order, first killing g, then f. The principal ideal (g) of $\mathbb{Z}[x]$ is the kernel of the homomorphism $\mathbb{Z}[x] \to \mathbb{Z}$ that sends $x \rightsquigarrow 2$. So when we kill $x - 2$ in $\mathbb{Z}[x]$, we obtain a ring isomorphic to $\mathbb{Z}$, in which the residue of x is 2. Then the residue of $f = x^2 + 1$ becomes 5. So we can also obtain $\overline{R}$ by killing 5 in $\mathbb{Z}$, and therefore $\overline{R} \approx \mathbb{F}_5$.

The rings we have mentioned are summed up in this diagram:

(11.4.6)

$$
\begin{array}{ccc}
 & \text{kill} & \\
 & x-2 & \\
\mathbb{Z}[x] & \longrightarrow & \mathbb{Z} \\
{\scriptstyle\text{kill}}\downarrow{\scriptstyle x^2+1} & \searrow & \downarrow{\scriptstyle\text{kill } 5} \\
\mathbb{Z}[i] & \longrightarrow & \mathbb{F}_5 \\
 & \text{kill} & \\
 & i-2 &
\end{array}
$$

$\square$

11.5   ADJOINING ELEMENTS

In this section we discuss a procedure closely related to that of adding relations: adjoining new elements to a ring. Our model for this procedure is the construction of the complex number field from the real numbers. That construction is completely formal: The complex number i has no properties other than its defining property: $i^2 = -1$. We will now describe the general principle behind this construction. We start with an arbitrary ring R, and consider the problem of building a bigger ring containing the elements of R and also a new element, which we denote by α. We will probably want α to satisfy some relation such as $\alpha^2 + 1 = 0$. A ring that contains another ring as a subring is called a *ring extension*. So we are looking for a suitable extension.

Sometimes the element α may be available in a ring extension R' that we already know. In that case, our solution is the subring of R' generated by R and α, the smallest subring containing R and α. The subring is denoted by $R[\alpha]$. We described this ring in Section 11.1 in the case $R = \mathbb{Z}$, and the description is no different in general: $R[\alpha]$ consists of the elements β of R' that have polynomial expressions

$$\beta = r_n\alpha^n + \cdots + r_1\alpha + r_0$$

with coefficients r_i in R.

But as happens when we construct $\mathbb{C}$ from $\mathbb{R}$, we may not yet have an extension containing α. Then we must construct the extension abstractly. We start with the polynomial ring $R[x]$. It is generated by R and x. The element x of satisfies no relations other than those implied by the ring axioms, and we will probably want our new element α to satisfy some relations. But now that we have the ring $R[x]$ in hand, we can add relations to it using the procedure explained in the previous section *on the polynomial ring $R[x]$*. The fact that R is replaced by $R[x]$ complicates the notation, but aside from this, nothing is different.

For example, we construct the complex numbers by introducing the relation $x^2 + 1 = 0$ into the ring $P = \mathbb{R}[x]$ of real polynomials. We form the quotient ring $\overline{P} = P/(x^2 + 1)$, and the residue of x becomes our element i. The relation $\overline{x}^2 + 1 = 0$ holds in $\overline{P}$ because the map $\pi : P \to \overline{P}$ is a homomorphism and because $x^2 + 1$ is in its kernel. So $\overline{P}$ is isomorphic to $\mathbb{C}$.

In general, say that we want to adjoin an element α to a ring R, and that we want α to satisfy the polynomial relation $f(x) = 0$, where

$$(11.5.1) \qquad f(x) = a_n x^n + a_{n-1} x^{n-1} + \cdots + a_1 x + a_0, \quad \text{with } a_i \text{ in } R.$$

The solution is $R' = R[x]/(f)$, where (f) is the principal ideal of $R[x]$ generated by f.

We let α denote the residue $\overline{x}$ of x in R'. Then because the map $\pi : R[x] \to R[x]/(f)$ is a homomorphism,

$$(11.5.2) \qquad \pi(f(x)) = \overline{f(x)} = \overline{a}_n \alpha^n + \cdots + \overline{a}_0 = 0.$$

Here $\overline{a}_i$ is the image in R' of the constant polynomial a_i. So, dropping bars, α satisfies the relation $f(\alpha) = 0$. The ring obtained in this way may be denoted by $R[\alpha]$ too.

An example: Let a be an element of a ring R. An inverse of a is an element α that satisfies the relation

$$(11.5.3) \qquad a\alpha - 1 = 0.$$

So we can adjoin an inverse by forming the quotient ring $R' = R[x]/(ax - 1)$.

The most important case is that our element α is a root of a monic polynomial:

$$(11.5.4) \qquad f(x) = x^n + a_{n-1} x^{n-1} + \cdots + a_1 x + a_0, \quad \text{with } a_i \text{ in } R.$$

We can describe the ring $R[\alpha]$ precisely in this case.

Proposition 11.5.5 Let R be a ring, and let $f(x)$ be a monic polynomial of positive degree n with coefficients in R. Let $R[\alpha]$ denote the ring $R[x]/(f)$ obtained by adjoining an element satisfying the relation $f(\alpha) = 0$.

(a) The set $(1, \alpha, \ldots, \alpha^{n-1})$ is a *basis* of $R[\alpha]$ over R: every element of $R[\alpha]$ can be written uniquely as a linear combination of this basis, with coefficients in R.

(b) Addition of two linear combinations is vector addition.

(c) Multiplication of linear combinations is as follows: Let β_1 and β_2 be elements of $R[\alpha]$, and let $g_1(x)$ and $g_2(x)$ be polynomials such that $\beta_1 = g_1(\alpha)$ and $\beta_2 = g_2(\alpha)$. One divides the product polynomial $g_1 g_2$ by f, say $g_1 g_2 = fq + r$, where the remainder $r(x)$, if not zero, has degree $<n$. Then $\beta_1 \beta_2 = r(\alpha)$.

The next lemma should be clear.

Lemma 11.5.6 Let f be a *monic* polynomial of degree n in a polynomial ring $R[x]$. Every nonzero element of (f) has degree at least n. $\qquad \square$

Proof of the proposition. **(a)** Since $R[\alpha]$ is a quotient of the polynomial ring $R[x]$, every element β of $R[\alpha]$ is the residue of a polynomial $g(x)$, i.e., $\beta = g(\alpha)$. Since f is monic, we can perform division with remainder: $g(x) = f(x)q(x) + r(x)$, where $r(x)$ is either zero or else has degree less than n (11.2.9). Then since $f(\alpha) = 0$, $\beta = g(\alpha) = r(\alpha)$. In this way, β is written as a combination of the basis. The expression for β is unique because the principal ideal (f) contains no element of degree $<n$. This also proves **(c)**, and **(b)** follows from the fact that addition in $R[x]$ is vector addition.   □

Examples 11.5.7 **(a)** The kernel of the substitution map $\mathbb{Z}[x] \to \mathbb{C}$ that sends $x \rightsquigarrow \gamma = \sqrt[3]{2}$ is the principal ideal $(x^3 - 2)$ of $\mathbb{Z}[x]$ (11.3.23). So $\mathbb{Z}[\gamma]$ is isomorphic to $\mathbb{Z}[x]/(x^3 - 2)$. The proposition shows that $(1, \gamma, \gamma^2)$ is a $\mathbb{Z}$-basis for $\mathbb{Z}[\gamma]$. Its elements are linear combinations $a_0 + a_1\gamma + a_2\gamma^2$, where a_i are integers. If $\beta_1 = (\gamma^2 - \gamma)$ and $\beta_2 = (\gamma^2 + 1)$, then

$$\beta_1\beta_2 = \gamma^4 - \gamma^3 + \gamma^2 - \gamma = f(\gamma)(\gamma - 1) + (\gamma^2 + \gamma - 2) = \gamma^2 + \gamma - 2.$$

(b) Let R' be obtained by adjoining an element δ to $\mathbb{F}_5$ with the relation $\delta^2 - 3 = 0$. Here δ becomes an abstract square root of 3. Proposition 11.5.5 tells us that the elements of R' are the 25 linear expressions $a + b\delta$ with coefficients a and b in $\mathbb{F}_5$.

We'll show that R' is a field of order 25 by showing that every nonzero element $a + b\delta$ of R' is invertible. To see this, consider the product $c = (a + b\delta)(a - b\delta) = (a^2 - 3b^2)$. This is is an element of $\mathbb{F}_5$, and because 3 isn't a square in $\mathbb{F}_5$, it isn't zero unless both a and b are zero. So if $a + b\delta \neq 0$, c is invertible in $\mathbb{F}_5$. Then the inverse of $a + b\delta$ is $(a - b\delta)c^{-1}$.

(c) The procedure used in **(b)** doesn't yield a field when it is applied to $\mathbb{F}_{11}$. The reason is that $\mathbb{F}_{11}$ already contains two square roots of 3, namely ± 5. If R' is the ring obtained by adjoining δ with the relation $\delta^2 - 3 = 0$, we are adjoining an abstract square root of 3, though $\mathbb{F}_{11}$ already contains two square roots. At first glance one might expect to get $\mathbb{F}_{11}$ back. We don't, because we haven't told δ to be equal to 5 or –5. We've told δ only that its square is 3. So $\delta - 5$ and $\delta + 5$ are not zero, but $(\delta + 5)(\delta - 5) = \delta^2 - 3 = 0$. This cannot happen in a field.   □

It is harder to analyze the structure of the ring obtained by adjoining an element when the polynomial relation isn't monic.

• There is a point that we have suppressed in our discussion, and we consider it now: When we adjoin an element α to a ring R with some relation $f(\alpha) = 0$, will our original R be a subring of the ring R' that we construct? We know that R is contained in the polynomial ring $R[x]$, as the subring of constant polynomials, and we also have the canonical map $\pi : R[x] \to R' = R[x]/(f)$. Restricting π to the constant polynomials gives us a homomorphism $R \to R'$, let's call it ψ. Is ψ injective? If it isn't injective, we cannot identify R with a subring of R'.

The kernel of ψ is the set of constant polynomials in the ideal:

(11.5.8) $\ker \psi = R \cap (f)$.

It is fairly likely that $\ker \psi$ is zero because f will have positive degree. There will have to be a lot of cancellation to make a polynomial multiple of f have degree zero. The kernel is zero when α is required to satisfy a monic polynomial relation. But it isn't always zero. For instance, let R be the ring $\mathbb{Z}/(6)$ of congruence classes modulo 6, and let f be the polynomial $2x + 1$ in $R[x]$. Then $3f = 3$. The kernel of the map $R \to R/(f)$ is not zero.

11.6   PRODUCT RINGS

The product $G \times G'$ of two groups was defined in Chapter 2. It is the product set, and the law of composition is componentwise: $(x, x')(y, y') = (xy, x'y')$. The analogous construction can be made with rings.

Proposition 11.6.1 Let R and R' be rings.

(a) The product set $R \times R'$ is a ring called the *product ring*, with component-wise addition and multiplication:
$$(x, x') + (y, y') = (x + y, x' + y') \quad \text{and} \quad (x, x')(y, y') = (xy, x'y'),$$

(b) The additive and multiplicative identities in $R \times R'$ are $(0, 0)$ and $(1, 1)$, respectively.

(c) The projections $\pi : R \times R' \to R$ and $\pi' : R \times R' \to R'$ defined by $\pi(x, x') = x$ and $\pi'(x, x') = x'$ are ring homomorphisms. The kernels of π and π' are the ideals $\{0\} \times R'$ and $R \times \{0\}$, respectively, of $R \times R'$.

(d) The kernel $R \times \{0\}$ of π' is a ring, with multiplicative identity $e = (1, 0)$. It is not a subring of $R \times R'$ unless R' is the zero ring. Similarly, $\{0\} \times R'$ is a ring with identity $e' = (0, 1)$. It is not a subring of $R \times R'$ unless R is the zero ring.

The proofs of these assertions are very elementary. We omit them, but see the next proposition for part **(d)**.   □

To determine whether or not a given ring is isomorphic to a product ring, one looks for the elements that in a product ring would be $(1, 0)$ and $(0, 1)$. They are idempotent elements.

• An *idempotent* element e of a ring S is an element of S such that $e^2 = e$.

Proposition 11.6.2 Let e be an idempotent element of a ring S.

(a) The element $e' = 1 - e$ is also idempotent, $e + e' = 1$, and $ee' = 0$.

(b) With the laws of composition obtained by restriction from S, the principal ideal eS is a ring with identity element e, and multiplication by e defines a ring homomorphism $S \to eS$.

(c) The ideal eS is not a subring of S unless e is the unit element 1 of S and $e' = 0$.

(d) The ring S is isomorphic to the product ring $eS \times e'S$.

Proof. **(a)** $e'^2 = (1 - e)^2 = 1 - 2e + e = e'$, and $ee' = e(1 - e) = e - e = 0$.

(b) Every ideal I of a ring S has the properties of a ring except for the existence of a multiplicative identity. In this case, e is an identity element for eS, because if a is in eS, say $a = es$, then $ea = e^2s = es = a$. The ring axioms show that multiplication by e is a homomorphism: $e(a + b) = ea + eb$, $e(ab) = e^2ab = (ea)(eb)$, and $e1 = e$.

(c) To be a subring of S, eS must contain the identity 1 of S. If it does, then e and 1 will both be identity elements of eS, and since the identity in a ring is unique, $e = 1$ and $e' = 0$.

(d) The rule $\varphi(x) = (ex, e'x)$ defines a homomorphism $\varphi : S \to eS \times e'S$, because both of the maps $x \rightsquigarrow ex$ and $x \rightsquigarrow e'x$ are homomorphisms and the laws of composition in the product ring are componentwise. We verify that this homomorphism is bijective. First, if $\varphi(x) = (0, 0)$, then $ex = 0$ and $e'x = 0$. If so, then $x = (e + e')x = ex + e'x = 0$ too. This shows that φ is injective. To show that φ is surjective, let (u, v) be an element of $eS \times e'S$, say $u = ex$ and $v = e'y$. Then $\varphi(u + v) = (e(ex + e'y), e'(ex + e'y)) = (u, v)$. So (u, v) is in the image, and therefore φ is surjective. $\qquad\square$

Examples 11.6.3 **(a)** We go back to the ring R' obtained by adjoining an abstract square root of 3 to $\mathbb{F}_{11}$. Its elements are the 11^2 linear combinations $a + b\delta$, with a and b in $\mathbb{F}_{11}$ and $\delta^2 = 3$. We saw in (11.5.7)**(c)** that this ring is not a field, the reason being that $\mathbb{F}_{11}$ already contains two square roots ± 5 of 3. The elements $e = \delta - 5$ and $e' = -\delta - 5$ are idempotents in R', and $e + e' = 1$. Therefore R' is isomorphic to the product $eR' \times e'R'$. Since the order of R' is 11^2, $|eR'| = |e'R'| = 11$. The rings eR' and $e'R'$ are both isomorphic to $\mathbb{F}_{11}$, and R' is isomorphic to the product ring $\mathbb{F}_{11} \times \mathbb{F}_{11}$.

(b) We define a homomorphism $\varphi : \mathbb{C}[x, y] \to \mathbb{C}[x] \times \mathbb{C}[y]$ from the polynomial ring in two variables to the product ring by $\varphi(f(x, y)) = (f(x, 0), f(0, y))$. Its kernel is the set of polynomials $f(x, y)$ divisible both by y and by x, which is the principal ideal of $\mathbb{C}[x, y]$ generated by xy. The map isn't quite surjective. Its image is the subring of the product consisting of pairs $(p(x), q(y))$ of polynomials with the same constant term. So the quotient $\mathbb{C}[x, y]/(xy)$ is isomorphic to that subring. $\qquad\square$

11.7   FRACTIONS

In this section we consider the use of fractions in rings other than the integers. For instance, a fraction p/q of polynomials p and q, with q not zero, is called a *rational function*.

Let's review the arithmetic of integer fractions. In order to apply the statements below to other rings, we denote the ring of integers by the neutral symbol R.

- A *fraction* is a symbol a/b, or $\frac{a}{b}$, where a and b are elements of R and b is not zero.
- Elements of R are viewed as fractions by the rule $a = a/1$.
- Two fractions a_1/b_1 and a_2/b_2 are equivalent, $a_1/b_1 \approx a_2/b_2$, if the elements of R that are obtained by "cross multiplying" are equal, i.e., if $a_1 b_2 = a_2 b_1$.
- Sums and products of fractions are given by $\quad \dfrac{a}{b} + \dfrac{c}{d} = \dfrac{ad + bc}{bd}, \quad$ and $\quad \dfrac{a}{b}\dfrac{c}{d} = \dfrac{ac}{bd}$.

We use the term "equivalent" in the third item because, strictly speaking, the fractions aren't actually *equal*.

A problem arises when one replaces the integers by an arbitrary ring R: In the definition of addition, the denominator of the sum is the product bd. Since denominators aren't allowed to be zero, bd had better not be zero. Since b and d are denominators, they aren't zero individually, but we need to know that the product of nonzero elements of R is nonzero. This turns out to be the only problem, but it isn't always true. For example, in the ring $\mathbb{Z}/(6)$ of congruence classes modulo 6, the classes 2 and 3 are not zero, but $2 \cdot 3 = 0$. Or, in a product $R \times R'$ of nonzero rings, the idempotents $(1, 0)$ and $(0, 1)$ are nonzero elements whose product is zero. One cannot work with fractions in those rings.

- An *integral domain* R, or just a *domain* for short, is a ring with this property: R is not the zero ring, and if a and b are elements of R whose product ab is zero, then $a = 0$ or $b = 0$.

Any subring of a field is a domain, and if R is a domain, the polynomial ring $R[x]$ is also a domain.

An element a of a ring is called a *zero divisor* if it is nonzero, and if there is another nonzero element b such that $ab = 0$. An integral domain is a nonzero ring which contains no zero divisors.

An integral domain R satisfies the *cancellation law*:

(11.7.1)   If $ab = ac$ and $a \neq 0$, then $b = c$.

For, from $ab = ac$ it follows that $a(b - c) = 0$. Then since $a \neq 0$ and since R is a domain, $b - c = 0$. $\qquad\square$

Theorem 11.7.2 Let F be the set of equivalence classes of fractions of elements of an integral domain R.

(a) With the laws defined as above, F is a field, called the *fraction field* of R.

(b) R embeds as a subring of F by the rule $a \rightsquigarrow a/1$.

(c) *Mapping Property:* If R is embedded as a subring of another field $\mathcal{F}$, the rule $a/b = ab^{-1}$ embeds F into $\mathcal{F}$ too.

The phrase "mapping property" is explained as follows: To write the property carefully, one should imagine that the embedding of R into $\mathcal{F}$ is given by an injective ring homomorphism $\varphi : R \rightarrow \mathcal{F}$. The assertion is then that the rule $\Phi(a/b) = \varphi(a)\varphi(b)^{-1}$ extends φ to an injective homomorphism $\Phi : F \rightarrow \mathcal{F}$.

The proof of Theorem 11.7.2 has many parts. One must verify that what we call equivalence of fractions is indeed an equivalence relation, that addition and multiplication are well-defined on equivalence classes, that the axioms for a field hold, and that sending $a \rightsquigarrow a/1$ is an injective homomorphism $R \rightarrow F$. Then one must check the mapping property. All of these verifications are straightfoward.

If we were the first people who wished to use fractions in a ring, we'd be nervous and would want to go carefully through each of the verifications. But they have been made many times. It seems sufficient to check a few of them to get a sense of what is involved.

Let us check that equivalence of fractions is a transitive relation. Suppose that $a_1/b_1 \approx a_2/b_2$ and also that $a_2/b_2 \approx a_3/b_3$ Then $a_1 b_2 = a_2 b_1$ and $a_2 b_3 = a_3 b_2$. We multiply by b_3 and b_1:

$$a_1 b_2 b_3 = a_2 b_1 b_3 \quad \text{and} \quad a_2 b_3 b_1 = a_3 b_2 b_1.$$

Therefore $a_1 b_2 b_3 = a_3 b_2 b_1$. Cancelling b_2, $a_3 b_1 = a_1 b_3$. Thus $a_1/b_1 \approx a_3/b_3$. Since we used the cancellation law, the fact that R is a domain is essential here.

Next, we show that addition of fractions is well-defined. Suppose that $a/b \approx a'/b'$ and $c/d \approx c'/d'$. We must show that $a/b + c/d \approx a'/b' + c'/d'$, and to do that, we cross multiply the expressions for the sums. We must show that $u = (ad + bc)(b'd')$ is equal to $v = (a'd' + b'c')(bd)$. The relations $ab' = a'b$ and $cd' = c'd$ show that

$$u = adb'd' + bcb'd' = a'dbd' + bc'b'd = v.$$

Verification of the mapping property is routine too. The only thing worth remarking is that, if R is contained in $\mathcal{F}$ and if a/b is a fraction, then $b \neq 0$, so the rule $a/b = ab^{-1}$ makes sense.

As mentioned above, a fraction of polynomials is called a *rational function*, and the fraction field of the polynomial ring $K[x]$, where K is a field, is called the *field of rational functions* in x, with coefficients in K. This field is usually denoted by $K(x)$:

(11.7.3) $K(x) = \left\{ \begin{array}{l} \text{equivalence classes of fractions } f/g, \text{ where } f \text{ and } g \\ \text{are polynomials, and } g \text{ is not the zero polynomial} \end{array} \right\}.$

The rational functions we define here are equivalence classes of fractions of the formal polynomials that were defined in Section 11.2. If $K = \mathbb{R}$, evaluation of a rational function $f(x)/g(x)$ defines an actual function on the real line, wherever $g(x) \neq 0$. But as with polynomials, we should distinguish the formally defined rational functions, which are fractions of formal polynomials, from the functions that they define.

11.8 MAXIMAL IDEALS

In this section we investigate the kernels of *surjective* homomorphisms

(11.8.1) $\varphi : R \to F$

from a ring R to a field F.

Let φ be such a map. The field F has just two ideals, the zero ideal (0) and the unit ideal (1) (11.3.19). The inverse image of the zero ideal is the kernel I of φ, and the inverse image of the unit ideal is the unit ideal of R. The Correspondence Theorem tells us that the only ideals of R that contain I are I and R. Because of this, I is called a maximal ideal.

• A *maximal ideal* M of a ring R is an ideal that isn't equal to R, and that isn't contained in any ideal other than M and R: If an ideal I contains M, then $I = M$ or $I = R$.

Proposition 11.8.2

(a) Let $\varphi : R \to R'$ be a surjective ring homomorphism, with kernel I. The image R' is a field if and only if I is a maximal ideal.

(b) An ideal I of a ring R is maximal if and only if $\overline{R} = R/I$ is a field.

(c) The zero ideal of a ring R is maximal if and only if R is a field.

Proof. (a) A ring is a field if it contains precisely two ideals (11.3.19), so the Correspondence Theorem asserts that the image of φ is a field if and only if there are two precisely ideals that contain its kernel I. This will be true if and only if I is a maximal ideal.

Parts (b) and (c) follow when (a) is applied to the canonical map $R \to R/I$.   □

Proposition 11.8.3 The maximal ideals of the ring $\mathbb{Z}$ of integers are the principal ideals generated by prime integers. □

Proof. Every ideal of $\mathbb{Z}$ is principal. Consider a principal ideal (n), with $n \geq 0$. If n is a prime, say $n = p$, then $\mathbb{Z}/(n) = \mathbb{F}_p$, a field. The ideal (n) is maximal. If n is not prime, there are three possibilities: $n = 0$, $n = 1$, or n factors. Neither the zero ideal nor the unit ideal is maximal. If n factors, say $n = ab$, with $1 < a < n$, then $1 \notin (a)$, $a \notin (n)$, and $n \in (a)$. Therefore $(n) < (a) < (1)$. The ideal (n) is not maximal. □

- A polynomial with coefficients in a field is called *irreducible* if it is not constant and if is not the product of two polynomials, neither of which is a constant.

Proposition 11.8.4

(a) Let F be a field. The maximal ideals of $F[x]$ are the principal ideals generated by the monic irreducible polynomials.

(b) Let $\varphi: F[x] \to R'$ be a homomorphism to an integral domain R', and let P be the kernel of φ. Either P is a maximal ideal, or $P = (0)$.

The proof of part **(a)** is analogous to the proof just given. We omit the proof of **(b)**. □

Corollary 11.8.5 There is a bijective correspondence between maximal ideals of the polynomial ring $\mathbb{C}[x]$ in one variable and points in the complex plane. The maximal ideal M_a that corresponds to a point a of $\mathbb{C}$ is the kernel of the substitution homomorphism $s_a: \mathbb{C}[x] \to \mathbb{C}$ that sends $x \rightsquigarrow a$. It is the principal ideal generated by the linear polynomial $x - a$.

Proof. The kernel M_a of the substitution homomorphism s_a consists of the polynomials that have a as a root, which are those divisible by $x - a$. So $M_a = (x - a)$. Conversely, let M be a maximal ideal of $\mathbb{C}[x]$. Then M is generated by a monic irreducible polynomial. The monic irreducible polynomials in $\mathbb{C}[x]$ are the polynomials $x - a$. □

The next theorem extends this corollary to polynomials rings in several variables.

Theorem 11.8.6 Hilbert's Nullstellensatz.[1] The maximal ideals of the polynomial ring $\mathbb{C}[x_1, \ldots, x_n]$ are in bijective correspondence with points of complex n-dimensional space. A point $a = (a_1, \ldots, a_n)$ of $\mathbb{C}^n$ corresponds to the kernel M_a of the substitution map $s_a: \mathbb{C}[x_1, \ldots, x_n] \to \mathbb{C}$ that sends $x_i \rightsquigarrow a_i$. The kernel M_a is generated by the n linear polynomials $x_i - a_i$.

Proof. Let a be a point of $\mathbb{C}^n$, and let M_a be the kernel of s_a. Since s_a is surjective and since $\mathbb{C}$ is a field, M_a is a maximal ideal. To verify that M_a is generated by the linear polynomials as asserted, we first consider the case that the point a is the origin $(0, \ldots, 0)$. We must show that the kernel of the map s_0 that evaluates a polynomial at the origin is generated by the variables $x_1, \ldots, x_n$. Well, $f(0, \ldots, 0) = 0$ if and only if the constant term of f is zero. If so, then every monomial that occurs in f is divisible by at least one of the variables, so f can

---

[1]The German word *Nullstellensatz* is a combination of three words whose translations are zero, places, theorem.

be written as a linear combination of the variables, with polynomial coefficients. The proof for an arbitrary point a can be made using the change of variable $x_i = x_i' + a_i$ to move a to the origin.

It is harder to prove that every maximal ideal has the form M_a. Let M be a maximal ideal, and let $\mathcal{F}$ denote the field $\mathbb{C}[x_1, \ldots, x_n]/M$. We restrict the canonical map (11.4.1) $\pi : \mathbb{C}[x_1, \ldots, x_n] \to \mathcal{F}$ to the subring $\mathbb{C}[x_1]$ of polynomials in in the first variable, obtaining a homomorphism $\varphi_1 : \mathbb{C}[x_1] \to \mathcal{F}$. Proposition 11.8.4 shows that the kernel of φ is either the zero ideal, or one of the maximal ideals $(x_1 - a_1)$ of $\mathbb{C}[x_1]$. We'll show that it cannot be the zero ideal. The same will be true when the index 1 is replaced by any other index, so M will contain linear polynomials of the form $x_i - a_i$ for each i. This will show that M contains one of the ideals M_a, and since M_a is maximal, M will be equal to that ideal.

In what follows, we drop the subscript from x_1. We suppose that $\ker \varphi = (0)$. Then φ maps $\mathbb{C}[x]$ isomorphically to its image, a subring of $\mathcal{F}$. The mapping property of fraction fields shows that this map extends to an injective map $\mathbb{C}(x) \to \mathcal{F}$, where $\mathbb{C}(x)$ is the field of rational functions – the field of fractions of the polynomial ring $\mathbb{C}[x]$. So $\mathcal{F}$ contains a field isomorphic to $\mathbb{C}(x)$. The next lemma shows that this is impossible. Therefore $\ker \varphi \neq (0)$.

Lemma 11.8.7

(a) Let R be a ring that contains the complex numbers $\mathbb{C}$ as a subring. The laws of composition on R can be used to make R into a complex vector space.

(b) As a vector space, the field $\mathcal{F} = \mathbb{C}[x_1, \ldots, x_n]/M$ is spanned by a countable set of elements.

(c) Let V be a vector space over a field, and suppose that V is spanned by a countable set of vectors. Then every independent subset of V is finite or countably infinite.

(d) When $\mathbb{C}(x)$ is made into a vector space over $\mathbb{C}$, the uncountable set of rational functions $(x - \alpha)^{-1}$, with α in $\mathbb{C}$, is independent.

Assume that the lemma has been proved. Then **(b)** and **(c)** show that every independent set in $\mathcal{F}$ is finite or countably infinite. On the other hand, $\mathcal{F}$ contains a subring isomorphic to $\mathbb{C}(x)$, so by **(d)**, $\mathcal{F}$ contains an uncountable independent set. This is a contradiction. $\square$

Proof of the Lemma. **(a)** For addition, one uses the addition law in R. Scalar multiplication ca of an element a of R by an element c of $\mathbb{C}$ is defined by multiplying these elements in R. The axioms for a vector space follow from the ring axioms.

(b) The surjective homomorphism $\pi : \mathbb{C}[x_1, \ldots, x_n] \to \mathcal{F}$ defines a map $\mathbb{C} \to \mathcal{F}$, by means of which we identify $\mathbb{C}$ as a subring of $\mathcal{F}$, and make $\mathcal{F}$ into a complex vector space. The countable set of monomials $x_1^{e_1} \cdots x_n^{e_n}$ forms a basis for $\mathbb{C}[x_1, \ldots, x_n]$, and since π is surjective, the images of these monomials span $\mathcal{F}$.

(c) Let S be a countable set that spans V, say $S = \{v_1, v_2, \ldots\}$. It could be finite or infinite. Let S_n be the subset $(v_1, \ldots, v_n)$ consisting of the first n elements of S, and let V_n be the span of S_n. If S is infinite, there will be infinitely many of these subspaces. Since S spans V, every element of V is a linear combination of finitely many elements of S, so it is in one of the spaces V_n. In other words, $\bigcup V_n = V$.

Let L be an independent set in V, and let $L_n = L \cap V_n$. Then L_n is a linearly independent subset of the space V_n, which is spanned by a set of n elements. So $|L_n| \le n$ (3.4.18). Moreover, $L = \bigcup L_n$ because $V = \bigcup V_n$. The union of countably many finite sets is finite or countably infinite.

(d) We must remember that linear combinations can involve only finitely many vectors. So we ask: Can we have a linear relation

$$\sum_{\nu=1}^{k} \frac{c_\nu}{x - \alpha_\nu} = 0,$$

where $\alpha_1, \dots, \alpha_k$ are distinct complex numbers and the coefficients c_ν aren't zero? No. Such a linear combination of formal rational functions defines a complex valued function except at the points $x = \alpha_\nu$. If the linear combination were zero, the function it defines would be identically zero. But $(x - \alpha_1)^{-1}$ takes on arbitrarily large values near α_1, while $(x - \alpha_\nu)^{-1}$ is bounded near α_1 for $\nu = 2, \dots, k$. So the linear combination does not define the zero function. $\qquad\square$

11.9   ALGEBRAIC GEOMETRY

A point $(a_1, \dots, a_n)$ of $\mathbb{C}^n$ is called a *zero* of a polynomial $f(x_1, \dots, x_n)$ of n variables if $f(a_1, \dots, a_n) = 0$. We also say that the polynomial f *vanishes* at such a point. The *common zeros* of a set $\{f_1, \dots, f_r\}$ of polynomials are the points of $\mathbb{C}^n$ at which all of them vanish – the solutions of the system of equations $f_1 = \cdots = f_r = 0$.

- A subset V of complex n-space $\mathbb{C}^n$ that is the set of common zeros of a finite number of polynomials in n variables is called an *algebraic variety*, or just a *variety*.

For instance, a *complex line* in the (x, y)-plane $\mathbb{C}^2$ is, by definition, the set of solutions of a linear equation $ax + by + c = 0$. This is a variety. So is a point. The point (a, b) of $\mathbb{C}^2$ is the set of common zeros of the two polynomials $x - a$ and $y - b$. The group $SL_2(\mathbb{C})$ is a variety in $\mathbb{C}^{2 \times 2}$. It is the set of zeros of the polynomial $x_{11}x_{22} - x_{12}x_{21} - 1$.

The Nullstellensatz provides an important link between algebra and geometry. It tells us that the maximal ideals in the polynomial ring $\mathbb{C}[x_1, \dots, x_n]$ correspond to points in $\mathbb{C}^n$. This correspondence also relates algebraic varieties to quotient rings of the polynomial ring.

Theorem 11.9.1   Let I be the ideal of $\mathbb{C}[x_1, \dots, x_n]$ generated by some polynomials $f_1, \dots f_r$, and let R be the quotient ring $\mathbb{C}[x_1, \dots, x_n]/I$. Let V be the variety of (common) zeros of the polynomials $f_1, \dots, f_r$ in $\mathbb{C}^n$. The maximal ideals of R are in bijective correspondence with the points of V.

Proof. The maximal ideals of R correspond to the maximal ideals of $\mathbb{C}[x_1, \dots, x_n]$ that contain I (Correspondence Theorem). An ideal of $\mathbb{C}[x_1, \dots, x_n]$ will contain I if and only if it contains the generators $f_1, \dots, f_r$ of I. Every maximal ideal of the ring $\mathbb{C}[x_1, \dots, x_n]$ is the kernel M_a of the substitution map that sends $x_i \rightsquigarrow a_i$ for some point $a = (a_1, \dots, a_n)$ of $\mathbb{C}^n$, and the polynomials $f_1, \dots, f_r$ are in M_a if and only if $f_1(a) = \cdots = f_r(a) = 0$, which is to say, if and only if a is a point of V. $\qquad\square$

As this theorem suggests, algebraic properties of the ring $R = \mathbb{C}[x]/I$ are closely related to geometric properties of the variety V. The analysis of this relationship is the field of mathematics called algebraic geometry.

A simple question one might ask about a set is whether or not it is empty. Is it possible for a ring to have no maximal ideals at all? This happens only for the zero ring.

Theorem 11.9.2 Let R be a ring. Every ideal I of R that is not R itself is contained in a maximal ideal.

To find a maximal ideal, one might try this procedure: If I is not maximal, choose a proper ideal I' that is larger than I. Replace I by I', and repeat. The proof follows this line of reasoning, but one may have to repeat the procedure many times, possibly uncountably often. Because of this, the proof requires the *Axiom of Choice*, or *Zorn's Lemma* (see the Appendix). The Hilbert Basis Theorem, which we will prove later (14.6.7), shows that for most rings that we study, the proof requires only a weak countable version of the Axiom of Choice. Rather than enter into a discussion of the Axiom of Choice here, we defer further discussion of the proof to Chapter 14. □

Corollary 11.9.3 The only ring R having no maximal ideals is the zero ring.

This follows from the theorem, because every nonzero ring R contains an ideal different from R: the zero ideal. □

Putting Theorems 11.9.1 and 11.9.2 together gives us another corollary:

Corollary 11.9.4 If a system of polynomial equations $f_1 = \cdots = f_r = 0$ in n variables has no solution in $\mathbb{C}^n$, then 1 is a linear combination $1 = \sum g_i f_i$ with polynomial coefficients g_i.

Proof. If the system has no solution, there is no maximal ideal that contains the ideal $I = (f_1, \ldots, f_r)$. So I is the unit ideal, and 1 is in I. □

Example 11.9.5 Most choices of three polynomials f_1, f_2, f_3 in two variables have no common solutions. For instance, the ideal of $\mathbb{C}[t, x]$ generated by

(11.9.6) $$f_1 = t^2 + x^2 - 2, \quad f_2 = tx - 1, \quad f_3 = t^3 + 5tx^2 + 1$$

is the unit ideal. This can be proved by showing that the equations $f_1 = f_2 = f_3 = 0$ have no solution in $\mathbb{C}^2$. □

It isn't easy to get a clear geometric picture of an algebraic variety in $\mathbb{C}^n$, but the general shape of a variety in $\mathbb{C}^2$ can be described fairly simply, and we do that here. We work with the polynomial ring in the two variables t and x.

Lemma 11.9.7 Let $f(t, x)$ be a polynomial, and let α be a complex number. The following are equivalent:

(a) $f(t, x)$ vanishes at every point of the locus $\{t = \alpha\}$ in $\mathbb{C}^2$,

(b) The one-variable polynomial $f(\alpha, x)$ is the zero polynomial,

(c) $t - \alpha$ divides f in $\mathbb{C}[t, x]$.

Proof. If f vanishes at every point of the locus $t = \alpha$, the polynomial $f(\alpha, x)$ is zero for every x. Then since a nonzero polynomial in one variable has finitely many roots, $f(\alpha, x)$ is the zero polynomial. This shows that **(a)** implies **(b)**.

A change of variable $t = t' + \alpha$ reduces the proof that **(b)** implies **(c)** to the case that $\alpha = 0$. If $f(0, x)$ is the zero polynomial, then t divides every monomial that occurs in f, and t divides f. Finally, the implication **(c)** implies **(a)** is clear. $\square$

Let $\mathcal{F}$ denote the field of rational functions $\mathbb{C}(t)$ in t, the field of fractions of the ring $\mathbb{C}[t]$. The ring $\mathbb{C}[t, x]$ is a subring of the one-variable polynomial ring $\mathcal{F}[x]$; its elements are polynomials in x,

(11.9.8) $$f(t, x) = a_n(t)x^n + \cdots + a_1(t)x + a_0(t),$$

whose coefficients $a_i(t)$ are rational functions in t. It can be helpful to begin by studying a problem about $\mathbb{C}[t, x]$ in the ring $\mathcal{F}[x]$, because its algebra is simpler. Division with remainder is available, and every ideal of $\mathcal{F}[x]$ is principal.

Proposition 11.9.9 Let $h(t, x)$ and $f(t, x)$ be nonzero elements of $\mathbb{C}[t, x]$. Suppose that h is not divisible by any polynomial of the form $t - \alpha$. If h divides f in $\mathcal{F}[x]$, then h divides f in $\mathbb{C}[t, x]$.

Proof. We divide by h in $\mathcal{F}[x]$, say $f = hq$, and we show that q is an element of $\mathbb{C}[t, x]$. Since q is an element of $\mathcal{F}[x]$, it is a polynomial in x whose coefficients are rational functions in t. We multiply both sides of the equation $f = hq$ by a monic polynomial in t to clear denominators in these coefficients. This gives us an equation of the form

$$u(t) f(t, x) = h(t, x)q_1(t, x),$$

where $u(t)$ is a monic polynomial in t, and q_1 is an element of $\mathbb{C}[t, x]$. We use induction on the degree of u. If u has positive degree, it will have a complex root α. Then $t - \alpha$ divides the left side of this equation, so it divides the right side too. This means that $h(\alpha, x)q_1(\alpha, x)$ is the zero polynomial in x. By hypothesis, $t - \alpha$ does not divide h, so $h(\alpha, x)$ is not zero. Since the polynomial ring $\mathbb{C}[x]$ is a domain, $q_1(\alpha, x) = 0$, and the lemma shows that $t - \alpha$ divides $q_1(t, x)$. We cancel $t - \alpha$ from u and q_1. Induction completes the proof. $\square$

Theorem 11.9.10 Two nonzero polynomials $f(t, x)$ and $g(t, x)$ in two variables have only finitely many common zeros in $\mathbb{C}^2$, unless they have a common nonconstant factor in $\mathbb{C}[t, x]$.

If the degrees of the polynomials f and g are m and n respectively, the number of common zeros is at most mn. This is known as the *Bézout bound*. For instance, two quadratic polynomials have at most four common zeros. (The analogue of this statement for real polynomials is that two conics intersect in at most four points.) It is harder to prove the Bézout bound than the finiteness. We won't need that bound, so we won't prove it.

Proof of Theorem 11.9.10. Assume that f and g have no common factor. Let I denote the ideal generated by f and g in $\mathcal{F}[x]$, where $\mathcal{F} = \mathbb{C}(t)$, as above. This is a principal ideal, generated by the (monic) greatest common divisor h of f and g in $\mathcal{F}[x]$.

If $h \neq 1$, it will be a polynomial whose coefficients may have denominators that are polynomials in t. We multiply by a polynomial in t to clear these denominators, obtaining a polynomial h_1 in $\mathbb{C}[t, x]$. We may assume that h_1 isn't divisible by any polynomial $t - \alpha$. Since the denominators are units in $\mathcal{F}$ and since h divides f and g in $\mathcal{F}[x]$, h_1 also divides f and g in $\mathcal{F}[x]$. Proposition 11.9.9 shows that h_1 divides f and g in $\mathbb{C}[t, x]$. Then f and g have a common nonconstant factor in $\mathbb{C}[t, x]$. We're assuming that this is not the case.

So the greatest common divisor of f and g in $\mathcal{F}[x]$ is 1, and $1 = rf + sg$, where r and s are elements of $\mathcal{F}[x]$. We clear denominators from r and s, multiplying both sides of the equation by a suitable polynomial $u(t)$. This gives us an equation of the form

$$u(t) = r_1(t, x) f(t, x) + s_1(t, x) g(t, x),$$

where all terms on the right are polynomials in $\mathbb{C}[t, x]$. This equation shows that if (t_0, x_0) is a common zero of f and g, then t_0 must be a root of u. But u is a polynomial in t, and a nonzero polynomial in one variable has finitely many roots. So at the common zeros of f and g, the variable t takes on only finitely many values. Similar reasoning shows that x takes on only finitely many values. This gives us only finitely many possibilities for the common zeros. $\qquad\square$

Theorem 11.9.10 suggests that the most interesting varieties in $\mathbb{C}^2$ are those defined as the locus of zeros of a single polynomial $f(t, x)$.

• The locus X of zeros in $\mathbb{C}^2$ of a polynomial $f(t, x)$ is called the *Riemann surface* of f.

It is also called a *plane algebraic curve* – a confusing phrase. As a topological space, the locus X has dimension two. Calling it an algebraic curve refers to the fact that the points of X depend only on one *complex* parameter. We give a rough description of a Riemann surface here. Let's assume that the polynomial f is *irreducible* – that it is not a product of two nonconstant polynomials, and also that it has positive degree in the variable x. Let

(11.9.11) $$X = \{(t, x) \in \mathbb{C}^2 \mid f(t, x) = 0\}$$

be its Riemann surface, and let T denote the complex t-plane. Sending $(t, x) \rightsquigarrow t$ defines a continuous map that we call a *projection*

(11.9.12) $$\pi : X \to T.$$

We will describe X in terms of this projection. However, our description will require that a finite set of "bad points" be removed from X. In fact, what is usually called the Riemann surface agrees with our definition only when suitable finite subsets are removed. The locus $\{f = 0\}$ may be "singular" at some points, and some other points of X may be "at infinity." The points at infinity are explained below (see (11.9.17)).

The simplest examples of singular points are *nodes*, at which the surface crosses itself, and *cusps*. The locus $x^2 = t^3 - t^2$ has a node at the origin, and the locus $x^2 = t^3$ has a cusp at the origin. The real points of these Riemann surfaces are shown here.

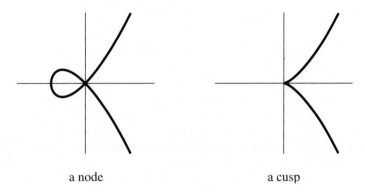

a node   a cusp

(11.9.13)   Some Singular Curves

To avoid repetition of the disclaimer "except on a finite set," we write X' for the complement of an unspecified finite subset of X, which is allowed to vary. Whenever a construction runs into trouble at some point, we simply delete that point. Essentially everything we do here and when we come back to Riemann surfaces in Chapter 15 will be valid only for X'. We keep X on hand for reference.

Our description of the Riemann surface will be as a branched covering of the complex t-plane T. The definition of covering space that we give here assumes that the spaces are Hausdorff spaces ([Munkres] p. 98). You can ignore this point if you don't know what it means. The sets in which we are interested are Hausdorff spaces because they are subsets of $\mathbb{C}^2$.

Definition 11.9.14   Let X and T be Hausdorff spaces. A continuous map $\pi: X \to T$ is an n-sheeted *covering space* if every fibre consists of n points, and if it has this property: Let x_0 be a point of X and let $\pi(x_0) = t_0$. Then π maps an open neigborhood U of x_0 in X homeomorphically to an open neighborhood V of t_0 in T.

A map π from X to the complex plane T is an n-sheeted *branched covering* if X contains no isolated points, the fibres of π are finite, and if there is a finite set Δ of points of T called *branch points*, such that the map $(X - \pi^{-1}\Delta) \to (T - \Delta)$ is an n-sheeted covering space. For emphasis, a covering space is sometimes called an *unbranched covering*.

Figure 11.9.15 below depicts the Riemann surface of the polynomial $x^2 - t$, a two-sheeted covering of T that is branched at the point $t = 0$. The figure has been obtained by writing t and x in terms of their real and imaginary parts, $t = t_0 + t_1 i$ and $x = x_0 + x_1 i$, and dropping the imaginary part x_1 of x, to obtain a surface in three-dimensional space. Its further projection to the plane is depicted using standard graphics.

The projected surface intersects itself along the negative t_0-axis, though the Riemann surface itself does not. Every negative real number t has two purely imaginary square roots. The real parts of these square roots are zero, and this produces the self-crossing in the projected surface.

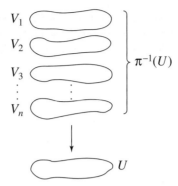

(11.9.14)   Part of an unbranched covering.

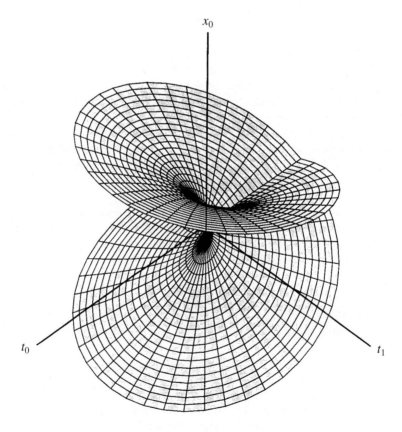

(11.9.15)   The Riemann surface $x^2 = t$.

Given a branched covering $X \to T$, we refer to the points in the set Δ as its *branch points*, though this is imprecise: The defining property continues to hold when we add any finite set of points to Δ. So we allow the possibility that some points of Δ don't need to be included – that they aren't "true" branch points.

Theorem 11.9.16 Let $f(t, x)$ be an irreducible polynomial in $\mathbb{C}[t, x]$ which has positive degree n in the variable x. The Riemann surface of f is an n-sheeted branched covering of the complex plane T.

Proof. The main step is to verify the first condition of (11.9.14), that the fibre $\pi^{-1}(t_0)$ consists of precisely n points except on a finite subset Δ.

The points of the fibre $\pi^{-1}(t_0)$ are the points (t_0, x_0) such that x_0 is a root of the one-variable polynomial $f(t_0, x)$. We must show that, except for a finite set of values $t = t_0$, this polynomial has n distinct roots. We write $f(t, x)$ as a polynomial in x whose coefficients are polynomials in t, say $f(x) = a_n(t)x^n + \cdots + a_0(t)$, and we denote $a_i(t_0)$ by a_i^0. The polynomial $f(t_0, x) = a_n^0 x^n + \cdots + a_1^0 x + a_0^0$ has degree at most n, so it has at most n roots. Therefore the fibre $\pi^{-1}(p)$ contains at most n points. It will have fewer than n points if either

(11.9.17)

(a) the degree of $f(t_0, x)$ is less than n, or
(b) $f(t_0, x)$ has a multiple root.

The first case occurs when t_0 is a root of $a_n(t)$. (If t_0 is a root of $a_n(t)$, one of the roots of $f(t_1, x)$ tends to infinity as $t_1 \to t_0$.) Since $a_n(t)$ is a polynomial, there are finitely many such values.

Consider the second case. A complex number x_0 is a multiple root of a polynomial $h(x)$ if $(x - x_0)^2$ divides $h(x)$, and this happens if and only if x_0 is a common root of $h(x)$ and its derivative $h'(x)$ (see Exercise 3.5). Here $h(x) = f(t_0, x)$. The first variable is fixed, so the derivative is the partial derivative $\frac{\partial f}{\partial x}$. Going back to the polynomial $f(t, x)$ in two variables, we see that the second case occurs at the points (t_0, x_0) that are common zeros of f and $\frac{\partial f}{\partial x}$. Now f cannot divide its partial derivative, which has lower degree in x. Since f is assumed to be irreducible, f and $\frac{\partial f}{\partial x}$ have no common nonconstant factor. Theorem 11.9.10 tells us that there are finitely many common zeros.

We now check the second condition of (11.9.14). Let t_0 be a point of T such that the fibre $\pi^{-1}(t_0)$ consists of n points, and let (t_0, x_0) be a point of X in the fibre. Then x_0 is a simple root of $f(t_0, x)$, and therefore $\frac{\partial f}{\partial x}$ is not zero at this point. The Implicit Function Theorem A.4.3 implies that one can solve for x as a function $x(t)$ of t in a neighborhood of t_0, such that $x(t_0) = x_0$. The neighborhood U referred to in the definition of covering space is the graph of this function. $\qquad\square$

To me algebraic geometry is algebra with a kick.

—Solomon Lefschetz

EXERCISES

Section 1   Definition of a Ring

1.1. Prove that $7 + \sqrt[3]{2}$ and $\sqrt{3} + \sqrt{-5}$ are algebraic numbers.

1.2. Prove that, for $n \neq 0$, $\cos(2\pi/n)$ is an algebraic number.

1.3. Let $\mathbb{Q}[\alpha, \beta]$ denote the smallest subring of $\mathbb{C}$ containing the rational numbers $\mathbb{Q}$ and the elements $\alpha = \sqrt{2}$ and $\beta = \sqrt{3}$. Let $\gamma = \alpha + \beta$. Is $\mathbb{Q}[\alpha, \beta] = \mathbb{Q}[\gamma]$? Is $\mathbb{Z}[\alpha, \beta] = \mathbb{Z}[\gamma]$?

1.4. Let $\alpha = \frac{1}{2}i$. Prove that the elements of $\mathbb{Z}[\alpha]$ are dense in the complex plane.

1.5. Determine all subrings of $\mathbb{R}$ that are discrete sets.

1.6. Decide whether or not S is a subring of R, when

    (a) S is the set of all rational numbers a/b, where b is not divisible by 3, and $R = \mathbb{Q}$,

    (b) S is the set of functions which are linear combinations with integer coefficients of the functions $\{1, \cos nt, \sin nt\}$, $n \in \mathbb{Z}$, and R is the set of all real valued functions of t.

1.7. Decide whether the given structure forms a ring. If it is not a ring, determine which of the ring axioms hold and which fail:

    (a) U is an arbitrary set, and R is the set of subsets of U. Addition and multiplication of elements of R are defined by the rules $A + B = (A \cup B) - (A \cap B)$ and $A \cdot B = A \cap B$.

    (b) R is the set of continuous functions $\mathbb{R} \to \mathbb{R}$. Addition and multiplication are defined by the rules $[f + g](x) = f(x) + g(x)$ and $[f \circ g](x) = f(g(x))$.

1.8. Determine the units in: **(a)** $\mathbb{Z}/12\mathbb{Z}$, **(b)** $\mathbb{Z}/8\mathbb{Z}$, **(c)** $\mathbb{Z}/n\mathbb{Z}$.

1.9. Let R be a set with two laws of composition satisfying all ring axioms except the commutative law for addition. Use the distributive law to prove that the commutative law for addition holds, so that R is a ring.

Section 2   Polynomial Rings

2.1. For which positive integers n does $x^2 + x + 1$ divide $x^4 + 3x^3 + x^2 + 7x + 5$ in $[\mathbb{Z}/(n)][x]$?

2.2. Let F be a field. The set of all formal power series $p(t) = a_0 + a_1 t + a_2 t^2 + \cdots$, with a_i in F, forms a ring that is often denoted by $F[[t]]$. By *formal* power series we mean that the coefficients form an arbitrary sequence of elements of F. There is no requirement of convergence. Prove that $F[[t]]$ is a ring, and determine the units in this ring.

Section 3   Homomorphisms and Ideals

3.1. Prove that an ideal of a ring R is a subgroup of the additive group R^+.

3.2. Prove that every nonzero ideal in the ring of Gauss integers contains a nonzero integer.

3.3. Find generators for the kernels of the following maps:

    (a) $\mathbb{R}[x, y] \to \mathbb{R}$ defined by $f(x, y) \rightsquigarrow f(0, 0)$,

    (b) $\mathbb{R}[x] \to \mathbb{C}$ defined by $f(x) \rightsquigarrow f(2 + i)$,

    (c) $\mathbb{Z}[x] \to \mathbb{R}$ defined by $f(x) \rightsquigarrow f(1 + \sqrt{2})$,

(d) $\mathbb{Z}[x] \to \mathbb{C}$ defined by $x \rightsquigarrow \sqrt{2} + \sqrt{3}$.

(e) $\mathbb{C}[x, y, z] \to \mathbb{C}[t]$ defined by $x \rightsquigarrow t, y \rightsquigarrow t^2, z \rightsquigarrow t^3$.

3.4. Let $\varphi : \mathbb{C}[x, y] \to \mathbb{C}[t]$ be the homomorphism that sends $x \rightsquigarrow t+1$ and $y \rightsquigarrow t^3 - 1$. Determine the kernel K of φ, and prove that every ideal I of $\mathbb{C}[x, y]$ that contains K can be generated by two elements.

3.5. The derivative of a polynomial f with coefficients in a field F is defined by the calculus formula $(a_n x^n + \cdots + a_1 x + a_0)' = n a_n x^{n-1} + \cdots + 1 a_1$. The integer coefficients are interpreted in F using the unique homomorphism $\mathbb{Z} \to F$.

(a) Prove the product rule $(fg)' = f'g + fg'$ and the chain rule $(f \circ g)' = (f' \circ g)g'$.

(b) Let α be an element of F. Prove that α is a multiple root of a polynomial f if and only if it is a common root of f and of its derivative f'.

3.6. An *automorphism* of a ring R is an isomorphism from R to itself. Let R be a ring, and let $f(y)$ be a polynomial in one variable with coefficients in R. Prove that the map $R[x, y] \to R[x, y]$ defined by $x \rightsquigarrow x + f(y), y \rightsquigarrow y$ is an automorphism of $R[x, y]$.

3.7. Determine the automorphisms of the polynomial ring $\mathbb{Z}[x]$ (see Exercise 3.6).

3.8. Let R be a ring of prime characteristic p. Prove that the map $R \to R$ defined by $x \rightsquigarrow x^p$ is a ring homomorphism. (It is called the *Frobenius map*.)

3.9. **(a)** An element x of a ring R is called *nilpotent* if some power is zero. Prove that if x is nilpotent, then $1 + x$ is a unit.

(b) Suppose that R has prime characteristic $p \neq 0$. Prove that if a is nilpotent then $1 + a$ is *unipotent*, that is, some power of $1 + a$ is equal to 1.

3.10. Determine all ideals of the ring $F[[t]]$ of formal power series with coefficients in a field F (see Exercise 2.2).

3.11. Let R be a ring, and let I be an ideal of the polynomial ring $R[x]$. Let n be the lowest degree among nonzero elements of I. Prove or disprove: I contains a monic polynomial of degree n if and only if it is a principal ideal.

3.12. Let I and J be ideals of a ring R. Prove that the set $I + J$ of elements of the form $x + y$, with x in I and y in J, is an ideal. This ideal is called the *sum* of the ideals I and J.

3.13. Let I and J be ideals of a ring R. Prove that the intersection $I \cap J$ is an ideal. Show by example that the set of products $\{xy \mid x \in I, y \in J\}$ need not be an ideal, but that the set of finite sums $\sum x_\nu y_\nu$ of products of elements of I and J is an ideal. This ideal is called the *product ideal*, and is denoted by IJ. Is there a relation between IJ and $I \cap J$?

Section 4   Quotient Rings

4.1. Consider the homomorphism $\mathbb{Z}[x] \to \mathbb{Z}$ that sends $x \rightsquigarrow 1$. Explain what the Correspondence Theorem, when applied to this map, says about ideals of $\mathbb{Z}[x]$.

4.2. What does the Correspondence Theorem tell us about ideals of $\mathbb{Z}[x]$ that contain $x^2 + 1$?

4.3. Identify the following rings: **(a)** $\mathbb{Z}[x]/(x^2 - 3, 2x + 4)$, **(b)** $\mathbb{Z}[i]/(2 + i)$, **(c)** $\mathbb{Z}[x]/(6, 2x - 1)$, **(d)** $\mathbb{Z}[x]/(2x^2 - 4, 4x - 5)$, **(e)** $\mathbb{Z}[x]/(x^2 + 3, 5)$.

4.4. Are the rings $\mathbb{Z}[x]/(x^2 + 7)$ and $\mathbb{Z}[x]/(2x^2 + 7)$ isomorphic?

Section 5   Adjoining Elements

5.1. Let $f = x^4 + x^3 + x^2 + x + 1$ and let α denote the residue of x in the ring $R = \mathbb{Z}[x]/(f)$. Express $(\alpha^3 + \alpha^2 + \alpha)(\alpha^5 + 1)$ in terms of the basis $(1, \alpha, \alpha^2, \alpha^3)$ of R.

5.2. Let a be an element of a ring R. If we adjoin an element α with the relation $\alpha = a$, we expect to get a ring isomorphic to R. Prove that this is true.

5.3. Describe the ring obtained from $\mathbb{Z}/12\mathbb{Z}$ by adjoining an inverse of 2.

5.4. Determine the structure of the ring R' obtained from $\mathbb{Z}$ by adjoining an element α satisfying each set of relations.

(a) $2\alpha = 6,\ 6\alpha = 15$,   (b) $2\alpha - 6 = 0,\ \alpha - 10 = 0$,   (c) $\alpha^3 + \alpha^2 + 1 = 0,\ \alpha^2 + \alpha = 0$.

5.5. Are there fields F such that the rings $F[x]/(x^2)$ and $F[x]/(x^2 - 1)$ are isomorphic?

5.6. Let a be an element of a ring R, and let R' be the ring $R[x]/(ax - 1)$ obtained by adjoining an inverse of a to R. Let α denote the residue of x (the inverse of a in R').

(a) Show that every element β of R' can be written in the form $\beta = \alpha^k b$, with b in R.

(b) Prove that the kernel of the map $R \to R'$ is the set of elements b of R such that $a^n b = 0$ for some $n > 0$.

(c) Prove that R' is the zero ring if and only if a is nilpotent (see Exercise 3.9).

5.7. Let F be a field and let $R = F[t]$ be the polynomial ring. Let R' be the ring extension $R[x]/(tx - 1)$ obtained by adjoining an inverse of t to R. Prove that this ring can be identified as the ring of *Laurent polynomials*, which are finite linear combinations of powers of t, negative exponents included.

Section 6   Product Rings

6.1. Let $\varphi: \mathbb{R}[x] \to \mathbb{C} \times \mathbb{C}$ be the homomorphism defined by $\varphi(x) = (1, i)$ and $\varphi(r) = (r, r)$ for r in $\mathbb{R}$. Determine the kernel and the image of φ.

6.2. Is $\mathbb{Z}/(6)$ isomorphic to the product ring $\mathbb{Z}/(2) \times \mathbb{Z}/(3)$? Is $\mathbb{Z}/(8)$ isomorphic to $\mathbb{Z}/(2) \times \mathbb{Z}/(4)$?

6.3. Classify rings of order 10.

6.4. In each case, describe the ring obtained from the field $\mathbb{F}_2$ by adjoining an element α satisfying the given relation:

(a) $\alpha^2 + \alpha + 1 = 0$,   (b) $\alpha^2 + 1 = 0$,   (c) $\alpha^2 + \alpha = 0$.

6.5. Suppose we adjoin an element α satisfying the relation $\alpha^2 = 1$ to the real numbers $\mathbb{R}$. Prove that the resulting ring is isomorphic to the product $\mathbb{R} \times \mathbb{R}$.

6.6. Describe the ring obtained from the product ring $\mathbb{R} \times \mathbb{R}$ by inverting the element $(2, 0)$.

6.7. Prove that in the ring $\mathbb{Z}[x]$, the intersection $(2) \cap (x)$ of the principal ideals (2) and (x) is the principal ideal $(2x)$, and that the quotient ring $R = \mathbb{Z}[x]/(2x)$ is isomorphic to the subring of the product ring $\mathbb{F}_2[x] \times \mathbb{Z}$ of pairs $(f(x), n)$ such that $f(0) \equiv n$ modulo 2.

6.8. Let I and J be ideals of a ring R such that $I + J = R$.

(a) Prove that $IJ = I \cap J$ (see Exercise 3.13).

(b) Prove the *Chinese Remainder Theorem*: For any pair a, b of elements of R, there is an element x such that $x \equiv a$ modulo I and $x \equiv b$ modulo J. (The notation $x \equiv a$ modulo I means $x - a \in I$.)

(c) Prove that if $IJ = 0$, then R is isomorphic to the product ring $(R/I) \times (R/J)$.

(d) Describe the idempotents corresponding to the product decomposition in **(c)**.

Section 7   Fractions

7.1. Prove that a domain of finite order is a field.

7.2. Let R be a domain. Prove that the polynomial ring $R[x]$ is a domain, and identify the units in $R[x]$.

7.3. Is there a domain that contains exactly 15 elements?

7.4. Prove that the field of fractions of the formal power series ring $F[[x]]$ over a field F can be obtained by inverting the element x. Find a neat description of the elements of that field (see Exercise 11.2.1).

7.5. A subset S of a domain R that is closed under multiplication and that does not contain 0 is called a *multiplicative set*. Given a multiplicative set S, define S-fractions to be elements of the form a/b, where b is in S. Show that the equivalence classes of S-fractions form a ring.

Section 8   Maximal Ideals

8.1. Which principal ideals in $\mathbb{Z}[x]$ are maximal ideals?

8.2. Determine the maximal ideals of each of the following rings:
(a) $\mathbb{R} \times \mathbb{R}$, **(b)** $\mathbb{R}[x]/(x^2)$, **(c)** $\mathbb{R}[x]/(x^2 - 3x + 2)$, **(d)** $\mathbb{R}[x]/(x^2 + x + 1)$.

8.3. Prove that the ring $\mathbb{F}_2[x]/(x^3 + x + 1)$ is a field, but that $\mathbb{F}_3[x]/(x^3 + x + 1)$ is not a field.

8.4. Establish a bijective correspondence between maximal ideals of $\mathbb{R}[x]$ and points in the upper half plane.

Section 9   Algebraic Geometry

9.1. Let I be the principal ideal of $\mathbb{C}[x, y]$ generated by the polynomial $y^2 + x^3 - 17$. Which of the following sets generate maximal ideals in the quotient ring $R = \mathbb{C}[x, y]/I$? $(x - 1, y - 4)$, $(x + 1, y + 4)$, $(x^3 - 17, y^2)$.

9.2. Let $f_1, \ldots, f_r$ be complex polynomials in the variables $x_1, \ldots, x_n$, let V be the variety of their common zeros, and let I be the ideal of the polynomial ring $R = \mathbb{C}[x_1, \ldots, x_n]$ that they generate. Define a homomorphism from the quotient ring $\overline{R} = R/I$ to the ring $\mathcal{R}$ of continuous, complex-valued functions on V.

9.3. Let $U = \{f_i(x_1, \ldots, x_m) = 0\}$, $V = \{g_j(y_1, \ldots, y_n) = 0\}$ be varieties in $\mathbb{C}^m$ and $\mathbb{C}^n$, respectively. Show that the variety defined by the equations $\{f_i(x) = 0, g_j(y) = 0\}$ in x, y-space $\mathbb{C}^{m+n}$ is the product set $U \times V$.

9.4. Let U and V be varieties in $\mathbb{C}^n$. Prove that the union $U \cup V$ and the intersection $U \cap V$ are varieties. What does the statement $U \cap V = \emptyset$ mean algebraically? What about the statement $U \cup V = \mathbb{C}^n$?

9.5. Prove that the variety of zeros of a set $\{f_1, \ldots, f_r\}$ of polynomials depends only on the ideal that they generate.

9.6. Prove that every variety in $\mathbb{C}^2$ is the union of finitely many points and algebraic curves.

9.7. Determine the points of intersection in $\mathbb{C}^2$ of the two loci in each of the following cases:
(a) $y^2 - x^3 + x^2 = 1$, $x + y = 1$, **(b)** $x^2 + xy + y^2 = 1$, $x^2 + 2y^2 = 1$,
(c) $y^2 = x^3$, $xy = 1$, **(d)** $x + y^2 = 0$, $y + x^2 + 2xy^2 + y^4 = 0$.

9.8. Which ideals in the polynomial ring $\mathbb{C}[x, y]$ contain $x^2 + y^2 - 5$ and $xy - 2$?

9.9. An *irreducible* plane algebraic curve C is the locus of zeros in $\mathbb{C}^2$ of an irreducible polynomial $f(x, y)$. A point p of C is a *singular point* of the curve if $f = \partial f/\partial x = \partial f/\partial y = 0$ at p. Otherwise p is a *nonsingular point*. Prove that an irreducible curve has only finitely many singular points.

9.10. Let L be the (complex) line $\{ax + by + c = 0\}$ in $\mathbb{C}^2$, and let C be the algebraic curve $\{f(x, y) = 0\}$, where f is an irreducible polynomial of degree d. Prove $C \cap L$ contains at most d points unless $C = L$.

9.11. Let C_1 and C_2 be the zeros of quadratic polynomials f_1 and f_2 respectively that don't have a common linear factor.

(a) Let p and q be distinct points of intersection of C_1 and C_2, and let L be the (complex) line through p and q. Prove that there are constants c_1 and c_2, not both zero, so that $g = c_1 f_1 + c_2 f_2$ vanishes identically on L. Prove also that g is the product of linear polynomials.

Hint: Force g to vanish at a third point of L.

(b) Prove that C_1 and C_2 have at most 4 points in common.

9.12. Prove in two ways that the three polynomials $f_1 = t^2 + x^2 - 2$, $f_2 = tx - 1$, $f_3 = t^3 + 5tx^2 + 1$ generate the unit ideal in $\mathbb{C}[x, y]$: by showing that they have no common zeros, and also by writing 1 as a linear combination of f_1, f_2, f_3, with polynomial coefficients.

*9.13. Let $\varphi : \mathbb{C}[x, y] \to \mathbb{C}[t]$ be a homomorphism that is the identity on $\mathbb{C}$ and sends $x \rightsquigarrow x(t)$, $y \rightsquigarrow y(t)$, and such that $x(t)$ and $y(t)$ are not both constant. Prove that the kernel of φ is a principal ideal.

Miscellaneous Exercises

M.1. Prove or disprove: If $a^2 = a$ for every a in a nonzero ring R, then R has characteristic 2.

M.2. A semigroup S is a set with an associative law of composition having an identity element. Let S be a commutative semigroup that satisfies the cancellation law: $ab = ac$ implies $b = c$. Prove that S can be embedded into a group.

M.3. Let R denote the set of sequences $a = (a_1, a_2, a_3, \ldots)$ of real numbers that are eventually constant: $a_n = a_{n+1} = \ldots$ for sufficiently large n. Addition and multiplication are componentwise, that is, addition is vector addition and multiplication is defined by $ab = (a_1 b_1, a_2 b_2, \ldots)$. Prove that R is a ring, and determine its maximal ideals.

M.4. (a) Classify rings R that contain $\mathbb{C}$ and have dimension 2 as vector space over $\mathbb{C}$.

(b) Do the same for rings that have dimension 3.

M.5. Define $\varphi : \mathbb{C}[x, y] \to \mathbb{C}[x] \times \mathbb{C}[y] \times \mathbb{C}[t]$ by $f(x, y) \rightsquigarrow (f(x, 0), f(0, y), f(t, t))$. Determine the image of this map, and find generators for the kernel.

M.6. Prove that the locus $y = \sin x$ in $\mathbb{R}^2$ doesn't lie on any algebraic curve in $\mathbb{C}^2$.

*M.7. Let X denote the closed unit interval $[0, 1]$, and let R be the ring of continuous functions $X \to \mathbb{R}$.

(a) Let $f_1, \ldots, f_n$ be functions with no common zero on X. Prove that the ideal generated by these functions is the unit ideal.

Hint: Consider $f_1^2 + \cdots + f_n^2$.

(b) Establish a bijective correspondence between maximal ideals of R and points on the interval.

CHAPTER 12

Factoring

You probably think that one knows everything about polynomials.

—Serge Lang

12.1 FACTORING INTEGERS

We study division in rings in this chapter, modeling our investigation on properties of the ring of integers, and we begin by reviewing those properties. Some have been used without comment in earlier chapters of the book, and some have been proved before.

A property from which many others follow is division with remainder: If a and b are integers and a is positive, there exist integers q and r so that

(12.1.1)
$$b = aq + r, \quad \text{and } 0 \leq r < a.$$

We've seen some of its important consequences:

Theorem 12.1.2

(a) Every ideal of the ring $\mathbb{Z}$ of integers is principal.

(b) A pair a, b of integers, not both zero, has a *greatest common divisor*, a positive integer d with these properties:

   (i) $\mathbb{Z}d = \mathbb{Z}a + \mathbb{Z}b$,

   (ii) d divides a and d divides b,

   (iii) if an integer e divides a and b, then e divides d.

   (iv) There are integers r and s such that $d = ra + sb$.

(c) If a prime integer p divides a product ab of integers, then p divides a or p divides b.

(d) *Fundamental Theorem of Arithmetic:* Every positive integer $a \neq 1$ can be written as a product $a = p_1 \cdots p_k$, where the p_i are positive prime integers, and $k > 0$. This expression is unique except for the ordering of the prime factors.

The proofs of these facts will be reviewed in a more general setting in the next section.

12.2   UNIQUE FACTORIZATION DOMAINS

It is natural to ask which rings have properties analogous to those of the ring of integers, and we investigate this question here. There are relatively few rings for which all parts of Theorem 12.1.2 can be extended, but polynomial rings over fields are important cases in which they do extend.

When discussing factoring, we assume that the ring R is an integral domain, so that the Cancellation Law 11.7.1 is available, and we exclude the element zero from consideration. Here is some terminology that we use:

(12.2.1) u is a *unit* if u has a multiplicative inverse in R.

a *divides* b if $b = aq$ for some q in R.

a is a *proper divisor* of b if $b = aq$ and neither a nor q is a unit.

a and b are *associates* if each divides the other, or if $b = ua$, and u is a unit.

a is *irreducible* if a is not a unit, and it has no proper divisor – its only divisors are units and associates.

p is a *prime element* if p is not a unit, and whenever p divides a product ab, then p divides a or p divides b.

These concepts can be interpreted in terms of the principal ideals generated by the elements. Recall that the principal ideal (a) generated by an element a consists of all elements of R that are are divisible by a. Then

(12.2.2) u is a unit $\quad \Leftrightarrow \quad (u) = (1)$.

a divides $b \quad \Leftrightarrow \quad (b) \subset (a)$.

a is a proper divisor of $b \quad \Leftrightarrow \quad (b) < (a) < (1)$.

a and b are associates $\quad \Leftrightarrow \quad (a) = (b)$.

a is irreducible $\quad \Leftrightarrow \quad (a) < (1)$, and there is no principal ideal (c) such that $(a) < (c) < (1)$.

p is a prime element $\quad \Leftrightarrow \quad ab \in (p)$ implies $a \in (p)$ or $b \in (p)$.

Before continuing, we note one of the simplest examples of a ring element that has more than one factorization. The ring is $R = \mathbb{Z}[\sqrt{-5}]$. It consists of all complex numbers of the form $a + b\sqrt{-5}$, where a and b are integers. We will use this ring as an example several times in this chapter and the next. In R, the integer 6 can be factored in two ways:

(12.2.3) $2 \cdot 3 = 6 = (1 + \sqrt{-5})(1 - \sqrt{-5}).$

It isn't hard to show that none of the four terms $2, 3, 1 + \sqrt{-5}, 1 - \sqrt{-5}$ can be factored further; they are irreducible elements of the ring.

We abstract the procedure of division with remainder first. To make sense of division with remainder, we need a measure of size of an element. A *size function* on an integral domain R can be any function σ whose domain is the set of nonzero elements of R, and whose range is the set of nonnegative integers. An integral domain R is a *Euclidean domain* if there is a size function σ on R such that division with remainder is possible, in the following sense:

(12.2.4)
> Let a and b be elements of R, and suppose that a is not zero.
> There are elements q and r in R such that $b = aq + r$,
> and either $r = 0$ or else $\sigma(r) < \sigma(a)$.

The most important fact about division with remainder is that r is zero, if and only if a divides b.

Proposition 12.2.5

(a) The ring $\mathbb{Z}$ of integers is a Euclidean domain, with size function $\sigma(\alpha) = |\alpha|$.
(b) A polynomial ring $F[x]$ in one variable over a field F is a Euclidean domain, with $\sigma(f) = $ degree of f.
(c) The ring $\mathbb{Z}[i]$ of Gauss integers is a Euclidean domain, with $\sigma(a) = |a|^2$.

The ring of integers and the polynomial rings were discussed in Chapter 11. We show here that the ring of Gauss integers is a Euclidean domain. The elements of $\mathbb{Z}[i]$ form a square lattice in the complex plane, and the multiples of a given nonzero element α form the principal ideal (α), which is a similar geometric figure. If we write $\alpha = re^{i\theta}$, then (α) is obtained from the lattice $\mathbb{Z}[i]$ by rotating through the angle θ and stretching by the factor r, as is illustrated below with $\alpha = 2 + i$:

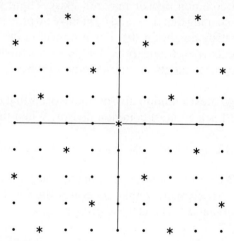

(12.2.6)   A Principal Ideal in the Ring of Gauss Integers.

For any complex number β, there is a point of the lattice (α) whose square distance from β is less than $|\alpha|^2$. We choose such a point, say $\gamma = \alpha q$, and let $r = \beta - \gamma$. Then $\beta = \alpha q + r$, and $|r|^2 < |\alpha|^2$, as required. Here q is in $\mathbb{Z}[i]$, and if β is in $\mathbb{Z}[i]$, so is r.

Division with remainder is not unique: There may be as many as four choices for the element γ. $\square$

• An integral domain in which every ideal is principal is called a *principal ideal domain*.

Proposition 12.2.7 A Euclidean domain is a principal ideal domain.

Proof. We mimic the proof that the ring of integers is a principal ideal domain once more. Let R be a Euclidean domain with size function σ, and let A be an ideal of R. We must show that A is principal. The zero ideal is principal, so we may assume that A is not the zero ideal. Then A contains a nonzero element. We choose a nonzero element a of A such that $\sigma(a)$ is as small as possible, and we show that A is the principal ideal (a) of multiples of a.

Because A is an ideal and a is in A, any multiple aq with q in R is in A. So $(a) \subset A$. To show that $A \subset (a)$, we take an arbitrary element b of A. We use division with remainder to write $b = aq + r$, where either $r = 0$, or $\sigma(r) < \sigma(a)$. Then b and aq are in A, so $r = b - aq$ is in A too. Since $\sigma(a)$ is minimal, we can't have $\sigma(r) < \sigma(a)$, and it follows that $r = 0$. This shows that a divides b, and hence that b is in the principal ideal (a). Since b is arbitrary, $A \subset (a)$, and therefore $A = (a)$. □

Let a and b be elements of an integral domain R, not both zero. A *greatest common divisor* d of a and b is an element with the following properties:

(a) d divides a and b.

(b) If an element e divides a and b, then e divides d.

Any two greatest common divisors d and d' are associate elements. The first condition tells us that both d and d' divide a and b, and then the second one tells us that d' divides d and also that d divides d'.

However, a greatest common divisor may not exist. There will often be a common divisor m that is maximal, meaning that a/m and b/m have no proper divisor in common. But this element may fail to satisfy condition **(b)**. For instance, in the ring $\mathbb{Z}[\sqrt{-5}]$ considered above (12.2.3), the elements $a = 6$ and $b = 2 + 2\sqrt{-5}$ are divisible both by 2 and by $1 + \sqrt{-5}$. These are maximal elements among common divisors, but neither one divides the other.

One case in which a greatest common divisor does exist is that a and b have no common factors except units. Then 1 is a greatest common divisor. When this is so, a and b are said to be *relatively prime*.

Greatest common divisors always exist in a principal ideal domain:

Proposition 12.2.8 Let R be a principal ideal domain, and let a and b be elements of R, which are not both zero. An element d that generates the ideal $(a, b) = Ra + Rb$ is a greatest common divisor of a and b. It has these properties:

(a) $Rd = Ra + Rb$,

(b) d divides a and b.

(c) If an element e of R divides both a and b, it also divides d.

(d) There are elements r and s in R such that $d = ra + sb$.

Proof. This is essentially the same proof as for the ring of integers. **(a)** restates that d generates the ideal (a, b). **(b)** states that a and b are in Rd, and **(d)** states that d is in the ideal $Ra + Rb$. For **(c)**, we note that if e divides a and b then a and b are elements of Re. In that case, Re contains $Ra + Rb = Rd$, so e divides d. □

Corollary 12.2.9 Let R be a principal ideal domain.

(a) If elements a and b of R are relatively prime, then 1 is a linear combination $ra + sb$.

(b) An element of R is irreducible if and only if it is a prime element.

(c) The maximal ideals of R are the principal ideals generated by the irreducible elements.

Proof. **(a)** This follows from Proposition 12.2.8**(d)**.

(b) In any integral domain, a prime element is irreducible. We prove this below, in Lemma 12.2.10. Suppose that R is a principal ideal domain and that an irreducible element q of R divides a product ab. We have to show that if q does not divide a, then q divides b. Let d be a greatest common divisor of a and q. Since q is irreducible, the divisors of q are the units and the associates of q. Since q does not divide a, d is not an associate of q. So d is a unit, q and a are relatively prime, and $1 = ra + sq$ with r and s in R. We multiply by b: $b = rab + sqb$. Both terms on the right side of this equation are divisible by q, so q divides the left side, b.

(c) Let q be an irreducible element. Its divisors are units and associates. Therefore the only principal ideals that contain (q) are (q) itself and the unit ideal (1) (see (12.2.2)). Since every ideal of R is principal, these are the only ideals that contain (q). Therefore (q) is a maximal ideal. Conversely, if an element b has a proper divisor a, then $(b) < (a) < (1)$, so (b) is not a maximal ideal. $\qquad \square$

Lemma 12.2.10 In an integral domain R, a prime element is irreducible.

Proof. Suppose that a prime element p is a product, say $p = ab$. Then p divides one of the factors, say a. But the equation $p = ab$ shows that a divides p too. So a and p are associates and b is a unit. The factorization is not proper. $\qquad \square$

What analogy to the Fundamental Theorem of Arithmetic 12.1.2**(d)** could one hope for in an integral domain? We may divide the desired statement of uniqueness of factorization into two parts. First, a given element should be a product of irreducible elements, and second, that product should be essentially unique.

Units in a ring complicate the statement of uniqueness. Unit factors must be disregarded and associate factors must be considered equivalent. The units in the ring of integers are ± 1, and in this ring it is natural to work with positive integers. Similarly, in the polynomial ring $F[x]$ over a field, it is natural to work with monic polynomials. But we don't have a reasonable way to normalize elements in an arbitrary integral domain; it is best not to try.

We say that factoring in an integral domain R is unique if, whenever an element a of R is written in two ways as a product of irreducible elements, say

(12.2.11)
$$p_1 \cdots p_m = a = q_1 \cdots q_n,$$

then $m = n$, and if the right side is rearranged suitably, q_i is an associate of p_i for each i. So in the statement of uniqueness, associate factorizations are considered equivalent.

For example, in the ring of Gauss integers,

$$(2 + i)(2 - i) = 5 = (1 + 2i)(1 - 2i).$$

These two factorizations of the element 5 are equivalent because the terms that appear on the left and right sides are associates: $-i(2 + i) = 1 - 2i$ and $i(2 - i) = 1 + 2i$.

It is neater to work with principal ideals than with elements, because associates generate the same principal ideal. However, it isn't too cumbersome to use elements and we will stay with them here. The importance of ideals will become clear in the next chapter.

When we attempt to write an element a as a product of irreducible elements, we always assume that it is not zero and not a unit. Then we attempt to factor a, proceeding this way: If a is irreducible, we stop. If not, then a has a proper factor, so it decomposes in some way as a product, say $a = a_1b_1$, where neither a_1 nor b_1 is a unit. We continue factoring a_1 and b_1, if possible, and we hope that this procedure terminates; in other words, we hope that after a finite number of steps all the factors are irreducible. We say that *factoring terminates* in R if this is always true, and we refer to a factorization into irreducible elements as an *irreducible factorization*.

An integral domain R is a *unique factorization domain* if it has these properties:

(12.2.12)
- Factoring terminates.
- The irreducible factorization of an element a is unique in the sense described above.

The condition that factoring terminates has a useful description in terms of principal ideals:

Proposition 12.2.13 Let R be an integral domain. The following conditions are equivalent:

- Factoring terminates.
- R does not contain an infinite strictly increasing chain $(a_1) < (a_2) < (a_3) < \cdots$ of principal ideals.

Proof. If the process of factoring doesn't terminate, there will be an element a_1 with a proper factorization such that the process fails to terminate for at least one of the factors. Let's say that the proper factorization is $a_1 = a_2b_2$, and that the process fails to terminate for the factor we call a_2. Since a_2 is a proper divisor of a_1, $(a_1) < (a_2)$ (see (12.2.2)). We replace a_1 by a_2 and repeat. In this way we obtain an infinite chain.

Conversely, if there is a strictly increasing chain $(a_1) < (a_2) < \cdots$, then none of the ideals (a_n) is the unit ideal, and therefore a_2 is a proper divisor of a_1, a_3 is a proper divisor of a_2, and so on (12.2.2). This gives us a nonterminating process. $\square$

We will rarely encounter rings in which factoring fails to terminate, and we will prove a theorem that explains the reason later (see (14.6.9)), so we won't worry much about it here. In practice it is the uniqueness that gives trouble. Factoring into irreducible elements will usually be possible, but it will not be unique, even when one takes into account the ambiguity of associate factors.

Going back to the ring $R = \mathbb{Z}[\sqrt{-5}]$, it isn't hard to show that all of the elements 2, 3, $1 + \sqrt{-5}$ and $1 - \sqrt{-5}$ are irreducible, and that the units of R are 1 and –1. So 2 is not an associate of $1 + \sqrt{-5}$ or of $1 - \sqrt{-5}$. Therefore $2 \cdot 3 = 6 = (1 + \sqrt{-5})(1 - \sqrt{-5})$ are essentially different factorizations: R is not a unique factorization domain.

Proposition 12.2.14

(a) Let R be an integral domain. Suppose that factoring terminates in R. Then R is a unique factorization domain if and only if every irreducible element is a prime element.

(b) A principal ideal domain is a unique factorization domain.

(c) The rings $\mathbb{Z}$, $\mathbb{Z}[i]$ and the polynomial ring $F[x]$ in one variable over a field F are unique factorization domains.

Thus the phrases *irreducible factorization* and *prime factorization* are synonymous in unique factorization domains, but most rings contain irreducible elements that are not prime. In the ring $\mathbb{Z}[\sqrt{-5}]$, the element 2 is irreducible. It is not prime because, though it divides the product $(1 + \sqrt{-5})(1 - \sqrt{-5})$, it does not divide either factor.

The converse of **(b)** is not true. We will see in the next section that the ring $\mathbb{Z}[x]$ of integer polynomials is a unique factorization domain, though it isn't a principal ideal domain.

Proof of Proposition (12.2.14). First of all, **(c)** follows from **(b)** because the rings mentioned in **(c)** are Euclidean domains, and therefore principal ideal domains.

(a) Let R be a ring in which every irreducible element is prime, and suppose that an element a factors in two ways into irreducible elements, say $p_1 \cdots p_m = a = q_1 \cdots q_n$, where $m \leq n$. If $n = 1$, then $m = 1$ and $p_1 = q_1$. Suppose that $n > 1$. Since p_1 is prime, it divides one of the factors $q_1, \ldots, q_n$, say q_1. Since q_1 is irreducible and since p_1 is not a unit, q_1 and p_1 are associates, say $p_1 = uq_1$, where u is a unit. We move the unit factor over to q_2, replacing q_1 by uq_1 and q_2 by $u^{-1}q_2$. The result is that now $p_1 = q_1$. Then we cancel p_1 and use induction on n.

Conversely, suppose that there is an irreducible element p that is not prime. Then there are elements a and b such that p divides the product $r = ab$, say $r = pc$, but p does not divide a or b. By factoring a, b, and c into irreducible elements, we obtain two inequivalent factorizations of r.

(b) Let R be a principal ideal domain. Since every irreducible element of R is prime (12.2.8), we need only prove that factoring terminates (12.2.14). We do this by showing that R contains no infinite strictly increasing chain of principal ideals. We suppose given an infinite weakly increasing chain

$$(a_1) \subset (a_2) \subset (a_3) \subset \ldots,$$

and we prove that it cannot be strictly increasing.

Lemma 12.2.15 Let $I_1 \subset I_2 \subset I_3 \subset \ldots$ be an increasing chain of ideals in a ring R. The union $J = \bigcup I_n$ is an ideal.

Proof. If u and v are in J, they are both in I_n for some n. Then $u + v$ and ru, for any r in R, are also in I_n, and therefore they are in J. This shows that J is an ideal.   □

We apply this lemma to our chain of principal ideals, with $I_\nu = (a_\nu)$, and we use the hypothesis that R is a principal ideal domain to conclude that the union J is a principal ideal, say $J = (b)$. Then since b is in the union of the ideals (a_n), it is in one of those ideals.

But if b is in (a_n), then $(b) \subset (a_n)$. On the other hand, $(a_n) \subset (a_{n+1}) \subset (b)$. Therefore $(b) = (a_n) = (a_{n+1})$. The chain is not strictly increasing.   □

One can decide whether an element a divides another element b in a unique factorization domain, in terms of their irreducible factorizations.

Proposition 12.2.16 Let R be a unique factorization domain.

(a) Let $a = p_1 \cdots p_m$ and $b = q_1 \cdots q_n$ be irreducible factorizations of two elements of R. Then a divides b in R if and only if $m \le n$ and, when the factors q_j are arranged suitably, p_i is an associate of q_i for $i = 1, \ldots, m$.

(b) Any pair of elements a, b, not both zero, has a greatest common divisor.

Proof. **(a)** This is very similar to the proof of Proposition 12.2.14**(a)**. The irreducible factors of a are prime elements. If a divides b, then p_1 divides b, and therefore p_1 divides some q_i, say q_1. Then p_1 and q_1 are associates. The assertion follows by induction when we cancel p_1 from a and q_1 from b. We omit the proof of **(b)**.   □

Note: Any two greatest common divisors of a and b are associates. But unless a unique factorization domain is a principal ideal domain, the greatest common divisor, though it exists, needn't have the form $ra + sb$. The greatest common divisor of 2 and x in the unique factorization domain $\mathbb{Z}[x]$ is 1, but we cannot write 1 as a linear combination of those elements with integer polynomials as coefficients.   □

We review the results we have obtained for the important case of a polynomial ring $F[x]$ over a field. The units in the polynomial ring $F[x]$ are the nonzero constants. We can factor the leading coefficient out of a nonzero polynomial to make it monic, and the only monic associate of a monic polynomial f is f itself. By working with monic polynomials, the ambiguity of associate factorizations can be avoided. With this taken into account, the next theorem follows from Proposition 12.2.14.

Theorem 12.2.17 Let $F[x]$ be the polynomial ring in one variable over a field F.

(a) Two polynomials f and g, not both zero, have a unique monic greatest common divisor d, and there are polynomials r and s such that $rf + sg = d$.

(b) If two polynomials f and g have no nonconstant factor in common, then there are polynomials r and s such that $rf + sg = 1$.

(c) Every irreducible polynomial p in $F[x]$ is a prime element of $F[x]$: If p divides a product fg, then p divides f or p divides g.

(d) *Unique factorization:* Every monic polynomial in $F[x]$ can be written as a product $p_1 \cdots p_k$, where p_i are monic irreducible polynomials in $F[x]$ and $k \ge 0$. This factorization is unique except for the ordering of the terms.   □

In the future, when we speak of the greatest common divisor of two polynomials with coefficients in a field, we will mean the unique monic polynomial with the properties **(a)** above. This greatest common divisor will sometimes be denoted by $\gcd(f, g)$.

The greatest common divisor $\gcd(f, g)$ of two polynomials f and g, not both zero, with coefficients in a field F can be found by repeated division with remainder, the process called the *Euclidean algorithm* that we mentioned in Section 2.3 for the ring of integers: Suppose that the degree of g is at least equal to the degree of f. We write $g = fq + r$ where the remainder r, if it is not zero, has degree less than that of f. Then $\gcd(f, g) = \gcd(f, r)$. If $r = 0$, $\gcd(f, g) = f$. If not, we replace f and g by r and f, and repeat the process. Since degrees are being lowered, the process is finite. The analogous method can be used to determine greatest common divisors in any Euclidean domain.

Over the complex numbers, every polynomial of positive degree has a root α, and therefore a divisor of the form $x - \alpha$. The irreducible polynomials are linear, and the irreducible factorization of a monic polynomial has the form

(12.2.18) $$f(x) = (x - \alpha_1) \cdots (x - \alpha_n),$$

where α_i are the roots of $f(x)$, with repetitions for multiple roots. The uniqueness of this factorization is not surprising.

When $F = \mathbb{R}$, there are two classes of irreducible polynomials: linear and quadratic. A real quadratic polynomial $x^2 + bx + c$ is irreducible if and only if its discriminant $b^2 - 4c$ is negative, in which case it has a pair of complex conjugate roots. The fact that every irreducible polynomial over the complex numbers is linear implies that no real polynomial of degree >2 is irreducible.

Proposition 12.2.19 Let α be a complex, not real, root of a real polynomial f. Then the complex conjugate $\overline{\alpha}$ is also a root of f. The quadratic polynomial $q = (x - \alpha)(x - \overline{\alpha})$ has real coefficients, and it divides f.   □

Factoring polynomials in the ring $\mathbb{Q}[x]$ of polynomials with rational coefficients is more interesting, because there exist irreducible polynomials in $\mathbb{Q}[x]$ of arbitrary degree. This is explained in the next two sections. Neither the form of the irreducible factorization nor its uniqueness are intuitively clear in this case.

For future reference, we note the following elementary fact:

Proposition 12.2.20 A polynomial f of degree n with coefficients in a field F has at most n roots in F.

Proof. An element α is a root of f if and only if $x - \alpha$ divides f (11.2.11). If so, we can write $f(x) = (x - \alpha)q(x)$, where $q(x)$ is a polynomial of degree $n - 1$. Let β be a root of f different from α. Substituting $x = \beta$, we obtain $0 = (\beta - \alpha)q(\beta)$. Since β is not equal to α, it must be a root of q. By induction on the degree, q has at most $n - 1$ roots in F. Putting those roots together with α, we see that f has at most n roots.   □

12.3   GAUSS'S LEMMA

Every monic polynomial $f(x)$ with rational coefficients can be expressed uniquely in the form $p_1 \cdots p_k$, where p_i are monic polynomials that are irreducible elements in the ring $\mathbb{Q}[x]$. But suppose that a polynomial $f(x)$ has integer coefficients, and that it factors in $\mathbb{Q}[x]$. Can it be factored without leaving the ring $\mathbb{Z}[x]$ of integer polynomials? We will see that it can, and also that $\mathbb{Z}[x]$ is a unique factorization domain.

Here is an example of an irreducible factorization in integer polynomials:

$$6x^3 + 9x^2 + 9x + 3 = 3(2x + 1)(x^2 + x + 1).$$

As we see, irreducible factorizations are slightly more complicated in $\mathbb{Z}[x]$ than in $\mathbb{Q}[x]$. Prime integers are irreducible elements of $\mathbb{Z}[x]$, and they may appear in the factorization of a polynomial. And, if we want to stay with integer coefficients, we can't require monic factors.

We have two main tools for studying factoring in $\mathbb{Z}[x]$. The first is the inclusion of the integer polynomial ring into the ring of polynomials with rational coefficients:

$$\mathbb{Z}[x] \subset \mathbb{Q}[x].$$

This can be useful because algebra in the ring $\mathbb{Q}[x]$ is simpler.

The second tool is reduction modulo some integer prime p, the homomorphism

(12.3.1) $$\psi_p : \mathbb{Z}[x] \to \mathbb{F}_p[x]$$

that sends $x \rightsquigarrow x$ (11.3.6). We'll often denote the image $\psi_p(f)$ of an integer polynomial by $\overline{f}$, though this notation is ambiguous because it doesn't mention p.

The next lemma should be clear.

Lemma 12.3.2 Let $f(x) = a_n x^n + \cdots + a_1 x + a_0$ be an integer polynomial, and let p be an integer prime. The following are equivalent:

- p divides every coefficient a_i of f in $\mathbb{Z}$,
- p divides f in $\mathbb{Z}[x]$,
- f is in the kernel of ψ_p. $\square$

The lemma shows that the kernel of ψ_p can be interpreted easily without mentioning the map. But the facts that ψ_p is a homomorphism and that its image $\mathbb{F}_p[x]$ is an integral domain make the interpretation as a kernel useful.

- A polynomial $f(x) = a_n x^n + \cdots + a_1 x + a_0$ with rational coefficients is called *primitive* if it is an integer polynomial of positive degree, the greatest commmon divisor of its coefficients $a_0, \ldots, a_n$ in the integers is 1, and its leading coefficient a_n is positive.

Lemma 12.3.3 Let f be an integer polynomial f of positive degree, with positive leading coefficient. The following conditions are equivalent:

- f is primitive,
- f is not divisible by any integer prime p,
- for every integer prime p, $\psi_p(f) \neq 0$. $\square$

Proposition 12.3.4

(a) An integer is a prime element of $\mathbb{Z}[x]$ if and only if it is a prime integer. So a prime integer p divides a product fg of integer polynomials if and only if p divides f or p divides g.

(b) *(Gauss's Lemma)* The product of primitive polynomials is primitive.

Proof. **(a)** It is obvious that an integer must be a prime if it is an irreducible element of $\mathbb{Z}[x]$. Let p be a prime integer. We use bar notation: $\bar{f} = \psi_p(f)$. Then p divides fg if and only if $\overline{fg} = 0$, and since $\mathbb{F}_p[x]$ is a domain, this is true if and only if $\bar{f} = 0$ or $\bar{g} = 0$, i.e., if and only if p divides f or p divides g.

(b) Suppose that f and g are primitive polynomials. Since their leading coefficients are positive, the leading coefficient of fg is also positive. Moreover, no prime p divides f or g, and by **(a)**, no prime divides fg. So fg is primitive.   □

Lemma 12.3.5 Every polynomial $f(x)$ of positive degree with rational coefficients can be written *uniquely* as a product $f(x) = c f_0(x)$, where c is a rational number and $f_0(x)$ is a primitive polynomial. Moreover, c is an integer if and only if f is an integer polynomial. If f is an integer polynomial, then the greatest common divisor of the coefficients of f is $\pm c$.

Proof. To find f_0, we first multiply f by an integer d to clear the denominators in its coefficients. This will give us a polynomial $df = f_1$ with integer coefficients. Then we factor out the greatest common divisor of the coefficients of f_1 and adjust the sign of the leading coefficient. The resulting polynomial f_0 is primitive, and $f = c f_0$ for some rational number c. This proves existence.

If f is an integer polynomial, we don't need to clear the denominator. Then c will be an integer, and up to sign, it is the greatest common divisor of the coefficients, as stated.

The uniqueness of this product is important, so we check it carefully. Suppose given rational numbers c and c' and primitive polynomials f_0 and f_0' such that $c f_0 = c' f_0'$. We will show that $f_0 = f_0'$. Since $\mathbb{Q}[x]$ is a domain, it will follow that $c = c'$.

We multiply the equation $c f_0 = c' f_0'$ by an integer and adjust the sign if necessary, to reduce to the case that c and c' are positive integers. If $c \neq 1$, we choose a prime integer p that divides c. Then p divides $c' f_0'$. Proposition 12.3.4**(a)** shows that p divides one of the factors c' or f_0'. Since f_0' is primitive, it isn't divisible by p, so p divides c'. We cancel p from both sides of the equation. Induction reduces us to the case that $c = 1$, and the same reasoning shows that then $c' = 1$. So $f_0 = f_0'$.   □

Theorem 12.3.6

(a) Let f_0 be a primitive polynomial, and let g be an integer polynomial. If f_0 divides g in $\mathbb{Q}[x]$, then f_0 divides g in $\mathbb{Z}[x]$.

(b) If two integer polynomials f and g have a common nonconstant factor in $\mathbb{Q}[x]$, they have a common nonconstant factor in $\mathbb{Z}[x]$.

Proof. **(a)** Say that $g = f_0 q$ where q has rational coefficients. We show that q has integer coefficients. We write $g = c g_0$, and $q = c' q_0$, with g_0 and q_0 primitive. Then $c g_0 = c' f_0 q_0$. Gauss's Lemma tells us that $f_0 q_0$ is primitive. Therefore by the uniqueness assertion of Lemma 12.3.5, $c = c'$ and $g_0 = f_0 q_0$. Since g is an integer polynomial, c is an integer. So $q = c q_0$ is an integer polynomial.

(b) If the integer polynomials f and g have a common factor h in $\mathbb{Q}[x]$ and if we write $h = c h_0$, where h_0 is primitive, then h_0 also divides f and g in $\mathbb{Q}[x]$, and by **(a)**, h_0 divides both f and g in $\mathbb{Z}[x]$.   □

Proposition 12.3.7

(a) Let f be an integer polynomial with positive leading coefficient. Then f is an irreducible element of $\mathbb{Z}[x]$ if and only if it is either a prime integer or a primitive polynomial that is irreducible in $\mathbb{Q}[x]$.

(b) Every irreducible element of $\mathbb{Z}[x]$ is a prime element.

Proof. Proposition 12.3.4**(a)** proves **(a)** and **(b)** for a constant polynomial. If f is irreducible and not constant, it cannot have an integer factor different from ± 1, so if its leading coefficient is positive, it will be primitive. Suppose that f is a primitive polynomial and that it has a proper factorization in $\mathbb{Q}[x]$, say $f = gh$. We write $g = cg_0$ and $h = c'h_0$, with g_0 and h_0 primitive. Then g_0h_0 is primitive. Since f is also primitive, $f = g_0h_0$. Therefore f has a proper factorization in $\mathbb{Z}[x]$ too. So if f is reducible in $\mathbb{Q}[x]$, it is reducible in $\mathbb{Z}[x]$. The fact that a primitive polynomial that is reducible in $\mathbb{Z}[x]$ is also reducible in $\mathbb{Q}[x]$ is clear. This proves **(a)**.

Let f be a primitive irreducible polynomial that divides a product gh of integer polynomials. Then f is irreducible in $\mathbb{Q}[x]$. Since $\mathbb{Q}[x]$ is a principal ideal domain, f is a prime element of $\mathbb{Q}[x]$ (12.2.8). So f divides g or h in $\mathbb{Q}[x]$. By (12.3.6) f divides g or h in $\mathbb{Z}[x]$. This shows that f is a prime element, which proves **(b)**. $\qquad\square$

Theorem 12.3.8 The polynomial ring $\mathbb{Z}[x]$ is a unique factorization domain. Every nonzero polynomial $f(x) \in \mathbb{Z}[x]$ that is not ± 1 can be written as a product

$$f(x) = \pm p_1 \cdots p_m q_1(x) \cdots q_n(x),$$

where p_i are integer primes and $q_j(x)$ are primitive irreducible polynomials. This expression is unique except for the order of the factors.

Proof. It is easy to see that factoring terminates in $\mathbb{Z}[x]$, so this theorem follows from Propositions 12.3.7 and 12.2.14. $\qquad\square$

The results of this section have analogues for the polynomial ring $F[t, x]$ in two variables over a field F. To set up the analogy, we regard $F[t, x]$ as the ring $F[t][x]$ of polynomials in x whose coefficients are polynomials in t. The analogue of the field $\mathbb{Q}$ will be the field $F(t)$ of rational functions in t, the field of fractions of $F[t]$. We'll denote this field by $\mathcal{F}$. Then $F[t, x]$ is a subring of the ring $\mathcal{F}[x]$ of polynomials

$$f = a_n(t)x^n + \cdots + a_1(t)x + a_0(t)$$

whose coefficients $a_i(t)$ are rational functions in t. This can be useful because every ideal of $\mathcal{F}[x]$ is principal.

The polynomial f is called *primitive* if it has positive degree, its coefficients $a_i(t)$ are polynomials in $F[t]$ whose greatest common divisor is equal to 1, and the leading coefficient $a_n(t)$ is monic. A primitive polynomial will be an element of the polynomial ring $F[t, x]$.

It is true again that the product of primitive polynomials is primitive, and that every element $f(t, x)$ of $\mathcal{F}[x]$ can be written in the form $c(t) f_0(t, x)$, where f_0 is a primitive polynomial in $F[t, x]$ and c is a rational function in t, both uniquely determined up to constant factor.

The proofs of the next assertions are almost identical to the proofs of Proposition 12.3.4 and Theorems 12.3.6 and 12.3.8.

Theorem 12.3.9 Let $F[t]$ be a polynomial ring in one variable over a field F, and let $\mathcal{F} = F(t)$ be its field of fractions.

(a) The product of primitive polynomials in $F[t, x]$ is primitive.

(b) Let f_0 be a primitive polynomial, and let g be a polynomial in $F[t, x]$. If f_0 divides g in $\mathcal{F}[x]$, then f_0 divides g in $F[t, x]$.

(c) If two polynomials f and g in $F[t, x]$ have a common nonconstant factor in $\mathcal{F}[x]$, they have a common nonconstant factor in $F[t, x]$.

(d) Let f be an element of $F[t, x]$ whose leading coefficient is monic. Then f is an irreducible element of $F[t, x]$ if and only if it is either an irreducible polynomial in t alone, or a primitive polynomial that is irreducible in $\mathcal{F}[x]$.

(e) The ring $F[t, x]$ is a unique factorization domain. □

The results about factoring in $\mathbb{Z}[x]$ also have analogues for polynomials with coefficients in any unique factorization domain R.

Theorem 12.3.10 If R is a unique factorization domain, the polynomial ring $R[x_1, \ldots, x_n]$ in any number of variables is a unique factorization domain.

Note: In contrast to the case of one variable, where every complex polynomial is a product of linear polynomials, complex polynomials in two variables are often irreducible, and therefore prime elements, of $\mathbb{C}[t, x]$. □

12.4   FACTORING INTEGER POLYNOMIALS

We pose the problem of factoring an integer polynomial

(12.4.1) $$f(x) = a_n x^n + \cdots + a_1 x + a_0,$$

with $a_n \neq 0$. Linear factors can be found fairly easily.

Lemma 12.4.2

(a) If an integer polynomial $b_1 x + b_0$ divides f in $\mathbb{Z}[x]$, then b_1 divides a_n and b_0 divides a_0.

(b) A primitive polynomial $b_1 x + b_0$ divides f in $\mathbb{Z}[x]$ if and only if the rational number $-b_0/b_1$ is a root of f.

(c) A rational root of a monic integer polynomial f is an integer.

Proof. (a) The constant coefficient of a product $(b_1 x + b_0)(q_{n-1} x^{n-1} + \cdots + q_0)$ is $b_0 q_0$, and if $q_{n-1} \neq 0$, the leading coefficient is $b_1 q_{n-1}$.

(b) According to Theorem 12.3.10(c), $b_1 x + b_0$ divides f in $\mathbb{Z}[x]$ if and only if it divides f in $\mathbb{Q}[x]$, and this is true if and only if $x + b_0/b_1$ divides f, i.e., $-b_0/b_1$ is a root.

(c) If $\alpha = a/b$ is a root, written with $b > 0$, and if $\gcd(a, b) = 1$, then $bx - a$ is a primitive polynomial that divides the monic polynomial f, so $b = 1$ and α is an integer.   ☐

The homomorphism $\psi_p : \mathbb{Z}[x] \to \mathbb{F}_p[x]$ (12.3.1) is useful for explicit factoring, one reason being that there are only finitely many polynomials in $\mathbb{F}_p[x]$ of each degree.

Proposition 12.4.3   Let $f(x) = a_n x^n + \cdots + a_0$ be an integer polynomial, and let p be a prime integer that does not divide the leading coefficient a_n. If the residue $\overline{f}$ of f modulo p is an irreducible element of $\mathbb{F}_p[x]$, then f is an irreducible element of $\mathbb{Q}[x]$.

Proof. We prove the contrapositive, that if f is reducible, then $\overline{f}$ is reducible. Suppose that $f = gh$ is a proper factorization of f in $\mathbb{Q}[x]$. We may assume that g and h are in $\mathbb{Z}[x]$ (12.3.6). Since the factorization in $\mathbb{Q}[x]$ is proper, both g and h have positive degree, and, if $\deg f$ denotes the degree of f, then $\deg f = \deg g + \deg h$.

Since ψ_p is a homomorphism, $\overline{f} = \overline{g}\overline{h}$, so $\deg \overline{f} = \deg \overline{g} + \deg \overline{h}$. For any integer polynomial p, $\deg \overline{p} \leq \deg p$. Our assumption on the leading coefficient of f tells us that $\deg \overline{f} = \deg f$. This being so we must have $\deg \overline{g} = \deg g$ and $\deg \overline{h} = \deg h$. Therefore the factorization $\overline{f} = \overline{g}\overline{h}$ is proper.   ☐

If p divides the leading coefficient of f, then $\overline{f}$ has lower degree, and using reduction modulo p becomes harder.

If we suspect that an integer polynomial is irreducible, we can try reduction modulo p for a small prime, $p = 2$ or 3 for instance, and hope that $\overline{f}$ turns out to be irreducible and of the same degree as f. If so, f will be irreducible too. Unfortunately, there exist irreducible integer polynomials that can be factored modulo every prime p. The polynomial $x^4 - 10x^2 + 1$ is an example. So the method of reduction modulo p may not work. But it does work quite often.

The irreducible polynomials in $\mathbb{F}_p[x]$ can be found by the "sieve" method. The *sieve of Eratosthenes* is the name given to the following method of determining the prime integers less than a given number n. We list the integers from 2 to n. The first one, 2, is prime because any proper factor of 2 must be smaller than 2, and there is no smaller integer on our list. We note that 2 is prime, and we cross out the multiples of 2 from our list. Except for 2 itself, they are not prime. The first integer that is left, 3, is a prime because it isn't divisible by any smaller prime. We note that 3 is a prime and then cross out the multiples of 3 from our list. Again, the smallest remaining integer, 5, is a prime, and so on.

$$2 \quad 3 \quad \cancel{4} \quad 5 \quad \cancel{6} \quad 7 \quad \cancel{8} \quad \cancel{9} \quad \cancel{10} \quad 11 \quad \cancel{12} \quad 13 \quad \cancel{14} \quad \cancel{15} \quad \cancel{16} \quad 17 \quad \cancel{18} \quad 19 \quad \cdots$$

The same method will determine the irreducible polynomials in $\mathbb{F}_p[x]$. We list the monic polynomials, degree by degree, and cross out products. For example, the linear polynomials in $\mathbb{F}_2[x]$ are x and $x + 1$. They are irreducible. The polynomials of degree 2 are x^2, $x^2 + x$, $x^2 + 1$, and $x^2 + x + 1$. The first three have roots in $\mathbb{F}_2$, so they are divisible by x or by $x + 1$. The last one, $x^2 + x + 1$, is the only irreducible polynomial of degree 2 in $\mathbb{F}_2[x]$.

(12.4.4)   The irreducible polynomials of degree ≤ 4 in $\mathbb{F}_2[x]$:

$$x, \quad x+1; \quad x^2+x+1; \quad x^3+x^2+1, \quad x^3+x+1;$$

$$x^4+x^3+1, \quad x^4+x+1, \quad x^4+x^3+x^2+x+1.$$

By trying the polynomials on this list, we can factor polynomials of degree at most 9 in $\mathbb{F}_2[x]$. For example, let's factor $f(x) = x^5 + x^3 + 1$ in $\mathbb{F}_2[x]$. If it factors, there must be an irreducible factor of degree at most 2. Neither 0 nor 1 is a root, so f has no linear factor. There is only one irreducible polynomial of degree 2, namely $p = x^2 + x + 1$. We carry out division with remainder: $f(x) = p(x)(x^3 + x^2 + x) + (x + 1)$. So p doesn't divide f, and therefore f is irreducible.

Consequently, the integer polynomial $x^5 - 64x^4 + 127x^3 - 200x + 99$ is irreducible in $\mathbb{Q}[x]$, because its residue in $\mathbb{F}_2[x]$ is the irreducible polynomial $x^5 + x^3 + 1$.

(12.4.5)   The monic irreducible polynomials of degree 2 in $\mathbb{F}_3[x]$:

$$x^2 + 1, \quad x^2 + x - 1, \quad x^2 - x - 1.$$

Reduction modulo p may help describe the factorization of a polynomial also when the residue is reducible. Consider the polynomial $f(x) = x^3 + 3x^2 + 9x + 6$. Reducing modulo 3, we obtain x^3. This doesn't look like a promising tool. However, suppose that $f(x)$ were reducible in $\mathbb{Z}[x]$, say $f(x) = (x+a)(x^2+bx+c)$. Then the residue of $x+a$ would divide x^3 in $\mathbb{F}_3[x]$, which would imply $a \equiv 0$ modulo 3. Similarly, we could conclude $c \equiv 0$ modulo 3. It is impossible to satisfy both of these conditions because the constant term ac of the product is supposed to be equal to 6. Therefore no such factorization exists, and $f(x)$ is irreducible.

The principle at work in this example is called the Eisenstein Criterion.

Proposition 12.4.6 Eisenstein Criterion. Let $f(x) = a_n x^n + \cdots + a_0$ be an integer polynomial and let p be a prime integer. Suppose that the coefficients of f satisfy the following conditions:

- p does not divide a_n;
- p divides all other coefficients $a_{n-1}, \ldots, a_0$;
- p^2 does not divide a_0.

Then f is an irreducible element of $\mathbb{Q}[x]$.

For example, the polynomial $x^4 + 25x^2 + 30x + 20$ is irreducible in $\mathbb{Q}[x]$.

Proof of the Eisenstein Criterion. Assume that f satisfies the conditions, and let $\overline{f}$ denote the residue of f modulo p. The hypotheses imply that $\overline{f} = \overline{a}_n x^n$ and that $\overline{a}_n \neq 0$. If f is reducible in $\mathbb{Q}[x]$, it will factor in $\mathbb{Z}[x]$ into factors of positive degree, say $f = gh$, where $g(x) = b_r x^r + \cdots + b_0$ and $h(x) = c_s x^s + \cdots + c_0$. Then $\overline{g}$ divides $\overline{a}_n x^n$, so $\overline{g}$ has the form $\overline{b}_r x^r$. Every coefficient of g except the leading coefficient is divisible by p. The same is true of h. The constant coefficient a_0 of f will be equal to $b_0 c_0$, and since p divides b_0 and c_0, p^2 must divide a_0. This contradicts the third condition. Therefore f is irreducible. $\square$

One application of the Eisenstein Criterion is to prove the irreducibility of the *cyclotomic polynomial* $\Phi(x) = x^{p-1} + x^{p-2} + \cdots + x + 1$, where p is a prime. Its roots are the pth roots of unity, the powers of $\zeta = e^{2\pi i/p}$ different from 1:

$$(12.4.7) \qquad\qquad (x - 1)\,\Phi(x) = x^p - 1.$$

Lemma 12.4.8 Let p be a prime integer. The binomial coefficient $\binom{p}{r}$ is an integer divisible exactly once by p for every r in the range $1 < r < p$.

Proof. The binomial coefficient $\binom{p}{r}$ is

$$\binom{p}{r} = \frac{p(p-1)\cdots(p-r+1)}{r(r-1)\cdots 1}.$$

When $r < p$, the terms in the denominator are all less than p, so they cannot cancel the single p that is in the numerator. Therefore $\binom{p}{r}$ is divisible exactly once by p. $\qquad\square$

Theorem 12.4.9 Let p be a prime. The cyclotomic polynomial $\Phi(x) = x^{p-1} + x^{p-2} + \cdots + x + 1$ is irreducible over $\mathbb{Q}$.

Proof. We substitute $x = y + 1$ into (12.4.7) and expand the result:

$$y\,\Phi(y+1) = (y+1)^p - 1 = y^p + \binom{p}{1}y^{p-1} + \cdots + \binom{p}{p-1}y + 1 - 1.$$

We cancel y. The lemma shows that the Eisenstein Criterion applies, and that $\Phi(y+1)$ is irreducible. It follows that $\Phi(x)$ is irreducible too. $\qquad\square$

Estimating the Coefficients

Computer programs factor integer polynomials by factoring modulo powers of a prime, usually the prime $p = 2$. There are fast algorithms, the *Berlekamp algorithms*, to do this. The simplest case is that f is a monic integer polynomial whose residue modulo p is the product of relatively prime monic polynomials, say $\overline{f} = \overline{g}\overline{h}$ in $\mathbb{F}_p[x]$. Then there will be a unique way to factor f modulo any power of p. (We won't take the time to prove this.) Let's suppose that this is so, and that we (or the computer) have factored modulo the powers $p, p^2, p^3, \ldots$ If f factors in $\mathbb{Z}[x]$, the coefficients of the factors modulo p^k will stabilize when they are represented by integers between $-p^k/2$ and $p^k/2$, and this will produce the integer factorization. If f is irreducible in $\mathbb{Z}[x]$, the coefficients of the factors won't stabilize. When they get too big, one can conclude that the polynomial is irreducible.

The next theorem of Cauchy can be used to estimate how big the coefficients of the integer factors could be.

Theorem 12.4.10 Let $f(x) = x^n + a_{n-1}x^{n-1} + \cdots + a_1 x + a_0$ be a monic polynomial with complex coefficients, and let r be the maximum of the absolute values $|a_i|$ of its coefficients. The roots of f have absolute value less than $r + 1$.

Proof of Theorem 12.4.10. The trick is to rewrite the expression for f in the form

$$x^n = f - \left(a_{n-1}x^{n-1} + \cdots + a_1 x + a_0\right)$$

and to use the triangle inequality:

$$(12.4.11) \qquad |x|^n \leq |f(x)| + |a_{n-1}||x|^{n-1} + \cdots + |a_1||x| + |a_0|$$

$$\leq |f(x)| + r\left(|x|^{n-1} + \cdots + |x| + 1\right) = |f(x)| + r\frac{|x|^n - 1}{|x| - 1}.$$

Let α be a complex number with absolute value $|\alpha| \geq r+1$. Then $\dfrac{r}{|\alpha| - 1} \leq 1$. We substitute $x = \alpha$ into (12.4.11):

$$|\alpha|^n \leq |f(\alpha)| + r\frac{|\alpha|^n - 1}{|\alpha| - 1} \leq |f(\alpha)| + |\alpha|^n - 1.$$

Therefore $|f(\alpha)| \geq 1$, and α is not a root of f. $\qquad\square$

We give two examples in which $r = 1$.

Examples 12.4.12   (a) Let $f(x) = x^6 + x^4 + x^3 + x^2 + 1$. The irreducible factorization modulo 2 is

$$x^6 + x^4 + x^3 + x^2 + 1 = (x^2 + x + 1)(x^4 + x^3 + x^2 + x + 1).$$

Since the factors are distinct, there is just one way to factor f modulo 2^2, and it is

$$x^6 + x^4 + x^3 + x^2 + 1 = (x^2 - x + 1)(x^4 + x^3 + x^2 + x + 1), \quad \text{modulo 4.}$$

The factorizations modulo 2^3 and modulo 2^4 are the same. If we had made these computations, we would guess that this is an integer factorization, which it is.

(b) Let $f(x) = x^6 - x^4 + x^3 + x^2 + 1$. This polynomial factors in the same way modulo 2. If f were reducible in $\mathbb{Z}[x]$, it would have a quadratic factor $x^2 + ax + b$, and b would be the product of two roots of f. Cauchy's theorem tells us that the roots have absolute value less than 2, so $|b| < 4$. Computing modulo 2^4,

$$x^6 - x^4 + x^3 + x^2 + 1 = (x^2 + x - 5)(x^4 - x^3 + 5x^2 + 7x + 3), \quad \text{modulo 16.}$$

The constant coefficient of the quadratic factor is -5. This is too big, so f is irreducible.

Note: It isn't necessary to use Cauchy's Theorem here. Since the constant coefficient of f is 1, the fact that $-5 \not\equiv \pm 1$ modulo 16 also proves that f is irreducible. $\qquad\square$

The computer implementations for factoring are interesting, but they are painful to carry out by hand. It is unpleasant to determine a factorization modulo 16 such as the one above by hand, though it can be done by linear algebra. We won't discuss computer methods further. If you want to pursue this topic, see [LL&L].

12.5   GAUSS PRIMES

We have seen that the ring $\mathbb{Z}[i]$ of Gauss integers is a Euclidean domain. Every element that is not zero and not a unit is a product of prime elements. In this section we describe these prime elements, called *Gauss primes*, and their relation to integer primes.

In $\mathbb{Z}[i]$, $5 = (2 + i)(2 - i)$, and the factors $2 + i$ and $2 - i$ are Gauss primes. On the other hand, the integer 3 doesn't have a proper factor in $\mathbb{Z}[i]$. It is itself a Gauss prime. These examples exhibit the two ways that prime integers can factor in the ring of Gauss integers.

The next lemma follows directly from the definition of a Gauss integer:

Lemma 12.5.1

- A Gauss integer that is a real number is an integer.

- An integer d divides a Gauss integer $a + bi$ in the ring $\mathbb{Z}[i]$ if and only if d divides both a and b in $\mathbb{Z}$. $\qquad\qquad\square$

Theorem 12.5.2

(a) Let π be a Gauss prime, and let $\bar{\pi}$ be its complex conjugate. Then $\bar{\pi}\pi$ is either an integer prime or the square of an integer prime.

(b) Let p be an integer prime. Then p is either a Gauss prime or the product $\bar{\pi}\pi$ of a Gauss prime and its complex conjugate.

(c) The integer primes p that are Gauss primes are those congruent to 3 modulo 4:
$$p = 3, 7, 11, 19, \ldots$$

(d) Let p be an integer prime. The following are equivalent:

  (i) p is the product of complex conjugate Gauss primes.

  (ii) p is congruent 1 modulo 4, or $p = 2$: $p = 2, 5, 13, 17, \ldots$

  (iii) p is the sum of two integer squares: $p = a^2 + b^2$.

  (iv) The residue of -1 is a square modulo p.

Proof of Theorem 12.5.2 **(a)** Let π be a Gauss prime, say $\pi = a + bi$. We factor the positive integer $\bar{\pi}\pi = a^2 + b^2$ in the ring of integers: $\bar{\pi}\pi = p_1 \cdots p_k$. This equation is also true in the Gauss integers, though it is not necessarily a prime factorization in that ring. We continue factoring each p_i if possible, to arrive at a prime factorization in $\mathbb{Z}[i]$. Because the Gauss integers have unique factorization, the prime factors we obtain must be associates of the two factors π and $\bar{\pi}$. Therefore k is at most two. Either $\bar{\pi}\pi$ is an integer prime, or else it is the product of two integer primes. Suppose that $\bar{\pi}\pi = p_1 p_2$, and say that π is an associate of the integer prime p_1, i.e., that $\pi = \pm p_1$ or $\pm i p_1$. Then $\bar{\pi}$ is also an associate of p_1, so is $\bar{\pi}$, so $p_1 = p_2$, and $\bar{\pi}\pi = p_1^2$.

(b) If p is an integer prime, it is not a unit in $\mathbb{Z}[i]$. (The units are $\pm 1, \pm i$.) So p is divisible by a Gauss prime π. Then $\bar{\pi}$ divides $\bar{p}$, and $\bar{p} = p$. So the integer $\bar{\pi}\pi$ divides p^2 in $\mathbb{Z}[i]$ and also in $\mathbb{Z}$. Therefore $\bar{\pi}\pi$ is equal to p or p^2. If $\bar{\pi}\pi = p^2$, then π and p are associates, so p is a Gauss prime.

Part **(c)** of the theorem follows from **(b)** and **(d)**, so we need not consider it further, and we turn to the proof of **(d)**. It is easy to see that **(d)(i)** and **(d)(iii)** are equivalent: If $p = \bar{\pi}\pi$

for some Gauss prime, say $\pi = a + bi$, then $p = a^2 + b^2$ is a sum of two integer squares. Conversely, if $p = a^2 + b^2$, then p factors in the Gauss integers: $p = (a - bi)(a + bi)$, and **(a)** shows that the two factors are Gauss primes. □

Lemma 12.5.3 below shows that **(d)(i)** and **(d)(iv)** are equivalent, because (12.5.3)**(a)** is the negation of **(d)(i)** and (12.5.3)**(c)** is the negation of **(d)(iv)**.

Lemma 12.5.3 Let p be an integer prime. The following statements are equivalent:

(a) p is a Gauss prime;
(b) the quotient ring $\overline{R} = \mathbb{Z}[i]/(p)$ is a field;
(c) $x^2 + 1$ is an irreducible element of $\mathbb{F}_p[x]$ (12.2.8)**(c)**.

Proof. The equivalence of the first two statements follows from the fact that $\mathbb{Z}[i]/(p)$ is a field if and only if the principal ideal (p) of $\mathbb{Z}[i]$ is a maximal ideal, and this is true if and only if p is a Gauss prime (see (12.2.9)).

What we are really after is the equivalence of **(a)** and **(c)**, and at a first glance these statements don't seem to be related at all. It is in order to obtain this equivalence that we introduce the auxiliary ring $\overline{R} = \mathbb{Z}[i]/(p)$. This ring can be obtained from the polynomial ring $\mathbb{Z}[x]$ in two steps: first killing the polynomial $x^2 + 1$, which yields a ring isomorphic to $\mathbb{Z}[i]$, and then killing the prime p in that ring. We may just as well introduce these relations in the opposite order. Killing the prime p first gives us the polynomial ring $\mathbb{F}_p[x]$, and then killing $x^2 + 1$ yields $\overline{R}$ again, as is summed up in the diagram below.

(12.5.4)
$$
\begin{array}{ccc}
\mathbb{Z}[x] & \xrightarrow{\text{kill } p} & \mathbb{F}_p[x] \\
{\scriptstyle\text{kill } x^2+1}\Big\downarrow & & \Big\downarrow{\scriptstyle\text{kill } x^2+1} \\
\mathbb{Z}[i] & \xrightarrow[\text{kill } p]{} & \overline{R}
\end{array}
\quad ,
$$

We now have two ways to decide whether or not $\overline{R}$ is a field. First, $\overline{R}$ will be a field if and only if the ideal (p) in the ring $\mathbb{Z}[i]$ is a maximal ideal, which will be true if and only if p is a Gauss prime. Second, $\overline{R}$ will be a field if and only if the ideal $(x^2 + 1)$ in the ring $\mathbb{F}_p[x]$ is a maximal ideal, which will be true if and only if $x^2 + 1$ is an irreducible element of that ring (12.2.9). This shows that **(a)** and **(c)** of Theorem 12.5.2 are equivalent. □

To complete the proof of equivalence of **(i)–(iv)** of Theorem 12.5.2**(d)**, it suffices to show that **(ii)** and **(iv)** are equivalent. It is true that -1 is a square modulo 2. We look at the primes different from 2. The next lemma does the job:

Lemma 12.5.5 Let p be an odd prime.

(a) The multiplicative group $\mathbb{F}_p^\times$ contains an element of order 4 if and only if $p \equiv 1$ modulo 4.
(b) The integer a solves the congruence $x^2 \equiv -1$ modulo p if and only if its residue $\overline{a}$ is an element of order 4 in the multiplicative group $\mathbb{F}_p^\times$.

Proof. **(a)** This follows from a fact mentioned before, that the multiplicative group $\mathbb{F}_p^\times$ is a cyclic group (see (15.7.3)). We give an ad hoc proof here. The order of an element divides the order of the group. So if $\bar{a}$ has order 4 in $\mathbb{F}_p^\times$, then the order of $\mathbb{F}_p^\times$, which is $p-1$, is divisible by 4. Conversely, suppose that $p-1$ is divisible by 4. We consider the homomorphism $\varphi : \mathbb{F}_p^\times \to \mathbb{F}_p^\times$ that sends $x \rightsquigarrow x^2$. The only elements of $\mathbb{F}_p^\times$ whose squares are 1 are ± 1 (see (12.2.20)). So the kernel of φ is $\{\pm 1\}$. Therefore its image, call it H, has even order $(p-1)/2$. The first Sylow Theorem shows that H contains an element of order 2. That element is the square of an element x of order 4.

(b) The residue $\bar{a}$ has order 4 if and only if $\bar{a}^2$ has order 2. There is just one element in $\mathbb{F}_p$ of order 2, namely the residue of -1. So $\bar{a}$ has order 4 if and only if $\bar{a}^2 = -\bar{1}$.   □

This competes the proof of Theorem 12.5.2.   □

> *You want to hit home run without going into spring training?*
>
> —Kenkichi Iwasawa

EXERCISES

Section 1   Factoring Integers

1.1. Prove that a positive integer n that is not an integer square is not the square of a rational number.

1.2. *(partial fractions)*

    (a) Write the fraction $7/24$ in the form $a/8 + b/3$.

    (b) Prove that if $n = uv$, where u and v are relatively prime, then every fraction $q = m/n$ can be written in the form $q = a/u + b/v$.

1.3. *(Chinese Remainder Theorem)*

    (a) Let n and m be relatively prime integers, and let a and b be arbitrary integers. Prove that there is an integer x that solves the simultaneous congruence $x \equiv a$ modulo m and $x \equiv b$ modulo n.

    (b) Determine all solutions of these two congruences.

1.4. Solve the following simultaneous congruences:

    (a) $x \equiv 3$ modulo 8, $x \equiv 2$ modulo 5,

    (b) $x \equiv 3$ modulo 15, $x \equiv 5$ modulo 8, $x \equiv 2$ modulo 7,

    (c) $x \equiv 13$ modulo 43, $x \equiv 7$ modulo 71.

1.5. Let a and b be relatively prime integers. Prove that there are integers m and n such that $a^m + b^n \equiv 1$ modulo ab.

Section 2 Unique Factorization Domains

2.1. Factor the following polynomials into irreducible factors in $\mathbb{F}_p[x]$.

(a) $x^3 + x^2 + x + 1$, $p = 2$, (b) $x^2 - 3x - 3$, $p = 5$, (c) $x^2 + 1$, $p = 7$

2.2. Compute the greatest common divisor of the polynomials $x^6 + x^4 + x^3 + x^2 + x + 1$ and $x^5 + 2x^3 + x^2 + x + 1$ in $\mathbb{Q}[x]$.

2.3. How many roots does the polynomial $x^2 - 2$ have, modulo 8?

2.4. Euclid proved that there are infinitely many prime integers in the following way: If $p_1, \ldots, p_k$ are primes, then any prime factor p of $(p_1 \cdots p_k) + 1$ must be different from all of the p_i. Adapt this argument to prove that for any field F there are infinitely many monic irreducible polynomials in $F[x]$.

2.5. (*partial fractions for polynomials*)

(a) Prove that every element of $\mathbb{C}(x)$ x can be written as a sum of a polynomial and a linear combination of functions of the form $1/(x - a)^i$.

(b) Exhibit a basis for the field $\mathbb{C}(x)$ of rational functions as vector space over $\mathbb{C}$.

2.6. Prove that the following rings are Euclidean domains.

(a) $\mathbb{Z}[\omega]$, $\omega = e^{2\pi i/3}$, (b) $\mathbb{Z}[\sqrt{-2}]$.

2.7. Let a and b be integers. Prove that their greatest common divisor in the ring of integers is the same as their greatest common divisor in the ring of Gauss integers.

2.8. Describe a systematic way to do division with remainder in $\mathbb{Z}[i]$. Use it to divide $4 + 36i$ by $5 + i$.

2.9. Let F be a field. Prove that the ring $F[x, x^{-1}]$ of Laurent polynomials (Chapter 11, Exercise 5.7) is a principal ideal domain.

2.10. Prove that the ring $\mathbb{R}[[t]]$ of formal power series (Chapter 11, Exercise 2.2) is a unique factorization domain.

Section 3 Gauss's Lemma

3.1. Let φ denote the homomorphism $\mathbb{Z}[x] \to \mathbb{R}$ defined by

(a) $\varphi(x) = 1 + \sqrt{2}$, (b) $\varphi(x) = \frac{1}{2} + \sqrt{2}$.

Is the kernel of φ a principal ideal? If so, find a generator.

3.2. Prove that two integer polynomials are relatively prime elements of $\mathbb{Q}[x]$ if and only if the ideal they generate in $\mathbb{Z}[x]$ contains an integer.

3.3. State and prove a version of Gauss's Lemma for Euclidean domains.

3.4. Let x, y, z, w be variables. Prove that $xy - zw$, the determinant of a variable 2×2 matrix, is an irreducible element of the polynomial ring $\mathbb{C}[x, y, z, w]$.

3.5. (a) Consider the map $\psi : \mathbb{C}[x, y] \to \mathbb{C}[t]$ defined by $f(x, y) \rightsquigarrow f(t^2, t^3)$. Prove that its image is the set of polynomials $p(t)$ such that $\frac{dp}{dt}(0) = 0$.

(b) Consider the map $\varphi : \mathbb{C}[x, y] \to \mathbb{C}[t]$ defined by $f(x, y) \rightsquigarrow (t^2 - t, t^3 - t^2)$. Prove that $\ker \varphi$ is a principal ideal, and find a generator $g(x, y)$ for this ideal. Prove that the image of φ is the set of polynomials $p(t)$ such that $p(0) = p(1)$. Give an intuitive explanation in terms of the geometry of the variety $\{g = 0\}$ in $\mathbb{C}^2$.

3.6. Let α be a complex number. Prove that the kernel of the substitution map $\mathbb{Z}[x] \to \mathbb{C}$ that sends $x \leadsto \alpha$ is a principal ideal, and describe its generator.

Section 4   Factoring Integer Polynomials

4.1. (a) Factor $x^9 - x$ and $x^9 - 1$ in $\mathbb{F}_3[x]$. **(b)** Factor $x^{16} - x$ in $\mathbb{F}_2[x]$.

4.2. Prove that the following polynomials are irreducible:
   (a) $x^2 + 1$, in $\mathbb{F}_7[x]$, **(b)** $x^3 - 9$, in $\mathbb{F}_{31}[x]$.

4.3. Decide whether or not the polynomial $x^4 + 6x^3 + 9x + 3$ generates a maximal ideal in $\mathbb{Q}[x]$.

4.4. Factor the integer polynomial $x^5 + 2x^4 + 3x^3 + 3x + 5$ modulo 2, modulo 3, and in $\mathbb{Q}$.

4.5. Which of the following polynomials are irreducible in $\mathbb{Q}[x]$?
   (a) $x^2 + 27x + 213$, **(b)** $8x^3 - 6x + 1$, **(c)** $x^3 + 6x^2 + 1$, **(d)** $x^5 - 3x^4 + 3$.

4.6. Factor $x^5 + 5x + 5$ into irreducible factors in $\mathbb{Q}[x]$ and in $\mathbb{F}_2[x]$.

4.7. Factor $x^3 + x + 1$ in $\mathbb{F}_p[x]$, when $p = 2, 3$, and 5.

4.8. How might a polynomial $f(x) = x^4 + bx^2 + c$ with coefficients in a field F factor in $F[x]$? Explain with reference to the particular polynomials $x^4 + 4x^2 + 4$ and $x^4 + 3x^2 + 4$.

4.9. For which primes p and which integers n is the polynomial $x^n - p$ irreducible in $\mathbb{Q}[x]$?

4.10. Factor the following polynomials in $\mathbb{Q}[x]$. **(a)** $x^2 + 2351x + 125$, **(b)** $x^3 + 2x^2 + 3x + 1$, **(c)** $x^4 + 2x^3 + 2x^2 + 2x + 2$, **(d)** $x^4 + 2x^3 + 3x^2 + 2x + 1$, **(e)** $x^4 + 2x^3 + x^2 + 2x + 1$, **(f)** $x^4 + 2x^2 + x + 1$, **(g)** $x^8 + x^6 + x^4 + x^2 + 1$, **(h)** $x^6 - 2x^5 - 3x^2 + 9x - 3$, **(j)** $x^4 + x^2 + 1$, **(k)** $3x^5 + 6x^4 + 9x^3 + 3x^2 - 1$, **(l)** $x^5 + x^4 + x^2 + x + 2$.

4.11. Use the sieve method to determine the primes <100, and discuss the efficiency of the sieve: How quickly are the nonprimes filtered out?

4.12. Determine:

   (a) the monic irreducible polynomials of degree 3 over $\mathbb{F}_3$,
   (b) the monic irreducible polynomials of degree 2 over $\mathbb{F}_5$,
   (c) the number of irreducible polynomials of degree 3 over the field $\mathbb{F}_5$.

4.13. *Lagrange interpolation formula*:

   (a) Let $a_0, \ldots, a_d$ be distinct complex numbers. Determine a polynomial $p(x)$ of degree n, which has $a_1, \ldots, a_n$ as roots, and such that $p(a_0) = 1$.
   (b) Let $a_0, \ldots, a_d$ and $b_0, \ldots, b_d$ be complex numbers, and suppose that the a_i are distinct. There is a unique polynomial g of degree $\leq d$ such that $g(a_i) = b_i$ for each $i = 0, \ldots, d$. Determine the polynomial g explicitly in terms of a_i and b_i.

4.14. By analyzing the locus $x^2 + y^2 = 1$, prove that the polynomial $x^2 + y^2 - 1$ is irreducible in $\mathbb{C}[x, y]$.

4.15. With reference to the Eisenstein criterion, what can one say when
   (a) $\bar{f}$ is constant, **(b)** $\bar{f} = x^n + \bar{b}x^{n-1}$?

4.16. Factor $x^{14} + 8x^{13} + 3$ in $\mathbb{Q}[x]$, using reduction modulo 3 as a guide.

4.17. Using congruence modulo 4 as an aid, factor $x^4 + 6x^3 + 7x^2 + 8x + 9$ in $\mathbb{Q}[x]$.

*4.18. Let $q = p^e$ with p prime, and let $r = p^{e-1}$. Prove that the cyclotomic polynomial $(x^q - 1)/(x^r - 1)$ is irreducible.

4.19. Factor $x^5 - x^4 - x^2 - 1$ modulo 2, modulo 16, and over $\mathbb{Q}$.

Section 5   Gauss Primes

5.1. Factor the following into primes in $\mathbb{Z}[i]$: **(a)** $1 - 3i$, **(b)** 10, **(c)** $6 + 9i$, **(d)** $7 + i$.

5.2. Find the greatest common divisor in $\mathbb{Z}[i]$ of **(a)** $11 + 7i, 4 + 7i$, **(b)** $11 + 7i, 8 + i$, **(c)** $3 + 4i, 18 - i$.

5.3. Find a generator for the ideal of $Z[i]$ generated by $3 + 4i$ and $4 + 7i$.

5.4. Make a neat drawing showing the primes in the ring of Gauss integers in a reasonable size range.

5.5. Let π be a Gauss prime. Prove that π and $\bar{\pi}$ are associates if and only if π is an associate of an integer prime, or $\bar{\pi}\pi = 2$.

5.6. Let R be the ring $\mathbb{Z}[\sqrt{-3}]$. Prove that an integer prime p is a prime element of R if and only if the polynomial $x^2 + 3$ is irreducible in $\mathbb{F}_p[x]$.

5.7. Describe the residue ring $\mathbb{Z}[i]/(p)$ for each prime p.

5.8. Let $R = \mathbb{Z}[\omega]$, where $\omega = e^{2\pi i/3}$. Make a drawing showing the prime elements of absolute value ≤ 10 in R.

*5.9. Let $R = \mathbb{Z}[\omega]$, where $\omega = e^{2\pi i/3}$. Let p be an integer prime $\neq 3$. Adapt the proof of Theorem 12.5.2 to prove the following:

   (a) The polynomial $x^2 + x + 1$ has a root in $\mathbb{F}_p$ if and only if $p \equiv 1$ modulo 3.
   (b) (p) is a maximal ideal of R if and only if $p \equiv -1$ modulo 3.
   (c) p factors in R if and only if it can be written in the form $p = a^2 + ab + b^2$, for some integers a and b.

5.10. [1]**(a)** Let α be a Gauss integer. Assume that α has no integer factor, and that $\bar{\alpha}\alpha$ is a square integer. Prove that α is a square in $\mathbb{Z}[i]$.
   (b) Let a, b, c be integers such that a and b are relatively prime and $a^2 + b^2 = c^2$. Prove that there are integers m and n such that $a = m^2 - n^2, b = 2mn$, and $c = m^2 + n^2$.

Miscellaneous Problems

M.1. Let S be a commutative semigroup – a set with a commutative and associative law of composition and with an identity element (Chapter 2, Exercise M.4). Suppose the Cancellation Law holds in S: If $ab = ac$ then $b = c$. Make the appropriate definitions and extend Proposition 12.2.14**(a)** to this situation.

M.2. Let $v_1, \dots, v_n$ be elements of $\mathbb{Z}^2$, and let S be the semigroup of all combinations $a_1 v_1 + \cdots + a_n v_n$ with non-negative integer coefficients a_i, the law of composition being *addition* (Chapter 2, Exercise M.4). Determine which of these semigroups has unique factorization **(a)** when the coordinates of the vectors v_i are nonnegative, and **(b)** in general.

   Hint: Begin by translating the terminology (12.2.1) into additive notation.

---

[1]Suggested by Nathaniel Kuhn.

M.3. Let p be an integer prime, and let A be an $n \times n$ integer matrix such that $A^p = I$ but $A \neq I$. Prove that $n \geq p - 1$. Give an example with $n = p - 1$.

*__M.4.__ (a) Let R be the ring of functions that are polynomials in $\cos t$ and $\sin t$, with real coefficients. Prove that R is isomorphic to $\mathbb{R}[x, y]/(x^2 + y^2 - 1)$.

  (b) Prove that R is not a unique factorization domain.

  (c) Prove that $S = \mathbb{C}[x, y]/(x^2 + y^2 - 1)$ is a principal ideal domain and hence a unique factorization domain.

  (d) Determine the units in the rings S and R.

  Hint: Show that S is isomorphic to a Laurent polynomial ring $\mathbb{C}[u, u^{-1}]$.

M.5. For which integers n does the circle $x^2 + y^2 = n$ contain a point with integer coordinates?

M.6. Let R be a domain, and let I be an ideal that is a product of distinct maximal ideals in two ways, say $I = P_1 \cdots P_r = Q_1 \cdots Q_s$. Prove that the two factorizations are the same, except for the ordering of the terms.

M.7. Let $R = \mathbb{Z}[x]$.

  (a) Prove that every maximal ideal in R has the form (p, f), where p is an integer prime and f is a primitive integer polynomial that is irreducible modulo p.

  (b) Let I be an ideal of R generated by two polynomials f and g that have no common factor other than ± 1. Prove that R/I is finite.

M.8. Let u and v be relatively prime integers, and let R' be the ring obtained from $\mathbb{Z}$ by adjoining an element α with the relation $v\alpha = u$. Prove that R' is isomorphic to $\mathbb{Z}\left[\frac{u}{v}\right]$ and also to $\mathbb{Z}\left[\frac{1}{v}\right]$.

M.9. Let R denote the ring of Gauss integers, and let W be the R-submodule of $V = R^2$ generated by the columns of a 2×2 matrix with coefficients in R. Explain how to determine the index $[V:W]$.

M.10. Let f and g be polynomials in $\mathbb{C}[x, y]$ with no common factor. Prove that the ring $R = \mathbb{C}[x, y]/(f, g)$ is a finite-dimensional vector space over $\mathbb{C}$.

M.11. (*Berlekamp's method*) The problem here is to factor efficiently in $\mathbb{F}_2[x]$. Solving linear equations and finding a greatest common divisor are easy compared with factoring. The derivative f' of a polynomial f is computed using the rule from calculus, but working modulo 2. Prove:

  (a) (*square factors*) The derivative f' is a square, and $f' = 0$ if and only if f is a square. Moreover, $\gcd(f, f')$ is the product of powers of the square factors of f.

  (b) (*relatively prime factors*) Let n be the degree of f. If $f = uv$, where u and v are relatively prime, the Chinese Remainder Theorem shows that there is a polynomial g of degree at most n such that $g^2 - g \equiv 0$ modulo f, and g can be found by solving a system of linear equations. Either $\gcd(f, g)$ or $\gcd(f, g - 1)$ will be a proper factor of f.

  (c) Use this method to factor $x^9 + x^6 + x^4 + 1$.

C H A P T E R   13

Quadratic Number Fields

Rien n'est beau que le vrai.

—Hermann Minkowski

In this chapter, we see how ideals substitute for elements in some interesting rings. We will use various facts about plane lattices, and in order not to break up the discussion, we have collected them together in Section 13.10 at the end of the chapter.

13.1 ALGEBRAIC INTEGERS

A complex number α that is the root of a polynomial with rational coefficients is called an *algebraic number*. The kernel of the substitution homomorphism $\varphi : \mathbb{Q}[x] \to \mathbb{C}$ that sends x to an algebraic number α is a principal ideal, as are all ideals of $\mathbb{Q}[x]$. It is generated by the monic polynomial of lowest degree in $\mathbb{Q}[x]$ that has α as a root. If α is a root of a product gh of polynomials, then it is a root of one of the factors. So the monic polynomial of lowest degree with root α is irreducible. We call this polynomial the *irreducible polynomial* for α over $\mathbb{Q}$.

• An algebraic number is an *algebraic integer* if its (monic) irreducible polynomial over $\mathbb{Q}$ has integer coefficients.

The cube root of unity $\omega = e^{2\pi i/3} = \frac{1}{2}(-1 + \sqrt{-3})$ is an algebraic integer because its irreducible polynomial over $\mathbb{Q}$ is $x^2 + x + 1$, while $\alpha = \frac{1}{2}(-1 + \sqrt{3})$ is a root of the irreducible polynomial $x^2 - x - \frac{1}{2}$ and is not an algebraic integer.

Lemma 13.1.1 A rational number is an algebraic integer if and only if it is an ordinary integer.

This is true because the irreducible polynomial over $\mathbb{Q}$ for a rational number a is $x - a$. $\square$

A *quadratic number field* is a field of the form $\mathbb{Q}[\sqrt{d}]$, where d is a fixed integer, positive or negative, which is not a square in $\mathbb{Q}$. Its elements are the complex numbers

$$(13.1.2) \qquad\qquad a + b\sqrt{d}, \quad \text{with } a \text{ and } b \text{ in } \mathbb{Q},$$

The notation $\sqrt{d}$ stands for the positive real square root if $d > 0$ and for the positive imaginary square root if $d < 0$. The field $\mathbb{Q}[\sqrt{d}]$ is a *real quadratic number field* if $d > 0$, and an *imaginary quadratic number field* if $d < 0$.

If d has a square integer factor, we can pull it out of the radical without changing the field. So we assume d *square-free*. Then d can be any one of the integers

$$d = -1, \pm 2, \pm 3, \pm 5, \pm 6, \pm 7, \pm 10, \ldots$$

We determine the algebraic integers in a quadratic number field $\mathbb{Q}[\sqrt{d}]$ now. Let δ denote $\sqrt{d}$, let $\alpha = a + b\delta$ be an element of $\mathbb{Q}[\delta]$ that is not in $\mathbb{Q}$, that is, with $b \neq 0$, and let $\alpha' = a - b\delta$. Then α and α' are roots of the polynomial

(13.1.3) $(x - \alpha')(x - \alpha) = x^2 - 2ax + (a^2 - b^2 d),$

which has rational coefficients. Since α is not a rational number, it is not the root of a linear polynomial. So this quadratic polynomial is irreducible over $\mathbb{Q}$. It is therefore the irreducible polynomial for α over $\mathbb{Q}$.

Corollary 13.1.4 A complex number $\alpha = a + b\delta$ with a and b in $\mathbb{Q}$ is an algebraic integer if and only if $2a$ and $a^2 - b^2 d$ are ordinary integers. $\square$

This corollary is also true when $b = 0$ and $\alpha = a$.

The possibilities for a and b depend on congruence modulo 4. Since d is assumed to be square free, we can't have $d \equiv 0$, so $d \equiv 1, 2,$ or 3 modulo 4.

Lemma 13.1.5 Let d be a square-free integer, and let r be a rational number. If $r^2 d$ is an integer, then r is an integer.

Proof. The square-free integer d cannot cancel a square in the denominator of r^2. $\square$

A *half integer* is a rational number of the form $m + \frac{1}{2}$, where m is an integer.

Proposition 13.1.6 The algebraic integers in the quadratic field $\mathbb{Q}[\delta]$, with $\delta^2 = d$ and d square free, have the form $\alpha = a + b\delta$, where:

- If $d \equiv 2$ or 3 modulo 4, then a and b are integers.
- If $d \equiv 1$ modulo 4, then a and b are either both integers, or both half integers.

The algebraic integers form a ring R, the *ring of integers* in F.

Proof. We assume that $2a$ and $a^2 - b^2 d$ are integers, and we analyze the possibilities for a and b. There are two cases: Either a is an integer, or a is a half integer.

Case 1: a is an integer. Then $b^2 d$ must be an integer. The lemma shows that b is an integer.

Case 2: $a = m + \frac{1}{2}$ is a half integer. Then $a^2 = m^2 + m + \frac{1}{4}$ will be in the set $\mathbb{Z} + \frac{1}{4}$. Since $a^2 - b^2 d$ is an integer, $b^2 d$ is also in $\mathbb{Z} + \frac{1}{4}$. Then $4b^2 d$ is an integer and the lemma shows that $2b$ is an integer. So b is a half integer, and then $b^2 d$ is in the set $\mathbb{Z} + \frac{1}{4}$ if and only if $d \equiv 1$ modulo 4.

The fact that the algebraic integers form a ring is proved by computation. $\square$

The imaginary quadratic case $d < 0$ is easier to handle than the real case, so we concentrate on it in the next sections. When $d < 0$, the algebraic integers form a lattice in the complex plane. The lattice is rectangular if $d \equiv 2$ or 3 modulo 4, and "isosceles triangular" if $d \equiv 1$ modulo 4.

When $d = -1$, R is the ring of Gauss integers, and the lattice is square. When $d = -3$, the lattice is equilateral triangular. Two other examples are shown below.

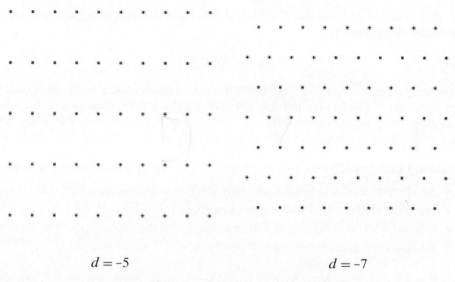

$$d = -5 \qquad\qquad\qquad d = -7$$

(13.1.7)   Integers in Some Imaginary Quadratic Fields.

Being a lattice is a very special property of the rings that we consider here, and the geometry of the lattices helps to analyze them.

When $d \equiv 2$ or 3 modulo 4, the integers in $\mathbb{Q}[\delta]$ are the complex numbers $a + b\delta$, with a and b integers. They form a ring that we denote by $\mathbb{Z}[\delta]$. A convenient way to write all the integers when $d \equiv 1$ modulo 4 is to introduce the algebraic integer

(13.1.8) $\eta = \tfrac{1}{2}(1 + \delta).$

It is a root of the monic integer polynomial

(13.1.9) $x^2 - x + h,$

where $h = (1 - d)/4$. The algebraic integers in $\mathbb{Q}[\delta]$ are the complex numbers $a + b\eta$, with a and b integers. The ring of integers is $\mathbb{Z}[\eta]$.

13.2   FACTORING ALGEBRAIC INTEGERS

The symbol R will denote the ring of integers in an imaginary quadratic number field $\mathbb{Q}[\delta]$. To focus your attention, it may be best to think at first of the case that d is congruent 2 or 3 modulo 4, so that the algebraic integers have the form $a + b\delta$, with a and b integers.

When possible, we denote ordinary integers by Latin letters $a, b, \ldots$, elements of R by Greek letters $\alpha, \beta, \ldots$, and ideals by capital letters $A, B, \ldots$ We work exclusively with nonzero ideals.

If $\alpha = a + b\delta$ is in R, its complex conjugate $\bar{\alpha} = a - b\delta$ is in R too. These are the roots of the polynomial $x^2 - 2ax + (a^2 - b^2 d)$ that was introduced in Section 13.1.

• The *norm* of $\alpha = a + b\delta$ is $N(\alpha) = \bar{\alpha}\alpha$.

The norm is equal to $|\alpha|^2$ and also to $a^2 - b^2 d$. It is a positive integer for all $\alpha \neq 0$, and it has the multiplicative property:

(13.2.1) $$N(\beta\gamma) = N(\beta)N(\gamma).$$

This property gives us some control of the factors of an element. If $\alpha = \beta\gamma$, then both terms on the right side of (13.2.1) are positive integers. To check for factors of α, it is enough to look at elements β whose norms divide the norm of α. This is manageable when $N(\alpha)$ is small. For one thing, it allows us to determine the units of R.

Proposition 13.2.2 Let R be the ring of integers in an imaginary quadratic number field.

• An element α of R is a unit if and only if $N(\alpha) = 1$. If so, then $\alpha^{-1} = \bar{\alpha}$.
• The units of R are $\{\pm 1\}$ unless $d = -1$ or -3.
• When $d = -1$, R is the ring of Gauss integers, and the units are the four powers of i.
• When $d = -3$, the units are the six powers of $e^{2\pi i/6} = \frac{1}{2}(1 + \sqrt{-3})$.

Proof. If α is a unit, then $N(\alpha)N(\alpha^{-1}) = N(1) = 1$. Since $N(\alpha)$ and $N(\alpha^{-1})$ are positive integers, they are both equal to 1. Conversely, if $N(\alpha) = \bar{\alpha}\alpha = 1$, then $\bar{\alpha}$ is the inverse of α, so α is a unit. The remaining assertions follow by inspection of the lattice R.   □

Corollary 13.2.3 Factoring terminates in the ring of integers in an imaginary quadratic number field.

This follows from the fact that factoring terminates in the integers. If $\alpha = \beta\gamma$ is a proper factorization in R, then $N(\alpha) = N(\beta)N(\gamma)$ is a proper factorization in $\mathbb{Z}$.   □

Proposition 13.2.4 Let R be the ring of integers in an imaginary quadratic number field. Assume that $d \equiv 3$ modulo 4. Then R is not a unique factorization domain except in the case $d = -1$, when R is the ring of Gauss integers.

Proof. This is analogous to what happens when $d = -5$. Suppose that $d \equiv 3$ modulo 4 and that $d < -1$. The integers in R have the form $a + b\delta$ which $a, b \in \mathbb{Z}$, and the units are ± 1. Let $e = (1 - d)/2$. Then

$$2e = 1 - d = (1 + \delta)(1 - \delta).$$

The element $1 - d$ factors in two ways in R. Since $d < -1$, there is no element $a + b\delta$ whose norm is equal to 2. Therefore 2, which has norm 4, is an irreducible element of R. If R were a unique factorization domain, 2 would divide either $1 + \delta$ or $1 - \delta$ in R, which it does not: $\frac{1}{2}(1 \pm \delta)$ is not an element of R when $d \equiv 3$ modulo 4.   □

There is a similar statement for the case $d \equiv 2$ modulo 4. (This is Exercise 2.2.) But note that the reasoning breaks down when $d \equiv 1$ modulo 4. In that case, $\frac{1}{2}(1 + \delta)$ is in R, and in fact there are more cases of unique factorization when $d \equiv 1$ modulo 4. A famous theorem enumerates these cases:

Theorem 13.2.5 The ring of integers R in the imaginary quadratic field $\mathbb{Q}[\sqrt{d}]$ is a unique factorization domain if and only if d is one of the integers $-1, -2, -3, -7, -11, -19, -43, -67, -163$.

Gauss proved that for these values of d, R has unique factorization. We will learn how to do this. He also conjectured that there were no others. This much more difficult part of the theorem was finally proved by Baker, Heegner, and Stark in the middle of the 20th century, after people had worked on it for more than 150 years. We won't be able to prove their theorem.

13.3 IDEALS IN $\mathbb{Z}[\sqrt{-5}]$

Before going to the general theory, we describe the ideals in the ring $R = \mathbb{Z}[\sqrt{-5}]$ as lattices in the complex plane, using an ad hoc method.

Proposition 13.3.1 Let R be the ring of integers in an imaginary quadratic number field. Every nonzero ideal of R is a sublattice of the lattice R. Moreover,

- If $d \equiv 2$ or 3 modulo 4, a sublattice A is an ideal if and only if $\delta A \subset A$.
- If $d \equiv 1$ modulo 4, a sublattice A is an ideal if and only if $\eta A \subset A$ (see (13.1.8)).

Proof. A nonzero ideal A contains a nonzero element α, and $(\alpha, \alpha\delta)$ is an independent set over $\mathbb{R}$. Also, A is discrete because it is a subgroup of the lattice R. Therefore A is a lattice (Theorem 6.5.5).

To be an ideal, a subset of R must be closed under addition and under multiplication by elements of R. Every sublattice A is closed under addition and multiplication by integers. If A is also closed under multiplication by δ, then it is closed under multiplication by an element of the form $a + b\delta$, with a and b integers. This includes all elements of R if $d \equiv 2$ or 3 modulo 4. So A is an ideal. The proof in the case $d \equiv 1$ modulo 4 is similar. $\qquad\square$

We describe ideals in the ring $R = \mathbb{Z}[\delta]$, when $\delta^2 = -5$.

Lemma 13.3.2 Let $R = \mathbb{Q}[\delta]$ with $\delta^2 = -5$. The lattice A of integer combinations of 2 and $1 + \delta$ is an ideal.

Proof. The lattice A is closed under multiplication by δ, because $\delta \cdot 2$ and $\delta \cdot (1 + \delta)$ are integer combinations of 2 and $1 + \delta$. $\qquad\square$

Figure 13.3.4 shows this ideal.

Theorem 13.3.3 Let $R = \mathbb{Z}[\delta]$, where $\delta = \sqrt{-5}$, and let A be a nonzero ideal of R. Let α be a nonzero element of A of minimal norm (or minimal absolute value). Then either

- The set $(\alpha, \alpha\delta)$ is a lattice basis for A, and A is the principal ideal (α), or
- The set $(\alpha, \frac{1}{2}(\alpha + \alpha\delta))$ is a lattice basis for A, and A is not a principal ideal.

This theorem has the following geometric interpretation: The lattice basis $(\alpha, \alpha\delta)$ of the principal ideal (α) is obtained from the lattice basis $(1, \delta)$ of the unit ideal R by multiplying by α. If we write α in polar coordinates $\alpha = re^{i\theta}$, then multiplication by α rotates the complex plane through the angle θ and stretches by the factor r. So all principal ideals are similar geometric figures. Also, the lattice with basis $(\alpha, \frac{1}{2}(\alpha + \alpha\delta))$ is obtained from the lattice $(2, 1 + \delta)$ by multiplying by $\frac{1}{2}\alpha$. All ideals of the second type are geometric figures similar to the one shown below (see also Figure 13.7.4).

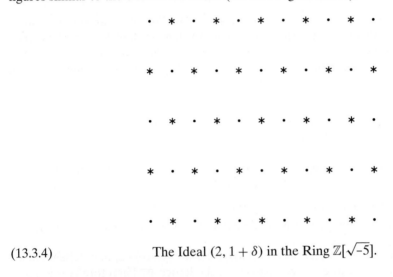

(13.3.4)   The Ideal $(2, 1 + \delta)$ in the Ring $\mathbb{Z}[\sqrt{-5}]$.

Similarity classes of ideals are called *ideal classes*, and the number of ideal classes is the *class number* of R. The theorem asserts that the class number of $\mathbb{Z}[\sqrt{-5}]$ is two. Ideal classes for other quadratic imaginary fields are discussed in Section 13.7.

Theorem 13.3.3 is based on the following simple lemma about lattices:

Lemma 13.3.5 Let A be a lattice in the complex plane, let r be the minimum absolute value among nonzero elements of A, and let γ be an element of A. Let n be a positive integer. The interior of the disk of radius $\frac{1}{n}r$ about the point $\frac{1}{n}\gamma$ contains no element of A other than the center $\frac{1}{n}\gamma$. The center may lie in A or not.

Proof. If β is an element of A in the interior of the disk, then $|\beta - \frac{1}{n}\gamma| < \frac{1}{n}r$, which is to say, $|n\beta - \gamma| < r$. Moreover, $n\beta - \gamma$ is in A. Since this is an element of absolute value less than the minimum, $n\beta - \gamma = 0$. Then $\beta = \frac{1}{n}\gamma$ is the center of the disk.   □

Proof of Theorem 13.3.3. Let α be a nonzero element of an ideal A of minimal absolute value r. Since A contains α, it contains the principal ideal (α), and if $A = (\alpha)$ we are in the first case.

Suppose that A contains an element β not in the principal ideal (α). The ideal (α) has the lattice basis $\mathbf{B} = (\alpha, \alpha\delta)$, so we may choose β to lie in the parallelogram $\Pi(\mathbf{B})$ of linear combinations $r\alpha + s\alpha\delta$ with $0 \leq r, s \leq 1$. (In fact, we can choose β so that $0 \leq r, s < 1$. See Lemma 13.10.2.) Because δ is purely imaginary, the parallelogram is a rectangle. How large the rectangle is, and how it is situated in the plane, depend on α, but the ratio of the side lengths is always $1:\sqrt{5}$. We'll be done if we show that β is the midpoint $\frac{1}{2}(\alpha + \alpha\delta)$ of the rectangle.

Figure 13.3.6 shows disks of radius r about the four vertices of such a rectangle, and also disks of radius $\frac{1}{2}r$ about three half lattice points, $\frac{1}{2}\alpha\delta$, $\frac{1}{2}(\alpha + \alpha\delta)$, and $\alpha + \frac{1}{2}\alpha\delta$. Notice that the interiors of these seven disks cover the rectangle. (It would be fussy to check this by algebra. Let's not bother. A glance at the figure makes it clear enough.)

According to Lemma 13.3.5, the only points of the interiors of the disks that can be elements of A are their centers. Since β is not in the principal ideal (α), it is not a vertex of the rectangle. So β must be one of the three half lattice points. If $\beta = \alpha + \frac{1}{2}\alpha\delta$, then since α is in A, $\frac{1}{2}\alpha\delta$ will be in A too. So we have only two cases to consider: $\beta = \frac{1}{2}\alpha\delta$ and $\beta = \frac{1}{2}(\alpha + \alpha\delta)$.

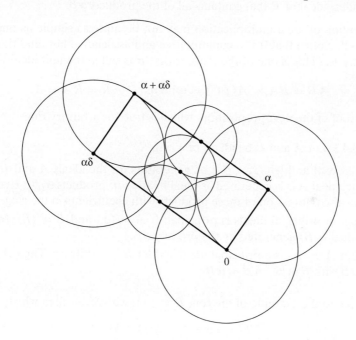

This exhausts the information we can get from the fact that A is a lattice. We now use the fact that A is an ideal. Suppose that $\frac{1}{2}\alpha\delta$ is in A. Multiplying by δ shows that $\frac{1}{2}\alpha\delta^2 = -\frac{5}{2}\alpha$ is in A. Then since α is in A, $\frac{1}{2}\alpha$ is in A too. This contradicts our choice of α as a nonzero element of minimal absolute value. So β cannot be equal to $\frac{1}{2}\alpha\delta$. The remaining possibility is that β is the center $\frac{1}{2}(\alpha + \alpha\delta)$ of the rectangle. If so, we are in the second case of the theorem. $\square$

13.4   IDEAL MULTIPLICATION

Let R be the ring of integers in an imaginary quadratic number field. As usual, the notation $A = (\alpha, \beta, \ldots, \gamma)$ means that A is the the ideal of R generated by the elements $\alpha, \beta, \ldots, \gamma$. It consists of all linear combinations of those elements, with coefficients in the ring.

Since a nonzero ideal A is a lattice, it has a lattice basis (α, β) consisting of two elements. Every element of A is an *integer* combination of α and β. We must be careful to distinguish between the concepts of a lattice basis and a generating set for an ideal. Any lattice basis generates the ideal, but the converse is false. For instance, a principal ideal is generated as an ideal by a single element, whereas a lattice basis has two elements.

Dedekind extended the notion of divisibility to ideals using the following definition of ideal multiplication:

- Let A and B be ideals in a ring R. The *product ideal* AB consists of all *finite sums of products*

$$(13.4.1) \qquad\qquad \sum_i \alpha_i \beta_i, \text{ with } \alpha_i \text{ in } A \text{ and } \beta_i \text{ in } B.$$

This is the smallest ideal of R that contains all of the products $\alpha\beta$.

The definition of ideal multiplication may not be quite as simple as one might hope, but it works well. Notice that it is a commutative and associative law, and that it has a unit element, namely R. (This is one of the reasons that R is called the unit ideal.)

$$(13.4.2) \qquad\qquad AB = BA, \quad A(BC) = (AB)C, \quad AR = RA = A.$$

We omit the proof of the next proposition, which is true for arbitrary rings.

Proposition 13.4.3 Let A and B be ideals of a ring $\mathcal{R}$.

(a) Let $\{\alpha_1, \dots, \alpha_m\}$ and $\{\beta_1, \dots, \beta_n\}$ be generators for the ideals A and B, respectively. The product ideal AB is generated as ideal by the mn products $\alpha_i \beta_j$: Every element of AB is a linear combination of these products with coefficients in the ring.

(b) The product of principal ideals is principal: If $A = (\alpha)$ and $B = (\beta)$, then AB is the principal ideal $(\alpha\beta)$ generated by the product $\alpha\beta$.

(c) Assume that $A = (\alpha)$ is a principal ideal and let B be arbitrary. Then AB is the set of products $\alpha\beta$ with β in B: $AB = \alpha B$. □

We go back to the example of the ring $R = \mathbb{Z}[\delta]$ with $\delta^2 = -5$, in which

$$(13.4.4) \qquad\qquad 2 \cdot 3 = 6 = (1 + \delta)(1 - \delta).$$

If factoring in R were unique, there would be an element γ in R dividing both 2 and $1 + \delta$, and then 2 and $1 + \delta$ would be in the principal ideal (γ). There is no such element. However, there is an *ideal* that contains 2 and $1 + \delta$, namely the ideal $(2, 1 + \delta)$ generated by these two elements, the one depicted in Figure 13.3.4.

We can make four ideals using the factors of 6:

$$(13.4.5) \qquad A = (2, 1 + \delta), \quad \overline{A} = (2, 1 - \delta), \quad B = (3, 1 + \delta), \quad \overline{B} = (3, 1 - \delta).$$

In each of these ideals, the generators that are given happen to form lattice bases. We denote the last of them by $\overline{B}$ because it is the complex conjugate of B:

$$(13.4.6) \qquad\qquad \overline{B} = \{\overline{\beta} \mid \beta \in B\}.$$

It is obtained by reflecting B about the real axis. The fact that $\overline{R} = R$ implies that the complex conjugate of an ideal is an ideal. The ideal $\overline{A}$, the complex conjugate of A, is equal to A. This accidental symmetry of the lattice A doesn't occur very often.

We now compute some product ideals. Proposition 13.4.3(a) tells us that the ideal $\overline{A}A$ is generated by the four products of the generators $(2, 1 - \delta)$ and $(2, 1 + \delta)$ of $\overline{A}$ and A:

$$\overline{A}A = (4, 2 + 2\delta, 2 - 2\delta, 6).$$

Each of the four generators is divisible by 2, so $\overline{A}A$ is contained in the principal ideal (2). (The notation (2) stands for the ideal $2R$ here.) On the other hand, 2 is an element of $\overline{A}A$ because $2 = 6 - 4$. Therefore $(2) \subset \overline{A}A$. This shows that $\overline{A}A = (2)$.

Next, the product AB is generated by four products:

$$AB = (6, 2 + 2\delta, 3 + 3\delta, (1 + \delta)^2).$$

Each of these four elements is divisible by $1 + \delta$, and $1 + \delta$ is the difference of two of them, so it is an element of AB. Therefore AB is equal to the principal ideal $(1 + \delta)$. One sees similarly that $\overline{A}\,B = (1 - \delta)$ and that $\overline{B}B = (3)$.

The principal ideal (6) is the product of four ideals:

(13.4.7) $(6) = (2)(3) = (\overline{A}A)(\overline{B}B) = (\overline{A}\,B)(AB) = (1 - \delta)(1 + \delta)$

Isn't this beautiful? The ideal factorization $(6) = \overline{A}A\overline{B}B$ has provided a common refinement of the two factorizations (13.4.4).

In the next section, we prove unique factorization of ideals in the ring of integers of any imaginary quadratic number field. The next lemma is the tool that we will need.

Lemma 13.4.8 Main Lemma. Let R be the ring of integers in an imaginary quadratic number field. The product of a nonzero ideal A of R and its conjugate $\overline{A}$ is a principal ideal, generated by a positive ordinary integer n: $\overline{A}A = (n) = nR$.

This lemma would be false for any ring smaller than R, for example, if one didn't include the elements with half integer coefficients, when $d \equiv 1$ modulo 4.

Proof. Let (α, β) be a lattice basis for the ideal A. Then $(\overline{\alpha}, \overline{\beta})$ is a lattice basis for $\overline{A}$. Moreover, $\overline{A}$ and A are generated as ideals by these bases, so the four products $\overline{\alpha}\alpha$, $\overline{\alpha}\beta$, $\overline{\beta}\alpha$, and $\overline{\beta}\beta$ generate the product ideal $\overline{A}A$. The three elements $\overline{\alpha}\alpha$, $\overline{\beta}\beta$, and $\overline{\beta}\alpha + \overline{\alpha}\beta$ are in $\overline{A}A$. They are algebraic integers equal to their complex conjugates, so they are rational numbers, and therefore ordinary integers (13.1.1). Let n be their greatest common divisor in the ring of integers. It is an integer combination of those elements, so it is also an element of $\overline{A}A$. Therefore $(n) \subset \overline{A}A$. If we show that n divides each of the four generators of $\overline{A}A$ in R, it will follow that $(n) = \overline{A}A$, and this will prove the lemma.

By construction, n divides $\overline{\alpha}\alpha$ and $\overline{\beta}\beta$ in $\mathbb{Z}$, hence in R. We have to show that n divides $\overline{\alpha}\beta$ and $\overline{\beta}\alpha$. How can we do this? There is a beautiful insight here. We use the definition of an algebraic integer. If we show that the quotients $\gamma = \overline{\alpha}\beta/n$ and $\overline{\gamma} = \overline{\beta}\alpha/n$ are algebraic integers, it will follow that they are elements of the ring of integers, which is R. This will mean that n divides $\overline{\alpha}\beta$ and $\overline{\beta}\alpha$ in R.

The elements γ and $\overline{\gamma}$ are roots of the polynomial $p(x) = x^2 - (\overline{\gamma} + \gamma)x + (\overline{\gamma}\gamma)$:

$$\overline{\gamma} + \gamma = \frac{\overline{\beta}\alpha + \overline{\alpha}\beta}{n}, \quad \text{and} \quad \overline{\gamma}\gamma = \frac{\overline{\beta}\alpha}{n}\frac{\overline{\alpha}\beta}{n} = \frac{\overline{\alpha}\alpha}{n}\frac{\overline{\beta}\beta}{n}.$$

By its definition, n divides each of the three integers $\overline{\beta}\alpha + \overline{\alpha}\beta$, $\overline{\alpha}\alpha$, and $\overline{\beta}\beta$. The coefficients of $p(x)$ are integers, so γ and $\overline{\gamma}$ are algebraic integers, as we hoped. (See Lemma 12.4.2 for the case that γ happens to be a rational number.) □

Our first applications of the Main Lemma are to divisibility of ideals. In analogy with divisibility of elements of a ring, we say that an ideal A *divides* another ideal B if there is an ideal C such that B is the product ideal AC.

Corollary 13.4.9 Let R be the ring of integers in an imaginary quadratic number field.

(a) *Cancellation Law*: Let A, B, C be nonzero ideals of R. Then $AB = AC$ if and only if $B = C$. Similarly, $AB \subset AC$, if and only if $B \subset C$, and $AB < AC$ if and only if $B < C$.

(b) Let A and B be nonzero ideals of R. Then $A \supset B$ if and only if A divides B, i.e., if and only if there is an ideal C such that $B = AC$.

Proof. (a) It is clear that if $B = C$, then $AB = AC$. If $AB = AC$, then $\overline{A}AB = \overline{A}AC$. By the Main Lemma, $\overline{A}A = (n)$, so $nB = nC$. Dividing by n shows that $B = C$. The other assertions are proved in the same way.

(b) We first consider the case that a principal ideal (n) generated by an ordinary integer n contains an ideal B. Then n divides every element of B in R. Let $C = n^{-1}B$ be the set of quotients, the set of elements $n^{-1}\beta$ with β in B. You can check that C is an ideal and that $nC = B$. Then B is the product ideal $(n)C$, so (n) divides B.

Now suppose that an ideal A contains B. We apply the Main Lemma again: $\overline{A}A = (n)$. Then $(n) = \overline{A}A$ contains $\overline{A}B$. By what has been shown, there is an ideal C such that $\overline{A}B = (n)C = \overline{A}AC$. By the Cancellation Law, $B = AC$.

Conversely, if A divides B, say $B = AC$, then $B = AC \subset AR = A$. □

13.5 FACTORING IDEALS

We show in this section that nonzero ideals in rings of integers in imaginary quadratic fields factor uniquely. This follows rather easily from the Main Lemma 13.4.8 and its Corollary 13.4.9, but before deriving it, we define the concept of a prime ideal. We do this to be consistent with standard terminology: the prime ideals that appear are simply the maximal ideals.

Proposition 13.5.1 Let $\mathcal{R}$ be a ring. The following conditions on an ideal $\mathcal{P}$ of $\mathcal{R}$ are equivalent. An ideal that satisfies these conditions is called a *prime ideal*.

(a) The quotient ring $\mathcal{R}/\mathcal{P}$ is an integral domain.

(b) $\mathcal{P} \neq \mathcal{R}$, and if a and b are elements of $\mathcal{R}$ such that $ab \in \mathcal{P}$, then $a \in \mathcal{P}$ or $b \in \mathcal{P}$.

(c) $\mathcal{P} \neq \mathcal{R}$, and if A and B are ideals of $\mathcal{R}$ such that $AB \subset \mathcal{P}$, then $A \subset \mathcal{P}$ or $B \subset \mathcal{P}$.

Condition (b) explains the term "prime." It mimics the important property of a prime integer, that if a prime p divides a product ab of integers, then p divides a or p divides b.

Proof. **(a)** $\Longleftrightarrow$ **(b)**: The conditions for R/P to be an integral domain are that $R/P \neq \{0\}$ and $\overline{ab} = 0$ implies $\overline{a} = 0$ or $\overline{b} = 0$. These conditions translate to $P \neq R$ and $ab \in P$ implies $a \in P$ or $b \in P$.

(b) $\Rightarrow$ **(c)**: Suppose that $ab \in P$ implies $a \in P$ or $b \in P$, and let A and B be ideals such that $AB \subset P$. If $A \not\subset P$, there is an element a in A that isn't in P. Let b be any element of B. Then ab is in AB and therefore in P. But a is not in P, so b is in P. Since b was an arbitrary element of B, $B \subset P$.

(c) $\Rightarrow$ **(b)**: Suppose that P has the property **(c)**, and let a and b be elements of R such that ab is in P. The principal ideal (ab) is the product ideal $(a)(b)$. If $ab \in P$, then $(ab) \subset P$, and so $(a) \subset P$ or $(b) \subset P$. This tells us that $a \in P$ or $b \in P$. $\qquad\square$

Corollary 13.5.2 Let R be a ring.

(a) The zero ideal of R is a prime ideal if and only if R is an integral domain.

(b) A maximal ideal of R is a prime ideal.

(c) A principal ideal (α) is a prime ideal of R if and only if α is a prime element of R.

Proof. **(a)** This follows from (13.5.1)**(a)**, because the quotient ring $R/(0)$ is isomorphic to R.

(b) This also follows from (13.5.1)**(a)**, because when M is a maximal ideal, R/M is a field. A field is an integral domain, so M is a prime ideal. Finally, **(c)** restates (13.5.1)**(b)** for a principal ideal. $\qquad\square$

This completes our discussion of prime ideals in arbitrary rings, and we go back to the ring of integers in an imaginary quadratic number field.

Corollary 13.5.3 Let R be the ring of integers in an imaginary quadratic number field, let A and B be ideals of R, and let P be a prime ideal of R that is not the zero ideal. If P divides the product ideal AB, then P divides one of the factors A or B.

This follows from (13.5.1)**(c)** when we use (13.4.9)**(b)** to translate inclusion into divisibility.$\square$

Lemma 13.5.4 Let R be the ring of integers in an imaginary quadratic number field, and let B be a nonzero ideal of R. Then

(a) B has finite index in R,

(b) there are finitely many ideals of R that contain B,

(c) B is contained in a maximal ideal, and

(d) B is a prime ideal if and only if it is a maximal ideal.

Proof. **(a)** is Lemma 13.10.3**(d)**, and **(b)** follows from Corollary 13.10.5

(c) Among the finitely many ideals that contain B, there must be at least one that is maximal.

(d) Let P be a nonzero prime ideal. Then by **(a)**, P has finite index in R. So R/P is a finite integral domain. A finite integral domain is a field. (This is Chapter 11, Exercise 7.1.) Therefore P is a maximal ideal. The converse is (13.5.2)**(b)**. $\qquad\square$

Theorem 13.5.5 Let R be the ring of integers in an imaginary quadratic field F. Every proper ideal of R is a product of prime ideals. The factorization of an ideal into prime ideals is unique except for the ordering of the factors.

Proof. If an ideal B is a maximal ideal, it is itself a prime ideal. Otherwise, there is an ideal A that properly contains B. Then A divides B, say $B = AC$. The cancellation law shows that C properly contains B too. We continue by factoring A and C. Since only finitely many ideals contain B, the process terminates, and when it does, all factors will be maximal and therefore prime.

If $P_1 \cdots P_r = Q_1 \cdots Q_s$, with P_i and Q_j prime, then P_1 divides $Q_1 \cdots Q_s$, and therefore P_1 divides one of the factors, say Q_1. Then P_1 contains Q_1, and since Q_1 is maximal, $P_1 = Q_1$. The uniqueness of factorization follows by induction when one cancels P_1 from both sides of the equation. $\square$

Note: This theorem extends to rings of algebraic integers in other number fields, but it is a very special property. Most rings do not admit unique factorization of ideals. The reason is that in most rings, $P \supset B$ does not imply that P divides B, and then the analogy between prime ideals and prime elements is weaker. $\square$

Theorem 13.5.6 The ring of integers R in an imaginary quadratic number field is a unique factorization domain if and only if it is a principal ideal domain, and this is true if and only if the class group $\mathcal{C}$ of R is the trivial group.

Proof. A principal ideal domain is a unique factorization domain (12.2.14). Conversely, suppose that R is a unique factorization domain. We must show that every ideal is principal. Since the product of principal ideals is principal and since every nonzero ideal is a product of prime ideals, it suffices to show that every nonzero prime ideal is principal.

Let P be a nonzero prime ideal of R, and let α be a nonzero element of P. Then α is a product of irreducible elements, and because R has unique factorization, they are prime elements (12.2.14). Since P is a prime ideal, P contains one of the prime factors of α, say π. Then P contains the principal ideal (π). But since π is a prime element, the principal ideal (π) is a nonzero prime ideal, and therefore a maximal ideal. Since P contains (π), $P = (\pi)$. So P is a principal ideal. $\square$

13.6   PRIME IDEALS AND PRIME INTEGERS

In Section 12.5, we saw how Gauss primes are related to integer primes. A similar analysis can be made for the ring R of integers in a quadratic number field, but we should speak of prime ideals rather than of prime elements. This complicates the analogues of some parts of Theorem 12.5.2. We consider only those parts that extend directly.

Theorem 13.6.1 Let R be the ring of integers in an imaginary quadratic number field.

(a) Let P be a nonzero prime ideal of R. Say that $\overline{P}P = (n)$ where n is a positive integer. Then n is either an integer prime or the square of an integer prime.

(b) Let p be an integer prime. The principal ideal $(p) = pR$ is either a prime ideal, or the product $\overline{P}P$ of a prime ideal and its conjugate.

(c) Assume that $d \equiv 2$ or 3 modulo 4. An integer prime p generates a prime ideal (p) of R if and only if d is not a square modulo p, and this is true if and only if the polynomial $x^2 - d$ is irreducible in $\mathbb{F}_p[x]$.

(d) Assume that $d \equiv 1$ modulo 4, and let $h = \frac{1}{4}(1 - d)$. An integer prime p generates a prime ideal (p) of R if and only if the polynomial $x^2 - x + h$ is irreducible in $\mathbb{F}_p[x]$.

Corollary 13.6.2 With the notation as in the theorem, any proper ideal strictly larger than (p) is a prime, and therefore a maximal, ideal. $\qquad\square$

• An integer prime p is said to *remain prime* if the principal ideal $(p) = pR$ is a prime ideal. Otherwise, the principal ideal (p) is a product $\overline{P}P$ of a prime ideal and its conjugate, and in this case the prime p is said to *split*. If in addition $\overline{P} = P$, the prime p is said to *ramify*.

Going back to the case $d = -5$, the prime 2 ramifies in $\mathbb{Z}[\sqrt{-5}]$ because $(2) = \overline{A}A$ and $\overline{A} = A$. The prime 3 splits. It does not ramify, because $(3) = \overline{B}B$ and $\overline{B} \neq B$ (see (13.4.5)).

Proof of Theorem 13.6.1. The proof follows that of Theorem 12.5.2 closely, so we omit the proofs of **(a)** and **(b)**. We discuss **(c)** in order to review the reasoning. Suppose $d \equiv 2$ or 3 modulo 4. Then $R = \mathbb{Z}[\delta]$ is isomorphic to the quotient ring $\mathbb{Z}[x]/(x^2 - d)$. A prime integer p remains prime in R if and only if $\tilde{R} = R/(p)$ is a field. (We are using a tilde here to avoid confusion with complex conjugation.) This leads to the diagram

(13.6.3)
$$
\begin{array}{ccc}
\mathbb{Z}[x] & \xrightarrow{\underset{(p)}{\text{kernel}}} & \mathbb{F}_p[x] \\[2pt]
{\scriptstyle\text{kernel}\atop (x^2 - d)}\Big\downarrow & & \Big\downarrow{\scriptstyle\text{kernel}\atop (x^2 - d)} \\[2pt]
\mathbb{Z}[\delta] & \xrightarrow[\underset{(p)}{\text{kernel}}]{} & \tilde{R}
\end{array}
$$

This diagram shows that $\tilde{R}$ is a field if and only if $x^2 - d$ is irreducible in $\mathbb{F}_p[x]$.

The proof of **(d)** is similar. $\qquad\square$

Proposition 13.6.4 Let A, B, C be nonzero ideals with $B \supset C$. The index $[B:C]$ of C in B is equal to the index $[AB:AC]$.

Proof. Since A is a product of prime ideals, it suffices to show that $[B:C] = [PB:PC]$ when P is a nonzero prime ideal. The lemma for an arbitrary ideal A follows when we multiply by one prime ideal at a time.

There is a prime integer p such that either $P = (p)$ or $\overline{P}P = (p)$ (13.6.1). If P is the principal ideal (p), the formula to be shown is $[B:C] = [pB:pC]$, and this is rather obvious (see (13.10.3)**(c)**).

Suppose that $(p) = \overline{P}P$. We inspect the chain of ideals $B \supset PB \supset \overline{P}PB = pB$. The cancellation law shows that the inclusions are strict, and $[B:pB] = p^2$. Therefore

$[B:PB] = p$. Similarly, $[C:PC] = p$ (13.10.3)**(b)**. The diagram below, together with the multiplicative property of the index (2.8.14), shows that $[B:C] = [PB:PC]$.

$$
\begin{array}{ccc}
B & \supset & C \\
\cup & & \cup \\
PB & \supset & PC
\end{array}
$$

$\square$

13.7 IDEAL CLASSES

As before, R denotes the ring of integers in an imaginary quadratic number field. We have seen that R is a principal ideal domain if and only if it is a unique factorization domain (13.5.6). We define an equivalence relation on nonzero ideals that is compatible with multiplication of ideals, and such that the principal ideals form one equivalence class.

- Two nonzero ideals A and A' of R are *similar* if, for some complex number λ,

(13.7.1) $$A' = \lambda A.$$

Similarity of ideals is an equivalence relation whose geometric interpretation was mentioned before: A and A' are similar if and only if, when regarded as lattices in the complex plane, they are similar geometric figures, by a similarity that is orientation-preserving. To see this, we note that a lattice looks the same at all of its points. So a geometric similarity can be assumed to relate the element 0 of A to the element 0 of A'. Then it will be described as a rotation followed by a stretching or shrinking, that is, as multiplication by a complex number λ.

- Similarity classes of ideals are called *ideal classes*. The class of an ideal A will be denoted by $\langle A \rangle$.

Lemma 13.7.2 The class $\langle R \rangle$ of the unit ideal consists of the principal ideals.

Proof. If $\langle A \rangle = \langle R \rangle$, then $A = \lambda R$ for some complex number λ. Since 1 is in R, λ is an element of A, and therefore an element of R. Then A is the principal ideal (λ). $\square$

We saw in (13.3.3) that there are two ideal classes in the ring $R = \mathbb{Z}[\delta]$, when $\delta^2 = -5$. Both of the ideals $A = (2, 1 + \delta)$ and $B = (3, 1 + \delta)$ represent the class of nonprincipal ideals. They are shown below, in Figure 13.7.4. Rectangles have been put into the figure to help you visualize the fact that the two lattices are similar geometric figures.

We see below (Theorem 13.7.10) that there are always finitely many ideal classes. The number of ideal classes in R is called the *class number* of R.

Proposition 13.7.3 The ideal classes form an abelian group $\mathcal{C}$, the *class group* of R, the law of composition being defined by multiplication of ideals: $\langle A \rangle \langle B \rangle = \langle AB \rangle$:

$$(\text{class of } A)(\text{class of } B) = (\text{class of } AB).$$

Proof. Suppose that $\langle A \rangle = \langle A' \rangle$ and $\langle B \rangle = \langle B' \rangle$, i.e., $A' = \lambda A$ and $B' = \gamma B$ for some complex numbers λ and γ. Then $A'B' = \lambda \gamma A B$, and therefore $\langle AB \rangle = \langle A'B' \rangle$. This shows that the law of composition is well defined. The law is commutative and associative because multiplication of ideals is commutative and associative, and the class $\langle R \rangle$ of the unit ideal is an identity element that we denote by 1, as usual. The only group axiom that isn't obvious is that every class $\langle A \rangle$ has an inverse. But this follows from the Main Lemma, which asserts that $\overline{A}A$ is a principal ideal (n). Since the class of a principal ideal is 1, $\langle \overline{A} \rangle \langle A \rangle = 1$ and $\langle \overline{A} \rangle = \langle A \rangle^{-1}$.   □

The class number is thought of as a way to quantify how badly unique factorization of elements fails. More precise information is given by the structure of $\mathcal{C}$ as a group. As we have seen, the class number of the ring $R = \mathbb{Z}[\sqrt{-5}]$ is two. The class group of R has order two. One consequence of this is that the product of any two nonprincipal ideals of R is a principal ideal. We saw several examples of this in (13.4.7).

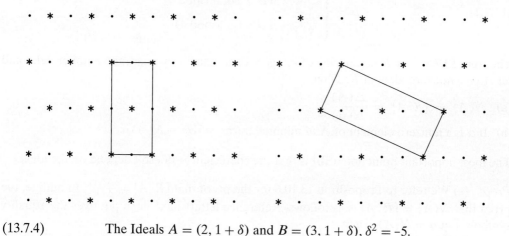

(13.7.4)   The Ideals $A = (2, 1 + \delta)$ and $B = (3, 1 + \delta)$, $\delta^2 = -5$.

Measuring an Ideal

The Main Lemma tells us that if A is a nonzero ideal, then $\overline{A}A = (n)$ is the principal ideal generated by a positive integer. That integer is defined to be the *norm* of A. It will be denoted by $N(A)$:

(13.7.5) $N(A) = n$, if n is the positive integer such that $\overline{A}A = (n)$.

The norm of an ideal is analogous to the norm of an element. As is true for norms of elements, this norm is multiplicative.

Lemma 13.7.6 If A and B are nonzero ideals, then $N(AB) = N(A)N(B)$. Moreover, the norm of the principal ideal (α) is equal to $N(\alpha)$, the norm of the element α.

Proof. Say that $N(A) = m$ and $N(B) = n$. This means that $\overline{A}A = (m)$ and $\overline{B}B = (n)$. Then $(\overline{AB})(AB) = (\overline{A}A)(\overline{B}B) = (m)(n) = (mn)$. So $N(AB) = mn$.

Next, suppose that A is the principal ideal (α), and let $n = N(\alpha)$ $(= \overline{\alpha}\alpha)$. Then $\overline{A}A = (\overline{\alpha})(\alpha) = (\overline{\alpha}\alpha) = (n)$, so $N(A) = n$ too.   □

We now have four ways to measure the size of an ideal A:

- the norm $N(A)$,
- the index $[R:A]$ of A in R,
- the area $\Delta(A)$ of the parallelogram spanned by a lattice basis for A,
- the minimum value taken on by the norm $N(\alpha)$, of the nonzero elements of A.

The relations among these measures are given by Theorem 13.7.8 below. To state that theorem, we need a peculiar number:

$$(13.7.7) \qquad \mu = \begin{cases} 2\sqrt{\frac{|d|}{3}} & \text{if } d \equiv 2 \text{ or } 3 \pmod 4 \\ \sqrt{\frac{|d|}{3}} & \text{if } d \equiv 1 \pmod 4. \end{cases}$$

Theorem 13.7.8 Let R be the ring of integers in an imaginary quadratic number field, and let A be a nonzero ideal of R. Then

(a) $N(A) = [R:A] = \dfrac{\Delta(A)}{\Delta(R)}.$

(b) If α is a nonzero element of A of minimal norm, $N(\alpha) \le N(A)\mu.$

The most important point about **(b)** is that the coefficient μ doesn't depend on the ideal.

Proof. **(a)** We refer to Proposition 13.10.6 for the proof that $[R:A] = \frac{\Delta(A)}{\Delta(R)}$. In outline, the proof that $N(A) = [R:A]$ is as follows. Reference letters have been put over the equality symbols. Let $n = N(A)$. Then

$$n^2 \overset{1}{=} [R:nR] \overset{2}{=} \left[R:\overline{A}A\right] \overset{3}{=} [R:A]\left[A:\overline{A}A\right] \overset{4}{=} [R:A]\left[R:\overline{A}\right] \overset{5}{=} [R:A]^2.$$

The equality labeled 1 is Lemma 13.10.3**(b)**, the one labeled 2 is the Main Lemma, which says that $nR = \overline{A}A$, and 3 is the multiplicative property of the index. The equality 4 follows from Proposition 13.6.4: $[A:\overline{A}A] = [RA:\overline{A}A] = [R:\overline{A}]$. Finally, the ring R is equal to its complex conjugate $\overline{R}$, and 5 comes down to the fact that $[\overline{R}:\overline{A}] = [R:A]$.

(b) When $d \equiv 2, 3$ modulo 4, R has the lattice basis $(1, \delta)$, and when $d \equiv 1$ modulo 4, R has the lattice basis $(1, \eta)$. The area $\Delta(R)$ of the parallelogram spanned by this basis is

$$(13.7.9) \qquad \Delta(R) = \begin{cases} \sqrt{|d|} & \text{if } d \equiv 2 \text{ or } 3 \text{ modulo } 4 \\ \frac{1}{2}\sqrt{|d|} & \text{if } d \equiv 1 \text{ modulo } 4. \end{cases}$$

So $\mu = \frac{2}{\sqrt{3}}\Delta(R)$. The length of the shortest vector in a lattice is estimated in Lemma 13.10.8: $N(\alpha) \le \frac{2}{\sqrt{3}}\Delta(A)$. We substitute $\Delta(A) = N(A)\Delta(R)$ from part **(a)** into this inequality, obtaining $N(\alpha) \le N(A)\mu$. $\square$

Theorem 13.7.10

(a) Every ideal class contains an ideal A with norm $N(A) \leq \mu$.

(b) The class group $\mathcal{C}$ is generated by the classes of prime ideals P whose norms are prime integers $p \leq \mu$.

(c) The class group $\mathcal{C}$ is finite.

Proof of Theorem 13.7.10. **(a)** Let A be an ideal. We must find an ideal C in the class $\langle A \rangle$ whose norm is at most μ. We choose a nonzero element α in A, with $N(\alpha) \leq N(A)\mu$. Then A contains the principal ideal (α), so A divides (α), i.e., $(\alpha) = AC$ for some ideal C, and $N(A)N(C) = N(\alpha) \leq N(A)\mu$. Therefore $N(C) \leq \mu$. Now since AC is a principal ideal, $\langle C \rangle = \langle A \rangle^{-1} = \langle \overline{A} \rangle$. This shows that the class $\langle \overline{A} \rangle$ contains an ideal, namely C, whose norm is at most μ. Then the class $\langle A \rangle$ contains $\overline{C}$, and $N(\overline{C}) = N(C) \leq \mu$.

(b) Every class contains an ideal A of norm $N(A) \leq \mu$. We factor A into prime ideals: $A = P_1 \cdots P_k$. Then $N(A) = N(P_1) \cdots N(P_k)$, so $N(P_i) \leq \mu$ for each i. The classes of prime ideals with norm $\leq \mu$ generate $\mathcal{C}$. The norm of a prime ideal P is either a prime integer p or the square p^2 of a prime integer. If $N(P) = p^2$, then $P = (p)$ (13.6.1). This is a principal ideal, and its class is trivial. We may ignore those primes.

(c) We show that there are finitely many ideals A with norm $N(A) \leq \mu$. If we write such an ideal as a product of prime ideals, $A = P_1 \cdots P_k$, and if $m_i = N(P_i)$, then $m_1 \cdots m_k \leq \mu$. There are finitely many sets of integers m_i, each a prime or the square of a prime, that satisfy this inequality, and there are at most two prime ideals with norms equal to a given integer m_i. So there are finitely many sets of prime ideals such that $N(P_1 \cdots P_k) \leq \mu$. $\square$

13.8 COMPUTING THE CLASS GROUP

The table below lists a few class groups. In the table, $\lfloor \mu \rfloor$ denotes the *floor* of μ, the largest integer $\leq \mu$. If n is an integer and if $n \leq \mu$, then $n \leq \lfloor \mu \rfloor$.

d	$\lfloor \mu \rfloor$	class group
-2	1	C_1
-5	2	C_2
-7	1	C_1
-14	4	C_4
-21	5	$C_2 \times C_2$
-23	2	C_3
-47	3	C_5
-71	4	C_7

(13.8.1)   Some Class Groups

To apply Theorem 13.7.10, we examine the prime integers $p \leq \lfloor \mu \rfloor$. If p splits (or ramifies) in R, we include the class of one of its two prime ideal factors in our set of generators for the class group. The class of the other prime factor is its inverse. If p remains prime, its class is trivial and we discard it.

Example 13.8.2 $d = -163$. Since $-163 \equiv 1$ modulo 4, the ring R of integers is $\mathbb{Z}[\eta]$, where $\eta = \frac{1}{2}(1 + \delta)$, and $\lfloor \mu \rfloor = 8$. We must inspect the primes $p = 2, 3, 5$, and 7. If p splits, we include one of its prime divisors as a generator of the class group. According to Theorem 13.6.1, an integer prime p remains prime in R if and only if the polynomial $x^2 - x + 41$ is irreducible modulo p. This polynomial happens to be irreducible modulo each of the primes 2, 3, 5, and 7. So the class group is trivial, and R is a unique factorization domain. $\square$

For the rest of this section, we consider cases in which $d \equiv 2$ or 3 modulo 4. In these cases, a prime p splits if and only if $x^2 - d$ has a root in $\mathbb{F}_p$. The table below tells us which primes need to be examined.

	$p \le \mu$
$-d \le 2$	
$-d \le 6$	2
$-d \le 17$	2, 3
$-d \le 35$	2, 3, 5
$-d \le 89$	2, 3, 5, 7
$-d \le 123$	2, 3, 5, 7, 11

(13.8.3)   Primes Less Than μ, When $d \equiv 2$ or 3 Modulo 4

If $d = -1$ or -2, there are no primes less than μ, so the class group is trivial, and R is a unique factorization domain.

Let's suppose that we have determined which of the primes that need to be examined split. Then we will have a set of generators for the class group. But to determine its structure we still need to determine the relations among these generators. It is best to analyze the prime 2 directly.

Lemma 13.8.4 Suppose that $d \equiv 2$ or 3 modulo 4. The prime 2 ramifies in R. The prime divisor P of the principal ideal (2) is

- $P = (2, 1 + \delta)$, if $d \equiv 3$ modulo 4,
- $P = (2, \delta)$, if $d \equiv 2$ modulo 4.

The class $\langle P \rangle$ has order two in the class group unless $d = -1$ or -2. In those cases, P is a principal ideal. In all cases, the given generators form a lattice basis of the ideal P.

Proof. Let P be as in the statement of the lemma. We compute the product ideal $\overline{P}P$. If $d \equiv 3$ modulo 4, $\overline{P}P = (2, 1 - \delta)(2, 1 + \delta) = (4, 2 + 2\delta, 2 - 2\delta, 1 - d)$, and if $d \equiv 2$ modulo 4, $\overline{P}P = (2, -\delta)(2, \delta) = (4, 2\delta, -d)$. In both cases, $\overline{P}P = (2)$. Theorem 15.10.1 tells us that the ideal (2) is either a prime ideal or the product of a prime ideal and its conjugate, so P must be a prime ideal.

We note also that $\overline{P} = P$, so 2 ramifies, $\langle P \rangle = \langle P \rangle^{-1}$, and $\langle P \rangle$ has order 1 or 2 in the class group. It will have order 1 if and only if it is a principal ideal. This happens when $d = -1$ or -2. If $d = -1$, $P = (1 + \delta)$, and if $d = -2$, $P = (\delta)$. When $d < -2$, the integer 2 has no proper factor in R, and then P is not a principal ideal. $\qquad\square$

Corollary 13.8.5 If $d \equiv 2$ or 3 modulo 4 and $d < -2$, the class number is even. $\qquad\square$

Example 13.8.6 $d = -26$. Table 13.8 tells us to inspect the primes $p = 2, 3$, and 5. The polynomial $x^2 + 26$ is reducible modulo 2, 3, and 5, so all of those primes split. Let's say that

$$(2) = \overline{P}P, \ (3) = \overline{Q}Q, \ \text{and} \ (5) = \overline{S}S.$$

We have three generators $\langle P \rangle$, $\langle Q \rangle$, $\langle S \rangle$ for the class group, and $\langle P \rangle$ has order 2. How can we determine the other relations among these generators? The secret method is to compute norms of a few elements, hoping to get some information. We don't have to look far: $N(1 + \delta) = 27 = 3^3$ and $N(2 + \delta) = 30 = 2 \cdot 3 \cdot 5$.

Let $\alpha = 1 + \delta$. Then $\overline{\alpha}\alpha = 3^3$. Since $(3) = \overline{Q}Q$, we have the ideal relation

$$(\overline{\alpha})(\alpha) = (\overline{Q}Q)^3.$$

Because ideals factor uniquely, the principal ideal (α) is the product of one half of the terms on the right, and $(\overline{\alpha})$ is the product of the conjugates of those terms. We note that 3 doesn't divide α in R. Therefore $\overline{Q}Q = (3)$ doesn't divide (α). It follows that (α) is either Q^3 or $\overline{Q}^3$. Which it is depends on which prime factor of (3) we label as Q.

In either case, $\langle Q \rangle^3 = 1$, and $\langle Q \rangle$ has order 1 or 3 in the class group. We check that 3 has no proper divisor in R. Then since Q divides (3), it cannot be a principal ideal. So $\langle Q \rangle$ has order 3.

Next, let $\beta = 2 + \delta$. Then $\overline{\beta}\beta = 2 \cdot 3 \cdot 5$, and this gives us the ideal relation

$$(\overline{\beta})(\beta) = \overline{P}P\overline{Q}Q\overline{S}S.$$

Therefore the principal ideal (β) is the product of one half of the ideals on the right and $(\overline{\beta})$ is the product of the conjugates of those ideals. We know that $\overline{P} = P$. If we don't care which prime factors of (3) and (5) we label as Q and S, we may assume that $(\beta) = PQS$. This gives us the relation $\langle P \rangle \langle Q \rangle \langle S \rangle = 1$.

We have found three relations:

$$\langle P \rangle^2 = 1, \ \langle Q \rangle^3 = 1, \ \text{and} \ \langle P \rangle \langle Q \rangle \langle S \rangle = 1.$$

These relations show that $\langle Q \rangle = \langle S \rangle^2$, $\langle P \rangle = \langle S \rangle^3$, and that $\langle S \rangle$ has order 6. The class group is a cyclic group of order 6, generated by a prime ideal divisor of 5.

The next lemma explains why the method of computing norms works.

Lemma 13.8.7 Let P, Q, S be prime ideals of the ring R of imaginary quadratic integers, whose norms are the prime integers p, q, s, respectively. Suppose that the relation

$\langle P \rangle^i \langle Q \rangle^j \langle S \rangle^k = 1$ holds in the class group $\mathcal{C}$. Then there is an element α in R with norm equal to $p^i q^j s^k$.

Proof. By definition, $\langle P \rangle^i \langle Q \rangle^j \langle S \rangle^k = \langle P^i Q^j S^k \rangle$. If $\langle P^i Q^j S^k \rangle = 1$, the ideal $P^i Q^j S^k$ is principal, say $P^i Q^j S^k = (\alpha)$. Then

$$(\overline{\alpha})(\alpha) = (\overline{P}P)^i (\overline{Q}Q)^j (\overline{S}S)^k = (p)^i (q)^j (s)^k = (p^i q^j s^k).$$

Therefore $N(\alpha) = \overline{\alpha}\alpha = p^i q^j s^k$. $\qquad\qquad\square$

We compute one more class group.

Example 13.8.8 $d = -74$. The primes to inspect are 2, 3, 5, and 7. Here 2 ramifies, 3 and 5 split, and 7 remains prime. Say that $(2) = \overline{P}P$, $(3) = \overline{Q}Q$, and $(5) = \overline{S}S$. Then $\langle P \rangle$, $\langle Q \rangle$, and $\langle S \rangle$ generate the class group, and $\langle P \rangle$ has order 2 (13.8.4). We note that

$$\begin{aligned}
N(1 + \delta) &= 75 = 3 \cdot 5^2 \\
N(4 + \delta) &= 90 = 2 \cdot 3^2 \cdot 5 \\
N(13 + \delta) &= 243 = 3^5 \\
N(14 + \delta) &= 270 = 2 \cdot 3^3 \cdot 5
\end{aligned}$$

The norm $N(13 + \delta)$ shows that $\langle Q \rangle^5 = 1$, so $\langle Q \rangle$ has order 1 or 5. Since 3 has no proper divisor in R, Q isn't a principal ideal. So $\langle Q \rangle$ has order 5. Next, $N(1 + \delta)$ shows that $\langle S \rangle^2 = \langle Q \rangle$ or $\langle \overline{Q} \rangle$, and therefore $\langle S \rangle$ has order 10. We eliminate $\langle Q \rangle$ from our set of generators. Finally, $N(4+\delta)$ gives us one of the relations $\langle P \rangle \langle Q \rangle^2 \langle S \rangle = 1$ or $\langle P \rangle \langle Q \rangle^2 \langle \overline{S} \rangle = 1$. Either one allows us to eliminate $\langle P \rangle$ from our list of generators. The class group is cyclic of order 10, generated by a prime ideal divisor of (5).

13.9 REAL QUADRATIC FIELDS

We take a brief look at real quadratic number fields, fields of the form $\mathbb{Q}[\sqrt{d}]$, where d is a square-free positive integer, and we use the field $\mathbb{Q}[\sqrt{2}]$ as an example. The ring of integers in this field is a unique factorization domain:

(13.9.1) $$R = \mathbb{Z}[\sqrt{2}] = \{a + b\sqrt{2} \mid a, b \in \mathbb{Z}\}.$$

It can be shown that unique factorization of ideals into prime ideals is true for the ring of integers in any real quadratic number field, and that the class number is finite [Cohn], [Hasse]. It is conjectured that there are infinitely many values of d for which the ring of integers has unique factorization.

When d is positive, $\mathbb{Q}[\sqrt{d}]$ is a subfield of the real numbers. Its ring of integers is not embedded as a lattice in the complex plane. However, we can represent R as a lattice in $\mathbb{R}^2$ by associating to the algebraic integer $a + b\sqrt{d}$ the point (u, v) of $\mathbb{R}^2$, where

(13.9.2) $$u = a + b\sqrt{d}, \quad v = a - b\sqrt{d}.$$

The resulting lattice is depicted below for the case $d = 2$. The reason that the hyperbolas have been put into the figure will be explained presently.

Recall that the field $Q[\sqrt{d}]$ is isomorphic to the abstractly constructed field

(13.9.3) $$F = \mathbb{Q}[x]/(x^2 - d).$$

If we replace $\mathbb{Q}[\sqrt{d}]$ by F and denote the residue of x in F by δ, then δ is an abstract square root of d rather than the positive real square root, and F is the set of elements $a + b\delta$, with a and b in $\mathbb{Q}$. The coordinates u, v represent the two ways that the abstractly defined field F can be embedded into the real numbers, namely, u sends $\delta \rightsquigarrow \sqrt{d}$ and v sends $\delta \rightsquigarrow -\sqrt{d}$.

For $\alpha = a + b\delta \in \mathbb{Q}[\delta]$, we denote by α' the "conjugate" element $a - b\delta$. The *norm* of α is

(13.9.4) $$N(\alpha) = \alpha'\alpha = a^2 - b^2 d.$$

If α is an algebraic integer, then $N(\alpha)$ is an ordinary integer. The norm is multiplicative:

(13.9.5) $$N(\alpha\beta) = N(\alpha)N(\beta).$$

However, $N(\alpha)$ is not necessarily positive. It isn't equal to $|\alpha|^2$.

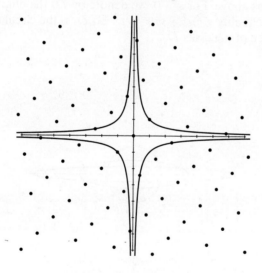

(13.9.6)   The Lattice $\mathbb{Z}[\sqrt{2}]$.

One significant difference between real and imaginary quadratic fields is that the ring of integers in a real quadratic field always contains infinitely many units. Since the norm of an algebraic integer is an ordinary integer, a unit must have norm ± 1, and if $N(\alpha) = \pm 1$, then the inverse of α is $\pm\alpha'$, so α is a unit. For example,

(13.9.7) $$\alpha = 1 + \sqrt{2}, \quad \alpha^2 = 3 + 2\sqrt{2}, \quad \alpha^3 = 7 + 5\sqrt{2}, \ldots$$

are units in the ring $R = \mathbb{Z}[\sqrt{2}]$. The element α has infinite order in the group of units.

The condition $N(\alpha) = a^2 - 2b^2 = \pm 1$ for units translates in (u, v)–coordinates to

(13.9.8) $$uv = \pm 1.$$

So the units are the points of the lattice that lie on one of the two hyperbolas $uv = 1$ and $uv = -1$, the ones depicted in Figure 13.9.6. It is remarkable that the ring of integers in a real quadratic field always has infinitely many units or, what amounts to the same thing, that the lattice always contains infinitely many points on these hyperbolas. This is far from obvious, either algebraically or geometrically, but a few such points are visible in the figure.

Theorem 13.9.9 Let R be the ring of integers in a real quadratic number field. The group of units in R is an infinite group.

We have arranged the proof as a sequence of lemmas. The first one follows from Lemma 13.10.8 in the next section.

Lemma 13.9.10 For every $\Delta_0 > 0$, there exists an $r > 0$ with the following property: Let L be a lattice in the (u, v)-plane P, let $\Delta(L)$ denote the area of the parallelogram spanned by a lattice basis, and suppose that $\Delta(L) \leq \Delta_0$. Then L contains a nonzero element γ with $|\gamma| < r$.   ☐

Let Δ_0 and r be as above. For $s > 0$, we denote by D_s the elliptical disk in the (u, v) plane defined by the inequality $s^{-2}u^2 + s^2v^2 \leq r^2$. So D_1 is the circular disk of radius r. The figure below shows three of the disks D_s.

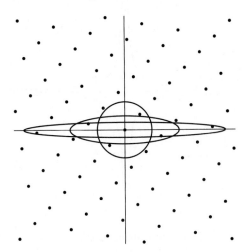

(13.9.11)   Elliptical Disks that Contain Points of the Lattice.

Lemma 13.9.12 With notation as above, let L be a lattice that contains no point on the coordinate axes except the origin, and such that $\Delta(L) \leq \Delta_0$.

(a) For any $s > 0$, the elliptical disk D_s contains a nonzero element of L.

(b) For any point $\alpha = (u, v)$ in the disk D_s, $|uv| \leq \frac{r^2}{2}$.

Proof. **(a)** The map $\varphi : \mathbb{R}^2 \to \mathbb{R}^2$ defined by $\varphi(x, y) = (sx, s^{-1}y)$ maps D_1 to D_s. The inverse image $L' = \varphi^{-1}L$ of L contains no point on the axes except the origin. We note that φ is an area-preserving map, because it multiplies one coordinate by s and the other by s^{-1}.

Therefore $\Delta(L') \leq \Delta_0$. Lemma 13.9.10 shows that the circular disk D_1 contains a nonzero element of L', say γ. Then $\alpha = \varphi(\gamma)$ is an element of L in the elliptical disk D_s.

(b) The inequality is true for the circular disk D_1. Let φ be the map defined above. If $\alpha = (u, v)$ is in D_s, then $\varphi^{-1}(\alpha) = (s^{-1}u, sv)$ is in D_1, so $|uv| = |(s^{-1}u)(sv)| \leq \frac{r^2}{2}$.   □

Lemma 13.9.13 With the hypotheses of the previous lemma, the lattice L contains infinitely many points (u, v) with $|uv| \leq \frac{r^2}{2}$.

Proof. We apply the previous lemma. For large s, the disk D_s is very narrow, and it contains a nonzero element of L, say α_s. The elements α_s cannot lie on the e_1-axis but they must become arbitrarily close to that axis as s tends to infinity. It follows that there are infinitely many points among them, and if $\alpha_s = (u_s, v_s)$, then $|u_s v_s| \leq \frac{r^2}{2}$.   □

Let R be the ring of integers in a real quadratic field, and let n be an integer. We call two elements β_i of R *congruent modulo n* if n divides $\beta_1 - \beta_2$ in R. When $d \equiv 2$ or 3 modulo 4 and $\beta_i = m_i + n_i\delta$, this simply means that $m_1 \equiv m_2$ and $n_1 \equiv n_2$ modulo n. The same is true when $d \equiv 1$ modulo 4, except that one has to write $\beta_i = m_i + n_i\eta$. In all cases, there are n^2 congruence classes modulo n.

Theorem 13.9.9 follows from the next lemma.

Lemma 13.9.14 Let R be the ring of integers in a real quadratic number field.

(a) There is a positive integer n such that the set S of elements of R with norm n is infinite. Moreover, there are infinitely many pairs of elements of S that are congruent modulo n.

(b) If two elements β_1 and β_2 of R with norm n are congruent modulo n, then β_2/β_1 is a unit of R.

Proof. **(a)** The lattice R contains no point on the axes other than the origin, because u and v aren't zero unless both a and b are zero. If α is an element of R whose image in the plane is the point (u, v), then $|N(\alpha)| = uv$. Lemma 13.9.13 shows that R contains infinitely many points with norm in a bounded interval. Since there are finitely many integers n in that interval, the set of elements of R with norm n is infinite for at least one of them. The fact that there are finitely many congruence classes modulo n proves the second assertion.

(b) We show that β_2/β_1 is in R. The same argument will show that β_1/β_2 is in R, hence that β_2/β_1 is a unit. Since β_1 and β_2 are congruent, we can write $\beta_2 = \beta_1 + n\gamma$, with γ in R. Let β_1' be the conjugate of β_1. So $\beta_1\beta_1' = n$. Then $\beta_2/\beta_1 = (\beta_1 + n\gamma)/\beta_1 = 1 + \beta_1'\gamma$. This is an element of R, as claimed.   □

13.10   ABOUT LATTICES

A lattice L in the plane $\mathbb{R}^2$ is *generated*, or *spanned* by a set S if every element of L can be written as an integer combination of elements of S. Every lattice L has a *lattice basis* $\mathbf{B} = (v_1, v_2)$ consisting of two elements. An element of L is an integer combination of the lattice basis vectors in exactly one way (see (6.5.5)).

Some notation:

(13.10.1)

$\Pi(\mathbf{B})$: the parallelogram of linear combinations $r_1 v_1 + r_2 v_2$ with $0 \le r_i \le 1$.
    Its vertices are 0, v_1, v_2, and $v_1 + v_2$.

$\Pi'(\mathbf{B})$: the set of linear combinations $r_1 v_1 + r_2 v_2$ with $0 \le r_i < 1$. It is obtained
    by deleting the edges $[v_1, v_1 + v_2]$ and $[v_2, v_1 + v_2]$ from $\Pi(\mathbf{B})$.

$\Delta(L)$: the area of $\Pi(\mathbf{B})$.

$[M : L]$: the index of a sublattice L of a lattice M – the number of additive cosets of L in M.

We will see that $\Delta(L)$ is independent of the lattice basis, so that notation isn't ambiguous. The other notation has been introduced before. For reference, we recall Lemma 6.5.8:

Lemma 13.10.2 Let $\mathbf{B} = (v_1, v_2)$ be a basis of $\mathbb{R}^2$, and let L be the lattice of integer combinations of $\mathbf{B}$. Every vector v in $\mathbb{R}^2$ can be written uniquely in the form $v = w + v_0$, with w in L and v_0 in $\Pi'(\mathbf{B})$. $\square$

Lemma 13.10.3 Let $K \subset L \subset M$ be lattices in the plane, and let $\mathbf{B}$ be a lattice basis for L. Then

(a) $[M : K] = [M : L][L : K]$.

(b) For any positive integer n, $[L : nL] = n^2$.

(c) For any positive real number r, $[M : L] = [rM : rL]$.

(d) $[M : L]$ is finite, and is equal to the number of points of M in the region $\Pi'(\mathbf{B})$.

(e) The lattice M is generated by L together with the finite set $M \cap \Pi'(\mathbf{B})$.

Proof. **(d),(e)** We can write an element x of M uniquely in the form $v + y$, where v is in L and y is in $\Pi'(\mathbf{B})$. Then v is in M, and so y is in M too. Therefore x is in the coset $y + L$. This shows that the elements of $M \cap \Pi'(\mathbf{B})$ are representative elements for the cosets of L in M. Since there is only one way to write $x = v + y$, these cosets are distinct. Since M is discrete and $\Pi'(\mathbf{B})$ is a bounded set, $M \cap \Pi'(\mathbf{B})$ is finite.

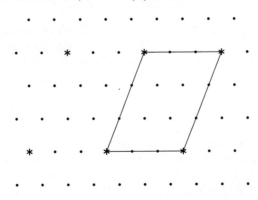

(13.10.4) $L = \{\cdot\}$ $3L = \{*\}$.

Formula **(a)** is the multiplicative property of the index (2.8.14). **(b)** follows from **(a)**, because the lattice nL is obtained by stretching L by the factor n, as is illustrated above for the case that $n = 3$. **(c)** is true because multiplication by r stretches both lattices by the same amount. $\square$

Corollary 13.10.5 Let $L \subset M$ be lattices in $\mathbb{R}^2$. There are finitely many lattices between L and M.

Proof. Let **B** be a lattice basis for L, and let N be a lattice with $L \subset N \subset M$. Lemma 13.10.3**(e)** shows that N is generated by L and by the set $N \cap \Pi'(\mathbf{B})$, which is a subset of the finite set $M \cap \Pi'(\mathbf{B})$. A finite set has finitely many subsets. $\square$

Proposition 13.10.6 If $L \subset M$ are lattices in the plane, $[M:L] = \dfrac{\Delta(L)}{\Delta(M)}$.

Proof. Say that **C** is the lattice basis (u_1, u_2) of M. Let n be a large positive integer, and let M_n denote the lattice with basis $\mathbf{C}_n = (\frac{1}{n}u_1, \frac{1}{n}u_2)$. Let Γ' denote the small region $\Pi'(\mathbf{C}_n)$. Its area is $\frac{1}{n^2}\Delta(M)$. The translates $x + \Gamma'$ of Γ' with x in M_n cover the plane without overlap, and there is exactly one element of M_n in each translate $x + \Gamma'$, namely x. (This is Lemma 13.10.2.)

Let B be a lattice basis for L. We approximate the area of $\Pi(\mathbf{B})$ in the way that one approximates a double integral, using translates of Γ'. Let $r = [M : L]$. Then $[M_n:L] = [M_n:M][M:L] = n^2 r$. Lemma 13.10.3**(d)** tells us that the region $\Pi'(\mathbf{B})$ contains $n^2 r$ points of the lattice M_n. Since the translates of Γ' cover the plane, the translates by these $n^2 r$ points cover $\Pi(\mathbf{B})$ approximately.

$$\Delta(L) \approx n^2 r \Delta(M_n) = r\Delta(M) = [M:L]\Delta(M).$$

The error in this approximation comes from the fact that $\Pi'(\mathbf{B})$ is not covered precisely along its boundary. One can bound this error in terms of the length of the boundary of $\Pi(\mathbf{B})$ and the diameter of Γ' (its largest linear dimension). The diameter tends to zero as $n \to \infty$, and so does the error. $\square$

Corollary 13.10.7 The area $\Delta(L)$ of the parallelogram $\Pi(\mathbf{B})$ is independent of the lattice basis **B**.

This follows when one sets $M = L$ in the previous proposition. $\square$

Lemma 13.10.8 Let v be a nonzero element of minimal length of a lattice L. Then $|v|^2 \leq \frac{2}{\sqrt{3}}\Delta(L)$.

The inequality becomes an equality for an equilateral triangular lattice.

Proof. We choose an element v_1 of L of minimal length. Then v_1 generates the subgroup $L \cap \ell$, where ℓ is the line spanned by v_1, and there is an element v_2 such that (v_1, v_2) is a lattice basis of L (see the proof of (6.5.5)). A change of scale changes $|v_1|^2$ and $\Delta(L)$ by the same factor, so we may assume that $|v_1| = 1$. We position coordinates so that $v_1 = (1, 0)^t$.

Say that $v_2 = (b_1, b_2)^t$. We may assume that b_2 is positive. Then $\Delta(L) = b_2$. We may also adjust v_2 by adding a multiple of v_1, to make $-\frac{1}{2} \le b_1 < \frac{1}{2}$, so that $b_1^2 \le \frac{1}{4}$. Since v_1 has minimal length among nonzero elements of L, $|v_2|^2 = b_1^2 + b_2^2 \ge |v_1|^2 = 1$. Therefore $b_2^2 \ge \frac{3}{4}$. Thus $\Delta(L) = b_2 \ge \frac{\sqrt{3}}{2}$, and $|v_1|^2 = 1 \le \frac{2}{\sqrt{3}}\Delta(L)$.   □

> *Nullum vero dubium nobis esse videtur,*
> *quin multa eaque egregia in hoc genere adhuc lateant*
> *in quibus alii vires suas exercere possint.*
>
> —Carl Friedrich Gauss

EXERCISES

Section 1   Algebraic Integers

1.1. Is $\frac{1}{2}(1 + \sqrt{5})$ an algebraic integer?

1.2. Prove that the integers in $\mathbb{Q}[\sqrt{d}]$ form a ring.

1.3. (a) Let α be a complex number that is the root of a monic integer polynomial, not necessarily an irreducible polynomial. Prove that α is an algebraic integer.

(b) Let α be an algebraic number that is the root of an integer polynomial $f(x) = a_n x^n + a_{n-1}x^{n-1} + \cdots + a_0$. Prove that $a_n\alpha$ is an algebraic integer.

(c) Let α be an algebraic integer that is the root of a monic integer polynomial $x^n + a_{n-1}n^{n-1} + \cdots + a_1 x + a_0$. Prove that α^{-1} is an algebraic integer if and only if $a_0 = \pm 1$.

1.4. Let d and d' be integers. When are the fields $\mathbb{Q}(\sqrt{d})$ and $\mathbb{Q}(\sqrt{d'})$ distinct?

Section 2   Factoring Algebraic Integers

2.1. Prove that 2, 3, and $1 \pm \sqrt{-5}$ are irreducible elements of the ring $R = \mathbb{Z}[\sqrt{-5}]$ and that the units of this ring are ± 1.

2.2. For which negative integers $d \equiv 2$ modulo 4 is the ring of integers in $\mathbb{Q}[\sqrt{d}]$ a unique factorization domain?

Section 3   Ideals in $\mathbb{Z}[\sqrt{-5}]$

3.1. Let α be an element of $R = \mathbb{Z}[\delta]$, $\delta = \sqrt{-5}$, and let $\gamma = \frac{1}{2}(\alpha + \alpha\delta)$. Under what circumstances is the lattice with basis (α, γ) an ideal?

3.2. Let $\delta = \sqrt{-5}$. Decide whether or not the lattice of integer combinations of the given vectors is an ideal: **(a)** $(5, 1 + \delta)$, **(b)** $(7, 1 + \delta)$, **(c)** $(4 - 2\delta, 2 + 2\delta, 6 + 4\delta)$.

3.3. Let A be an ideal of the ring of integers R in an imaginary quadratic field. Prove that there is a lattice basis for A, one of whose elements is an ordinary positive integer.

3.4. For each ring R listed below, use the method of Proposition 13.3.3 to describe the ideals in R. Make a drawing showing the possible shapes of the lattices in each case.

(a) $R = \mathbb{Z}[\sqrt{-3}]$, (b) $R = \mathbb{Z}[\frac{1}{2}(1 + \sqrt{-3})]$, (c) $R = \mathbb{Z}[\sqrt{-6}]$,

(d) $R = \mathbb{Z}[\frac{1}{2}(1 + \sqrt{-7})]$, (e) $R = \mathbb{Z}[\sqrt{-10}]$

Section 4   Ideal Multiplication

4.1. Let $R = \mathbb{Z}[\sqrt{-6}]$. Find a lattice basis for the product ideal AB, where $A = (2, \delta)$ and $B = (3, \delta)$.

4.2. Let R be the ring $\mathbb{Z}[\delta]$, where $\delta = \sqrt{-5}$, and let A denote the ideal generated by the elements **(a)** $3+5\delta, 2+2\delta$, **(b)** $4+\delta, 1+2\delta$. Decide whether or not the given generators form a lattice basis for A, and identify the ideal $\overline{A}A$.

4.3. Let R be the ring $\mathbb{Z}[\delta]$, where $\delta = \sqrt{-5}$, and let A and B be ideals of the form $A = (\alpha, \frac{1}{2}(\alpha + \alpha\delta))$, $B = (\beta, \frac{1}{2}(\beta + \beta\delta))$. Prove that AB is a principal ideal by finding a generator.

Section 5   Factoring Ideals

5.1. Let $R = \mathbb{Z}[\sqrt{-5}]$.

(a) Decide whether or not 11 is an irreducible element of R and whether or not (11) is a prime ideal of R.

(b) Factor the principal ideal (14) into prime ideals in $\mathbb{Z}[\delta]$.

5.2. Let $\delta = \sqrt{-3}$ and $R = \mathbb{Z}[\delta]$. This is not the ring of integers in the imaginary quadratic number field $\mathbb{Q}[\delta]$. Let A be the ideal $(2, 1 + \delta)$.

(a) Prove that A is a maximal ideal, and identify the quotient ring R/A.

(b) Prove that $\overline{A}A$ is not a principal ideal, and that the Main Lemma is not true for this ring.

(c) Prove that A contains the principal ideal (2) but that A does not divide (2).

5.3. Let $f = y^2 - x^3 - x$. Is the ring $\mathbb{C}[x, y]/(f)$ an integral domain?

Section 6   Prime Ideals and Prime Integers

6.1. Let $d = -14$. For each of the primes $p = 2, 3, 5, 7, 11$, and 13, decide whether or not p splits or ramifies in R, and if so, find a lattice basis for a prime ideal factor of (p).

6.2. Suppose that d is a negative integer, and that $d \equiv 1$ modulo 4. Analyze whether or not 2 remains prime in R in terms of congruence modulo 8.

6.3. Let R be the ring of integers in an imaginary quadratic field.

(a) Suppose that an integer prime p remains prime in R. Prove that $R/(p)$ is a field with p^2 elements.

(b) Prove that if p splits but does not ramify, then $R/(p)$ is isomorphic to the product ring $\mathbb{F}_p \times \mathbb{F}_p$.

6.4. When d is congruent 2 or 3 modulo 4, an integer prime p remains prime in the ring of integers of $\mathbb{Q}[\sqrt{d}]$ if the polynomial $x^2 - d$ is irreducible modulo p.

(a) Prove that this is also true when $d \equiv 1$ modulo 4 and $p \neq 2$.

(b) What happens to $p = 2$ when $d \equiv 1$ modulo 4?

6.5. Assume that d is congruent 2 or 3 modulo 4.

(a) Prove that a prime integer p ramifies in R if and only if $p = 2$ or p divides d.

(b) Let p be an integer prime that ramifies, and say that $(p) = P^2$. Find an explicit lattice basis for P. In which cases is P a principal ideal?

6.6. Let d be congruent to 2 or 3 modulo 4. An integer prime might be of the form $a^2 - b^2 d$, with a and b in $\mathbb{Z}$. How is this related to the prime ideal factorization of (p) in the ring of integers R?

6.7. Suppose that $d \equiv 2$ or 3 modulo 4, and that a prime $p \neq 2$ does not remain prime in R. Let a be an integer such that $a^2 \equiv d$ modulo p. Prove that $(p, a + \delta)$ is a lattice basis for a prime ideal that divides (p).

Section 7   Ideal Classes

7.1. Let $R = \mathbb{Z}[\sqrt{-5}]$, and let $B = (3, 1 + \delta)$. Find a generator for the principal ideal B^2.

7.2. Prove that two nonzero ideals A and A' in the ring of integers in an imaginary quadratic field are similar if and only if there is a nonzero ideal C such that both AC and $A'C$ are principal ideals.

7.3. Let $d = -26$. With each of the following integers n, decide whether n is the norm of an element α of R. If it is, find α: $n = 75,\ 250,\ 375,\ 5^6$.

7.4. Let $R = \mathbb{Z}[\delta]$, where $\delta^2 = -6$.

(a) Prove that the lattices $P = (2, \delta)$ and $Q = (3, \delta)$ are prime ideals of R.

(b) Factor the principal ideal (6) into prime ideals explicitly in R.

(c) Determine the class group of R.

Section 8   Computing the Class Group

8.1. With reference to Example 13.8.6, since $\langle P \rangle = \langle S \rangle^3$ and $\langle Q \rangle = \langle S \rangle^2$, Lemma 13.8.7 predicts that there are elements whose norms are $2 \cdot 5^3$ and $3^2 \cdot 5^2$. Find such elements.

8.2. With reference to Example 13.8.8, explain why $N(4 + \delta)$ and $N(14 + \delta)$ don't lead to contradictory conclusions.

8.3. Let $R = \mathbb{Z}[\delta]$, with $\delta = \sqrt{-29}$. In each case, compute the norm, explain what conclusions one can draw about ideals in R from the norm computation, and determine the class group of R: $N(1 + \delta),\ N(4 + \delta),\ N(5 + \delta),\ N(9 + 2\delta),\ N(11 + 2\delta)$.

8.4. Prove that the values of d listed in Theorem 13.2.5 have unique factorization.

8.5. Determine the class group and draw the possible shapes of the lattices in each case:

(a) $d = -10$, (b) $d = -13$, (c) $d = -14$, (d) $d = -21$.

8.6. Determine the class group in each case:

(a) $d = -41$, (b) $d = -57$, (c) $d = -61$, (d) $d = -77$, (e) $d = -89$.

Section 9   Real Quadratic Fields

9.1. Prove that $1 + \sqrt{2}$ is an element of infinite order in the group of units of $\mathbb{Z}[\sqrt{2}]$.

9.2. Determine the solutions of the equation $x^2 - y^2 d = 1$ when d is a positive integer.

9.3. (a) Prove that the size function $\sigma(\alpha) = |N(\alpha)|$ makes the ring $\mathbb{Z}[\sqrt{2}]$ into a Euclidean domain, and that this ring has unique factorization.

  (b) Make a sketch showing the principal ideal $(\sqrt{2})$ of $R = \mathbb{Z}[\sqrt{2}]$, in the embedding depicted in Figure 13.9.6.

9.4. Let R be the ring of integers in a real quadratic number field. What structures are possible for the group of units in R?

9.5. Let R be the ring of integers in a real quadratic number field, and let U_0 denote the set of units of R that are in the first quadrant in the embedding (13.9.2).

  (a) Prove that U_0 is an infinite cyclic subgroup of the group of units.

  (b) Find a generator for U_0 when $d = 3$ and when $d = 5$.

  (c) Draw a figure showing the hyperbolas and the units in a reasonable size range for $d = 3$.

Section 10   About Lattices

10.1. Let M be the integer lattice in $\mathbb{R}^2$, and let L be the lattice with basis $((2, 3)^t, (3, 6)^t)$. Determine the index $[M : L]$.

10.2. Let $L \subset M$ be lattices with bases $\mathbf{B}$ and $\mathbf{C}$, respectively, and let A be the integer matrix such that $\mathbf{B}A = \mathbf{C}$. Prove that $[M : L] = |\det A|$.

Miscellaneous Problems

M.1. Describe the subrings S of $\mathbb{C}$ that are lattices in the complex plane.

*M.2. Let $R = \mathbb{Z}[\delta]$, where $\delta = \sqrt{-5}$, and let p be a prime integer.

  (a) Prove that if p splits in R, say $(p) = \overline{P}P$, then exactly one of the ellipses $x^2 + 5y^2 = p$ or $x^2 + 5y^2 = 2p$ contains an integer point.

  (b) Find a property that determines which ellipse has an integer point.

M.3. Describe the prime ideals in   **(a)** the polynomial ring $\mathbb{C}[x, y]$ in two variables,

  (b) the ring $\mathbb{Z}[x]$ of integer polynomials.

M.4. Let L denote the integer lattice $\mathbb{Z}^2$ in the plane $\mathbb{R}^2$, and let P be a polygon in the plane whose vertices are points of L. *Pick's Theorem* asserts that the area $\Delta(P)$ is equal to $a + b/2 - 1$, where a is the number of points of L in the interior of P, and b is the number of points of L on the boundary of P.

  (a) Prove Pick's Theorem.

  (b) Derive Proposition 13.10.6 from Pick's Theorem.

CHAPTER 14

Linear Algebra in a Ring

Be wise! Generalize!

—Picayune Sentinel

Solving linear equations is a basic problem of linear algebra. We consider systems $AX = B$ when the entries of A and B are in a ring R here, and we ask for solutions $X = (x_1, \ldots, x_n)^t$ with x_i in R. This becomes difficult when the ring R is complicated, but we will see how it can be solved when R is the ring of integers or a polynomial ring over a field.

14.1 MODULES

The analog for a ring R of a vector space over a field is called a module.

• Let R be a ring. An *R-module* V is an abelian group with a law of composition written $+$, and a scalar multiplication $R \times V \to V$, written $r, v \rightsquigarrow rv$, that satisfy these axioms:

(14.1.1) $1v = v$, $(rs)v = r(sv)$, $(r+s)v = rv + sv$, and $r(v + v') = rv + rv'$,

for all r and s in R and all v and v' in V.

These are precisely the axioms for a vector space (3.1.2). However, the fact that elements of a ring needn't be invertible makes modules more complicated.

Our first examples are the modules R^n of *R-vectors*, column vectors with entries in the ring. They are called *free modules*. The laws of composition for R-vectors are the same as for vectors with entries in a field:

$$\begin{bmatrix} a_1 \\ \vdots \\ a_n \end{bmatrix} + \begin{bmatrix} b_1 \\ \vdots \\ b_n \end{bmatrix} = \begin{bmatrix} a_1 + b_1 \\ \vdots \\ a_n + b_n \end{bmatrix} \quad \text{and} \quad r \begin{bmatrix} a_1 \\ \vdots \\ a_n \end{bmatrix} = \begin{bmatrix} ra_1 \\ \vdots \\ ra_n \end{bmatrix}$$

But when R isn't a field, it is no longer true that they are the only modules. There will be modules that aren't isomorphic to any free module, though they are spanned by a finite set.

An abelian group V, its law of composition written additively, can be made into a module over the integers in exactly one way. The distributive law forces us to set $2v = (1+1)v = v + v$, and so on:

$$nv = v + \cdots + v = \text{``}n \text{ times } v\text{''}$$

and $(-n)v = -(nv)$, for any positive integer n. It is intuitively plausible this makes V into a $\mathbb{Z}$-module, and also that it is the only way to do so. Let's not bother with a formal proof.

Conversely, any $\mathbb{Z}$-module has the structure of an abelian group, given by keeping only the addition law and forgetting about its scalar multiplication.

(14.1.2)   Abelian group and $\mathbb{Z}$ – module are equivalent concepts.

We must use additive notation in the abelian group in order to make this correspondence seem natural, and we do so throughout the chapter.

Abelian groups provide examples to show that modules over a ring needn't be free. Since $\mathbb{Z}^n$ is infinite when n is positive, no finite abelian group except the zero group is isomorphic to a free module.

A *submodule* W of an R-module V is a nonempty subset that is closed under addition and scalar multiplication. The laws of composition on V make a submodule W into a module. We've seen submodules in one case before, namely submodules of the ring R, when it is thought of as the free R-module R^1.

Proposition 14.1.3 The submodules of the R-module R are the ideals of R.

By definition, an ideal is a nonempty subset of R that is closed under addition and under multiplication by elements of R. $\square$

The definition of a *homomorphism* $\varphi: V \to W$ of R-modules copies that of a linear transformation of vector spaces. It is a map compatible with the laws of composition:

(14.1.4) $\varphi(v + v') = \varphi(v) + \varphi(v')$ and $\varphi(rv) = r\varphi(v),$

for all v and v' in V and r in R. An *isomorphism* is a bijective homomorphism. The *kernel* of a homomorphism $\varphi: V \to W$, the set of elements v in V such that $\varphi(v) = 0$, is a submodule of the domain V, and the *image* of φ is a submodule of the range W.

One can extend the quotient construction to modules. Let W be a submodule of an R-module V. The quotient module $\overline{V} = V/W$ is the group of additive cosets $\overline{v} = [v + W]$. It is made into an R-module by the rule

(14.1.5) $r\overline{v} = \overline{rv}.$

The main facts about quotient modules are collected together below.

Theorem 14.1.6 Let W be a submodule of an R-module V.

(a) The set $\overline{V}$ of additive cosets of W in V is an R-module, and the canonical map $\pi: V \to \overline{V}$ sending $v \rightsquigarrow \overline{v} = [v + W]$ is a surjective homomorphism of R-modules whose kernel is W.

(b) *Mapping property*: Let $f: V \to V'$ be a homomorphism of R-modules whose kernel K contains W. There is a unique homomorphism: $\overline{f}: \overline{V} \to V'$ such that $f = \overline{f} \circ \pi$.

(c) *First Isomorphism Theorem*: Let $f : V \to V'$ be a surjective homomorphism of R-modules whose kernel is equal to W. The map $\bar{f}$ defined in **(b)** is an isomorphism.

(d) *Correspondence Theorem*: Let $f : V \to \mathcal{V}$ be a surjective homomorphism of R-modules, with kernel W. There is a bijective correspondence between submodules of $\mathcal{V}$ and submodules of V that contain W. This correspondence is defined as follows: If $\mathcal{S}$ is a submodule of $\mathcal{V}$, the corresponding submodule of V is $S = f^{-1}(\mathcal{S})$ and if S is a submodule of V that contains W, the corresponding submodule of W is $\mathcal{S} = f(S)$. If S and $\mathcal{S}$ are corresponding modules, then V/S is isomorphic to $\mathcal{V}/\mathcal{S}$.

We have seen the analogous facts for rings and ideals, and for groups and normal subgroups. The proofs follow the pattern set previously, so we omit them. $\square$

14.2   FREE MODULES

Free modules form an important class, and we discuss them here. Beginning in Section 14.5, we look at other modules.

- Let R be a ring. An *R-matrix* is a matrix whose entries are in R. An *invertible R-matrix* is an R-matrix that has an inverse that is also an R-matrix. The $n \times n$ invertible R-matrices form a group called the *general linear group over R:*

$$(14.2.1) \qquad GL_n(R) = \{n \times n \text{ invertible } R\text{-matrices}\}.$$

The *determinant* of an R-matrix $A = (a_{ij})$ can be computed by any one of the rules described in Chapter 1. The complete expansion (1.6.4), for example, exhibits $\det A$ as a polynomial in the n^2 matrix entries, with coefficients ± 1.

$$(14.2.2) \qquad \det A = \sum_p \pm a_{1,p1} \cdots a_{n,pn}.$$

As before, the sum is over all permutations p of the indices $\{1, \dots, \mathbf{n}\}$, and the symbol $\pm$ stands for the sign of the permutation. When we evaluate this formula on an R-matrix, we obtain an element of R. Rules for the determinant, such as

$$(\det A)(\det B) = \det (AB),$$

continue to hold. We have proved this rule when the matrix entries are in a field (1.4.10), and we discuss the reason that such properties are true for R-matrices in the next section. Let's assume for now that they are true.

Lemma 14.2.3 Let R be a ring, not the zero ring.

(a) A square R-matrix A is invertible if and only if it has either a left inverse or a right inverse, and also if and only if its determinant is a unit of the ring.

(b) An invertible R-matrix is square.

Proof. **(a)** If A has a left inverse L, the equation $(\det L)(\det A) = \det I = 1$ shows that $\det A$ has an inverse in R, so it is a unit. Similar reasoning shows that $\det A$ is a unit if A has a right inverse.

If A is an R-matrix whose determinant δ is a unit, Cramer's Rule: $A^{-1} = \delta^{-1}\text{cof}(A)$, where $\text{cof}(A)$ is the cofactor matrix (1.6.7), shows that there is an inverse with coefficients in R.

(b) Suppose that an $m \times n$ R-matrix P is invertible, i.e., that there is an $n \times m$ R-matrix Q such that $PQ = I_m$ and also $QP = I_n$. Interchanging P and Q if necessary, we may suppose that $m \geq n$. If $m \neq n$, we make P and Q square by adding zeros:

$$\begin{bmatrix} P & | & 0 \end{bmatrix} \begin{bmatrix} Q \\ \hline 0 \end{bmatrix} = I_m.$$

This does not change the product PQ, but the determinants of these square matrices are zero, so they are not invertible. Therefore $m = n$. $\square$

When R has few units, the fact that the determinant of an invertible matrix must be a unit is a strong restriction. For instance, if R is the ring of integers, the determinant must be ± 1. Most integer matrices are invertible when thought of as real matrices, so they are in $GL_n(\mathbb{R})$. But unless the determinant is ± 1, the entries of the inverse matrix won't be integers: they won't be elements of $GL_n(\mathbb{Z})$. Nevertheless, when $n > 1$, there are many invertible $n \times n$ R-matrices. The elementary matrices $E = I + ae_{ij}$, with $i \neq j$ and a in R, are invertible, and they generate a large group.

We return to the discussion of modules. The concepts of basis and independence (Section 3.4) are carried over from vector spaces. An ordered set $(v_1, \ldots, v_k)$ of elements of a module V is said to *generate* V, or to *span* V if every element v is a linear combination:

(14.2.4) $$v = r_1 v_1 + \cdots + r_k v_k,$$

with coefficients in R. If this is true, the elements v_i are called *generators*. A module V is *finitely generated* if there exists a finite set of generators. Most of the modules we study will be finitely generated.

A set of elements $(v_1, \ldots, v_n)$ of a module V is *independent* if, whenever a linear combination $r_1 v_1 + \cdots + r_n v_n$ with r_i in R is zero, all of the coefficients r_i are zero. A set $(v_1, \ldots, v_n)$ that generates V and is independent is a *basis*. As with vector spaces, the set $(v_1, \ldots, v_n)$ is a basis if every v in V is a linear combination (14.2.4) in a unique way. The *standard basis* $\mathbf{E} = (e_1, \ldots, e_k)$ is a basis of R^n.

We may also speak of linear combinations and independence of infinite sets, using the terminology of Section 3.7. Even when S is infinite, a linear combination can involve only finitely many terms.

If we denote an ordered set $(v_1, \ldots, v_n)$ of elements of V by $\mathbf{B}$, as in Chapter 3. Then multiplication by $\mathbf{B}$,

$$\mathbf{B}X = (v_1, \ldots, v_n) \begin{bmatrix} x_1 \\ \vdots \\ x_n \end{bmatrix} = v_1 x_1 + \cdots + v_n x_n,$$

defines a homomorphism of modules that we may also denote by **B**:

(14.2.5)
$$R^n \xrightarrow{\ \mathbf{B}\ } V.$$

As before, the scalars have migrated to the right side. This homomorphism is surjective if and only if **B** generates V, injective if and only if **B** is independent, and bijective if and only if **B** is a basis. Thus a module V has a basis if and only if it is isomorphic to one of the free modules R^k, and if so, it is called a *free module* too. A module is free if and only if it has a basis.

Most modules have no basis.

A free $\mathbb{Z}$-module is also called a *free abelian group.* Lattices in $\mathbb{R}^2$ are free abelian groups, while finite, nonzero abelian groups are not free.

    Computation with bases of free modules is done in the same way as with bases of vector spaces. If **B** is a basis of a free module V, the *coordinate vector* of an element v, with respect to **B**, is the unique column vector X such that $v = \mathbf{B}X$. If two bases $\mathbf{B} = (v_1, \ldots v_m)$ and $\mathbf{B}' = (v_1', \ldots, v_n')$ for the same free module V are given, the basechange matrix is obtained as in Chapter 3, by writing the elements of the new basis as linear combinations of the old basis: $\mathbf{B}' = \mathbf{B}P$.

Proposition 14.2.6 Let R be a ring that is not the zero ring.

(a) The matrix P of a change of basis in a free module is an invertible R-matrix.

(b) Any two bases of the same free module over R have the same cardinality.

    The proof of **(a)** is the same as the proof of Proposition 3.5.9, and **(b)** follows from **(a)** and from Lemma 14.2.3.   □

    The number of elements of a basis for a free module V is called the *rank* of V. The rank is analogous to the dimension of a vector space. (Many concepts have different names when used for modules over rings.)

    As is true for vector spaces, every homomorphism f between free modules R^n and R^m is given by left multiplication by an R-matrix A:

(14.2.7)
$$R^n \xrightarrow{\ A\ } R^m.$$

The jth column of A is $f(e_j)$. Similarly, if $\varphi : V \to W$ is a homomorphism of free R-modules with bases $\mathbf{B} = (v_1, \ldots, v_n)$ and $\mathbf{C} = (w_1, \ldots, w_m)$, respectively, the matrix of the homomorphism with respect to **B** is defined to be $A = (a_{ij})$, where

(14.2.8)
$$\varphi(v_j) = \sum_i w_i a_{ij}.$$

If X is the coordinate vector of a vector v, i.e., if $v = \mathbf{B}X$ then $Y = AX$ is the coordinate vector of its image, i.e., $\varphi(v) = \mathbf{C}Y$.

(14.2.9)

$$
\begin{array}{ccc}
R^n \xrightarrow{\ A\ } R^m & & X \rightsquigarrow Y \\
\Big\downarrow{\scriptstyle \mathbf{B}} \quad \Big\downarrow{\scriptstyle \mathbf{C}} & & \\
V \xrightarrow{\ \varphi\ } W & & v \rightsquigarrow \varphi(v)
\end{array}
$$

As is true for linear transformations, a change of the bases **B** and **C** by invertible R-matrices P and Q changes the matrix of φ to $A' = Q^{-1}AP$.

14.3 IDENTITIES

In this section we address the following question: Why do certain properties of matrices with entries in a field continue to hold when the entries are in a ring? Briefly, they continue to hold if they are *identities*, which means that they are true when the matrix entries are variables. To be specific, suppose that we want to prove a formula such as the multiplicative property of the determinant, $(\det A)(\det B) = \det(AB)$, or Cramer's Rule. Suppose we have already proved the formula for matrices with complex entries. We don't want to do the work again, and besides, we may have used special properties of $\mathbb{C}$, such as the field axioms, to check the formula there. We did use the properties of a field to prove the ones mentioned, so the proofs we gave will not work for rings. We show here how to deduce such formulas for all rings, once they have been shown for the complex numbers.

The principle is quite general, but in order to focus attention, we consider the multiplicative property $(\det A)(\det B) = \det(AB)$, using the complete expansion (14.2.2) of the determinant as its definition. We replace the matrix entries by variables. Denoting by X and Y indeterminate $n \times n$ matrices, the variable identity is $(\det X)(\det Y) = \det(XY)$. Let's write

(14.3.1) $$f(X, Y) = (\det X)(\det Y) - \det(XY).$$

This is a polynomial in the $2n^2$ variable matrix entries x_{ij} and $y_{k\ell}$, an element of the ring $\mathbb{Z}[\{x_{ij}\}, \{y_{k\ell}\}]$ of integer polynomials in those variables.

Given matrices $A = (a_{ij})$ and $B = (b_{k\ell})$ with entries in a ring R, there is a unique homomorphism

(14.3.2) $$\varphi : \mathbb{Z}[\{x_{ij}\}, \{y_{k\ell}\}] \to R,$$

the substitution homomorphism, that sends $x_{ij} \rightsquigarrow a_{ij}$ and $y_{k\ell} \rightsquigarrow b_{k\ell}$.

Referring back to the definition of the determinant, we see that because φ is a homomorphism, it will send

$$f(X, Y) \rightsquigarrow f(A, B) = (\det A)(\det B) - \det(AB).$$

To prove the multiplicative property for matrices in an arbitrary ring, it suffices to prove that f is the zero element in the polynomial ring $\mathbb{Z}[\{x_{ij}\}, \{y_{k\ell}\}]$. That is what it means to say that the formula is an identity. If so, then since $\varphi(0) = 0$, it will follow that $f(A, B) = 0$ for any matrices A and B in any ring.

Now: If we were to expand f and collect terms, to write it as a linear combination of monomials, all coefficients would be zero. However, we don't know how to do this, nor do we want to. To illustrate this point, we look at the 2×2 case. In that case,

$$f(X, Y) = ((x_{11}x_{22} - x_{12}x_{21})(y_{11}y_{22} - y_{12}y_{21}))$$
$$- (x_{11}y_{11} + x_{12}y_{21})(x_{21}y_{12} + x_{22}y_{22})$$
$$+ (x_{11}y_{12} + x_{12}y_{22})(x_{21}y_{11} + x_{22}y_{22}).$$

This is the zero polynomial, but it isn't obvious that it is zero, and we wouldn't want to make the computation for larger matrices.

Instead, we reason as follows: Our polynomial determines a function on the space of $2n^2$ complex variables $\{x_{ij}, y_{k\ell}\}$ by evaluation: If A and B are complex matrices and if we evaluate f at $\{a_{ij}, b_{k\ell}\}$, we obtain $f(A, B) = (\det A)(\det B) - \det(AB)$. We know that $f(A, B)$ is equal to zero because our identity is true for complex matrices. So the function that f determines is identically zero. The only (formal) polynomial that defines the zero function is the zero polynomial. Therefore f is equal to zero.

It is possible to formalize this discussion and to prove a general theorem about the validity of identities in an arbitrary ring. However, even mathematicians occasionally feel that formulating a general theorem isn't worthwhile – that it is easier to consider each case as it comes along. This is one of those occasions.

14.4   DIAGONALIZING INTEGER MATRICES

We consider the problem mentioned at the beginning of the chapter: Given an $m \times n$ *integer matrix A* (a matrix whose entries are integers) and a integer column vector B, find the integer solutions of the system of linear equations

$$(14.4.1) \hspace{3cm} AX = B.$$

Left multiplication by the integer matrix A defines a map $\mathbb{Z}^n \xrightarrow{A} \mathbb{Z}^m$. Its kernel is the set of integer solutions of the homogeneous equation $AX = 0$, and its image is the set of integer vectors B such that the equation $AX = B$ has a solution in integers. As usual, all solutions of the inhomogeneous equation $AX = B$ can be obtained from a particular one by adding solutions of the homogeneous equation.

When the coefficients are in a field, row reduction is often used to solve linear equations. These operations are more restricted here: We should use them only when they are given by *invertible integer matrices* – integer matrices that have integer matrices as their inverses. The invertible integer matrices form the *integer general linear group* $GL_n(\mathbb{Z})$.

The best results will be obtained when we use both row and column operations to simplify a matrix. So we allow these operations:

(14.4.2)

- add an integer multiple of one row to another, or add an integer multiple of one column to another;
- interchange two rows or two columns;
- multiply a row or column by –1.

Any such operation can be made by multiplying A on the left or right by an *elementary integer matrix* – an elementary matrix that is an invertible integer matrix. The result of a sequence of operations will have the form

$$(14.4.3) \hspace{3cm} A' = Q^{-1}AP,$$

where Q and P are invertible integer matrices of the appropriate sizes.

Over a field, any matrix can be brought into the block form

$$A' = \begin{bmatrix} I & \\ & 0 \end{bmatrix}$$

by row and column operations (4.2.10). We can't hope for such a result when working with integers: We can't do it for 1×1 matrices. But we can diagonalize.

An example:

(14.4.4)
$$A = \begin{bmatrix} 1 & 2 & 3 \\ 4 & 6 & 6 \end{bmatrix} \xrightarrow[\text{oper}]{\text{row}} \begin{bmatrix} 1 & 2 & 3 \\ 0 & -2 & -6 \end{bmatrix} \xrightarrow[\text{opers}]{\text{col}} \begin{bmatrix} 1 & 0 & 0 \\ 0 & -2 & -6 \end{bmatrix}$$

$$= \begin{bmatrix} 1 & 0 & 0 \\ 0 & -2 & -6 \end{bmatrix} \xrightarrow[\text{oper}]{\text{row}} \begin{bmatrix} 1 & 0 & 0 \\ 0 & 2 & 6 \end{bmatrix} \xrightarrow[\text{oper}]{\text{col}} \begin{bmatrix} 1 & 0 & 0 \\ 0 & 2 & 0 \end{bmatrix} = A'$$

The matrix obtained has the form $A' = Q^{-1}AP$, where Q and P are invertible integer matrices:

(14.4.5)
$$Q^{-1} = \begin{bmatrix} 1 & \\ 4 & -1 \end{bmatrix} \quad \text{and} \quad P = \begin{bmatrix} 1 & -2 & 3 \\ & 1 & -3 \\ & & 1 \end{bmatrix}$$

(It is easy to make a mistake when computing these matrices. To compute Q^{-1}, the elementary matrices that produce the row operations multiply in reverse order, while to compute P one must multiply in the order that the operations are made.)

Theorem 14.4.6 Let A be an integer matrix. There exist products Q and P of elementary integer matrices of appropriate sizes, so that $A' = Q^{-1}AP$ is diagonal, say

$$A' = \begin{bmatrix} \begin{bmatrix} d_1 & & \\ & \ddots & \\ & & d_k \end{bmatrix} & \\ & 0 \end{bmatrix},$$

where the diagonal entries d_i are positive, and each one divides the next: $d_1 \mid d_2 \mid \cdots \mid d_k$.

Note that the diagonal will not lead to the bottom right corner unless A is a square matrix, and if k is less than both m and n, the diagonal will have some zeros at the end.

We can sum up the information inherent in the four matrices that appear in the theorem by the diagram

(14.4.7)
$$\begin{array}{ccc} \mathbb{Z}^n & \xrightarrow{A'} & \mathbb{Z}^m \\ {\scriptstyle P}\downarrow & & \downarrow{\scriptstyle Q} \\ \mathbb{Z}^n & \xrightarrow[A]{} & \mathbb{Z}^m \end{array}$$

where the maps are labeled by the matrices that are used to define them.

Proof. We assume $A \neq 0$. The strategy is to perform a sequence of operations, so as to end up with a matrix

(14.4.8)
$$\begin{bmatrix} d_1 & 0 & \cdots & 0 \\ 0 & & & \\ \vdots & & M & \\ 0 & & & \end{bmatrix}$$

in which d_1 divides every entry of M. When this is done, we work on M. We describe a systematic method, though it may not be the quickest way to proceed. The method is based on repeated division with remainder.

Step 1: By permuting rows and columns, we move a nonzero entry with smallest absolute value to the upper left corner. We multiply the first row by -1 if necessary, so that this upper left entry a_{11} becomes positive.

Next, we try to clear out the first column. Whenever an operation produces a nonzero entry in the matrix whose absolute value is smaller than a_{11}, we go back to Step 1 and start the whole process over. This will spoil the work we have done, but progress is made because a_{11} decreases. We won't need to return to Step 1 infinitely often.

Step 2: If the first column contains a nonzero entry a_{i1} with $i > 1$, we divide by a_{11}:

$$a_{i1} = a_{11}q + r,$$

where q and r are integers, and the remainder r is in the range $0 \leq r < a_{11}$. We subtract $q(\text{row } 1)$ from $(\text{row } i)$. This changes a_{i1} to r. If $r \neq 0$, we go back to Step 1. If $r = 0$, we have produced a zero in the first column.

Finitely many repetitions of Steps 1 and 2 result in a matrix in which $a_{i1} = 0$ for all $i > 1$. Similarly, we may use column operations to clear out the first row, eventually ending up with a matrix in which the only nonzero entry in the first row and the first column is a_{11}.

Step 3: Assume that a_{11} is the only nonzero entry in the first row and column, but that some entry b of M is not divisible by a_{11}. We add the column of A that contains b to column 1. This produces an entry b in the first column. We go back to Step 2. Division with remainder produces a smaller nonzero matrix entry, sending us back to Step 1. $\square$

We are now ready to solve the integer linear system $AX = B$.

Proposition 14.4.9 Let A be an $m \times n$ matrix, and let P and Q be invertible integer matrices such that $A' = Q^{-1}AP$ has the diagonal form described in Theorem 14.4.6.

(a) The integer solutions of the homogeneous equation $A'X' = 0$ are the integer vectors X' whose first k coordinates are zero.

(b) The integer solutions of the homogeneous equation $AX = 0$ are those of the form $X = PX'$, where $A'X' = 0$.

(c) The image W' of multiplication by A' consists of the integer combinations of the vectors $d_1e_1, \ldots, d_ke_k$.

(d) The image W of multiplication by A consists of the vectors $Y = QY'$, where Y' is in W'.

Proof. **(a)** Because A' is diagonal, the equation $A'X' = 0$ reads

$$d_1x'_1 = 0, \;\; d_2x'_2 = 0, \ldots, \;\; d_kx'_k = 0.$$

In order for X' to solve the diagonal system $A'X' = 0$, we must have $x'_i = 0$ for $i = 1, \ldots, r$, and x'_i can be arbitrary if $i > k$,

(c) The image of the map A' is generated by the columns of A', and because A' is diagonal, the columns are especially simple: $A'_j = d_je_j$ if $j \leq k$, and $A'_j = 0$ if $j > k$.

(b),(d) We regard Q and P as matrices of changes of basis in $\mathbb{Z}^n$ and $\mathbb{Z}^m$, respectively. The vertical arrows in the diagram 14.4.7 are bijective, so P carries the kernel of A' bijectively to the kernel of A, and Q carries the image of A' bijectively to the image of A. $\quad\square$

We go back to example (14.4.4). Looking at the matrix A' we see that the solutions of $A'X' = 0$ are the integer multiples of e_3. So the solutions of $AX = 0$ are the integer multiples of Pe_3, which is the third column $(3, -3, 1)^t$ of P. The image of A' consists of integer combinations of the vectors e_1 and $2e_2$, and the image of A is obtained by multiplying these vectors by Q. It happens in this example that $Q = Q^{-1}$. So the image consists of the integer combinations of the columns of the matrix

$$QA' = \begin{bmatrix} 1 & 0 \\ 4 & -1 \end{bmatrix} \begin{bmatrix} 1 & 0 \\ 0 & 2 \end{bmatrix} = \begin{bmatrix} 1 & 0 \\ 4 & -2 \end{bmatrix}.$$

Of course, the image of A is also the set of integer combinations of the columns of A, but those columns do not form a $\mathbb{Z}$-basis.

The solution we have found isn't unique. A different sequence of row and column operations could produce different bases for the kernel and image. But in our example, the kernel is spanned by one vector, so that vector is unique up to sign.

Submodules of Free Modules

The theorem on diagonalization of integer matrices can be used to describe homomorphisms between free abelian groups.

Corollary 14.4.10 Let $\varphi : V \to W$ be a homomorphism of free abelian groups. There exist bases of V and W such that the matrix of the homomorphism has the diagonal form (14.4.6). $\quad\square$

Theorem 14.4.11 Let W be a free abelian group of rank m, and let U be a subgroup of W. Then U is a free abelian group, and its rank is less than or equal to m.

Proof. We begin by choosing a basis $\mathbf{C} = (w_1, \ldots, w_m)$ for W and a set of generators $\mathbf{B} = (u_1, \ldots, u_n)$ for U. We write $u_j = \sum_i w_ia_{ij}$, and we let $A = (a_{ij})$. The columns of the matrix A are the coordinate vectors of the generators u_j, when computed with respect to the basis $\mathbf{C}$ of W. We obtain a commutative diagram of homomorphisms of abelian groups

(14.4.12)

$$
\begin{array}{ccc}
\mathbb{Z}^n & \xrightarrow{\ A\ } & \mathbb{Z}^m \\
\downarrow{\scriptstyle \mathbf{B}} & & \downarrow{\scriptstyle \mathbf{C}} \\
U & \xrightarrow{\ i\ } & W
\end{array}
$$

where i denotes the inclusion of U into W. Because $\mathbf{C}$ is a basis, the right vertical arrow is bijective, and because $\mathbf{B}$ generates U, the left vertical arrow is surjective.

We diagonalize A. With the usual notation $A' = Q^{-1}AP$, we interpret P as the matrix of a change of basis for $\mathbb{Z}^n$, and Q as the matrix of a change of basis in $\mathbb{Z}^m$. Let the new bases be $\mathbf{C}'$ and $\mathbf{B}'$. Since our original choices of basis $\mathbf{C}$ and the generating set $\mathbf{B}$ were arbitrary, we may replace $\mathbf{C}, \mathbf{B}$ and A by $\mathbf{C}', \mathbf{B}'$ and A' in the above diagram. So we may assume that the matrix A has the diagonal form given in (14.4.6). Then $u_j = d_j w_j$ for $j = 1, \ldots, k$.

Roughly speaking, this is the proof, but there are still a few points to consider. First, the diagonal matrix A may contain columns of zeros. A column of zeros corresponds to a generator u_j whose coordinate vector with respect to the basis $\mathbf{C}$ of W is the zero vector. So u_j is zero too. This vector is useless as a generator, so we throw it out. When we have done this, all diagonal entries will be positive, and we will have $k = n$ and $n \le m$.

If W is the zero subgroup, we will end up throwing out all the generators. As with vector spaces, we must agree that the empty set is a basis for the zero module, or else mention this exceptional case in the statement of the theorem.

We assume that the $m \times n$ matrix A is diagonal, with positive diagonal entries $d_1, \ldots, d_n$ and with $n \le m$, and we show that the set $(u_1, \ldots, u_n)$ is a basis of U. Since this set generates U, what has to be proved is that it is independent. We write a linear relation $a_1 u_1 + \cdots + a_n u_n = 0$ in the form $a_1 d_1 w_1 + \cdots + a_n d_n w_n = 0$. Since $(w_1, \ldots, w_m)$ is a basis, $a_i d_i = 0$ for each i, and since $d_i > 0$, $a_i = 0$.

The final point is more serious: We needed a finite set of generators of U to get started. How do we know that there is such a set? It is a fact that every subgroup of a finitely generated abelian group is finitely generated. We prove this in Section 14.6. For the moment, the theorem is proved only with the additional hypothesis that U is finitely generated. $\quad\square$

Suppose that a lattice L in $\mathbb{R}^2$ with basis $\mathbf{B} = (v_1, v_2)$ is a sublattice of the lattice M with the basis $\mathbf{C} = (u_1, u_2)$, and let A be the integer matrix such that $\mathbf{B} = \mathbf{C}A$. If we change bases in L and M, the matrix A will be changed to a matrix $A' = Q^{-1}AP$, where P and Q are invertible integer matrices. According to Theorem 14.4.6, bases can be chosen so that A is diagonal, with positive diagonal entries d_1 and d_2. Suppose that this has been done. Then if $\mathbf{B} = (v_1, v_2)$ and $\mathbf{C} = (u_1, u_2)$, the equation $\mathbf{B} = \mathbf{C}A$ reads $v_1 = d_1 u_1$ and $v_2 = d_2 u_2$.

Example 14.4.13 Let $Q = \begin{bmatrix} 1 & 1 \\ 3 & 1 \end{bmatrix}$, $A = \begin{bmatrix} 2 & -1 \\ 1 & 2 \end{bmatrix}$, $P = \begin{bmatrix} 1 & 1 \\ 1 & 2 \end{bmatrix}$, $A' = Q^{-1}AP = \begin{bmatrix} 1 & \\ & 5 \end{bmatrix}$.

Let M be the integer lattice with its standard basis $\mathbf{C} = (e_1, e_2)$, and let L be the lattice with basis $\mathbf{B} = (v_1, v_2) = ((2, 1)^t, (-1, 2)^t)$. Its coordinate vectors are the columns of A. We interpret P as the matrix of a change of basis in L, and Q as the matrix of change of basis in M. In coordinate vector form, the new bases are $\mathbf{C}' = (e_1, e_2)Q = ((1, 3)^t, (0, 1)^t)$ and $\mathbf{B}' = (v_1, v_2)P = ((1, 3)^t, (0, 5)^t)$.

The left-hand figure below shows the squares spanned by the two original bases, and the figure on the right shows the parallelograms spanned by the two new bases. The parallelogram spanned by the new basis for L is filled precisely by five translates of the shaded parallelogram, which is the parallelogram spanned by the new basis for M. The index is 5. Note that there are five lattice points in the region $\Pi'(v_1, v_2)$. This agrees with Proposition 13.10.3(**d**). The figure on the right also makes it clear that the ratio $\Delta(L)/\Delta(M)$ is 5. $\square$

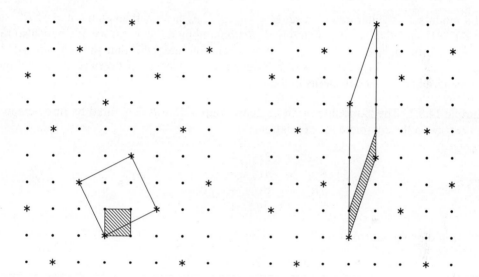

(14.4.14)   Diagonalization, Applied to a Sublattice.

14.5   GENERATORS AND RELATIONS

In this section we turn our attention to modules that are not free. We show how to describe a large class of modules by means of matrices called *presentation matrices*.

Left multiplication by an $m \times n$ R-matrix A defines a homomorphism of R-modules $R^n \xrightarrow{A} R^m$. Its image consists of all linear combinations of the columns of A with coefficients in the ring, and we may denote the image by AR^n. We say that the quotient module $V = R^m/AR^n$ is *presented* by the matrix A. More generally, we call any isomorphism $\sigma: R^m/AR^n \to V$ a *presentation* of a module V, and we say that the matrix A is a *presentation matrix* for V if there is such an isomorphism.

For example, the cyclic group C_5 of order 5 is presented as a $\mathbb{Z}$-module by the 1×1 integer matrix [5], because C_5 is isomorphic to $\mathbb{Z}/5\mathbb{Z}$.

We use the canonical map $\pi: R^m \to V = R^m/AR^n$ (14.1.6) to interpret the quotient module $V = R^m/AR^n$, as follows:

Proposition 14.5.1

(a) V is generated by a set of elements $\mathbf{B} = (v_1, \ldots, v_m)$, the images of the standard basis elements of R^m.

(b) If $Y = (y_1, \ldots, y_m)^t$ is a column vector in R^m, the element $\mathbf{B}Y = v_1 y_1 + \cdots + v_m y_m$ of V is zero if and only if Y is a linear combination of the columns of A, with coefficients in R – if and only if there exists a column vector X with entries in R such that $Y = AX$.

Proof. The images of the standard basis elements generate V because the map π is surjective. Its kernel is the submodule $A R^n$. This submodule consists precisely of the linear combinations of the columns of A. $\square$

- If a module V is generated by a set $\mathbf{B} = (v_1, \ldots, v_m)$, we call an element Y of R^m such that $\mathbf{B}Y = 0$ a *relation vector*, or simply a *relation* among the generators. We may also refer to the equation $v_1 y_1 + \cdots + v_m y_m = 0$ as a relation, meaning that the left side yields 0 when it is evaluated in V. A set S of relations is a *complete set* if every relation is a linear combination of S with coefficients in the ring.

Example 14.5.2 The $\mathbb{Z}$-module or an abelian group V that is generated by three elements v_1, v_2, v_3 with the complete set of relations

(14.5.3)
$$
\begin{aligned}
3v_1 &+ 2v_2 &+ v_3 &= 0 \\
8v_1 &+ 4v_2 &+ 2v_3 &= 0 \\
7v_1 &+ 6v_2 &+ 2v_3 &= 0 \\
9v_1 &+ 6v_2 &+ v_3 &= 0
\end{aligned}
$$

is presented by the matrix

(14.5.4)
$$
A = \begin{bmatrix} 3 & 8 & 7 & 9 \\ 2 & 4 & 6 & 6 \\ 1 & 2 & 2 & 1 \end{bmatrix}.
$$

Its columns are the coefficients of the relations (14.5.3):

$$(v_1, v_2, v_3) A = (0, 0, 0, 0).$$ $\square$

We now describe a theoretical method of finding a presentation of an R-module V. The method is very simple: We choose a set of generators $\mathbf{B} = (v_1, \ldots, v_m)$ for V. These generators provide us with a surjective homomorphism $R^m \to V$ that sends a column vector Y to the linear combination $\mathbf{B}Y = v_1 y_1 + \cdots + v_m y_m$. Let us denote the kernel of this map by W. It is the *module of relations*; its elements are the relation vectors.

We repeat the procedure, choosing a set of generators $\mathbf{C} = (w_1, \ldots, w_m)$ for W, and we use these generators to define a surjective map $R^n \to W$. But here the generators w_j are elements of R^m. They are column vectors. We assemble the coordinate vectors A_j of w_j into an $m \times n$ matrix

(14.5.5)
$$
A = \begin{bmatrix} | & & | \\ A_1 & \cdots & A_m \\ | & & | \end{bmatrix}.
$$

Then multiplication by A defines a map

$$R^n \xrightarrow{\;A\;} R^m$$

that sends $e_j \rightsquigarrow A_j = w_j$. It is the composition of the map $R^n \to W$ with the inclusion $W \subset R^m$. By construction, W is its image, and we denote it by AR^n.

Since the map $R^m \to V$ is surjective, the First Isomorphism Theorem tells us that V is isomorphic to $R^m/W = R^m/AR^n$. Therefore the module V is presented by the matrix A. Thus the presentation matrix A for a module V is determined by

(14.5.6)

- a set of generators for V, and
- a set of generators for the module of relations W.

Unless the set of generators forms a basis of V, in which case A is empty, the number of generators will be equal to the number of rows of A.

This construction depends on two assumptions: We must assume that our module V has a finite set of generators. Fair enough: We can't expect to describe a module that is too big, such as an infinite dimensional vector space, in this way. We must also assume that the module W of relations has a finite set of generators. This is a less desireable assumption because W is not given; it is an auxiliary module that was obtained in the course of the construction. We need to examine this point more closely, and we do this in the next section (see (14.6.5)). But except for this point, we can now speak of generators and relations for a finitely generated R-module V.

Since the presentation matrix depends on the choices (14.5.6), many matrices present the same module, or isomorphic modules. Here are some rules for manipulating a matrix A without changing the isomorphism class of the module it presents:

Proposition 14.5.7 Let A be an $m \times n$ presentation matrix for a module V. The following matrices A' present the same module V:

(i) $A' = Q^{-1}A$, with Q in $GL_m(R)$;

(ii) $A' = AP$, with P in $GL_n(R)$;

(iii) A' is obtained by deleting a column of zeros;

(iv) the jth column of A is e_i, and A' is obtained from A by deleting (*row i*) and (*column j*).

The operations **(iii)** and **(iv)** can also be done in reverse. One can add a column of zeros, or one can add a new row and column with 1 as their common entry, all other entries being zero.

Proof. We refer to the map $R^n \xrightarrow{\;A\;} R^m$ defined by the matrix.

(i) The change of A to $Q^{-1}A$ corresponds to a change of basis in R^m.

(ii) The change of A to AP corresponds to a change of basis in R^n.

(iii) A column of zeros corresponds to the trivial relation, which can be omitted.

(iv) A column of A equal to e_i corresponds to the relation $v_i = 0$. The zero element is useless as a generator, and its appearance in any other relation is irrelevant. So we may delete v_i from the generating set and from the relations. Doing so changes the matrix A by deleting the ith row and jth column. $\qquad\square$

It may be possible to simplify a matrix quite a lot by these rules. For instance, our original example of the integer matrix (14.5.4) reduces as follows:

$$A = \begin{bmatrix} 3 & 8 & 7 & 9 \\ 2 & 4 & 6 & 6 \\ 1 & 2 & 2 & 1 \end{bmatrix} \rightarrow \begin{bmatrix} 0 & 2 & 1 & 6 \\ 0 & 0 & 2 & 4 \\ 1 & 2 & 2 & 1 \end{bmatrix} \rightarrow \begin{bmatrix} 2 & 1 & 6 \\ 0 & 2 & 4 \end{bmatrix} \rightarrow \begin{bmatrix} 2 & 1 & 6 \\ -4 & 0 & -8 \end{bmatrix} \rightarrow$$

$$\rightarrow [-4 \ \ -8] \rightarrow [4 \ \ 8] \rightarrow [4 \ \ 0] \rightarrow [4].$$

Thus A presents the abelian group $\mathbb{Z}/4\mathbb{Z}$.

By definition, an $m \times n$ matrix presents a module by means of m generators and n relations. But as we see from this example, the numbers m and n depend on choices; they are not uniquely determined by the module.

Another example: The 2×1 matrix $\begin{bmatrix} 4 \\ 0 \end{bmatrix}$ presents an abelian group V by means of two generators (v_1, v_2) and one relation $4v_1 = 0$. We can't simplify this matrix. The abelian group that it presents is the direct sum $\mathbb{Z}/4\mathbb{Z} \oplus \mathbb{Z}$ of a cyclic group of order four and an infinite cyclic group (see Section 14.7). On the other hand, as we saw above, the matrix $[4 \ \ 0]$ presents a group with one generator v_1 and two relations, the second of which is the trivial relation. It is a cyclic group of order 4.

14.6 NOETHERIAN RINGS

In this section we discuss finite generation of the module of relations. For modules over a nasty ring, the module of relations needn't be finitely generated, though V is. Fortunately this doesn't occur with the rings we have been studying, as we show here.

Proposition 14.6.1 The following conditions on an R-module V are equivalent:

(i) Every submodule of V is finitely generated;

(ii) *ascending chain condition*: There is no infinite strictly increasing chain

$W_1 < W_2 < \cdots$ of submodules of V.

Proof. Assume that V satisfies the ascending chain condition, and let W be a submodule of V. We select a set of generators of W in the following way: If $W = 0$, then W is generated by the empty set. If not, we start with a nonzero element w_1 of W, and we let W_1 be the span of (w_1). If $W_1 = W$ we stop. If not, we choose an element w_2 of W not in W_1, and we let W_2 be the span of (w_1, w_2). Then $W_1 < W_2$. If $W_2 < W$, we choose an element w_3 not in W_2, etc. In this way we obtain a strictly increasing chain $W_1 < W_2 < \cdots$ of submodules of W. Since V satisfies the ascending chain condition, this chain cannot be continued indefinitely. Therefore some W_k is equal to W, and then $(w_1, \ldots, w_k)$ generates W.

The proof of the converse is similar to the proof of Proposition 12.2.13, which states that factoring terminates in a domain if and only if it has no strictly increasing chain of principal ideals. Assume that every submodule of V is finitely generated, and let $W_1 \subset W_2 \subset \ldots$ be an infinite weakly increasing chain of submodules of V. We show that this chain is not strictly increasing. Let U denote the union of these submodules. Then U is a submodule of V. The proof is the same as the one given for ideals (12.2.15). So U is finitely generated. Let $(u_1, \ldots, u_r)$ be a set of generators for U. Each u_ν is in one of the modules W_i and since the chain is increasing, there is an i such that W_i contains all of the elements $u_1, \ldots, u_r$. Then W_i contains the module U generated by $(u_1, \ldots, u_k)$: $U \subset W_i \subset W_{i+1} \subset U$. This shows that $U = W_i = W_{i+1} = U$, and that the chain is not strictly increasing.   □

Definition 14.6.2 A ring R is *noetherian* if every ideal of R is finitely generated.

Corollary 14.6.3 A ring is noetherian if and only if it satisfies the ascending chain condition: There is no infinite strictly increasing chain $I_1 < I_2 < \cdots$ of ideals of R.   □

Principal ideal domains are noetherian because every ideal in such a ring is generated by one element. So the rings $\mathbb{Z}$, $\mathbb{Z}[i]$, and $F[x]$, with F a field, are noetherian.

Corollary 14.6.4 Let R be a noetherian ring. Every proper ideal I of R is contained in a maximal ideal.

Proof. If I is not maximal itself, then it is properly contained in a proper ideal I_2, and if I_2 is not maximal, it is properly contained in a proper ideal I_3, and so on. By the ascending chain condition (14.6.1), the chain $I < I_2 < I_3 \cdots$ must be finite. Therefore I_k is maximal for some k.   □

The relevance of the concept of a noetherian ring to the problem of finite generation of a submodule is shown by the following theorem:

Theorem 14.6.5 Let R be a noetherian ring. Every submodule of a finitely generated R-module V is finitely generated.

Proof. Case 1: $V = R^m$. We use induction on m. A submodule of R^1 is an ideal of R (14.1.3). Since R is noetherian, the theorem is true when $m = 1$. Suppose that $m > 1$. We consider the projection

$$\pi : R^m \to R^{m-1}$$

given by dropping the last entry: $\pi(a_1, \ldots, a_m) = (a_1, \ldots, a_{m-1})$. Its kernel is the set of vectors of R^m whose first $m - 1$ coordinates are zero. Let W be a submodule of R^m, and let $\varphi : W \to R^{m-1}$ be the restriction of π to W. The image $\varphi(W)$ is a submodule of R^{m-1}. It is finitely generated by induction. Also, $\ker \varphi = (W \cap \ker \pi)$ is a submodule of $\ker \pi$, which is a module isomorphic to R^1. So $\ker \varphi$ is finitely generated. Lemma 14.6.6 shows that W is finitely generated.

Case 2: The general case. Let V be a finitely generated R-module. Then there is a surjective map $\varphi : R^m \to V$ from a free module to V. Given a submodule W of V, the Correspondence Theorem tells us that $U = \varphi^{-1}(W)$ is a submodule of the module R^m, so it is finitely generated, and $W = \varphi(U)$. Therefore W is finitely generated (14.6.6)(a).   □

Lemma 14.6.6 Let $\varphi : V \to V'$ be a homomorphism of R-modules.

(a) If V is finitely generated and φ is surjective, then V' is finitely generated.

(b) If the kernel and the image of φ are finitely generated, then V is finitely generated.

(c) Let W be a submodule of an R-module V. If both W and $\overline{V} = V/W$ are finitely generated, then V is finitely generated. If V is finitely generated, so is $\overline{V}$.

Proof. **(a)** Suppose that φ is surjective and let $(v_1, \ldots, v_n)$ be a set of generators for V. The set $(v'_1, \ldots, v'_n)$, with $v'_i = \varphi(v_i)$, generates V'.

(b) We follow the proof of the dimension formula for linear transformations (4.1.5). We choose a set of generators $(u_1, \ldots, u_k)$ for the kernel and a set of generators $(v'_1, \ldots, v'_m)$ for the image. We also choose elements v_i of V such that $\varphi(v_i) = v'_i$, and we show that the set $(u_1, \ldots, u_k; v_1, \ldots, v_m)$ generates V. Let v be any element of V. Then $\varphi(v)$ is a linear combination of $(v'_1, \ldots, v'_m)$, say $\varphi(v) = a_1 v'_1 + \cdots + a_m v'_m$. Let $x = a_1 v_1 + \cdots + a_m v_m$. Then $\varphi(x) = \varphi(v)$, hence $v - x$ is in the kernel of φ. So $v - x$ is a linear combination of $(u_1, \ldots, u_k)$, say $v - x = b_1 u_1 + \cdots + b_k u_k$, and

$$v = a_1 v_1 + \cdots + a_m v_m + b_1 u_1 + \cdots + b_k u_k.$$

Since v was arbitrary, the set $(u_1, \ldots, u_k; v_1, \ldots, v_m)$ generates.

(c) This follows from **(b)** and **(a)** when we replace φ by the canonical homomorphism $\pi : V \to \overline{V}$. $\square$

This theorem completes the proof of Theorem 14.4.11.

Since principal ideal domains are noetherian, submodules of finitely generated modules over these rings are finitely generated. In fact, most of the rings that we have been studying are noetherian. This follows from another of Hilbert's theorems:

Theorem 14.6.7 Hilbert Basis Theorem. Let R be a noetherian ring. The polynomial ring $R[x]$ is noetherian.

The proof of this theorem is below. It shows by induction that the polynomial ring $R[x_1, \ldots, x_n]$ in several variables over a noetherian ring R is noetherian. Therefore the rings $\mathbb{Z}[x_1, \ldots, x_n]$ and $F[x_1, \ldots, x_n]$, with F a field, are noetherian. Also, quotients of noetherian rings are noetherian:

Proposition 14.6.8 Let R be a noetherian ring, and let I be an ideal of R. Any ring that is isomorphic to the quotient ring $\overline{R} = R/I$ is noetherian.

Proof. Let $\overline{J}$ be an ideal of $\overline{R}$, and let $\pi : R \to \overline{R}$ be the canonical map. Let $J = \pi^{-1}(\overline{J})$ be the corresponding ideal of R. Since R is noetherian, J is finitely generated, and it follows that $\overline{J}$ is finitely generated (14.6.6)**(a)**. $\square$

Corollary 14.6.9 Let P be a polynomial ring in a finite number of variables over the integers or over a field. Any ring R that is isomorphic to a quotient ring P/I is noetherian. $\square$

We turn to the proof of the Hilbert Basis Theorem now.

Lemma 14.6.10 Let R be a ring and let I be an ideal of the polynomial ring $R[x]$. The set A whose elements are the leading coefficients of the nonzero polynomials in I, together with the zero element of R, is an ideal of R, the *ideal of leading coefficients*.

Proof. We must show that if α and β are in A, then $\alpha+\beta$ and $r\alpha$ are also in A. If any one of the three elements α, β, or $\alpha+\beta$ is zero, then $\alpha+\beta$ is in A, so we may assume that these elements are not zero. Then α is the leading coefficient of an element f of I, and β is the leading coefficient of an element g of I. We multiply f or g by a suitable power of x so that their degrees become equal. The polynomial we get is also in I. Then $\alpha+\beta$ is the leading coefficient of $f+g$. Since I is an ideal, $f+g$ is in I and $\alpha+\beta$ is in A. The proof that $r\alpha$ is in A is similar. $\square$

Proof of the Hilbert Basis Theorem. We suppose that R is a noetherian ring, and we let I be an ideal in the polynomial ring $R[x]$. We must show that there is a finite subset S of I that generates this ideal – a subset such that every element of I can be expressed as a linear combination of its elements, with polynomial coefficients.

Let A be the ideal of leading coefficients of I. Since R is noetherian, A has a finite set of generators, say $(\alpha_1, \ldots, \alpha_k)$. We choose for each $i = 1, \ldots, k$ a polynomial f_i in I with leading coefficient α_i, and we multiply these polynomials by powers of x as necessary, so that their degrees become equal, say to n.

Next, let P denote the set consisting of the polynomials in $R[x]$ of degree less than n, together with 0. This is a free R-module with basis $(1, x, \ldots, x^{n-1})$. The subset $P \cap I$, which consists of the polynomials of degree less than n that are in I together with zero, is an R-submodule of P. Let's call this submodule W. Since P is a finitely generated R-module and since R is noetherian, W is a finitely generated R-module. We choose generators $(h_1, \ldots, h_\ell)$ for W. Every polynomial in I of degree less than n is a linear combination of $(h_1, \ldots, h_\ell)$, with coefficients in R.

We show now that the set $(f_1, \ldots, f_k; h_1, \ldots, h_\ell)$ generates the ideal I. We use induction on the degree d of g.

Case 1: $d < n$. In this case, g is an element of W, so it is a linear combination of $(h_1, \ldots, h_\ell)$ with coefficients in R. We don't need polynomial coefficients here.

Case 2: $d \geq n$. Let β be the leading coefficient of g, so $g = \beta x^d + (lower\ degree\ terms)$. Then β is an element of the ideal A of leading coefficients, so it is a linear combination $\beta = r_1\alpha_1 + \cdots + r_k\alpha_k$ of the leading coefficients α_i of f_i, with coefficients in R. The polynomial

$$q = \sum r_i x^{d-n} f_i$$

is in the ideal generated by $(f_1, \ldots, f_k)$. It has degree d, and its leading coefficient is β. Therefore the degree of $g - q$ is less than d. By induction, $g - q$ is a polynomial combination of $(f_1, \ldots, f_k; h_1, \ldots, h_\ell)$. Then $g = q + (g - q)$ is also such a combination. $\square$

14.7   STRUCTURE OF ABELIAN GROUPS

The Structure Theorem for abelian groups, which is below, asserts that a finite abelian group V is a direct sum of cyclic groups. The work of the proof has been done. We know that there exists a diagonal presentation matrix for V. What remains to do is to interpret the meaning of this matrix for the group.

The definition of a direct sum of modules is the same as that of a direct sum of vector spaces.

- Let $W_1, \ldots, W_k$ be submodules of an R-module V. Their *sum* is the submodule that they generate. It consists of all elements that are sums:

(14.7.1) $$W_1 + \cdots + W_k = \{v \in V \mid v = w_1 + \cdots + w_k, \text{ with } w_i \text{ in } W_i\}.$$

We say that V is the *direct sum* of the submodules $W_1, \ldots, W_k$, and we write $V = W_1 \oplus \cdots \oplus W_k$, if

(14.7.2)
- they *generate*: $V = W_1 + \cdots + W_k$, and
- they are *independent*: If $w_1 + \cdots + w_k = 0$, with w_i in W_i, then $w_i = 0$ for all i.

Thus V is the direct sum of the submodules W_i if every element v in V can be written uniquely in the form $v = w_1 + \cdots + w_k$, with w_i in W_i. As is true for vector spaces, a module V is the direct sum $W_1 \oplus W_2$ of two submodules W_1 and W_2 if and only if $W_1 + W_2 = V$ and $W_1 \cap W_2 = 0$ (see (3.6.6)).

The same definitions are used for abelian groups. An abelian group V is the direct sum $W_1 \oplus \cdots \oplus W_k$ of the subgroups $W_1, \ldots, W_k$ if:

- Every element v of V can be written as a sum $v = w_1 + \cdots + w_k$ with w_i in W_i, i.e., $V = W_1 + \cdots + W_k$.
- If a sum $w_1 + \cdots + w_k$, with w_i in W_i is zero, then $w_i = 0$ for all i.

Theorem 14.7.3 Structure Theorem for Abelian Groups. A finitely generated abelian group V is a direct sum of cyclic subgroups $C_{d_1}, \ldots, C_{d_k}$ and a free abelian group L:

$$V = C_{d_1} \oplus \cdots \oplus C_{d_k} \oplus L,$$

where the order d_i of C_{d_i} is greater than 1, and d_i divides d_{i+1} for $i = 1, \ldots, k-1$.

Proof of the Structure Theorem. We choose a presentation matrix A for V, determined by a set of generators and a complete set of relations. We can do this because V is finitely generated and because $\mathbb{Z}$ is a Noetherian ring. After a suitable change of generators and relations, A will have the diagonal form given in Theorem 14.4.6. We may eliminate any diagonal entry that is equal to 1, and any column of zeros (see (14.5.7)). The matrix A will then have the shape

(14.7.4)
$$A = \begin{bmatrix} d_1 & & \\ & \ddots & \\ & & d_k \\ & 0 & \end{bmatrix}$$

with $d_1 > 1$ and $d_1 | d_2 | \cdots | d_k$. It will be an $m \times k$ matrix, $0 \le k \le m$. The meaning of this for our abelian group is that V is generated by a set of m elements $\mathbf{B} = (v_1, \ldots, v_m)$, and that

(14.7.5) $$d_1 v_1 = 0, \ldots, d_k v_k = 0$$

forms a complete set of relations among these generators.

Let C_j denote the cyclic subgroup generated by v_j, for $j = 1, \ldots, m$. For $j \le k$, C_j is cyclic of order d_j, and for $j > k$, C_j is infinite cyclic. We show that V is the direct sum of these cyclic groups. Since **B** generates, $V = C_1 + \cdots + C_m$. Suppose given a relation $w_1 + \cdots + w_m = 0$ with w_j in C_j. Since v_j generates C_j, $w_j = v_j y_j$ for some integer y_j. The relation is $\mathbf{B}Y = v_1 y_1 + \cdots + v_m y_m = 0$. Since the columns of A form a complete set of relations, $Y = AX$ for some integer vector X, which means that y_j is a multiple of d_j if $j \le k$ and $y_j = 0$ if $j > k$. Since $v_j d_j = 0$ if $j \le k$, $w_j = 0$ if $j \le k$. The relation is trivial, so the cyclic groups C_j are independent. The direct sum of the infinite cyclic groups C_j with $j > k$ is the free abelian group L. $\square$

A finite abelian group is finitely generated, so as stated above, the Structure Theorem decomposes a finite abelian group into a direct sum of finite cyclic groups, in which the order of each summand divides the next. The free summand will be zero.

It is sometimes convenient to decompose the cyclic groups further, into cyclic groups of prime power order. This decomposition is based on Proposition 2.11.3: If a and b are relatively prime integers, the cyclic group C_{ab} of order ab is isomorphic to the direct sum $C_a \oplus C_b$ of cyclic subgroups of orders a and b. Combining this with the Structure Theorem yields the following:

Corollary 14.7.6 Structure Theorem (Alternate Form). Every finite abelian group is a direct sum of cyclic groups of prime power orders. $\square$

It is also true that the orders of the cyclic subgroups that occur are uniquely determined by the group. If the order of V is a product of distinct primes, there is no problem. For example, if the order is 30, then V must be isomorphic to $C_2 \oplus C_3 \oplus C_5$ and to C_{30}. But is $C_2 \oplus C_2 \oplus C_4$ isomorphic to $C_4 \oplus C_4$? It isn't difficult to show that it is not, by counting elements of orders 1 or 2. The group $C_4 \oplus C_4$ contains four such elements, while $C_2 \oplus C_2 \oplus C_4$ contains eight of them. This counting method always works.

Theorem 14.7.7 Uniqueness for the Structure Theorem. Suppose that a finite abelian group V is a direct sum of cyclic groups of prime power orders $d_j = p_j^{r_j}$. The integers d_j are uniquely determined by the group V.

Proof. Let p be one of the primes that appear in the direct sum decomposition of V, and let c_i denote the number of cyclic groups of order p^i in the decomposition. The set of elements whose orders divide p^i is a subgroup of V whose order is a power of p, say p^{ℓ_i}. Let k be the largest index such that $c_k > 0$. Then

$$\ell_1 = c_1 + \;\; c_2 + \;\; c_3 + \cdots + \;\; c_k$$
$$\ell_2 = c_1 + 2c_2 + 2c_3 + \cdots + 2c_k,$$
$$\ell_3 = c_1 + 2c_2 + 3c_3 + \cdots + 3c_k$$
$$\cdots$$
$$\ell_k = c_1 + 2c_2 + 3c_3 + \cdots + kc_k.$$

The exponents ℓ_i determine the integers c_i. $\square$

The integers d_i are also uniquely determined when they are chosen, as in Theorem 14.7.3, so that $d_1 | \cdots | d_k$.

14.8   APPLICATION TO LINEAR OPERATORS

The classification of abelian groups has an analogue for the polynomial ring $R = F[t]$ in one variable over a field F. Theorem 14.4.6 about diagonalizing integer matrices carries over because the key ingredient in the proof of Theorem 14.4.6, the division algorithm, is available in $F[t]$. And since the polynomial ring is noetherian, any finitely generated R-module V has a presentation matrix (14.2.7).

Theorem 14.8.1   Let $R = F[t]$ be a polynomial ring in one variable over a field F and let A be an $m \times n$ R-matrix. There are products Q and P of elementary R-matrices such that $A' = Q^{-1}AP$ is diagonal, each nonzero diagonal entry d_i of A' is a monic polynomial, and $d_1 \,|\, d_2 \,|\, \ldots \,|\, d_k$.   □

Example 14.8.2   Diagonalization of a matrix of polynomials:

$$A = \begin{bmatrix} t^2 - 3t + 1 & t - 2 \\ (t-1)^3 & t^2 - 3t + 2 \end{bmatrix} \xrightarrow{\text{row}} \begin{bmatrix} t^2 - 3t + 1 & t - 2 \\ t^2 - t & 0 \end{bmatrix} \xrightarrow{\text{col}}$$

$$\xrightarrow{\text{col}} \begin{bmatrix} -1 & t - 2 \\ t^2 - t & 0 \end{bmatrix} \xrightarrow{\text{col}} \begin{bmatrix} -1 & 0 \\ t^2 - t & t^3 - 3t + 2t \end{bmatrix} \xrightarrow{\text{row}} \begin{bmatrix} 1 & 0 \\ 0 & t^3 - 3t^2 + 2t \end{bmatrix}.$$

Note: It is not surprising that we ended up with 1 in the upper left corner in this example. This will happen whenever the greatest common divisor of the matrix entries is 1.   □

As is true for the ring of integers, Theorem 14.8.1 provides us with a method to determine the polynomial solutions of a system $AX = B$, when the entries of A and B are polynomial matrices (see Proposition 14.4.9).

We extend the structure theorem to polynomial rings next. To carry along the analogy with abelian groups, we define a *cyclic R-module C*, where R is any ring, to be a module that is generated by a single element v. Then there is a surjective homomorphism $\varphi: R \to C$ that sends $r \rightsquigarrow rv$. The kernel of φ, the module of relations, is a submodule of R, an ideal I. By the First Isomorphism Theorem, C is isomorphic to the R-module R/I.

When $R = F[t]$, the ideal I will be principal, and C will be isomorphic to $R/(d)$ for some polynomial d. The module of relations will be generated by a single element.

Theorem 14.8.3 Structure Theorem for Modules over Polynomial Rings. Let $R = F[t]$ be the ring of polynomials in one variable with coefficients in a field F.

(a) Let V be a finitely generated module over R. Then V is a direct sum of cyclic modules $C_1, C_2, \ldots, C_k$ and a free module L, where C_i is isomorphic to $R/(d_i)$, the elements $d_1, \ldots, d_k$ are monic polynomials of positive degree, and $d_1 \,|\, d_2 \,|\, \ldots \,|\, d_k$.

(b) The same assertion as **(a)**, except that the condition that d_i divides d_{i+1} is replaced by: Each d_i is a power of a monic irreducible polynomial.   □

It is also true that the prime powers occurring in (**b**) are unique, but we won't take the time to prove this.

For example, let $R = \mathbb{R}[t]$, and the *R-module* V presented by the matrix A of Example 14.8.2. It is also presented by the diagonal matrix

$$A' = \begin{bmatrix} 1 & 0 \\ 0 & t^3 - 3t^2 + 2t \end{bmatrix}.$$

and we can drop the first row and column from this matrix (14.5.7). So V is presented by the 1×1 matrix $[g]$, where $g(t) = t^3 - 3t^2 + 2t = t(t-1)(t-2)$. This means that V is a cyclic module, isomorphic to $C = R/(g)$. Since g has three relatively prime factors, V can be further decomposed. It is isomorphic to a direct sum of cyclic R-modules:

(14.8.4) $R/(g) \approx \big(R/(t)\big) \oplus \big(R/(t-1)\big) \oplus \big(R/(t-2)\big).$

We now apply the theory we have developed to study linear operators on vector spaces over a field. This application provides a good example of how abstraction can lead to new insights. The method developed for abelian groups is extended formally to modules over polynomial rings, and is then applied in a concrete new situation. This was not the historical development. The theories for abelian groups and for linear operators were developed independently and were tied together later. But it is striking that the two cases, abelian groups and linear operators, can end up looking so different when the same theory is applied to them.

The key observation that allows us to proceed is that if we are given a linear operator

(14.8.5) $T : V \to V$

on a vector space over a field F, we can use this operator to make V into a module over the polynomial ring $F[t]$. To do so, we must define multiplication of a vector v by a polynomial $f(t) = a_n t^n + \cdots + a_1 t + a_0$. We set

(14.8.6) $f(t)v = a_n T^n(v) + a_{n-1} T^{n-1}(v) + \cdots + a_1 T(v) + a_0 v$

The right side could also be written as $[f(T)](v)$, where $f(T)$ denotes the linear operator $a_n T^n + a_{n-1} T^{n-1} + \cdots + a_1 T + a_0 I$. (The brackets have been added to make it clear that it is the operator $f(T)$ that acts on v.) With this notation, we obtain the formulas

(14.8.7) $tv = T(v)$ and $f(t)v = [f(T)](v).$

The fact that rule (14.8.6) makes V into an $F[t]$-module is easy to verify, and the formulas (14.8.7) may appear tautological. They raise the question of why we need a new symbol t. But $f(t)$ is a polynomial, while $f(T)$ is a linear operator.

Conversely, if V is an $F[t]$-module, scalar multiplication of elements of V by a polynomial is defined. In particular, we are given a rule for multiplying by the constant polynomials, the elements of F. If we keep the rule for multiplying by constants but forget for the moment about multiplication by nonconstant polynomials, then the axioms for a module show that V becomes a vector space over F (14.1.1). Next, we can multiply elements of V by the polynomial t. Let us denote the operation of multiplication by t on V as T. So T is the map

(14.8.8) $V \xrightarrow{T} V$,   defined by $T(v) = tv$.

This map is a linear operator when V is considered as a vector space over F. By the distributive law, $t(v + v') = tv + tv'$, therefore $T(v + v') = T(v) + T(v')$. If c is a scalar, then $tcv = ctv$, and therefore $T(cv) = cT(v)$. So an $F[t]$-module V provides us with a linear operator on a vector space. The rules we have described, going from linear operators to modules and back, are inverse operations.

(14.8.9)   Linear operator on an F-vector space and
                   $F[t]$-module are equivalent concepts.

We will want to apply this observation to finite-dimensional vector spaces, but we note in passing the linear operator that corresponds to the free $F[t]$-module of rank 1. When $F[t]$ is considered as a vector space over F, the monomials $(1, t, t^2, \ldots)$ form a basis, and we can use this basis to identify $F[t]$ with the infinite-dimensional space Z, the space of infinite row vectors $(a_0, a_1, a_2, \ldots)$ with finitely many entries different from zero that was defined in (3.7.2). Multiplication by t on $F[t]$ corresponds to the *shift operator* T:

$$(a_0, a_1, a_2, \ldots) \rightsquigarrow (0, a_0, a_1, a_2, \ldots).$$

The shift operator on the space Z corresponds to the free $F[t]$-module of rank 1.

We now begin our application to linear operators. Given a linear operator T on a vector space V over F, we may also view V as an $F[t]$-module. We suppose that V is finite-dimensional as a vector space, say of dimension n. Then it is finitely generated as a module, and it has a presentation matrix. There is some danger of confusion here, because there are two matrices around: the presentation matrix for the module V, and the matrix of the linear operator T. The presentation matrix is an $r \times s$ matrix with polynomial entries, where r is the number of chosen generators for the module and s is the number of relations. The matrix of the linear operator is an $n \times n$ matrix whose entries are scalars, where n is the dimension of V. Both matrices contain the information needed to describe the module and the linear operator.

Regarding V as an $F[t]$-module, we can apply Theorem 14.8.3 to conclude that V is a direct sum of cyclic submodules, say

$$V = W_1 \oplus \cdots \oplus W_k,$$

where W_i is isomorphic to $F[t]/(f_i)$, f_i being a monic polynomial in $F[t]$. When V is finite-dimensional, the free summand is zero.

To interpret the meaning of the direct sum decomposition for the linear operator T, we choose bases $\mathbf{B}_i$ for the subspaces W_i. Then with respect to the basis $\mathbf{B} = (\mathbf{B}_1, \ldots, \mathbf{B}_k)$, the matrix of T has a block form (4.4.4), where the blocks are the matrices of T restricted to the invariant subspaces W_i. Perhaps it will be enough to examine the operator that corresponds to a cyclic module.

Let W be a cyclic $F[t]$-module, generated as a module by a single element that we label as w_0. Since every ideal of $F[t]$ is principal, W will be isomorphic to $F[t]/(f)$, where $f = t^n + a_{n-1}t^{n-1} + \cdots + a_1 t + a_0$ is a monic polynomial in $F[t]$. The isomorphism $F[t]/(f) \to W$ will send $1 \rightsquigarrow w_0$. The set $(1, t, \ldots, t^{n-1})$ is a basis of $F[t]/(f)$ (11.5.5), so the set $(w_0, tw_0, t^2 w_0, \ldots t^{n-1} w_0)$ is a basis of W as vector space.

The corresponding linear operator $T : W \to W$ is multiplication by t. Written in terms of T, the basis of W is $(w_0, w_1, \ldots w_{n-1})$, with $w_j = T^j w_0$. Then

$$T(w_0) = w_1, \quad T(w_1) = w_2 \ , \ldots, \quad T(w_{n-2}) = w_{n-1}, \text{ and}$$

$$[f(T)]w_0 = T^n w_0 + a_{n-1}T^{n-1} w_0 + \cdots + a_1 T w_0 + a_0 w_0 = 0.$$

$$= T w_{n-1} + a_{n-1} w_{n-1} + \cdots + a_1 w_1 + a_0 w_0 = 0.$$

This determines the matrix of T. It has the form illustrated below for small values of n:

(14.8.10)
$$[-a_0], \begin{bmatrix} 0 & -a_0 \\ 1 & -a_1 \end{bmatrix}, \begin{bmatrix} 0 & 0 & -a_0 \\ 1 & 0 & -a_1 \\ 0 & 1 & -a_2 \end{bmatrix}, \ldots,$$

The characteristic polynomial of this matrix is $f(t)$.

Theorem 14.8.11   Let T be a linear operator on a finite-dimensional vector space V over a field F. There is a basis for V with respect to which the matrix of T is made up of blocks of the type shown above. $\quad\square$

This form for the matrix of a linear operator is called a *rational canonical form*. It is the best available for an arbitrary field.

Example 14.8.12   Let $F = \mathbb{R}$. The matrix A shown below is in rational canonical form. Its characteristic polynomial is $t^3 - 1$. Since this is a product of relatively prime polynomials: $t^3 - 1 = (t - 1)(t^2 + t + 1)$, the cyclic $\mathbb{R}[t]$-module that it presents is a direct sum of cyclic modules. The matrix A' is another rational canonical form that describes the same module. Over the complex numbers, A is diagonalizable. Its diagonal form is A'', where $\omega = e^{2\pi i/3}$.

(14.8.13)
$$A = \begin{bmatrix} 0 & 0 & 1 \\ 1 & 0 & 0 \\ 0 & 1 & 0 \end{bmatrix}, \ A' = \left[\begin{array}{c|cc} 1 & & \\ \hline & 0 & -1 \\ & 1 & -1 \end{array}\right], \ A'' = \begin{bmatrix} 1 & & \\ & \omega & \\ & & \omega^2 \end{bmatrix}$$

$\quad\square$

Various relations between properties of an $F[t]$-module and the corresponding linear operator are summed up in the table below.

(14.8.14)

$F[t]$-**module**	**Linear operator** T
multiplication by t	operation of T
free module of rank 1	shift operator
submodule	T-invariant subspace
direct sum of submodules	direct sum of T-invariant subspaces
cyclic module generated by w	subspace spanned by w, $T(w)$, $T^2(w)$, $\ldots$

14.9   POLYNOMIAL RINGS IN SEVERAL VARIABLES

Modules over a ring become increasingly complicated with increasing complication of the ring, and it can be difficult to determine whether or not an explicitly presented module is free. In this section we describe, without proof, a theorem that characterizes free modules over polynomial rings in several variables. This theorem was proved by Quillen and Suslin in 1976.

Let $R = \mathbb{C}[x_1, \ldots, x_k]$ be the polynomial ring in k variables, and let V be a finitely generated R-module. Let A be a presentation matrix for V. The entries of A will be polynomials $a_{ij}(x)$, and if A is an $m \times n$ matrix, then V is isomorphic to the cokernel $R^m / A R^n$ of multiplication by A on R-vectors.

When we evaluate the matrix entries $a_{ij}(x)$ at a point $(c_1, \ldots, c_k)$ of $\mathbb{C}^k$, we obtain a complex matrix $A(c)$ whose i, j-entry is $a_{ij}(c)$.

Theorem 14.9.1   Let V be a finitely generated module over the polynomial ring $\mathbb{C}[x_1, \ldots, x_k]$, and let A be an $m \times n$ presentation matrix for V. Denote by $A(c)$ the evaluation of A at a point c of $\mathbb{C}^k$. Then V is a free module of rank r if and only if the matrix $A(c)$ has rank $m - r$ at every point c.

The proof of this theorem requires too much background to give here. However, we can use it to determine whether or not a given module is free. For example, let V be the module over $\mathbb{C}[x, y]$ presented by the 4×2 matrix

(14.9.2)
$$A = \begin{bmatrix} 1 & x \\ y & x+3 \\ x & y \\ x^2 & y^2 \end{bmatrix}$$

So V has four generators, say $v_1, \ldots, v_4$, and two relations:

$$v_1 + yv_2 + xv_3 + x^2v_4 = 0 \quad \text{and} \quad xv_1 + (x+3)v_2 + yv_3 + y^2v_4 = 0.$$

It isn't very hard to show that $A(c)$ has rank 2 for every point c in $\mathbb{C}^2$. Theorem 14.9.1 tells us that V is a free module of rank 2.

One can get an intuitive understanding for this theorem by considering the vector space $W(c)$ spanned by the columns of the matrix $A(c)$. It is a subspace of $\mathbb{C}^m$. As c varies in the space $\mathbb{C}^k$, the matrix $A(c)$ varies continuously. Therefore the subspace $W(c)$ will also vary continuously, provided that its dimension does not jump around. Continuous families of vector spaces of constant dimension, parametrized by a topological space $\mathbb{C}^k$, are called *vector bundles* over $\mathbb{C}^k$. The module V is free if and only if the family of vector spaces $W(c)$ forms a vector bundle.

> *"Par une déformation coutumière aux mathématiciens,*
> *je me'en tenais au point de vue trop restreint.*
>
> —Jean-Louis Verdier

EXERCISES

Section 1 Modules

1.1. Let R be a ring, and let V denote the R-module R. Determine all homomorphisms $\varphi : V \to V$.

1.2. Let V be an abelian group. Prove that if V has a structure of $\mathbb{Q}$-module with its given law of composition as addition, then that structure is uniquely determined.

1.3. Let $R = \mathbb{Z}[\alpha]$ be the ring generated over $\mathbb{Z}$ by an algebraic integer α. Prove that for any integer m, R/mR is finite, and determine its order.

1.4. A module is called *simple* if it is not the zero module and if it has no proper submodule.

   (a) Prove that any simple R-module is isomorphic to an R-module of the form R/M, where M is a maximal ideal.

   (b) Prove *Schur's Lemma*: Let $\varphi : S \to S'$ be a homomorphism of simple modules. Then φ is either zero, or an isomorphism.

Section 2 Free Modules

2.1. Let $R = \mathbb{C}[x, y]$, and let M be the ideal of R generated by the two elements x and y. Is M a free R-module?

2.2. Prove that a ring R having the property that every finitely generated R-module is free is either a field or the zero ring.

2.3. Let A be the matrix of a homomorphism $\varphi : \mathbb{Z}^n \to \mathbb{Z}^m$ of free $\mathbb{Z}$-modules.

   (a) Prove that φ is injective if and only if the rank of A, as a real matrix, is n.

   (b) Prove that φ is surjective if and only if the greatest common divisor of the determinants of the $m \times m$ minors of A is 1.

2.4. Let I be an ideal of a ring R.

   (a) Under what circumstances is I a free R-module?

   (b) Under what circumstances is the quotient ring R/I a free R-module?

Section 3 Identities

3.1. Let $\tilde{f}$ denote the function on $\mathbb{C}^n$ defined by evaluation of a (formal) complex polynomial $f(x_1, \ldots, x_n)$. Prove that if $\tilde{f}$ is the zero function, then f is the zero polynomial.

3.2. It might be convenient to verify an identity only for the real numbers. Would this suffice?

3.3. Let A and B be $m \times m$ and $n \times n$ R-matrices, respectively. Use permanence of identities to prove that trace of the linear operator $f(M) = AMB$ on the space $R^{m \times n}$ is the product $(\text{trace } A)(\text{trace } B)$.

3.4. In each case, decide whether or not permanence of identities allows the result to be carried over from the complex numbers to an arbitrary commutative ring.

   (a) the associative law for matrix multiplication,

   (b) the Cayley-Hamilton Theorem,

(c) Cramer's Rule,

(d) the product rule, quotient rule, and chain rule for differentiation of polynomials,

(e) the fact that a polynomial of degree n has at most n roots,

(f) Taylor expansion of a polynomial.

Section 4   Diagonalizing Integer Matrices

4.1. (a) Reduce each matrix to diagonal form by integer row and column operations.

$$\begin{bmatrix} 3 & 1 \\ -1 & 2 \end{bmatrix}, \quad \begin{bmatrix} 4 & 7 & 2 \\ 2 & 4 & 6 \end{bmatrix}, \quad \begin{bmatrix} 3 & 1 & -4 \\ 2 & -3 & 1 \\ -4 & 6 & -2 \end{bmatrix}$$

(b) For the first matrix, let $V = \mathbb{Z}^2$ and let $L = AV$. Draw the sublattice L, and find bases of V and L that exhibit the diagonalization.

(c) Determine integer matrices Q^{-1} and P that diagonalize the second matrix.

4.2. Let $d_1, d_2, \ldots$ be the integers referred to in Theorem 14.4.6. Prove that d_1 is the greatest common divisor of the entries a_{ij} of A.

4.3. Determine all integer solutions to the system of equations $AX = 0$, when $A = \begin{bmatrix} 4 & 7 & 2 \\ 2 & 4 & 6 \end{bmatrix}$. Find a basis for the space of integer column vectors B such that $AX = B$ has a solution.

4.4. Find a basis for the $\mathbb{Z}$-module of integer solutions of the system of equations $x + 2y + 3z = 0, x + 4y + 9z = 0$.

4.5. Let α, β, γ be complex numbers. Under what conditions is the set of integer linear combinations $\{\ell\alpha + m\beta + n\gamma \mid \ell, m, n, \in \mathbb{Z}\}$ a lattice in the complex plane?

4.6. Let $\varphi : \mathbb{Z}^k \to \mathbb{Z}^k$ be a homomorphism given by multiplication by an integer matrix A. Show that the image of φ is of finite index if and only if A is nonsingular and that if so, then the index is equal to $|\det A|$.

4.7. Let $A = (a_1, \ldots, a_n)^t$ be an integer column vector, and let d be the greatest common divisor of $a_1, \ldots, a_n$. Prove that there is a matrix $P \in GL_n(\mathbb{Z})$ such that $PA = (d, 0, \ldots, 0)^t$.

4.8. Use invertible row and column operations in the ring $\mathbb{Z}[i]$ of Gauss integers to diagonalize the matrix $\begin{bmatrix} 3 & 2+i \\ 2-i & 9 \end{bmatrix}$.

4.9. Use diagonalization to prove that if $L \subset M$ are lattices, then $[M:L] = \dfrac{\Delta(L)}{\Delta(M)}$.

Section 5   Generators and Relations

5.1. Let $R = \mathbb{Z}[\delta]$, where $\delta = \sqrt{-5}$. Determine a presentation matrix as R-module for the ideal $(2, 1 + \delta)$.

5.2. Identify the abelian group presented by the matrix $\begin{bmatrix} 3 & 1 & 2 \\ 1 & 1 & 1 \\ 2 & 3 & 6 \end{bmatrix}$.

Section 6   Noetherian Rings

6.1. Let $V \subset \mathbb{C}^n$ be the locus of common zeros of an infinite set of polynomials $f_1, f_2, f_3, \ldots$. Prove that there is a finite subset of these polynomials whose zeros define the same locus.

6.2. Find an example of a ring R and an ideal I of R that is not finitely generated.

Section 7   Structure of Abelian Groups

7.1. Find a direct sum of cyclic groups isomorphic to the abelian group presented by the matrix
$$\begin{bmatrix} 2 & 2 & 2 \\ 2 & 2 & 0 \\ 2 & 0 & 2 \end{bmatrix}.$$

7.2. Write the abelian group generated by x and y, with the relation $3x + 4y = 0$ as a direct sum of cyclic groups.

7.3. Find an isomorphic direct product of cyclic groups, when V is the abelian group generated by x, y, z, with the given relations.

(a) $3x + 2y + 8z = 0, 2x + 4z = 0$
(b) $x + y = 0, 2x = 0, 4x + 2z = 0, 4x + 2y + 2z = 0$
(c) $2x + y = 0, x - y + 3z = 0$
(d) $7x + 5y + 2z = 0, 3x + 3y = 0, 13x + 11y + 2z = 0$

7.4. In each case, identify the abelian group that has the given presentation matrix:
$$\begin{bmatrix} 2 \\ 1 \end{bmatrix}, \begin{bmatrix} 0 \\ 5 \end{bmatrix}, [2 \quad 0 \quad 0], \begin{bmatrix} 1 & 0 \\ 0 & 1 \\ 0 & 0 \end{bmatrix}, \begin{bmatrix} 2 & 3 \\ 1 & 2 \end{bmatrix}, \begin{bmatrix} 2 & 4 \\ 1 & 4 \end{bmatrix}, \begin{bmatrix} 2 & 4 \\ 6 & 4 \end{bmatrix}, \begin{bmatrix} 4 & 6 \\ 2 & 3 \end{bmatrix}.$$

7.5. Determine the number of isomorphism classes of abelian groups of order 400.

7.6. (a) Let a and b be relatively prime positive integers. By manipulating the diagonal matrix with diagonal entries a and b, prove that the cyclic group C_{ab} is isomorphic to the product $C_a \oplus C_b$.

(b) What can you say if the assumption that a and b are relatively prime is dropped?

7.7. Let $R = \mathbb{Z}[i]$ and let V be the R-module generated by elements v_1 and v_2 with relations $(1 + i)v_1 + (2 - i)v_2 = 0, 3v_1 + 5iv_2 = 0$. Write this module as a direct sum of cyclic modules.

7.8. Let $F = \mathbb{F}_p$. For which prime integers p does the additive group F^1 have a structure of $\mathbb{Z}[i]$-module? How about F^2?

7.9. Show that the following concepts are equivalent:

- R-module, where $R = \mathbb{Z}[i]$,
- abelian group V, with a homomorphism $\varphi: V \to V$ such that $\varphi \circ \varphi = -identity$.

Section 8   Application to Linear Operators

8.1. Let T be the linear operator on $\mathbb{C}^2$ whose matrix is $\begin{bmatrix} 2 & 1 \\ 0 & 1 \end{bmatrix}$. Is the corresponding $\mathbb{C}[t]$-module cyclic?

8.2. Let M be a $\mathbb{C}[t]$-module the form $\mathbb{C}[t]/(t-\alpha)^n$. Show that there is a $\mathbb{C}$-basis for M, such that the matrix of the corresponding linear operator is a Jordan block.

8.3. Let $R = F[x]$ be the polynomial ring in one variable over a field F, and let V be the R-module generated by an element v that satisfies the relation $(t^3 + 3t + 2)v = 0$. Choose a basis for V as F-vector space, and determine the matrix of the operator of multiplication by t with respect to this basis.

8.4. Let V be an $F[t]$-module, and let $\mathbf{B} = (v_1, \ldots, v_n)$ be a basis for V as F-vector space. Let B be the matrix of T with respect to this basis. Prove that $A = tI - B$ is a presentation matrix for the module.

8.5. Prove that the characteristic polynomial of the matrix (14.8.10) is $f(t)$.

8.6. Classify finitely generated modules over the ring $\mathbb{C}[\epsilon]$, where $\epsilon^2 = 0$.

Section 9   Polynomial Rings in Several Variables

9.1. Determine whether or not the modules over $\mathbb{C}[x, y]$ presented by the following matrices are free.

$$\text{(a)}\begin{bmatrix} x^2+1 & x \\ x^2y+x+y & xy+1 \end{bmatrix}, \quad \text{(b)}\begin{bmatrix} xy-1 \\ x^2-y^2 \\ y \end{bmatrix}, \quad \text{(c)}\begin{bmatrix} x-1 & x \\ y & y+1 \\ x & y \\ x^2 & 2y \end{bmatrix}$$

9.2. Prove that the module presented by (14.9.2) is free by exhibiting a basis.

9.3. Following the model of the polynomial ring in one variable, describe modules over the ring $\mathbb{C}[x, y]$ in terms of complex vector spaces with additional structure.

9.4. Prove the easy half of the theorem of Quillen and Suslin: If V is free, then the rank of $A(c)$ is constant.

9.5. Let $R = \mathbb{Z}[\sqrt{-5}]$, and let V be the module presented by the matrix $A = \begin{bmatrix} 2 \\ 1+\delta \end{bmatrix}$. Prove that the residue of A in R/P has rank 1 for every prime ideal P of R, but that V is not a free module.

Miscellaneous Problems

M.1. In how many ways can the additive group $\mathbb{Z}/5\mathbb{Z}$ be given the structure of a module over the Gauss integers?

M.2. Classify finitely generated modules over the ring $\mathbb{Z}/(6)$.

M.3. Let A be a finite abelian group, and let $\varphi : A \to \mathbb{C}^\times$ be a homomorphism that is not the trivial homomorphism. Prove that $\sum_{a \in A} \varphi(a) = 0$.

M.4. When an integer 2×2 matrix A is diagonalized by $Q^{-1}AP$, how unique are the matrices P and Q?

M.5. Which matrices A in $GL_2(\mathbb{R})$ stabilize some lattice L in $\mathbb{R}^2$?

M.6. (a) Describe the orbits of right multiplication by $G = GL_2(\mathbb{Z})$ on the space of 2×2 integer matrices.

(b) Show that for any integer matrix A, there is an invertible integer matrix P such that AP has the following *Hermitian normal form*:

$$\begin{bmatrix} d_1 & 0 & 0 & 0 & \cdots \\ a_2 & d_2 & 0 & 0 & \\ a_3 & b_3 & d_3 & 0 & \\ \vdots & & & & \ddots \end{bmatrix},$$

where the entries are nonnegative, $a_2 < d_2, a_3, b_3 < d_3$, etc.

M.7. Let S be a subring of the polynomial ring $R = \mathbb{C}[t]$ that contains $\mathbb{C}$ and is not equal to $\mathbb{C}$. Prove that R is a finitely generated S-module.

*M.8. (a) Let α be a complex number, and let $\mathbb{Z}[\alpha]$ be the subring of $\mathbb{C}$ generated by α. Prove that α is an algebraic integer if and only if $\mathbb{Z}[\alpha]$ is a finitely generated abelian group.

(b) Prove that if α and β are algebraic integers, then the subring $\mathbb{Z}[\alpha, \beta]$ of $\mathbb{C}$ that they generate is a finitely generated abelian group.

(c) Prove that the algebraic integers form a subring of $\mathbb{C}$.

*M.9. Consider the Euclidean space $\mathbb{R}^k$, with dot product $(v \cdot w)$. A *lattice* L in V is a discrete subgroup of V^+ that contains k independent vectors. If L is a lattice, define $L^* = \{w \mid (v \cdot w) \in \mathbb{Z} \text{ for all } v \in L\}$.

(a) Show that L has a lattice basis $\mathbf{B} = (v_1, \ldots, v_k)$, a set of k vectors that spans L as $\mathbb{Z}$-module.

(b) Show that L^* is a lattice, and describe how one can determine a lattice basis for L^* in terms of $\mathbf{B}$.

(c) Under what conditions is L a sublattice of L^*?

(d) Suppose that $L \subset L^*$. Find a formula for the index $[L^*:L]$.

*M.10. (a) Prove that the multiplicative group $\mathbb{Q}^\times$ of rational numbers is isomorphic to the direct sum of a cyclic group of order 2 and a free abelian group with countably many generators.

(b) Prove that the additive group $\mathbb{Q}^+$ of rational numbers is not a direct sum of two proper subgroups.

(c) Prove that the quotient group $\mathbb{Q}^+/\mathbb{Z}^+$ is not a direct sum of cyclic groups.

C H A P T E R   15

Fields

Our difficulty is not in the proofs, but in learning what to prove.

—Emil Artin

15.1 EXAMPLES OF FIELDS

Much of the theory of fields has to do with a pair $F \subset K$ of fields, one contained in the other. Given such a pair, K is called a *field extension* of F, or an *extension field*. The notation K/F will indicate that K is a field extension of F.

Here are the three most important classes of fields.

Number Fields

A number field K is a subfield of $\mathbb{C}$.

Any subfield of $\mathbb{C}$ contains the field $\mathbb{Q}$ of rational numbers, so it is a field extension of $\mathbb{Q}$. The number fields most commonly studied are *algebraic number fields*, all of whose elements are algebraic numbers. We studied quadratic number fields in Chapter 13.

Finite Fields

A finite field is a field that contains finitely many elements.

A finite field contains one of the prime fields $\mathbb{F}_p$, and therefore it is an extension of that field. Finite fields are described in Section 15.7.

Function Fields

Extensions of the field $F = \mathbb{C}(t)$ of rational functions are called function fields.

A function field can be defined by an equation $f(t, x) = 0$, where f is an irreducible complex polynomial in the variables t and x, such as $f(t, x) = x^2 - t^3 + t$, for example. We may use the equation $f(t, x) = 0$ to define x "implicitly" as a function $x(t)$ of t, as we learn to do in calculus. In our example, this function is $x(t) = \sqrt{t^3 - t}$. The corresponding function field K consists of the combinations $p + q\sqrt{t^3 - t}$, where p and q are rational functions in t. One can work in this field just as one would in a field such as $\mathbb{Q}(\sqrt{-5})$. For most polynomials $f(t, x)$, there won't be an explicit expression for the implicitly defined function $x(t)$, but by definition, it satisfies the equation $f(t, x(t)) = 0$. We will see in Section 15.9 that $x(t)$ defines an extension field of F.

15.2   ALGEBRAIC AND TRANSCENDENTAL ELEMENTS

Let K be an extension of a field F, and let α be an element of K. By analogy with the definition of algebraic numbers (11.1), α is *algebraic over* F if it is a root of a monic polynomial with coefficients in F, say

(15.2.1) $$f(x) = x^n + a_{n-1}x^{n-1} + \cdots + a_0, \text{ with } a_i \text{ in } F,$$

and $f(\alpha) = 0$. An element is *transcendental over* F if it is not algebraic over F – if it is not a root of any such polynomial.

These properties, algebraic and transcendental, depend on F. The complex number $2\pi i$ is algebraic over the field of real numbers but transcendental over the field of rational numbers. Every element α of a field K is algebraic over K, because it is the root of the polynomial $x - \alpha$, which has coefficients in K.

The two possibilities for α can be described in terms of the substitution homomorphism

(15.2.2) $$\varphi: F[x] \to K, \text{ defined by } x \rightsquigarrow \alpha.$$

An element α is transcendental over F if φ is injective, and algebraic over F if φ is not injective, that is, if the kernel of φ is not zero. We won't have much to say about the case that α is transcendental.

Suppose that α is algebraic over F. Since $F[x]$ is a principal ideal domain, the kernel of φ is a principal ideal, generated by a monic polynomial $f(x)$ with coefficients in F. This polynomial can be described in various ways.

Proposition 15.2.3 Let α be an element of an extension field K of a field F that is algebraic over F. The following conditions on a monic polynomial f with coefficients in F are equivalent. The unique monic polynomial that satisfies these conditions is called the *irreducible polynomial for α over F* .

- f is the monic polynomial of lowest degree in $F[x]$ that has α as a root.
- f is an irreducible element of $F[x]$, and α is a root of f.
- f has coefficients in F, α is a root of f, and the principal ideal of $F[x]$ that is generated by f is a maximal ideal.
- α is a root of f, and if g is any polynomial in $F[x]$ that has α as a root, then f divides g.   □

The degree of the irreducible polynomial for α over F is called the *degree of α over F*.

It is important to keep in mind that the irreducible polynomial f depends on F as well as on α, because irreducibility of a polynomial depends on the field. The irreducible polynomial for $\sqrt{i}$ over $\mathbb{Q}$ is $x^4 + 1$, but this polynomial factors in the field $\mathbb{Q}(i)$. The irreducible polynomial for $\sqrt{i}$ over $\mathbb{Q}(i)$ is $x^2 - i$. When there are several fields around, it is ambiguous to say that a polynomial is irreducible. It is better to say that f is *irreducible over* F, or that it is an *irreducible element of* $F[x]$.

Let K be an extension field of F. The subfield of K generated by an element α of K will be denoted by $F(\alpha)$:

(15.2.4) $F(\alpha)$ is the smallest subfield of K that contains F and α.

Similarly, if $\alpha_1, \ldots, \alpha_k$ are elements of an extension field K of F, the notation $F(\alpha_1, \ldots, \alpha_k)$ will stand for the smallest subfield of K that contains these elements and F.

As in Chapter 11, we denote the *ring* generated by α over F by $F[\alpha]$. It is the image of the map $\varphi : F[x] \to K$ defined above, and it consists of the elements β of K that can be expressed as polynomials in α with coefficients in F:

(15.2.5) $$\beta = b_n \alpha^n + \cdots + b_1 \alpha + b_0, \quad b_i \text{ in } F.$$

The field $F(\alpha)$ is isomorphic to the field of fractions of $F[\alpha]$. Its elements are ratios of elements of the form (15.2.5) (see Section 11.7).

Similarly, if $\alpha_1, \ldots, \alpha_k$ are elements of K, the smallest subring of K that contains F and these elements is denoted by $F[\alpha_1, \ldots, \alpha_k]$. It consists of the elements β of K that can be expressed as polynomials in the α_i with coefficients in F. The field $F(\alpha_1, \ldots, \alpha_k)$ is the field of fractions of the ring $F[\alpha_1, \ldots, \alpha_k]$.

If an element α of F is transcendental over F, the map $F[x] \to F[\alpha]$ is an isomorphism. In that case $F(\alpha)$ is isomorphic to the field $F(x)$ of rational functions. The field extensions $F(\alpha)$ are isomorphic for all transcendental elements α.

Things are different when α is algebraic:

Proposition 15.2.6 Let α be an element of an extension field K/F which is algebraic over F, and let f be the irreducible polynomial for α over F.

(a) The canonical map $F[x]/(f) \to F[\alpha]$ is an isomorphism, and $F[\alpha]$ is a field. Thus $F[\alpha] = F(\alpha)$.

(b) More generally, let $\alpha_1, \ldots, \alpha_k$ be elements of an extension field K/F, which are algebraic over F. The ring $F[\alpha_1, \ldots, \alpha_k]$ is equal to the field $F(\alpha_1, \ldots, \alpha_k)$.

Proof. **(a)** Let $\varphi : F[x] \to K$ be the map (15.2.2). Since the ideal (f) is maximal, $f(x)$ generates the kernel, and $F[x]/(f)$ is isomorphic to the image of φ, which is $F[\alpha]$. Moreover, $F[x]/(f)$ is a field, and therefore $F[\alpha]$ is a field. Since $F(\alpha)$ is the fraction field of $F[\alpha]$, it is equal to $F[\alpha]$.

(b) This follows by induction:

$$F[\alpha_1, \ldots, \alpha_k] = F[\alpha_1, \ldots, \alpha_{k-1}][\alpha_k] = F(\alpha_1, \ldots, \alpha_{k-1})[\alpha_k] = F(\alpha_1, \ldots, \alpha_n). \quad \square$$

The next proposition is a special case of Proposition 11.5.5.

Proposition 15.2.7 Let α be an algebraic element over F, and let $f(x)$ be the irreducible polynomial for α over F. If $f(x)$ has degree n, i.e., if α has degree n over F, then $(1, \alpha, \ldots, \alpha^{n-1})$ is a basis for $F(\alpha)$ as a vector space over F. $\square$

For instance, the irreducible polynomial for $\omega = e^{2\pi i/3}$ over $\mathbb{Q}$ is $x^2 + x + 1$. The degree of ω over $\mathbb{Q}$ is 2, and $(1, \omega)$ is a basis for $\mathbb{Q}(\omega)$ over $\mathbb{Q}$.

It may not be easy to tell whether two algebraic elements α and β generate isomorphic field extensions, though Proposition 15.2.7 provides a *necessary* condition: They must have the same degree over F, because the degree of α over F is the dimension of $F(\alpha)$ as an F-vector space. This is obviously not a sufficient condition. All of the imaginary quadratic fields studied in Chapter 13 are obtained by adjoining elements of degree 2 over $\mathbb{Q}$, but they aren't isomorphic.

On the other hand, if α is a complex root of $x^3 - x + 1$, then $\beta = \alpha + 1$ is a root of $x^3 - 3x^2 + 2x + 1$. The fields $\mathbb{Q}(\alpha)$ and $\mathbb{Q}(\beta)$ are the same. If we were presented only with the two polynomials, it might take some time to notice how they are related.

What we can describe easily are the circumstances under which there is an isomorphism $F(\alpha) \rightarrow F(\beta)$ that fixes F and sends α to β. The next proposition, though very simple, is fundamental to our understanding of field extensions.

Proposition 15.2.8 Let F be a field, and let α and β be elements of field extensions K/F and L/F. Suppose that α and β are algebraic over F. There is an isomorphism of fields $\sigma: F(\alpha) \rightarrow F(\beta)$ that is the identity on F and that sends $\alpha \rightsquigarrow \beta$ if and only if the irreducible polynomials for α and β over F are equal.

Proof. Since α is algebraic over F, $F[\alpha] = F(\alpha)$, and similarly, $F[\beta] = F(\beta)$. Suppose that the irreducible polynomials for α and for β over F are both equal to f. Proposition 15.2.6 tells us that there are isomorphisms

$$F[x]/(f) \overset{\varphi}{\rightarrow} F[\alpha] \quad \text{and} \quad F[x]/(f) \overset{\psi}{\rightarrow} F[\beta].$$

The composed map $\sigma = \psi\varphi^{-1}$ is the required isomorphism $F(\alpha) \rightarrow F(\beta)$. Conversely, if there is an isomorphism σ that is the identity on F and that sends α to β, and if $f(x)$ is a polynomial with coefficients in F such that $f(\alpha) = 0$, then $f(\beta) = 0$ too. (See Proposition 15.2.10 below.) So the irreducible polynomials for the two elements are equal. ☐

For instance, let α_1 denote the real cube root of 2, and let $\omega = e^{2\pi i/3}$ be a complex cube root of 1. The three complex roots of $x^3 - 2$ are α_1, $\alpha_2 = \omega\alpha$ and $\alpha_3 = \omega^2\alpha$. Therefore there is an isomorphism $\mathbb{Q}(\alpha_1) \overset{\sim}{\rightarrow} \mathbb{Q}(\alpha_2)$ that sends α_1 to α_2. In this case the elements of $\mathbb{Q}(\alpha_1)$ are real numbers, but α_2 is not a real number. To understand this isomorphism, we must look only at the internal algebraic structure of the fields.

Definition 15.2.9 Let K and K' be extensions of the same field F. An isomorphism $\varphi: K \rightarrow K'$ that restricts to the identity on the subfield F is called an *F-isomorphism*, or an *isomorphism of field extensions*. If there exists an F-isomorphism $\varphi: K \rightarrow K'$, K and K' are *isomorphic* extension fields.

The next proposition was proved for complex conjugation before (12.2.19).

Proposition 15.2.10 Let $\varphi: K \rightarrow K'$ be an isomorphism of field extensions of F, and let f be a polynomial with coefficients in F. Let α be a root of f in K, and let $\alpha' = \varphi(\alpha)$ be its image in K'. Then α' is also a root of f.

Proof. Say that $f(x) = a_n x^n + \cdots + a_1 x + a_0$. Since φ is an F-isomorphism and since a_i are in F, $\varphi(a_i) = a_i$. Since φ is a homomorphism,

$$0 = \varphi(0) = \varphi(f(\alpha)) = \varphi(a_n \alpha^n + \cdots + a_1 \alpha + a_0)$$
$$= \varphi(a_n)\varphi(\alpha)^n + \cdots + \varphi(a_1)\varphi(\alpha) + \varphi(a_0) = a_n \alpha'^n + \cdots + a_1 \alpha' + a_0.$$

Therefore α' is a root of f. $\square$

15.3   THE DEGREE OF A FIELD EXTENSION

A field extension K of F can always be regarded as an F-vector space. Addition is the addition law in K, and scalar multiplication of an element of K by an element of F is obtained by multiplying these two elements in K. The dimension of K, when regarded as an F-vector space, is called the *degree* of the field extension. This degree, which is denoted by $[K:F]$, is a basic property of a field extension.

(15.3.1) $[K:F]$ is the dimension of K, as an F-vector space.

For example, $\mathbb{C}$ has the $\mathbb{R}$-basis $(1, i)$, so the degree $[\mathbb{C}:\mathbb{R}]$ is 2.

A field extension K/F is a *finite extension* if its degree is finite. Extensions of degree 2 are *quadratic* extensions, those of degree 3 are *cubic* extensions, and so on.

Lemma 15.3.2

(a) A field extension K/F has degree 1 if and only if $F = K$.
(b) An element α of a field extension K has degree 1 over F if and only if α is an element of F.

Proof. **(a)** If the dimension of K as vector space over F is 1, any nonzero element of K, including 1, will be an F-basis, and if 1 is a basis, every element of K is in F.

(b) By definition, the degree of α over F is the degree of the (monic) irreducible polynomial for α over F. If α has degree 1, then this polynomial must be $x - \alpha$, and if $x - \alpha$ has coefficients in F, then α is in F. $\square$

Proposition 15.3.3 Assume that the field F does not have characteristic 2, that is, $1 + 1 \neq 0$ in F. Then any extension K of degree 2 over F can be obtained by adjoining a square root: $K = F(\delta)$, where $\delta^2 = d$ is an element of F. Conversely, if δ is an element of a field extension of F, and if δ^2 is in F but δ is not in F, then $F(\delta)$ is a quadratic extension of F.

It is not true that all cubic extensions can be obtained by adjoining a cube root. We learn more about this point in the next chapter (see Section 16.11).

Proof. We first show that every quadratic extension K can be obtained by adjoining a root of a quadratic polynomial $f(x)$ with coefficients in F. We choose an element α of K that is not in F. Then $(1, \alpha)$ is a linearly independent set over F. Since K has dimension 2 as a vector space over F, this set is a basis for K. It follows that α^2 is a linear combination of

$(1, \alpha)$ with coefficients in F. We write this linear combination as $\alpha^2 = -b\alpha - c$. Then α is a root of $f(x) = x^2 + bx + c$, and since α is not in F, this polynomial is irreducible over F. This much is also true when the characteristic is 2.

The discriminant of the quadratic polynomial f is $D = b^2 - 4c$. In a field of characteristic not 2, the quadratic formula $\frac{1}{2}(-b + \sqrt{D})$ solves the equation $x^2 + bx + c = 0$. This is proved by substituting into the polynomial. There are two choices for the square root, let δ be one of them. Then δ is in K, δ^2 is in F, and because α is in the field $F(\delta)$, δ generates K over F. Conversely, if δ^2 is in F but δ is not in F, then $(1, \delta)$ will be an F-basis for $F(\delta)$, so $[F(\delta):F] = 2$. $\square$

The term *degree* comes from the case that K is generated by one algebraic element α: $K = F(\alpha)$. This is the first important property of the degree:

Proposition 15.3.4

(a) If an element α of an extension field is algebraic over F, the degree $[F(\alpha):F]$ of $F(\alpha)$ over F is equal to the degree of α over F.

(b) An element α of an extension field is algebraic over F if and only if the degree $[F(\alpha):F]$ is finite.

Proof. If α is algebraic over F, then by definition, its degree over F is equal to the degree of its irreducible polynomial f over F. And if f has degree n, then $F(\alpha)$ has the F-basis $(1, \alpha, \ldots, \alpha^{n-1})$ (Proposition 15.2.7), so $[F(\alpha):F] = n$. If α is not algebraic, then $F[\alpha]$ and $F(\alpha)$ have infinite dimension over F. $\square$

The second important property relates degrees in chains of field extensions.

Theorem 15.3.5 Multiplicative Property of the Degree. Let $F \subset K \subset L$ be fields. Then $[L:F] = [L:K][K:F]$. Therefore both $[L:K]$ and $[K:F]$ divide $[L:F]$.

Proof. Let $\mathbf{B} = (\beta_1, \ldots, \beta_n)$ be a basis for L as a K-vector space, and let $\mathbf{A} = (\alpha_1, \ldots, \alpha_m)$ be a basis for K as F-vector space. So $[L:K] = n$ and $[K:F] = m$. To prove the theorem, we show that the set of mn products $\mathbf{P} = \{\alpha_i \beta_j\}$ is a basis of L as F-vector space. The reasoning in case one of the degrees is infinite is similar.

Let γ be an element of L. Since $\mathbf{B}$ is a basis for L over K, γ can be expressed uniquely as a linear combination $b_1 \beta_1 + \cdots + b_n \beta_n$, with coefficients b_j in K. Since $\mathbf{A}$ is a basis for K over F, each b_j can be expressed uniquely as a linear combination $a_{1j}\alpha_1 + \cdots + a_{mj}\alpha_m$, with coefficients a_{ij} in F. Then $\gamma = \sum_{i,j} a_{ij}\alpha_i\beta_j$. This shows that $\mathbf{P}$ spans L as an F-vector space. If a linear combination $\sum_{i,j} a_{ij}\alpha_i\beta_j$ is zero, then because $\mathbf{B}$ is a basis for L as K-vector space, the coefficient $\sum_i a_{ij}\alpha_i$ of β_j is zero for every j. This being so, a_{ij} is zero for every i and every j because $\mathbf{A}$ is a basis for K over F. Therefore $\mathbf{P}$ is independent, and hence it is a basis for L over F. $\square$

Corollary 15.3.6

(a) Let $F \subset K$ be a finite field extension of degree n, and let α be an element of K. Then α is algebraic over F, and its degree over F divides n.

(b) Let $F \subset F' \subset L$ be fields. If an element α of L is algebraic over F, it is algebraic over F'. If α has degree d over F, its degree over F' is at most d.

(c) A field extension K that is generated over F by finitely many algebraic elements is a finite extension. A finite extension is generated by finitely many elements.

(d) If K is an extension field of F, the set of elements of K that are algebraic over F is a subfield of K.

Proof. **(a)** The element α generates an intermediate field $F \subset F(\alpha) \subset K$, and the multiplicative property states that $[K:F] = [K:F(\alpha)][F(\alpha):F]$. Therefore $[F(\alpha):F]$ is finite, and it divides $[K:F]$.

(b) Let f denote the irreducible polynomial for α over F. Since $F \subset F'$, f is also an element of $F'[x]$. Since α is a root of f, the irreducible polynomial g for α over F' divides f. So the degree of g is at most equal to the degree of f.

(c) Let $\alpha_1, \ldots, \alpha_k$ be elements that generate K and are algebraic over F, and let F_i denote the field $F(\alpha_1, \ldots, \alpha_i)$ generated by the first i of the elements. These fields form a chain $F = F_0 \subset F_1 \subset \cdots \subset F_k = K$. Since α_i is algebraic over F, it is also algebraic over the larger field F_{i-1}. Therefore the degree $[F_i : F_{i-1}]$ is finite for every i. By the multiplicative property, $[K:F]$ is finite. The second assertion is obvious.

(d) We must show that if α and β are elements of K that are algebraic over F, then $\alpha + \beta$, $\alpha\beta$, etc., are algebraic over F. This follows from **(a)** and **(c)** because they are elements of the field $F(\alpha, \beta)$. $\square$

Corollary 15.3.7 Let K be an extension field of F of prime degree p. If an element α of K is not in F, then α has degree p over F and $K = F(\alpha)$. $\square$

Corollary 15.3.8 Let $\mathcal{K}$ be an extension field of a field F, let K and F' be subfields of $\mathcal{K}$ that are finite extensions of F, and let K' denote the subfield of $\mathcal{K}$ generated by the two fields K and F' together. Let $[K':F] = N$, $[K:F] = m$ and $[F':F] = n$. Then m and n divide N, and $N \leq mn$.

Proof. The multiplicative property shows that m and n divide N. Next, suppose that F' is generated over F by one element: $F' = F(\beta)$ for some element β. Then $K' = K(\beta)$. Corollary 15.3.6**(b)** shows that the degree of β over K, which is equal to $[K':K]$, is at most equal to the degree of β over F, which is n. The multiplicative property shows that $N \leq mn$. The case that F is generated by several elements follows by induction, when one adjoins one element at a time. $\square$

The diagram below sums up the corollary:

(15.3.9)

$$
\begin{array}{ccc}
 & K' & \\
{\scriptstyle \leq n}\diagup & \big| & \diagdown{\scriptstyle \leq m} \\
K & N & F' \\
{\scriptstyle m}\diagdown & \big| & \diagup{\scriptstyle n} \\
 & F &
\end{array}
$$

It follows from the corollary that the degree N of K' over F is divisible by the least common multiple of m and n, and that if m and n are relatively prime, then $N = mn$.

It might be tempting to guess that N divides mn, but this isn't always true.

Examples 15.3.10

(a) The three complex roots of $x^3 - 2$ are $\alpha_1 = \alpha$, $\alpha_2 = \omega\alpha$, and $\alpha_3 = \omega^2\alpha$, where $\alpha = \sqrt[3]{2}$ and $\omega = e^{2\pi i/3}$. Each of the roots α_i has degree 3 over $\mathbb{Q}$, but $\mathbb{Q}(\alpha_1, \alpha_2) = \mathbb{Q}(\alpha, \omega)$, and since ω has degree 2 over $\mathbb{Q}$, $[\mathbb{Q}(\alpha_1, \alpha_2):\mathbb{Q}] = 6$.

(b) Let $\alpha = \sqrt[3]{2}$ and let β be a root of the irreducible polynomial $x^4 + x + 1$ over $\mathbb{Q}$. Because 3 and 4 are relatively prime, $\mathbb{Q}(\alpha, \beta)$ has degree 12 over $\mathbb{Q}$. Therefore α is not in the field $\mathbb{Q}(\beta)$. On the other hand, since i has degree 2 over $\mathbb{Q}$, it is not so easy to decide whether or not i is in $\mathbb{Q}(\beta)$. (It is not.)

(c) Let $K = \mathbb{Q}(\sqrt{2}, i)$ be the field generated over $\mathbb{Q}$ by $\sqrt{2}$ and i. Both i and $\sqrt{2}$ have degree 2 over $\mathbb{Q}$, and since i is complex, it is not in $\mathbb{Q}(\sqrt{2})$. So $[\mathbb{Q}(\sqrt{2}, i):\mathbb{Q}] = 4$. Therefore the degree of i over $\mathbb{Q}(\sqrt{2})$ is 2. Since $\sqrt{-2}$ and i also generate K, i is not in the field $\mathbb{Q}[\sqrt{-2}]$ either. $\square$

15.4   FINDING THE IRREDUCIBLE POLYNOMIAL

Let γ be an element of an extension field K of F, and assume that γ is algebraic over F. There are two general methods to find the irreducible polynomial $f(x)$ for γ over F. One is to compute the powers of γ and to look for a linear relation among them. Sometimes, though not very often, one can guess the other roots of f, say $\gamma_1, \dots, \gamma_k$, with $\gamma = \gamma_1$. Then expanding the product will $(x - \gamma_1) \cdots (x - \gamma_k)$ produce the polynomial. We'll give an example to illustrate the two methods, in which F is the field $\mathbb{Q}$ of rational numbers.

Example 15.4.1 Let $\gamma = \sqrt{2} + \sqrt{3}$. We compute powers of γ, and simplify when possible: $\gamma^2 = 5 + 2\sqrt{6}$, $\gamma^4 = 49 + 20\sqrt{6}$. We won't need the other powers because we can eliminate $\sqrt{6}$ from these two equations, obtaining the relation $\gamma^4 - 10\gamma^2 + 1 = 0$. Thus γ is a root of the polynomial $g(x) = x^4 - 10x^2 + 1$. $\square$

Two important elementary observations are implicit here:

Lemma 15.4.2

(a) A linear dependence relation $c_n\gamma^n + \cdots + c_1\gamma + c_0 = 0$ among powers of an element γ means that γ is a root of the polynomial $c_n x^n + \cdots + c_1 x + c_0$.

(b) Let α and β be algebraic elements of an extension field of F, and let their degrees over F be d_1 and d_2, respectively. The $d_1 d_2$ monomials $\alpha^i \beta^j$, with $0 \le i < d_1$ and $0 \le j < d_2$, span $F(\alpha, \beta)$ as F-vector space.

Proof. Though important, **(a)** is trivial. To prove **(b)**, we note that because α and β are algebraic over F, $F(\alpha, \beta) = F[\alpha, \beta]$ (15.2.6). The monomials listed span $F[\alpha, \beta]$. $\square$

Example 15.4.3 The alternate approach to Example 15.4.1 is to guess that the roots of g might be $\gamma_1 = \sqrt{2} + \sqrt{3}$, $\gamma_2 = -\sqrt{2} - \sqrt{3}$, $\gamma_3 = -\sqrt{2} + \sqrt{3}$, and $\gamma_4 = \sqrt{2} - \sqrt{3}$. Expanding the polynomial with these roots, we find

$$(x - \gamma_1)(x - \gamma_2)(x - \gamma_3)(x - \gamma_4)$$
$$= (x^2 - (\sqrt{2} + \sqrt{3})^2)(x^2 - (\sqrt{2} - \sqrt{3})^2) = x^4 - 10x^2 + 1.$$

This is the polynomial that we obtained before. □

The lemma shows that one can always produce a polynomial having an element such as $\gamma = \alpha + \beta$ as a root, provided that the irreducible polynomials for α and β are known. Say that α and β have degrees d_1 and d_2 over F, respectively. Given any element γ of $F(\alpha, \beta)$, we write its powers $1, \gamma, \gamma^2, \ldots, \gamma^n$ as linear combinations of the monomials $\alpha^i \beta^j$ with $0 \le i < d_1$ and $0 \le j < d_2$. When $n = d_1 d_2$, we get $n + 1$ powers γ^ν that are linear combinations of n monomials. So the powers are linearly dependent. A linear dependence relation determines a polynomial with coefficients in F with γ as a root. However, there is a point that complicates matters. The polynomial with root γ that we find in this way may be reducible. The irreducible polynomial for γ over F is the *lowest degree* polynomial with root γ. To determine it by this method, we would need a *basis* for K over F.

Examples 15.4.4

(a) In Example 15.4.1, where $\alpha = \sqrt{2}$, $\beta = \sqrt{3}$ and $d_1 = d_2 = 2$, the elements $\alpha^i \beta^j$ with $i, j < 2$ are $1, \sqrt{2}, \sqrt{3}$, and $\sqrt{6}$. These elements do form a basis of K over $\mathbb{Q}$. The polynomial $x^4 - 10x^2 + 1$ is irreducible.

(b) We go back to Example 15.3.10**(a)**, in which the three roots of the polynomial $x^3 - 2$ are labeled α_i, $i = 1, 2, 3$. Let $F = \mathbb{Q}$, $L = \mathbb{Q}(\alpha_1)$ and $K = \mathbb{Q}(\alpha_1, \alpha_2)$. Each of the roots α_i has degree 3 over F. According to the lemma, the nine monomials $\alpha_1^i \alpha_2^j$ with $0 \le i, j < 3$ span K over F. However, these monomials aren't independent. Since f has a root α_1 in the field L, it factors in $L[x]$, say $f(x) = (x - \alpha_1)q(x)$. Then α_2 is a root of $q(x)$, so α_2 has degree at most 2 over L. The set $(1, \alpha_2)$ is a basis for K over the field L, so the six monomials $\alpha_1^i \alpha_2^j$ with $0 \le i < 3$ and $0 \le j < 2$ form a basis for K over F. If we want a basis of monomials, we should use this one. □

15.5   RULER AND COMPASS CONSTRUCTIONS

Famous theorems assert that certain geometric constructions cannot be done with ruler and compass alone. To illustrate these theorems, we use the concept of degree of a field extension to prove the impossibility of trisection of an angle.

Here are the rules for ruler and compass construction:

(15.5.1)

- Two points in the plane are given to start with. These points are *constructed*.

- If two points p_0, p_1 have been constructed, we may draw the line through them, or draw a circle with center at p_0 and passing through p_1. Such lines and circles are then *constructed*.

- The points of intersection of constructed lines and circles are *constructed*.

Points, lines, and circles will be called *constructible* if they can be obtained in finitely many steps, using these rules.

Notice that our ruler may be used only to draw lines through constructed points. We are not allowed to use it for measurement. Sometimes the ruler is referred to as a "straight-edge" to emphasize this point.

We begin with some familiar constructions. In each figure, the lines and circles are to be drawn in the order indicated. The first two constructions make use of a point q on ℓ whose only restriction is that it is not on the perpendicular. Whenever we need an arbitrary point, we will construct a particular one for the purpose. We can do this because a constructed line ℓ contains infinitely many points that can be constructed.

Construction 15.5.2 Construct a line through a constructed point p and perpendicular to a constructed line ℓ.

Case 1: $p \notin \ell$

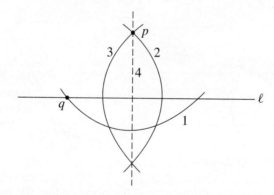

Case 2: $p \in \ell$

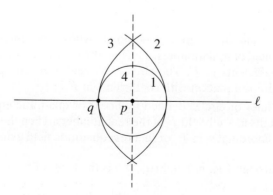

Construction 15.5.3 Construct a line parallel to a constructed line ℓ and passing through a constructed point p.

Apply Cases 1 and 2 above:

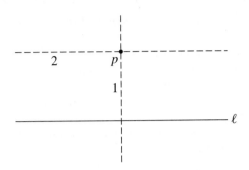

Construction 15.5.4 Mark off a length defined by two points onto a constructed line ℓ, with endpoint p.

Use the construction of parallels:

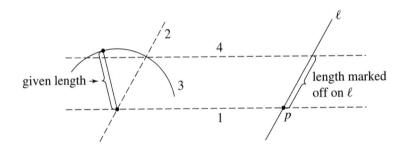

We introduce Cartesian coordinates into the plane so that the points that are given at the start have coordinates $(0, 0)$ and $(1, 0)$.

Proposition 15.5.5

(a) Let $p_0 = (a_0, b_0)$ and $p_1 = (a_1, b_1)$ be points whose coordinates a_i and b_i are in a subfield F of the field of real numbers. The line through p_0 and p_1 is defined by a linear equation with coefficients in F. The circle with center p_0 and passing through p_1 is defined by a quadratic equation with coefficients in F.

(b) Let A and B be lines or circles defined by linear or quadratic equations, respectively, that have coefficients in a subfield F of the real numbers. Then the points of intersection of A and B have coordinates in F, or in a real quadratic field extension F' of F.

Proof. **(a)** The line through (a_0, b_0) and (a_1, b_1) is the locus of the linear equation

$$(a_1 - a_0)(y - b_0) = (b_1 - b_0)(x - a_0).$$

The circle with center (a_0, b_0) and passing through (a_1, b_1) is the locus of the quadratic equation

$$(x - a_0)^2 + (y - b_0)^2 = (a_1 - a_0)^2 + (b_1 - b_0)^2.$$

The coefficients of these equations are in F.

(b) The point of intersection of two lines is found by solving two linear equations with coefficients in F, so its coordinates are in F. To find the intersection of a line and a circle, we use the equation of the line to eliminate one variable from the equation of the circle, obtaining a quadratic equation in one unknown. This quadratic equation has solutions in the field $F' = F(\sqrt{D})$, where D is its discriminant. The discriminant is an element of F. If $F' \neq F$, then the degree of F' over F is 2. If D is negative, there is no real solution to the equations. Then the line and circle do not intersect.

Consider the intersection of two circles, say

$$(x - a_1)^2 + (y - b_1)^2 = c_1 \quad \text{and} \quad (x - a_2)^2 + (y - b_2)^2 = c_2,$$

where a_i, b_i, and c_i are in F. In general, the solution of a pair of quadratic equations in two variables requires solving an equation of degree 4. In this case we are lucky: The difference of the two quadratic equations is a linear equation. We can use that linear equation to eliminate one variable, as before. The lucky event reflects the fact that, whereas a pair of conics may intersect in four points, two circles intersect in at most two points. □

Theorem 15.5.6 Let p be a constructible point. For some integer n, there is a chain of fields

$$\mathbb{Q} = F_0 \subset F_1 \subset F_2 \subset \cdots \subset F_n = K, \quad \text{such that}$$

- K is a subfield of the field of real numbers;
- the coordinates of p are in K;
- for each $i = 0, \ldots, n - 1$, the degree $[F_{i+1} : F_i]$ is equal to 2.

Therefore the degree $[K : \mathbb{Q}]$ is a power of 2.

Proof. We introduced coordinates so that the points originally given are $(0, 0)$ and $(1, 0)$. These points have coordinates in $\mathbb{Q}$. The process of constructing the point p involves a sequence of steps, each one of which draws a line or a circle.

Suppose that all points constructed by the time we are at the kth step have coordinates in a field F. The next step constructs a line or circle through two of these points, and according to Proposition 15.5.5**(a)**, the line or circle has an equation with coefficients in F. The field does not change. Then according to Proposition 15.5.5**(b)**, any point of intersection of the lines and circles constructed so far will have coordinates, either in F, or in a real quadratic extension of F. The assertion follows by induction from Proposition 15.5.5 and from the multiplicative property of the degree. □

- We call a real number a *constructible* if the point $(a, 0)$ is constructible. Since we can construct perpendiculars, this is the same thing as saying that a is the x-coordinate of a constructible point. And since we can mark off lengths, a positive real number a is constructible if and only if there is a pair p, q of constructible points whose distance apart is a.

Corollary 15.5.7 Let a be a constructible real number. Then a is an algebraic number, and its degree over $\mathbb{Q}$ is a power of 2.

Since a is in a field K that is the end of a chain of fields as in the theorem, and since $[K:\mathbb{Q}]$ is a power of 2, the degree of a is also a power of 2 (15.3.6). $\qquad\square$

The converse of this corollary is false. There exist real numbers of degree 4 over $\mathbb{Q}$ that aren't constructible. Galois theory provides a way to understand this. (This is Exercise 9.17 of Chapter 16.)

We can now prove the impossibility of certain geometric constructions. The method is to show that if a certain construction were possible, then it would also be possible to construct an algebraic number whose degree over $\mathbb{Q}$ is not a power of 2. This would contradict the corollary. Our example is the impossibility of trisection of the angle, which asks for a construction of the angle $\frac{1}{3}\theta$ when θ is given. Now many angles, $45°$ for instance, can be trisected. The trisection problem asks for a general method of construction that will work for any "given" angle.

Since it is easy to construct an angle of $60°$, we can give this angle to ourselves, using ruler and compass constructions. If trisection were possible, we could construct an angle of $20°$. We will show that it is impossible to construct that particular angle, and therefore that there is no general method of trisection.

We'll say that an angle θ is constructible if it is possible to construct a pair of lines meeting with angle θ. If we mark off a unit length on one of the lines and drop a perpendicular to the other line, we will have constructed the real number $\cos\theta$. Conversely, if $\cos\theta$ is a constructible real number, we can reverse this process to construct a pair of lines meeting with angle θ.

The next lemma shows that $20° = \pi/9$ cannot be constructed.

Lemma 15.5.8 The real number $\cos 20°$ is algebraic over $\mathbb{Q}$ and its degree over $\mathbb{Q}$ is 3. Therefore $\cos 20°$ is not a constructible number.

Proof. Let $\alpha = 2\cos\theta = e^{i\theta} + e^{-i\theta}$, where $\theta = \pi/9$. Then $e^{3i\theta} + e^{-3i\theta} = 2\cos(\pi/3) = 1$, and

$$\alpha^3 = (e^{i\theta} + e^{-i\theta})^3 = e^{3i\theta} + 3e^{i\theta} + 3e^{-i\theta} + e^{-3i\theta} = 1 + 3\alpha,$$

so α is a root of the polynomial $x^3 - 3x - 1$. This polynomial is irreducible over $\mathbb{Q}$ because it has no integer root. It is therefore the irreducible polynomial for α over $\mathbb{Q}$. So α has degree 3 over $\mathbb{Q}$, and so does $\cos\theta$. $\qquad\square$

One more example: The regular 7-gon cannot be constructed. This is similar to the above problem, because constructing $20°$ is equivalent to constructing the 18-gon. We'll vary the approach slightly. Let $\theta = 2\pi/7$ and let $\zeta = e^{i\theta}$. Then ζ is a seventh root of unity, a root of the irreducible polynomial equation $x^6 + x^5 + \cdots + 1$ (Theorem 12.4.9), so ζ has degree 6 over $\mathbb{Q}$. If the 7-gon were constructible, then $\cos\theta$ and $\sin\theta$ would be constructible numbers. They would lie in a real field extension K whose degree over $\mathbb{Q}$ is a power of 2, say 2^k. Call this field K, and consider the extension $K(i)$ of K. This extension has degree 2, so $[K(i):\mathbb{Q}] = 2^{k+1}$. But $\zeta = \cos\theta + i\sin\theta$ is in $K(i)$. This contradicts the fact that the degree of ζ is 6.

The argument we have used is not special to the number 7. It applies to any prime integer p, provided that $p - 1$, the degree of the irreducible polynomial $x^{p-1} + \cdots + x + 1$, is not a power of 2.

Corollary 15.5.9 Let p be a prime integer. If the regular p-gon can be constructed with ruler and compass, then $p = 2^r + 1$ for some integer r.   □

Gauss proved the converse: If a prime has the form $2^r + 1$, then the regular p-gon can be constructed. The regular 17-gon, for example, can be constructed by ruler and compass. We will learn how to prove this in the next chapter (see Corollary 16.10.5).

To complete the discussion, we prove a converse to Theorem 15.5.6.

Theorem 15.5.10 Let $\mathbb{Q} = F_0 \subset F_1 \subset \cdots \subset F_n = K$ be a chain of subfields of the field $\mathbb{R}$ of real numbers with the property that for each $i = 0, \ldots, n-1$, $[F_{i+1} : F_i] = 2$. Then every element of K is constructible.

Since any extension of degree 2 can be obtained by adjoining a square root, the theorem follows from the next lemma.

Lemma 15.5.11

(a) The constructible numbers form a subfield of $\mathbb{R}$.

(b) If a is a positive constructible number, then so is $\sqrt{a}$.

Proof. **(a)** We must show that if a and b are positive constructible numbers, then $a + b, -a$, ab, and a^{-1} (if $a \neq 0$) are also constructible. The closure in case a or b is negative follows easily. Addition and subtraction are done by marking lengths on a line. For multiplication and division, we use similar right triangles.

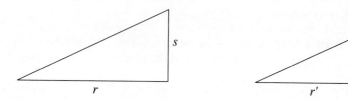

Given one triangle and one side of a second triangle, the second triangle can be constructed by parallels. To construct the product ab, we take $r = 1, s = a$, and $r' = b$. Then $s' = ab$. To construct a^{-1}, we take $r = a, s = 1$, and $r' = 1$. Then $s' = a^{-1}$.

(b) We use similar triangles again. We must construct them so that $r = a, r' = s$, and $s' = 1$. Then $s = \sqrt{a}$. How to make the construction is less obvious this time, but we can use inscribed triangles in a circle. A triangle inscribed into a circle, with a diameter as its hypotenuse, is a right triangle. This is a theorem of high school geometry, and it can be checked using the equation for a circle and Pythagoras's theorem. So we construct a circle whose diameter is $1 + a$ and proceed as in the figure below.

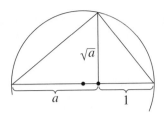

$\square$

15.6   ADJOINING ROOTS

Up to this point, we have used subfields of the complex numbers as our examples. Abstract constructions are not needed to create these fields, except that the construction of the complex number field as an extension of the real number field is abstract. We simply adjoin complex numbers to the rational numbers as desired, and work with the subfield they generate. But finite fields and function fields are not subfields of a familiar, all-encompassing field analogous to $\mathbb{C}$, so these fields must be constructed. The fundamental tool for their construction is the adjunction of elements to a ring, which was described in Chapter 11. It is applied here to the case that the ring we start with is a field.

We review the construction. Given a polynomial $f(x)$ with coefficients in a field F, we may adjoin a root of f to F. The procedure is to form the quotient ring

(15.6.1)
$$K = F[x]/(f)$$

of the polynomial ring $F[x]$. This construction always yields a ring K and a homomorphism $F \to K$, such that the residue $\overline{x}$ of x satisfies the relation $f(\overline{x}) = 0$ (11.5.2). However, we want to construct not only a ring, but a field. Here the theory of polynomials over a field comes into play. It tells us that the principal ideal (f) in the polynomial ring $F[x]$ is a maximal ideal if and only if f is an irreducible polynomial (12.2.8). Therefore K will be a field if and only if f is irreducible (11.8.2).

Lemma 15.6.2 Let F be a field, and let f be an irreducible polynomial in $F[x]$. Then the ring $K = F[x]/(f)$ is an extension field of F, and the residue $\overline{x}$ of x is a root of $f(x)$ in K.

Proof. The ring K is a field because (f) is a maximal ideal, and the homomorphism $F \to K$, which sends the elements of F to the residues of the constant polynomials, is injective because F is a field (11.3.20). So we may identify F with its image, a subfield of K. The field K becomes an extension of F by means of this identification. Finally, $\overline{x}$ satisfies the equation $f(\overline{x}) = 0$. It is a root of f (see (11.5.2)). $\square$

- A polynomial f *splits completely* in a field K if it factors into linear factors in K.

Proposition 15.6.3 Let F be a field, and let $f(x)$ be a monic polynomial in $F[x]$ of positive degree. There exists a field extension K of F such that $f(x)$ splits completely in K.

Proof. We use induction on the degree of f. The first case is that f has a root α in F, so that $f(x) = (x - \alpha)q(x)$ for some polynomial q. If so, we replace f by q, and we are done by induction. Otherwise, we choose an irreducible factor g of f. By Lemma 15.6.2, there is a field extension F_1 of F in which g has a root α. Then α is a root of f too. We replace F by F_1, and this reduces us to the first case.   ☐

As we see, the polynomial ring $F[x]$ is an important tool for studying extensions of a field F. When we are working with field extensions, there is an interplay between the polynomial rings over the fields. This interplay doesn't present serious difficulties, but instead of scattering the points that should be mentioned about in the text, we have collected them into the next proposition.

Proposition 15.6.4 Let f and g be polynomials with coefficients in a field F, with $f \neq 0$, and let K be an extension field of F.

(a) The polynomial ring $K[x]$ contains $F[x]$ as subring, so computations made in the ring $F[x]$ are also valid in $K[x]$.

(b) Division with remainder of g by f gives the same answer, whether carried out in $F[x]$ or in $K[x]$.

(c) f divides g in $K[x]$ if and only if f divides g in $F[x]$.

(d) The (monic) greatest common divisor d of f and g is the same, whether computed in $F[x]$ or in $K[x]$.

(e) If f and g have a common root in K, they are not relatively prime in $F[x]$. If f and g are not relatively prime in $F[x]$, there exists an extension field in which they have a common root.

(f) If f is an irreducible element of $F[x]$ and if f and g have a common root in K, then f divides g in $F[x]$.

Proof. **(a)** This is obvious.

(b) Carry out the division in $F[x]$: $g = fq + r$. This equation remains true in the bigger ring $K[x]$, and since division with remainder in $K[x]$ is unique, carrying the division out in $K[x]$ leads to the same result.

(c) This is **(b)** in the case that the remainder is zero.

(d) Let d and d' denote the greatest common divisors of f and g in $F[x]$ and $K[x]$, respectively. Then d is a common divisor in $K[x]$, and since d' is the greatest common divisor in $K[x]$, d divides d'. In addition, we know that d has the form $d = pf + qg$, for some elements p and q in $F[x]$. Since d' divides f and g, d' divides d. Thus d and d' are associates in $K[x]$, and since they are monic polynomials, they are equal.

(e) Let α be a common root of f and g in K. Then $x - \alpha$ is a common divisor of f and g in $K[x]$. So the greatest common divisor of f and g in $K[x]$ isn't 1. By **(d)**, it isn't 1 in $F[x]$ either. Conversely, if f and g have a common divisor d of positive degree, there is an extension field of F in which d has a root. This root will be a common root of f and g.

(f) If f is irreducible, its only monic divisors in $F[x]$ are 1 and f. Part **(e)** tells us that the greatest common divisor of f and g in $F[x]$ isn't 1. Therefore it is f.   ☐

The final topic of this section is the derivative $f'(x)$ of a polynomial $f(x)$. The derivative is computed using the rules from calculus for differentiating polynomial functions.

In other words, if $f(x) = a_n x^n + a_{n-1} x^{n-1} + \cdots + a_1 x + a_0$, then

(15.6.5) $$f'(x) = na_n x^{n-1} + (n-1)a_{n-1} x^{n-2} + \cdots + a_1.$$

The integer coefficients in this formula are interpreted as the elements $1 + \cdots + 1$ of F. So if f has coefficients in a field F, its derivative does too. It can be shown that familiar rules of differentiation, such as the product rule, hold. (This is Exercise 3.5.)

The derivative can be used to recognize multiple roots of a polynomial.

Lemma 15.6.6 Let f be a polynomial with coefficients in a field F. An element α in an extension field K of F is a *multiple root*, meaning that $(x - \alpha)^2$ divides f, if and only if it is a root of f and also a root of f'.

Proof. If α is a root of f, then $x - \alpha$ divides f, say $f(x) = (x - \alpha)g(x)$. Then α is a multiple root of f if and only if it is a root of g. By the product rule for differentiation,

$$f'(x) = (x - \alpha)g'(x) + g(x).$$

Substituting $x = \alpha$, one sees that $f'(\alpha) = 0$ if and only if $g(\alpha) = 0$. $\quad\square$

Proposition 15.6.7 Let $f(x)$ be a polynomial with coefficients in F. There exists a field extension K of F in which f has a multiple root if and only if f and f' are not relatively prime.

Proof. If f has a multiple root in K, then f and f' have a common root in K, so they are not relatively prime in K or in F. Conversely, if f and f' are not relatively prime, then they have a common root in some field extension K, hence f has a multiple root there. $\quad\square$

Here is one of the most important applications of the derivative to field theory:

Proposition 15.6.8 Let f be an *irreducible* polynomial in $F[x]$.

(a) f has no multiple root in any field extension of F unless the derivative f' is the zero polynomial.

(b) If F is a field of characteristic zero, then f has no multiple root in any field extension of F.

Proof. (a) We must show that f and f' are relatively prime unless f' is the zero polynomial. Since it is irreducible, f will have a nonconstant factor in common with another polynomial g only if f divides g. And if f divides g, then unless $g = 0$, the degree of g will be at least as large as the degree of f. If the derivative f' isn't zero, its degree is less than the degree of f, and then f and f' have no common nonconstant factor.

(b) In a field of characteristic zero, the derivative of a nonconstant polynomial isn't zero. $\quad\square$

The derivative of a nonconstant polynomial f may be zero when F has prime characteristic p. This happens when the exponent of every monomial that occurs in f is divisible by p. A typical polynomial whose derivative is zero in characteristic 5 is

$$f(x) = x^{15} + ax^{10} + bx^5 + c,$$

where a, b, c can be any elements of F. Since the derivative of this polynomial is identically zero, all of its roots in an extension field will be multiple roots.

15.7   FINITE FIELDS

In this section, we describe the fields of finite order. The characteristic of a finite field K cannot be zero, so it is a prime integer (3.2.10), and therefore K will contain one of the prime fields $F = \mathbb{F}_p$. Since K is finite, it will be finite-dimensional when considered as a vector space over this field.

Let r denote the degree $[K : F]$. As an F-vector space, K is isomorphic to the space F^r of column vectors, which contains p^r elements. So the *order* of a finite field, the number of its elements, is a power of a prime. It is customary to use the letter q for this order:

$$(15.7.1) \qquad\qquad |K| = p^r = q.$$

In this section, q will denote a positive power of a prime integer p. Fields of order q are often denoted by $\mathbb{F}_q$. We are going to show that all finite fields of order q are isomorphic, so this notation isn't too ambiguous, though when $r > 1$ the isomorphism between two of them will not be unique.

The simplest example of a finite field other than a prime field is the field $\mathbb{F}_4$ of order 4. Let K denote this field, and let $F = \mathbb{F}_2$. There is just one irreducible polynomial of degree 2 in $F[x]$, namely $x^2 + x + 1$ (12.4.4), and K is obtained by adjoining a root α of this polynomial to F:

$$K \approx F[x]/(x^2 + x + 1).$$

Because the element α, the residue of x, has degree 2, the set $(1, \alpha)$ forms a basis of K over F (15.2.7). The elements of K are the four linear combinations of the basis, with coefficients modulo 2:

$$(15.7.2) \qquad\qquad K = \{0, 1, \alpha, 1 + \alpha\}.$$

The element $1 + \alpha$ is the other root of $f(x)$ in K. Computation in $\mathbb{F}_4$ is made using the relations $1 + 1 = 0$ and $\alpha^2 + \alpha + 1 = 0$.

Try not to confuse the field $\mathbb{F}_4$ with the ring $\mathbb{Z}/(4)$, which isn't a field.

Here are the main facts about finite fields:

Theorem 15.7.3 Let p be a prime integer, and let $q = p^r$ be a positive power of p.

(a) Let K be a field of order q. The elements of K are roots of the polynomial $x^q - x$.

(b) The irreducible factors of the polynomial $x^q - x$ over the prime field $F = \mathbb{F}_p$ are the irreducible polynomials in $F[x]$ whose degrees divide r.

(c) Let K be a field of order q. The multiplicative group $K^\times$ of nonzero elements of K is a cyclic group of order $q - 1$.

(d) There exists a field of order q, and all fields of order q are isomorphic.

(e) A field of order p^r contains a subfield of order p^k if and only if k divides r.

Corollary 15.7.4 For every positive integer r, there exists an irreducible polynomial of degree r over the prime field $\mathbb{F}_p$.

Proof. According to **(d)**, there is a field K of order $q = p^r$. Its degree $[K:F]$ over $F = \mathbb{F}_p$ is r. According to **(c)**, the multiplicative group $K^\times$ is cyclic. It is obvious that a generator α for this cyclic group will generate K as extension field, i.e., that $K = F(\alpha)$. Since $[K:F] = r$, α has degree r over F. So α is the root of an irreducible polynomial of degree r.   □

We examine a few examples in which q is a power of 2. The irreducible polynomials of degree at most 4 over $\mathbb{F}_2$ are listed in (12.4.4).

Examples 15.7.5

(i) The field $\mathbb{F}_4$ has degree 2 over $\mathbb{F}_2$. Its elements are the roots of the polynomial

(15.7.6) $$x^4 - x = x(x-1)(x^2 + x + 1).$$

Note that the factors of $x^2 - x$ appear, because $\mathbb{F}_4$ contains $\mathbb{F}_2$.
Since we are working in characteristic 2, signs are irrelevant: $x - 1 = x + 1$.

(ii) The field $\mathbb{F}_8$ of order 8 has degree 3 over the prime field $\mathbb{F}_2$. Its elements are the eight roots of the polynomial $x^8 - x$. The factorization of this polynomial in $\mathbb{F}_2$ is

(15.7.7) $$x^8 - x = x(x-1)(x^3 + x + 1)(x^3 + x^2 + 1).$$

The cubic factors are the two irreducible polynomials of degree 3 in $\mathbb{F}_2[x]$.
To compute in the field $\mathbb{F}_8$, we choose an element β in that field, a root of one of the irreducible cubic factors, say of $x^3 + x + 1$. It will have degree 3 over $\mathbb{F}_2$. Then $(1, \beta, \beta^2)$ is a basis of $\mathbb{F}_8$ as a vector space over $\mathbb{F}_2$. The elements of $\mathbb{F}_8$ are the eight linear combinations of this basis with coefficients 0 and 1:

(15.7.8) $$\mathbb{F}_8 = \{0, 1, \beta, 1+\beta, \beta^2, 1+\beta^2, \beta+\beta^2, 1+\beta+\beta^2\}.$$

Computation in $\mathbb{F}_8$ is done using the relations $1 + 1 = 0$ and $\beta^3 + \beta + 1 = 0$.
Note that $x^2 + x + 1$ is not a factor of $x^8 - x$, and therefore $\mathbb{F}_8$ does not contain $\mathbb{F}_4$. It couldn't, because $[\mathbb{F}_8:\mathbb{F}_2] = 3$, $[\mathbb{F}_4:\mathbb{F}_2] = 2$, and 2 does not divide 3.

(iii) The field $\mathbb{F}_{16}$ of order 16 has degree 4 over $\mathbb{F}_2$. Its elements are roots of the polynomial $x^{16} - x$. This polynomial factors in $\mathbb{F}_2[x]$ as

(15.7.9) $$x^{16}-x = x(x-1)(x^2+x+1)(x^4+x^3+x^2+x+1)(x^4+x^3+1)(x^4+x+1)$$

The three irreducible polynomials of degree 4 in $\mathbb{F}_2[x]$ appear here. The factors of $x^4 - x$ are also among the factors, because $\mathbb{F}_{16}$ contains $\mathbb{F}_4$.   □

We now begin the proof of Theorem (15.7.3). We let F denote the prime field $\mathbb{F}_p$.

Proof of Theorem 15.7.3(a). *(the elements of K are roots of $x^q - x$)* Let K be a field of order q. The multiplicative group $K^\times$ has order $q - 1$. Therefore the order of any element α of $K^\times$

divides $q - 1$, so $\alpha^{(q-1)} - 1 = 0$, which means that α is a root of the polynomial $x^{(q-1)} - 1$. The remaining element of K, zero, is the root of the polynomial x. So every element of K is a root of $x(x^{(q-1)} - 1) = x^q - x$. $\qquad\square$

Proof of Theorem 15.7.3(c). *(the multiplicative group is cyclic)* The proof is based on the Structure Theorem 14.7.3 for abelian groups, which tells us that $K^\times$ is a direct sum of cyclic groups.

The Structure Theorem was stated with additive notation: A finite abelian group V is a direct sum $C_1 \oplus \cdots \oplus C_k$ of cyclic subgroups of orders $d_1, \ldots, d_k$, such that each d_i divides the next: $d_1 | d_2 | \cdots | d_k$. Let $d = d_k$. If w_i is a generator for C_i, then $d_i w_i = 0$, and since d_i divides d, $dw_i = 0$. Therefore $dv = 0$ for every element v of V. The order of every element of V divides d.

Going over to multiplicative notation, $K^\times$ is a direct sum of cyclic subgroups, say $H_1 \oplus \cdots \oplus H_k$, where H_i has order d_i, and $d_1 | d_2 | \cdots | d_k$. With $d = d_k$ as before, the order of every element α of $K^\times$ divides d, which means that $\alpha^d = 1$. Therefore every element of $K^\times$ is a root of the polynomial $x^d - 1$. This polynomial has at most d roots in K (12.2.20), and therefore $|K^\times| = q - 1 \le d$. On the other hand, $|K^\times| = |H_1 \oplus \cdots \oplus H_k| = d_1 \ldots d_k$. So $d_1 \ldots d_k = |K^\times| = q - 1 \le d$. Since $d = d_k$, the only possiblility is that $k = 1$ and $q - 1 = d$. Therefore $K^\times = H_1$, and $K^\times$ is cyclic. $\qquad\square$

Proof of Theorem 15.7.3(d). *(existence of a field with q elements)* Since we have proved **(a)**, we know that the elements of a field of order q will be roots of the polynomial $x^q - x$. There exists a field extension L of F in which this polynomial splits completely (15.6.3). The natural thing to try is to take such a field L and hope for the best, that the roots of $x^q - x$ in L form the subfield K that we are looking for. This is shown by Lemma 15.7.11 below.

Lemma 15.7.10 Let F be a field of prime characteristic p, and let $q = p^r$ be a positive power of p.

(a) The polynomial $x^q - x$ has no multiple root in any field extension of F.

(b) In the polynomial ring $F[x, y]$, $(x + y)^q = x^q + y^q$.

Proof. **(a)** The derivative of $x^q - x$ is $qx^{(q-1)} - 1$. In characteristic p, the coefficient q is equal to 0, so the derivative is -1. Since the constant polynomial -1 has no root, $x^q - x$ and its derivative have no common root, and therefore $x^q - x$ has no multiple root (Lemma 15.6.6).

(b) We expand $(x + y)^p$ in $\mathbb{Z}[x, y]$:

$$(x + y)^p = x^p + \binom{p}{1}x^{p-1}y + \binom{p}{2}x^{p-2}y^2 + \cdots + \binom{p}{p-1}xy^{p-1} + y^p.$$

Lemma 12.4.8 tells us that the binomial coefficients $\binom{p}{r}$ are divisible by p for r in the range $1 < r < p$. Since F has characteristic p, the map $\mathbb{Z}[x, y] \to F[x, y]$ sends these coefficients to zero, and $(x + y)^p = x^p + y^p$ in $F[x, y]$. The fact that $(x + y)^q = x^q + y^q$ when $q = p^r$ follows by induction. $\qquad\square$

Lemma 15.7.11 Let p be a prime and let $q = p^r$ be a positive power of p. Let L be a field of characteristic p, and let K be the set of roots of $x^q - x$ in L. Then K is a subfield of L.

Proof. Let α and β be roots of the polynomial $x^q - x$ in L. We have to show that $\alpha + \beta, -\alpha,$ $\alpha\beta, \alpha^{-1}$ (if $\alpha \neq 0$), and 1 are roots of the same polynomial. So we assume that $\alpha^q = \alpha$ and $\beta^q = \beta$. The proofs that $\alpha\beta, \alpha^{-1}$, and 1 are roots are obvious enough that we omit them. Substitution into Lemma 15.7.10(b) shows that $(\alpha + \beta)^q = \alpha^q + \beta^q = \alpha + \beta$.

Finally, we verify that -1 is a root of $x^q - x$. Since products of roots are roots, it will follow that $-\alpha$ is a root. If $p \neq 2$, then q is an odd integer, and it is true that $(-1)^q = -1$. If $p = 2$, q is even, and $(-1)^q = 1$. But in this case, the characteristic of L is 2, so $1 = -1$ in L.   □

We must still show that two fields K and K' of the same order $q = p^r$ are isomorphic. Let α be a generator for the cyclic group $K^\times$. Then $K = F(\alpha)$, so the irreducible polynomial f for α over F has degree equal to $[K:F] = r$. Then f generates the ideal of polynomials in $F[x]$ with root α, and since α is also a root of $x^q - x$, f divides $x^q - x$. Since $x^q - x$ splits completely in K', f has a root α' in K' too. Then $F(\alpha)$ and $F(\alpha')$ are both isomorphic to $F[x]/(f)$, hence to each other. Counting degrees shows that $F(\alpha') = K'$, so K and K' are isomorphic.   □

Proof of Theorem 15.7.3(e). (subfields of $\mathbb{F}_q$) Let $q = p^r$ and $q' = p^k$. Then $[\mathbb{F}_q : \mathbb{F}_p] = r$ and $[\mathbb{F}_{q'} : \mathbb{F}_p] = k$, we can't have $\mathbb{F}_p \subset \mathbb{F}_{q'} \subset \mathbb{F}_q$ unless k divides r. Suppose that k divides r, say $r = ks$. Substitution of $y = p^k$ into the equation $y^s - 1 = (y - 1)(y^{s-1} + \cdots + y + 1)$ shows that $q' - 1$ divides $q - 1$. Since the multiplicative group $K^\times$ is cyclic of order $q - 1$, and since $q' - 1$ divides $q - 1$, $K^\times$ contains an element β of order $q' - 1$. The $q' - 1$ powers of this element are roots of $x^{(q'-1)} - 1$ in K. Therefore $x^{q'} - x$ splits completely in K. Lemma 15.7.11 shows that the roots form a field of order q'.   □

Proof of Theorem 15.7.3(b). (the irreducible factors of $x^q - x$) Let g be an irreducible polynomial over F of degree k. The polynomial $x^q - x$ factors into linear factors in K because it has q roots in K. If g divides $x^q - x$, it will also factor into linear factors, so it will have a root β in K. The degree of β over F divides $[K:F] = r$, and is equal to k. So k divides r. Conversely, suppose that k divides r. Let β be a root of g in an extension field of F. Then $[F(\beta):F] = k$, and by (e), K contains a subfield isomorphic to $F(\beta)$. Therefore g has a root in K, and so g divides $x^q - x$.

This completes the proof of Theorem 15.7.3.   □

15.8 PRIMITIVE ELEMENTS

Let K be a field extension of a field F. An element α that generates K/F, i.e., such that $K = F(\alpha)$, is called a *primitive element* for the extension. Primitive elements are useful because computation in $F(\alpha)$ can be done easily, provided that the irreducible polynomial for α over F is known.

Theorem 15.8.1 Primitive Element Theorem. Every finite extension K of a field F of characteristic zero contains a primitive element.

The statement is true also when F is a finite field, though the proof is different. For an infinite field of characteristic $p \neq 0$, the theorem requires an additional hypothesis. Since we won't be studying such fields, we won't consider that case.

Proof of the Primitive Element Theorem. Since the extension K/F is finite, K is generated by a finite set. For example, a basis for K as F-vector space will generate K over F. Say that $K = F(\alpha_1, \ldots, \alpha_k)$. We use induction on k. There is nothing to prove when $k = 1$. For $k > 1$, induction allows us to assume the theorem true for the field $K_1 = F(\alpha_1, \ldots, \alpha_{k-1})$ generated by the first $k - 1$ elements α_i. So we may assume that K_1 is generated by a single element β. Then K will be generated by the two elements α_k and β. The proof of the theorem is thereby reduced to the case that K is generated by two elements. The next lemma takes care of this case. $\square$

Lemma 15.8.2 Let F be a field of characteristic zero, and let K be an extension field that is generated over F by two elements α and β. For all but finitely many c in F, $\gamma = \beta + c\alpha$ is a primitive element for K over F.

Proof. Let $f(x)$ and $g(x)$ be the irreducible polynomials for α and β, respectively, over F, and let $\mathcal{K}$ be a field extension of K in which f and g split completely. Call their roots $\alpha_1, \ldots, \alpha_m$ and $\beta_1, \ldots, \beta_n$, respectively, with $\alpha = \alpha_1$ and $\beta = \beta_1$.

Since the characteristic is zero, the roots α_i are distinct, as are the roots β_j (15.6.8)**(b)**. Let $\gamma_{ij} = \beta_j + c\alpha_i$, with $i = 1, \ldots, m$ and $j = 1, \ldots, n$. When $(i, j) \neq (k, \ell)$, the equation $\gamma_{ij} = \gamma_{k\ell}$ holds for at most one c. So for all but finitely many elements c of F, the γ_{ij} will be distinct. We will show that if c avoids these "bad" values, then $\gamma_{11} = \beta_1 + c\alpha_1$ will be a primitive element. We drop the subscript, and write $\gamma = \beta_1 + c\alpha_1$.

Let $L = F(\gamma)$. To show that γ is a primitive element, it will be enough to show that α_1 is in L. Then $\beta_1 = \gamma - c\alpha_1$ will be in L too, and therefore L will be equal to K. To begin with, α_1 is a root of $f(x)$. The trick is to use g to cook up a second polynomial with α_1 as a root, namely $h(x) = g(\gamma - cx)$. This polynomial doesn't have coefficients in F, but because g is in $F[x]$, c is in F, and γ is in L, the coefficients of g are in L.

We inspect the greatest common divisor d of f and h. It is the same, whether computed in $L[x]$ or in the extension field $\mathcal{K}[x]$ (15.6.4). Since $f(x) = (x - \alpha_1) \cdots (x - \alpha_m)$ in $\mathcal{K}$, d is the product of the factors $x - \alpha_i$ that also divide h, i.e., those such that α_i is a common root of h and f. One common root is α_1. If we show that this is the only common root, it will follow that $d = x - \alpha_1$, and because the greatest common divisor is an element of $L[x]$ (15.6.4)(d), that α_1 is an element of L.

So all we have to do is check that α_i is not a root of h when $i > 1$. We substitute: $h(\alpha_i) = g(\gamma - c\alpha_i)$. The roots of g are $\beta_1, \ldots, \beta_n$, so we must check that $\gamma - c\alpha_i \neq \beta_j$ for any j, or that $\beta_1 + c\alpha_1 \neq \beta_j + c\alpha_i$. This is true because c has been chosen so that the elements γ_{ij} are distinct. $\square$

15.9   FUNCTION FIELDS

In this section we look at *function fields*, the third class of field extensions mentioned at the beginning of the chapter. The field $\mathbb{C}(t)$ of rational functions in t will be denoted by F. Its elements are fractions p/q of complex polynomials, with $q \neq 0$. Function fields are finite field extensions of F.

Let α be a primitive element for an extension field K of F degree n, and let f be the irreducible polynomial for α over F, so that $K = F(\alpha)$ is isomorphic to the field $F[x]/(f)$, with α corresponding to the residue of x. By clearing denominators, we make f into a primitive polynomial that we write as a polynomial in x:

$$(15.9.1) \qquad f(t, x) = a_n(t)x^n + \cdots + a_1(t)x + a_0(t).$$

The hypothesis that f is a primitive polynomial means that the coefficients $a_i(t)$ are polynomials in t with greatest common divisor 1, and that $a_n(t)$ is monic (12.3.9). The *Riemann surface* X of such a polynomial was defined in Section 11.9, as the locus of zeros $\{f = 0\}$ in complex (t, x)-space $\mathbb{C}^2$. It was shown there that X is an n-sheeted branched covering of the complex t-plane T (11.9.16). The branch points are the points $t = t_0$ of T at which the one-variable polynomial $f(t_0, x)$ has fewer than n roots, which happens when $f(t_0, x)$ has a multiple root, or when t_0 is a root of the leading coefficient $a_n(t)$ of f (11.9.17).

As before, we use the notation X' for a set obtained by deleting an unspecified finite subset from X, and instead of saying that some statement is true except at a finite set of points of X, we will say that it is true on X'.

An *isomorphism of extension fields* K and L of F was defined in (15.2.9). It is an isomorphism of fields $\varphi: K \to L$ that restricts to the identity on F:

$$(15.9.2) \qquad \begin{array}{ccc} K & \xrightarrow{\ \varphi\ } & L \\ \uparrow & & \uparrow \\ F & =\!\!=\!\!= & F \end{array}$$

The vertical arrows in this diagram are the inclusions of F as a subfield into K and L, and the long equality symbol stands for the identity map.

• An *isomorphism of branched coverings* X and Y of T is a continuous, bijective map $\eta : X' \to Y'$ that is compatible with the projections of these surfaces to T:

$$(15.9.3) \qquad \begin{array}{ccc} X' & \xrightarrow{\ \eta\ } & Y' \\ \downarrow & & \downarrow \\ T' & =\!\!=\!\!= & T'. \end{array}$$

The primes indicate that we expect to delete finite sets of points from X and Y in order that the map η be defined and bijective.

Speaking a bit loosely, we call a branched covering $\pi : X \to T$ *path connected* if X' is path connected, by which we mean that for every finite subset Δ of X, the set $X - \Delta$ is path connected.

The object of this section is to explain the next theorem, which describes function fields in terms of their Riemann surfaces.

Theorem 15.9.4 Riemann Existence Theorem. There is a bijective correspondence between isomorphism classes of function fields of degree n over F and isomorphism classes of connected, n-sheeted branched coverings of T, such that the class of the field extension K defined by an irreducible polynomial $f(t, x)$ corresponds to the class of its Riemann surface X.

This theorem gives us a way to decide when two polynomials of the same degree in x define isomorphic field extensions. A simple criterion that can often be used is that the branch points of their Riemann surfaces must match up. However, the theorem fails to tell us how to find a polynomial with a given branched cover as its Riemann surface. It cannot do this. Many polynomials define isomorphic field extensions, and finding something is difficult when there are many choices.

The proof of the theorem is too long to include, but one part is rather easy to verify:

Proposition 15.9.5 Let $f(t, x)$ and $g(t, y)$ be irreducible polynomials in $\mathbb{C}[t, x]$ and $\mathbb{C}[t, y]$, respectively. Let $K = F[x]/(f)$ and $L = F[y]/(g)$ be the field extensions they define, and let X and Y be the Riemann surfaces $\{f = 0\}$ and $\{g = 0\}$. If K/F and L/F are isomorphic field extensions, then X and Y are isomorphic branched coverings of T.

Proof. The residue of y in $L = F[y]/(g)$, let's call it β, is a root of g, i.e., $g(t, \beta) = 0$, and an F-isomorphism $\varphi: K \to L$ gives us a root of g in K, namely $\gamma = \varphi^{-1}(\beta)$. So $g(t, \gamma) = 0$. As is true for any element of $K = F[x]/(f)$, γ can be represented as the residue modulo (f) of an element of $F[x]$. We let u be such an element, and we define the isomorphism $\eta: X \to Y$ by $\eta(t, x) = (t, u(t, x))$.

We must show that if (t, x) is a point of X, then (t, u) is a point of Y. Since $g(t, \gamma) = 0$ in K and since u is an element of $F[x]$ that represents γ, $g(t, u)$ is in the ideal (f). There is an element h of $F[x]$ such that

$$g(t, u) = fh.$$

If (t, x) is a point of X, then $f(t, x) = 0$, and so $g(t, u) = 0$ too. Therefore (t, u) is indeed a point of Y. However, since u and h are elements of $F[x]$, their coefficients are rational functions in t that may have denominators. So η may be undefined at a finite set of points.

The inverse function to η is obtained by interchanging the roles of K and L. $\square$

Cut and Paste

"Cut and paste" is a procedure to construct or deconstruct a branched covering.

We go back to our example of the Riemann surface X of the polynomial $x^2 - t$, and write $x = x_0 + x_1 i$ as before. If we cut X open along the double locus of Figure 11.9.15, the negative real t-axis, it decomposes into the two parts $x_0 > 0$ and $x_0 < 0$. Each of these parts projects bijectively to T, provided that we disregard what happens along the cut.

Turning this procedure around, we can construct a branched covering isomorphic to X in the following way: We stack two copies S_1, S_2 of the complex plane over T and cut them open along the negative real axis. These copies of T will be called *sheets*. Then we glue *side A* of the cut on S_1 to *side B* of the cut on S_2 and vice versa. (This cannot be done in three-dimensional space.)

$$\underline{\qquad\qquad \textit{side A} \qquad\qquad}\bullet$$
$$\textit{side B}$$

(15.9.6)   Sides A and B.

Suppose we are given an n-sheeted branched covering $X \to T$, and let $\Delta = \{p_1, \dots, p_k\}$ be the set of its branch points in T. For $\nu = 1, \dots, k$, we choose nonintersecting half lines C_ν that lead from p_ν to infinity. We cut T open along these half lines, and we also cut X open at all points that lie over them.

We should be specific about what we mean by cutting. Let's agree that cutting T open means removing all points of the half lines C_ν, including p_ν, and that cutting X open means removing all points that lie over those half lines.

Lemma 15.9.7 When X is cut open above the half lines C_ν, it decomposes as a union of n "sheets" $S_1, \dots, S_n$, which can be numbered arbitrarily. Each sheet projects bijectively to the cut plane T.

This is true because the cut surface X is an unbranched covering space of the cut plane T, which is a simply-connected set: Any loop in the cut plane can be contracted continuously to a point. It is intuitively plausible that every unbranched covering of a simply connected space decomposes completely. The sheet that contains a point p of X consists of all points that can be joined to p by a path without crossing the cuts. (This is an exercise in [Munkres], p. 342). $\square$

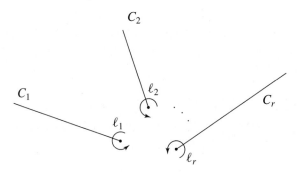

(15.9.8)   The Cut Plane T.

Now to reconstruct the surface X we take n copies of the cut plane T, we call them "sheets" and label them as $S_1, \dots, S_n$. We stack them up over T. Except for the cuts, the union of these sheets is our branched covering. We must describe the rule for gluing the sheets back together along the cuts. On T, we make a loop ℓ_ν that circles a branch point p_ν in the counterclockwise direction, and we call the side of C_ν we pass through before crossing C_ν as "$\textit{side A}$" and the side we pass through after crossing as "$\textit{side B}$." We label the corresponding sides of the sheet S_i as $\textit{side } A_i$ and $\textit{side } B_i$, respectively. Then the rule for gluing X amounts to instructions that *side A_i is glued to side B_j* for some j. This rule is described by the permutation σ_ν of the indices $\mathbf{1}, \ldots, \mathbf{n}$ that sends $\mathbf{i} \rightsquigarrow \mathbf{j}$.

It seems clear that we can construct a covering using an arbitrary set of permutations σ_ν, except that what should happen above the branch points themselves is not clear. To avoid ambiguity, we simply delete all branch points and all points that lie over them.

- *Branching Data:* For $\nu = 1, \ldots, r$, a permutation σ_ν of the indices $\mathbf{1}, \ldots, \mathbf{n}$.
- *Gluing Instructions:* If $\sigma_\nu(\mathbf{i}) = \mathbf{j}$, glue *side A_i* to *side B_j* along the cut C_ν.

When the gluing is done no cuts remain, and the union of the sheets is our covering. As is true of the Riemann surface depicted in Figure 11.9.15, four dimensions will be needed to do the gluing without self crossings.

If σ_ν is the trivial permutation, then each sheet is glued to itself above C_ν. Then that cut isn't needed, and we say that p_ν is not a *true* branch point.

The next corollary restates the above discussion.

Lemma 15.9.9 Every n-sheeted branched covering $X \to T$ is isomorphic to one constructed by the cut-and-paste process. □

Note: The numbering of the sheets is arbitrary, and the concept of a "top sheet" has no intrinsic meaning for a Riemann surface. If there were a top sheet, one could define x as a single valued function of t by choosing the value on that sheet. One can do this only after the Riemann surface has been cut open. Wandering around on X leads from one sheet to another. □

Except for the arbitrary numbering of the sheets, the permutations σ_ν are uniquely determined by the branched covering X. A change of numbering by a permutation ρ will change each σ_ν to the conjugate $\rho^{-1}\sigma_\nu\rho$.

Lemma 15.9.10 Let X and Y be branched coverings constructed by cut and paste, using the same points p_ν and half lines C_ν. Let the permutations defining their gluing data be σ_ν and τ_ν, respectively. Then X and Y are isomorphic branched coverings if and only if there is a permutation ρ such that $\tau_\nu = \rho^{-1}\sigma_\nu\rho$ for each ν. □

Lemma 15.9.11 The branched covering X constructed by cut and paste is path connected if and only if the permutations $\sigma_1, \ldots, \sigma_r$ generate a subgroup H of the symmetric group that operates transitively on the indices $\mathbf{1}, \ldots, \mathbf{n}$.

Proof. Each sheet is path connected. If the permutation σ_ν sends the index $\mathbf{i}$ to $\mathbf{j}$, the sheets S_i and S_j are glued together along the cut C_ν. Then there will be a short path across the cut that leads from a point of S_i to a point of S_j, and because the sheets themselves are path connected, all points of $S_i \cup S_j$ can be connected by paths. So X is path connected if and only if, for every pair of indices $\mathbf{i}, \mathbf{j}$, there is a sequence of the permutations σ_ν that carries $\mathbf{i} = \mathbf{i}_0 \rightsquigarrow \mathbf{i}_1 \rightsquigarrow \cdots \rightsquigarrow \mathbf{i}_d = \mathbf{j}$. This will be true if and only if H operates transitively. □

Example 15.9.12 The simplest k-sheeted path connected branched coverings of T are branched at a single point. Let Y be such a covering, branched only at the origin $t = 0$. The branching data for Y consists of a single permutation σ, the one that corresponds to a loop around the origin. The previous lemma tells us that, since Y is path connected, σ must operate transitively on the k indices, and the only permutations that operate transitively are the cyclic permutations of order k. So with suitable numbering of the sheets, $\sigma = (\mathbf{12} \cdots \mathbf{k})$. There is, up to isomorphism, exactly one k-sheeted branched covering branched only at the origin. The Riemann Existence Theorem tells us that there is, up to isomorphism, a unique field extension with this Riemann surface. It is not hard to guess this field extension: it is the one defined by the polynomial $y^k - t$, i.e., $K = F(y)$, where $y = \sqrt[k]{t}$. The Riemann surface Y has k sheets. It is branched only at the origin because each t different from zero has k complex kth roots.

There are two more things to be said here. First, the theorem asserts that this is the *only* field extension of degree k branched at the single point $t = 0$. This isn't obvious. Second, the same field extension $K = F(y)$ can be generated by many elements. For most choices of generators, it would not be obvious that there is only one *true* branch point. $\square$

Computing the Permutations

Given a polynomial $f(t, x)$, one wishes to determine the permutations σ_ν that define the gluing data of its Riemann surface. Two problems present themselves. First, the "local problem:" At each branch point p one must determine the permutation σ of the sheets that occurs when one circles that point. As we have seen, σ depends on the numbering of the sheets. Second, one must take care to use the same numbering for each branch point. This is the more difficult problem. A computer has no problem with it, but except in very simple cases, it is difficult to do by hand.

To compute the permutations, the computer chooses a "base point" b in the cut plane T and computes the n roots of the polynomial $f(b, x)$ numerically, with a suitable accuracy. It numbers these roots arbitrarily, say $\gamma_1, \ldots, \gamma_n$, and labels the sheets by calling S_i the sheet that contains the root γ_i. Then it walks to a point b_ν in the vicinity of a branch point p_ν, taking care not to cross any of the cuts. The roots γ_i vary continuously, and the computer can follow this variation by recomputing roots every time it takes a small step. This tells it how to label the sheets at the point b_ν. Then to determine the permutation σ_ν, the computer follows a counterclockwise loop ℓ_ν around p_ν, again recomputing roots as it goes along. Because the loop crosses the cut C_ν, the roots will have been permuted by σ_ν when the path returns to b_ν. In this way, the computer determines σ_ν. And because the numbering has been established at the base point b, it will be the same for all of the branch points.

Needless to say, doing this by hand is incredibly tedious. We find ways to get around the problem in the examples we present below.

The local problem can be solved by analytic methods, and we give an incomplete analysis here. The method is to relate the Riemann surface to one that we know, namely to the Riemann surface Y of the polynomial $y^k - t$. Let t_0 be a branch point of the Riemann surface X: $\{f(t, x) = 0\}$, where f is a polynomial of the form (15.9.1). Substituting $t = t_0$, we obtain the one-variable polynomial $f^0(x) = f(t_0, x)$.

Lemma 15.9.13 Let x_0 be a root of $f^0(x)$. Suppose that

- x_0 is a k-fold root of $f^0(x)$, and
- the partial derivative $\frac{\partial f}{\partial t}$ is not zero at the point (t_0, x_0).

Then the permutation of the sheets at the point t_0 contains a k-cycle.

Proof. We change variables to move the point (t_0, x_0) to the origin $(0, 0)$, so that $f^0(x) = f(0, x)$, and we write $f(t, x) = f^0(x) - tv(t, x)$. Then $\frac{\partial f}{\partial t}(0, 0) = -v(0, 0)$. Our hypotheses tell us that $v(0, 0) \neq 0$. Also, since $x = 0$ is a k-fold root of $f^0(x)$, that polynomial has the form $x^k u(x)$ where $u(x)$ is a polynomial in x and $u(0) \neq 0$. Then $f(t, x) = x^k u(x) - tv(t, x)$. Let $c = u(0)/v(0, 0)$. We replace t by $c^{-1}t$. The result is that now $u(0)/v(0, 0) = 1$.

We restrict attention to a small neighborhood U of the origin $(0, 0)$ in (t, x)-space, and write the equation $f = 0$ as

$$x^k u/v = t.$$

For (t, x) in U, u/v is near to 1. Among the kth roots of u/v, one will be near to 1, and that root, call it w, depends continuously on the point (t, x) in U. The other kth roots will be $\zeta^\nu w$, where $\zeta = e^{2\pi i/k}$.

Let $y = xw$. Then in our neighborhood U, the equation $f(t, x) = 0$ is equivalent with $y^k = t$. Therefore there are k sheets of our Riemann surface X that intersect U, and when we make a loop around the point $t = 0$, those k sheets will be permuted in the same way as the sheets of the Riemann surface Y, i.e., cyclically. $\square$

We now describe the branching data for a few simple polynomials. We take polynomials that are monic in x. The branch points will be the points t_0 at which $f(t_0, x)$ has multiple roots – the points at which $f(t_0, x)$ and $\frac{\partial f}{\partial x}(t_0, x)$ have a common root. Proposition 15.9.13 will be our main tool.

Examples 15.9.14 (a) $f(t, x) = x^2 - t^3 + t$, $\frac{\partial f}{\partial x} = 2x$, $\frac{\partial f}{\partial t} = -3t^2 + 1$.

Here X is a two-sheeted covering of T. There are three branch points $t = 0$, $t = 1$, and $t = -1$, and $\frac{\partial f}{\partial t} \neq 0$ at all of them. So the permutation of the sheets at each of these points contains a two-cycle. Since there are two sheets, each of the permutations is the transposition $(\mathbf{12})$. We don't need to be careful about the numbering when there are two sheets.

(b) We ask for a path connected, three-sheeted branched covering X of T branched at two points p_1 and p_2, and such that the permutation σ_i at the point p_i is a transposition.

We may label the sheets so that $\sigma_1 = (\mathbf{12})$. Then because X is path connected, the permutation σ_2 must be either $(\mathbf{23})$ or $(\mathbf{13})$ (15.9.11). Switching the sheets called S_1 and S_2 doesn't affect σ_1, but it interchanges the two other transpositions, so with suitable numbering of the sheets, $\sigma_1 = (\mathbf{1}\,\mathbf{2})$ and $\sigma_2 = (\mathbf{2}\,\mathbf{3})$. There is just one isomorphism class of such coverings.

The Riemann Existence theorem tells us that there is, up to isomorphism, a unique field extension K of F with this covering as its Riemann surface. Of course K will depend on the location of the two branch points but they can be moved to any position by a linear change of variable in t.

How do we find a polynomial $f(t, x)$ whose Riemann surface has this form? There is no general method, so one has to guess, and this case is simple enough that it can be guessed fairly easily. Since there is very minimal branching, we look for a very simple polynomial that is cubic in x. It takes a bit of courage to start looking, but one of the first attempts might be a polynomial of the form $x^3 + x + t$. This will work, but let's take $f(t, x) = x^3 - 3x + t$ instead. Then $\frac{\partial f}{\partial x} = 3x^2 - 3$ and $\frac{\partial f}{\partial t} = 1$. Substituting the roots $x = \pm 1$ of $\frac{\partial f}{\partial x}$ into f, one finds that the branch points are the points $t = \pm 2$. Since $\frac{\partial f}{\partial t}$ is nowhere zero, Proposition 15.9.13 applies.

There is a double root at the point $p_1 = (2, -1)$. So σ_1 contains 2-cycle, a transposition. Similarly, σ_2 is a transposition. So apart from the location of the two branch points, the Riemann surface X of the polynomial $f = x^3 - 3x + t$ has the desired properties, and $F[x]/(f)$ defines the field extension with that branching.

(c) $f(t, x) = x^3 - t^3 + t^2$, $\frac{\partial f}{\partial x} = 3x^2$, $\frac{\partial f}{\partial t} = -3t^2 + t$.

Here X is a three-sheeted covering of T. The branch points are at $t = 0$ and $t = 1$, and both $f(0, x)$ and $f(1, x)$ have triple roots. Let σ_0 and σ_1 denote the permutations of the sheets at the branch points. The partial derivative $\frac{\partial f}{\partial t}$ is not zero at $t = 1$, so the three sheets are permuted cyclically there. With suitable numbering, σ_1 will be $(\mathbf{1}\,\mathbf{2}\,\mathbf{3})$.

The point $t = 0$ presents problems. First, $\frac{\partial f}{\partial t}$ vanishes there. Second, how can we make sure to use the same numbering of the sheets at the two points? In the previous example, knowing that the Riemann surface must be path connected was enough to determine the branching. This fact gives us no information here because σ_1 operates transitively on the sheets by itself.

We use a trick that works only in the simplest cases. That is to compute the permutation that we get by walking around a large circle Γ. A large circular path will cross each of the cuts once (see Figure 15.9.8), so the sheets will be permuted by the product permutation $\sigma_0\sigma_1$, or by $\sigma_1\sigma_0$, depending on where we start. If we can determine that permutation, then since we know σ_1, we will be able to recover σ_0.

The substitution $t = u^{-1}$ maps T bijectively to the complex u-plane U, except that it is undefined at the points $t = 0$ and $u = 0$. Because $u \to 0$ as $t \to \infty$, the point $u = 0$ of U is called the *point at infinity* of T. Our large circle Γ in T corresponds to a small circle, we'll call it L, that circles the origin in U. However, a counterclockwise walk around Γ corresponds to a clockwise walk around L: If $t = re^{i\theta}$, then $u = r^{-1}e^{-i\theta}$.

We make the substitution $t = u^{-1}$ into the polynomial $f = x^3 - t^3 + t^2$ and clear denominators, obtaining $x^3u^3 - 1 + u$. When analyzing such a substitution, one usually has to substitute for x as well. It seems clear here that we should set $y = ux$. This gives us

$$y^3 - 1 + u.$$

Let's call this polynomial $g(u, y)$. The Riemann surfaces X and $Y : \{g = 0\}$ correspond via the substitution $(x, t) \leftrightarrow (y, u)$, which is defined and invertible except above the origins in the planes T and U. Therefore the permutation of sheets of X defined by a counterclockwise walk around Γ will be the same as the permutation of sheets of Y defined by a clockwise walk around L. That permutation is trivial, because the Riemann surface Y is not branched at $u = 0$. Therefore $\sigma_0 \sigma_1 = 1$, and since $\sigma_1 = (\mathbf{1\,2\,3})$, $\sigma_0 = (\mathbf{3\,2\,1})$. $\square$

15.10   THE FUNDAMENTAL THEOREM OF ALGEBRA

A field F is *algebraically closed* if every polynomial of positive degree with coefficients in F has a root in F. The Fundamental Theorem of Algebra asserts that the field of complex numbers is algebraically closed.

Theorem 15.10.1 Fundamental Theorem of Algebra. Every nonconstant polynomial with complex coefficients has a complex root.

There are several proofs of this theorem, and one of them is particularly appealing. We present it in outline. We must prove that a nonconstant polynomial

$$(15.10.2) \qquad f(x) = x^n + a_{n-1}x^{n-1} + \cdots + a_1 x + a_0$$

with complex coefficients has a complex root. If $a_0 = 0$, then 0 is a root, so we may assume that $a_0 \neq 0$.

The rule $y = f(x)$ defines a function from the complex x-plane to the complex y-plane. Let C_r denote a circle of radius r about the origin in the complex x-plane, parametrized as $x = re^{i\theta}$, with $0 \leq \theta < 2\pi$. We inspect the image $f(C_r)$ of C_r.

To warm up, we consider the function defined by the polynomial $y = x^n = r^n e^{ni\theta}$. As θ runs from 0 to 2π, the point x travels once around the circle of radius r. At the same time, $n\theta$ runs from 0 to $2\pi n$. The point y winds n times around the circle of radius r^n.

Now let f be the polynomial (15.10.2). For sufficiently large r, x^n is the dominant term in $f(x)$. To make this precise, let M be the maximum absolute value of the coefficients a_i of f. Then if $|x| = r \geq 10nM$,

$$\left| f(x) - x^n \right| = \left| a_{n-1}x^{n-1} + \cdots + a_1 x + a_0 \right| \leq nM|x|^{n-1} \leq \tfrac{1}{10}r^n.$$

It follows from this inequality that, as θ runs from 0 to 2π and x^n winds n times around the circle of radius r^n, $f(x)$ also winds around the origin n times. A good way to visualize this conclusion is with the dog-on-a-leash model. If someone walks a dog n times around a large circular path, the dog also goes around n times, though perhaps following a different path. This will be true provided that the leash is shorter than the radius of the path. Here x^n represents the position of the person at the time θ, and $f(x)$ represents the position of the dog. The radius of the path is r^n and the length of the leash is $\tfrac{1}{10}r^n$.

We vary the radius r. Since f is a continuous function, the image $f(C_r)$ will vary continuously with r. When the radius r is very small, $f(C_r)$ makes a small loop around the constant term a_0 of f. This small loop won't wind around the origin at all. But as we just saw, $f(C_r)$ winds n times around the origin if r is large enough. The only explanation for this is that for some intermediate radius r', $f(C_{r'})$ passes through the origin. This means that for some point α on the circle $C_{r'}$, $f(\alpha) = 0$. Then α is a root of f.

> *I don't consider this algebra,*
> *but this doesn't mean that algebraists can't do it.*
>
> —Garrett Birkhoff

EXERCISES

Section 1   Examples of Fields

1.1. Let R be an integral domain that contains a field F as subring and that is finite-dimensional when viewed as vector space over F. Prove that R is a field.

1.2. Let F be a field, not of characteristic 2, and let $x^2 + bx + c = 0$ be a quadratic equation with coefficients in F. Prove that if δ is an element of F such that $\delta^2 = b^2 - 4c$, $x = (-b + \delta)/2a$ solves the quadratic equation in F. Prove also that if the discriminant $b^2 - 4c$ is not a square, the polynomial has no root in F.

1.3. Which subfields of $\mathbb{C}$ are dense subsets of $\mathbb{C}$?

Section 2   Algebraic and Transcendental Elements

2.1. Let α be a complex root of the polynomial $x^3 - 3x + 4$. Find the inverse of $\alpha^2 + \alpha + 1$ in the form $a + b\alpha + c\alpha^2$, with a, b, c in $\mathbb{Q}$.

2.2. Let $f(x) = x^n - a_{n-1}x^{n-1} + \cdots \pm a_0$ be an irreducible polynomial over F, and let α be a root of f in an extension field K. Determine the element α^{-1} explicitly in terms of α and of the coefficients a_i.

2.3. Let $\beta = \omega\sqrt[3]{2}$, where $\omega = e^{2\pi i/3}$, and let $K = \mathbb{Q}(\beta)$. Prove that the equation $x_1^2 + \cdots + x_k^2 = -1$ has no solution with x_i in K.

Section 3   The Degree of a Field Extension

3.1. Let F be a field, and let α be an element that generates a field extension of F of degree 5. Prove that α^2 generates the same extension.

3.2. Prove that the polynomial $x^4 + 3x + 3$ is irreducible over the field $\mathbb{Q}[\sqrt[3]{2}]$.

3.3. Let $\zeta_n = e^{2\pi i/n}$. Prove that $\zeta_5 \notin \mathbb{Q}(\zeta_7)$.

3.4. Let $\zeta_n = e^{2\pi i/n}$. Determine the irreducible polynomial over $\mathbb{Q}$ and over $\mathbb{Q}(\zeta_3)$ of
(a) ζ_4, (b) ζ_6, (c) ζ_8, (d) ζ_9, (e) ζ_{10}, (f) ζ_{12}.

3.5. Determine the values of n such that ζ_n has degree at most 3 over $\mathbb{Q}$.

3.6. Let a be a positive rational number that is not a square in $\mathbb{Q}$. Prove that $\sqrt[4]{a}$ has degree 4 over $\mathbb{Q}$.

3.7. (a) Is i in the field $\mathbb{Q}(\sqrt[4]{-2})$? **(b)** Is $\sqrt[3]{5}$ in the field $\mathbb{Q}(\sqrt[3]{2})$?

3.8. Let α and β be complex numbers. Prove that if $\alpha + \beta$ and $\alpha\beta$ are algebraic numbers, then α and β are also algebraic numbers.

3.9. Let α and β be complex roots of irreducible polynomials $f(x)$ and $g(x)$ in $\mathbb{Q}[x]$. Let $K = \mathbb{Q}(\alpha)$ and $L = \mathbb{Q}(\beta)$. Prove that $f(x)$ is irreducible in $L[x]$ if and only if $g(x)$ is irreducible in $K[x]$.

3.10. A field extension K/F is an *algebraic extension* if every element of K is algebraic over F. Let K/F and L/K be algebraic field extensions. Prove that L/F is an algebraic extension.

Section 4   Finding the Irreducible Polynomial

4.1. Let $K = \mathbb{Q}(\alpha)$, where α is a root of $x^3 - x - 1$. Determine the irreducible polynomial for $\gamma = 1 + \alpha^2$ over $\mathbb{Q}$.

4.2. Determine the irreducible polynomial for $\alpha = \sqrt{3} + \sqrt{5}$ over the following fields.
(a) $\mathbb{Q}$,   **(b)** $\mathbb{Q}(\sqrt{5})$,   **(c)** $\mathbb{Q}(\sqrt{10})$,   **(d)** $\mathbb{Q}(\sqrt{15})$.

4.3. With reference to Example 15.4.4**(b)**, determine the irreducible polynomial for $\gamma = \alpha_1 + \alpha_2$ over $\mathbb{Q}$.

Section 5   Constructions with Ruler and Compass

5.1. Express $\cos 15°$ in terms of real square roots.

5.2. Prove that the regular pentagon can be constructed by ruler and compass
(a) by field theory,   **(b)** by finding an explicit construction.

5.3. Decide whether or not the regular 9-gon is constructible by ruler and compass.

5.4. Is it possible to construct a square whose area is equal to that of a given triangle?

5.5. Referring to the proof of Proposition 15.5.5, suppose that the discriminant D is negative. Determine the line that appears at the end of the proof geometrically.

5.6. Thinking of the plane as the complex plane, describe the set of constructible points as complex numbers.

Section 6   Adjoining Roots

6.1. Let F be a field of characteristic zero, let f' denote the derivative of a polynomial f in $F[x]$, and let g be an irreducible polynomial that is a common divisor of f and f'. Prove that g^2 divides f.

6.2. (a) Let F be a field of characteristic zero. Determine all square roots of elements of F that a quadratic extension of the form $F(\sqrt{a})$ contains.
(b) Classify quadratic extensions of $\mathbb{Q}$.

6.3. Determine the quadratic number fields $\mathbb{Q}[\sqrt{d}]$ that contain a primitive nth root of unity, for some integer n.

Section 7   Finite Fields

7.1. Identify the group $\mathbb{F}_4^+$.

7.2. Determine the irreducible polynomial of each of the elements of $\mathbb{F}_8$ in the list 15.7.8

7.3. Find a 13th root of 2 in the field $\mathbb{F}_{13}$.

7.4. Determine the number of irreducible polynomials of degree 3 over $\mathbb{F}_3$ and over $\mathbb{F}_5$.

7.5. Factor $x^9 - x$ and $x^{27} - x$ in $\mathbb{F}_3$.

7.6. Factor the polynomial $x^{16} - x$ over the fields $\mathbb{F}_4$ and $\mathbb{F}_8$:

7.7. Let K be a finite field. Prove that the product of the nonzero elements of K is -1.

7.8. The polynomials $f(x) = x^3 + x + 1$ and $g(x) = x^3 + x^2 + 1$ are irreducible over $\mathbb{F}_2$. Let K be the field extension obtained by adjoining a root of f, and let L be the extension obtained by adjoining a root of g. Describe explicitly an isomorphism from K to L, and determine the number of such isomorphisms.

7.9. Work this problem without appealing to Theorem (15.7.3). Let $F = \mathbb{F}_p$.

   (a) Determine the number of monic irreducible polynomials of degree 2 in $F[x]$.

   (b) Let $f(x)$ be an irreducible polynomial of degree 2 in $F[x]$. Prove that $K = F[x]/(f)$ is a field of order p^2, and that its elements have the form $a + b\alpha$, where a and b are in F and α is a root of f in K. Moreover, every such element with $b \neq 0$ is the root of an irreducible quadratic polynomial in $F[x]$.

   (c) Show that every polynomial of degree 2 in $F[x]$ has a root in K.

   (d) Show that all the fields K constructed as above for a given prime p are isomorphic.

*7.10. Let F be a finite field, and let $f(x)$ be a nonconstant polynomial whose derivative is the zero polynomial. Prove that f cannot be irreducible over F.

7.11. Let $f = ax^2 + bx + c$ with a, b, c in a ring R. Show that the ideal of the polynomial ring $R[x]$ that is generated by f and f' contains the discriminant, the constant polynomial $b^2 - 4ac$.

7.12. Let p be a prime integer, and let $q = p^r$ and $q' = p^k$. For which values of r and k does $x^{q'} - x$ divide $x^q - x$ in $\mathbb{Z}[x]$?

7.13. Prove that a finite subgroup of the multiplicative group of any field F is a cyclic group.

7.14. Find a formula in terms of the Euler ϕ function for the number of irreducible polynomials of degree n over the field $\mathbb{F}_p$.

Section 8   Primitive Elements

8.1. Prove that every finite extension of a finite field has a primitive element.

8.2. Determine all primitive elements for the extension $K = \mathbb{Q}(\sqrt{2}, \sqrt{3})$ of $\mathbb{Q}$.

Section 9   Function Fields

9.1. Let $f(x)$ be a polynomial with coefficients in a field F. Prove that if there is a rational function $r(x)$ such that $r^2 = f$, then r is a polynomial.

9.2. Determine the branch points and the gluing data for the Riemann surfaces of the following polynomials.

(a) $x^2 - t^2 + 1$, (b) $x^4 - t - 1$, (c) $x^3 - 3tx - 4t$, (d) $x^3 - 3x^2 - t$,

(e) $x^3 - t(t-1)$, (f) $x^3 - 3tx^2 + t$, (g) $x^4 + 4x + t$, (h) $x^3 - 3tx - t - t^2$.

9.3. (a) Determine the number of isomorphism classes of function fields K of degree 3 over $F = \mathbb{C}(t)$ that are ramified only at the points 1 and -1.

(b) Describe gluing data for the Riemann surface corresponding to each isomorphism class of fields as a pair of permutations.

(c) For each isomorphism class, find a polynomial $f(t, x)$ such that $K = F[t]/(f)$ represents the isomorphism class.

*9.4. Prove the Riemann Existence Theorem for quadratic extensions.

Hint: Show that up to isomorphism, a quadratic extension of F is described by the finite set $\{p_1, \ldots, p_k\}$ of its true branch points.

*9.5. Write a computer program that determines the branch points p_ν and the permutations σ_ν for the Riemann surface of a given polynomial.

Section 10   The Fundamental Theorem of Algebra

10.1. Prove that the subset of $\mathbb{C}$ consisting of the algebraic numbers is algebraically closed.

10.2. Construct an algebraically closed field that contains the prime field $\mathbb{F}_p$.

*10.3. With notation as at the end of the section, a comparison of the images $f(C_r)$ for varying radii shows another interesting geometric feature: For large r, the curve $f(C_r)$ makes n loops around the origin. Its total curvature is $2\pi n$. Assuming that the coefficient a_1 is not zero, the linear term $a_1 z + a_0$ dominates $f(z)$ for small z. Then for small r, $f(C_r)$ makes a single loop around a_0. Its total curvature is only 2π. Something happens to the loops as r varies. Explain.

*10.4. Write a computer program to illustrate the variation of $f(C_r)$ with r.

Miscellaneous Exercises

M.1. Let $K = F(\alpha)$ be a field extension generated by a transcendental element α, and let β be an element of K that is not in F. Prove that α is algebraic over the field $F(\beta)$.

M.2. Factor $x^7 + x + 1$ in $\mathbb{F}_7[x]$.

*M.3. Let $f(x)$ be an irreducible polynomial of degree 6 over a field F, and let K be a quadratic extension of F. What can be said about the degrees of the irreducible factors of f in $K[x]$?

M.4. **(a)** Let p be an odd prime. Prove that exactly half of the elements of $\mathbb{F}_p^\times$ are squares and that if α and β are nonsquares, then $\alpha\beta$ is a square.

(b) Prove the same assertion for any finite field of odd order.

(c) Prove that in a finite field of even order, every element is a square.

(d) Prove that the irreducible polynomial for $\gamma = \sqrt{2} + \sqrt{3}$ over $\mathbb{Q}$ is reducible modulo p for every prime p.

*M.5. Prove that any element of $GL_2(\mathbb{Z})$ of finite order has order 1, 2, 3, 4, or 6

(a) by using field theory.

(b) by applying the Crystallographic Restriction.

*M.6. **(a)** Prove that a rational function $f(t)$ that generates the field $\mathbb{C}(t)$ of all rational functions defines a bijective map $T' \to T'$.

(b) Prove a rational function $f(x)$ generates the field of rational functions $\mathbb{C}(x)$ if and only if it is of the form $(ax + b)/(cx + d)$, with $ad - bc \neq 0$.

(c) Identify the group of automorphisms of $\mathbb{C}(x)$ that are the identity on $\mathbb{C}$.

*M.7. Prove that the homomorphism $SL_2(\mathbb{Z}) \to SL_2(\mathbb{F}_p)$ obtained by reducing the matrix entries modulo p is surjective.

CHAPTER 16

Galois Theory

En un mot les calculs sont impraticables.

—Evariste Galois

We have seen that computation in an extension field generated by a single algebraic element α can be made simply, by identifying it with the formally constructed field $F[x]/(f)$, where f is the irreducible polynomomial for α over F. But suppose that f factors into linear factors in an extension field K. It isn't clear how to compute with all of the roots at the same time. To do that we need to know how they are related, and that depends on the particular case. The fundamental discovery that arose through the work of several people, especially of Lagrange and Galois, is that the relationships between the roots are best understood indirectly, in terms of symmetry. That symmetry is the topic of this chapter.

Beginning in Section 16.4, we assume that the fields we are working with have characteristic zero. The most important consequences of this assumption are:

- The roots of an irreducible polynomial over a field F are distinct (15.6.8).
- A finite extension field K/F has a primitive element (15.8.1).

16.1 SYMMETRIC FUNCTIONS

Let $R[u]$ denote the polynomial ring $R[u_1, \ldots, u_n]$ in n variables over a ring R. A permutation σ of the indices $\{1, \ldots, n\}$ operates on polynomials by permuting the variables:

(16.1.1) $$f = f(u_1, \ldots, u_n) \rightsquigarrow f(u_{\sigma 1}, \ldots, u_{\sigma n}) = \sigma(f).$$

In this way, σ defines an automorphism of $R[u]$ that we denote by σ too. Because σ acts as the identity on the constant polynomials, it is called an *R-automorphism*. The symmetric group S_n operates by R-automorphisms on the polynomial ring. A *symmetric* polynomial is one that is fixed by every permutation. The symmetric polynomials form a subring of the polynomial ring $R[u]$.

A polynomial g is symmetric if two monomials that are in the same orbit, such as $u_1 u_2^2$ and $u_2 u_3^2$, have the same coefficient in g. We call the sum of the monomials in an orbit an *orbit sum*. The orbit sums form a basis for the space of symmetric polynomials. The orbit sums of degree at most 3 in three variables are

$$1, \quad \boldsymbol{u_1 + u_2 + u_3}, \quad u_1^2 + u_2^2 + u_3^2, \quad \boldsymbol{u_1 u_2 + u_1 u_3 + u_2 u_3},$$
$$u_1^3 + u_3^3 + u_3^3, \quad u_1 u_2^2 + u_2 u_1^2 + u_1 u_3^2 + u_3 u_1^2 + u_2 u_3^2 + u_3 u_2^2, \quad \boldsymbol{u_1 u_2 u_3}.$$

The *elementary symmetric functions* are some special symmetric polynomials. When there are n variables, they are

$$s_1 = \sum_i u_i \qquad\qquad = u_1 + u_2 + \cdots + u_n$$

$$s_2 = \sum_{i<j} u_i u_j \qquad = u_1 u_2 + u_1 u_3 + \cdots$$

$$s_3 = \sum_{i<j<k} u_i u_j u_k \quad = u_1 u_2 u_3 + \cdots$$

$$\vdots \qquad\qquad \vdots \qquad\qquad\quad \vdots$$

$$s_n = \quad u_1 u_2 \cdots u_n \qquad = u_1 u_2 \cdots u_n.$$

Indices have been chosen so that s_i is the orbit sum of the monomial $u_1 u_2 \cdots u_i$. The elementary symmetric functions in three variables are shown above in boldface.

The elementary symmetric functions are the coefficients of the polynomial with variable roots $u_1, \ldots, u_n$:

$$P(x) = (x - u_1)(x - u_2) \cdots (x - u_n)$$
$$= x^n - s_1 x^{n-1} + s_2 x^{n-2} - \cdots \pm s_n.$$

(16.1.2)

When $n = 2$,

$$P(x) = (x - u_1)(x - u_2) = x^2 - (u_1 + u_2)x + (u_1 u_2),$$

and when $n = 3$,

$$P(x) = x^3 - (u_1 + u_2 + u_3)\,x^2 + (u_1 u_2 + u_1 u_3 + u_2 u_3)\,x - (u_1 u_2 u_3).$$

The order of the indices in (16.1.2) is the reverse of the one we have used for the coefficients of a polynomial previously, and the signs alternate. Because of the way these indices and signs appear, we will label undetermined coefficients of a polynomial in the analogous form in this chapter:

(16.1.3) $$f(x) = x^n - a_1 x^{n-1} + a_2 x^{n-2} - \cdots \pm a_n.$$

As before, we say that a polynomial f *splits completely* in a field K if it factors into linear factors, say

(16.1.4) $$f(x) = (x - \alpha_1) \cdots (x - \alpha_n),$$

with α_i in K. If so, then substituting $u_i = \alpha_i$ shows that the coefficients of f are obtained by evaluating the symmetric functions.

Lemma 16.1.5 If (16.1.4) is a factorization of the polynomial (16.1.3), then $a_i = s_i(\alpha_1, \ldots, \alpha_n)$.   □

Theorem 16.1.6 Symmetric Functions Theorem. Every symmetric polynomial $g(u_1, \ldots, u_n)$ with coefficients in a ring R can be written in a unique way as a polynomial in the elementary symmetric functions $s_1, \ldots, s_n$.

To be precise: If $g(u)$ is a symmetric polynomial, there is a unique polynomial $G(z_1, \ldots, z_n)$ with coefficients in R in another set of variables $z_1, \ldots, z_n$, such that $g(u)$ is obtained by the substitution $z_i \rightsquigarrow s_i$: $g(u_1, \ldots, u_n) = G(s_1, \ldots, s_n)$.

We prove the theorem below, but first, some examples:

Examples 16.1.7 (a) The symmetric polynomial $u_1^2 + \cdots + u_n^2$, because it has degree 2, is a linear combination $c_1 s_1^2 + c_2 s_2$. One can use special values of the variables to determine the coefficients. Substituting $u = (1, 0, \ldots, 0)$ shows that $c_1 = 1$, and substituting $u = (1, -1, 0, \ldots, 0)$ shows that $c_2 = -2$:

(16.1.8)
$$u_1^2 + \cdots + u_n^2 = s_1^2 - 2s_2.$$

(b) We use a different method for the symmetric polynomial

(16.1.9)
$$g(u) = u_1 u_2^2 + u_2 u_1^2 + u_1 u_3^2 + u_3 u_1^2 + u_2 u_3^2 + u_3 u_2^2$$

in the three variables u_1, u_2, u_3. The first step is to set $u_3 = 0$. We obtain the symmetric polynomial $g^\circ = u_1^2 u_2 + u_2^2 u_1$ in the remaining variables. Let s_i° denote the elementary symmetric functions in u_1, u_2: $s_1^\circ = u_1 + u_2$ and $s_2^\circ = u_1 u_2$. We notice that $g^\circ = s_1^\circ s_2^\circ$.

The second step is to compare the polynomial g with the three-variable symmetric polynomial $s_1 s_2$:
$$s_1 s_2 = (u_1 + u_2 + u_3)(u_1 u_2 + u_1 u_3 + u_2 u_3).$$

We won't expand the right side explicitly. Instead, we note that the expansion has nine terms, one of which is $u_1^2 u_2$. Since $s_1 s_2$ is symmetric, the orbit sum g of $u_1^2 u_2$, which has six terms, appears. The three remaining terms are equal to $u_1 u_2 u_3 = s_3$:

(16.1.10)
$$g = s_1 s_2 - 3s_3.$$

This computation is an example of a systematic method, and the proof of the Symmetric Functions Theorem, which we explain next, is based on that method.   □

Proof of the Symmetric Functions Theorem. There is nothing to show when $n = 1$, because $u_1 = s_1$ in that case. Proceeding by induction, we assume the theorem proved for symmetric functions in $n - 1$ variables. Given a symmetric polynomial g in $u_1, \ldots, u_n$, we consider the polynomial g° obtained by substituting zero for the last variable: $g^\circ(u_1, \ldots, u_{n-1}) = g(u, \ldots, u_{n-1}, 0)$. We note that g° is symmetric in $u_1, \ldots, u_{n-1}$. So by the induction hypothesis, g° may be written as a polynomial in the elementary symmetric functions in $u_1, \ldots, u_{n-1}$, which we label as $s_1^\circ, \ldots, s_{n-1}^\circ$:

$$s_1^\circ = u_1 + u_2 + \cdots + u_{n-1}, \text{ etc.}$$

There is a symmetric polynomial $Q(z_1, \ldots, z_{n-1})$ such that $g^\circ = Q(s_1^\circ, \ldots, s_{n-1}^\circ)$.

Lemma 16.1.11 Let g be a symmetric polynomial of degree d in the variables $u_1, \ldots, u_n$, and suppose that $g^\circ = Q(s_1^\circ, \ldots, s_{n-1}^\circ)$. Then $g = Q(s_1, \ldots, s_{n-1}) + s_n h$, where h is a symmetric polynomial in $u_1, \ldots, u_n$ of degree $d - n$.

Proof. Let $p(u_1, \ldots, u_n) = g(u_1, \ldots, u_n) - Q(s_1, \ldots, s_{n-1})$. This is a difference of symmetric polynomials, so it is symmetric, and if we set $u_n = 0$, we obtain $p(u_1, \ldots, u_{n-1}, 0) = g^\circ - Q(s^\circ) = 0$. Therefore u_n divides p. Because p is symmetric, every u_i divides p, and therefore s_n divides p. Writing $p = s_n h$, the polynomial h is symmetric. This gives us an equation of the form claimed by the lemma. $\square$

We go back to the proof of the Symmetric Functions Theorem. The lemma tells us that $g = Q(s) + s_n h$, where h is symmetric. A second induction, this time on the degree of a symmetric polynomial, allows us to conclude that h is a polynomial in the symmetric functions. Then so is g.

One can show that G is uniquely determined by going over this proof. $\square$

We give one more example of the systematic method. Let g be the orbit sum of the monomial $u_1 u_2^2$, but this time in four variables $u_1, \ldots, u_4$. Let $s_1, \ldots, s_4$ denote the elementary symmetric functions in four variables. We set $u_4 = 0$, and obtain formula (16.1.10), written now as $g^\circ = s_1^\circ s_2^\circ - 3s_3^\circ$. Then as in the above lemma,

$$g = s_1 s_2 - 3s_3 + s_4 h.$$

Since g has degree 3, $h = 0$. Formula 16.1.10 remains valid when g is the orbit sum of $u_1^2 u_2$ in any number $n \geq 3$ of variables.

Here is an important consequence of the Symmetric Functions Theorem:

Corollary 16.1.12 Suppose that a polynomial $f(x) = x^n - a_1 x^{n-1} + \cdots \pm a_n$ has coefficients in a field F, and that it splits completely in an extension field K, with roots $\alpha_1, \ldots, \alpha_n$. Let $g(u_1, \ldots, u_n)$ be a symmetric polynomial in $u_1, \ldots, u_n$ with coefficients in F. Then $g(\alpha_1, \ldots, \alpha_n)$ is an element of F.

For instance, $\alpha_1^k + \alpha_2^k + \cdots + \alpha_n^k$ will be an element of F.

Proof. The Symmetric Functions Theorem tells us that g is a polynomial in the elementary symmetric functions. Say that $g(u_1, \ldots, u_n) = G(s_1, \ldots, s_n)$, where $G(z)$ is a polynomial with coefficients in F. When we evaluate at $u = \alpha$, we obtain $s_i(\alpha) = a_i$ (16.1.5). So

(16.1.13) $$g(\alpha_1, \ldots, \alpha_n) = G(a_1, \ldots, a_n).$$

Because $a_1, \ldots, a_n$ are in F and G has coefficients in F, $G(a)$ is in F. $\square$

The next proposition provides a way to construct symmetric polynomials, starting with any polynomial.

Proposition 16.1.14 Let $p_1 = p_1(u_1, \ldots, u_n)$ be a polynomial, let $\{p_1, \ldots, p_k\}$ be its orbit for the operation of the symmetric group on the variables, and let $w = w_1, \ldots, w_k$ be another set of variables, where k is the number of polynomials in the orbit of p_1. (So k divides the order $n!$ of the symmetric group.) If $h(w_1, \ldots, w_k)$ is a symmetric polynomial in w, then $h(p_1, \ldots, p_k)$ is a symmetric polynomial in u.

Proof. Except that it is slightly confusing, this is nearly trivial. A permutation of the variables $u_1, \ldots, u_n$ permutes the set $\{p_1, \ldots, p_k\}$ because that set is an orbit. And because h is a symmetric polynomial, a permutation of $p_1, \ldots, p_k$ carries $h(p_1, \ldots, p_k)$ to itself. $\square$

Example 16.1.15 There are three variables u_1, u_2, u_3 and $p_1 = u_1^2 + u_2 u_3$. The orbit of p_1 consists of three polynomials:

$$p_1 = u_1^2 + u_2 u_3, \quad p_2 = u_2^2 + u_3 u_1, \quad p_3 = u_3^2 + u_1 u_2.$$

We substitute $w = p$ into the symmetric polynomial $w_1 w_2 + w_1 w_3 + w_2 w_3$, obtaining a symmetric polynomial in u:

$$\overset{\text{3 terms}}{} \quad \overset{\text{6 terms}}{} \quad \overset{\text{3 terms}}{}$$
$$p_1 p_2 + p_2 p_3 + p_3 p_1 = (u_1^2 u_2^2 + \cdots) + (u_1^3 u_3 + \cdots) + (u_1 u_2 u_3^2 + \cdots).$$ $\square$

16.2 THE DISCRIMINANT

The most important symmetric polynomial, aside from the elementary symmetric functions, is the *discriminant* of the polynomial

$$P(x) = x^n - s_1 x^{n-1} + s_2 x^{n-2} - \cdots \pm s_n$$

with the variable roots $u_1, \ldots, u_n$. By definition, the discriminant is

$$(16.2.1) \qquad D(u) = (u_1 - u_2)^2 (u_1 - u_3)^2 \cdots (u_{n-1} - u_n)^2 = \prod_{i<j} (u_i - u_j)^2.$$

Its main properties are:

- $D(u)$ is a symmetric polynomial with integer coefficients.
- If $\alpha_1, \ldots, \alpha_n$ are elements of a field, then $D(\alpha) = 0$ if and only if two of the elements α_i are equal.

The Symmetric Functions Theorem tells us that the discriminant D can be written uniquely as an integer polynomial in the elementary symmetric functions. Let

$$(16.2.2) \qquad\qquad \Delta(z) = \Delta(z_1, \ldots, z_n)$$

be that polynomial, so that $D(u) = \Delta(s)$. When $n = 2$,

$$(16.2.3) \qquad D = (u_1 - u_2)^2 = s_1{}^2 - 4s_2, \quad \text{and} \quad \Delta(z) = z_1^2 - 4z_2.$$

This is the familiar formula for the discriminant of the quadratic polynomial $x^2 - s_1 x + s_2$, though the fact that D is the square of the difference of the roots wasn't emphasized when I was in school.

Unfortunately, D and Δ are very complicated when n is larger. I don't know what they are when $n > 3$. The discriminant of the general cubic polynomial

$$(16.2.4) \qquad\qquad P(x) = x^3 - s_1 x^2 + s_2 x - s_3$$

is already too complicated to remember:

$$(16.2.5) \qquad \begin{aligned} D &= (u_1 - u_2)^2 (u_1 - u_3)^2 (u_2 - u_3)^2 \\ &= -4s_1^3 s_3 + s_1^2 s_2^2 + 18 s_1 s_2 s_3 - 4 s_2^3 - 27 s_3^2, \end{aligned}$$

$$\Delta = -4z_1^3 z_3 + z_1^2 z_2^2 + 18 z_1 z_2 z_3 - 4 z_2^3 - 27 z_3^2.$$

These formulas remain true when substitutions are made for the variables u_i. If we are given particular elements $\alpha_1, \dots, \alpha_n$ in a ring R, and if

$$(x - \alpha_1)(x - \alpha_2) \cdots (x - \alpha_n) = x^n - a_1 x^{n-1} + a_2 x^{n-2} - \cdots \pm a_n,$$

then, substituting α_i for u_i,

$$D(\alpha_1, \dots, \alpha_n) = \prod_{i<j} (\alpha_i - \alpha_j)^2 = \Delta(a_1, \dots, a_n).$$

Whether or not a polynomial $f(x) = x^n - a_1 x^{n-1} + a_2 x^{n-2} - \cdots \pm a_n$ is a product of linear factors, its *discriminant* is defined to be the element $\Delta(a_1, \dots, a_n)$, where $\Delta(z)$ is the polynomial (16.2.2). If f has coefficients in a field F, then $\Delta(z)$ has coefficients in F and $\Delta(a)$ is an element of F.

The discriminant of a cubic becomes simpler when the coefficient of x^2 in $f(x)$ is zero. Provided that the characteristic is not 3, the quadratic term in the general polynomial (16.2.4) can be eliminated by a substitution analogous to completing squares, called a *Tschirnhausen transformation*,

$$(16.2.6) \qquad\qquad x = y + s_1/3.$$

If we write a cubic whose quadratic term vanishes as

$$(16.2.7) \qquad\qquad f(x) = x^3 + px + q,$$

the discriminant is obtained by substituting into (16.2.5):

$$(16.2.8) \qquad\qquad \Delta(0, p, -q) = -4p^3 - 27q^2.$$

Since the elementary symmetric function s_i has degree i in the variables u, it is convenient to assign the *weight* i to the variable z_i, and to define the *weighted degree* of a monomial $z_1^{e_1} z_2^{e_2} \cdots z_n^{e_n}$ to be $e_1 + 2e_2 + \cdots + ne_n$. Substitution of s_i for z_i into a monomial of weighted degree d in z yields a polynomial of ordinary degree d in $u_1, \dots, u_n$. For instance, $z_1 z_2$ has weighted degree 3, and $s_1 s_2 = (u_1 + \cdots)(u_1 u_2 + \cdots)$ has degree 3. If $g(u)$ is a symmetric polynomial of degree d, and if $G(z)$ is the polynomial such that $g(u) = G(s)$, then G will have weighted degree d in z.

The discriminant of the cubic (16.2.4) is a homogeneous polynomial of degree 6 in u. There are seven monomials in z_1, z_2, z_3 of weighted degree 6:

$$(16.2.9) \qquad z_1^6, \ z_1^4 z_2, \ z_1^2 z_2^2, \ z_2^3, \ z_1^3 z_3, \ z_1 z_2 z_3, \ z_3^2,$$

and Δ is an integer combination of those monomials. We'll determine the coefficients of the first four of these monomials using the systematic method: We set $u_3 = 0$ in $D = (u_1 - u_2)^2 (u_1 - u_3)^2 (u_2 - u_3)^2$, obtaining the symmetric polynomial $(u_1 - u_2)^2 u_1^2 u_2^2 = (s_1^{\circ 2} - 4s_2^{\circ}) s_2^{\circ 2}$ in u_1, u_2. Therefore $D = s_1^2 s_2^2 - 4 s_2^3 + s_3 h$, where h is a symmetric cubic polynomial. The coefficients of s_1^6 and $s_1^4 s_2$ are zero. I don't know an easy way to determine the remaining three coefficients of Δ, but one way is to assign some special values to the variables u_1, u_2, u_3.

16.3   SPLITTING FIELDS

Let f be a polynomial with coefficients in a field F, not necessarily an irreducible polynomial. A *splitting field* for f over F is an extension field K/F such that

- f splits completely in K, say $f(x) = (x - \alpha_1) \cdots (x - \alpha_n)$ with α_i in K, and
- K is generated by the roots: $K = F(\alpha_1, \ldots, \alpha_n)$.

The second condition implies that, for every element β of K, there is a polynomial $p(u_1, \ldots, u_n)$ with coefficients in F, such that $p(\alpha_1, \ldots, \alpha_n) = \beta$. In fact there will be many such polynomials: Since the roots are algebraic over F, some polynomials evaluate to zero.

If our field F is a subfield of the complex numbers $\mathbb{C}$, a splitting field K can be obtained simply by adjoining the complex roots of f to F, and we may refer to K as *the* splitting field of f. But if F is not a subfield of $\mathbb{C}$, we have to construct a splitting field abstractly, as was explained in the last chapter (Section 15.6).

Lemma 16.3.1

(a) If $F \subset L \subset K$ are fields, and if K is a splitting field of a polynomial f over F, then K is also a splitting field of the same polynomial over L.

(b) Every polynomial $f(x)$ in $F[x]$ has a splitting field.

(c) A splitting field is a finite extension of F, and every finite extension is contained in a splitting field.

Proof. **(a)** This is obvious.

(b) Given a polynomial f with coefficients in F, there is a field extension K' of F in which f splits completely (15.6.3). The subfield of K' generated by the roots of f will be a splitting field.

(c) A splitting field is generated by finitely many elements that are algebraic over F, so it is a finite extension of F. Conversely, a finite extension L/F is generated by finitely many elements, say $\gamma_1, \ldots, \gamma_k$, each of which is algebraic over F. Let g_i be the irreducible polynomial for γ_i over F, and let f be the product $g_1 \cdots g_k$. We may extend the field L to a splitting field K of f over L, and then K will be a splitting field over F too. $\qquad \square$

We now use symmetric functions to prove an amazing fact:

Theorem 16.3.2 Splitting Theorem. Let K be an extension of a field F that is a splitting field of a polynomial $f(x)$ with coefficients in F. If an irreducible polynomial $g(x)$ with coefficients in F has one root in K, then it splits completely in K.

This theorem provides a characteristic property of splitting fields. A splitting field K over F is a finite field extension with this property:

> *An irreducible polynomial over F with one root in K splits completely in K.*

Which polynomial is used to define K as a splitting field is not important.

Proof of the Splitting Theorem. Let f and g be as in the statement of the theorem. We are given a root β_1 of g in K, and we must show that g splits completely in K. Since g is irreducible, it is the irreducible polynomial for β_1 over F.

The splitting field K is generated over F by the roots $\alpha_1, \ldots, \alpha_n$ of f. Every element of K can be written as a polynomial in α, with coefficients in F. We choose a polynomial $p_1(u_1, \ldots, u_n)$ such that $p_1(\alpha) = \beta_1$.

Let $\{p_1, \ldots, p_k\}$ be the orbit of $p_1(u)$ for the operation of the symmetric group S_n on the polynomial ring $F[u_1, \ldots, u_n]$, and let $\beta_j = p_j(\alpha)$. So $\beta_1, \ldots, \beta_k$ are elements of K. We will prove the splitting theorem by showing that the polynomial

$$h(x) = (x - \beta_1) \cdots (x - \beta_k)$$

has coefficients in F. Suppose that this has been proved. Then since β_1 is a root of h, it will follow that the irreducible polynomial for β_1 over F, which is g, divides h, and since h splits completely in K, g does too.

Say that $h(x) = x^k - b_1 x^{k-1} + b_2 x^{k-2} - \cdots \pm b_k$. The coefficients $b_1, \ldots, b_k$ are obtained by evaluating elementary symmetric functions at $\beta = \beta_1, \ldots, \beta_k$. But these are the elementary symmetric functions in k variables. We introduce new variables $w_1, \ldots, w_k$, and we label the elementary symmetric functions in these variables as $s'_1(w), \ldots, s'_k(w)$, using a prime to remind us that the variables are the new ones. Then $b_j = s'_j(\beta)$.

We evaluate s'_j in two steps: First, we substitute $w = p$, i.e., $w_j = p_j(u)$. Because $s'_j(w)$ is symmetric in w, $s'_j(p)$ is a symmetric polynomial in u (16.1.14). Next, we substitute $u_i = \alpha_i$. Because $s'_j(p(u))$ is symmetric in u, $s'_j(p(\alpha))$ is in the field F (16.1.12). On the other hand, $s'_j(p(\alpha)) = s'_j(\beta) = b_j$. The coefficients b_j are in F.   □

16.4   ISOMORPHISMS OF FIELD EXTENSIONS

For the rest of the chapter, we assume that our fields have *characteristic zero*, and we won't mention this assumption again. The field extensions that we consider will be finite extensions. We need a few definitions:

• Let K and K' be extension fields of F. The concept of an *F-isomorphism* $\sigma : K \to K'$ was introduced before (see (15.2.9)). It is an isomorphism whose restriction to the subfield F is the identity map. An *F-automorphism* of an extension field K is an F-isomorphism from K to itself. The F-automorphisms of K are the symmetries of the field extension.

- The F-automorphisms of a finite extension K form a group called the *Galois group* of K over F, which is often denoted by $G(K/F)$.

- A finite extension K/F is a *Galois extension* if the order of its Galois group $G(K/F)$ is equal to the degree of the extension: $|G(K/F)| = [K:F]$.

   We will see below (16.6.2) that the order of the Galois group always divides the degree of the extension.

Example 16.4.1   The complex number field $\mathbb{C}$ is a Galois extension of the field $\mathbb{R}$ of real numbers. The Galois group $G(\mathbb{C}/\mathbb{R})$ is a cyclic group of order two, generated by the automorphism of complex conjugation. There is an analogous statement for any quadratic extension K/F. A quadratic extension is obtained by adjoining a square root, say that $K = F(\alpha)$, where $\alpha^2 = a$ is in F. The Galois group G of K/F has order two, and the element τ of G different from the identity interchanges the two square roots α and $-\alpha$. For instance, if $F = \mathbb{Q}$ and $K = \mathbb{Q}(\sqrt{2})$, there is an F-automorphism τ of K that sends $a + b\sqrt{2} \rightsquigarrow a - b\sqrt{2}$. We have seen this automorphism before.   □

Lemma 16.4.2   Let K and K' be extensions of a field F.

(a) Let $f(x)$ be a polynomial with coefficients in F, and let σ be an F-isomorphism from K to K'. If α is a root of f in K, then $\sigma(\alpha)$ is a root of f in K'.

(b) Suppose that K is generated over F by some elements $\alpha_1, \ldots, \alpha_n$. Let σ and σ' be F-isomorphisms $K \to K'$. If $\sigma(\alpha_i) = \sigma'(\alpha_i)$ for $i = 1, \ldots, n$, then $\sigma = \sigma'$. If an F-automorphism σ of K fixes all of the generators, it is the identity map.

(c) Let f be an irreducible polynomial with coefficients in F, and let α and α' be roots of f in K and K', respectively. There is a unique F-isomorphism $\sigma: F(\alpha) \to F(\alpha')$ that sends α to α'. If $F(\alpha) = F(\alpha')$, then σ is an F-automorphism.

Proof. **(a)** was proved in the last chapter (15.2.10). We omit the proof of **(b)**. In **(c)**, the existence of σ was proved in the last chapter (15.2.8), and **(b)** shows that σ is unique.   □

Proposition 16.4.3

(a) Let f be a polynomial with coefficients in F. An extension field L/F contains at most one splitting field of f over F.

(b) Let f be a polynomial with coefficients in F. Any two splitting fields of f over F are isomorphic extension fields.

Proof. **(a)** If L contains a splitting field of f, then f splits completely in L, say $f = (x - \alpha_1) \cdots (x - \alpha_n)$ with α_i in L. If β is any root of f in L, substitution into this product shows that $\beta = \alpha_i$ for some i. So f has no other roots in L, and the only splitting field of f that is contained in L is $F(\alpha_1, \ldots, \alpha_n)$.

(b) Let K_1 and K_2 be two splitting fields of f over F. The first splitting field K_1 is a finite extension of F, so it has a primitive element γ. Let g be the irreducible polynomial for γ over F. We choose an extension L of the second field K_2 in which g has a root γ', and we let K' denote the subfield $F(\gamma')$ of L generated by γ'. There is an F-isomorphism $\varphi: K_1 \to K'$

that sends γ to γ', and because K' is F-isomorphic to the splitting field K_1, it is also a splitting field of f. Then both K' and K_2 are splitting fields contained in the field L, and **(a)** shows that they are equal. Therefore φ is an F-isomorphism from K_1 to K_2. $\square$

16.5   FIXED FIELDS

Let H be a group of automorphisms of a field K. The *fixed field* of H, which is often denoted by K^H, is the set of elements of K that are fixed by every group element:

$$(16.5.1) \qquad\qquad K^H = \{\alpha \in K \mid \sigma(\alpha) = \alpha \text{ for all } \sigma \text{ in } H\}.$$

It is easy to verify that K^H is a subfield of K, and that H is a subgroup of the Galois group $G(K/K^H)$. The Fixed Field Theorem below shows that, in fact, H is equal to $G(K/K^H)$.

Theorem 16.5.2 Let H be a finite group of automorphisms of a field K and let F denote the fixed field K^H. Let β_1 be an element of K, and let $\{\beta_1, \ldots \beta_r\}$ be the H-orbit of β_1.

(a) The irreducible polynomial for β_1 over F is $g(x) = (x - \beta_1) \cdots (x - \beta_r)$.

(b) β_1 is algebraic over F, and its degree over F is equal to the order of its orbit. Therefore the degree of β_1 over F divides the order of H.

Proof. Part **(b)** of the theorem follows from **(a)**. We prove **(a)**. Say that

$$g(x) = (x - \beta_1) \cdots (x - \beta_r) = x^r - b_1 x^{r-1} + \cdots \pm b_r.$$

The coefficients of g are symmetric functions of the orbit $\{\beta_1, \ldots, \beta_r\}$ (16.1.5). Since the elements of H permute the orbit, they fix the coefficients. Therefore g has coefficients in the fixed field.

Let h be a polynomial with coefficients in F that has β_1 as a root. For $i = 1, \ldots, r$, there is an element σ of H such that $\sigma(\beta_1) = \beta_i$. Because the elements of H are F-automorphisms of K and because h has coefficients in F, β_i is also a root of h (16.4.2)**(a)**. So $x - \beta_i$ divides f. Since this is true for every i, g divides f in $K[x]$ and in $F[x]$ (15.6.4)**(b)**. This shows that g generates the principal ideal of polynomials in $F[x]$ with root β_1, and that g is the irreducible polynomial for β_1 over F (15.2.3). $\square$

An extension field K/F is called *algebraic* if every element of K is algebraic over F.

Lemma 16.5.3 Let K be an algebraic extension of a field F that is not a finite extension of F. There exist elements in K whose degrees over F are arbitrarily large.

Proof. We form a chain of intermediate fields $F < F_1 < F_2 < \cdots$ as follows: We choose an element α_1 of K that is not in F, and we let $F_1 = F(\alpha_1)$. Then α_1 is algebraic over F, so $[F_1 : F] < \infty$, and therefore $F_1 < K$. Next, we choose an element α_2 of K that is not in F_1, and we let $F_2 = F(\alpha_1, \alpha_2)$. Then $[F_2 : F] < \infty$ and $F_1 < F_2 < K$. We choose α_3 in K, not in F_2, etc. This chain of fields gives us a strictly increasing chain of finite extensions of F. The degrees $[F_i:F]$ become arbitrarily large, while remaining finite. Each extension F_i/F has a primitive element γ_i, and the degrees of γ_i over F become arbitrarily large too. $\square$

Theorem 16.5.4 Fixed Field Theorem. Let H be a finite group of automorphisms of a field K, and let $F = K^H$ be its fixed field. Then K is a finite extension of F, and its degree $[K:F]$ is equal to the order $|H|$ of the group.

Proof. Let $F = K^H$ and let n be the order of H. Theorem 16.5.2 shows that the extension K/F is algebraic, and that the degree over F of any element β of K divides n. Therefore the degree $[K:F]$ is finite (16.5.3). Let γ be a primitive element for this extension. Every element σ of H is the identity on F, so if σ also fixes γ, it will be the identity map – the identity element of H. Therefore the stabilizer of γ is the trivial subgroup $\{1\}$ of H, and the orbit of γ has order n. Theorem 16.5.2 shows that γ has degree n over F. Since $K = F(\gamma)$, the degree $[K:F]$ is equal to n too. $\square$

Automorphisms of the field $\mathbb{C}(t)$ of rational functions in one variable provide examples that illustrate the Fixed Field Theorem and Theorem 16.5.2.

Example 16.5.5 Let $K = \mathbb{C}(t)$, and let σ and τ be the automorphisms of K that are the identity on $\mathbb{C}$ and such that $\sigma(t) = it$ and $\tau(t) = t^{-1}$. Then $\sigma^4 = 1$, $\tau^2 = 1$, and $\tau\sigma = \sigma^{-1}\tau$. Therefore σ and τ generate a group of automorphisms H that is isomorphic to the dihedral group D_4.

Lemma 16.5.6 The rational function $u = t^4 + t^{-4}$ is transcendental over $\mathbb{C}$.

Proof. Let $g(x) = x^d + c_{d-1}x^{d-1} + \cdots + c_0$ be a monic polynomial of degree d with complex coefficients. Then $t^{4d}g(u)$ is a monic polynomial of degree $8d$ in t. Since t is transcendental, $t^{4d}g(u) \neq 0$, and therefore $g(u) \neq 0$. $\square$

It follows from the lemma that the field $\mathbb{C}(u)$ is isomorphic to a field of rational functions in one variable. We show that it is the fixed field K^H. We note that u is fixed by σ and τ. So it is in the fixed field K^H, and therefore $\mathbb{C}(u) \subset K^H$. Theorem 16.5.2 tells us that the irreducible polynomial for t over K^H is the polynomial whose roots form its orbit. The orbit of t is

$$\left\{ t,\ it,\ -t,\ -it,\ t^{-1},\ -it^{-1},\ -t^{-1},\ it^{-1} \right\}$$

and the polynomial whose roots are the elements of this orbit is

$$(x^4 - t^4)(x^4 - t^{-4}) = x^8 - ux^4 + 1.$$

So t is a root of a polynomial of degree 8 with coefficients in $\mathbb{C}(u)$, and therefore the degree $[K:\mathbb{C}(u)]$ is at most 8. The Fixed Field Theorem asserts that $[K:K^H] = 8$. Since $\mathbb{C}(u) \subset K^H$, it follows that $\mathbb{C}(u) = K^H$. $\square$

This example illustrates a famous theorem:

Theorem 16.5.7 Lüroth's Theorem. Let F be a subfield of the field $\mathbb{C}(t)$ of rational functions that contains $\mathbb{C}$ and is not $\mathbb{C}$ itself. Then F is isomorphic to a field $\mathbb{C}(u)$ of rational functions.   □

16.6 GALOIS EXTENSIONS

We come now to the main topic of the chapter: Galois theory.

• If K is an extension field of F, an *intermediate field* L is a field such that $F \subset L \subset K$. An intermediate field is *proper* if it is neither F nor K.

If L is an intermediate field, then every L-automorphism of K will be an F-automorphism, and therefore

(16.6.1) $$G(K/L) \subset G(K/F).$$

Lemma 16.6.2

(a) The Galois group G of a finite field extension K/F is a finite group whose order divides the degree $[K:F]$ of the extension.

(b) Let H be a finite group of automorphisms of a field K. Then K is a Galois extension of its fixed field K^H, and H is the Galois group of K/K^H.

Proof. **(a)** By definition of F-automorphism, the elements of G act trivially on F, so F is contained in the fixed field K^G. Then $F \subset K^G \subset K$, so $[K:K^G]$ divides $[K:F]$. By the Fixed Field Theorem, $|G| = [K:K^G]$.

(b) By definition of K^H, the elements of H are K^H-automorphisms. Therefore H is a subgroup of the Galois group $G(K/K^H)$. Since $|G(K/K^H)|$ divides $[K:K^H]$ and $|H| = [K:K^H]$, the two groups are equal, and K is a Galois extension of K^H.   □

Lemma 16.6.3 Let γ_1 be a primitive element for a finite extension K of a field F and let $f(x)$ be the irreducible polynomial for γ_1 over F. Let $\gamma_1, \ldots, \gamma_r$ be the roots of f that are in K. There is a unique F-automorphism σ_i of K such that $\sigma_i(\gamma_1) = \gamma_i$. These are all of the F-automorphisms of K, so $G(K/F)$ has order r.

Proof. There is a unique F-isomorphism $\sigma_i : F(\gamma_1) \to F(\gamma_i)$ that sends $\gamma_1 \rightsquigarrow \gamma_i$ (16.4.2)**(c)**. We are given that $K = F(\gamma_1)$, and since $F(\gamma_i)$ has the same degree over F, $K = F(\gamma_i)$ too. Therefore σ_i is an F-automorphism of K. Every F-automorphism of K sends γ_1 to a root of f, so it is one of the automorphisms σ_i.   □

Theorem 16.6.4 Characteristic Properties of Galois Extensions. Let K/F be a finite extension and let G be its Galois group. The following are equivalent:

(a) K/F is a Galois extension, i.e., $|G| = [K:F]$,

(b) The fixed field K^G is equal to F,

(c) K is a splitting field over F.

Part **(b)** of the theorem can be used to show that an element of a Galois extension K is actually in the field F, and **(c)** can be used to show that an extension is Galois.

Proof of the Theorem. **(a)** ⇔ **(b)**: By the Fixed Field Theorem, $|G| = [K : K^G]$. Since $F \subset K^G \subset K$, $|G| = [K:F]$ if and only if $F = K^G$.

(a) ⇔ **(c)**: Let $n = [K : F]$. We choose a primitive element γ_1 for K over F. Let f be its irreducible polynomial over F. Since γ_1 is a primitive element, the degree of f is n. Let $\gamma_1, \ldots, \gamma_r$ be the roots of f that are in K. Lemma 16.6.3 tells us that $|G| = r$. So $|G| = [K:F]$, i.e., the extension is Galois, if and only if f splits completely in K. Because K is generated over F by γ_1, it is also generated by the set of all the roots of f, so K is a splitting field over F if and only if f splits completely in K. ☐

If K is the splitting field of a polynomial f over F, we may also refer to the Galois group $G(K/F)$ of the extension K/F also as the *Galois group of* f.

Corollary 16.6.5

(a) Every finite extension K/F is contained in a Galois extension.

(b) If K/F is a Galois extension, and if L is an intermediate field, then K is also a Galois extension of L, and the Galois group $G(K/L)$ is a subgroup of the Galois group $G(K/F)$.

Proof. Theorem 16.6.4 allows us to replace the phrase "Galois extension" by "splitting field." Then the Corollary follows from Lemmas 16.3.1 and 16.6.2. ☐

Theorem 16.6.6 Let K/F be a Galois extension with Galois group G, and let g be a polynomial with coefficients in F that splits completely in K. Let its roots in K be $\beta_1, \ldots, \beta_r$.

(a) The group G operates on the set of roots $\{\beta_i\}$.

(b) If K is a splitting field of g over F, the operation on the roots is faithful, and by its operation on the roots, G embeds as a subgroup of the symmetric group S_r.

(c) If g is irreducible over F, the operation on the roots is transitive.

(d) If K is a splitting field of g over F and g is irreducible over F, then G embeds as a transitive subgroup of S_r.

Proof. **(a)** is (16.4.2)**(a)** and **(b)** is (16.4.2)**(b)**. If g is irreducible, it is the irreducible polynomial for β_1 over F. Since F is the fixed field of G, Theorem 16.5.2 tells us that the roots β_i of g form the G-orbit of β_1. So the operation is transitive, as **(c)** asserts. Finally, **(d)** is the combination of **(b)** and **(c)**. ☐

This theorem is useful, though it doesn't suffice to determine the Galois group. Both the integer r and the embedding into S_r depend on f, not only on the Galois extension K. Also, the symmetric group S_r has several transitive subgroups when $r > 2$.

16.7 THE MAIN THEOREM

One of the most important parts of Galois theory is the determination of the intermediate fields. The Main Theorem of Galois theory asserts that when K/F is a Galois extension, the intermediate fields are in bijective correspondence with the subgroups of the Galois group. It will not be immediately clear why this fact is important; we will have to see it used to understand that.

Theorem 16.7.1 Main Theorem. Let K be a Galois extension of a field F, and let G be its Galois group. There is a bijective correspondence between subgroups of G and intermediate fields:

$$\{\text{subgroups}\} \longleftrightarrow \{\text{intermediate fields}\}.$$

This correspondence associates to a subgroup H its fixed field, and to an intermediate field L the Galois group of K over L. The maps

$$H \rightsquigarrow K^H \quad \text{and} \quad L \rightsquigarrow G(K/L).$$

are inverse functions.

Proof. We must show that the composition of the two maps in either order is the identity map, and the work has been done. Let H be a subgroup of G and let L be its fixed field. The Fixed Field Theorem tells us that $G(K/L) = H$. In the other order, let L be an intermediate field and let H be the Galois group of K over L. Then K is a Galois extension of L (Corollary 16.6.5(b)). Theorem 16.6.4 tells us that the fixed field of H is L. $\qquad\square$

Corollary 16.7.2 **(a)** The correspondence given by the Main Theorem reverses inclusions: If L and L' are intermediate fields and if H and H' are the corresponding subgroups, then $L \subset L'$ if and only if $H \supset H'$.

(b) The subgroup that corresponds to the field F is the whole group $G(K/F)$, and the subgroup that corresponds to K is the trivial subgroup $\{1\}$.

(c) If L corresponds to H, then $[K:L] = |H|$ and $[L:F] = [G:H]$.

   In **(c)**, the first equality follows from the facts that K is a Galois extension of L and that $H = G(K/L)$. Then the second equality follows, because

$$|G| = [K:F] = [K:L][L:F] \quad \text{and also} \quad |G| = |H|[G:H]. \qquad\square$$

Corollary 16.7.3 A finite field extension K/F has finitely many intermediate fields $F \subset L \subset K$.

Proof. This follows from the Main Theorem when K/F is a Galois extension, because a finite group has finitely many subgroups. Since we can embed any finite extension into a Galois extension, it is true for any finite extension. $\qquad\square$

Example 16.7.4 Let F be the field of rational numbers, and let $\alpha = \sqrt{3}$ and $\beta = \sqrt{5}$, so that $\alpha\beta = \sqrt{15}$. The splitting field $K = F(\alpha, \beta)$ of the polynomial $(x^2 - 3)(x^2 - 5)$ is a Galois extension of F of degree 4. Its Galois group G has order 4, so it is either the Klein four group or a cyclic group. It is easy to find three intermediate fields of degree 2 over F, namely $F(\alpha)$, $F(\beta)$, and $F(\alpha\beta)$. These three intermediate fields correspond to three proper subgroups of G. Therefore G is the Klein four group, which has three elements of order 2, hence three subgroups of order 2. The cyclic group of order 4 has only one subgroup of order 2.

The subgroups of order 2 are the only proper subgroups of G, so the Main Theorem tells us that there are no proper intermediate fields other than the three we have found. Consequently, an element $\gamma = a + b\alpha + c\beta + d\alpha\beta$ of K, with a, b, c, d in F, has degree 4 over F unless it is in one of the three proper intermediate fields, and this happens only when at least two of the coefficients b, c, d are zero. $\square$

Suppose that we are given a chain of fields $F \subset L \subset K$, and that K is a Galois extension of F. Then K is also a Galois extension of L. However, L needn't be a Galois extension of F. To complete the picture, we show that the intermediate fields L that are Galois extensions of F correspond to normal subgroups of G.

Theorem 16.7.5 Let K/F be a Galois extension with Galois group G, and let L be the fixed field L of a subgroup H of G. The extension L/F is a Galois extension if and only if H is a normal subgroup of G. If so, then the Galois group $G(L/F)$ is isomorphic to the quotient group G/H.

$$
\begin{array}{ll}
G = G(K/F) & \left.K\right\} \quad \begin{array}{l} H = G(K/L) \\ \textit{operates on } K \\ \textit{fixing } L \end{array} \\
\textit{operates on } K \left\{ \; L \right. \\
\textit{fixing } F & \left.\begin{array}{l} \\ \textit{If } H \textit{ is normal,} \\ \textit{then } G/H = G(L/F) \\ F \end{array}\right\} \; \textit{operates here}
\end{array}
$$

Proof. Let ϵ_1 be a primitive element for the extension L/F, and let g be the irreducible polynomial for ϵ_1 over F. This polynomial splits completely in the splitting field K; let its roots be $\epsilon_1, \ldots, \epsilon_r$. We have the following facts to work with:

• L/F is a Galois extension if and only if it is a splitting field, which happens when all of the roots ϵ_i are in L.

• If a root ϵ_i is in L, then $L = F(\epsilon_i)$, because ϵ_i and ϵ_1 have the same degree over F and $L = F(\epsilon_1)$.

• An element σ of G is the identity on L if and only if it fixes ϵ_1. So the stabilizer of ϵ_1 is equal to H.

• The operation of G on the set $\{\epsilon_1, \ldots, \epsilon_r\}$ is transitive: For any $i = 1, \ldots, r$, there is an element σ of G such that $\sigma(\epsilon_1) = \epsilon_i$ (16.4.2)**(c)**.

Let σ be an element of G, and say that $\sigma(\epsilon_1) = \epsilon_i$. Then $F(\epsilon_i) = L$ if and only if ϵ_i is in L, and if so, the stabilizer of ϵ_i will be equal to H. On the other hand, the stabilizer of $\sigma(\epsilon_1)$ is the conjugate group $\sigma H \sigma^{-1}$. Therefore K/F is a Galois extension if and only if $\sigma H \sigma^{-1} = H$ for all σ, i.e., if and only if H is a normal subgroup.

Suppose that L is a Galois extension of F. Then the roots ϵ_i are in L. An element σ of the Galois group G will map ϵ_1 to another root ϵ_i, and therefore it will map $L = F(\epsilon_1)$ to $F(\epsilon_i) = L$. So restricting σ to L defines an F-automorphism of L. This restriction gives us a homomorphism $\varphi: G \to G(L/F)$. The kernel of φ is the set of σ that restrict to the identity on L, which is H. Moreover, $|G/H| = [G:H] = |G(L/F)|$. The First Isomorphism Theorem tells us that G/H is isomorphic to $G(L/F)$. $\square$

In the next sections, we examine some of the most important situations in which Galois theory can be used.

16.8 CUBIC EQUATIONS

Let $f(x) = x^3 - a_1 x^2 + a_2 x - a_3$ be an irreducible polynomial over F, and let K be a splitting field of f over F. Say that the roots of f in K are $\alpha_1, \alpha_2, \alpha_3$. Then in $K[x]$,

(16.8.1) $$f(x) = (x - \alpha_1)(x - \alpha_2)(x - \alpha_3).$$

Since a_1 is in F and $a_1 = \alpha_1 + \alpha_2 + \alpha_3$, the third root α_3 is in the field generated by the first two roots. So we have a chain of extension fields

$$F \subset F(\alpha_1) \subset F(\alpha_1, \alpha_2) \quad \text{and} \quad F(\alpha_1, \alpha_2) = F(\alpha_1, \alpha_2, \alpha_3) = K.$$

Let L denote the field $F(\alpha_1)$. Since f is irreducible over F, $[L:F] = 3$. And since α_1 is in L, the polynomial f factors in $L[x]$:

(16.8.2) $$f(x) = (x - \alpha_1)q(x),$$

where q is the quadratic polynomial whose roots are α_2 and α_3. So K is obtained from L by adjoining a root of a quadratic polynomial. There are two cases: If q is irreducible over L, then $[K:L] = 2$ and $[K:F] = 6$. If q is reducible over L, then α_2 and α_3 are in L, $L = K$, and $[K:F] = 3$.

Examples 16.8.3 (a) $f(x) = x^3 + 3x + 1$ is irreducible over $\mathbb{Q}$, and its derivative is nowhere zero on the real line. Therefore f defines an increasing function of the real variable x that takes the value zero exactly once: f has one real root. This root does not generate the splitting field K, which also contains two complex roots. So $[K:\mathbb{Q}] = 6$.

(b) $f(x) = x^3 - 3x + 1$ is also irreducible over $\mathbb{Q}$. In this case, it happens that if α_1 is a root of f, then $\alpha_2 = \alpha_1^2 - 2$ is another root. This can be checked by substituting into f. So the splitting field K is equal to $\mathbb{Q}(\alpha_1)$ and $[K:\mathbb{Q}] = 3$. $\square$

We go back to an arbitrary irreducible cubic. By its operation on the roots, the Galois group G of K/F becomes a transitive subgroup of the symmetric group S_3 (16.4.2)(c). The transitive subgroups are S_3 and A_3 – a cyclic group of order 3. If $[K:F] = 3$, then $G = A_3$, and if $[K:F] = 6$, then $G = S_3$. To distinguish these two cases, we need to decide whether or not the quadratic polynomial $q(x)$ that appears in (16.8.2) is irreducible over the field $L = F(\alpha_1)$. Working in the field L is painful. We would rather make a computation in the field F. Fortunately, there is an element that makes it possible to decide, the square root δ of the discriminant (16.2.5) of f:

(16.8.4) $$\delta = (\alpha_1 - \alpha_2)(\alpha_1 - \alpha_3)(\alpha_2 - \alpha_3).$$

Its main properties are:

- δ is an element of K,
- $\delta \neq 0$ (because the roots α_i are distinct), and
- a permutation of the roots multiplies δ by the sign of the permutation.

Theorem 16.8.5 Galois Theory for a Cubic. Let K be the splitting field of an irreducible cubic polynomial f over a field F, let D be the discriminant of f, and let G be the Galois group of K/F.

- If D is a square in F, then $[K:F] = 3$ and G is the alternating group A_3.
- If D is not a square in F, then $[K:F] = 6$ and G is the symmetric group S_3.

The discriminant of $x^3 + 3x + 1$ is $-5 \cdot 3^3$, not a square, while the discrminant of $x^3 - 3x + 1$ is 3^4, a square (see 16.2.8)). This agrees with the discussion of the examples above.

Proof of Theorem 16.8.5. A permutation of the roots multiplies δ by the sign of the permutation. If δ is in F, it is fixed by every element of G. In that case odd permutations can't be in G, and therefore $G = A_3$ and $[K:F] = 3$. If δ isn't in F then it isn't fixed by G, so G contains an odd permutation. In that case, $G = S_3$ and $[K:F] = 6$. $\square$

The alternating group has no proper subgroups. So if $G = A_3$ there are no proper intermediate fields. This is obvious, because $[K:F] = 3$ is a prime. The symmetric group S_3 has four proper subgroups. With the usual notation, they are the three groups $\langle y \rangle, \langle xy \rangle, \langle x^2 y \rangle$ of order 2, and the group $\langle x \rangle$ of order 3, which is A_3. The Main Theorem tells us that when $G = S_3$, there are four proper intermediate fields. They are $F(\alpha_3), F(\alpha_2), F(\alpha_1)$, and $F(\delta)$.

16.9   QUARTIC EQUATIONS

Let $f(x)$ be an irreducible quartic polynomial with coefficients in F, and let the roots of f in a splitting field K over F be $\alpha_1, \alpha_2, \alpha_3, \alpha_4$. By its operation on the roots, the Galois group $G = G(K/F)$ is represented as a transitive subgroup of S_4 (16.6.6). The transitive subgroups are easy to determine because S_4 is isomorphic to the octahedral group, a rotation group. Any subgroup will be a rotation group too, so it will be one of the groups listed in Theorem 6.12.1. The transitive subgroups of S_4 are

$$(16.9.1) \qquad\qquad S_4, \ A_4, \ D_4, \ C_4, \ D_2.$$

There are three conjugate subgroups isomorphic to D_4, and three conjugate subgroups isomorphic to C_4. The subgroup D_2, the Klein four group, consists of the identity and the three products of disjoint transpositions. It is a normal subgroup of S_4 that we have seen before (2.5.15). (Some other subgroups of S_4 are isomorphic to D_2, but they aren't transitive.) Notice that the order of G, which is equal to the degree $[K:F]$, distinguishes all of these groups except the last two. Unfortunately, it isn't very easy to determine the degree.

We begin with a type of quartic polynomial that can be analyzed concretely. I learned this from Susan Landau [Landau].

Examples 16.9.2 Here F denotes the field $\mathbb{Q}$ of rational numbers.

(a) Let α be the "nested" square root $\alpha = \sqrt{4 + \sqrt{5}}$. To determine the irreducible polynomial for α over F, we guess that its roots might be $\pm\alpha$ and $\pm\alpha'$, where $\alpha' = \sqrt{4 - \sqrt{5}}$. Having made this guess, we expand the polynomial

$$f(x) = (x - \alpha)(x + \alpha)(x - \alpha')(x + \alpha') = x^4 - 8x^2 + 11.$$

It isn't very hard to show that this polynomial is irreducible over F. We'll leave the proof as an exercise. So it is the irreducible polynomial for α over F. Let K be the splitting field of f. Then

$$F \subset F(\alpha) \subset F(\alpha, \alpha') \quad \text{and} \quad F(\alpha, \alpha') = K.$$

Since f is irreducible, $[F(\alpha):F] = 4$ and since $\sqrt{5}$ is in $F(\alpha)$, $\alpha' = \sqrt{4 - \sqrt{5}}$ has degree at most 2 over $F(\alpha)$. We don't yet know whether or not α' is in the field $F(\alpha)$. In any case, $[K:F]$ is 4 or 8. The Galois group G of K/F also has order 4 or 8, so it is D_4, C_4, or D_2.

Which of the conjugate subgroups D_4 might operate depends on how we number the roots. Let's number them this way:

$$\alpha_1 = \alpha, \quad \alpha_2 = \alpha', \quad \alpha_3 = -\alpha, \quad \alpha_4 = -\alpha'.$$

With this ordering, an automorphism that sends $\alpha_1 \rightsquigarrow \alpha_i$ also sends $\alpha_3 \rightsquigarrow -\alpha_i$. The permutations with this property form the dihedral group D_4 generated by

(16.9.3) $\sigma = (1\,2\,3\,4)$ and $\tau = (2\,4)$.

Our Galois group is a subgroup of this group. It can be the whole group D_4, the cyclic group C_4 generated by σ, or the dihedral group D_2 generated by σ^2 and τ.

Note: We must be careful: Every element of this group D_4 permutes the roots, but we don't yet know which of these permutations come from automorphisms of K. A permutation that doesn't come from an automorphism tells us nothing about K. □

There is one permutation, $\rho = \sigma^2 = (1\,3)(2\,4)$, that is in all three of the groups D_4, C_4, and D_2, so it extends to an F-automorphism of K that we denote by ρ too. This automorphism generates a subgroup N of G of order 2.

To compute the fixed field K^N, we look for expressions in the roots that are fixed by ρ. It isn't hard to find some: $\alpha^2 = 4 + \sqrt{5}$ and $\alpha\alpha' = \sqrt{11}$. So K^N contains the field $L = F(\sqrt{5}, \sqrt{11})$. We inspect the chain of fields $F \subset L \subset K^N \subset K$. We have $[K:F] \leq 8$, $[L:F] = 4$, and $[K:K^N] = 2$ (Fixed Field Theorem). It follows that $L = K^N$, that $[K:F] = 8$, and that G is the dihedral group D_4.

(b) Let $\alpha = \sqrt{2 + \sqrt{2}}$. The irreducible polynomial for α over F is $x^4 - 4x^2 + 2$. Its roots are α, $\alpha' = \sqrt{2 - \sqrt{2}}$, $-\alpha$, $-\alpha'$ as before. Here $\alpha\alpha' = \sqrt{2}$, which is in the field $F(\alpha)$. Therefore α' is also in that field. The degree $[K:F]$ is 4, and G is either C_4 or D_2.

Because the operation of G on the roots is transitive, there is an element σ' of G that sends $\alpha \rightsquigarrow \alpha'$. Since $\alpha^2 = 2 + \sqrt{2}$ and $\alpha'^2 = 2 - \sqrt{2}$, σ' sends $\sqrt{2} \rightsquigarrow -\sqrt{2}$ and $\alpha\alpha' \rightsquigarrow -\alpha\alpha'$.

This implies that $\alpha' \rightsquigarrow -\alpha$. So $\sigma' = \sigma$. The Galois group is the cyclic group C_4.

(c) Let $\alpha = \sqrt{4 + \sqrt{7}}$. Its irreducible polynomial over F is $x^4 - 8x^2 + 9$. Here $\alpha\alpha' = 3$. Again, α' is in the field $F(\alpha)$, and the degree $[K:F]$ is 4. If an automorphism σ' sends $\alpha \rightsquigarrow \alpha'$, then since $\alpha\alpha' = 3$, it must send $\alpha' \rightsquigarrow \alpha$. The Galois group is D_2.

One can analyze any quartic polynomial of the form $x^4 + bx^2 + c$ in this way.   □

It is harder to analyze a general quartic

$$(16.9.4) \qquad\qquad f(x) = x^4 - a_1 x^3 + a_2 x^2 - a_3 x + a_4,$$

because its roots $\alpha_1, \ldots, \alpha_4$ can rarely be written explicitly in a useful way. The main method is to look for expressions in the roots that are fixed by some, but not all, of the permutations in S_4. The square root of the discriminant D is the first such expression:

$$\delta = \prod_{i<j}(\alpha_i - \alpha_j) = (\alpha_1 - \alpha_2)(\alpha_1 - \alpha_3)(\alpha_1 - \alpha_4)(\alpha_2 - \alpha_3)(\alpha_2 - \alpha_4)(\alpha_3 - \alpha_4).$$

Because the roots are distinct, δ isn't zero, and as is true for cubic equations (16.8.4), a permutation σ of the roots multiplies δ by the sign of the permutation. Even permutations fix δ and odd permutations do not fix δ.

Proposition 16.9.5 Let G be the Galois group of an irreducible quartic polynomial f. The discriminant D of f is a square in F if and only if G contains no odd permutation. Therefore

- If D is a square in F, then G is A_4 or D_2.
- If D is not a square in F, then G is S_4, D_4, or C_4.

Proof. D is a square in F if and only if δ is in F, which happens when every element of G fixes δ. The permutations that fix δ are the even permutations. The last statements are proved by looking at the list (16.9.1) of transitive subgroups of S_4.   □

There is an analogous statement for splitting fields of a polynomial of any degree.

Proposition 16.9.6 Let K be a splitting field over F of an irreducible polynomial f of degree n in $F[x]$, and let D be the discriminant of f. The Galois group $G(K/F)$ is a subgroup of the alternating group A_n if and only if D is a square in F.   □

Lagrange found another useful expression in the roots α_i, one that is special to quartic polynomials. Let

$$(16.9.7) \qquad \beta_1 = \alpha_1\alpha_2 + \alpha_3\alpha_4, \quad \beta_2 = \alpha_1\alpha_3 + \alpha_2\alpha_4, \quad \beta_3 = \alpha_1\alpha_4 + \alpha_2\alpha_3,$$

and let

$$g(x) = (x - \beta_1)(x - \beta_2)(x - \beta_3).$$

This polynomial is called the *resolvent cubic* of f. Every permutation of the roots α_i permutes the elements β_j, so the coefficients of g are symmetric functions in the roots. They are elements of F that can be computed when needed.

By a lucky accident, the fact that the roots of an irreducible quartic are distinct implies that the elements β_i are also distinct. For instance,

$$\beta_1 - \beta_2 = \alpha_1\alpha_2 + \alpha_3\alpha_4 - \alpha_1\alpha_3 - \alpha_2\alpha_4 = (\alpha_1 - \alpha_4)(\alpha_2 - \alpha_3).$$

Since the α_ν are distinct, $\beta_1 - \beta_2$ isn't zero. The discriminants of the polynomials f and g are actually *equal*.

Whether or not the resolvent cubic has a root in F gives us more information about the Galois group G.

Proposition 16.9.8 Let G be the Galois group of an irreducible quartic polynomial f over F, and let g be the resolvent cubic of f. Then g is irreducible if and only if the order of G is divisible by 3. Moreover,

- If g splits completely in F, then $G = D_2$.
- If g has one root in F, then $G = D_4$ or C_4.
- If g is irreducible over F, then $G = S_4$ or A_4.

Proof. The proof of the proposition is simple, but the fact that the three elements β_i are distinct is an essential point that could easily be overlooked. Let B denote the set $\{\beta_1, \beta_2, \beta_3\}$. It has order 3. The operation of the symmetric group S_4 on the roots α_ν defines a transitive operation on B, and the associated permutation representation is a homomorphism $\varphi \colon S_4 \to S_3$ that we have seen before (2.5.13). Its kernel is the subgroup D_2. If g splits completely in F, the Galois group operates trivially on B, and therefore $G = D_2$.

If g is irreducible over F, G operates transitively on B (16.6.6), so its order is divisible by three. Conversely, if $|G|$ is divisible by three, then G contains an element of order 3, say ρ. Since the kernel of φ is D_2, ρ does not operate trivially on B. It permutes the three elements cyclically. Therefore G operates transitively on B, and g is irreducible.

The rest of the proposition follows by looking back at the list (16.9.1). $\square$

Thus the polynomials $x^2 - D$, where D is the discriminant, and the resolvent cubic $g(x)$ nearly suffice to describe the Galois group. The results are summed up in this table:

(16.9.9)

	D a square	D not a square
g reducible	$G = D_2$	$G = D_4$ or C_4
g irreducible	$G = A_4$	$G = S_4$

Unfortunately, there is no simple expression in the roots that removes the remaining ambiguity (see Exercise M.11).

Note: The proof of Proposition 16.9.8 makes use of the particular formulas (16.9.7) to define a permutation of the set B in terms of a permutation of the roots α_ν. If a permutation of the roots comes from an F-automorphism, the permutation of B will be given by that automorphism. However, if the permutation doesn't come from an F-automorphism, the permutation of B defined using the formulas has no meaning for the field.

For example, let K be the splitting field of the polynomial $x^4 - 2$ over $\mathbb{Q}$. We index the roots from 1 to 4 in the order $\alpha_1 = \alpha$, $\alpha_2 = i\alpha$, $\alpha_3 = -\alpha$, $\alpha_4 = -i\alpha$, where α is the positive real fourth root of 2. Then $\beta_1 = 2i\sqrt{2}$, $\beta_2 = 0$, $\beta_3 = -2i\sqrt{2}$. The transposition $\epsilon = (12)$ isn't an element of the Galois group. When we use the formulas 16.9.7 to define how ϵ permutes the set B, the operation we obtain switches β_2 and β_3. Since $\beta_2 = 0$ and $\beta_3 \neq 0$, this permutation makes no sense algebraically. $\square$

16.10   ROOTS OF UNITY

In this section, F denotes the field $\mathbb{Q}$ of rational numbers. The subfield of the complex numbers generated over F by an nth root of unity $\zeta_n = e^{2\pi i/n}$ is called a *cyclotomic field*. We'll assume that n is a prime integer p. The irreducible polynomial for $\zeta = e^{2\pi i/p}$ over the rational numbers is

(16.10.1) $$f(x) = x^{p-1} + \cdots + x + 1$$

(Theorem 12.4.9). Its roots are the powers $\zeta, \zeta^2, \ldots, \zeta^{p-1}$, so ζ generates the splitting field of f, and therefore $K = F(\zeta)$ is a Galois extension of F of degree $p - 1$.

Proposition 16.10.2

(a) Let p be a prime, and let $\zeta = e^{2\pi i/p}$. The Galois group of $\mathbb{Q}(\zeta)$ over $\mathbb{Q}$ is a cyclic group of order $p - 1$. It is isomorphic to the multiplicative group $\mathbb{F}_p^\times$ of nonzero elements of the prime field $\mathbb{F}_p$.

(b) For any subfield F' of $\mathbb{C}$, the Galois group of $F'(\zeta)$ over F' is a cyclic group.

Proof. **(a)** With $F = \mathbb{Q}$, let G be the Galois group of $F(\zeta)$ over F. An element σ of G is determined by the image $\sigma(\zeta)$, which can be any one of the $p - 1$ roots of f. Let's call σ_i the element such that $\sigma_i(\zeta) = \zeta^i$. The exponent i is determined as a nonzero residue modulo p because $\zeta^p = 1$. So sending $\sigma_i \rightsquigarrow i$ defines a bijective map $\epsilon : G \to \mathbb{F}_p^\times$. The computation

$$\sigma_i \sigma_j(\zeta) = \sigma_i(\zeta^j) = \sigma_i(\zeta)^j = \zeta^{ij}$$

shows that ϵ is a homomorphism, and therefore an isomorphism. The fact that $\mathbb{F}_p^\times$ is cyclic is a part of Theorem 15.7.3.

The element σ_ν that sends $\zeta \rightsquigarrow \zeta^\nu$ generates G if and only if ν is a primitive root modulo p, a generator for the cyclic group $\mathbb{F}_p^\times$.

(b) An element σ of the Galois group $G' = G(F'(\zeta)/F')$ will also send ζ to a power ζ^ν. The proof above shows that G' is isomorphic to a subgroup of the cyclic group $\mathbb{F}_p^\times$. Therefore it is a cyclic group too. $\square$

Example 16.10.3 $p = 17$ and $\zeta = e^{i\theta}$, where $\theta = 2\pi/17$.

The residue of 3 is a primitive root modulo 17, so the Galois group $G = G(K/F)$ is a cyclic group of order 16, generated by the automorphism σ that sends $\zeta \rightsquigarrow \zeta^3$. There are five subteps, of orders 16, 8, 4, 2, and 1, generated by σ, σ^2, σ^4, σ^8, and 1, respectively. Let the fixed fields of the subgroups be $F = L_0 = K^{\langle\sigma\rangle}$, $L_1 = K^{\langle\sigma^2\rangle}$, $L_2 = K^{\langle\sigma^4\rangle}$, $L_3 = K^{\langle\sigma^8\rangle}$, and $L_4 = K$. They form a chain of fields $L_0 \subset L_1 \subset L_2 \subset L_3 \subset L_4$, where the degree of each extension L_i/L_{i-1} is 2. The Main Theorem tells us that these are the only intermediate fields.

Lemma 16.10.4 The field L_3 defined above is generated by $\cos\theta$, and it has degree 8 over F.

Proof. Let $L' = F(\cos\theta)$. Since $\zeta + \zeta^{-1} = 2\cos\theta$, $\cos\theta$ is in $K = F(\zeta)$. Moreover, ζ is a root of the quadratic polynomial $(x - \zeta)(x - \zeta^{-1}) = x^2 - 2(\cos\theta)x + 1$, which has coefficients in L', so $[K:L'] \leq 2$ and $[L':F] \geq 8$. Therefore L' is either L_3 or K, and since L' is a subfield of $\mathbb{R}$ but K is not, $L' = L_3$. $\square$

Corollary 16.10.5 The regular 17-gon can be constructed with ruler and compass.

Proof. The chain $F \subset L_1 \subset L_2 \subset L_3$ shows that we can reach the field L_3, which contains $\cos\theta$, by a sequence of three successive square root adjunctions, and since L_3 is a subfield of $\mathbb{R}$, these square roots are real. (See (15.5.10).) $\square$

The next lemma is useful for describing the quadratic extension L_1 of F:

Lemma 16.10.6 Let $\alpha = c_1\zeta + c_2\zeta^2 + \cdots + c_{p-2}\zeta^{p-2} + c_{p-1}\zeta^{p-1}$ be a linear combination with rational coefficients c_i, where $\zeta = e^{2\pi i/p}$ and p is prime. If α is a rational number, then $c_1 = c_2 = \ldots = c_{p-1}$, and $\alpha = -c_1$.

Proof. Since ζ is a root of f (16.10.1), we can solve for ζ^{p-1} and rewrite the given linear combination as $\alpha = (-c_{p-1})1 + (c_1 - c_{p-1})\zeta + \cdots + (c_{p-2} - c_{p-1})\zeta^{p-2}$. Because the powers $1, \zeta, \ldots, \zeta^{p-2}$ form a basis for K over F, this combination is a rational number only if all coefficients except $-c_{p-1}$ are equal to zero. If so, then $c_i = c_{p-1}$ for every i and $\alpha = -c_1$, as asserted. $\square$

Example 16.10.7 The case $p = 17$, continued.

The powers of the primitive root 3 modulo 17, listed in order, and with representatives for the congruence classes taken between -8 and 8, are

$$(16.10.8) \qquad 1, 3, -8, -7, -4, 5, -2, -6, -1, -3, 8, 7, 4, -5, 2, 6.$$

The automorphism σ of $K = F(\zeta)$ that sends ζ to ζ^3 generates the Galois group G, and it runs through the powers of ζ in the corresponding order:

$$(16.10.9) \qquad \zeta \rightsquigarrow \zeta^3 \rightsquigarrow \zeta^{-8} \rightsquigarrow \zeta^{-7} \rightsquigarrow \cdots$$

The G-orbit of ζ consists of the 16 powers of ζ different from 1.

Let H denote the subgroup $\langle\sigma^2\rangle$ of order 8. The G-orbit of ζ splits into two H-orbits that are obtained by taking every other term in the sequence of powers (16.10.9):

$$\{\zeta, \zeta^{-8}, \zeta^{-4}, \cdots\} \quad \text{and} \quad \{\zeta^3, \zeta^{-7}, \zeta^5, \cdots\}.$$

Let α_1 and α_2 denote the sums over these two orbits, respectively: $\alpha_1 = \zeta + \zeta^{-8} + \cdots$. The set $\{\alpha_1, \alpha_2\}$ is a G-orbit. Theorem 16.5.2 tells us that the elements α_i have degree 2 over the fixed field of G, which is F, and that the irreducible polynomial for α_i over F is $(x - \alpha_1)(x - \alpha_2)$. To determine this polynomial, we need to compute the two symmetric functions $s_1(\alpha) = \alpha_1 + \alpha_2$ and $s_2(\alpha) = \alpha_1\alpha_2$.

To begin with, we note that $s_1(\alpha)$ is the sum of all powers of ζ different from 1, so $s_1(\alpha) = -1$ (16.10.6). Next,

$$s_2(\alpha) = \alpha_1\alpha_2 = (\zeta + \zeta^{-8} + \cdots)(\zeta^3 + \zeta^{-7} + \cdots).$$

Writing α_i requires writing ζ many times, so we use a shorthand. We write

(16.10.10) $\qquad \alpha_1 = [1, -8, -4, -2, -1, 8, 4, 2]$, $\quad \alpha_2 = [3, -7, 5, -6, -3, 7, -5, 6]$.

This notation indicates that α_1 is the sum of the powers of ζ whose exponents are in the first bracketed string. To compute $s_2(\alpha)$, we must add each of the eight terms in the first string to those in the second string, modulo p, obtaining 64 exponents. Then $s_2(\alpha)$ will be the sum of the corresponding powers of ζ. Let's not do this explicitly. Since $s_2(\alpha)$ is a rational number, all powers different from $\zeta^0 = 1$ must occur the same number of times (16.10.6). We notice that we won't get any zeros when we do the addition, because a residue and its negative are in the same bracketed sequence. So the 64 terms must include four of each of the 16 nonzero exponents. Therefore $s_2(\alpha) = -4$. The irreducible polynomial for α_i over F is

(16.10.11) $\qquad\qquad (x - \alpha_1)(x - \alpha_2) = x^2 + x - 4.$

Its discriminant is 17, so $L_1 = F(\sqrt{17})$. $\qquad\qquad\qquad\qquad\qquad\qquad\qquad\qquad$ □

One can determine the extension field of degree 2 over F that is contained in the cyclotomic field $F(\zeta_p)$ for any odd prime p in the same way.

Theorem 16.10.12 Let p be a prime different from 2, and let L be the unique quadratic extension of $\mathbb{Q}$ contained in the cyclotomic field $\mathbb{Q}(\zeta_p)$. If $p \equiv 1$ modulo 4, then $L = \mathbb{Q}(\sqrt{p})$, and if $p \equiv 3$ modulo 4, then $L = \mathbb{Q}(\sqrt{-p})$.

This seems to be an occasion for "proof by example." The case that $p \equiv 1$ modulo 4 is illustrated by the prime 17, and the computation is analogous for any such prime. We'll illustrate the case $p \equiv 3$ modulo 4 by the prime 11. The residue of 2 is a primitive root modulo 11. Its powers list the nonzero residue classes modulo 11 in the order

$$1, 2, 4, -3, 5, -1, -2, -4, 3, -5.$$

Let $\zeta = \zeta_{11}$ and let σ be the automorphism that sends $\zeta \rightsquigarrow \zeta^2$. With shorthand notation as above, the orbit sums of σ^2 are

$$\alpha_1 = [1, 4, 5, -2, 3], \quad \alpha_2 = [2, -k, -1, -4, -5].$$

Here if k is in the list of exponents for the sum α_1, then $-k$ is in the list for α_2. Therefore zero occurs five times among the 25 terms in the list of exponents for $\alpha_1\alpha_2$, and this contributes 5 to $\alpha_1\alpha_2$. Since $\alpha_1\alpha_2$ is in $\mathbb{Q}$, the 20 remaining terms must consist of two of each of the 10 nonzero congruence classes modulo 11. The sum of these terms contributes -2. Therefore $\alpha_1\alpha_2 = 3$. The irreducible polynomial for α_i is $x^2 + x + 3$. Its discriminant is -11. □

Theorem (16.10.12) is a special case of a beautiful theorem of algebraic number theory.

Theorem 16.10.13 Kronecker-Weber Theorem. Every Galois extension of the field $\mathbb{Q}$ of rational numbers whose Galois group is abelian is contained in one of the cyclotomic fields $\mathbb{Q}(\zeta_n)$. $\qquad\square$

16.11 KUMMER EXTENSIONS

This section is devoted to the following theorem:

Theorem 16.11.1 Let F be a subfield of $\mathbb{C}$ that contains the pth root of unity $\zeta = e^{2\pi i/p}$, p prime, and let K/F be a Galois extension of degree p. Then K is obtained by adjoining a pth root. In other words, K is generated over F by an element β, with β^p in F.

Extensions of this type are often called *Kummer extensions*. The Galois group of a Kummer extension is a cyclic group of prime order.

The theorem is familiar for $p = 2$: Every extension of degree 2 can be obtained by adjoining a square root. But suppose that $p = 3$ and that F contains the cube root of unity $\omega = e^{2\pi i/3}$. If the discriminant of the irreducible cubic polynomial f (16.2.7) is a square in F, then the splitting field of f has degree 3 (16.8.5). The theorem asserts that the splitting field has the form $F(\sqrt[3]{b})$, for some b in F. This isn't obvious. If the discriminant is not a square, the roots cannot be obtained by adjoining a cube root. (This is Exercise 11.1.)

The next proposition completes the picture. Suppose that β is the pth root of a nonzero element b of F in an extension field K. Then it will be a root of the polynomial $g(x) = x^p - b$, and if ζ is in F, the roots of f in K will be $\zeta^\nu \beta$ for $\nu = 0, 1, \ldots, p-1$. So β will generate the splitting field of g over F.

Proposition 16.11.2 Let p be a prime, let F be a field that contains the pth root of unity $\zeta = e^{2\pi i/p}$, and let b be a nonzero element of F. The polynomial $g(x) = x^p - b$ is either irreducible over F, or else it splits completely.

Proof. Let K be a splitting field of g over F, and suppose that some root β of g is not in F. Then the degree $[K:F]$ will be greater than 1, so the Galois group $G = G(K/F)$ will contain an element σ different from the identity. Since β generates K over F, $\sigma(\beta)$ cannot be equal to β. So $\sigma(\beta) = \zeta^\nu \beta$ for some ν with $0 < \nu < p$. We also have $\sigma(\zeta) = \zeta$. Therefore $\sigma^2(\beta) = \zeta^\nu(\zeta^\nu \beta) = \zeta^{2\nu}\beta$, and in general, $\sigma^k(\beta) = \zeta^{k\nu}\beta$. Since $0 < \nu < p$ and p is prime, the multiples of ν run through all residues modulo p. This shows that G operates transitively on the p roots of g. Therefore g is irreducible over F. $\qquad\square$

Proof of Theorem (16.11.1). The proof is nice. We view K as a vector space over F, and we verify that an element σ of the Galois group G is a linear operator on K: If α and β are in K and c is in F, then $\sigma(c) = c$. Since σ is an automorphism,

$$\sigma(\alpha + \beta) = \sigma(\alpha) + \sigma(\beta) \quad \text{and} \quad \sigma(c\alpha) = \sigma(c)\sigma(\alpha) = c\sigma(\alpha),$$

We choose a generator σ for the cyclic Galois group G. Then $\sigma^p = 1$, so any eigenvalue λ of σ must satisfy the relation $\lambda^p = 1$, which means that λ is a power of ζ. These eigenvalues are in the field F by hypothesis. Moreover, a linear operator of order p has at least one eigenvalue different from 1. This is because, over the complex numbers, the matrix of σ is diagonalizable (see Theorem 4.7.14 or Corollary (10.3.9)). Its eigenvalues are the entries of the corresponding diagonal matrix Λ. If σ is not the identity, then $\Lambda \neq I$, so some diagonal entry must be different from 1.

Let β be an eigenvector of σ with eigenvalue $\lambda \neq 1$, and let $b = \beta^p$. Then $\sigma(\beta) = \lambda\beta$, hence $\sigma(b) = (\lambda\beta)^p = b$. Since σ generates G, b is in the fixed field, which is F, while β is not in F. Since $[K:F]$ is prime, $F(\beta) = K$. $\square$

With notation as in Theorem 16.11.1, say that K is the splitting field over F of an irreducible polynomial f of degree p. There is a simple expression in the roots of f that often yields an eigenvector for the operator σ. The permutation of the roots $\alpha_1, \ldots, \alpha_p$ of f that is defined by σ will be cyclic, so if we number the roots appropriately, σ will be the permutation $(\mathbf{1\,2} \cdots \mathbf{p})$. Let λ be an eigenvalue of σ, and let

(16.11.3) $$\beta = \alpha_1 + \lambda\alpha_2 + \cdots + \lambda^{p-1}\alpha_p.$$

Then $\sigma(\beta) = \alpha_2 + \lambda\alpha_3 \cdots + \lambda^{p-2}\alpha_{p-1} + \lambda^{p-1}\alpha_1 = \lambda^{-1}\beta$. So unless β happens to be zero, it will be an eigenvector with eigenvalue λ^{-1}.

Example 16.11.4 Kummer's theorem leads to a formula for the roots of a cubic polynomial that was discovered in the sixteenth century by Cardano and Tartaglia. The derivation that we outline here isn't as short as Cardano's, but it is easier to remember because it is systematic. We suppose that the quadratic coefficient of the cubic is zero, and to avoid denominators in the solution, we write it as

$$f(x) = x^3 + 3px + 2q.$$

Then $s_1 = 0$, $s_2 = 3p$, $s_3 = -2q$, and the discriminant is $D = -2^2 3^3 (q^2 + p^3)$.

Let the roots be u_1, u_2, u_3, numbered arbitrarily. With $\omega = e^{2\pi i/3}$, the elements

$$z = u_1 + \omega u_2 + \omega^2 u_3 \text{ and } z' = u_1 + \omega^2 u_2 + \omega u_3$$

are eigenvectors for the cyclic permutation $\sigma = (\mathbf{1\,2\,3})$. Since $1 + \omega + \omega^2 = 0$,

$$z + z' = s_1 + z + z' = u_1.$$

The cubes z^3 and z'^3 are fixed by σ, so according to Kummer's Theorem and Theorem 16.8.5, they can be written in terms of $p, q, \delta = \sqrt{D}$, and ω. When the cubes are written in this way, $u_1 = z + z'$ will be expressed as a sum of cube roots.

One makes the following computations. Let

$$A = u_1^2 u_2 + u_2^2 u_3 + u_3^2 u_1,$$
$$B = u_2^2 u_1 + u_3^2 u_2 + u_1^2 u_3.$$

Then

$$A - B = (u_1 - u_2)(u_1 - u_3)(u_2 - u_3) = \delta,$$
$$A + B = s_1 s_2 - 3s_3 = 6q.$$

Also, $u_1^3 + u_2^3 + u_3^3 = s_1^3 + 3s_1 s_2 + 3s_3 = -6q$.

One solves for A, B and expands z^3 and z'^3. The result of this computation is Cardano's formula:

(16.11.5) $$u_1 = \sqrt[3]{-q + \sqrt{q^2 + p^3}} + \sqrt[3]{-q - \sqrt{q^2 + p^3}}.$$

For instance, if $f(x) = x^3 + 3x + 2$, then $x = \sqrt[3]{-1 + \sqrt{2}} - \sqrt[3]{-1 - \sqrt{2}}$.

However, the formula is ambiguous. In the term $\sqrt[3]{-q + \sqrt{q^2 + p^3}}$, the square root can take two values, and when a square root is chosen, there are three possible values for the cube root, giving six ways to read that term. There are also six ways to read the other term. But f has only three roots. $\square$

16.12 QUINTIC EQUATIONS

The main motivation behind Galois's work was the problem of solving fifth-degree equations. A short time earlier, Abel had shown that the quintic equation

$$(16.12.1) \qquad x^5 - a_1 x^4 + a_2 x^3 - a_3 x^2 + a_4 x - a_5 = 0$$

with variable coefficients a_i couldn't be solved by radicals, but no equation with integer coefficients that couldn't be solved was known. Anyhow, the problem was over 200 years old, and it continued to interest people. In the meantime Galois's ideas have turned out to be much more important than the problem that motivated them. It is amazing that Galois was able to do what he did before the concept of a group was developed.

Proposition 16.12.2 Let F be a subfield of the complex numbers. The following two conditions on a complex number α are equivalent, and α is called *solvable* over F if it satisfies either one of them:

(a) There is a chain of subfields $F = F_0 \subset F_1 \subset \ldots \subset F_r = K$ of $\mathbb{C}$ such that α is in K, and

- $j = 1, \ldots, r$, $F_j = F_{j-1}(\beta_j)$, where a power of β_j is in F_{j-1}.

(b) There is a chain of subfields $F = F_0 \subset F_1 \subset \ldots \subset F_s = K$ of $\mathbb{C}$ such that α is in K, and

- for $j = 1, \ldots, r$, F_{j+1} is a Galois extension of F_j of prime degree.

The proof of the proposition isn't difficult, but it doesn't have much intrinsic interest, so we defer it to the end of the section. We need condition **(b)** in order to be able to use Galois theory. It is the more important characterization of solvability, and one can avoid the technicality of the proposition by accepting it as the definition.

Condition **(a)** means that F_j is generated over F_{j-1} by an nth root for some integer n (that depends on j). It is similar to the description of the real numbers that can be constructed by ruler and compass. In that description, only square roots of positive real numbers are allowed. Theoretically, one could unravel the extensions to write a solvable element α using a succession of nested roots. But as with Cardano's solution of the cubic equation, there is a great deal of ambiguity in a formula involving radicals, because there are n choices for an nth root. It is useless to write a root explicitly as a complicated expression in radicals. Indeed, Cardano's formula is useless.

Proposition 16.12.3 If α is a root of a polynomial of degree at most four with coefficients in a field F, then α is solvable over F.

Proof. For quadratic polynomials, the quadratic formula proves this. For cubics, Cardano's formula 16.11.7 gives the solution. If $f(x)$ is quartic, we begin by adjoining the square root δ of D. Then we use Cardano's formula to solve for a root of the resolvent cubic $g(x)$, and we adjoin it. At this point, Table 16.9.9 shows that the Galois group of f over the field that we obtain is a subgroup of the Klein four group. Therefore f can be solved by a sequence of at most two more square root extensions. $\square$

Theorem 16.12.4 Let f be an irreducible polynomial of degree 5 over a subfield F of the complex numbers, whose Galois group G is either the alternating group A_5 or the symmetric group S_5. Then the roots of f are not solvable over F.

Proof. If $G = S_5$, we replace F by the quadratic extension $F(\delta)$, where δ is the square root of the discriminant. If we can solve over F, we can solve over the larger field $F(\delta)$. So we may assume that G is the alternating group A_5, a simple group (7.5.4).

Our strategy is as follows: We consider a Galois extension of F'/F of prime degree p, with Galois group G', a cyclic group of order p, and we show that no progress toward solving the equation $f = 0$ is made when one replaces F by F'. We do this by showing that the Galois group of f over F' is again the alternating group A_5. Because A_5 contains an element of order 5, it cannot be the Galois group of a reducible polynomial of degree 5. So f remains irreducible over F'. Therefore there is no chain of type (16.12.2)**(b)**, and the roots of f are not solvable.

We choose such an extension F', and then we have two Galois extensions. The first, K/F, is the splitting field of the quintic polynomial f over F. Its Galois group is $G = A_5$. The second, F'/F, has a cyclic Galois group G' of order p, and since it is a Galois extension, it is the splitting field of some irreducible polynomial g over F.

Let K' be the splitting field over F of the product polynomial fg. It is generated by the complex roots $\alpha_1, \ldots, \alpha_5$ and $\beta_1, \ldots, \beta_p$ of f and g, respectively. The roots α_i generate the splitting field K of f, and the roots β_j generate the splitting field F' of g. The inclusions among the four fields are shown in the diagram below. Each of the extension fields is a Galois extension, and the Galois groups have been labeled in the diagram.

$$
\begin{array}{ccc}
 & K' & \\
H' \nearrow & & \nwarrow H \\
K & \mathcal{G} & F' \\
G \nwarrow & & \nearrow G' \\
 & F &
\end{array}
$$

Since K is a Galois extension of F, G is isomorphic to the quotient group $\mathcal{G}/H'$, and since F' is a Galois extension of F, G' is isomorphic to the quotient group $\mathcal{G}/H$ (16.7.5). Our plan is to show that H is isomorphic to G, i.e., that H is the alternating group A_5.

The group H' consists of the F-automorphisms of K' that fix the roots α_i, and H consists of the F-automorphisms that fix the roots β_j. If an F-automorphism of K' fixes the roots α_i and also the roots β_j, then since these roots generate K', it is the identity. Therefore $H \cap H'$ is the trivial group.

We restrict the canonical map $\mathcal{G} \to \mathcal{G}/H \approx G'$ to the subgroup H'. The kernel of this restriction is the trivial group $H \cap H'$, so the restriction is injective. It maps H' isomorphically to a subgroup of G'. By hypothesis, G' is cyclic of prime order p. So there are only two possibilities: either H' is the trivial group, or else H' is cyclic of order p.

Case 1: H' is the trivial group. Then the surjective map from $\mathcal{G}$ to the quotient group $\mathcal{G}/H' \approx G$ is an isomorphism, and $\mathcal{G}$ is isomorphic to the simple group $G = A_5$. This makes the existence of a surjective map from $\mathcal{G}$ to the cyclic quotient group $\mathcal{G}/H \approx G'$ impossible. So this case is ruled out.

Case 2: H' is cyclic of order p. Then $|\mathcal{G}| = |G||H'| = p|G|$ and also $|\mathcal{G}| = |G'||H| = p|H|$. Therefore G and H have the same order, 60. We restrict the canonical map $\mathcal{G} \to \mathcal{G}/H' \approx G$ to the subgroup H. The kernel of this restriction is the trivial group $H \cap H'$, so the restriction is injective. It maps H isomorphically to a subgroup of G. Since both groups have order 60, the restriction is an isomorphism, and $H \approx G = A_5$. $\qquad\square$

We now exhibit an irreducible polynomial of degree 5 over $\mathbb{Q}$, whose Galois group is S_5. The facts that 5 is a prime integer and that the Galois group G acts transitively on the roots $\alpha_1, \ldots, \alpha_5$ limit the possible Galois groups. Since the action is transitive, $|G|$ is divisible by 5. Thus G contains an element of order 5. The only elements of order 5 in S_5 are the 5-cycles. We leave the next lemma as an exercise.

Lemma 16.12.5 If a subgroup G of S_5 contains a 5-cycle and also a transposition, then $G = S_5$. $\qquad\square$

Corollary 16.12.6 Let $f(x)$ be an irreducible polynomial of degree 5 over $\mathbb{Q}$. If f has exactly three real roots, its Galois group G is the symmetric group, and hence its roots are not solvable.

Proof. Let the roots be $\alpha_1, \ldots, \alpha_5$, with $\alpha_1, \alpha_2, \alpha_3$ real and α_4, α_5 complex, and let K be the splitting field of f. The only permutations of the roots that fix the first three roots are the identity and the transposition $(\mathbf{45})$. Since $F(\alpha_1, \alpha_2, \alpha_3) \neq K$, that transposition must be in G. Since G operates transitively on the roots, it contains an element of order 5, a 5-cycle. So $G = S_5$. $\qquad\square$

Example 16.12.7 The polynomial $x^5 - 16x = x(x^2 - 4)(x^2 + 4)$ has three real roots. Of course it is reducible, but we we can add a small constant without changing the number of real roots. This is seen by looking at the graph of the polynomial. For instance, $x^5 - 16x + 2$ also has three real roots, and it is irreducible over $\mathbb{Q}$. Its roots are not solvable over $\mathbb{Q}$. $\qquad\square$

We now prove Proposition 16.12.2.

Lemma 16.12.8 Let K/F be a Galois extension whose Galois group G is abelian. There is a chain of intermediate fields $F = F_0 \subset F_1 \subset \cdots \subset F_m = K$ such that F_i/F_{i-1} is a Galois extension of prime degree for each i.

Proof. The abelian group G contains a subgroup H of prime order. This subgroup corresponds to an intermediate field L, and K is a Galois extension of L with group H. Because G is abelian, H is a normal subgroup, and therefore L is a Galois extension of F with abelian Galois group $\tilde{G} = G/H$. Since $\tilde{G}$ has smaller order than G, induction completes the proof. $\qquad\square$

Proof of Proposition 16.12.2. (a) $\Rightarrow$ (b) We begin with the chain of fields (a), and we add more extensions and more fields to the chain to arrive at a chain having the properties

(b). First, since $\sqrt[rs]{a} = \sqrt[r]{\sqrt[s]{a}}$, we can, at the cost of adding intermediate fields, suppose that all the roots that occur in our chain are pth roots for various primes p. We make a note of the primes $p_1, \ldots, p_k$ that occur, and set this chain aside for the moment.

We go back to the field F, and to start, we adjoin the p_νth roots of unity for $\nu = 1, \ldots, k$, one after the other. Each of these extensions is Galois, with a cyclic Galois group (Proposition 16.10.2(**b**)). Lemma 16.12.8 shows that each of them contains a chain whose layers are Galois extensions of prime degree.

Let F' be the field we obtain. We continue by adjoining the roots that we were given, but to F'. By Kummer theory, each of these root adjunctions will now be a Galois extension with a cyclic Galois group of prime order, unless it becomes a trivial extension. The field K' that we obtain at the end of our new chain will contain the last field K of the chain given to start, so α will be an element of K'. Therefore this new chain is one of the form **(b)**.

(b)$\Rightarrow$(a) Suppose that we are given a chain **(b)**, and consider one of the extensions in the chain, say $F_{i-1} \subset F_i$. It is a Galois extension of prime degree, say degree p. Theorem 16.11.1 shows that this extension is obtained by adjoining a pth root, provided that the pth roots of unity are in F_{i-1}. So we enlarge the chain, beginning by adjoining the required pth roots of unity to F. The enlarged chain will satisfy condition **(a)**. $\qquad\square$

> *Il parait après cela qu'il n'y a aucun fruit à tirer de la solution que nous proposons.*
>
> —Evariste Galois

EXERCISES

Section 1   Symmetric Functions

1.1. Determine the orbit of the polynomial below. If the polynomial is symmetric, write it in terms of the elementary symmetric functions.

(a) $u_1^2 u_2 + u_2^2 u_3 + u_3^2 u_1$ $(n = 3)$,
(b) $(u_1 + u_2)(u_2 + u_3)(u_1 + u_3)$ $(n = 3)$,
(c) $(u_1 - u_2)(u_2 - u_3)(u_1 - u_3)$ $(n = 3)$,
(d) $u_1^3 u_2 + u_2^3 u_3 + u_3^3 u_1 - u_1 u_2^3 - u_2 u_3^3 - u_3 u_1^3$ $(n = 3)$,
(e) $u_1^3 + u_2^3 + \cdots + u_n^3$.

1.2. Find two bases for the ring of symmetric polynomials, as a module over the ring R.

*1.3. Let $w_k = u_1^k + \cdots + u_n^k$.

(a) Prove *Newton's identities*: $w_k - s_1 w_{k-1} + \cdots \pm s_{k-1} w_1 \mp k s_k = 0$.
(b) Do $w_1, \ldots, w_n$ generate the ring of symmetric functions?

Section 2   The Discriminant

2.1. Prove that the discriminant is a symmetric function.

2.2. (a) Prove that the discriminant of a real cubic is non-negative if and only if the cubic has three real roots.

(b) Suppose that a real quartic polynomial has a positive discriminant. What can you say about the number of real roots?

2.3. (a) Prove that the Tschirnhausen substitution (16.2.6) does not change the discriminant of a cubic polynomial.

(b) Determine the coefficients p and q in (16.2.7) that are obtained from the general cubic (16.2.4) by the Tschirnhausen substitution.

2.4. Use undetermined coefficents to determine the discriminant of the polynomial

(a) $x^3 + px + q$, **(b)** $x^4 + px + q$, **(c)** $x^5 + px + q$.

2.5. Use the systematic method on the discriminant in four variables, to determine the coefficients in $\Delta(s_1, \ldots, s_4)$ of all monomials not divisible by s_4.

2.6. Let $u'_i = u_i + t, i = 1, 2, 3$. Compute the derivatives $\frac{d}{dt} s_i(u')$ and $\frac{d}{dt} \Delta(u')$, and use your results to verify Formula 16.2.5 for the discriminant of a cubic.

2.7. There are n variables. Let $m = u_1 u_2^2 u_3^3 \cdots u_{n-1}^{n-1}$ and let $p(u) = \sum_{\sigma \in A_n} \sigma(m)$. The S_n-orbit of $p(u)$ contains two elements, p and another polynomial q. Prove that $(p - q)^2 = D(u)$.

Section 3   Splitting Fields

3.1. Let f be a polynomial of degree n with coefficients in F and let K be a splitting field for f over F. Prove that $[K:F]$ divides $n!$.

3.2. Determine the degrees of the splitting fields of the following polynomials over $\mathbb{Q}$:

(a) $x^3 - 2$, **(b)** $x^4 - 1$, **(c)** $x^4 + 1$.

3.3. Let $F = \mathbb{F}_2(u)$ be the field of rational functions over the prime field $\mathbb{F}_2$. Prove that the polynomial $x^2 - u$ is irreducible over F, and that it has a double root in a splitting field.

Section 4   Isomorphisms of Field Extensions

4.1. (a) Determine all automorphisms of the field $\mathbb{Q}(\sqrt[3]{2})$, and of the field $\mathbb{Q}(\sqrt[3]{2}, \omega)$, where $\omega = e^{2\pi i/3}$.

(b) Let K be the splitting field over $\mathbb{Q}$ of $f(x) = (x^2 - 2x - 1)(x^2 - 2x - 7)$. Determine all automorphisms of K.

Section 5   Fixed Fields

5.1. For each of the following sets of automorphisms of the field of rational functions $\mathbb{C}(t)$, determine the group of automorphisms that they generate, and determine the fixed field explicitly.

(a) $\sigma(t) = t^{-1}$, **(b)** $\sigma(t) = it$, **(c)** $\sigma(t) = -t$, $\tau(t) = t^{-1}$,

(d) $\sigma(t) = \omega t$, $\tau(t) = t^{-1}$, where $\omega = e^{2\pi i/3}$.

5.2. Show that the automorphisms $\sigma(t) = \dfrac{t+i}{t-i}$ and $\tau(t) = \dfrac{it-i}{t+1}$ of $\mathbb{C}(t)$ generate a group isomorphic to the alternating group A_4, and determine the fixed field of this group.

5.3. Let $F = \mathbb{C}(t)$ be the field of rational functons in t. Prove that every element of F that is not in $\mathbb{C}$ is transcendental over $\mathbb{C}$.

Section 6   Galois Extensions

6.1. Let α be a complex root of the polynomial $x^3 + x + 1$ over $\mathbb{Q}$, and let K be a splitting field of this polynomial over $\mathbb{Q}$. Is $\sqrt{-31}$ in the field $\mathbb{Q}(\alpha)$? Is it in K?

6.2. Let $K = \mathbb{Q}(\sqrt{2}, \sqrt{3}, \sqrt{5})$. Determine $[K:\mathbb{Q}]$, prove that K is a Galois extension of $\mathbb{Q}$, and determine its Galois group.

6.3. Let $K \supset L \supset F$ be a chain of extension fields of degree 2. Show that K can be generated over F by the root of an irreducible quartic polynomial of the form $x^4 + bx^2 + c$.

Section 7   The Main Theorem

7.1. Determine the intermediate fields of an extension field of the form $F(\sqrt{a}, \sqrt{b})$ without appealing to the Main Theorem.

7.2. Let K/F be a Galois extension such that $G(K/F) \approx C_2 \times C_{12}$. How many intermediate fields L are there with **(a)** $[L:F] = 4$, **(b)** $[L:F] = 9$, **(c)** $G(K/L) \approx C_4$?

7.3. How many intermediate fields L with $[L:F] = 2$ are there when K/F is a Galois extension with Galois group **(a)** the alternating group A_4, **(b)** the dihedral group D_4?

7.4. Let $F = \mathbb{Q}$ and $K = \mathbb{Q}(\sqrt{2}, \sqrt{3}, \sqrt{5})$. Determine all intermediate fields.

7.5. Let $f(x)$ be an irreducible cubic polynomial over $\mathbb{Q}$ whose Galois group is S_3. Determine the possible Galois groups of the polynomial $(x^3 - 1) f(x)$.

7.6. Let K/F be a Galois extension whose Galois group is the symmetric group S_3. Is K the splitting field of an irreducible cubic polynomial over F?

7.7. **(a)** Determine the irreducible polynomial for $i + \sqrt{2}$ over $\mathbb{Q}$.

(b) Prove that the set $(1, i, \sqrt{2}, i\sqrt{2})$ is a basis for $\mathbb{Q}(i, \sqrt{2})$ over $\mathbb{Q}$.

7.8. Let α denote the positive real fourth root of 2. Factor the polynomial $x^4 - 2$ into irreducible factors over each of the fields $\mathbb{Q}$, $\mathbb{Q}(\sqrt{2})$, $\mathbb{Q}(\sqrt{2}, i)$, $\mathbb{Q}(\alpha)$, $\mathbb{Q}(\alpha, i)$.

7.9. Let $\zeta = e^{2\pi i/5}$. Prove that $K = \mathbb{Q}(\zeta)$ is a splitting field for the polynomial $x^5 - 1$ over $\mathbb{Q}$, and determine the degree $[K:\mathbb{Q}]$. Without using Theorem 16.7.1, prove that K is a Galois extension of $\mathbb{Q}$, and determine its Galois group.

7.10. Let K/F be a Galois extension with Galois group G, and let H be a subgroup of G. Prove that there exists an element $\beta \in K$ whose stabilizer is equal to H.

7.11. Let $\alpha = \sqrt[3]{2}$, $\beta = \sqrt{3}$, and $\gamma = \alpha + \beta$. Let L be the field $\mathbb{Q}(\alpha, \beta)$, and let K be the splitting field of the polynomial $(x^3 - 2)(x^2 - 3)$ over $\mathbb{Q}$.

(a) Determine the irreducible polynomial f for γ over $\mathbb{Q}$, and its roots in $\mathbb{C}$.

(b) Determine the Galois group of $k/\mathbb{Q}$.

Section 8   Cubic Equations

8.1. Let K/F be a Galois extension whose group G is the Klein four group D_2. Prove that K can be obtained by adjoining two square roots to F, and explain how G acts on K.

8.2. Determine the Galois groups of the following polynomials over $\mathbb{Q}$:
(a) $x^3 - 2$,   (b) $x^3 + 3x + 14$,   (c) $x^3 - 3x^2 + 1$,   (d) $x^3 - 21x + 7$,
(e) $x^3 + x^2 - 2x - 1$,   (f) $x^3 + x^2 - 2x + 1$.

8.3. Determine the quadratic polynomial $q(x)$ that appears in (16.8.2) explicitly, in terms of α_1 and the coefficients of f.

8.4. Let $K = \mathbb{Q}(\alpha)$, where α is a root of the polynomial $x^3 + 2x + 1$, and let $g(x) = x^3 + x + 1$. Does $g(x)$ have a root in K?

8.5. Let α_i be the roots of a cubic polynomial $f(x) = x^3 + px + q$. Find a formula for a second root α_2 in terms of the elements α_1, δ, and the coefficients of f.

Section 9   Quartic Equations

9.1. Let K be a Galois extension of F whose Galois group is the symmetric group S_4. Which integers occur as degrees of elements of K over F?

9.2. With reference to Example 16.9.2(a), write the element $\alpha + \alpha'$ as a nested square root. What other nested square roots does K contain?

9.3. Can $\sqrt{4 + \sqrt{7}}$ be written in the form $\sqrt{a} + \sqrt{b}$, with rational numbers a and b?

9.4. (a) Prove that the polynomial $x^4 - 8x^2 + 11$ is irreducible over $\mathbb{Q}$ in two ways: using the methods of Chapter 12 and computing with its roots.
(b) Do the same for the polynomial $x^4 - 8x^2 + 9$.
(c) Determine all intermediate fields when K is the splitting field of $x^4 - 8x^2 + 11$ over $\mathbb{Q}$.

9.5. Consider a nested square root $\alpha = \sqrt{r + \sqrt{t}}$ with r and t in a field F. Assume that α has degree 4 over F, let f be the irreducible polynomial of α over F, and let K be a splitting field of f over F.

(a) Compute the irreducible polynomial $f(x)$ for α over F. Prove that $G(K/F)$ is one of the groups D_4, C_4, or D_2.
(b) Explain how to determine the Galois group in terms of the element $r^2 - t$.
(c) Assume that the Galois group of K/F is the dihedral group D_4. Determine generators for all intermediate fields $F \subset L \subset K$.

9.6. Compute the discriminant of the quartic polynomial $x^4 + 1$, and determine its Galois group over $\mathbb{Q}$.

9.7. Assume that an extension field K/F has the form $K = F(\sqrt{a}, \sqrt{b})$. Determine all nested square roots $\sqrt{r + \sqrt{t}}$ that are in K, with r and t in F.

9.8. Determine whether or not the following nested radicals can be written in terms of unnested square roots, and if so, find an expression.

(a) $\sqrt{2 + \sqrt{11}}$, (b) $\sqrt{10 + 5\sqrt{2}}$, (c) $\sqrt{11 + 6\sqrt{2}}$, (d) $\sqrt{6 + \sqrt{11}}$, (e) $\sqrt{11 + \sqrt{6}}$.

9.9. (a) Determine the discriminant and the resolvent cubic of a polynomial of the form $f(x) = x^4 + rx + s$.
(b) Determine the Galois groups of $x^4 + 8x + 12$ and $x^4 + 8x - 12$ over $\mathbb{Q}$.
(c) Can the roots of the polynomial $x^4 + x - 5$ be constructed by ruler and compass?

9.10. (a) What are the possible Galois groups of an irreducible quartic polynomial over $\mathbb{Q}$ that has exactly two real roots?

(b) What are the possible Galois groups over $\mathbb{Q}$ of an irreducible quartic polynomial $f(x)$ whose discriminant is negative?

9.11. Let $F = \mathbb{Q}$, and let K be the splitting field of the polynomial $f(x) = x^4 - 2$ over F. The roots are $\alpha, -\alpha, i\alpha, -i\alpha$, with $\alpha = \sqrt[4]{2}$.

(a) Determine the Galois group $G = G(K/F)$, and the subgroup $H = G(K/F(i))$.

(b) Show how each element of H permutes the roots of f.

(c) Find all intermediate fields.

9.12. Determine the Galois groups of the following polynomials over $\mathbb{Q}$.

(a) $x^4 + 4x^2 + 2$, (b) $x^4 + 2x^2 + 4$, (c) $x^4 + 1$,

(d) $x^4 + x + 1$, (e) $x^4 + x^3 + x^2 + x + 1$, (f) $x^4 + x^2 + 1$.

9.13. Let K be the splitting field over $\mathbb{Q}$ of the polynomial $x^4 - 2x^2 - 1$. Determine the Galois group G of $K/\mathbb{Q}$, find all intermediate fields, and match them up with the subgroups of G.

*9.14. Let $F = \mathbb{Q}(\omega)$, where $\omega = e^{2\pi i/3}$. Determine the Galois group over F of the splitting field of (a) $\sqrt[3]{2 + \sqrt{2}}$, (b) $\sqrt{2 + \sqrt[3]{2}}$.

*9.15. Let K be the splitting field of an irreducible quartic polynomial $f(x)$ over F, and let the roots of $f(x)$ in K be $\alpha_1, \alpha_2, \alpha_3, \alpha_4$. Assume that the resolvent cubic $g(x)$ has a root $\beta_1 = \alpha_1\alpha_2 + \alpha_3\alpha_4$ in F. Express the root α_1 explicitly in terms of nested square roots.

9.16. Determine the resolvent cubic of the general quartic polynomial (16.9.4).

9.17. Determine the real numbers α of degree 4 over $\mathbb{Q}$ that can be constructed with ruler and compass, in terms of the Galois groups of their irreducible polynomials.

9.18. Prove that any Galois extension whose Galois group is the dihedral group D_4 is the splitting field of a polynomial of the form $x^4 + bx^2 + c$.

Section 10   Roots of Unity

10.1. Determine the degree of ζ_7 over the field $\mathbb{Q}(\zeta_3)$.

10.2. Let $\zeta = \zeta_{17}$. Find generators for the intermediate field L_2 described in Example 16.10.3.

10.3. Let $\zeta = \zeta_7$. Determine the degree of the following elements over $\mathbb{Q}$.

(a) $\zeta + \zeta^5$, (b) $\zeta^3 + \zeta^4$, (c) $\zeta^3 + \zeta^5 + \zeta^6$.

10.4. Let $\zeta = \zeta_{13}$. Determine the degrees of the following elements over $\mathbb{Q}$.

(a) $\zeta + \zeta^{12}$, (b) $\zeta + \zeta^2$, (c) $\zeta + \zeta^5 + \zeta^8$, (d) $\zeta^2 + \zeta^5 + \zeta^6$, (e) $\zeta + \zeta^5 + \zeta^8 + \zeta^{12}$,

(f) $\zeta + \zeta^2 + \zeta^5 + \zeta^{12}$, (g) $\zeta + \zeta^3 + \zeta^4 + \zeta^9 + \zeta^{10} + \zeta^{12}$.

10.5. Let $K = \mathbb{Q}(\zeta_p)$. Determine explicitly all intermediate fields when

(a) $p = 5$, (b) $p = 7$, (c) $p = 11$, (d) $p = 13$.

10.6. (a) Carry out the proof of Theorem 16.10.12.

(b) Prove the Kronecker-Weber Theorem for quadratic extensions.

10.7. Let $\zeta_n = e^{2\pi i/n}$ and let $K = \mathbb{Q}(\zeta_n)$.

    (a) Prove that K is a Galois extension of $\mathbb{Q}$.

    (b) Define an injective homomorphism $G(K/\mathbb{Q}) \to U$ to the group U of units in the ring $\mathbb{Z}/(n)$.

    (c) Prove that this homomorphism is bijective when $n = 6, 8, 12$. (In fact, this map is always bijective.)

10.8. Determine the Galois groups of the polynomials $x^8 - 1$, $x^{12} - 1$, $x^9 - 1$.

10.9. Let $f(x) = (x - \alpha_1) \cdots (x - \alpha_n)$.

    (a) Prove that the discriminant of f is $\pm f'(\alpha_1) \cdots f'(\alpha_n)$, where f' is the derivative of f, and determine the sign.

    (b) Use the formula to compute the discriminant of the polynomial $x^p - 1$, and use it to give another proof of Theorem 16.10.12.

10.10. With regard to the eigenvector γ described at the end of Section 16.11, show that at least one of the elements $\gamma_i = \alpha_1 + \zeta^i \alpha_2 + \cdots + \zeta^{(p-1)i} \alpha_p$ isn't zero.

Section 11   Kummer Extensions

11.1. Prove that if the discriminant of an irreducible cubic polynomial in $F[x]$ is not a square in F, then the roots cannot be obtained by adoining a cube root to F.

11.2. **(a)** Prove Proposition 16.11.2 without using Galois theory.

    (b) With F arbitrary, prove if $x^p - a$ is reducible in $F[x]$, then it has a root in F.

*11.3. Let F be a subfield of $\mathbb{C}$ that contains i, and let K be a Galois extension of F whose group is C_4. Is it true that K has the form $F(\alpha)$, with α^4 in F?

11.4. Carry out the computation to arrive at Cardano's formula (16.13.3).

11.5. **(a)** How does Cardano's formula (16.13.3) express the roots of the polynomials $x^3 + 3x$, $x^3 + 2$, $x^3 - 3x + 2$ and $x^3 - 3x + 2$?

    (b) What are the correct choices of roots in Cardano's formula?

Section 12   Quintic Equations

12.1. Is every Galois extension of degree 10 solvable?

12.2. Determine the transitive subgroups of S_5.

12.3. Let G be the Galois group of an irreducible quintic polynomial. Show that if G contains an element of order 3, then G is either S_5 or A_5.

12.4. Let $s_1, \ldots, s_n$ be the elementary symmetric functions in variables $u_1, \ldots, u_n$, and let F be a field.

    (a) Prove that the field $F(u)$ of rational functions in $u_1, \ldots, u_n$ is a Galois extension of the field $F(s_s, \ldots, s_n)$, and that its Galois group is the symmetric group S_n.

    (b) Suppose that $n = 5$, and let $w = u_1 u_2 + u_2 u_3 + u_3 u_4 + u_4 u_5 + u_5 u_1$. Determine the Galois group of $F(u)$ over the field $F(s, w)$.

    (c) Let G be a finite group. Prove that there exists a field F and a Galois extension K of F whose Galois group is G.

12.5. Let K be a Galois extension of $\mathbb{Q}$ whose degree is a power of 2, and such that $K \subset \mathbb{R}$. Prove that the elements of K can be constructed by ruler and compass.

12.6. Prove that if the Galois group of a polynomial f is a nonabelian simple group, then the roots are not solvable.

12.7. Find a polynomial of degree 7 over $\mathbb{Q}$ whose Galois group is S_7.

12.8. Let p be a prime. Prove that the symmetric group S_p is generated by any p-cycle together with any transposition.

Miscellaneous Problems

M.1. Let $F_1 \subset F_2$ be a field extension, and let f be a polynomial with coefficients in F_1. A splitting field K_2 of f over F_2 will contain a splitting field K_1 of f over F_1. What is the relation between the Galois groups $G(K_1/F_1)$ and $G(K_2/F_2)$?

M.2. Let L/F and K/L be Galois extensions. Is K/F necessarily a Galois extension?

M.3. (*Vandermonde determinant*)

    (a) Prove that the determinant of the matrix

$$\begin{bmatrix} 1 & u_1 & u_1{}^2 & \cdots & u_1^{n-1} \\ 1 & u_2 & & & u_2^{n-1} \\ \vdots & \vdots & & & \vdots \\ 1 & u_n & \cdots & \cdots & u_n^{n-1} \end{bmatrix}$$

    is a constant multiple of the square root of the discriminant $\delta(u) = \prod_{i<j}(u_i - u_j)$.

    (b) Determine the constant.

M.4. **(a)** The non-negative real numbers are those having a real square root. Use this fact to prove that the field $\mathbb{R}$ has no automorphism except the identity.

    *(b) Prove that $\mathbb{C}$ has no *continuous* automorphisms other than complex conjugation and the identity.

M.5. Let $K = \mathbb{F}_q$, where $q = p^r$.

    (a) Prove that the *Frobenius* map φ defined by $\varphi(x) = x^p$ is an automorphism of $F = \mathbb{F}_p$.

    (b) Prove that the Galois group $G(K/F)$ is a cyclic group of order r that is generated by the Frobenius map φ.

    (c) Prove that the Main Theorem of Galois theory is true for the extension K/F.

M.6. [1]Let K be a subfield of $\mathbb{C}$, and let G be its group of automorphisms. We can view G as acting on the point set K in the complex plane. The action will probably be discontinuous, but nevertheless, we can define an action on line segments $[\alpha, \beta]$ whose endpoints are in K by defining $g[\alpha, \beta] = [g\alpha, g\beta]$. Then G also acts on polygons whose vertices are in K.

    (a) Let $K = \mathbb{Q}(\zeta)$ where ζ is a primitive fifth root of 1. Find the G-orbit of the regular pentagon whose vertices are $1, \zeta, \zeta^2, \zeta^3, \zeta^4$.

    (b) Let α be the side length of the pentagon of **(a)**. Show that α^2 is in K, and find the irreducible equation for α over $\mathbb{Q}$.

---

[1]In memory of Bruce Renshaw.

*M.7. A polynomial f in $F[u_1, \ldots, u_n]$ is $\frac{1}{2}$-symmetric if $f(u_{\sigma 1}, \ldots u_{\sigma n}) = f(u_1, \ldots, u_n)$ for every even permutation σ, and skew-symmetric if $f(u_{\sigma 1}, \ldots, u_{\sigma n}) = (\text{sign } \sigma)$ $f(u_1, \ldots, u_n)$ for every permutation σ.

    (a) Prove that the square root of the discriminant $\delta = \prod_{i<j}(u_i - u_j)$ is skew-symmetric.

    (b) Prove that every $\frac{1}{2}$-symmetric polynomial has the form $f + g\delta$, where f, g are symmetric polynomials.

*M.8. [2]With variables u_0, u_1, u_2, u_3, let $p_i = (u_i - u_{i+1})(u_i - u_{i+2})(u_{i+1} - u_{i+2})$, indices read modulo 4. Determine

    (a) $\sum_{i=0}^{3} \frac{u_i}{p_{i+1}}$,   **(b)** $\sum_{i=0}^{3} \frac{u_i^3}{p_{i+1}}$.

*M.9. Let $f(t, x)$ be an irreducible polynomial in $\mathbb{C}[t, x]$ that is monic and cubic when regarded as a polynomial in x. Assume that for some t_0, the polynomial $f(t_0, x)$ has one simple root and one double root. Prove that the splitting field K of $f(x)$ over $\mathbb{C}(t)$ has degree 6.

*M.10. Let K be a finite extension of a field F, and let $f(x)$ be in $K[x]$. Prove that there is a nonzero polynomial $g(x)$ in $K[x]$ such that the product $f(x)g(x)$ is in $F[x]$.

*M.11. Let $f(x)$ be an irreducible quartic polynomial in $F[x]$ and let $\alpha_1, \alpha_2, \alpha_3, \alpha_4$ be its roots in a splitting field K. Assume that the resolvent cubic has a root $\beta = \alpha_1\alpha_2 + \alpha_3\alpha_4$ in F, but that the discriminant D is not a square in F. According to (16.9.9), the Galois group of K/F is either C_4 or D_4.

    (a) Determine the subgroup H of the group S_4 of permutations of the roots α_i, which stabilizes β explicitly. Don't forget to prove that no permutations other than those you list fix β.

    (b) Let $\gamma = \alpha_1\alpha_2 - \alpha_3\alpha_4$ and $\epsilon = \alpha_1 + \alpha_2 - \alpha_3 - \alpha_4$. Prove that γ^2 and ϵ^2 are in F.

    (c) Let δ be the square root of the discriminant. Prove that if $\gamma \neq 0$, then $\delta\gamma$ is a square in F if and only if $G = C_4$. Similarly, prove that if $\epsilon \neq 0$, then $\delta\epsilon$ is a square in F if and only if $G = C_4$.

    (d) Prove that γ and ϵ can't both be zero.

*M.12. A finite group G is *solvable* if it contains a chain of subgroups $G = H_0 \subset H_1 \subset \cdots \subset H_k = \{1\}$ such that for every $i = 1, \ldots, k$, H_i is a normal subgroup of H_{i-1}, and the quotient group H_i/H_{i+1} is a cyclic group. Let f be an irreducible polynomial over a field F, and let G be its Galois group. Prove that the roots of f are solvable over F if and only if G is a solvable group.

*M.13. [3]Let K/F be a Galois extension with Galois group G. If we think of K as an F-vector space, we obtain a representation of G on K. Let χ denote the character of this representation. Show that if F contains enough roots of unity, then χ is the character of the regular representation.

Wie weit diese Methoden reichen werden, muss erst
die Zukunft zeigen.

—Emmy Noether

---

[2]Suggested by Harold Stark.
[3]Suggested by Galyna Dobrovolska.

APPENDIX

Background Material

*Historically speaking, it is of course quite untrue
that mathematics is free from contradiction;
non-contradiction appears as a goal to be achieved,
not as a God-given quality that has been granted us once for all.*

—Nicolas Bourbaki

A.1 ABOUT PROOFS

What mathematicians consider an appropriate way to present a proof is not easy to make clear. One cannot give proofs that are complete in the sense that every step consists in applying a rule of logic to the previous step. Writing such a proof would take too long, and the main points wouldn't be emphasized. On the other hand, all difficult steps of the proof are supposed to be included. Someone reading the proof should be able to fill in as many details as needed to understand it. How to write a proof is a skill that can be learned only by experience.

Three general methods used to construct a proof are *dichotomy*, *induction*, and *contradiction*.

The word *dichotomy* means division into two parts. It is used to subdivide a problem into smaller, more easily managed pieces. Other names for this procedure are *case analysis* and *divide and conquer*.

Here is an example of dichotomy: By definition, the *binomial coefficient* $\binom{n}{k}$ (read n choose k) is the number of subsets of order k in the set of indices $\{1, 2, \ldots, n\}$. For example, $\binom{4}{2} = 6$. The set $\{1, 2, 3, 4\}$ has six subsets of order 2.

Proposition A.1.1 For every integer n and every $k \le n$, $\quad \binom{r}{k} = \binom{r-1}{k} + \binom{r-1}{k-1}$.

Proof. Let S be a subset of $\{1, 2, \ldots, n\}$ of order k. Then either $\mathbf{n}$ is in S or $\mathbf{n}$ is not in S. This is our dichotomy.

Case 1: $\mathbf{n}$ is not in S. In this case, S is actually a subset of $\{1, 2, \ldots, \mathbf{n}-1\}$. By definition, there are $\binom{n-1}{k}$ of these subsets.

Case 2: **n** is in S. Let $S' = S - \{\mathbf{n}\}$ be the set obtained by deleting the index **n** from the set S. Then S' is a subset of $\{1, 2, \ldots, \mathbf{n} - 1\}$, of order $n - 1$. There are $\binom{n-1}{k-1}$ such sets S'. Hence there are $\binom{n-1}{k-1}$ subsets of order k that contain n.

This gives us $\binom{n-1}{k} + \binom{n-1}{k-1}$ subsets of order k altogether. $\qquad\square$

The remarkable power of the method of dichotomy is shown here: In each of the two cases, $\mathbf{n} \in S$ and $\mathbf{n} \notin S$, we have an additional fact about our set S. This additional fact can be used in the proof.

Often a proof will require sorting through several possibilities, examining each in turn. This is dichotomy, or case analysis. It is analogous to the way Gray's *Manual of Botany* is used to determine the species of a plant. The procedure in Gray's Manual leads through a sequence of dichotomies. A typical one is "leaves opposite," or "leaves alternate." Classification of mathematical structures will also proceed through a sequence of dichotomies. They need not be spelled out formally in simple cases, but when one is dealing with a complicated range of possibilities, careful sorting is needed.

Induction is the main method for proving a sequence of statements P_n, indexed by positive integers n. To prove P_n for all n, the principle of induction requires us to do two things:

(A.1.2)

  (i) prove that P_1 is true, and

  (ii) prove that if, for some integer $k > 1$, P_k is true, then P_{k+1} is also true.

Sometimes it is more convenient to prove that if, for some integer $k \geq 0$, P_{k-1} is true, then P_k is true. This is just a change of the index.

Here are some examples of induction. If n is a positive integer, then $n!$ ("n factorial") is the product $1 \cdot 2 \cdots n$ of the integers from 1 to n. Also, 0! is defined to be 1.

Proposition A.1.3 $\dbinom{n}{k} = \dfrac{n!}{k!(n-k)!}.$

Proof. Let P_r be the statement that $\binom{r}{\ell} = \frac{r!}{\ell!(r-\ell)!}$ for all $\ell = 1, \ldots, r$. You will be able to check that P_1 is true. Assume that P_{r-1} is true. Then the formula is true when we substitute $n = r - 1$ and $\ell = k$ and is also true when we substitute $n = r - 1$ and $\ell = k - 1$:

$$\binom{r-1}{k} = \frac{(r-1)!}{k!(r-1-k)!} \quad \text{and} \quad \binom{r-1}{k-1} = \frac{(r-1)!}{(k-1)!(r-k)!}.$$

According to Proposition (A.1.1),

$$\binom{r}{k} = \binom{r-1}{k} + \binom{r-1}{k-1} = \frac{(r-1)!}{k!(r-k-1)!} + \frac{(r-1)!}{(k-1)!(r-k)!}$$

$$= \frac{(r-k)(r-1)!}{k!(r-k)!} + \frac{k(r-1)!}{k!(r-k)!} = \frac{r!}{k!(r-k)!}.$$

This shows that P_r is true. $\qquad\square$

As another example, let us prove the "pigeonhole principle." Here $|S|$ denotes the order, the number of elements, of a set S.

Proposition A.1.4 If $\varphi: S \to T$ is an injective map between finite sets, then T contains at least as many elements as are in S: $|S| \leq |T|$.

Proof. We use induction on $n = |S|$. The assertion is true if $n = 0$, that is, if S is empty. We suppose that the theorem has been proved for $n = k - 1$, and we proceed to check it for $n = k$, where $k > 0$. We suppose that $|S| = k$, and we choose an element s of S. Let $t = \varphi(s)$ be the image of s in T. Since φ is injective, s is the only element whose image is t. Therefore φ maps the set $S' = S - \{s\}$ obtained by removing s injectively to the set $T' = T - \{t\}$. Obviously, $|S'| = |S| - 1 = k - 1$ and $|T'| = |T| - 1$. By the induction assumption, $|S'| \leq |T'|$, and so $|S| \leq |T|$. $\qquad\square$

There is a variant of the principle of induction, called *complete induction*. Here again, we wish to prove a statement P_n for each positive integer n. The principle of complete induction asserts that it is enough to prove the following statement:

> If n *is a positive integer, and if* P_k *is true for every positive integer* k < n, *then* P_n *is true.*

When $n = 1$, there are no positive integers $k < n$. The hypothesis in the statement is automatically satisfied when $n = 1$. So a proof using complete induction must include a proof of P_1.

The principle of complete induction is used when there is a procedure to reduce P_n to P_k for some smaller integers k, but not necessarily to P_{n-1}. Here is an example:

Theorem A.1.5 Every integer n greater than 1 is a product of prime integers.

Proof. Let P_n be the statement that n is a product of primes. We assume that P_k is true for all $k < n$, and we must prove that P_n is true, i.e., that n is a product of primes. If n is prime itself, then it is the product of one prime. Otherwise, n can be written as a product $n = ab$ of positive integers neither of which is equal to 1. Then a and b are less than n, so the induction hypothesis tells us that P_a and P_b are both true, that is, a and b are products of primes. Putting these products side by side gives us the required factorization of n. $\qquad\square$

Proofs by contradiction proceed by assuming that the desired conclusion is false and deriving a contradiction from this assumption. The conclusion must therefore be true. Such proofs are often fakes, in the sense that the argument by contradiction is easily turned into a direct proof. Here is an example:

Proposition A.1.6 Let $\varphi: S \to T$ be an injective map between finite sets. If φ is bijective, then $|S| = |T|$.

Proof. Since we are given that φ is injective, φ will be bijective if and only if it is surjective. We assume that $|S| = |T|$, but that φ is not surjective. Then there is an element t in T, which is not in the image of S. This being so, φ actually maps S injectively to the set $T' = T - \{t\}$. Then Proposition A.1.4 tells us that $|S| \le |T'| = |T| - 1$ and this contradicts $|S| = |T|$. $\square$

Try not to arrange proofs this way. The assumption made in the proof that $|S| = |T|$ is irrelevant. Put positively, the argument shows that if an injective map φ isn't bijective, then $|S| < |T|$.

If X stands for some statement, we let *not X* stand for the statement that X is false. The assertion "if *not B*, then *not A*" is the *contrapositive* of the assertion "if A, then B," and is logically equivalent with it. The argument presented above proves the contrapositive of the assertion of the proposition.

It isn't easy to find very simple examples of good proofs by contradiction, but there are some in the text.

A.2   THE INTEGERS

We learn elementary properties of addition and multiplication of integers in elementary school, but let us look again, to see what would be required in order to prove some of the properties, such as the associative and distributive laws. Complete proofs require a fair amount of writing, and we will only make a start here. It is customary to begin by defining addition and multiplication for positive integers. Negative numbers are introduced later. This means that several cases have to be treated as one goes along, which is boring, or else a clever notation has to be found to avoid such a case analysis. We will content ourselves with a description of the operations on positive integers. Positive integers are also called *natural numbers*.

The set $\mathbb{N}$ of natural numbers is characterized by these properties:

Peano's Axioms

- The set $\mathbb{N}$ contains a particular element 1.
- *Successor function*: There is a map $\sigma \colon \mathbb{N} \to \mathbb{N}$ that sends an integer to another integer, called the *successor* or *next integer*. This map is injective, and for every n in $\mathbb{N}$, $\sigma(n) \ne 1$.
- *Induction axiom*: Suppose that a subset S of $\mathbb{N}$ has these properties:

   (i) 1 is an element of S, and

   (ii) if n is in S, then $\sigma(n)$ is in S.

   Then S contains every natural number: $S = \mathbb{N}$.

The successor $\sigma(n)$ will turn into $n + 1$ when addition is defined. At this stage the notation $n + 1$ could be confusing. It is better to use a neutral notation, and we will denote the successor by n' for now. The successor function allows us to use the natural numbers for counting, which is the basis of arithmetic.

The induction property can be described intuitively by saying that the natural numbers are obtained from 1 by repeatedly taking the next integer:

$$\mathbb{N} = \{1, 1', 1'', \ldots\} \quad (= \{1, 2, 3, \ldots\}).$$

In other words, counting runs through all natural numbers. This property is the basis of induction proofs.

Peano's axioms can also be used to make recursive definitions. The phrases *recursive definition*, or *inductive definition*, refer to the definition of a sequence of objects C_n indexed by the natural numbers, in which each object is defined in terms of the preceding one. For instance, a recursive definition of the function x^n is

$$x^1 = x \quad \text{and} \quad x^{n'} = x^n x.$$

The important points are:

(A.2.1) C_1 is defined, and a rule is given for determining $C_{n'}(= C_{n+1})$ from C_n.

It is intuitively clear that these properties determine the sequence C_n uniquely, though to give a quick proof of this fact from Peano's axioms isn't easy. We won't carry the proof out.

Given the set of positive integers and the ability to make recursive definitions, we can define addition and multiplication of positive integers as follows:

(A.2.2)
Addition: $m + 1 = m'$ and $m + n' = (m + n)'$.
Multiplication: $m \cdot 1 = m$ and $m \cdot n' = m \cdot n + m$.

In these definitions, we take an arbitrary integer m and define addition and multiplication for that integer m and for every n recursively. In this way, $m + n$ and $m \cdot n$ are defined for all m and n.

The proofs of the associative, commutative, and distributive laws for the integers are exercises in induction that might be called "Peano playing." We will carry out one of the verifications here as a sample.

Proof of the associative law for addition. We are to prove that for all a, b, and n in $\mathbb{N}$, $(a + b) + n = a + (b + n)$. We first check the case $n = 1$ for all a and b. Three applications of the definition give

$$(a + b) + 1 = (a + b)' = a + b' = a + (b + 1).$$

Next, assume the associative law true for a particular value of n and for all a, b. Then we verify it for n' as follows:

$$
\begin{aligned}
(a + b) + n' &= (a + b) + (n + 1) & &\text{(definition)} \\
&= ((a + b) + n) + 1 & &\text{(case } n = 1) \\
&= (a + (b + n)) + 1 & &\text{(induction hypothesis)} \\
&= a + ((b + n) + 1) & &\text{(case } n = 1) \\
&= a + (b + (n + 1)) & &\text{(case } n = 1) \\
&= a + (b + n') & &\text{(definition)}. \qquad \square
\end{aligned}
$$

The proofs of other properties of addition and multiplication follow similar lines.

A.3 ZORN'S LEMMA

At a few places in the text, we refer to Zorn's Lemma, a tool for handling infinite sets. We now describe it.

• A *partial ordering* of a set S is a relation $s \le s'$, which may hold between certain elements and which satisfies the following axioms for all s, s', s'' in S:

(A.3.1)

 (i) $s \le s$;
 (ii) if $s \le s'$ and $s' \le s''$, then $s \le s''$;
(iii) if $s \le s'$ and $s' \le s$, then $s = s'$.

A partial ordering is called a *total ordering* if, in addition,

(iv) for all s, s' in S, either $s \le s'$ or $s' \le s$.

For example, let S be a set whose elements are sets. If A, B are in S, we may define $A \le B$ if A is a subset of B: $A \subset B$. This is a partial ordering on S, called the *ordering by inclusion*. Whether or not it is a total ordering depends on the particular case.

An element m of a partially ordered set S is a *maximal element* if there is no element s in S with $m \le s$, except for m itself. A partially ordered set S may contain many different maximal elements. For example, a subset V of a set U is a *proper* subset if V is neither the empty set, nor the whole set U. The set of all proper subsets of the set $\{\mathbf{1}, \dots, \mathbf{n}\}$, ordered by inclusion, contains n maximal elements, one of which is $\{\mathbf{2}, \mathbf{3}, \mathbf{4}, \dots, \mathbf{n}\}$.

A nonempty finite partially ordered set S contains at least one maximal element, but an infinite partially ordered set, such as the set of integers, may contain no maximal element at all. A totally ordered set contains at most one maximal element.

• If A is a subset of a partially ordered set S, then an *upper bound* for A is an element b in S such that for all a in A, $a \le b$. A partially ordered set S is *inductive* if every totally ordered subset T of S has an upper bound.

A finite totally ordered set contains a unique maximal element, and is inductive.

Lemma A.3.2 Zorn's Lemma. An inductive partially ordered set S has at least one maximal element.

Zorn's Lemma is equivalent with the *axiom of choice*, which is known to be independent of the basic axioms of set theory. We won't enter into a further discussion of this equivalence, but we will show how Zorn's Lemma can be used to show that every vector space has a basis.

Proposition A.3.3 Every vector space V over a field F has a basis.

Proof. Let S be the set whose elements are the linearly independent subsets of V, partially ordered by inclusion. We show that S is inductive: Let T be a totally ordered subset of S.

Then we claim that the union of the sets making up T is also linearly independent. This will show that it is in S. To verify this, let

$$B = \bigcup_{A \in T} A$$

be the union. By definition, a relation of linear dependence on B is finite, so it can be written in the form

(A.3.4) $$c_1 v_1 + \cdots + c_n v_n = 0,$$

with v_i in B. Since B is a union of the sets in T, each v_i is contained in one of these subsets, call it A_i. The collection $\{A_1, \ldots, A_n\}$ of these subsets is a finite, totally ordered subset of T. It has a unique maximal element A. Then v_i is in A for every $i = 1, \ldots, n$. But since A is in S, it is a linearly independent set. Therefore (A.3.4) is the trivial relation. This shows that B is linearly independent, hence that it is an element of S.

We have verified the hypothesis of Zorn's Lemma. So S contains a maximal element M, and we claim that M is a basis. By definition of S, M is linearly independent. Let $W = \text{Span}(M)$. If $W < V$, then we choose an element v in V, which is not in W. The set $M \cup \{v\}$ will be linearly independent. This contradicts the maximality of M and shows that $W = V$, hence that M is a basis. $\square$

A similar argument proves Theorem (11.9.2) of Chapter 11:

Proposition A.3.5 Let R be a ring. Every ideal $I \neq R$ is contained in a maximal ideal. $\square$

A.4   THE IMPLICIT FUNCTION THEOREM

The Implicit Function Theorem for complex polynomial functions is used a few times in this book, and for lack of a reference, we derive it here from the theorem for real valued functions that we state below. The theorem for real valued functions can be found in [Rudin], Theorem 9.27.

Theorem A.4.1 Implicit Function Theorem. Let $f_1(x, y), \ldots, f_r(x, y)$ be functions of $n + r$ real variables $x_1, \ldots, x_m, y_1 \ldots, y_r$, which have continuous partial derivatives in an open set of $\mathbb{R}^{n+r}$ containing the point (a, b). Assume that the Jacobian determinant

$$\det \begin{bmatrix} \dfrac{\partial f_1}{\partial y_1} & \cdots & \dfrac{\partial f_1}{\partial y_r} \\ \vdots & & \vdots \\ \dfrac{\partial f_r}{\partial y_1} & \cdots & \dfrac{\partial f_r}{\partial y_r} \end{bmatrix}$$

is not zero at the point (a, b). There is a neighborhood U of the point a in $\mathbb{R}^n$ such that there are unique continuously differentiable functions $Y_1(x), \ldots, Y_r(x)$ on U satisfying

$$f_i(x, Y(x)) = 0 \quad \text{for} \quad i = 1, \cdots, r, \quad \text{and} \quad Y(a) = b. \qquad \square$$

The partial derivatives of a complex polynomial $f(x, y)$ are defined using the rules of calculus. But we can also write everything in terms of the real and imaginary parts, say $x = x_0 + x_1 i$, $y = y_0 + y_1 i$, where x_0, x_1, y_0, y_1 are real variables, and $f = f_0 + f_1 i$, where $f_i = f_i(x_0, x_1, y_0, y_1)$ is a real-valued function of the four real variables. Since f is a polynomial in x and y, the real functions f_i are polynomials in the real variables x_i and y_i. So they have continuous partial derivatives.

Lemma A.4.2   Let $f(x, y)$ be a polynomial in two variables with complex coefficients. Then with notation as above,

(a) $\dfrac{\partial f}{\partial y} = \dfrac{\partial f_0}{\partial y_0} + \dfrac{\partial f_1}{\partial y_0} i$, and

(b)   (*Cauchy-Riemann equations*) $\dfrac{\partial f_0}{\partial y_0} = \dfrac{\partial f_1}{\partial y_1}$, and $-\dfrac{\partial f_0}{\partial y_1} = \dfrac{\partial f_1}{\partial y_0}$.

Proof. One can use the product rule to verify these formulas. Suppose that $f = gh$. Then $f_0 = g_0 h_0 - g_1 h_1$ and $f_1 = g_0 h_1 + g_1 h_0$. If the formulas are true for g and h, they follow for f. So it is enough to verify the lemma for the functions $f = y$ and $f = x$, for which they are obvious. $\qquad \square$

Theorem A.4.3 Implicit Function Theorem for Complex Polynomials. Let $f(x, y)$ be a complex polynomial. Suppose that for some (a, b) in $\mathbb{C}^2$, $f(a, b) = 0$ and $\frac{\partial f}{\partial y}(a, b) \neq 0$. There is a neighborhood U of x in $\mathbb{C}$ on which a unique continuous function $Y(x)$ exists having the properties

$$f(x, Y(x)) = 0 \quad \text{and} \quad Y(a) = b.$$

Proof. We reduce the theorem to the real Implicit Function Theorem A.4.1. The same argument will apply when there are more variables.

With notation as above, we are to solve the pair of equations $f_0 = f_1 = 0$ for y_0 and y_1 as functions of x_0 and x_1. To do this, we show that the Jacobian determinant

$$\det \begin{bmatrix} \dfrac{\partial f_0}{\partial y_0} & \dfrac{\partial f_0}{\partial y_1} \\[2mm] \dfrac{\partial f_1}{\partial y_0} & \dfrac{\partial f_1}{\partial y_1} \end{bmatrix}$$

is not zero at (a, b). By hypothesis, $f_i(a_0, a_1, b_0, b_1) = 0$. Also, since $\frac{\partial f}{\partial y}(a, b) \neq 0$, Lemma A.4.2**(a)** tells us that $\frac{\partial f_0}{\partial y_0} = d_0$ and $\frac{\partial f_1}{\partial y_0} = d_1$, are not both zero. Part **(b)** of the lemma shows that the Jacobian determinant is

$$\det \begin{bmatrix} d_0 & -d_1 \\ d_1 & d_0 \end{bmatrix} = d_0{}^2 + d_1{}^2 > 0.$$

This shows that the hypotheses of the Implicit Function Theorem (A.4.1) are satisfied. $\qquad \square$

EXERCISES

Section A.1   About Proofs

A.1. Use induction to find a closed form for each of the following expressions.

    (a) $1 + 3 + 5 + \cdots + (2n + 1)$
    (b) $1^2 + 2^2 + 3^2 + \cdots + n^2$

A.2. Prove that $1^3 + 2^3 + \cdots + n^3 = (n(n + 1))^2/4$.

A.3. Prove that $1/(1 \cdot 2) + 1/(2 \cdot 3) + \cdots + 1/(n(n + 1)) = n/(n + 1)$.

A.4. Let $\varphi: S \to T$ be a surjective map between finite sets. Prove by induction that $|S| \geq |T|$ and that if $|S| = |T|$, then φ is bijective.

A.5. Let n be a positive integer. Show that if $2^n - 1$ is a prime number, then n is prime.

A.6. Let $a_n = 2^{2^n} + 1$. Prove that $a_n = a_0 a_1 \ldots a_{n-1} + 2$.

A.7. A nonconstant polynomial with rational coefficients is called irreducible if it is not a product of two nonconstant polynomials with rational coefficients. Prove that every polynomial with rational coefficients can be written as a product of irreducible polynomials.

Section A.2   The Integers

A.8. Prove that every natural number n except 1 has the form m' for some natural number m.

A.9. Prove the following laws for the natural numbers.

    (a) the commutative law for addition,
    (b) the associative law for multiplication,
    (c) the distributive law,
    (d) the cancellation law for addition: if $a + b = a + c$, then $b = c$.

A.10. The relation $<$ on $\mathbb{N}$ can be defined by the rule $a < b$ if $b = a + n$ for some n. Assume that properties of addition have been proved.

    (a) Prove that if $a < b$, then $a + n < b + n$ for all n.
    (b) Prove that the relation $<$ is transitive.
    (c) Prove that if a and b are natural numbers, then $a < b$, or $a = b$, or $b < a$.

A.11. Assume that basic properties of the relation $<$ on $\mathbb{N}$ are known (see Exercise A.10). Prove the principle of *complete induction*: A subset S of $\mathbb{N}$ is equal to $\mathbb{N}$ if it has the following property: If n is an element of $\mathbb{N}$ such that m is in S for every $m < n$, then n is in S.

Section A.3   Zorn's Lemma

A.12. Let S be a partially ordered set.

    (a) Prove that if S contains an upper bound b, then b is unique, and also b is a maximal element.

    (b) Prove that if S is totally ordered, then a maximal element m is an upper bound for S.

A.13. Use Zorn's Lemma to prove that every ideal I of a ring R that is not R itself is contained in a maximal ideal.

Section A.4   The Implicit Function Theorem

A.14. Prove Lemma (A.4.2).

A.15. Let $f(x, y)$ be a complex polynomial. Assume that the equations

$$f = 0, \quad \frac{\partial f}{\partial x} = 0, \quad \frac{\partial f}{\partial y} = 0,$$

have no common solution in $\mathbb{C}^2$. Prove that the locus $f = 0$ is a manifold of dimension 2.

Bibliography

GENERAL ALGEBRA TEXTS

- G. Birkhoff and S. MacLane, *A Survey of Modern Algebra*, 3rd ed., Macmillan, New York, 1965.
- I. N. Herstein, *Topics in Algebra*, 2nd ed., Wiley, New York, 1975.
- N. Jacobson, *Basic Algebra I, II*, Freeman, San Francisco, 1974, 1980.
- S. Lang, *Algebra*, 2nd ed., Addison Wesley, Reading, MA, 1965.
- B. L. van der Waerden, *Modern Algebra*, Ungar, New York, 1970.

LINEAR ALGEBRA

- P. D. Lax, *Linear Algebra and Its Applications*, 2nd ed., Wiley, Hoboken, NJ, 2007.
- G. Strang, *Linear Algebra and Its Applications*, 3rd ed., Harcourt Brace Jovanovich, San Diego, 1988.

ANALYSIS AND TOPOLOGY

- A. P. Mattuck, *Introduction to Analysis*, Prentice-Hall, Upper Saddle, River, N.J., 1999.
- J. R. Munkres, Topology; *A First Course*, 2nd ed., Prentice Hall, Englewood Cliffs, NJ, 2000.
- W. Rudin, *Principles of Mathematical Analysis*, 3rd ed., McGraw-Hill, New York, 1976.

NUMBER THEORY

- H. Cohn, *A Second Course in Member Theory*, John Wiley & Sons, New York-London, 1962.
- K. F. Gauss, *Disquisitiones Arithmeticae*, Leipzig, 1801.
- H. Edwards, *Galois Theory*, Springer-Verlag, New York, 1984.
- H. Hasse, *Number Theory*, Springer-Verlag, New York, 1980.
- J.-P. Serre, *A Course in Arithmetic*, Springer-Verlag, New York, 1973.
- J. H. Silverman, *The Arithmetic of Elliptic Curves*, Springer-Verlag, New York, 1992.
- H. Stark, *An Introduction to Number Theory*, M.I.T. Press, Cambridge, MA, 1978.

GROUPS

- M. R. Sepanski, *Compact Lie Groups*, Springer-Verlag, New York, 2009.
- J.-P. Serre, *Linear Representations of Finite Groups*, Springer-Verlag, New York, 1977.
- H. Weyl, *The Classical Groups*, Princeton University Press, Princeton, N.J., 1946.

GEOMETRY

- G. A. Bliss, *Algebraic Functions*, AMS Colloquium Publications XVI, New York, 1933.
- H. S. M. Coxeter, *Introduction to Geometry*, Wiley, New York, 1961.
- D. Schwarzenbach, *Crystallography*, Wiley, Chichester, U.K., 1993.
- M. Senechal, *Quasicrystals*, Cambridge University Press, Cambridge, U.K., 1996.

HISTORY OF MATHEMATICS

- N. Bourbaki, *Elements d'histoire des mathematiques*, Hermann, Paris, 1974.
- M. Kline, *Mathematical Thought from Ancient to Modern Times*, Oxford, New York, 1972.
- E. Landau, *Foundations of Analysis*, AMS Chelsea, New York, 2001.
- B. L. van der Waerden, *A History of Algebra*, Springer-Verlag, Berlin, New York, 1985.

JOURNAL ARTICLES

- A. F. Filippov, *A short proof of the theorem on reduction of a matrix to jordan form*, Vestnik Mosk. Univ. Ser. I Mat. Meh. 26 (1971) 18–19.
- R. Howe, *Very Basic Lie Theory,* Math Monthly 90 (1983) 600–623.
- S. Landau, *How to tangle with a nested radical*, Math. Intelligencer 16 (1994) 49–55.
- A.K. Lenstra, H. W. Lenstra, and L. Lovász, *Factoring polynomials with rational coefficients* Math. Annalen 261 (1982) 515–534.
- J. Milnor, *Analytic proofs of the "hairy ball theorem" and the Brouwer fixed-point theorem*, Amer. Math. Monthly (1978) 521–524.
- J. Stillwell, *The word problem and the isomorphism problem for groups*, Bull. Amer. Math. Soc. 6 (1982) 33–56.
- J. A. Todd and H. S. M. Coxeter, *A practical method for enumerating cosets of a finite abstract group*, Proc. Edinburg Math. Soc, II Ser. 5 (1936) 26–34.

Notation

$\langle A \rangle$ the class of the ideal A (13.7.2)

A^{t} the transpose of the matrix A (1.3.1)

A_n the alternating group (2.5.6)

$\mathbb{C}$ the field of complex numbers (2.2.2)

C_n the cyclic group of order n (6.4.1)

$C(x)$ the conjugacy class of the element x (7.2.3)

$\mathrm{cof}(A)$ the cofactor matrix of the matrix A (1.6.7)

D_n the dihedral group (6.4.1)

$\det A$ the determinant of the matrix A (1.4.1)

e_i, e_{ij} a standard basis vector (1.1.24), a matrix unit (1.1.21)

F^n the space of n-dimensional column vectors with entries in F (3.3.6)

$F^{m \times n}$ the space of $m \times n$ matrices with coefficients in F (3.3.6)

$\mathbb{F}_p$ the field of integers modulo p (3.2.4)

GL_n the general linear group (2.2.4)

I, I the identity matrix (1.1.11), the icosahedral group (6.12.1)

$\mathrm{im}\,\varphi$ the image of the map φ (2.5.4)

$\ker\varphi$ the kernel of the homomorphism φ (2.5.5), (4.1.5)

K^G a fixed field (16.5.1)

ℓ^∞ the space of bounded sequences (3.7.2)

M, M_n the group of isometries of the plane, of n-space (Section 6.2)

$\mathbb{N}$ the set of positive integers, also called *natural numbers* (A.2.1)

$N(H)$ the normalizer of the subgroup H (7.6.1)

$n!$ n factorial: the product of the integers $1, 2, \ldots, n$.

$\binom{n}{k}$ a binomial coefficient (A.1.1)

O_n the orthogonal group (6.7.3), (9.1.2)

$O_{3,1}$ the Lorentz group (9.1.5)

PSL_n the projective group (9.8.1)

$\mathbb{R}$	the field of real numbers (2.2.2)
R^+	the additive group of R (2.1.1)
$R^\times$	the multiplicative group of invertible elements of R (2.1.1)
$S_n,$	the symmetric group (2.2.5)
$\mathbb{S}^n$	the n-dimensional sphere (Section 9.2)
SL_n	the special linear group (2.2.11), (9.1.3)
SO_n	the special orthogonal group (5.1.11), (9.1.3)
SP_{2n}	the symplectic group (9.1.4)
SU_n	the special unitary group (9.1.3)
T	the tetrahedral group (6.12.1)
U_n	the unitary group (8.3.14), (9.1.3)
$\langle x \rangle$	the subgroup generated by the element x (2.4.1)
Z	the center of a group (2.5.12)
$\mathbb{Z}$	the ring of integers (2.2.2)
$Z(x)$	the centralizer of the element x (7.2.2)
ζ_n	the nth root of unity $e^{2\pi i/n}$ (12.4.7)
$\lfloor \mu \rfloor$	the largest integer $\leq \mu$: the *floor* of μ (13.7.7)
ω	the cube root of unity $e^{2\pi i/3}$ (10.4.14)
$\approx$	indicates that two structures are isomorphic, as in $G \approx G'$ (2.6.3)
$\equiv$	congruence, as in $a \equiv b$ modulo n (2.9.1), see also (2.8.2), (2.7.14)
$*$	If A is a complex matrix, then A^* is the adjoint matrix $\overline{A}^{\mathrm{t}}$ (8.3.5). In a matrix display, $*$ denotes an undetermined entry. The starred exercises are some of the more difficult ones.
$\oplus$	direct sum (3.6.5), (14.7.2)

If S and T are sets, we use the following notation:

$\lvert S \rvert$	the number of elements, the order, of the set S
$[S]$	the subset S, when it is regarded as an element of a set of subsets (2.7.8)
$s \in S$	s is an element of S.
$S \subset T$	S is a subset of T, or S is contained in T. In other words, every element of S is also an element of T.

$T \supset S$ T contains S, which is the same as $S \subset T$.

$S < T$ S is a proper subset of T, meaning that it is a subset, and T contains an element that is not a member of S.

$T > S$ This is the same as $S < T$.

$S \cap T$ the *intersection* of the sets: the set of all elements in common to S and T.

$S \cup T$ the *union* of the sets: the set of all elements that are contained in at least one of the sets S or T.

$S \times T$ the *product* set. Its elements are ordered pairs (s, t), with s in S and t in T.

$\varphi : S \to T$ a *map* φ from S to T, a function whose domain is S and whose range is T.

$s \rightsquigarrow t$ This wiggly arrow indicates that the map under consideration sends the element s to the element t, i.e., that $\varphi(s) = t$.

$\square$ This symbol indicates that a digression in the text, such as a proof or an example, has ended, and that the text returns to the main thread. $\square$

Index

A

Abelian groups, 40, 81, 412–13, 421
    finite, 431
    free, 225
    infinite, 41
    Structure Theorem for, 429–30
Abstract symmetry, 176–78, 190–91
Addition
    of matrices, 2
    of relations, 337–38
    vector, 78
Adjoint matrix, 233
Adjoint operator, 242
Adjoint representation, 289
Affine group, 288
Algebraically closed field, 471
Algebraic element, 443–46, 472
Algebraic extension, 473
Algebraic geometry, 347–53
Algebraic integers, 383–85, 408
    factoring, 385–87
Algebraic number, 383
Algebraic number field, 442
Algebraic variety, 347
Alternating group, 49, 63
Angle
    of rotation, 171
    between vectors, 242
Antipodal point, 269
Ascending chain condition, 426
Associative law, 5, 68, 176
    for addition, 517
    for congruence classes, 61
    for scalar multiplication, 90
Augmented matrix, 12
Automorphism, 52, 176
    F-automorphism, 484
    inner, 193

    R-automorphism, 477
    of ring, 355
Averaging, over a group, 294
Axiom of choice, 98, 348, 518. *See also*
    Zorn's Lemma
Axis of rotation, 134

B

Basechange matrix, 93–94
Base point, 468
Bases, 86–91, 99–100
    change of, 93–95
    computing with, 90–91, 100
    defined, 88
    infinite, 98
    lattice, 169, 405
    of module, 415
    orthogonal, 252
    orthonormal, 133, 240, 252
    standard, 88, 415
Berlekamp algorithms, 374, 382
Bézout bound, 349
Bilateral symmetry, 154
Bilinear form, 229–60
    Euclidean space, 241–42
    Hermitian form, 232–35
    Hermitian space, 241–42
    orthogonality, 235–41
    skew-symmetric form, 249–52
    spectral theorem and, 242–45
    symmetric form, 231–32
Binomial coefficient, 513
Block multiplication, 8–9
Branched covering, 351
    cut and paste, 465–68
    isomorphism of, 464
Branch points, 351, 353
Burnside's formula, 194

C

Cancellation law, 41–43, 82–83, 343, 392
Canonical map, 66, 335, 423
Cardano's formula, 501
Cartesian coordinates, 452
Case analysis, 513
Cauchy-Riemann equations, 520
Cauchy's Theorem, 375
Cayley-Hamilton theorem, 140
Cayley's theorem, 195
Celestial sphere, 264
Center
    of group, 196
    of p-group, 197
Center of gravity, 166
Centroid, 166. *See also* Center of gravity
Change of basis, 93–95
Character, 291, 298–303
    dimension of, 299
    Hermitian product on, 299
    irreducible, 299
    one-dimensional, 303–4
    table, 302
Characteristic polynomial, 113–16
    of linear operator, 115
Characteristic subgroup, 225
Characteristic zero, 83, 484
Chinese Remainder Theorem, 73, 356, 378
Circle group, 262, 320
Circulant, 258
Class
    congruence, 60
    ideal, 388, 396–99, 410
Class equation, 195–97
    of icosahedral group, 198–200
Class function, 300
Class group, 399–402, 410
Class number, 396
Closure in subgroups, 42–43
Cofactor matrix, 29–31
Column index, 1
Column rank, 108
Column space, 87, 104
Column vector, 2
Combination, linear, 7, 79, 86, 97

Common zeros, 347
Commutative law, 5–6
    for congruence classes, 61
Commutative diagram, 105
Commutator subgroup, 225
Compact groups, 311
Complete expansion, of determinants, 29
Complete induction, 515, 521
Complete of relations, 215, 424
Complex algebraic group, 282
Complex line, 347
Complex representations, 293
Congruence, 60
Conics, 245–49
    degenerate, 245
    nondegenerate, 246
Conjugacy class, 196
Conjugate representation, 293
Conjugate subgroups, 72, 178, 203
Conjugation, 52, 195
    in symmetric group, 200–203
Connected component, 76
Constructible point, line, circle, 451–54
Construction, ruler and compass, 450–55
Continuity, proof by, 138–40
Contradiction, proofs by, 515
Coordinates, 90
    change of, 158–59
Coordinate system, 159
Coordinate vectors, 78, 93, 94, 105, 416
Correspondence Theorem, 61–64, 336–37, 414
    proof of, 63–64, 336
Coset, 56–59
    double, 76
    left, 49, 56
    operation on, 178–80
    right, 58–59, 216
Counting formula, 57, 58, 62, 180–81, 185
Covering space, 351
Cramer's Rule, 415, 417
Crystallographic group, 187
Crystallographic restriction, 171–72
Cubic, resolvent, 496
Cubic equations, 492–93, 507–8
Cubic extensions, 446

Cusp, 351
Cut and paste, 465–68
Cycle notation, 24
Cyclic group, 46–47, 163, 183, 208
    generator for, 84
    infinite, 47
    of order n, 46
Cyclic R-module, 432
Cyclotomic polynomial, 374

D

Defining relations, 212
Degenerate conic, 245
Degree
    of field extension, 446–49
    of a monomial, 327
    multiplicative property of, 447
    total, 327
    weighted, 482
Determinant homomorphism, 49, 56, 62
Determinant, 7, 18–24
    complete expansion of, 29
    formulas for, 27–31
    multiplicative property of, 21–24
    of permutation matrix, 27
    recursive definition of, 20
    of R-matrix, 414
    uniqueness of, 20–21
    Vandermonde, 511
Diagonal entries, 6
Diagonal form, 116–19
Diagonalizable matrix, 117
Diagonalizable operator, 119
Diagonal matrix, 6
Dichotomy, 513
Differential equations, 141–45, 151
Dihedral group, 163, 183, 316
Dimension, 86–91
    of character, 299
    of vector space, 90
    of linear group, 262
Dimension formula, 102–4
Direct sums, 95–96, 295
    of modules, 429
    of submodules, 430

Discrete group, 167–72
Discrete subgroup, 168
Discriminant, 481–83
Distinct, 17
Distributive law, 5, 81, 324
    for congruence classes, 61
    for matrix multiplication, 147
    for vector spaces, 84
Divide and conquer, 513
Divisor
    greatest common, 44–45, 334, 359,
        362
    zero, 343
Domain
    Euclidean, 361, 376
    factorization, 360–67, 379, 400
    integral, 343
    principal ideal, 361
    unique factorization, 364
Dot product, 132, 229
Double coset, 76

E

Eigenspace, 126
    generalized, 131
Eigenvalue, 111, 113, 114, 116, 234
Eigenvectors, 110–13, 116, 124
    generalized, 120
    positive, 112
Eisenstein criterion, 373–74
Elementary integer matrix, 418
Elementary matrix, 10–12, 77
Elementary row operation, 10
Elementary symmetric function, 478
Elements
    adjoining, 338–41
    algebraic, 443–46
    inverse image of, 55
    irreducible, 444
    maximal, 518
    norm of, 386
    prime, 360
    primitive, 462–63
    relatively prime, 362
    representative, 55

Elements (*continued*)
  solvable, 502
  stabilizer of, 177–78
  transcendental, 443–46
  zero, 417
Ellipse, 246
Ellipsoid, 248, 269
Equation, 4
  Cauchy-Riemann, 520
  class, 195–97
  cubic, 492–93
  differential, 141–45
  homogeneous, 15, 88, 92
  quartic, 493–97
  quintic, 502–5
Equator, 265, 267
Equivalence relation, 52–56
  defined, 53
  defined by a map, 55–56
  reflexive, 53
  symmetric, 53
  transitive, 53
Euclidean Algorithm, 45, 367
Euclidean domain, 361, 376
Euclidean space, 241–42
  standard, 241
Euler's theorem, 137–38
Exceptional group, 283
Expansion by minors, 19, 28
  on the i*th* row, 28
Extension
  algebraic, 472
  cubic, 446
  field, 442
  finite, 446
  Galois, 485, 488–89
  Kummer, 500–502
  ring, 338

F

Factoring, 359–82
  algebraic integers, 385–87
  Gauss primes, 376–78
  Gauss's lemma, 367–71
  ideals, 392–94, 409
    integer polynomials, 371–75, 380–81
    integers, 359, 378
    unique factorization domains, 360–67
Factorization
  ideal, 391
  irreducible, 364, 365
  prime, 365
Faithful operation, 182
Faithful representation, 291
F-automorphism, 484
Fermat's theorem, 99
Fibonacci numbers, 152
Field extension, 442
  algebraic, 486
  degree of, 446–49
  isomorphism of, 445, 484–86
Fields, 80–84, 98–99, 442–76
  adjoining roots, 456–59
  algebraically closed, 471
  algebraic and transcendental elements, 443–46
  characteristic of, 83
  finding irreducible polynomials, 449–50
  finite, 442, 459–62
  fixed, 486–88
  function, 442–43, 463–71
  intermediate, 488
  number, 442
  quadratic number, 383–411
  of rational functions, 344
  real quadratic, 402–5
  ruler and compass constructions, 450–55
  splitting, 483–84
  tangent vector, 280
Finite abelian group, 431
Finite-dimensional vector space, 89
  dimension of, 90
  subspaces of, 95
Finite extension, 446
Finite field, 442, 459–62
  order of, 459
Finite group, 41
  homomorphism of, 58
  of orthogonal operators on plane, 163–67

Finitely generated module, 415
Finite simple group, 283
Finite subgroups of rotation group, 183–87
First Isomorphism Theorem, 68–69, 215,
       335, 414, 432, 492
Fixed field, 486–88
Fixed Field Theorem, 487–88
Fixed point theorem, 166, 198
Fixed vector, 111
Form
       Hermitian, 232–35
       Killing, 289
       Lorentz, 231
       matrix of, 230
       nondegenerate, 236, 252
       quadratic, 246
       rational canonical, 435
       signature of, 240
       skew-symmetric, 230, 249–52
       symmetric, 230
Fourier matrix, 260
Fractions, 342–44
Free abelian group, 225
Free group, 210–11
       mapping property of, 214
Free modules, 412, 437
       submodules of, 421–23
Frobenius map, 355, 511
Frobenius reciprocity, 321
Function field, 442–43, 463–71
       cut and paste, 465–68
Functions
       rational, 487
       successor, 516
       symmetric, 477–81
Fundamental domain, 193
Fundamental Theorem
       of Algebra, 471
       of Arithmetic, 359, 363

G

Galois extension, 485, 488–89
       characteristic properties of, 488–89
Galois group, 485
       of a polynomial, 489

Galois theory, 477–512
       for a cubic, 493
       cubic equations, 492–93
       discriminant, 481–83
       fixed fields, 486–88
       isomorphisms and field extensions,
              484–86
       Kummer extensions, 500–502
       Main Theorem, 489–92
       quartic equations, 493–97
       quintic equations, 502–5
       roots of unity, 497–500
       splitting fields, 483–84
       symmetric functions and, 477–81
Gauss integer, 323, 386
Gauss prime, 376–78, 394
Gauss's lemma, 367–71
Generalized eigenspace, 131
Generalized eigenvector, 120
General linear group, 8, 41
       integer, 418
       over R, 414
Generators, 212–16, 225–26, 423–26, 438
       Jordan, 122
       of a module, 415
Geometry, algebraic, 347–53, 357–58
Glide reflection, 160
Glide symmetry, 155
Gram-Schmidt procedure, 241
Greatest common divisor, 44, 334, 359, 362
Group homomorphism, 48
Group operation, 176–78
Group representation, 290–322
Groups, 37–77
       abelian, 40, 81, 412–13, 421
       affine, 288
       alternating, 49, 63
       averaging over, 294
       center of, 50, 196
       circle, 262
       compact, 311
       complex algebraic, 282
       correspondence theorem, 61–64
       cosets, 56–59
       crystallographic, 187
       cyclic, 46–47, 64, 163, 183

Groups (*continued*)
  defined, 40
  defining relations for, 42
  dihedral, 163, 183
  discrete, 167–72
  equivalence relations and partitions, 52–56
  exceptional, 283
  finite, 41, 163–67
  finite simple, 283
  free, 210–11
  free abelian, 225
  Galois, 485
  general linear, 41
  homomorphisms, 47–51
  homophonic, 77
  icosahedral, 183
  infinite, 41
  isomorphic, 51
  isomorphism of, 51–52
  laws of composition, 37–40
  linear, 261–89
  Lorentz, 262
  Mathieu, 283
  modular arithmetic, 60–61
  multiplicative, 84
  nonabelian, 222
  octahedral, 183
  one-parameter, 272–75
  operation of, 293
  opposite, 70
  order of, 40
  orthogonal, 134, 261
  p-groups, 197–98
  plane crystallographic, 172–76
  point, 170–71
  product group, 64–66
  protective, 280
  quotient, 66–69, 74–75
  representation of, 292
  rotation, 137, 269–72
  simple, 199
  special linear, 43, 50
  spin, 269
  sporadic, 283
  surjective, 62

  symmetric, 41, 50, 197
  symplectic, 261
  tetrahedral, 183
  translation, 168–70
  translation in, 277–80
  triangle, 226
  two-dimensional crystallographic, 172
  unitary, 235, 261

H

Half integer, 384
Half space, 259
Hausdorff space, 351
Hermitian form, 232–35, 254
  standard, 232
Hermitian matrix, 233
Hermitian operator, 257
Hermitian product, 299
Hermitian space, 241–42, 256
  standard, 241
Hermitian symmetry, 233
Hilbert Basis Theorem, 428–29
Hilbert Nullstellensatz, 345
Homeomorphism, 262
Homogeneity in a group, 277
Homogeneous linear equation, 15, 88, 92
Homogeneous polynomial, 328
Homomorphism, 47–51, 158
  determinant, 49, 56, 62
  group, 48
  image of, 48–49
  kernel of, 49, 56, 62, 69, 331, 413
  restriction of, 61
  of modules, 413
  of rings, 328–34
  of R-modules, 427
  spin, 269
  trivial, 48
Homophonic group, 77
Hyperbola, 246
Hyperplane, 259
Hypervector, 86

I

Icosahedral group, 183
   class equation of, 198–200
Ideal, 331, 387
   factorization, 391–94
   generated by a set, 332
   of leading coefficients, 428
   maximal, 344–47, 394
   prime, 392, 394–96
   principal, 331
   product, 355, 390
   proper, 331
   unit, 331
   zero, 331
Ideal class, 388, 396–99
Ideal multiplication, 389–92
Idempotent, 341
Identities, 5, 417–18
   Newton, 505
Identity element, 42
Identity matrix, 6
Image, of homomorphism, 413
Imaginary quadratic number field,
   383
Implicit Function Theorem, 522
Inclusion, ordering by, 518
Inclusion map, 48
Indefinite form, 231
Independence, 87, 95, 97, 415
Independent subspaces, 95
Index, multiplicative property of, 58
Induced law, 42
Induced representation, 321
Induction, 513–516
Inductive definition, 517
Inductive set, 518
Infinite basis, 98
Infinite cyclic group, 47
Infinite-dimensional space, 96–98
Infinite group, 41
Infinite order, 47
Infinite set, span of, 97
Inner automorphism, 193
Integer general linear group, 418
Integer matrix
   diagonalizing, 418–23

elementary, 418
invertible, 418
Integer polynomials, factoring,
   371–75
Integers, 390, 516–17
   algebraic, 383–85
   factoring, 378
   Gauss, 323, 386
   half, 384
   modulo, 66
   next, 516
   norm of, 397
   prime, 64, 394–96
   ring of, 384
   square-free, 384
   subgroups of additive group of,
     43
   successor, 516
Integral domain, 343
Intermediate field, 488
Intersection, 527
Invariant
   form, 297
   operator, 307
   subspace, 110, 294
   vector, 294
Inverse, 7, 40
Inverse image, 55
   left, right, 7
Invertible integer matrix, 418
Invertible matrix, 7, 15
Invertible operator, 109
Irreducible character, 299
Irreducible element, 444
Irreducible factorization, 364
Irreducible polynomial, 350, 383,
   443, 458
   finding, 449–50
Irreducible representation, 294–96
Isometrix, 156–59
   discrete group of, 167–72
   fixed point of, 162
   orientation-preserving, 160
   orientation-reversing, 160
   of the plane, 159–63
Isomorphic groups, 51

Isomorphism, 51–52
  of branched coverings, 464
  of field extensions, 445, 464, 484–86
  of groups, 51–52
  modules and, 413
  of representations, 293, 307
  of rings, 328
  of vector spaces, 85, 91
Isomorphism class of a group, 52

J

Jacobi identity, 276
Jordan block, 121, 148
Jordan form, 120–25, 148
Jordan generators, 122

K

Kaleidoscope principle, 167
Kernel
  of homomorphism, 49, 56, 62, 413
  of ring homomorphism, 331
Killing form, 289
Klein Four Group, 47, 65, 490, 493, 503
Kronecker delta, 133
Kronecker-Weber Theorem, 500
Kummer extensions, 500–502

L

Lagrange interpolation formula, 17,
    380
Lagrange's theorem, 57
Latitude, 265–66
Lattice, 403, 405–8
Lattice basis, 169, 405
Laurent polynomials, 356
Law of composition, 37–40
  associative, 37
  commutative, 38
  identity for, 39
Law of cosines, 242
Leading coefficients, 325
  ideal of, 428
Left coset, 49, 56
Left multiplication, 195, 277–78

  by G, 177
Left translation, 277
Lie algebra, 275–77, 286
Lie bracket, 276
Linear algebra, in ring, 412–41
  free modules, 414–17
  generators and relations, 423–26
  linear operators and, 432–35
  modules, 412–14
  noetherian rings, 426–29
  polynomial rings in several variables,
      436
  structure of abelian groups, 429–32
Linear combination, 9, 79, 86, 97
Linear equation, homogeneous, 15, 88, 91
Linear group, 261–89
  classical groups, 261–62
  dimension of, 262
  integer general, 418
  Lie algebra, 275–77
  normal subgroups of SL_2, 280–83
  one-parameter groups, 272–75
  rotation group SO_3, 269–72
  special unitary group SU_2, 266–69
  spheres and, 263–66
  translation in group, 277–80
Linear operator, 102–31, 293, 432–35
  applications of, 132–53
  characteristic polynomial of, 113–16,
      115
  defined, 108–10
  dimension formula, 102–4
  eigenvectors, 110–13
  Jordan form, 120–25
  left shift, right shift, 109
  triangular and diagonal form,
      116–19
Linear relation, 103
  among vectors, 87
Linear transformation, 102
  matrix of, 104–8
Longitude, 265–66
Lorentz form, 231
Lorentz group, 262
Lorentz transformation, 262
Lüroth's Theorem, 488

M

Main Lemma, 392
Main Theorem of Galois theory,
    489–92
Manifold, 278
Mapping property
    of free groups, 214
    of quotient groups, 214
    of quotient modules, 413
    of quotient rings, 335, 343
Maps
    canonical, 66, 335, 423
    equivalence relation defined by,
        55–56
    Frobenius, 355
    surjective, 54
    well defined, 180
    zero, 328
Maschke's theorem, 296, 298
Mathieu group, 283
Matrix, 1–36
    addition of, 2
    adjoint, 233
    augmented, 12
    basechange, 94
    block multiplication, 8–9
    cofactor, 29–31
    determinant of, 7, 18–24
    diagonal, 6, 117, 146
    diagonal entries in, 6
    diagonalizable, 117, 124
    elementary, 10–12
    elementary integer, 418
    Fourier, 260
    Hermitian, 233
    identity, 6
    integer, 418–23
    invertible, 7, 15
    of linear transformation, 104–8
    multiplication of, 2–3, 78
    nonzero, 9
    normal, 242
    orthogonal, 132–38
    permutation, 24–27, 51
    of polynomials, 432

    positive, 112
    presentation, 423
    R-matrix, 414
    rotation, 108, 134
    row echelon, 13–15
    row reduction of, 10–17
    scalar multiplication of, 2
    self-adjoint, 233
    skew-Hermitian, 267
    square, 2, 8
    unitary, 235, 244–45
    upper triangular, 6
    zero, 6
Matrix entries, 1
Matrix exponential, 145–50,
    278
Matrix multiplication, 2–4
Matrix notation, 4, 86
Matrix of form, 230
Matrix of transformation, 105
Matrix product, 3
Matrix representation, 290
Matrix transpose, 17–18
Matrix units, 9–10
Maximal element, 518
Maximal ideal, 344–47, 394
Minors, 19
    expansion by, 19
Modular arithmetic, 60–61
Modules, 412–14
    basis of, 415
    direct sum of, 429
    finitely generated, 415
    free, 412, 414–17
    generators of, 415
    homomorphism, 413
    isomorphism, 413
    rank of, 416
    of relations, 424
    R-module, 412
    Structure Theorem for,
        432–35
Monic polynomial, 325, 340
Monomial, 325, 327
Multi-index, 327
Multiple root, 458

Multiplication
  block, 8–9
  ideal, 389–92
  left, 177, 195, 277–78
  of matrices, 78
  matrix, 2–4
  right, 216
  scalar, 2, 5, 78, 90
  table, 38
Multiplicative group, structure of, 84
Multiplicative property
  of degree, 447
  of index, 58
  of the determinant, 21–24
Multiplicative set, 357

N

Natural number, 516
n-dimensional sphere (n–sphere),
    263
Negative definite, 231
Negative semidefinite, 231
Newton's identities, 505
Nilpotent, 122, 127, 355
Node, 351
Noetherian ring, 426–29
Nonabelian group, 222
Noncommutative ring, 324
Nondegeneracy on a subspace, 252
Nondegenerate form, 236, 252
Nonsingular point, 358
Nonzero, 9
Norm
  of an element, 386, 403
  of an ideal, 397
Normalizer, 203
Normal matrix, 242
Normal subgroup, 66
  generated by a set, 212
North pole, 263, 264
Notation
  cycle, 24
  fraction, 40, 343–44
  matrix, 4, 86
  power, 40

sigma, 4
  summation, 5, 28
Nullity, 103
Nullspace, 79, 103
Null vector, 236, 252
Number field, 442
  algebraic, 442

O

Octahedral group, 183
One-dimensional character, 303–4
One-parameter group, 272–75
Operation
  on cosets, 178–80
  faithful, 182
  of a group, 176–78, 293
  partial, 217, 218
  on subsets, 181
Operator
  adjoint, 242
  determinant of, 118
  diagonalizable, 117
  Hermitian, 244
  invertible, 109
  linear, 110, 293, 432–35
  normal, 242
  nilpotent, 122, 127
  orientation-preserving, 159
  orientation-reversing, 159
  orthogonal, 134, 162, 245
  self-adjoint, 243
  shift, 109, 434
  singular, 109
  symmetric, 245
  trace of, 118
  unitary, 242
Opposite group, 70
Orbit, 166, 177, 185
Orbit sum, 477
Order
  of finite field, 459
  of group, 40, 208–10
  by inclusion, 518
  partial, 518
  total, 518

Ordered set, 86
Orientation, 159
Orientation-preserving isometry, 160
Orientation-reversing isometry, 160
Orthogonal basis, 252
Orthogonal group, 134, 261
Orthogonality, 235–41, 254–56
Orthogonality relations, 300
    proof of, 309–11
Orthogonal matrix, 132–38
Orthogonal operator, 134, 245
Orthogonal projection, 238–41
Orthogonal representation, 269
Orthogonal space, 236
    to a subspace, 252
Orthogonal sum, 237
Orthogonal vectors, 252
Orthonormal basis, 133, 240

P

Parabola, 246
Parallelogram law, 256
    for vector addition, 112
Partial operation, 217, 220
Partial ordering, 518
Partition, 52–56, 57
Peano's axioms, 516–17
Permutation matrix, 26, 51
    determinant of, 27
Permutation representation, 181–83,
    304
Permutation, 24–27, 41, 50, 201
    cycle notation, 24
    representation, 181–83, 192
    symmetric group, 24
    transposition, 25
p-group, 197–98
Pick's Theorem, 411
Plane algebraic curve, 350
Plane crystallographic group, 172–76, 189–90
Point group, 170–71
Point, 163
    base, 468
    branch, 351, 353
Polar decomposition, 259, 287

Pole, 184, 186
    north, 263, 264
Polynomial ring, 325–28, 432–35
    in several variables, 436, 440
Polynomial, 85, 327
    characteristic, 113–16, 197
    complex, 520
    constant, 325
    cyclotomic, 374
    discriminant of, 481–83
    homogeneous, 328
    integer, 380–81
    irreducible, 350, 383, 443, 449–50, 458
    Laurent, 356
    matrix of, 432
    monic, 325, 340
    paths of, 101
    primitive, 368, 371
    quadratic, 247
    quartic, 495
    ring, 325–328
    roots of, 116
    symmetric, 477
Positive combination, 259
Positive definite, 229, 231, 232, 234
Positive eigenvector, 112
Positive matrix, 112
Power notation, 40
Presentation matrix, 423
Prime
    Gauss, 376–78, 381, 394
    ramified, 395
    split, 395
Prime element, 360
Prime factorization, 365
Prime ideal, 392, 394–96
Prime integer, 64, 394–96
Primitive element, 462–63
Primitive Element Theorem, 462–63
Primitive polynomial, 368, 371
Primitive root, 84
Principal ideal, 331
Principal ideal domain, 361
Product group, 64–66, 74
Product ideal, 355, 390
Product matrix, 3

Product permutation, 24
Product ring, 341–42
Product rule, 142
Product set, 67, 527
Projection, 64
    orthogonal, 238–41
    stereographic, 263
Projective group, 280
Proper ideal, 331
Proper subgroup, 43
Proper subspace, 79
Pythagoras' theorem, 133

Q

Quadratic form, 246
Quadratic number field, 383–411
    algebraic integer, 383–85
    class group, 396–99
    factoring algebraic integers, 385–87
    factoring ideals, 392–94
    ideal class, 396–99
    ideal multiplication, 389–92
    ideals, 387–89
    imaginary, 383
    lattices and, 405–8
    real, 402–5
Quadric, 245–49
Quartic equation, 493–97
Quartic polynomial, 495
Quaternion algebra, 266, 288
Quaternion group H, 47
Quintic equation, 502–5
Quotient group, 66–69, 74–75
    mapping property of, 214–15
Quotient ring, 334–38
    mapping property of, 335, 343

R

Ramified prime, 395
Rank, 103
    of a free module, 416
Rational canonical form, 435
Rational function, 342, 344, 487
    field of, 344
R-automorphism, 477

Real quadratic field, 402–5
Recursive definition, 517
    of the determinant, 20
Reducible representation, 295
Reflection, 134, 160
    glide, 160
Regular representation, 304–7
Relations, 212–16, 423–26
    adding, 337–38
    complete set of, 215
    defining, 212
    module of, 424
    orthogonality, 309–11
Relation vector, 424
Relatively prime elements, 362
Representation
    adjoint, 289
    complex, 293
    conjugate, 293
    faithful, 291
    of a group, 290–92
    induced, 321
    irreducible, 294–96
    isomorphism of, 293, 307
    matrix, 290
    orthogonal, 269
    permutation, 181–83, 304
    reducible, 295
    regular, 304–7
    sign, 291
    standard, 291
    of SU_2, 311–14
    trivial, 291
    unitary, 296–98
Representative element, 55
Residue, 330, 335
Resolvent cubic, 496
Restriction, 110, 181
    crystallographic, 171–72
    of homomorphism, 61
Riemann Existence Theorem, 465
Riemann surface, 350, 352, 464
Right coset, 58–59, 216
Right inverse, 7
Right multiplication, 216
Right shift operator, 109

Rings, 323–58
   automorphism of, 355
   characteristic of, 334
   extension of, 338
   homomorphism of, 328–34
   ideals in, 328–34, 387–89
   of integers, 384
   linear algebra in, 412–41
   noetherian, 426–29
   noncommutative, 324
   polynomial, 325–28, 339, 432–35, 436
   product, 341–42
   quotient, 334–38
   unit of, 325
   zero, 324, 414
R-matrix, 414
   determinant of, 414
R-module, 412
   homomorphism of, 427
Root
   adjoining, 456–59
   multiple, 458
Root of unity, 497–500
Rotation, 134, 160
   axis of, 134
Rotational symmetry, 154
Rotation group, 137
   finite subgroups of, 183–87
   SO_3, 269–72
Rotation matrix, 108, 134
Row echelon matrix, 13–15
Row index, 1
Row operation, 10
   elementary, 10
Row rank, 108
Row reduction, 10–17
Row vector, 2, 97, 108

S

Scalar multiplication, 2, 5, 78, 84, 90
   associative law for, 90
Scalars, 2
Schur's lemma, 307–9
Schwartz inequality, 256
Second Isomorphism Theorem, 227

Self-adjoint matrix, 233
Self-adjoint operator, 243
Semigroup, 75
Sets
   independent, 87, 95, 97, 415
   inductive, 518
   ordered, 86
   product, 527
Sheets, 465
Shift operator, 434
Sieve of Eratosthenes, 372
Sigma notation, 4
Signature of a form, 240
Sign representation, 291
Simple groups, 199
Singular operator, 109
Singular point, 358
Size function, 360
Skew-Hermitian matrix, 267
Skew-symmetric form, 230, 249–52
Solvable element, 502
Space
   covering, 351
   Euclidean, 241–42
   Hermitian, 241–42
Span, 86
   defined, 91
   of infinite set, 97
Special linear group, 43, 50
Spectral theorem, 242–45, 253
   for Hermitian operators, 244
   for normal operators, 244
   for symmetric operators, 245
   for unitary matrices, 244–45
Sphere, 263–66
   celestial, terrestrial, 264
Spin group, homomorphism, 269
Split prime, 395
Splitting field, 483–84
Splitting Theorem, 484
Sporadic group, 283
Square-free integer, 384
Square matrix, 2, 8
Square system, 16–17
Stabilizer, of element, 177–78
Standard basis, 88, 415

Standard representation, 291
Stereographic projection, 263
Structure Theorem
    for abelian groups, 429–30
    for modules, 432–35
    uniqueness for, 431–32
Subfield, 80
Subgroup, 42
    of additive group of integers,
        43–46
    characteristic, 225
    commutator, 225
    conjugate, 72, 178, 203
    discrete, 168
    finite, 183–87
    normal, 66
    proper, 43
    of SL_2, 280–83
    Sylow p-subgroups, 203
    trivial, 43
    zero, 422
Submodule, 413
    direct sum of, 430
    of free modules, 421–23
Subring, 323, 324
Subsets, operation on, 181
Subspace, 78–80, 85
    independent, 95
    linear transformation and, 102
    nondegenerate on a, 236
    orthogonal space to, 252
    proper, 79
    sum of, 95
Substitution Principle, 329
Successor function, 516
Summation notation, 5, 28
Surjective map, 54
Sylow p-subgroups, 203
Sylow theorems, 195, 203–7
Sylvester's law, 240, 256, 258
Symbolic notation, 55
Symmetric form, 229, 230
Symmetric function, 477–81
    elementary, 478
Symmetric Functions Theorem,
    479–81

Symmetric group, 24, 41, 50, 197
    conjugation in, 200–203
Symmetric operator, 245
    spectral theorem for, 245
Symmetric polynomial, 477
Symmetry, 154–94
    abstract, 176–78
    bilateral, 154
    glide, 155
    Hermitian, 233
    of plane figures, 154–56
    rotational, 154
    translational, 155
Symplectic group, 261
System, 4
    coordinate, 159
    square, 16–17

T

Tangent vector field, 280
Terrestrial sphere, 264
Tetrahedral group, 183
Third Isomorphism Theorem, 227
T-invariant, 110
Todd-Coxeter Algorithm, 206, 216–20
Total ordering, 518
Trace, 116
Transcendental element, 443–46
Transformation
    Lorentz, 262
    Tschirnhausen, 482
Translation, 156, 160
    in a group, 277–80, 286–87
    left, 277
Translation group, 168–70
Translation vector, 163
Translational symmetry, 155
Transpose, matrix, 17–18
Transposition, 25
Triangle group, 226
Triangular form, 116–19
Trivial homomorphism, 48
Trivial representation, 291
Trivial subgroup, 43
Truncated polyhedron, 186

Tschirnhausen transformation, 482
Two-dimensional crystallographic group, 172

U

Unbranched covering, 351
Union, 527
Unipotent, 355
Unique factorization domain, 364
Uniqueness of the determinant, 20–21
Unit, of a ring, 325
Unitary group, 235, 261
    SU_2, 266–69, 284
Unitary matrix, 235
    spectral theorem for, 244–45
Unitary representations, 296–98
Unit ball, 264
Unit ideal, 331
Unit vector, 133
Unity, root of, 497–500
Upper bound, 518
Upper triangular matrix, 6

V

Vandermonde determinant, 511
Variety, 347
Vector
    angle between, 242
    column, 2
    coordinate, 78, 90, 416
    fixed, 111
    length of, 242
    nonzero, 113
    null, 236, 252
    orthogonal, 252

    relation, 424
    tangent, 280
    translation, 163
    unit, 133
Vector addition, 78
Vector bundle, 436
Vector space, 78–101, 99
    bases and dimension, 86–91
    computing with bases, 91–95
    defined, 84–86
    direct sum, 95–96
    fields, 80–84
    finite-dimensional, 89
    infinite-dimensional, 96–98
    isomorphism of, 85, 91
    subspace, 78–80

W

Weight, weighted degree, 482
Well-defined, 180
Wilson's theorem, 99
Word problem, 213

Z

Zero
    characteristic, 83, 484
    common, 347
Zero divisor, 343
Zero element, 417
Zero ideal, 331
Zero map, 328
Zero matrix, 6
Zero ring, 324, 414
Zero vector, 126
Zorn's Lemma, 98, 348, 518–19